ABBREVIATED TABLE OF CONVERSIONS (See Appendix also)

Length

1 meter = 39.4 in. = 3.28 ft
1 in. = 2.54 cm (exactly)
1 mi = 1.61 km = 5280 ft

Time

1 day = 86,400 s; 1 yr = 365 days = 3.15×10^7 s

Mass

1 kg = 2.21 lb
1 slug = 32.2 lb = 14.6 kg

Velocity

1 mi/hr = 0.447 m/s = 1.47 ft/s

Angle

1 rad = 57.3° = 0.159 rev

Force

1 N = 10^5 dyne = 0.225 lb (force)

Pressure

1 N/m² = 1.45×10^{-4} lb/in.² = 9.87×10^{-6} atm

Energy and power

1 J = 0.738 ft · lb; 1 cal = 4.18 J; 1 eV = 1.60×10^{-19} J
1 hp = 746 W = 550 ft · lb/s

Physics

Chris Zafiratos

University of Colorado

John Wiley & Sons, Inc.

New York   London   Sydney   Toronto

Cover: The beam of light from a metal-vapor laser dispersed into its constituent colors by a diffraction grating. (*Photo by Fritz Goro*)

Library of Congress Cataloging in Publication Data

Zafiratos, Chris.
   Physics.

   Includes index.
    1.   Physics.   I.   Title.
QC21.2.Z3   530   75–14034
ISBN 0–471–98104–4

Printed in the United States of America

10 9 8 7 6 5 4 3 2 1

Preface

This book grew out of my experience teaching the standard calculus-based physics course at the University of Colorado. Though many of the students had a previous calculus course, calculus was not a prequisite but rather a corequisite requirement. It was tempting to blame insufficiencies of some students upon their lack of a previous calculus course, but this did not seem to be the case. Instead, the most frequent difficulty of the beginning students was an inability to describe an everyday setting in the analytical language of physics, and to project that language back into their reality. Alternatively put, translation from the English language into the language of mathematics, and the inverse process, is a skill which most beginning physics students lack.

The transition from "hunting for a formula which gives the answer," to a reasoning approach that draws upon several facets of knowledge is probably the most difficult step in the development of a physics student. If new ideas are presented too rapidly this transition often is not made at all by the average student. Even a simple notational change, e.g. the dot product, is another fact to be absorbed and digested and can be a distraction. Accordingly, I have tried to pace reasonably the introduction of new concepts. I have also tried to explain more of the reasoning behind many concepts because my students appear to profit from this kind of help.

I have encouraged students to draw sketches and to use mathematical reasoning by example and by explicit direction and discussion. Each new block of material is followed by simple exercises which can be solved readily. These exercises help the student retain the basic ideas and calculational steps utilized in discussing a given phenomenon. When the student is then asked to combine several ideas he is better prepared to do so.

Since many of my students are more interested in dynamics rather than statics, I have begun the text with a discussion of motion and Newton's laws of motion. The concept of momentum is used in a fundamental way and both the differential and integral form of Newton's second law are discussed immediately. The early introduction of calculus does not seem to be a special source of difficulty since most calculus courses are still stressing foundations of calculus at the time a parallel physics course with *any* ordering of topics has need for differentiation and integration. To aid the student enrolled in a corequisite calculus course I have included some math where appropriate. In fact, Chapter Two is a brief introduction to calculus in the setting of kinematics. Mathematical topics are subsequently discussed as they are needed. They are not discussed in a general way, as they would be in a mathematics course, but only in the limited context required in our coverage of physics. This not only serves to prevent overburdening the student with too much knowledge too quickly, but also frequently helps the student who later encounters the same topics in all their appropriate generality in a mathematics course. Most of this supplemental mathematical material is arranged so that it can be deleted by the well-prepared reader.

The first half of the book is taken up with mechanics and thermodynamics. The second half begins with wave motion and leads directly into light as an example of a wave phenomenon. Physical and geometrical optics are discussed. Electricity and magnetism then follow, including a chapter which deals with Maxwell's prediction of electromagnetic waves and the inclusion of light waves in this phenomenon. The final chapter introduces the student to quantum physics. It is historical in its organization and emphasizes wave-particle duality. The overwhelming emphasis of the book is on classical physics, but underlying atomic phenomena are discussed wherever they are appropriate.

In recent years introductory physics courses have concentrated on a physicist's view of physics rather than that of the *user* of physics. Thus, topics such as musical scales, motors, alternating-current circuits and optical instruments have been dropped or de-emphasized by some texts. I have tried to steer a middle course, including those topics which seemed of interest or value to many of the variety of students encountered in introductory physics.

The International System of Units (SI) is the primary system utilized in this book. However, the wide variety of units encountered in *applications* of basic physics tends to put off the student who has never made conversions from one set of units to another. Further, it is still true that most readers have a better mental image of the foot, pound, quart and mile than of their SI analogs. Accordingly, some use is made of English units in early sections of the book. Other units are mentioned and occasionally used in examples when their use seems appropriate, but SI units are emphasized throughout.

I am indebted to many people for their help during the preparation of this book. John Taylor deserves very special thanks for his thorough and constructive criticism after teaching from the entirety of the mimeographed notes. Many other colleagues at the University of Colorado have helped with discussion and William Campbell helped by teaching from the first half of the notes at the University of Nebraska. David Cannell of the University of California, Santa Barbara, Dorothy Wollum of California State, Fullerton, and Jack Wollum of California State, Los Angeles all provided valuable suggestions. Helpful comments were also given by Haywood Blum of Drexel University, Lowell Wood of the University of Houston and Thomas Erber of the Illinois Institute of Technology. Several students have made useful comments and suggestions, most notably Molly Rothenberg, Mark Utlaut and Larry Zanetti; and many have offered encouragement. Several fine photographs (those not otherwise credited) resulted from the joint efforts of John Groft and Robert Stoller. George Thomsen's early support and persistent encouragement were most welcome. Finally, for the typing and proofreading of endless chapters and revisions, and for her constant support, I thank my wife Joellen.

Chris D. Zafiratos
February 1975

About the Author

Chris Zafiratos is a Professor of Physics at the University of Colorado. He received his B.S. at Lewis and Clark College in 1957 and his Ph.D. at the University of Washington in 1962. He continued with post-doctoral research in Nuclear Physics at the Los Alamos Scientific Laboratory and taught at Oregon State University before coming to the University of Colorado in 1966. He is a member of the American Physical Society and publishes research articles in experimental nuclear physics. He has been involved with the evolution of intro-ductory physics courses at the University of Colorado and currently (1974–76) serves as chairman of the Nuclear Physics Laboratory of the University of Colorado.

Contents

Omit

Omit

Omit

Chapter *1*

Introduction

1-1 The Content of Physics   Many of the physical sciences imply their content in their name: astronomy, biology, and oceanography are examples. Physics, however, does not. Indeed, the subject matter of physics is so broad that it underlies and runs through all of the physical sciences. Because of the tremendous revolutions in relativity and quantum physics that occurred around 1900, it is convenient to divide physics into classical physics (pre-1900) and modern physics. Some topics included in these two areas of physics are tabulated below.

TABLE 1-1   Some Topics of Classical and Modern Physics

Classical Physics	Modern Physics
Mechanics	Quantum mechanics
Gravitation	Relativity
Heat	Atoms
Sound	Nuclei
Light	Elementary particles
Electricity	Condensed matter
Magnetism	

Mechanics is the name given to the theory of motion. It gives a precise relationship between motion and the forces that cause it. By the beginning of this century it was already clear that heat and sound phenomena could be understood in the framework of classical mechanics, the mechanics originated by Newton in the seventeenth century. It was also clear that electricity and magnetism formed a unified topic and, further, that light was simply an electromagnetic wave. The physicist of the 1890s had every reason to be smug. The way in which electric charges moved could be understood by combining electricity and magnetism with mechanics, and the way in which planets and stars moved could be understood by combining gravitation and mechanics. Since matter is composed of atoms, which themselves contain electric charges and gravitating mass, complete understanding of the physical universe seemed at hand.

Unfortunately, classical mechanics failed badly when it was applied to the internal motion of atoms. As an example of the magnitude of the failure, consider the lifetime of isolated atoms. Classical physics predicts that such atoms will collapse in 10^{-8} seconds, yet atoms are known to be perfectly stable for periods of time enormously longer than that.

Over a period of years, culminating in the 1930s, the new theory of quantum mechanics was developed. It dealt, correctly, with questions of atomic and subatomic motion and structure. It also showed that the older classical physics was correct for objects much larger than atoms, as long as the velocities involved were much smaller than that of light.

For motion involving velocities near that of light (3×10^8 meters/second), classical mechanics fails even for objects much larger than atomic dimensions. Einstein's *special theory of relativity* leads to a *relativistic mechanics* that is part of modern physics and that can cope with all possible velocities. Einstein's *general theory of relativity* replaces Newton's classical theory of gravitation. General relativity is required, for example, to understand the gravitational effects of ultradense stars, which are currently of great interest in astronomy. However, for the sort of densities and velocities encountered in our solar system, classical gravitational theory is for the most part sufficient.

Most of the phenomena of an everyday nature and a good many of those encountered in modern technology can be understood most simply and elegantly with the aid of classical physics. Furthermore, the groundwork covered in a study of classical physics is required in any thorough study of modern physics. Some ideas of modern physics will be touched upon in this text but we will concentrate heavily upon classical physics.

Having briefly described the major topics of physics and the portion covered in this text, we ask, what about the work carried on by the contemporary physicist? A good many physicists, along with their colleagues in engineering, are occupied with applications of physical knowledge for benefit of consumers, corporations, governments, etc. These physicists collectively utilize the full range of classical and modern knowledge. Obviously a good many physicists teach physics. Finally, many contemporary physicists pursue basic research aimed at the unanswered questions of physics. Of course, many individual physicists fall into more than one category. We can obtain some idea of the unanswered questions that attract the attention of research physicists by listing the major categories of the papers that are published in current research journals.

TABLE 1-2 Major Categories in Current Physics Journals

Atoms, molecules and related topics
Condensed matter (solids and fluids)
Electromagnetic, gravitational, and quantum fields
Elementary particles
Nuclear reactions and nuclear structure
Quantum mechanics
Statistical mechanics and thermodynamics

What may be the most important modern development in science is perhaps too close in time to be clearly recognized. This development is the wide-ranging unification of all the physical sciences, including engineering and biology. This unification is an ongoing process that has enormous benefits in increasing our intellectual power with an economy of required basic concepts. The contemporary physicist finds more and more of the language and methods his predecessors developed appearing in all the physical sciences. The increasing unity in the panorama of the sciences may be a reflection of a basically simple and unified universe. Or we may be in for a rude awakening in the near future. In any event, these are lively times for the intellectually curious person.

1-2 The Fundamental Quantities of Mechanics

Many quantities enter into mechanics: speed, force, mass, acceleration, length, time, direction, and weight are examples that come to mind. Most of the above quantities can be defined in terms of one another. Speed, for example, is simply the ratio of the distance covered (a length) to the time required to do so. However, a certain bare minimum of the above quantities must be *fundamental quantities,* which are not derived from others. The commonly used fundamental quantities in mechanics are *length, mass,* and *time.* These quantities are defined by *operational definitions.*

An operational definition gives a clear procedure for measurement and defines a *standard* unit of the quantity. As an example, we could say that time intervals will be measured in *seconds* and that the standard second is the time required for a simple pendulum 3.26 feet long to swing from one side to the other. (The actual, more precise definition of the second is given in the next section.) In this manner the romantic, philosophical question, "What *is* time?" is ignored in favor of a definition so that we can get on with a study of motion.

1-3   Measurement and Standards

The precision of measurements of the fundamental quantities has grown steadily with various technical advances. Frequently these advances in precision have been prompted by a technological need. At other times such advances have been made purely for the sake of the improvement itself. On some occasions entirely new physical phenomena have been discovered in the process of refining measurement techniques.

As the accuracy of measurements improve, more and more precise standards must be chosen to keep pace with the state of the art of measurement. Thus, one of the English length standards was once the foot and was literally the length of a human foot. Subsequently the yard was standardized by Henry II and a standard yardstick was preserved as the primary standard. The French Academy of Science then adopted (1889) the metric system, with a unit of length called the *meter.* The meter was defined to be one ten millionth of the distance from the north pole to the equator along the meridian running near Dunkirk, Paris, and Barcelona. In fact, however, the standard meter was taken to be the distance between two scribed marks on a platinum-iridium bar, which has been kept under carefully controlled conditions at the International Bureau of Weights and Measures in Paris since 1889. Of course, the standard meter was made as close to one ten-millionth of the pole-equator distance as was possible at the time, but nonetheless the actual primary standard was the meter bar itself.

Techniques have since been developed that allow the measurement of distances in units of light wavelengths. The wavelength of light is quite short, about half a millionth of a meter, and distances can be measured to an accuracy of about one hundredth of a wave-

Fig. 1-1. The sensitivity of a commercial laser interferometer is demonstrated by its reading of 1 microinch—the deflection produced in the 4-ft aluminum I beam by the weight of one penny. This instrument directly uses the modern length standardization in terms of the length of light waves. (Courtesy *Hewlett-Packard Journal*)

length. Thus, lengths near one meter in magnitude can be determined to one part in one hundred million. Commercial instruments utilizing this method are now readily available (see Fig. 1-1). The length standard has been changed to keep pace with these new developments; in 1960 the meter was defined by the International Committee of Weights and Measures to be 1,670,767.73 wavelengths in vacuum of the light emitted by a specified transition in krypton-86 atoms. Light from lasers has a technical advantage (discussed in Chapter 24) in these measurements and light from a specified type of laser may well replace the krypton light source in a future standard. The instrument that measures length in terms of light wavelengths is called an *interferometer* and is discussed in Section 24-10.

At first glance, the desire for ever more accurate measurements and standards may seem pointless; why add another decimal place to a measured number? Besides their use in fundamental scientific research, precision measurements of length have many practical applications. One such application is in the production of microscopic electronic integrated circuits. Another is the possibility of

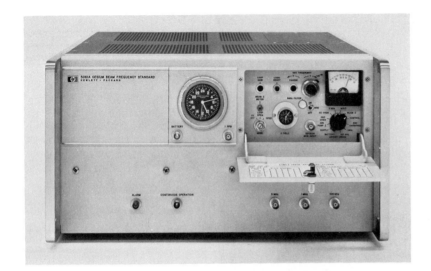

Fig. 1-2. This commercial atomic clock is small enough to be portable and can be powered by batteries. It has an accuracy of one part in 10^{11}. (By permission of Hewlett-Packard)

earthquake prediction by use of extremely accurate long-term measurements of earth movement near faults.

Present day capability of measuring time intervals to one part in one hundred billion with portable instruments has application in collision avoidance for jet aircraft. One collision avoidance system, for example, utilizes radio signals transmitted by all aircraft at precisely the same time once every three seconds. These signals propagate at approximately 1000 feet per one millionth of a second. All aircraft monitor the arrival time of signals from other aircraft and a computer detects potentially dangerous encounters. In order to operate successfully, the primary signals from all aircraft must be synchronized to within one millionth of a second of each other. An atomic clock (see Fig. 1-2) has the requisite stability to maintain this synchronization over the period of a day. In fact, it is this type of clock that is now used for the primary time standard. An excellent account of the development of present standards is given in "Standards of Measurement" by Allen V. Astin in the June 1968 issue of *Scientific American.*

1-4   Systems of Units

Physical quantities are measured in terms of units; the centimeter, foot, and kilometer are examples of units of length, and the second and day are units of time. We have already indicated that mechanics requires three fundamental quantities: length, mass, and time. There are several conventional *systems of units* in which these quantities are expressed. The centimeter-gram-second system (cgs) and the meter-kilogram-second system (mks) utilize metric system units. English units in use in the United States employ the foot, slug (defined in Chapter 4), and second or the foot, pound, and second as units. These U.S. units are actually defined in terms of metric unit standards by international agreement.

The metric system utilizes a decimal system to obtain multiples of standard units. The multiplying factor is always a power of ten and is indicated by a prefix, as in Table 1-3. The units of the metric system are abbreviated m for meter, g for gram, and s for second. Other units are obtained by combining prefixes with these units. Thus the millimeter (mm) is one thousandth of a meter and the kilogram (kg) is one thousand grams.

The generally accepted mks system has been augmented by four additional fundamental quantities (and their standards) to form a complete system of fundamental units. All other quantities in physics can be derived from these seven fundamental quantities. This system of units is called the International System of Units and is abbreviated SI (for Système International).* The four additional units with their symbols in parentheses are the ampere (A) for electric current, the kelvin (K) for thermodynamic temperature and for temperature intervals, the candela (cd) for luminous intensity, and the mole (mol) for quantity of substance. Supplementary units and derived units that are formed from the base units are tabulated at the end of this book. Conversions among various units are also tabulated there.

The definition of each fundamental SI quantity involves the choice of a standard. The three fundamental quantities of mechanics have the following standards.

Length: The meter is the length equal to 1,650,673.73 wavelengths in vacuum of the light corresponding to the transition between two specified levels ($2p_{10}$ and $5d_5$) of the krypton-86 atom. The U.S. units are related to the meter by defining one inch equal to 2.54 cm exactly.

Mass: The kilogram is still defined by the prototype kilogram of 1889. This prototype was fabricated to have a mass equal

TABLE 1-3   Prefixes in the Metric System

Power of Ten	Prefix	Symbol
10^{12}	tera	T
10^{9}	giga	G
10^{6}	mega	M
10^{3}	kilo	k
10^{2}	hecto	h
10^{1}	deka	da
10^{-1}	deci	d
10^{-2}	centi	c
10^{-3}	milli	m
10^{-6}	micro	μ
10^{-9}	nano	n
10^{-12}	pico	p
10^{-15}	femto	f
10^{-18}	atto	a

*SI units and their historical background are concisely discussed and defined in a publication from the U.S. Government Printing Office: "The International System of Units (SI)," edited by Chester H. Page and Paul Vigoureaux, National Bureau of Standards Special Publication 330, 1972 Edition. For sale by the Superintendent of Documents, U.S. Government Printing Office, Washington, D.C. 20402 (Order by SD Catalog No. C 13.10:330/2) for 30 cents.

to that of one thousand cubic centimeters of water at its temperature of maximum density. It was subsequently determined that an error of 28 parts in one million was made in constructing the standard. Nonetheless, the metal prototype kilogram is retained as the standard.

Time: Originally, the unit of time, the second, was defined as 1/86,400 of the mean solar day. Slight irregularities in the rotation of the earth, about one part in one hundred million, can be detected by modern timekeeping techniques and affect the length of the mean solar day. The second is now defined by atomic vibrations to be the duration of 9,192,631,770 periods of the radiation corresponding to the transition between the two hyperfine levels of the ground state of the cesium-133 atom.

The standards for the remaining four fundamental quantities of physics will be discussed at appropriate points in the book.

The reader will proceed to find small groups of exercises after most blocks of new material. These exercises are intended to be simple, as they are designed only to test the reader's grasp of the elementary points in the preceding section. Answers are given in the back of the text. Problems ranging in difficulty beyond the exercises are collected at the ends of the chapters. Problems that are unusually difficult or that involve lengthy calculations are marked with an asterisk.

Exercises

1. The length of a bolt is found to be 16 mm. Express this length in (*a*) cm, (*b*) m, and (*c*) km.

2. A time-interval name that appeared in the jargon of World War II research in nuclear weapons was the "shake," where 1 shake $= 10^{-8}$ s. Express the shake in units of μs and ns.

1-5   Dimensions and Algebraic Operations

We recognize a logical inconsistency in adding different entities. We accomplish nothing by adding seven apples to three bicycles because the answer is not ten applecycles but seven apples and three bicycles. Similarly the primary quantities of physics cannot be summed to form new derived quantities. It is possible, however, to multiply and divide primary quantities to obtain new derived quantities. A familiar example is that of *velocity*, which may be expressed in miles per hour (mi/hr), feet per second (ft/s), or meters per second (m/s). All means of expressing velocity contain a length interval divided by a time interval. We say that velocity has the *dimensions* of length divided by time, often abbreviated lt^{-1}. Similarly the dimensions of area are l^2 and of volume, l^3.

When we write an equation containing magnitudes of physical quantities, each term in the equation must have the same dimensions. As an example, consider the distance from town, s, of a car that began M miles from town and moved away steadily at a velocity of N miles per hour. This distance is given by the equation

$$s = M \text{ mi} + \left(N \frac{\text{mi}}{\text{hr}} \times t \text{ hr} \right)$$

where t is the number of hours elapsed since the motion began. The units in parentheses cancel to produce

$$s = M \text{ mi} + \left(N \frac{\text{mi}}{\text{hr}} \times t \text{ hr} \right)$$

$$= M \text{ mi} + Nt \text{ mi} = (M + Nt) \text{ mi}$$

Thus we see that the equation is dimensionally consistent. Each term has the dimensions of length. Alternatively, the cancellation can be carried out when values are substituted. Thus if the original distance is 10 mi, the velocity is 20 mi/hr, and the elapsed time is 3 hr, we have

$$s = 10 \text{ mi} + 20 \frac{\text{mi}}{\text{hr}} \cdot 3 \text{ hr} = 10 \text{ mi} + 60 \text{ mi} = 70 \text{ mi}$$

We will often use a single symbol to represent both the numerical value and the units of a quantity. In the preceding example, the distance could be represented by s, the starting distance by s_0, the velocity by v, and time by t. We then have the simple statement

$$s = s_0 + vt$$

The dimensional consistency of this equation can be checked by substituting the dimensions of each term:

$$l = l + (lt^{-1})t = l + l$$

The equation is seen to be consistent since all terms in the equation have the dimensions of length.

Conversions from one set of units to another are made by multiplying by a dimensionless ratio:

$$5 \text{ mi} = 5 \text{ mi} \cdot 5280 \text{ ft/mi} = 26400 \text{ ft}$$

where the quantity 5280 ft/mi is a ratio of length units and, hence, has no dimension. Dimensionless ratios also can be inserted into equations to convert the units of one term to be consistent with others.

Example   A convenient mnemonic for velocity conversions is

$$60 \text{ mi/hr} = 88 \text{ ft/s}$$

Show that this is correct.

Solution   Insert the conversion from miles to feet into the numerator and the conversion from hours to seconds into the denominator:

$$60 \, \frac{\text{mi}}{\text{hr}} \cdot \frac{5280 \text{ ft/mi}}{3600 \text{ s/hr}} = \frac{316800 \text{ ft}}{3600 \text{ s}} = 88 \, \frac{\text{ft}}{\text{s}}$$

The ratio is exact so this particular mnemonic has not been rounded off to two significant figures.

Exercises

3. Convert 60 mi/hr to m/s and km/hr. The required conversions are 1 mi = 1610 m, 1 mi = 1.61 km, and 1 hr = 3600 s.

4. The following formula gives the time required for an object moving with an initial velocity v and accelerating with an acceleration a to reach a distance s from its original position.

$$t = \frac{-v \pm \sqrt{v^2 + 2as}}{a}$$

The dimensions of v are lt^{-1} and those of a are lt^{-2}. Is the formula dimensionally consistent?

1-6   Rounding off Numbers

When someone tells us that a desk is two meters long, we cannot tell whether he means that it is precisely two meters in length (surely there must be a limit to the precision of any such statement) or only that its length is closer to two meters than it is to 1.9 or 2.1 meters. To remove this ambiguity we write as many figures in the number as are *significant*.

Usually, the *precision* of our knowledge of a number limits the number of significant figures. Thus, if the desk's length is measured with a scale on which rulings are spaced every mm, we write the length of the desk as 2.000 m to indicate the number of significant figures.

In this day of widely available electronic calculators, the meaning of significant figures is often lost. For example, a calculator with eight-figure capability might be used to compute the area of a desk top with dimensions

$$L = 2.47 \text{ m}, \qquad W = 1.03 \text{ m}$$

and find

$$A = L \cdot W = 2.5441000 \text{ m}^2$$

Surely the area of the desk top is not known *this* well. All we know is that the length of the desk top lies between 2.465 m and 2.475 m, while the width lies between 1.025 m and 1.035 m. If, for example, the "true" length were 2.4750000 m and the width 1.0350000 m, the actual area would be

$$A = L \cdot W = 2.561625 \text{ m}^2$$

which differs from our earlier result in the third significant figure.

A general rule we will follow is to round off our answers to multiplication and division operations to the number of significant figures of our least precise number or, at most, to one more significant figure than the least precise number.

Example
$$2.2 \cdot 4.553 = 10$$
$$2.200 \cdot 4.5530 = 10.02$$
$$0.0442 \cdot 3.21 = 0.142$$
$$0.0022/1.555556 = 0.0014$$

Note in this example that leading zeros are not significant figures since they serve only to indicate division by precise powers of ten. Trailing zeros, on the other hand, are used by convention to indicate the precision of a number that happens to "come out even." This point is more easily seen when scientific notation is used:

$$(4.42 \times 10^{-2}) \cdot 3.21 = 1.42 \times 10^{-1}$$

$$(2.2 \times 10^{-3})/1.555556 = 1.4 \times 10^{-3}$$

When numbers are added and subtracted, the decimal location of the least significant figure determines the number of figures that should appear in the result.

Example
$$\begin{array}{r} 1.23456 \\ + 1.10000 \\ \hline 2.33456 \end{array}$$

However, if one of these numbers is poorly known we have

$$\begin{array}{r} 1.23456 \\ + 1.1 \\ \hline 2.3 \end{array}$$

Since we *do not know* the values of the number 1.1 beyond the tenths column in the last example above, we should not express the answer beyond the tenths column. To do so would indicate that we know the result with a greater accuracy than warranted by the given information.

Exercises

5. Round off the answers to the following calculations:

$453 \cdot 22 =$
$0.0015 \cdot 30000 =$
$6.6666/2.000 =$
$6.6666/2.0 =$

6. Round off the answers for the following:

$55.01 + 4.3 =$
$55.01 + 4.322 =$
$55.01 + 0.00004 =$
$55.01 - 10 =$

Chapter *2*

Introduction to Calculus

2-1 Introduction Calculus is a branch of modern mathematics that deals with change. Since motion involves change in a fundamental way, it is not surprising that much of the development of calculus occurred simultaneously with the growth of the first complete theory of motion. This theory of motion is now called classical mechanics and rests heavily upon the work of Sir Isaac Newton (1642–1727). The ideas developed by Newton for handling changing quantities, now incorporated in what we call calculus, were developed independently at about the same time by Wilhelm Leibnitz (1646–1716), a German philosopher and mathematician. Though physics owes its greatest debt to Newton, modern writing has largely adopted the notation used by Leibnitz.

Retracing the development of concepts of motion is an excellent first introduction to calculus. In this chapter we will discuss the most elementary ideas of differential and integral calculus, primarily as applied to motion. We will limit this discussion of calculus to include only functions that can be expressed in graphical form or as simple polynomials. Those readers already familiar with elementary calculus may want to treat this chapter as a review or to omit it entirely.

2-2 Motion of a Particle

The study of motion can be quite complex if we consider an extended object that spins or tumbles while it moves along a curved path. Often, however, the essential features of an object's motion can be understood without regard to its orientation. For example, the motion of the earth in its yearly orbit about the sun is not affected by its own daily rotation about its own axis. The earth could just as well be a particle with the same absence of extension possessed by a mathematical point; it would follow the same yearly orbit. In this and many other applications, the fiction of an idealized point particle is useful and we will begin by considering the motion of point particles.

In order to describe the motion of a particle, we must first be able to define the position of the particle; it is changes in position that we will call motion. A conventional Cartesian coordinate system is most useful: an example of its use is shown in Fig. 2-1. Note that the z coordinate of the point P is given by the distance up from the x-y plane, the x coordinate by the distance out from the y-z plane, and the y coordinate by the distance from the x-z plane. Of course, the coordinate system itself may be in motion if we take a larger view. This possibility will be discussed subsequently and, for now, we will consider only a single coordinate system as our *frame of reference*. The problem of describing the motion of a point in that frame of reference now becomes that of giving its $x, y,$ and z coordinates at *specified times*. It is only if one or more coordinates vary as time elapses that we can say there is motion, so we will want to examine carefully the dependence of the coordinates upon time.

Let us begin our discussion by considering only one coordinate. The dependence of a coordinate, say x, on the time t is aptly described by saying that x is a *function* of t. This statement is often written symbolically as

$$x = f(t)$$

where the symbol in parentheses is called the *argument* of the function. The functional dependence of x upon t is sometimes expressed by saying that x is the *dependent* variable, while t is the *independent* variable. We may also indicate that x is a function of t by writing $x(t)$, with only the parentheses indicating the functional dependence. An example of such a function is the algebraic function $x = 10t^2 + 3t - 7$ or, more generally, $x = at^2 + bt + c$. Another example is given by the graph of Fig. 2-4. There x is given as a function of t but the function cannot conveniently be written down in any algebraic form.

If an object, confined to only one dimension of a coordinate system, is in motion, then its distance from the origin of a coordinate system must change as time elapses. Let us call that distance the *position*, symbolized by the letter x, and let us take as an example a lecturer pacing back and forth in front of a classroom. A sketch

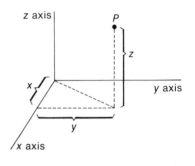

Fig. 2-1. Cartesian coordinates of a point P.

of the situation is shown in Fig. 2-2, where x is taken to be the lecturer's distance from the left wall of the sketch as he paces back and forth. The lecturer is certainly not a point particle, but we can fix our attention upon one point on the lecturer.

A graph of the changing x coordinate is shown in Fig. 2-3. Each point gives the value of x, the lecturer's distance from the left wall, at one instant of time with each instant taken to be one second later than the preceding. We see that the lecturer begins one meter away from the wall, then moves to 2 meters away from the wall during the first 4 seconds of elapsed time. He is stationary for the next 4 seconds (until $t = 8$ seconds), then once again increases his distance from the wall to a value of 4 meters at $t = 12$ seconds. He now must turn around and retrace his steps, since we see that x steadily *decreases* for the next 7 or 8 seconds out to $t = 20$ seconds, where our graph ends. We could, if we wish, record our lecturer's position more frequently than once every second, with the results shown in Fig. 2-4.

Modern laboratory apparatus can measure position every nanosecond (billionth of a second) without great difficulty, in which case

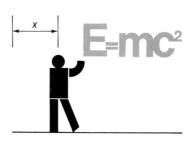

Fig. 2-2.   The position x of the lecturer varies as he paces back and forth.

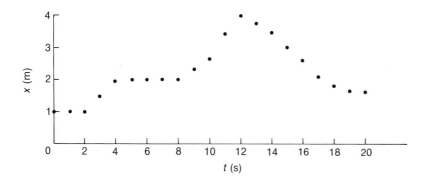

Fig. 2-3.   Graph of x vs t for the lecturer of Fig. 2-2.

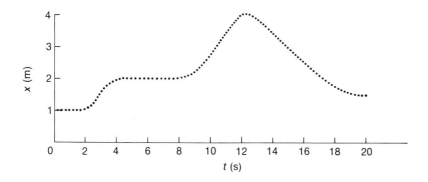

Fig. 2-4.   Graph of x vs t as in Fig. 2-3 but with more frequent data collection.

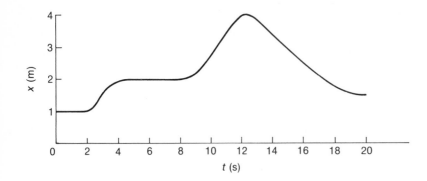

Fig. 2-5.   A graph of x as a continuous function of t.

we could no longer distinguish the individual data points of Fig. 2-4. Since we can, in principle, measure an object's location with almost arbitrary rapidity, it is quite common in discussions of motion to assume that positions such as x are *continuous* functions of time.* A graph showing x as a continuous function of time is shown in Fig. 2-5.

Now let us examine Fig. 2-5 and find those times when the lecturer is moving most rapidly. First note that during the first two seconds of elapsed time, the position of the lecturer has not changed, that is, he is stationary. The same condition occurs between $t = 4$ and $t = 8$ seconds and for a fleeting instant near $t = 12$ seconds. The x coordinate is increasing steadily between $t = 2$ and $t = 4$ seconds and between $t = 8$ and $t = 12$ seconds; thus the lecturer is moving away from the left wall of the lecture hall during these times. Between $t = 12$ and $t = 20$ seconds, the x coordinate is steadily decreasing; the lecturer is moving back toward the wall. Let us formalize these ideas by defining *velocity*, a quantity that will be zero when there is no motion, small when there is a slow change of position, and large when the motion is rapid.

A quantity that has such properties is distance traveled divided by the time required to cover that distance. Illustrating with units that are common to American highway travel, we all know that an automobile that is covering 60 miles during every hour is moving with a velocity of 60 miles per hour. Velocity is thus the *change* in position divided by the change in time that occurred while the position was changing. The reader has perhaps already noted a difficulty with this concept of velocity. What if the automobile moves

---

*This assumption will need closer examination when we discuss the motion of objects as small as atoms. For such small objects the very act of measurement cannot help but change the motion that one intends to measure. This point will be discussed further in Chapter 40.

at 120 miles per hour for one-half hour and then is stationary for one half hour? We would then say that the automobile *averaged* 60 miles per hour over the one hour interval. In general, the *average velocity* is defined by

$$\bar{v} = \frac{\Delta x}{\Delta t} \qquad (2\text{-}1)$$

where Δx is the change in position and Δt is the time interval in which the change occurred. The quantity Δx is called the *displacement* and is given by

$$\Delta x = x_2 - x_1$$

where x_1 is the position at the beginning of the time interval and x_2 is the position at the end of the time interval. The bar over the symbol for velocity in Eq. (2-1) is our notation to indicate an averaged quantity. Use of this definition is illustrated in Fig. 2-6 for a 2-second time interval centered about $t = 10$ s. The displacement during this time is seen to be equal to 1.2 m. The average velocity is thus

$$\bar{v} = \frac{\Delta x}{\Delta t} = \frac{1.2 \text{ m}}{2 \text{ s}} = 0.6 \text{ m/s}$$

Note that the dimensions of velocity are length divided by time, meters/second (read meters *per* second) in our example. Other commonly used units are feet/second, kilometers/hour, and miles/hour. Our definition of average velocity is identical with a mathematician's definition of the *slope* of x as a function of t. When the slope of the x-vs-t curve is steep, Δx divided by Δt gives a larger result than when the slope is shallow.

Because it is an average over a time interval, the average velocity does not contain information about details of the motion during

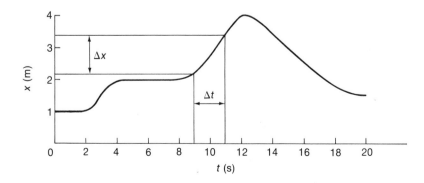

Fig. 2-6.   The average velocity for the 2-s interval between $t = 9$ s and $t = 11$ s is found by use of Eq. (2-1) with $\Delta x = x_2 - x_1 = 3.4$ m $- 2.2$ m $= 1.2$ m and $\Delta t = t_2 - t_1$ $= 11$ s $- 9$ s $= 2$ s.

that time interval. In Fig. 2-7, for example, the average velocity for the chosen 2-second time interval is given by

$$\bar{v} = \frac{\Delta x}{\Delta t} = \frac{30 \text{ m} - 20 \text{ m}}{4 \text{ s} - 2 \text{ s}} = 5 \text{ m/s}$$

The motion could have followed the path indicated by the dotted straight line with the same result. The ratio $\Delta x/\Delta t$ is the *slope* of that straight line. Yet we see that there is a brief period of time during that 2-second interval in which the velocity is negative. This is indicated in Fig. 2-8, where, for the shorter time interval chosen, we have

$$\bar{v} = \frac{\Delta x}{\Delta t} = \frac{22 \text{ m} - 23 \text{ m}}{2.9 \text{ s} - 2.8 \text{ s}} = -10 \text{ m/s}$$

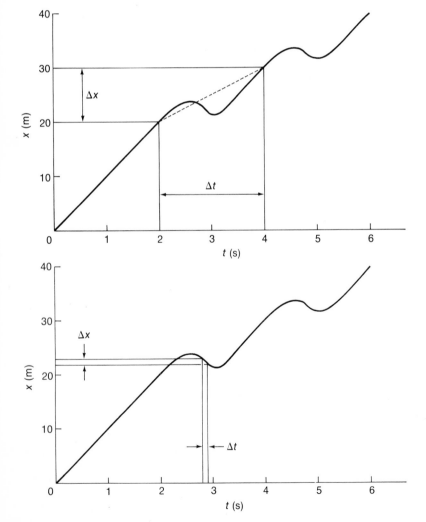

Fig. 2-7. The average velocity over the chosen interval is positive, though there is a short period of time within that interval in which the velocity is negative. The ratio $\Delta x/\Delta t$ defines the *slope* of the dotted line shown.

Fig. 2-8. The average velocity is negative over a shorter time span than that of Fig. 2-7.

Thus there can be details of motion "overlooked" by an average velocity taken over a long time interval. In the comparison between Figs. 2-9 and 2-10, we see the increased detail visible when the average velocity is obtained for very short time intervals.

Exercises

1. The position of an automobile varies as shown in the table.

x (meters)	5	5	6	7$\frac{1}{2}$	10	14	18	22	24	25	25
t (seconds)	0	1	2	3	4	5	6	7	8	9	10

Find its average velocity between the following times: (*a*) 0–10 seconds, (*b*) 0–1 seconds, (*c*) 2–5 seconds, (*d*) 4–5 seconds, (*e*) 5–6 seconds, (*f*) 6–7 seconds, (*g*) 7–8 seconds, (*h*) 8–9 seconds.

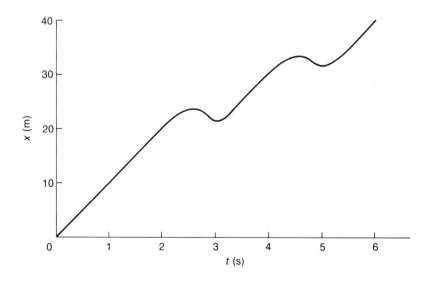

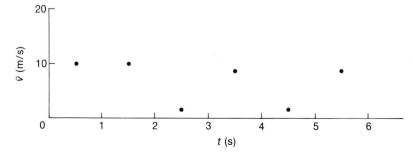

Fig. 2-9.   During each 1-s time interval, the average velocity is computed for the motion in the upper graph. The results are indicated below by points centered in each 1-s time interval.

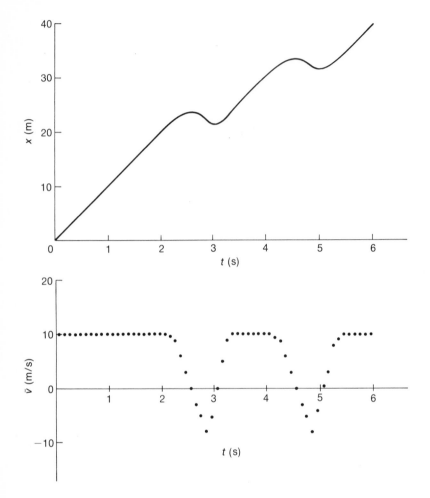

2. Let the coordinate x stand for the height of a ball above the floor. A graph of x against time is given in Fig. 2-11, from the time the ball is dropped until it rebounds to its original height. Find the average velocity of the ball between (a) $t = 0$ and $t = 0.6$ seconds, (b) $t = 0$ and $t = 0.2$ seconds, (c) $t = 0.4$ and $t = 0.5$ seconds, (d) $t = 0.5$ and $t = 0.6$ seconds.

3. Let x as a function of t be given by the equation $x = ct$, where c is a constant. Show that the average velocity between *any* two times t_1 and t_2 is equal to c.

2-3   Instantaneous Velocity

In the preceding section we saw that average velocities obtained for very short time intervals revealed details of motion that were

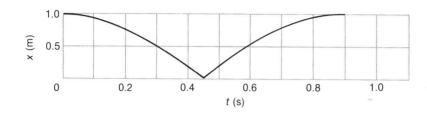

Fig. 2-11.

hidden when the velocity was averaged over a longer time interval. By choosing extremely short time intervals, the average velocity for a given interval is more nearly appropriate to a unique instant of time. We can, at least in thought, consider reducing the time interval indefinitely. The effect of ever-shortening time intervals on a graph with constant slope is indicated in Fig. 2-12. Consider the period between $t = 2$ s and $t = 6$ s:

$$\bar{v} = \frac{\Delta x}{\Delta t} = \frac{20 \text{ m}}{4 \text{ s}} = 5 \text{ m/s}$$

Similarly, for the intervals between 2 s and 4 s and then between 2 s and 3 s, we have

$$\bar{v} = \frac{10 \text{ m}}{2 \text{ s}} = 5 \text{ m/s}$$

$$\bar{v} = \frac{5 \text{ m}}{1 \text{ s}} = 5 \text{ m/s}$$

Though each time interval in this example is only half as large as the preceding one, the distance covered is also half as large, so that their ratio is unchanged. We can continue indefinitely to reduce the time interval Δt, and since the corresponding displacement Δx will reduce in proportion, we always obtain the same result for the veloc-

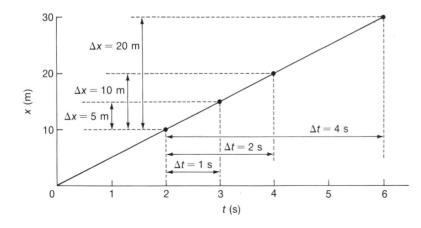

Fig. 2-12.   The effect of reducing time intervals when the slope is constant.

ity. If we let the time interval diminish to an *arbitrarily short* time, we call the velocity so obtained the *instantaneous* velocity. To make our definition of instantaneous velocity more concise we write:

$$v = \lim_{\Delta t \to 0} \frac{\Delta x}{\Delta t} \qquad (2\text{-}2)$$

The symbol v without a bar over it is our notation for the instantaneous velocity and the expression

$$\lim_{\Delta t \to 0} \frac{\Delta x}{\Delta t}$$

is our notation for the above-described process of reduction to an arbitrarily short time interval. The notation is read, "the limit of Δx divided by Δt as Δt goes to zero."

In the event that x as a function of time does *not* have constant slope, the limiting process we have described leads to the concept of the slope of a curve *at a point*. A curve with varying slope is illustrated in Fig. 2-13. If we find the average slope (hence the average velocity) between the times t_1 and t_4, it is equal to the slope of the straight line A that connects those two points. Finding the average velocity between times t_1 and t_3 corresponds to finding the slope of the straight line labeled B in Fig. 2-13. As the time interval beginning at t_1 is made shorter and shorter, the slope of the straight line segment shown approaches more and more closely to the in-

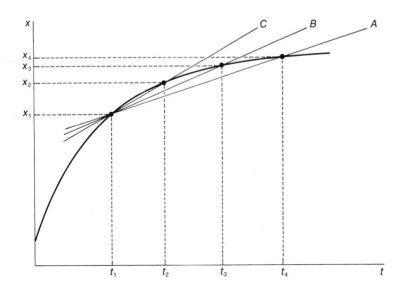

Fig. 2-13. The time intervals $t_4 - t_1$, $t_3 - t_1$, and $t_2 - t_1$ all begin at t_1. Each successively smaller interval defines an average velocity that is closer to the value of the instantaneous velocity at the point x_1, t_1.

stantaneous velocity at point x_1,t_1. In the limiting case, as Δt approaches zero, the slope so defined is that of a straight line *tangent* to the curve at point x_1,t_1. This tangent is shown in Fig. 2-14. The length of the tangent drawn is immaterial since the slope of a straight line is independent of its length.

Let us review the preceding sections. First, we defined the average velocity to be

$$\bar{v} = \frac{\Delta x}{\Delta t} = \frac{x_2 - x_1}{t_2 - t_1} \tag{2-3}$$

The value of $\bar{v}$ also equals the *slope* of the straight line connecting the two points x_1,t_1 and x_2,t_2. Second, we defined the instantaneous velocity to be

$$v = \lim_{\Delta t \to 0} \frac{\Delta x}{\Delta t} \tag{2-4}$$

This limit is equal to the slope of the straight line that is tangent to the x-vs-t curve at the point of interest. This concept of slope at a point is the heart of differential calculus.

The statement of (2-4) is usually abbreviated by the notation

$$\frac{dx}{dt} = \lim_{\Delta t \to 0} \frac{\Delta x}{\Delta t}$$

so that (2-4) can be rewritten

$$v = \frac{dx}{dt} \tag{2-5}$$

Equation (2-5) is read, "v equals the *derivative* of x with respect to t."

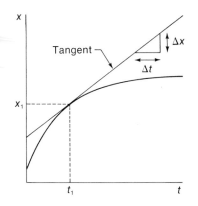

Fig. 2-14. The straight line above is tangent to the curve shown in Fig. 2-13 at the point x_1,t_1. The slope of this tangent is equal to the value of the instantaneous velocity at the point x_1,t_1. This slope is determined from the ratio $\Delta x/\Delta t$ for any portion of the tangent line.

2-4 Graphical Differentiation

The velocity at a given time is equal to the derivative of x with respect to t at that time. Since the value of this derivative equals the slope of the tangent to the curve, we can simply lay a ruler tangent to the curve and measure the slope of the ruler to obtain the velocity.

Example Consider the motion of an automobile between two stop signs 100 meters apart. Let the x coordinate be the distance of the auto from the first stop sign. A graph of the x coordinate versus time is shown in Figure 2-15. First, let us look over the graph of x vs t to find its main features. These are:

1. The graph begins, at $t = 0$, at the first stop sign ($x = 0$ meters) and ends, at $t = 10$ seconds, at the second stop sign ($x = 100$ meters).

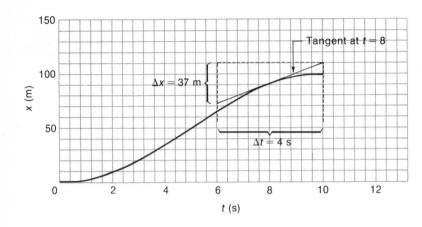

Fig. 2-15. The position of an automobile as a function of time. An example of graphical determination of velocity is shown at $t = 8$ s. Note that the length of the tangent line is immaterial, as the ratio $\Delta x/\Delta t$ for that tangent line determines the velocity.

2. Since 100 meters are covered in 10 seconds, the *average* velocity over the entire interval is 10 meters/second.

3. The slope of the curve is zero at the beginning ($t = 0$) and end ($t = 10$ s) of the graph, hence the velocity is zero at those two times.

4. The slope of the curve is steepest between $t = 5$ and 6 seconds, hence this is the time when the car is moving most rapidly.

Now we will find the instantaneous velocity at a sufficient number of points to see how the velocity varies during the 10-second time interval. As an example, consider the point at $t = 8$ seconds. As seen in Fig. 2-15, the *tangent* at $t = 8$ s has a slope given by

$$\frac{\Delta x}{\Delta t} = \frac{37 \text{ m}}{4 \text{ s}} = 9.25 \text{ m/s}$$

This slope is the value of the derivative of x with respect to t at that point and hence is equal to the instantaneous velocity there. A graph of the results of this procedure for many points is shown in Fig. 2-16. With a smoothly varying function, such as the example we are considering, the number of points shown in Fig. 2-16 is sufficient to guide

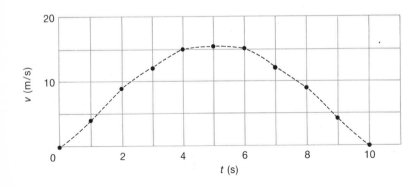

Fig. 2-16. Velocity derived from the position-vs-time curve of Fig. 2-15.

us in drawing a complete curve for *v* as a function of *t*. The dotted curve is a guess at the shape of this curve for our example. If more information is desired between the points we obtained, the derivative can be determined at a greater number of points. The accuracy of the velocities we obtain in this manner is limited by our ability to estimate whether or not a straightedge is tangent to the curve.

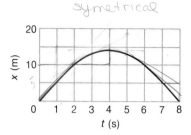

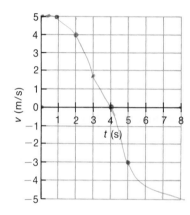

Exercises

4. A displacement-vs-time graph is given in Fig. 2-17. Use a straightedge (a folded piece of paper will do) as a tangent to determine the instantaneous velocity at $t = 1, 2, 3, 4, 5, 6$, and 7 seconds. Plot the points that you obtain on the lower graph in Fig. 2-17.

5. Two displacement-vs-time curves are graphed in Fig. 2-18. Use a straightedge to measure the slope of these curves at a sufficient number of points to enable you to sketch a complete graph of the velocity as a function of time for each case. Be sure to indicate the units and numerical scale of your graphs. Can you understand why the same result was obtained for both cases?

2-5 Algebraic Functions

If we are given an algebraic statement of the dependence of *x* on *t*, such as $x = t^2$, we can use the foregoing graphical method to obtain the derivative of the function. However, it is usually more convenient to apply the definition of the derivative in such a case. Since the definition of the derivative involves letting Δt approach zero,

Fig. 2-17.

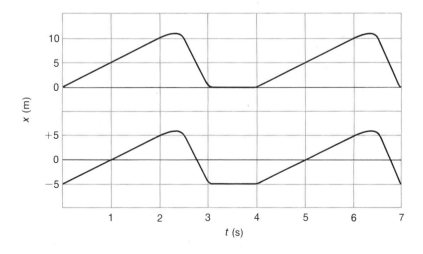

Fig. 2-18.

we want to write $\Delta x/\Delta t$ in such a way that we can find the limit of this ratio as Δt approaches zero.

Let us proceed as follows: Suppose $x = t^2$; then we can in principle find x at any time t we wish, simply by squaring t. To find the instantaneous velocity at some particular time t, we can calculate the value x will have at the slightly later time, $t + \Delta t$, where Δt is a small time interval, and subtract the value x had at the original time t. The amount by which x changed, during the time interval Δt, is then given by

$$\begin{aligned}
\Delta x &= x(t + \Delta t) - x(t) \\
&= (t + \Delta t)^2 - t^2 \\
&= t^2 + 2t\Delta t + (\Delta t)^2 - t^2 \\
&= 2t\Delta t + (\Delta t)^2
\end{aligned}$$

Remember that $x(t)$ means the value of x at time t, so $x(t + \Delta t)$ means the value of x at time $t + \Delta t$.

Next we form the ratio $\Delta x/\Delta t$ by dividing through by Δt:

$$\frac{\Delta x}{\Delta t} = 2t + \Delta t$$

Now it is clear that

$$\lim_{\Delta t \to 0} \frac{\Delta x}{\Delta t} = 2t$$

So we have the result that when $x = t^2$, $dx/dt = 2t$.* Look at the graph of this function $x = t^2$ in Fig. 2-19 to see if this result is reasonable. The solid curve is the function $x = t^2$ and, indeed, it does have a negative slope for negative values of t, zero slope at $t = 0$, and an increasing positive slope as t increases in the positive direction. Our result for the derivative is shown as a dashed line and exhibits all of these properties.

Suppose in the preceding discussion we had considered the function $x = Bt^2$, where B is a constant, rather than $x = t^2$. The reader can quickly verify, by repeating a similar calculation, that we now obtain:

$$dx/dt = 2Bt$$

The effect of such a multiplying constant is illustrated in the comparison of Figs. 2-19 and 2-20. In Fig. 2-20 the function is $x = 5t^2$, so that its derivative is $dx/dt = 10t$. Generally, a constant factor

---

*Although the result obtained for the limit as Δt approaches zero in this case is obvious, such is not always the case. A considerable portion of a calculus course is justifiably spent in developing the *general* behavior of limits. For our present purposes, elementary ideas will suffice.

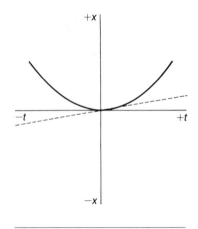

Fig. 2-19.   A graph of the function $x = t^2$ is shown as a solid curve. The dashed line is a graph of $2t$, which is the derivative of x with respect to t.

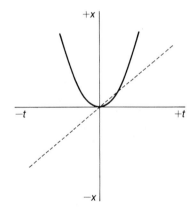

Fig. 2-20.   The solid curve is a graph of $x = 5t^2$, which is plotted on the same scale as Fig. 2-19. Multiplication by the constant factor of 5 has made the function steeper at any given value of t. The derivative, $dx/dt = 10t$, is shown as a dashed line and appropriately has larger values than the derivative in Fig. 2-19.

that multiplies a function simply appears as a multiplicative factor in the derivative as well. The general rule is

$$\frac{d}{dt} \text{constant} \cdot f(t) = \text{constant} \cdot \frac{df(t)}{dt} \qquad (2\text{-}6)$$

In either example above, note that the process of differentiation produces a new function, the *derived* function. If our original function is represented by $f(t)$, it is common notation to represent the derived function by $f'(t)$. Equation (2-6) can thus be rewritten

$$\frac{d}{dt} \text{constant} \cdot f(t) = \text{constant} \cdot f'(t) \qquad (2\text{-}7)$$

In an introductory calculus course it would be shown that the result $dx/dt = 2t$ when $x = t^2$ is a special case of a more general rule:

If $x = t^n$ then $dx/dt = nt^{n-1}$. Further, applying (2-7) we find

$$\frac{dx}{dt} = nBt^{n-1} \qquad \text{when} \qquad x = Bt^n \qquad (2\text{-}8)$$

Application of this rule to the function

$$x = 0.5t^3$$

gives the result

$$\frac{dx}{dt} = 3(0.5)t^2 = 1.5t^2$$

and is graphed in Fig. 2-21.

The derivative of a constant is zero, since the derivative of a function has a nonzero value only where the function is *changing*. If we recall that $t^0 = 1$, we note that we can write a constant, say C, as $C = Ct^0$. Applying (2-8) we find

$$\frac{dC}{dt} = 0t^{-1} = 0$$

One more useful result that can be obtained by application of the limiting process is that *the derivative of a sum is simply the sum of the derivatives*. More briefly,

$$\frac{d}{dt}[f(t) + g(t)] = f'(t) + g'(t) \qquad (2\text{-}9)$$

Example   Suppose the value of x is given by the equation

$$x = a + bt + ct^2 + et^3$$

where a, b, c, and e are all constants. Find v.

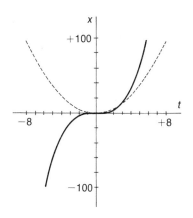

Fig. 2-21.   The function $x = 0.5t^3$ is shown as a solid curve. Its derivative, according to Eq. (2-8), is $dx/dt = 1.5t^2$ and is shown as a dashed curve. Note that this function is negative for negative t but the *slope* is positive there.

Solution   We wish to find v, which is given by

$$v = \frac{dx}{dt}$$

Using the relations (2-8) and (2-9) above:

$$\frac{dx}{dt} = \frac{d(a)}{dt} + \frac{d(bt)}{dt} + \frac{d(ct^2)}{dt} + \frac{d(et^3)}{dt}$$

$$= 0 + b + 2ct + 3et^2$$

In this example, differentiation of a polynomial of third degree reduced it to a second-degree polynomial. Does this make sense? A glance at the graphs of these functions tells us it does. In Fig. 2-22 we see a graph of a third-degree polynomial. The number of "bends" made by such a function can be deduced if we recall that a third-order polynomial equation has up to three real roots. Note that the slope of this function is positive at the left side of the graph, then negative near the center, and finally positive for large positive values of t. Thus, the derivative of this third-order polynomial should look like the function graphed as a dashed line—a parabola that has two real roots and is the graph of a second-degree polynomial.

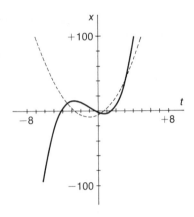

Fig. 2-22.   The function $x = -2 - 5t + 3t^2 + t^3$ is shown as a solid curve. The derivative, $dx/dt = -5 + 6t + 3t^2$, is shown as a dashed curve. Note that this derivative passes through zero value at both of the times where a straightedge tangent to the solid curve has zero slope.

Exercises

6. Use the limiting technique of Section 2-5 to find the derivative of the function $x = ct$, where c is a constant.

7. Use the limiting technique of Section 2-5 to find the derivative of the function $x = -kt^2$, where k is a positive constant. Sketch a crude graph of this function and another of your result for its derivative. Does the graphical procedure for differentiation agree with your result? What are the dimensions of the constant k?

8. Let $x = 16t^5$. Apply (2-8) to find dx/dt.

9. If $x = a + bt + ct^2$, find dx/dt.

10. The position of an automobile is given by

$$x = 10 \text{ mi} + 5 \frac{\text{mi}}{\text{s}^2} (t^2)$$

where t is given in seconds. Differentiate to obtain v as a function of t. Find v at $t = 3$ s.

2-6   Motion with Constant Velocity

Motion in which the velocity is constant is particularly simple. A constant velocity means that the displacement is changing uniformly. This means that no matter what time interval we consider, and no matter how long a time interval we consider, the ratio $\Delta x/\Delta t$ will be constant. The value of that ratio is equal to the velocity and has the units length divided by time, such as meters/second or miles/hour. This situation can be illustrated with a graph, as in Fig. 2-23.

The particular time interval marked off in Fig. 2-23 is two hours long and the displacement changes by 50 miles during that time. The velocity is thus 50 miles/2 hours = 25 mi/hr. The reader should convince himself that any other time interval could have been chosen with the same result. In particular, a vanishingly small time interval could be chosen with no change in the ratio $\Delta x/\Delta t$, hence the instantaneous velocity, dx/dt, and the average velocity, $\Delta x/\Delta t$, are the same for this special case of constant velocity (and *only* for this special case).

Motion with constant velocity can also be described algebraically. Writing the equation of the line graphed in Fig. 2-23, we have

$$x = v_0 t + x_0 \qquad (2\text{-}10)$$

where v_0 is a constant whose value gives the slope of the line (and hence is equal to the velocity) and x_0 is a constant equal to the value of the displacement at $t = 0$. For the case of Fig. 2-23 the constants have the values $v_0 = 25$ mi/hr and $x_0 = 10$ mi.

Example The displacement of an object from some fixed origin is given by the equation $x = v_0 t + x_0$, where $v_0 = -15$ m/s and $x_0 = 30$ m. (a) At what time is the displacement equal to zero? (b) Graph the displacement as a function of time.

Solution (a) At $t = 0$, v_0 gives the slope, which is negative, and x_0 gives the displacement, which is positive. As time increases from zero,

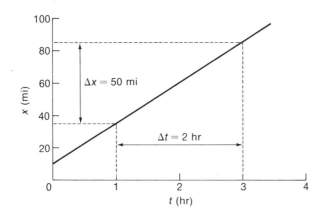

Fig. 2-23. Position-vs-time graph for constant velocity.

the displacement *decreases* (negative slope) and hence must pass through zero at some time. Let us call that time t_0. In general, $x = v_0 t + x_0$. Hence, since $x = 0$ at the time t_0,

$$0 = v_0 t_0 + x_0$$

Thus,

$$t_0 = \frac{-x_0}{v_0} = \frac{-30 \text{ m}}{-15 \text{ m/s}} = 2 \text{ s}$$

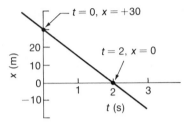

(*b*) We now know that at $t = 0$, $x = 30$ m and at $t = +2$ seconds, $x = 0$. These two points suffice to determine the straight line shown in Fig. 2-24. This straight line represents the displacement as a function of time as long as the velocity remains constant.

Fig. 2-24.

   Thus far, we have proceeded from a known position-vs-time relationship to find the corresponding velocity. What about the inverse question? Given the velocity (as for example, from an automobile speedometer), what can we say about the position? To find the displacement we can use the definition of average velocity:

$$\bar{v} = \frac{\Delta x}{\Delta t}$$

Transposing, $\Delta x = \bar{v} \, \Delta t$. So knowledge of the average velocity allows us to compute the changes of position (displacement) for a given time interval. In the case of motion with constant velocity, the average velocity and the velocity are equal.

Example   An automobile moves with constant velocity, $v = 30$ miles/hour. At $t = 0$ the auto is at a position of $x = x_0 = -10$ miles. What is its position at $t = 2$ hours?

Solution   The change in position, Δx, is given by $\Delta x = \bar{v} \, \Delta t = 30$ mi/hr $\cdot$ 2 hr $= +60$ mi. Thus $x = x_0 + \Delta x = -10$ mi $+ 60$ mi $= +50$ mi.

Exercises

11.   An automobile moves along the x axis of a coordinate system with constant velocity. Write an equation that gives x as a function of t for each of the following cases. Sketch a crude graph of x vs t for each case. (*a*) Its velocity equals 10 mi/hr and its position at $t = 0$ is $x = 0$. (*b*) Its velocity equals 10 mi/hr and its position at $t = 0$ is $x = -10$ mi. (*c*) Its velocity equals -10 mi/hr and its position at $t = 0$ is $x = 10$ mi.

12. A table of position versus time is given:

x (feet)	t (seconds)
−11	0
− 7	$\frac{1}{2}$
− 3	1
+ 1	$1\frac{1}{2}$
+ 5	2
+ 9	$2\frac{1}{2}$
+13	3
+17	$3\frac{1}{2}$

(*a*) On the basis of this information, is the velocity uniform? (*b*) Is it possible that the motion is not that of uniform velocity? Explain. (*c*) If the motion is that of uniform velocity, what is the value of that velocity?

13. The positions of two bicycles are given as follows:

$$\text{bicycle 1: } x_1 = x_0 + v_1 t$$
$$\text{bicycle 2: } x_2 = v_2 t$$

where v_1 and v_2 are constant velocities. (*a*) Find an expression for the time t_p at which one bicycle passes the other. This is most easily done by setting x_1 equal to x_2 and solving for t. (*b*) For $x_0 = 10$ km, $v_1 = 10$ km/hr, and $v_2 = 20$ km/hr, what is the value of t_p in hours?

2-7 The Antiderivative or Indefinite Integral

Given x as a function of t, we can differentiate this function to find the velocity as a function of t. In more compact form we could write:

$$\text{Given} \quad x = f(t)$$

$$\text{then} \quad v = f'(t)$$

But what about the inverse question: Given velocity as a function of time, how do we find the position?

First, let us note that knowledge of velocity alone cannot completely determine the position. If we set out at $t = 0$ and drive at 50 mi/hr for 2 hours, all we know is that the *change* in position is 100 mi. We must be given the starting point as well as the velocity to obtain our final position. This point may seem trivial, yet its mathematical consequences are not. Knowledge of x as a function of time completely determines dx/dt. Knowledge of dx/dt as a function of time does *not* determine x. We require an additional piece of information, the initial value of x.

As an illustration, consider

$$dx/dt = At^n$$

Recalling the rule for differentiation of this function (Eq. (2-8)), it does not require a great deal of guessing to deduce that

$$x = \frac{A}{n+1} t^{n+1}$$

would give

$$dx/dt = At^n$$

upon differentiation (except for the troublesome case of $n = -1$). But so would

$$x = \frac{A}{n+1} t^{n+1} + C$$

where C is a constant of any value whatsoever. The following notation is generally used to describe all of this: if

$$v = dx/dt \quad \text{then} \quad x = \int v \, dt + C$$

where the symbol $\int (\ ) \, dt$ stands for the operation that is the inverse of differentiation. This operation is called *integration,* and the term

$$\int v \, dt$$

is called the *antiderivative* of v or the *indefinite integral* of v.

Let us use this notation to describe the preceding example. In that example we were given v in the form of a power law and found:

$$x = \int v \, dt + C = \frac{A}{n+1} t^{n+1} + C \qquad (2\text{-}11)$$

when $\quad v = At^n \quad (n \neq -1)$

When x and v are position and velocity, the constant C is simply the initial position, while the indefinite integral gives the *change* in position. In general:

$$x = \int v \, dt + x_0 \qquad (2\text{-}12)$$

Example   Given that $v = At^2$ and the initial position is x_0, find x as a function of time. For the specific case of $A = 6$ m/s³ and $x_0 = -200$ m, find x at $t = +10$ s.

Solution $$x = \int v \, dt + x_0 = \int (At^2) \, dt + x_0$$

By use of (2-11) we find

$$x = \frac{A}{3} t^3 + x_0$$

which gives x as a function of time. Now for the specific case we simply substitute the values given:

$$x(t = +10 \text{ s}) = \frac{6 \text{ m/s}^3}{3} \cdot (10 \text{ s})^3 - 200 \text{ m}$$

$$= 2000 \text{ m} - 200 \text{ m} = 1800 \text{ m}$$

Exercises

14. If $f'(t) = t^3$, what is $f(t)$?

15. Given $v = At^2 + Bt^3$, can you guess the correct expression for $\int v \, dt$? Check by differentiating your result to see if you obtain the given expression for the velocity.

16. The readings of an automobile speedometer are found to obey a simple law:

$$v = 10 \, \frac{\text{km}}{\text{hr}^2} \, t$$

where t is given in hours. Find the position of the auto at $t = 3$ hr, given $x_0 = 20$ km.

2-8   The Definite Integral

The notation that we use for the indefinite integral seems arbitrary and cumbersome. It has its origin in the idea of the definite integral. We can understand the notation and gain insight into the meaning of the integral if we examine the graph of v vs t shown in Fig. 2-25. Since v is not constant over the time interval between t_1 and t_6, we *cannot* simply write

$$\Delta x = v \Delta t = v(t_6 - t_1)$$

If we break this time interval into short intervals, the velocity will not vary greatly during any one of these short intervals. Thus the increment Δx that is gained during a short time Δt is approximately

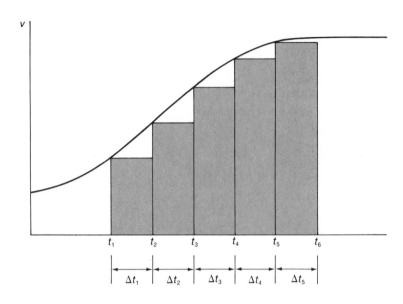

Fig. 2-25.   On this plot of v vs t, the net displacement between t_1 and t_6 is approximately given by the stairstep shaded area. Note that the *dimensions* of the area on this graph are velocity multiplied by time, i.e., length.

equal to the velocity at the beginning of the time interval multiplied by the length of the time interval:

$$\Delta x_1 \approx v_1 \Delta t_1$$

$$\Delta x_2 \approx v_2 \Delta t_2$$

and so on. The total distance covered is then approximately equal to the sum of all these increments:

$$\Delta x \approx v_1 \Delta t_1 + v_2 \Delta t_2 + v_3 \Delta t_3 + v_4 \Delta t_4 + v_5 \Delta t_5$$

The value of $v_1 \Delta t_1$ is equal to the *area* of the rectangle directly over Δt_1, as shown in the graph of Fig. 2-25. Our approximate result for Δx is thus equal to the sum of the areas of all the rectangular strips in Fig. 2-25.

How can we make our approximation more exact? The error in our approximation is caused by the fact that v is *not* constant over each interval Δt. Let us make Δt *smaller* so that the velocity can vary less during Δt. In Fig. 2-26, intervals of half the original size are chosen and the stairstep area enclosed by the rectangles is closer to the exact distance covered between t_1 and t_{12}. If we imagine ever smaller increments Δt, it is clear that our approximation becomes more and more exact, since v changes so little during any one Δt. However, the number of strips whose area must be summed becomes very large as the number of subdivisions is increased. In the limit of *arbitrarily small* intervals Δt, the stairstep area becomes exactly the entire area bounded by the v-vs-t curve,

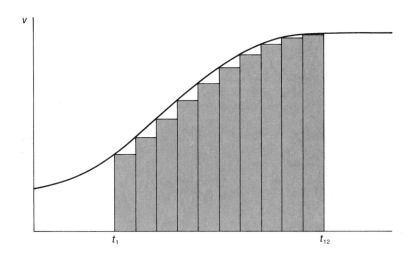

t_1 t_{12}

Fig. 2-26. A closer approximation to the total displacement is
given by halving the increments
Δt that were used in Fig. 2-25.
The approximation can be further
improved by making each Δt
even smaller.

the t axis, and the vertical lines at t_{initial} and t_{final} as shown in Fig.
2-27. In this limit of vanishingly small time intervals, our approxima-
tion becomes exact. This is written as follows:

$$\Delta x = \lim_{\Delta t \to 0} \sum_{i=1}^{N} v_i \Delta t_i \qquad (2\text{-}13)$$

The notation

$$\sum_{i=1}^{N} v_i \Delta t_i$$

stands for

$$v_1 \Delta t_1 + v_2 \Delta t_2 + v_3 \Delta t_3 + \cdots + v_N \Delta t_N$$

The number N is given by

$$N = \frac{t_{\text{final}} - t_{\text{initial}}}{\Delta t}$$

if each small increment is of the same size. As $\Delta t \to 0$, the number
N grows without bound, i.e., $N \to \infty$. Standard notation for an
increment Δt that is allowed to become arbitrarily small is dt and is
called an *infinitesimal*. Thus the modern notation that has evolved is

$$\Delta x = \lim_{\Delta t \to 0} \sum_{i=1}^{N} v_i \Delta t_i = \int_{t_{\text{initial}}}^{t_{\text{final}}} v\, dt \qquad (2\text{-}14)$$

where an integral between two specific endpoints is called a *definite
integral*. The two endpoints, t_{final} and t_{initial}, are called the *limits*
of the definite integral.

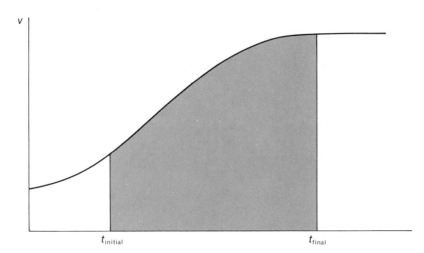

Fig. 2-27. The net displace-
ment between $t_{initial}$ and t_{final} is
given exactly by the complete
shaded area under the velocity-
time curve.

We immediately see two methods of calculating the value of
this definite integral. One method, which generally involves the use
of a computer because of the tedium involved, is actually to break
up the time interval between the two limits into a very large number
of intervals and carry out the sum (2-14) for a value of N that is
not infinite, but large enough for the desired accuracy. This technique is called *numerical integration*. The second method is to
measure the area under the graph in Fig. 2-27. There are many
ingenious ways to do this, but perhaps one illustration will suffice.
We can simply cut out the area with a pair of scissors and weigh it
on an analytical balance. Comparison of this weight to the weight
of a standard square of the same type of graph paper with a known
area then gives us the desired result. Problem 2-21 at the end of the
chapter illustrates this method.

Comparison of (2-12) and (2-14) leads to a third method, which
is perhaps the most exciting possibility of all and is of sufficient
importance to be called *the fundamental theorem of calculus:*

$$x = \int v\, dt + x_0 \qquad (2\text{-}12)$$

$$\Delta x = \int_{t_{initial}}^{t_{final}} v\, dt \qquad (2\text{-}14)$$

Comparison of (2-12) and (2-14) can be facilitated by using (2-12)
to find the same value of Δx given by (2-14). We do this by finding
both x_{final} (at t_{final}) and $x_{initial}$ (at $t_{initial}$) by use of (2-12); then we
find Δx as follows:

$$x_{\text{final}} = \int v\, dt \,(\text{evaluated at } t_{\text{final}}) + x_0$$

$$x_{\text{initial}} = \int v\, dt \,(\text{evaluated at } t_{\text{initial}}) + x_0$$

Finally, for the difference $x_f - x_i$ we have

$$\Delta x = x_f - x_i = \int v\, dt \,(\text{at } t_f) - \int v\, dt \,(\text{at } t_i) \qquad (2\text{-}15)$$

But we also know that Δx is given by the definite integral given in (2-14). Equating Δx from (2-14) to that from (2-15), we find

$$\int_{t_i}^{t_f} v\, dt = \int v\, dt \,(\text{at } t_f) - \int v\, dt \,(\text{at } t_i) \qquad (2\text{-}16)$$

The value of the *definite* integral of v on the left side of (2-16) equals the difference in the value of the *indefinite* integrals of v evaluated at the initial and final limits of the definite integral.

This connection between definite and indefinite integrals is the fundamental theorem of calculus. Its importance lies in the fact that indefinite integrals are antiderivatives, so that we now have a powerful means for computing definite integrals of algebraic functions.

Example   Suppose that a rocket moves according to the law

$$v = Kt^2 \qquad (2\text{-}17)$$

where K is a constant. (a) How far does it move between the two times t_i and t_f? (b) If $K = 24$ ft/s³, $t_i = 2$ s, and $t_f = 4$ s, how far does it move?

Solution   As we have seen, the distance it moves is given by the definite integral

$$\Delta x = \int_{t_i}^{t_f} v\, dt = \int_{t_i}^{t_f} Kt^2\, dt$$

By the fundamental theorem of calculus, Eq. (2-16), we have

$$\Delta x = \int v\, dt \,(\text{at } t = t_f) - \int v\, dt \,(\text{at } t = t_i)$$

Since $\int v\, dt$ is the antiderivative of v, we can easily deduce it:

$$\int Kt^2\, dt = \frac{K}{3} t^3 + C \qquad (2\text{-}18)$$

since $(K/3)t^3 + C$ does give Kt^2 upon differentiation. (a) Now we can evaluate (2-16) by use of (2-18):

$$\Delta x = \left[\frac{K}{3} t_f^3 + C \right] - \left[\frac{K}{3} t_i^3 + C \right] = \frac{K}{3} [t_f^3 - t_i^3]$$

(b) Substituting the values given into the answer to part (a), we find $\Delta x = 24 \text{ ft/s}^3 [(4 \text{ s})^3 - (2 \text{ s})^3] = 1344 \text{ ft}$. The area that corresponds to the value of this definite integral is shown in Fig. 2-28.

Example   The rate at which water is delivered to a reservoir is graphed in Fig. 2-29. Find the total quantity of water, in gallons, delivered to the reservoir between $t = 1$ and $t = 3$ hours.

Solution   If the flow were constant at R gal/hr, then $R \cdot \Delta t$ would give the total number of gallons delivered. But the flow varies, just as the velocity varied in our discussion of the definite integral. Thus, finding the total quantity of water supplied between $t = 1$ and $t = 3$ hours amounts to finding the area under the curve between those limits. Note that the dimensions of this area (hr × gal/hr = gal) are those of a quantity of water. The integral shown in Fig. 2-29 can be approximately evaluated with an extremely simple method. We simply count the squares of graph paper that comprise the shaded area. Of course, some of the squares are partly inside and partly outside the area, so our result will not be completely accurate, but the error can be made smaller by making the graph paper squares smaller. In this particular case the shaded area contains approximately 382 small squares. Each square has a value of 1000 gal/hr × 1/10 hr = 100 gal (note the ordinate is marked off in 10^4 gal/hr major divisions). Thus the total quantity delivered between $t = 1$ and $t = 3$ hours is 38,200 gal.

Example   Find the antiderivative of the function $v = c_1 t + c_2$, where c_1 and c_2 are constants.

Solution   What function, when differentiated, gives the above expression? A little educated guessing is required. We know, for example, that if we differentiate $c_1 t^2$ we obtain $2c_1 t$, not $c_1 t$, as is desired. Hence

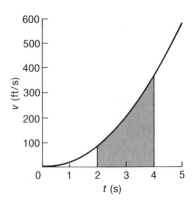

Fig. 2-28.   The equation $v = 24t^2$ is graphed above. The shaded area is equal to the value of the definite integral, which gives the displacement between $t = 2$ s and $t = 4$ s. This value is found by algebraic means in the accompanying example.

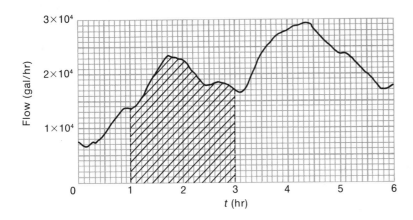

Fig. 2-29.   Flow rate for a water supply system.

the antiderivative we are after must contain the term $\frac{1}{2}c_1 t^2$. Proceeding in this fashion, we find

$$x = \tfrac{1}{2}c_1 t^2 + c_2 t + c_3$$

will do the job. The reader can verify this by differentiation, using (2-8) and (2-9).

Exercises

17. The velocity of a bicycle (for a short time) is given by

$$v = Kt \qquad\qquad (2\text{-}19)$$

where K is equal to 3 ft/s² when t is given in seconds. Find the change in position of the bicycle between $t = 1$ and $t = 5$ seconds by finding the antiderivative of (2-19) and using it to evaluate

$$\Delta x = \int_{t_i}^{t_f} v\, dt$$

18. Fig. 2-30 gives a velocity as a function of time. Find Δx between $t = 1$ and $t = 5$ seconds by counting squares to obtain the required area. Can you see an easier way to find the area involved by using the procedure of Exercise 17?

2-9   Summary

The position of an object in three dimensions is conveniently given by its x, y, and z coordinates in a Cartesian coordinate system. If

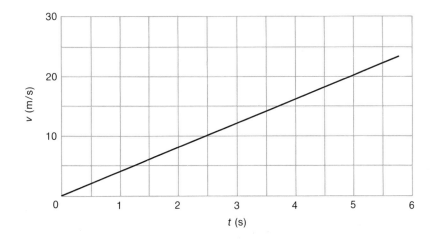

Fig. 2-30.

any of these coordinates are changing with time, we say the object is in motion. In one dimension, say x, we define the velocity thus:

$$v = dx/dt$$

The derivative, dx/dt, can be obtained in two ways.

1. Graphically: The function of interest, $x = f(t)$, can be graphed. The slope of a tangent to the graphed curve is the value of the derivative of the function at the point of contact between the tangent and the curve.

2. Algebraically: The fundamental definition of a derivative,

$$\frac{dx}{dt} = \lim_{\Delta t \to 0} \frac{\Delta x}{\Delta t},$$

may be used if the function can be written in algebraic form so that the manipulative power of algebra may be used.

The process that is inverse to differentiation is integration. Symbolically we may say

If $F = \displaystyle\int f(t)\, dt$ then $dF/dt = f(t)$.

The indefinite integral (or antiderivative) contains an arbitrary constant, since the derivative from which it is obtained contains information concerning only the *slope* of a function, not the actual function value. The fundamental theorem of calculus tells us that a definite integral is simply the difference between two indefinite integrals evaluated at two specific values of the independent variable. These two values are called the limits of the integral. A definite integral is equal to the area under the $f(t)$-vs-t curve between the limits t_i and t_f.

$$\text{area} = \int_{t_i}^{t_f} f(t)\, dt = \int f(t)\, dt \ [\text{at } t_f] - \int f(t)\, dt \ [\text{at } t_i]$$

Indefinite integrals can be obtained by educated guessing, i.e., "What function, when differentiated, gives back my original function?" Definite integrals can be evaluated by directly determining the appropriate area on a graph or by taking the difference of the indefinite integrals evaluated at the limits.

Problems

2-1. World records for running several different distances are given below. Calculate the average velocity over the entire time interval for each event.

Distance (m)	50	60	100	200	400	800	1000
Time (s)	5.5	6.5	9.9	19.5	44.5	104.3	136.2

2-2. The distance of a falling object from its starting point is tabulated below.

Distance (m)	0.0	4.9	19.6	44.1	78.4	122.5
Time (s)	0	.1	2	3	4	5

Find the average velocity, $\bar{v}$, during the time intervals $t = 0$ to $t = 1$ s, $t = 1$ s to $t = 2$ s, $t = 2$ s to $t = 3$ s, $t = 3$ s to $t = 4$ s, $t = 4$ s to $t = 5$ s, and $t = 0$ to $t = 5$ s.

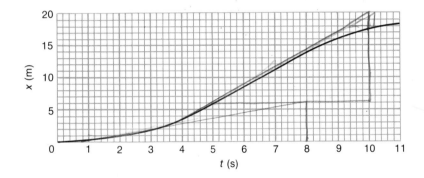

Fig. 2-31.

2-3. The position of an object is graphed in the sketch of Fig. 2-31. What is its average velocity in the time intervals $t = 2$ to $t = 3$ seconds, and $t = 5$ to $t = 7$ seconds?

2-4. A graph of x vs t is shown in Fig. 2-32. Find $\bar{v}$ between (*a*) 0 and 2 s, (*b*) 0 and 4 s, (*c*) 6 and 12 s, (*d*) 0 and 10 s.

2-5. A formula for x as a function of t is given by the equation $x = kt^2$, where k is a constant. Find the average velocity between the times (*a*) 0 and 1 second, (*b*) 1 and 2 seconds, (*c*) 2 and 3 seconds. (*d*) What dimensions should we associate with the constant k so that the equation $x = kt^2$ will balance dimensionally?

2-6. The graphs in Fig. 2-33 give position as a function of time. Sketch a qualitative graph of the velocity associated with each. Before proceeding, it is helpful to note those points where the slope of the position-vs-time curve is zero, where

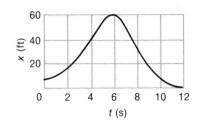

Fig. 2-32.

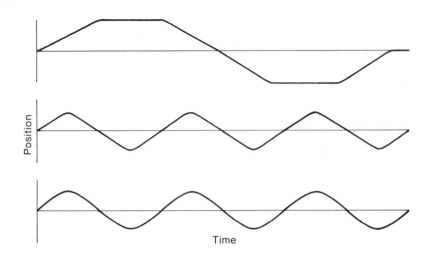

Fig. 2-33.

Position

Time

it has large positive values, and where it has large negative values.

2-7.   A function that has a maximum and minimum value within the given interval is graphed in Fig. 2-34. By placing a straightedge tangent to the curve at those two points, we see that the derivative of the function is zero there, since both tangents are horizontal. Can you sketch a function that has either a maximum or a minimum within an interval and does *not* have a vanishing derivative at the point of maximum or minimum?

2-8.   An object's position is always on the x axis and is given by $x = At^3 + Bt^2 + Ct + D$. (a) Find the times at which $v = 0$ (i.e., find the *roots of the equation* for v). (b) What conditions do the coefficients A, B, and C satisfy if: (1) there are two such times; (2) there is only one such time; (3) there are no such times.

2-9.   The position of a projectile is given by

$$x = v_0 t - Kt^2$$

where v_0 equals 250 m/s and t is measured in seconds. Given that the velocity of this projectile falls to a value of 150 m/s at $t = 5$ s, find the value of K. What are the units of K?

2-10.   Given that

$$x = At^3 - Bt^2$$

(a) differentiate to find an expression for the velocity. (b) If $A = 2$ ft/s^3 and $x = 200$ ft at $t = 5$ s, find the value of v at $t = 4$ s.

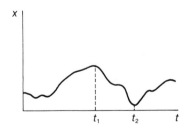

Fig. 2-34.

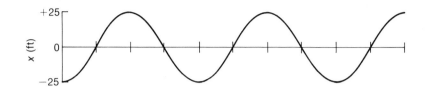

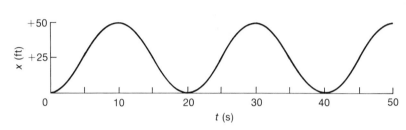

Fig. 2-35.

2-11. Two functions are graphed in Fig. 2-35. Use a straightedge to measure the slope of these curves at a sufficient number of points to enable you to sketch a complete graph of the instantaneous velocity versus time for each case. Why are your results the same in both cases?

2-12. An observer located near $x = 0$ sees two cars moving toward him from opposite directions along the x axis. At $t = 0$ one car is at $x = -500$ m and moving with a velocity of 10 m/s. The other car is at $x = 100$ m and has a velocity $v = -5$ m/s. If both velocities remain constant, (a) write equations for the positions, x_1 and x_2, of each car as a function of time. (b) When will the two cars collide? (c) Where will the two cars collide?

2-13. Two automobiles move along the x axis of a coordinate system with constant velocity. Car A has a velocity of $+20$ mi/hr and an initial position (the position at $t = 0$) of $+10$ mi. Car B has an initial position of -20 mi and a velocity of $+30$ mi/hr. At what time will car B pass car A?

2-14. Two vehicles are each moving with constant velocity, as shown in Fig. 2-36. (a) When will A overtake B? (b) How

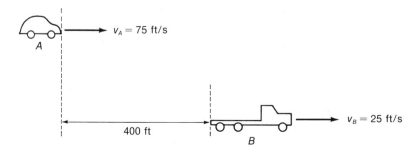

Fig. 2-36.

many feet will B have moved before being overtaken by A?
(c) When will A be 100 feet ahead of B?

2-15. Given $f'(t) = At + B$, find its antiderivative, $f(t) = \int f'(t)\, dt$.

2-16. Given that $f'(t) = 9\, t^2 + 4t$, find its antiderivative,

$$f(t) = \int f'(t)\, dt.$$

2-17. For the function $f(t)$ of the preceding problem, find the value of the definite integral

$$\int_{t=1}^{t=4} f'(t)\, dt.$$

2-18. Find the indefinite integral of the expression

$$v = (30 \text{ m/s}^3)t^2$$

2-19. Find the indefinite integral of

$$v = c_1 + c_2 t + c_3 t^2 + c_4 t^3$$

2-20. A graph of v as a function of time is given in Fig. 2-37. Assuming that $x_0 = 0$, integrate $v(t)$ by any means (counting squares under the curve between $t = 0$ and the times $t = 1$ s, $t = 2$ s, $t = 3$ s, etc., is probably the simplest) to obtain x at a sufficient number of points to sketch $x(t)$ on the lower part of the graph.

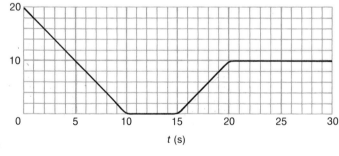

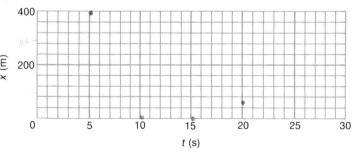

Fig. 2-37.

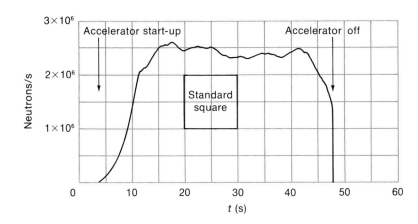

Fig. 2-38.

2-21.   A radio-chemist irradiates a target with neutrons in order to produce a radioactive isotope. The accelerator that produces the neutrons is not precisely stable, so with a neutron detector he monitors the number of neutrons impinging upon his target per second. The chart recording made by this instrument is shown in Fig. 2-38. The total number of neutrons during the entire bombardment is equal to the area under the curve. In order to obtain the value of this area, the chemist cuts out the area with a pair of scissors and weighs it on an analytical balance with a result of 345 mg. The standard square on the graph paper is also cut out and weighed. Its weight is 38 mg. Find the total number of neutrons that struck the target.

2-22.   The velocity of a rocket sled is closely approximated by the expression

$$v = 196 \,\frac{m}{s^2} \cdot t$$

for the first few seconds of its motion. If it starts from rest, find its displacement from the starting point at $t = \frac{1}{2}$ s, 1 s, $1\frac{1}{2}$ s, and 2 s.

Chapter **3**

Kinematics in One Dimension

3-1 Introduction Kinematics is that part of mechanics that describes motion, while dynamics "explains" motion. These words come from the Greek: *kinesis*, meaning motion, and *dynami*, meaning strength. In this chapter we will first describe motion in one dimension, and then Chapter 4 will introduce the dynamics of one-dimensional motion. The reader with previous preparation in calculus may choose to omit Chapter 2 and proceed directly to this chapter. The reader who has just completed Chapter 2 can skip over Sections 3-2 and 3-3 and go directly to Section 3-4.

3-2 Velocity

The position of an object lying on the x axis will be denoted by x. A change in position, $\Delta x = x_2 - x_1$, is called a *displacement*. If a displacement Δx occurs in a time $\Delta t = t_2 - t_1$, we define the average velocity as follows:

$$\bar{v} = \frac{\Delta x}{\Delta t}$$

Suppose that we are given the position as a function of time:

$$x = f(t)$$

The instantaneous velocity is then defined to be the derived function

$$v = f'(t) \tag{3-1}$$

Alternative notation for this definition is

$$v = \lim_{\Delta t \to 0} \frac{\Delta x}{\Delta t} = \frac{dx}{dt} \tag{3-2}$$

where dx/dt is the derivative of x with respect to t. Note that $f(t)$ may be given in a familiar algebraic form such as (a) below, or as a transcendental function such as (b) below, or in graphical form as in (c) below.

(a) $x = 10t^3 - 2t^2 + t$

(b) $x = A \sin(kt)$

(c)

In all these cases, the derivative, and hence the velocity, is well defined.* For case (c) above, the graphical procedure discussed in Section 2-4 is required to find the derivative.

If we are given the velocity as a function of time and wish to find the position, we perform *integration*, the operation that is inverse to differentiation:

$$x(t) = \int v(t) \, dt + \text{constant}$$

In this case the constant of integration is simply the value of x at $t = 0$, called x_0, while the antiderivative of v, $\int v(t) \, dt$, evaluated at times greater than $t = 0$ gives the additional displacement caused by the velocity. Thus, if we simply observe the speedometer of an automobile and continually integrate its reading, we can deduce how far we have moved. We do not know where we are, however, until we know our original position, x_0.

$$x(t) = \int v \, dt + x_0 \tag{3-3}$$

---

*It is certainly possible to devise pathological functions that are so discontinuous that it is not possible to define their derivative. Such functions are not encountered in our discussion of motion.

Alternatively, once we are given the velocity, we can phrase the process of finding the displacement in terms of the definite integral:

$$\Delta x = x_2 - x_1 = \int_{t_1}^{t_2} v\, dt \qquad (3\text{-}4)$$

Exercises

1. The position of a body varies according to the function $x = bt + ct^3$. (a) Find an expression for its velocity as a function of time. (b) If $b = 10$ m/s and $c = 3$ m/s^3, find its velocity at $t = 0$ and $t = 3$ s.

2. The velocity of a projectile is found to vary as $v = 40$ m/s $- (2$ m/s$^3)t^2$ for times between $t = 0$ and $t = 4$ s. Find the displacement, Δx, that occurs between $t = 1$ s and $t = 4$ s. *Suggestion:* Denote by b the value 40 m/s and by c the value 2 m/s^3 in your evaluation of the integral given by (3-4). Only as a last step substitute the values of b, c, and t into your solution.

3-3   Motion with Constant Velocity

The words "constant velocity" describe the derivative of position with respect to time. Expressed as an equation, $v = v_0 =$ constant. To find the behavior of the position itself, we simply integrate the velocity:

$$x = \int v\, dt = \int v_0\, dt = v_0 t + \text{constant}$$

In this case it is clear that the constant is simply the starting position, x_0:

$$x = x_0 + v_0 t \qquad (3\text{-}5)$$

As a check, we differentiate (3-5):

$$x = x_0 + v_0 t$$

$$v = \frac{dx}{dt} = 0 + v_0$$

Thus we obtain the originally assumed constant velocity.

3-4   Acceleration

In a given physical situation, the velocity of an object may or may not be constant. If its velocity is changing with time, we say the object is *accelerating*. A definition of acceleration that makes the foregoing precise is:

$$a = \frac{dv}{dt}$$ (3-6)

where a is the acceleration and dv/dt is the derivative of v with respect to t. We can also make the inverse statement:

$$v = \int a \, dt + C$$

usually written in the form

$$v = v_0 + \int a \, dt$$ (3-7)

Since $v(t)$ is the time derivative of x, we can rewrite the definition of acceleration:

$$a = \frac{dv}{dt} = \frac{d}{dt}\left(\frac{dx}{dt}\right) = \frac{d^2x}{dt^2}$$

where d^2x/dt^2 is read "the second derivative of x with respect to t." Alternative notation is

$$a = f''(t)$$

when $x = f(t)$.

As was the case for velocity, an average acceleration $\bar{a}$ can be defined for a specified time interval Δt:

$$\bar{a} = \frac{\Delta v}{\Delta t}$$

Example Find the acceleration of a body whose displacement is given in the upper part of Fig. 3-1.

Solution The slope of x vs t is zero during the first 1.6 seconds, then becomes positive between 1.6 and 6.4 seconds. The slope is again zero between 6.4 and 11.6 seconds, then becomes negative between 11.6 and 16 seconds. The resulting velocity-vs-time graph is shown in the second graph in Fig. 3-1. Now, to find the acceleration we find the slope of the velocity-vs-time graph. We see that the velocity is constant (slope = 0) between 0 and 1.6 seconds, between 2.4 and 5.6 seconds, and between 6.4 and 11.6 seconds. During these times the acceleration is zero. Where the slope of the v-vs-t graph is nonzero, the value of the acceleration is given by the value of that slope. The acceleration is shown in the lower portions of Fig. 3-1. Its value is constant during each of the three 0.8-s intervals in which the velocity was changing.

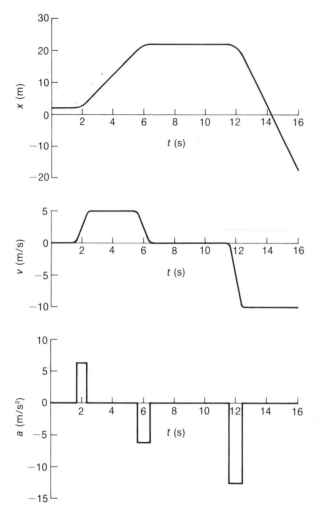

Fig. 3-1. The velocity-vs-time graph in the center was deduced from the slope of the displacement-vs-time graph above. The acceleration-vs-time graph at the bottom was similarly deduced from the slope of the velocity-vs-time graph.

Note that the units of acceleration are those of velocity divided by time, since it is defined to be a change in velocity divided by the corresponding time interval. These units in our example are meters/second/second or, more briefly, m/s².

Notice that near $t = 2$ seconds, even though the object is not yet moving very rapidly, it must be gaining speed *to the right*. Near $t = 6$ seconds, the object has a large positive velocity, i.e., it is moving rapidly to the right, and yet it has a negative acceleration because it is rapidly losing its positive velocity as it stops. Near $t = 12$ seconds it has a negative acceleration because it is gaining negative velocity, i.e., beginning to move to the *left*. The other possibility that is not shown in the figure is that of an object that is moving to the left but slowing down. In this case it is losing negative velocity, which is equivalent to gaining positive velocity, i.e., its acceleration is *positive*. You should spend time thinking about many different situations and determining whether the acceleration is positive or negative, as in Problem 3-2.

Example If $x = kt^3$, where k is a constant, find the acceleration as a function of time.

Solution First we find the velocity by differentiation:

$$v = \frac{dx}{dt} = 3kt^2$$

Then we differentiate the expression for the velocity to obtain the acceleration:

$$a = \frac{dv}{dt} = 6kt$$

Note that the dimensions of k must be those of lt^{-3} to balance dimensionally the expression given for x. These same dimensions for k also dimensionally balance the derived expressions for velocity and acceleration.

Exercises

3. A displacement-vs-time graph is shown in Fig. 3-2. From this graph construct a velocity-vs-time graph and an acceleration-vs-time graph.

4. A displacement as a function of time is given by the expression $x = kt^4$, where k is a constant equal to 3 m/s^4. Find the value of the acceleration at $t = 2$ s and $t = 5$ s.

3-5   Motion with Constant Acceleration

An object with constant acceleration is said to be accelerating *uniformly*. As we will see in the following chapter, uniform acceleration occurs when a constant force is applied to a mass. This situation occurs often enough to deserve special consideration.

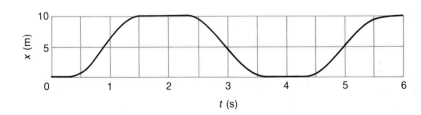

Fig. 3-2.

A constant acceleration implies a velocity-vs-time dependence that is of constant slope, i.e., v is a linear function of t. In Fig. 3-3 we see three v-vs-t graphs that all lead to the same value of acceleration: $\frac{1}{3}$ m/s². An equation for the velocity in each of these three cases would have the same form:

$$v = v_0 + at \qquad (3\text{-}8)$$

where a is $\frac{1}{3}$ m/s² in each case and v_0 equals 0, -5, and -10 m/s from top to bottom in order in Fig. 3-3. Equation (3-8), with arbitrary values of v_0 and a, is the most general possible behavior of velocity as a function of time that gives constant acceleration. Equation (3-8) can be deduced by integration of a constant acceleration:

$$v = \int a \, dt = at + \text{constant}$$

(as long as a is of constant value). The value of the integration constant is easily found by considering the time $t = 0$ and calling the velocity at that time v_0:

$$v_0 = 0 + \text{constant}$$

Hence $v_0 = \text{constant}$ and we conclude that $v = at + \text{constant} = v_0 + at$.

To find the variation of x with t that occurs under uniform acceleration, we integrate a second time:

$$x = \int v \, dt + C = \int (v_0 + at) \, dt + C$$

$$= \int v_0 \, dt + \int at \, dt + C = v_0 t + \tfrac{1}{2} at^2 + C \qquad (3\text{-}9)$$

The value of this integration constant is simply the initial position, x_0. Rewriting (3-9) we have

$$x = x_0 + v_0 t + \tfrac{1}{2} at^2 \qquad \text{(for constant } a\text{)} \qquad (3\text{-}10)$$

Checking to see that (3-10) is correct, we differentiate (3-10):

$$v = \frac{dx}{dt} = 0 + v_0 + at$$

and see that we again obtain (3-8), as we should. Equation (3-10) thus gives the dependence of position upon time for a constant acceleration.

A graph of (3-10) is shown in Fig. 3-4. This curve is a parabola in general, since (3-10) is a quadratic equation. The particular parabola shown intercepts the x axis above zero, corresponding to a positive value of x_0. It also corresponds to a positive value of

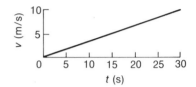

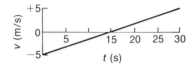

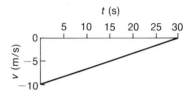

Fig. 3-3. Each of these three velocity-vs-time graphs yield the same value of acceleration: $\frac{1}{3}$ ms⁻².

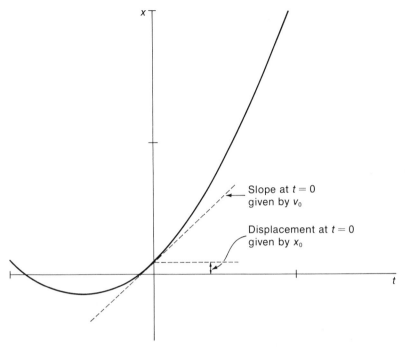

Fig. 3-4.   A graph of Eq. (3-10) with both x_0 and v_0 positive.

Slope at $t = 0$ given by v_0

Displacement at $t = 0$ given by x_0

v_0, since the slope is positive at $t = 0$. The slope of the parabola continually grows steeper, indicating a positive acceleration.

If the acceleration were negative, for example with both x_0 and v_0 equal to zero, we would have the situation graphed in Fig. 3-5. This is the situation frequently depicted in a "Tom and Jerry" cartoon in which the cat has a rubber band attached to his tail, so that as he approaches the mouse ($x = 0$) his velocity continually decreases until, finally, he disappears back toward negative x with large negative velocity.

Example   An object accelerates uniformly along a straight track with acceleration $a = 10$ m/s². At $t = 0$ it is at $x = -8$ m and moving with a velocity of 3 m/s. (a) What is the position of the object at $t = 3$ s? (b) What is the velocity of the object at $t = 3$ s?

Solution   (a) From the given information we have $x_0 = -8$ m and $v_0 = 3$ m/s. Thus

$$x = x_0 + v_0 t + \tfrac{1}{2}at^2$$

$$= -8 \text{ m} + (3 \text{ m/s})\, t + \tfrac{1}{2}(10 \text{ m/s}^2)\, t^2$$

Setting $t = 3$ s,

$$x = -8 \text{ m} + (3 \text{ m/s}) \cdot 3 \text{ s} + (5 \text{ m/s}^2) \cdot 9 \text{ s}^2$$

$$= -8 \text{ m} + 9 \text{ m} + 45 \text{ m} = 46 \text{ m}$$

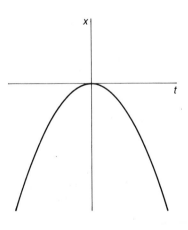

Fig. 3-5.   Graph of displacement versus time for constant negative acceleration.

Note that the dimensions of each term in the above equation cancel to leave an answer in meters, as must be the case if the equation is logical and no algebraic errors were made. Checking the dimensions of a result is an excellent method for catching errors in a calculation.

(b) $v = v_0 + at = 3 \text{ m/s} + (10 \text{ m/s}^2) \cdot 3 \text{ s}$

$= 3 \text{ m/s} + 30 \text{ m/s} = 33 \text{ m/s}$

Example   Car 1 and car 2 have initial positions of $+x_0$ and $-x_0$, respectively. Their initial velocities are zero and they accelerate with accelerations a (car 1) and $2a$ (car 2), respectively. Since car 2 starts behind car 1 but has larger acceleration, it will pass car 1. Find the time at which the positions of the cars coincide.

Solution   Let the position of car 1 be given by x_1 and the position of car 2 by x_2. Then

$$x_1 = +x_0 + \tfrac{1}{2}at^2$$
$$x_2 = -x_0 + \tfrac{1}{2}(2a)\,t^2$$

Note that $v_0 = 0$ in both cases. Coincidence of the cars means that $x_1 = x_2$. If it occurs at time t_p, we have

$$x_0 + \tfrac{1}{2}at_p{}^2 = -x_0 + at_p{}^2$$

Now, solving for t_p,

$$\tfrac{1}{2}at_p{}^2 = 2x_0$$
$$t_p{}^2 = 4x_0/a$$
$$t_p = \pm 2\sqrt{x_0/a}$$

Note the interesting result that there are *two* times at which the cars pass one another. This can be understood by plotting the parabolas of their displacement-vs-time graphs, as in Fig. 3-6. The velocities are large and negative before $t = 0$ because the assumed constant positive acceleration can lead to $v = 0$ at $t = 0$ only if v is negative before $t = 0$. If there were no positive acceleration before $t = 0$, we could cut off the left half of the graph and find only one time ($t = +\sqrt{2x_0/a}$) of coincidence. Taking numbers appropriate to actual automobiles, let $x_0 = 50$ meters (thus the cars are initially separated by roughly the length of a football field) and $a = 2 \text{ m/s}^2$ (the second car accelerating at $2a$ will be accelerating almost as rapidly as passenger cars can). We then find

$$t_p = 2\sqrt{\frac{50 \text{ m}}{2 \text{ m/s}^2}} = 2\sqrt{25 \text{ s}^2} = 10 \text{ s}$$

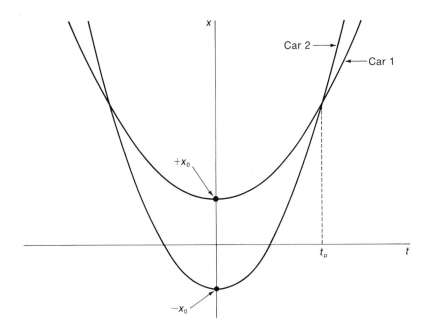

Fig. 3-6. Graph of both positions versus time for the car-passing problem of the example.

Exercises

5. In the last example, how fast was car 2 moving when it passed car 1?

6. An object starts from rest with a constant acceleration of 5 m/s^2 at $t = 0$. (*a*) How fast is it moving at $t = 10$ seconds? (*b*) How far has it moved in those first 10 seconds?

7. Can a vehicle have negative acceleration and positive velocity? Can you think of an everyday example?

The two integrations that led to (3-10) can also be evaluated by a graphical procedure, as shown in Fig. 3-7. The first graph shows that the area between $t = 0$ and any time t equals at. A constant of integration is required since the initial velocity is undetermined by integration. We call this initial velocity v_0. The second graph shows the resulting velocity-vs-time graph. The area under this curve, up to a time t, is the sum of the lower rectangular area, given by $v_0 t$, and the triangular area. The triangle has base t and height at, giving an area of $\frac{1}{2}at^2$. The constant of integration acquired in this integration is called x_0, so we obtain

$$x = x_0 + v_0 t + \tfrac{1}{2}at^2$$

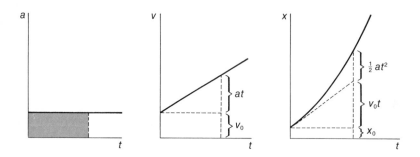

Fig. 3-7. The kinematics of constant acceleration obtained by successive graphical integrations.

The final result is shown in the third graph of Fig. 3-7, where we see the contribution of each of the three terms.

3-6   Free Fall

If a falling object is acted upon by no influences other than earth's gravity, we say the object is in *free fall.* Air resistance generally prevents the occurrence of true free fall, yet if objects are reasonably dense, air resistance is very nearly negligible for falls of a few feet or so.* Free fall baffled men for centuries, partly because of the confusing and variable nature of air resistance and partly because of the difficulty of the notion that velocity changes *continuously,* never lingering at any one value for a finite instant of time. Galileo Galilei (1564–1642) is generally credited for the understanding that free fall involves continual acceleration of the simplest kind, constant acceleration. By studying the motion of objects on inclined planes, he essentially slowed the effects of gravity sufficiently to enable an accurate study with the crude timing methods then available.** Though others studied free fall before Galileo, their views were not widely propagated, partly because of the resistance of the medieval church. The published work of Galileo† showed that the distance covered by a falling object starting from rest is proportional to the square of the elapsed time, and that this is the

---

*This statement is necessarily vague. A feather, for example, exhibits free fall behavior for only the first few cm of its fall, while a stone does so even when dropped from a tall building.

**In these studies he usually used a constant flow of water filling a vessel. The elapsed time from the start to the stop of the flow was proportional to the quantity of water accumulated in the container.

†See for example, "Galileo's Discovery of the Law of Free Fall" by Stillman Drake *Scientific American,* May 1973.

behavior expected for uniform acceleration, as we have already seen.

The magnitude of the acceleration of a freely falling object is usually symbolized by g, where

$$g = 9.8 \text{ m/s}^2 \quad \text{or} \quad g = 32.2 \text{ ft/s}^2$$

The value of g actually varies slightly at different points on the earth's surface, as discussed in Chapter 17.

If we consider "up" to be the positive y direction, we can rewrite Eq. (3-10) for an object in free fall:

$$y = y_0 - \tfrac{1}{2}gt^2 \tag{3-11}$$

where y_0 is the initial height of the object and the initial velocity is zero. The minus sign on the second term reflects the fact that the acceleration is directed down toward smaller values of y. The acquired velocity is given by

$$v = -gt \tag{3-12}$$

where the minus sign indicates a downward velocity.

In Fig. 3-8 we see the coordinate y and another coordinate, d, which represents the distance covered by the falling object measured from its starting point. In terms of this coordinate d, we have simply

$$d = \tfrac{1}{2}gt^2 \tag{3-13}$$

since $d = 0$ at $t = 0$. Again, no initial velocity is assumed.

Equation (3-13) can be used to determine a person's reaction time by a simple distance measurement. Suppose we wish to determine a driver's reaction time in applying the brakes of a car in the most favorable possible circumstance. Fig. 3-9 shows the subject's foot poised near a wall as if over a brake pedal; another person is holding a card near the wall, with the bottom of the card positioned at the point where the subject's foot will strike the wall. The card is released with no advance warning, while the "driver" tries to stop it by pinching it against the wall with his toe. The time lag t is found easily by measuring d, how far from the bottom of the card the "driver" catches it.

Since $d = \tfrac{1}{2}gt^2$,

$$t = \sqrt{2d/g}$$

For the short distances involved, the card will move in free fall. (You can test this hypothesis by repeating the experiment over and over, using first heavy cardboard and progressing to tissue paper,

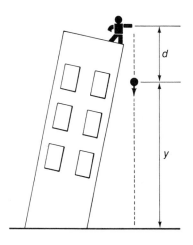

Fig. 3-8. Coordinates useful in describing free fall.

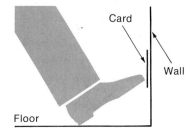

Fig. 3-9. A simple reaction-time measurement.

for which air resistance is much more important.) An excellent reaction time of $\frac{1}{10}$ s corresponds to

$$d = \frac{1}{2} \cdot 32 \, \frac{\text{ft}}{\text{s}} \cdot \left(\frac{1}{10 \, \text{s}}\right)^2$$

$$= \frac{16}{100} \, \text{ft} = 1.9 \, \text{in.}$$

Example   A flower pot is released from the top of a building with a height of 128 ft. When does it pass a window 32 ft above the ground? Assume $g = 32$ ft/s² for simplicity.

Solution   The most direct approach is to substitute 128 ft for y_0 and 32 ft for y in (3-11). We can then solve for the value of t satisfying this equation. However, it is always less cumbersome first to solve algebraically for the unknown, then to substitute numerical values as a last step:

$$y = y_0 - \tfrac{1}{2}gt^2 \qquad\qquad (3\text{-}11)$$

$$t^2 = 2(y_0 - y)/g$$

$$t = \pm\sqrt{2(y_0 - y)/g} \qquad\qquad (3\text{-}14)$$

At this point we make a dimensional check of our result. The dimensions of (3-14) are

$$t = \sqrt{\frac{l}{l/t^2}} = \sqrt{t^2} = t$$

The dimensions of the right side cancel to produce t, so (3-14) is dimensionally correct. Note that if we had substituted only numerical values at the outset, this simple check of consistency would have been impossible. For our numerical result we substitute values into (3-14):

$$t = \pm\sqrt{\frac{2 \cdot 96 \, \text{ft}}{32 \, \text{ft/s}^2}} = \pm\sqrt{6} \, \text{s}$$

The flower pot passes the window at $t = \sqrt{6}$ s $= 2.45$ s after being dropped. The time $t = -2.45$ s, also obtained above, is the time *before* $t = 0$ at which the pot *would have passed* the window if it were thrown up, so that it would be just fleetingly at rest at $y = 128$ ft at $t = 0$. In our problem we are interested only in times after $t = 0$ and so we simply discard the negative solution.

Exercises

8. How would you modify Eq. (3-11) to include an initial velocity, v_0, directed upward along y?

9. Light propagates through empty space with a velocity of
    3×10^8 m/s. How long would a ship, with an acceleration always
    equal to g, take to reach one-tenth the speed of light? Take
    $g = 10$ m/s^2 and round off the number of seconds in a year to
    3×10^7 s.

3-7   An Important Derived Equation

For the case of constant acceleration, we know that

$$v = v_0 + at \qquad (3\text{-}15)$$

and
$$x = x_0 + v_0 t + \tfrac{1}{2}at^2$$

However, we often wish to know the velocity of an object moving
with constant acceleration, *not* for a known time, but for a known
distance, d. Since this distance is $x - x_0$, we may write

$$d = v_0 t + \tfrac{1}{2}at^2 \qquad (3\text{-}16)$$

In order to obtain a relation between v and d in which we need
not know the time elapsed, we must eliminate t from Eq. (3-16).
We do this solving for t from (3-15):

$$v = v_0 + at \qquad (3\text{-}15)$$

$$at = v - v_0$$

$$t = \frac{v - v_0}{a}$$

Now, substituting this result in (3-16),

$$d = v_0\left(\frac{v - v_0}{a}\right) + \tfrac{1}{2}a\left(\frac{v - v_0}{a}\right)^2$$

$$= \frac{v_0 v}{a} - \frac{v_0^2}{a} + \frac{v^2}{2a} - \frac{v_0 v}{a} + \frac{v_0^2}{2a}$$

$$= \frac{v^2}{2a} - \frac{v_0^2}{2a}$$

Rearranging terms we obtain

$$v^2 = v_0^2 + 2ad \qquad (3\text{-}17)$$

for constant acceleration. Equation (3-17) gives us the desired connection between v and d. In the special case of zero initial velocity,
it reduces to $v = \sqrt{2ad}$.

$V = \sqrt{2ad}$ at $V_0 = 0$

Example   The acceleration due to the earth's gravity is 32 ft/s^2. If a
flower pot falls out a fifth-floor window, what is its velocity when it

passes windows on the fourth floor and the third floor? The distance between floors is eight feet.

Solution   Since the initial velocity is zero, we can write $v = \sqrt{2ad}$. At the fourth floor, the pot has fallen eight feet, so

$$v = \sqrt{2\,(32\text{ ft/s}^2)\,8\text{ ft}} = \sqrt{512\text{ ft}^2/\text{s}^2}$$

$$= 22.6\text{ ft/s}$$

At the third floor, $s = 16$ ft:

$$v = \sqrt{2 \cdot 32 \cdot 16\text{ ft}^2/\text{s}^2} = 32\text{ ft/s}$$

Example   An automobile is initially moving at 30 mi/hr. It accelerates at 10 ft/s² for one city block ($d = 250$ ft). What is its final velocity?

Solution   We must make the units consistent, so we can either convert the distance to miles or change mi/hr into ft/s. We choose the latter. Since 60 mi/hr corresponds to 88 ft/s, the initial velocity of 30 mi/hr is equal to 44 ft/s. Now, using Eq. (3-17),

$$v^2 = v_0{}^2 + 2ad = 44^2\text{ ft}^2/\text{s}^2 + 2\,(32\text{ ft/s}^2)\,250\text{ ft}$$

$$= 1936\text{ ft}^2/\text{s}^2 + 16000\text{ ft}^2/\text{s}^2 = 17936\text{ ft}^2/\text{s}^2$$

Hence $$v = 134\,\frac{\text{ft}}{\text{s}} \cdot \frac{60\text{ mi/hr}}{88\text{ ft/s}} = 91\tfrac{1}{2}\,\frac{\text{mi}}{\text{hr}}$$

Thus we see that an acceleration of about 1/3 that of gravity is quite large for an automobile.

Exercises

10. How does Eq. (3-17) change if the acceleration is directed opposite to the velocity?

11. A uniformly accelerating car starting from rest reaches 60 mi/hr in 15 s. At this acceleration, in how many seconds can it cover a quarter mile, starting from rest?

12. A uniformly accelerating object is observed to change its velocity from 10 m/s to 20 m/s in a distance of 15 meters. What is the value of the acceleration? Is the result changed if v_0 is 20 m/s and the final velocity 30 m/s?

3-8   Motion with Variable Acceleration

The equation

$$x = x_0 + v_0 t + \tfrac{1}{2}at^2$$

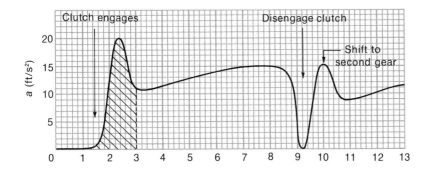

Fig. 3-10. A chart recording of acceleration versus time for a test automobile. The shaded area shown equals the velocity gained in the first three seconds.

was obtained by two integrations explicitly for the case of constant acceleration. If the acceleration varies, we must use Eqs. (3-7) and (3-3) directly in order to find the velocity and displacement. In Figure 3-10 we see a chart recording obtained by placing an accelerometer in an automobile being tested.* Since

$$\Delta v = \int_{t_1}^{t_2} a(t) \, dt$$

finding the change in velocity between any two times t_1 and t_2 involves simply finding the definite integral of $a(t)$ between those times. That integral is equal to the area under the curve in Fig. 3-10 that lies between t_1 and t_2. The shaded area shown there equals the velocity change between the beginning of the acceleration ($t=0$) and $t = 3$ s. Since the test run starts at zero velocity, Δv and v are equal in this case. Fig. 3-11 shows the velocity obtained by integration between $t = 0$ s and successively larger times in Fig. 3-10. This integration was carried out crudely by simply counting squares in Fig. 3-10 to evaluate the area up to any given time. An example is indicated for the point obtained at $t = 3$ s.

Finding the displacement requires integration of the curve in Fig. 3-11. Proceeding as before, we obtain the graph of Fig. 3-12. Note that the formulas developed for constant acceleration, such as (3-10), were not applicable in this case. Instead we applied the

---

*As we will see in the next chapter, an accelerated mass is subject to a force that is proportional to the acceleration. A small mass attached to a spring serves as an accelerometer, since the deflection of the spring indicates the force on the mass and hence the acceleration.

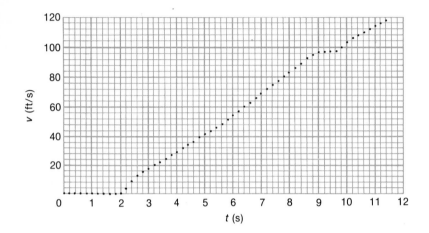

Fig. 3-11. Velocity versus time obtained by graphical integration of the curve in Fig. 3-10. The point at $t = 3$ s, for example, was found by estimating 90 squares ($= 18$ ft/s) in the shaded area of Fig. 3-10.

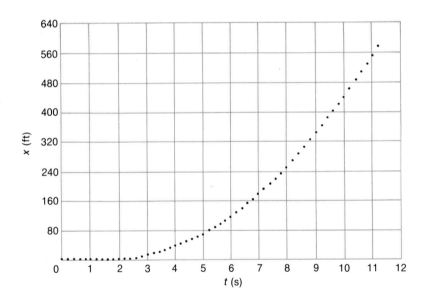

Fig. 3-12. Graph of displacement for the test auto; derived from the graph of Fig. 3-11.

fundamental relations of (3-3) and (3-7). Two successive integrations of an accelerometer recording finally gave us the displacement.

An inertial guidance system utilizes this same method to measure the displacement of a vehicle without reference to external guides such as the stars or radio signals. Submarines can be navigated under the polar ice cap by such devices. In practice, inertial guidance systems use three accelerometers to measure the acceleration along the x, y, and z coordinates. Two electronic integrations performed on each of these accelerations then continuously gives

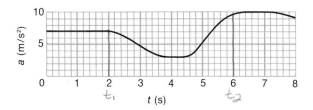

Fig. 3-13.

the displacement in each of the three dimensions from the initial point of the voyage.*

Exercises

13. The acceleration of an object is graphed in Fig. 3-13. If the velocity at $t = 2$ s is zero, find the velocity at $t = 6$ s.

14. The acceleration is given by $a = bt$, where $b = 5$ m/s³. If $v = 10$ m/s at $t = 0$ s, find the displacement that occurs between $t = 0$ and $t = 15$ s.

3-9   Summary

The formal definitions of velocity and acceleration in one dimension are given by

$$v = \frac{dx}{dt}$$

$$a = \frac{dv}{dt} = \frac{d^2x}{dt^2}$$

For variable acceleration we can write

$$\Delta v = v_2 - v_1 = \int_{t_1}^{t_2} a\, dt \qquad (3\text{-}18)$$

with the change in position then being given by

$$\Delta x = x_2 - x_1 = \int_{t_1}^{t_2} v\, dt \qquad (3\text{-}19)$$

For the case of constant acceleration starting at $t = 0$, these calculations are easily carried out explicitly, giving

$$v = v_0 + at$$

$$x = x_0 + v_0 t + \tfrac{1}{2}at^2$$

---

*Integration by means of electric circuits is briefly discussed in Section 37-4.

An object in free fall undergoes constant (downward) acceleration of magnitude

$$g = 9.8 \text{ m/s}^2 = 32.2 \text{ ft/s}^2$$

If we consider the distance fallen from $t = 0$ to be a positive quantity called d, we have

$$d = \tfrac{1}{2}gt^2$$

for free fall with zero initial velocity. If instead we wish to consider the distance y upward from some reference elevation (such as street level) as a positive quantity, we write

$$y = y_0 - \tfrac{1}{2}gt^2$$

where y_0 is the height from which free fall begins.

For constant acceleration we can show that

$$v^2 = v_0^2 + 2ad$$

where v is the velocity obtained by an object moving a distance d with an initial velocity v_0.

If the acceleration of an object is not uniform, the velocity and subsequently the displacement can be found only by integration, as indicated by Eqs. (3-18) and (3-19).

Problems

3-1. Let x as a function of t be given by the equation $x = x_0 + v_0 t + \tfrac{1}{2}at^2$ with the values $x_0 = 10$ m, $v_0 = 10$ m/s, and $a = 10$ m/s². Graph this function for the interval between $t = -1$ s and $t = +5$ s under the following conditions: (a) x_0 negative, v_0 and a positive; (b) x_0 and v_0 negative, a positive; (c) x_0 and v_0 positive, a negative.

3-2. Consider cars travelling north on Broadway to be travelling in the $+x$ direction. Consider the zero of the x axis to be at a given crosswalk. What are the signs ($+$ or $-$) of x, v, and a under the following conditions: (a) the car is travelling north, is approaching the crosswalk, and has its brakes on; (b) the car is travelling north, has passed the crosswalk, and is speeding up; (c) the car is travelling south, has passed the crosswalk, and is slowing down.

3-3. An object is dropped from a height of 400 meters at $t = 0$. (a) When does the object strike the ground? (b) How fast is it moving when it strikes the ground?

(a)

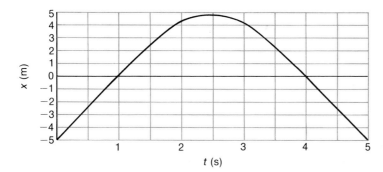

Fig. 3-14.

(b)

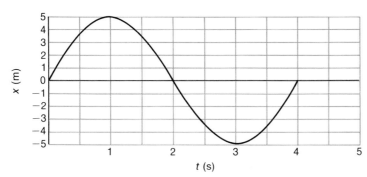

(c)

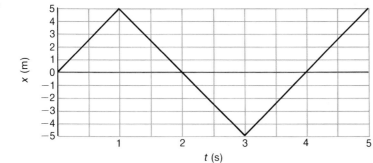

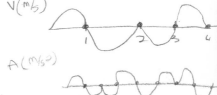

3-4. A free-falling stone starting from rest at the top of a cliff strikes the ground with a velocity of 30 m/s. If it is thrown downward with an initial velocity of 40 m/s, what is its velocity when it strikes the ground?

3-5. An object's position is always on the y axis and given by

$$y = At^3 + Bt^2 + Ct + D$$

(*a*) Find the times at which $v = 0$ (i.e., find the zeros of the expression for dy/dt). (*b*) What conditions do the coefficients A, B, and C satisfy if: (1) there are two such times; (2) there is only one such time; (3) there are no such times?

3-6.   Three different graphs of displacement versus time are given in Fig. 3-14. Sketch a graph of the acceleration as a function of time for each case. The sketches can be rough, but indicate an approximate numerical scale for your results. Hint: In part c, the sharp corners will cause difficulty. Think of rounding the corners off slightly and consider the limiting case as the corners become sharper and sharper.

3-7.   Given that $x = At^3 + Bt^2 + Ct + D$, (a) find expressions for the velocity and the acceleration. (b) What are the conditions on the coefficients so that there is one time at which both the velocity and the acceleration are zero?

3-8.   In the expression given in the preceding problem, let $A = 1$ m/s^3, $b = 2$ m/s^2, $C = 1$ m/s and $D = 5$ m. (a) At what time (or times) is the acceleration equal to zero? (b) What is the value of the acceleration when $v = 0$?

3-9.   A police car at rest observes a car approaching from behind that radar determines to be moving at 90 mi/hr. At the moment the speeding car passes the police car, the police give chase with a constant acceleration of 18 ft/s^2. (a) When and where does the police car overtake the speeding car if the latter maintains 90 mi/hr? (b) If this occurs, how fast is each car moving at the moment of passing?

3-10.   Repeat Problem 3-9, except with a somewhat more realistic acceleration for the police car. Assume that

$$a = a_0 - bt$$

where $a_0 = 18$ ft/s^2 and $b = \frac{3}{4}$ ft/s^3. (a) When does the police car overtake the speeding car? (b) What is the speed of the police car at the moment of overtaking?

3-11.   An acceleration is given by $a = bt^2 + c$. If the velocity at t_1 is equal to v_1, find the displacement that occurs between t_2 and t_3. Your answer will contain the given parameters b, c, t_1, t_2, and t_3.

3-12.   A model rocket starts from rest and moves with an acceleration given by $a = 100$ m/s$^2 - (4$ m/s$^4)t^2$ between $t = 0$ and $t = 5$ s. Find its *velocity* at $t = 1$ s, 2 s, 3 s, 4 s, and 5 s.

3-13.   The engineer of a fast freight with a velocity of 30 m/s suddenly sees a slow freight ahead that is moving in the same direction but at a velocity of only 10 m/s. He quickly applies the brakes, causing constant acceleration equal to -2 m/s². At the moment he applies the brakes, the front of his engine is 99 m from the back of the slow train. The slow train continues at a constant 10 m/s. (*a*) Does a collision occur? (*b*) If so, where does it occur?

3-14. A physics student is standing 48 ft from a building when he notices a flower pot accidentally dislodged from the building's balcony, 64 ft above ground. He dashes toward the building and catches the pot just before it strikes the ground. What is the student's average velocity?

3-15. A bicycle has an initial velocity of 3 m/s and experiences a constant negative acceleration of $\frac{1}{2}$ m/s². How far does the bicycle move during the first 3 s after the velocity begins to decrease?

3-16. A car has an initial velocity of 44 ft/s and slows uniformly to 22 ft/s with $a = -11$ m/s². How far does it move in the process?

3-17. Add a term to Eq. (3-11) to include an initial upward-directed velocity. Now consider a stone, thrown vertically upward from the ground, that passes a window 64 ft above ground one second later. When will it pass that window again?

3-18. One car follows another along a freeway. They are separated by a distance d and both move with the same initial velocity v_0. Suppose it takes the second driver a time τ to apply his brakes and both cars have equally effective brakes. *Not on exam* (*a*) Find the minimum value of d required to avoid a collision in the event the lead car suddenly applies its brakes, eventually coming to rest. Assume a constant negative acceleration. (*b*) Sketch on a single graph the position-vs-time curves for both cars under the conditions of your solution. (*c*) Find the numerical value of d for $\tau = \frac{1}{4}$ s and a typical freeway speed.

3-19. The record for the quarter-mile "drag race" listed in the *1974 Guinness Book of World Records* is 243.90 mi/hr at the end of a 440-yd run starting from rest. The elapsed time was 6.175 s. Assuming a constant value of acceleration, find the magnitude of a (in ft/s²) first using Eq. (3-15), then using (3-17). How do you explain the discrepancy between these two results?

Chapter **4**

Introduction to Dynamics

4-1 Introduction   The preceding two chapters gave us a language well suited to a description of motion but they contained no physics. The description of motion was there but the "why" of motion was not. We need to develop a *theory* of motion, a theory that allows us to predict what kind of motion will occur in any given physical situation. We will develop the concepts of inertia, momentum, and force in this chapter and give a theory of motion for the simplest case: the motion of mass points along one dimension. The general topic of motion and its causes is called *dynamics*. We usually reserve the word *mechanics* for the entirety of physical theories of motion, including kinematics, statics, and dynamics, for rotating as well as translating systems.

4-2 The Law of Inertia: Newton's First Law

One simple question that we can ask is "Why do objects move?" The "why" of this question seems appropriate; after all, most motion we observe has an obvious cause—the jet engine that moves an airplane, the muscle that moves the arm, or the force of gravity that propels a skier. Until the time of the Renaissance, it was commonly thought that motion was not possible without a causal agent. Galileo's work led to the discovery of the law of inertia, which he clearly stated in his "Discourse on Two New Sciences," published in 1638. Careful observation of the motion of smooth objects on nearly frictionless inclined planes led him to the conclusion that an object, once in motion, would continue in a straight line with unchanging velocity if no external influences were acting upon it. This law was difficult to discover because most objects in motion *are* acted upon by external influences: the jet airliner feels a tremendous drag due to air resistance and will begin to slow immediately when the engines are stopped; the skier who has acquired a high velocity on a steep hill will, upon encountering a flat area, coast for some distance, but the friction of his skis on the snow must bring him to a stop eventually. Galileo observed that objects lost their velocity only gradually if friction was reduced to a minimum, hence he concluded that in the limit of no external influences, a moving object would maintain its motion undiminished.

Galileo argued that external influences, more precisely called *forces,* are required to *change* the motion of a body but in their absence, a moving body moves in a straight line with unchanging velocity. He could set a block into motion by pushing it, but once moving it would coast indefinitely if friction could be avoided. This idea is called the *law of inertia,* where the word inertia refers to a body's resistance to changes in its state of motion.

Sir Isaac Newton (1642–1727) adopted Galileo's law of inertia and made it the first of his three laws of motion. Newton wrote in Latin, the language of scholars in his day, but a modern translation would read:

> Every body persists in its state of rest or of uniform motion in a straight line unless it is compelled to change that state by forces impressed on it.

All of this can be put more concisely as follows:

$$v = \text{constant} \quad \text{if} \quad F = 0$$

where v is the velocity of the body and F the net force. The words "net force" are used here because it is possible, for example, for two equal and opposite forces to cancel one another's effects, thus producing no *net* force. We will later discuss this point more fully. Note that $v = 0$ (the state of rest) is but one particular case of the above equation.

Before going on to Newton's second law of motion, let us mention an important point contained in the first law. We already have defined what we mean by the word *force:* it is anything that can cause a body's velocity to change. Examples are the force of gravity, the forces exerted by our muscles, and the force due to wind pressure. In general, those things we identify as a push or pull are forces.

4-3   Inertial Mass: Introduction

The law of inertia applies to all objects, yet we know that there is a difference in behavior between "heavy" objects and "light" objects. The most obvious difference between them is the strength of the force that the earth exerts upon them—the force of gravity. But there is another quality associated with heavy objects. The word "massive" or "ponderous" relates to this quality. Once a heavy object is moving, it is more difficult to stop than a light object moving similarly. Conversely, once stopped, it is more difficult to set a heavy object into motion. This quality of objects is called their *mass*, or, more specifically, their *inertial mass*. Inertial mass is a measure of how difficult it is to change the velocity of the object. Thus inertial mass is a *measure* of the inertia of an object.

4-4   Inertial Mass: Definition

Our definition of mass is an operational definition. We adopt a simple (in thought, at least) procedure that utilizes an inertial balance.* Fig. 4-1 illustrates the procedure. Two objects, m_1 and m_2, are attached to the ends of a beam, as shown. This apparatus is placed on a surface as nearly frictionless as possible and then set into motion by a pull applied to the center of the beam, as shown in part *a* of the figure. If, for example, m_1 has greater mass than m_2, it will be more difficult to set into motion and will lag behind m_2, as shown in part *b* of the figure. This device will be used to *define* which of two objects is the more massive. We can then pick one object, for example a one-kilogram standard mass, and make several more objects with the same mass by carefully adding to or subtracting from their mass until they balance on the inertial balance with our one-kilogram standard. These masses are then secondary standards. Having done this, we can fashion masses that are multiples or fractions of the standard. Units of $\frac{1}{10}$ kg, for example, are checked by balancing 10 of them against the 1-kg standard. Thus a system of units and standards can be established by defining mass in terms of the inertial properties of matter.

The reader may object at this point, recalling that masses are really compared to one another in a standards laboratory, *not* by pulling on a connecting beam and watching to see if one lags, but simply by placing them on a balance. Actually, both methods agree with one another, as we will see later in this chapter. We have described the inertial-balance method to emphasize that *mass is a measure of inertia.*

*A simple inertial balance can be fabricated with two roller skates and a meterstick.

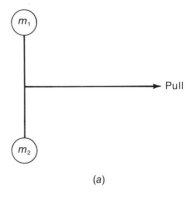

(a)

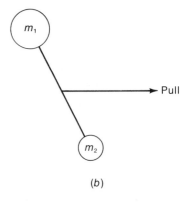

(b)

Fig. 4-1. An inertial balance viewed from overhead. The beam connecting the masses is pulled at its center, and if one mass lags behind the other, as in (b), we conclude it has greater inertial mass.

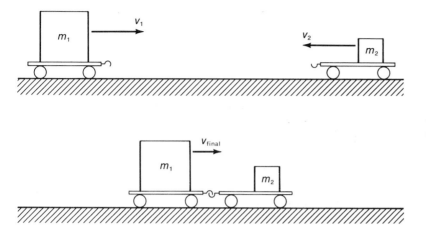

Fig. 4-2. Two carts are sent toward one another with equal but oppositely directed velocities. When they collide, a coupler keeps them locked together. The final common velocity of the two carts is in the same direction as the initial motion of the cart that was most difficult to stop.

4-5   Momentum

Having defined what we mean by the inertial mass, we are now in a position to observe a new experimental fact. Suppose two carts of unequal mass are set into motion toward one another, as in Fig. 4-2, and then allowed to collide and stick together. The final direction of motion of the pair of carts will be in the direction of the cart that was most difficult to stop. In conformity with the idea of inertial mass, we find that when the magnitudes of the initial velocities are equal, the final direction of motion is that of the most massive cart.

If we now use two carts of equal mass but unequal velocity, the final direction of motion is that of the fastest moving cart. The difficulty of stopping a moving body is thus seen to depend *both* on its inertial mass and on its velocity. This agrees with our own experience. For example, we know the degree of difficulty of stopping a running man depends both on his mass and his speed.

The simplest hypothesis that can be made that incorporates these experimental results is that it is the *product* of the mass, m, and the velocity, v, that determines the degree of difficulty in stopping a moving body. This product, mv, is called the *momentum* and is designated by the letter p. More briefly,

$$\text{momentum} = p = mv$$

This definition is consistent with the English language usage of momentum, where a movement (political, historical, or physical) that has a great deal of momentum is one that tends to keep going and can be stopped only with difficulty.

We have now reached the same degree of understanding of motion that existed during Galileo's time, partly due to his experiments and studies.

1. Inertia is the tendency of objects to move with unchanging velocity.

2. Inertial mass is a measure of difficulty in changing the velocity of an object.

3. Both the mass and velocity of an object enter into its momentum. The momentum of an object is a measure of the difficulty of arresting its motion.

These three ideas are correct, as far as they go. Galileo had completed his work in the first half of the seventeenth century. In the last half of the seventeenth century, the importance of momentum was firmly established in a series of experiments concerning collisions. These studies were made by Huygens, Wallis, and Wren in response to a proposal of the Royal Society (London) to "solve the problem of inpact." The discovery they made was that of the law of *conservation of momentum*. Today, most physicists consider this law to be a cornerstone of both classical and quantum mechanics. Since the experiments of Huygens, Wallis, and Wren, this law has been tested in innumerable ways, which include such diverse examples as the motion of massive rockets and the collisions of subatomic particles.

Let us review one of the types of experiments that first revealed momentum conservation as a law. Fig. 4-3 shows a cart collision experiment. The carts are fitted with elastic spring steel bumpers, their masses are adjustable, and they roll as nearly without friction as possible. Cart 1 of mass m_1 is moving toward cart 2 of mass m_2. Cart 1 has velocity v_1 and cart 2 is at rest. If the two masses m_1 and m_2 are equal, a simple result occurs. After the collision, cart 1 is at rest and cart 2 is found to be moving in the same direction and with the same velocity, v_1, that cart 1 originally had. If we modify this experiment by putting some sticky substance on the bumpers so that the carts must stick together after their collision, the result of the experiment is, of course, different. Now we will find, after

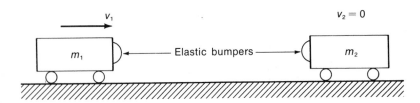

Fig. 4-3. A cart-collision experiment.

the collision, both carts stuck together and moving in the same direction but with one-half the initial velocity. If we perform the experiment with a small explosive on the bumpers, so that the carts both recoil away from their point of collision, their momenta still obey a simple rule. Results are shown in Table 4-1 for three different collisions.

TABLE 4-1

	Before Collision			After Collision		
	m_1v_1	m_2v_2	$m_1v_1 + m_2v_2$	m_1v_1	m_2v_2	$m_1v_1 + m_2v_2$
Elastic bumpers	mv	0	mv	0	mv	mv
Sticky bumpers	mv	0	mv	mv/2	mv/2	mv
Explosive collision	mv	0	mv	−0.2 mv	1.2 mv	mv

The total momentum before the collision, $m_1v_1 + m_2v_2$, is mv in all cases. The startling result is that the total momentum of this system (consisting of the two carts) is absolutely unchanged after the collision, even though the three collisions occurred under different conditions. In each of the three cases, the final total momentum is still mv. The constancy of the total momentum of a system, even though collisions occur among its parts, is the meaning of the law of conservation of momentum. It is true even when many objects are involved in collisions that are elastic, sticky, explosive, or soft.

The total momentum of a system of N masses is given by

$$p_{total} = m_1v_1 + m_2v_2 + \cdots + m_Nv_N = \sum_{i=1}^{N} m_iv_i \qquad (4\text{-}1)$$

Note that some of the velocities may be negative, as for a cart bouncing backward in a collision experiment. The several parts of the system may interact with one another (as when carts collide with one another), but no net *external* force may act if momentum is to be conserved. A concise statement of the law of conservation of momentum is:

$$p_{total} = \text{constant} \quad \text{if net external force} = 0 \qquad (4\text{-}2)$$

Example   A 1-kg mass moving without friction at 15 m/s strikes and sticks to a 4-kg mass that was initially at rest. What is the velocity of the stuck pair after the collision?

Solution Let us designate the 1-kg mass by m_1 and the 4-kg mass by m_2. Let v_1 refer to the velocity of m_1 before the collision and v_{final}, or v_f, refer to their common velocity after the collision. Then:

$$\text{total momentum before collision} = p_i = m_1 v_1 + 0$$

$$\text{total momentum after collision} = p_f = (m_1 + m_2)v_f$$

By the law of conservation of momentum, $p_f = p_i$, so

$$m_1 v_1 = (m_1 + m_2)v_f$$

We solve for v_f, substituting the numerical values and units of these values:

$$v_f = \frac{m_1}{m_1 + m_2} v_1 = \frac{1 \text{ kg}}{5 \text{ kg}} \cdot 15 \text{ m/s} = 3 \text{ m/s}$$

Note that the units balance correctly.

   An important point of this example is the reduction of the problem to a solution in terms of general quantities, such as m_1 and m_2, *before* numbers are substituted to obtain the desired numerical result. This procedure allows us to discover logical errors more readily, to obtain an answer that has a wider range of validity than the particular numerical problem being calculated, and to check the dimensions of the answer. Further, it enables us to communicate to others our thought process in solving the problem. As an example of the opposite approach consider this "solution":

$$1 \times 15 = 5v \qquad \text{so} \qquad v = 3$$

There is little chance to catch logical errors in such a presentation, since most of the reasoning was done mentally with only the result being placed on paper. The significance of algebraic symbols is clear; that of bare numbers is not.

Example Two frictionless carts initially at rest are placed together with a small explosive charge between them. The explosive is fired and the two carts recoil away from one another. If the ratio of their masses is given by $m_1/m_2 = 2$, what is the ratio of their recoil velocities?

Solution The momentum is initially zero. Further no *external* forces act upon this two-cart system. Therefore, each cart must have the same (though oppositely directed) momentum finally, so that the sum of their momenta is still zero. Since one cart is twice as massive as the other, it must be moving half as fast as the other. More formally:

$$p_{initial} = m_1(0) + m_2(0) = 0$$

By conservation of momentum we know that

$$p_{final} = m_1 v_1 + m_2 v_2 = 0$$

Hence $m_1v_1 = -m_2v_2$, or

$$\frac{v_1}{v_2} = -\frac{m_2}{m_1} = -\frac{1}{2}$$

so that $v_1 = -\frac{1}{2}v_2$. That is, cart 1 is moving half as fast as cart 2 and in the opposite direction. Since this experiment measures the ratio of masses, it can be used to define a scale of inertial masses just as well as the inertial balance described earlier.*

Example   A 2-kg cart moving from left to right at 15 m/s strikes an 8-kg cart that is at rest. After the collision, the 8-kg cart is seen to be moving to the right at 5 m/s. What is the final velocity of the 2-kg cart?

Solution   Let us designate the initial momentum by p:

$$p = m_1 v_{\text{initial}}$$

Then the total momentum after the collision must still equal p. So after the collision we may write

$$m_1v_1 + m_2v_2 = p$$

where m_1 refers to the 2-kg cart and m_2 to the 8-kg cart. Solving for v_1:

$$v_1 = \frac{p - m_2v_2}{m_1}$$

Substituting values,

$$v_1 = \frac{2 \text{ kg} \cdot 15 \text{ m/s} - 8 \text{ kg} \cdot 5 \text{ m/s}}{2 \text{ kg}}$$

$$= -5 \text{ m/s}$$

Thus, the lighter cart has bounced back toward the left (its velocity is negative) at 5 m/s.

Exercises

1. An ice skater whose mass is 80 kg and a 2-kg stone are initially stationary on the ice. The ice skater pushes the rock, which moves off at 20 m/s with respect to the ice. The ice skater recoils backward as he pushes the rock. What is his velocity with respect to the ice?

---

*This type of mass measurement is well suited to atoms and nuclei, while the inertial balance is not. In actual practice, such recoil effects *are* used to measure masses of atomic and subatomic particles. However, the most *accurate* measurements of atomic masses apply Newton's second law $(F = ma)$, where the known force is electromagnetic.

2. A 15-g bullet is fired with a muzzle velocity of 350 m/s. In order to measure its velocity accurately, a wooden block is placed on wheels so that when the bullet strikes it and sticks in it, the block acquires a small velocity that is easy to measure. If we wish the final velocity of the block (initially at rest) to be about 4 m/s, what should the mass of the block be?

3. A hunter fires a 10-g bullet at 450 m/s. If the hunter and gun together weigh 75 kg, what is his recoil velocity in the absence of friction?

4-6 Force

Up to now we have considered only systems in which there was no friction, so that there would be no external influences acting on them. Now we wish to understand the effect of external influences. We will call such external influences *forces* and define precisely what we mean by a force.

Since the momentum of an object or collection of objects is unchanged in the absence of external influences, it seems reasonable to *define* a force as that which changes momentum. In other words, if the momentum of an object changes, a force must be acting. The magnitude of the force will be given by the rate at which it can change the momentum. That is:

$$F = \frac{dp}{dt} \tag{4-3}$$

This equation was first proposed by Newton and is called Newton's second law of motion. If the momentum is constant, as it would be for a freely coasting object, then the time derivative of the momentum (dp/dt) is zero; that is, no forces are acting. If, on the other hand, the momentum is changing with time, the quantity dp/dt will be nonzero and the force, by definition, is equal in magnitude to the value of dp/dt. The faster the momentum is changing, the larger is dp/dt, and hence the greater the force that is acting. Intuitively, we know that pushing and pulling on objects can speed them up or slow them down. Equation (4-3) formalizes our intuitive notion and makes a precise definition of "pushes and pulls." They are *forces,* and a measure of a force is the rapidity with which it can change momentum.

There is more to Newton's second law than a simple definition. The law can be used to make predictions that are susceptible to experimental test. For example, using (4-3) we can calibrate a standard force produced by a stretched spring when applied to a

2-kg mass. The rate of change of momentum of the mass gives the magnitude of the standard force. We now apply this *same* standard force to a *different* mass, say a 4-kg mass. We find that the velocity of this 4-kg mass increases at only half the rate obtained for the 2-kg mass. Since momentum is the product of mass and velocity, we see by this second experiment that, indeed, the rate of momentum change produced by a *given* force is not dependent upon the particular body to which it is applied. A wide variety of observations, including the motion of distant stars, indicates that Newton's second law is indeed a universal law.

If the mass of the object described by (4-3) is not changing, then the only way the momentum can change is if the velocity of the object changes. But a changing velocity implies acceleration, so as long as the mass is constant, we can rewrite (4-3) in terms of acceleration as follows:

$$F = \frac{dp}{dt} = \frac{d}{dt}\,(mv) = m\frac{dv}{dt} = ma$$

The last equality follows from the definition of acceleration. Hence, if the mass of an object is unchanging, we may write Newton's second law in this form:

$$F = ma \qquad\qquad (4\text{-}4)$$

If m is given in kg and a in m/s², the units of force are seen to be kg·m/s². The force that produces a 1-m/s² acceleration when acting on a 1-kg mass is called a *newton*. The dimension of the newton can always be replaced by kg·m/s², from its definition. In the international system of units, the newton is abbreviated N.

N – newton

kg · m/s²

Example   A 2-kg object accelerates at 10 m/s². How large is the force acting upon it?

Solution $F = ma = 2 \text{ kg} \cdot 10 \text{ m/s}^2$

$$= 20 \text{ kg} \cdot \text{m/s}^2 = 20 \text{ N}$$

An experimental observation that is often required in application of Newton's second law is the following: if more than one force acts upon an object, it is the *net* force that determines the acceleration. Two equal forces acting in opposite directions upon an object produce no acceleration. Such is the case for a book resting on a table: the force of gravity pulls down on the book and the table pushes up on the book. The net force, and hence the acceleration, is zero. When we rest in a chair, certain of our nerve endings are quite aware of the upward force provided by the chair.

Remove the chair and the unopposed force of gravity causes us to accelerate downward. For forces that are parallel to one another, the net force is simply the algebraic sum of forces.* If the forces act horizontally with respect to the observer, it is customary to consider forces (and acceleration) to his right to be positive and those to his left, negative. Similarly, for forces acting vertically, upward-directed forces are customarily considered positive, downward forces negative.

Example   An automobile of 1000-kg mass is acted upon by an accelerating force, F_1, as shown in Fig. 4-4. A retarding force, F_2, due to air resistance and friction acts in the opposite direction. If the magnitudes of F_1 and F_2 are 1000 N and 500 N, respectively, find the acceleration.

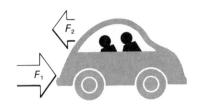

Fig. 4-4.   An automobile acted upon by an accelerating force, F_1, and a retarding force, F_2, due to friction and air resistance.

Solution   Considering forces to the right to be positive and those to the left negative, we write:

$$F_1 = +1000 \text{ N} \qquad F_2 = -500 \text{ N}$$

$$F_{net} = F_1 + F_2$$

$$a = \frac{F_{net}}{m} = \frac{F_1 + F_2}{m} = \frac{1000 \text{ N} - 500 \text{ N}}{1000 \text{ kg}} = \frac{1}{2} \frac{\text{m}}{\text{s}^2}$$

The acceleration is directed in the positive direction, toward the right. The units of our answer follow easily from the fact that the dimensions, $\text{kg} \cdot \text{m/s}^2$, can always be substituted for N. If Fig. 4-4 is unchanged but we reverse our sign convention (consider forces and accelerations to the left to be positive), will we conclude that the car accelerates toward the left? No, for then we would write

$$a = \frac{-1000 \text{ N} + 500 \text{ N}}{1000 \text{ kg}} = -\frac{1}{2} \frac{\text{m}}{\text{s}^2}$$

Since accelerations to the left are now considered positive, our negative result once again indicates that the car accelerates to the right. Physics does not depend on sign conventions. As long as we are consistent with our choice of signs, we will obtain consistent results.

Exercises

4. A body with a mass of 10 kg is observed to accelerate uniformly at the rate of 4 m/s². What is the magnitude of the net force acting upon it?

---

*The net force that results from nonparallel forces is discussed in Chapter 8.

5.   An engineer determines that a driving force of 10^3 N is required
     to maintain a test car in motion with a constant velocity of
     10 m/s. At a steady speed of 30 m/s, the required force is
     9×10^3 N. What is the magnitude of the retarding force caused
     by friction and air resistance at each of these two speeds?

4-7   Gravitation, Weight, and Inertial Mass

A phenomenon that we frequently observe is that of objects in
free fall. They are apparently acted upon by some force, since they
are observed to accelerate downward. We call this the *force of
gravity;* we are most aware of it when we hold an object so that it
cannot accelerate downward. In this case we supply an upward
force so that the *net* force on the object is zero. The magnitude of
this gravitational force is called the *weight* of an object.

The weight of an object is easily inferred from its mass (and
vice versa) because of an important observation of Galileo: he
performed a wide variety of experiments and concluded that all
falling objects experience the same acceleration in the absence
of air resistance. A feather normally stops accelerating downward
shortly after its release, because the upward force of the air resistance on the feather equals the force of gravity. However, a startling
demonstration of Galileo's conclusion was performed originally
by Newton: if a feather and a lead weight are simultaneously released inside a glass vessel from which the air has been pumped,
the feather will fall as swiftly as the weight.

The conclusion of many measurements is that all objects, when
acted upon *only* by the force of gravity, accelerate downward
(toward the center of the earth) with the same acceleration. The
magnitude of this acceleration is 9.8 m/s² (32.2 ft/s²) and is symbolized by the letter *g*.* Since objects with any mass *m* experience
the same acceleration, *g*, when acted upon by their weight *W*, we
may write Newton's second law for any object in free fall:

$$W = mg \qquad\qquad (4\text{-}5)$$

We see that the weight of an object (a force) is *g* times its mass.
This exact proportionality between the weight of an object and its
inertial mass is what allows us to use an ordinary gravity balance
to determine whether or not two *masses* are equal. The gravity
balance, strictly speaking, determines only equality of *weight*, but
we now see that it implies equality of *mass*, as well. Whenever an

[handwritten margin note: All objects when acted upon only by the force of gravity, accelerate downward with the same acceleration]

---

*The value of *g* varies slightly over the surface of the earth, as discussed in Chapter
17.

object is acted upon only by its weight, it accelerates with $a = g$. This leads to a system of units widely used by aircraft engineers and pilots but rarely discussed in textbooks. The units of force are pounds, the units of mass are *also* pounds, and the units of acceleration are g's. Thus a 1-lb force acting on a 1-lb mass (precisely the case when we release a 1-lb weight) produces an acceleration of 1 g. A 2-lb force acting on the 1-lb mass would produce an acceleration of 2 g's.

Example   A 10-lb rocket initially at rest is acted upon by a 50-lb thrust for two seconds. What is its velocity at the end of two seconds?

Solution
$$a = F/m = \frac{50 \text{ lb}}{10 \text{ lb}} = 5 \text{ } g\text{'s}$$

In terms of feet and seconds, we know that $g = 32$ ft/s², so $5g = 160$ ft/s². For an object initially at rest and subjected to a constant acceleration, we also know that

$$v = at = (160 \text{ ft/s}^2) \cdot 2 \text{ s} = 320 \text{ ft/s} \text{ } (= 218 \text{ mi/hr})$$

Often we wish to use familiar units for acceleration, such as m/s² or ft/s². In that case we must choose our units of force and mass properly if we wish to use $F = ma$. Three systems of units that accomplish this are tabulated below.

System of Units	Force	Mass	Acceleration
mks (SI)	newtons	kg	m/s²
cgs	dynes	g	cm/s²
British engineering	pounds	slugs	ft/s²

The force unit of the mks system (the newton) is chosen so that a one-newton force acting on a one-kilogram object produces an acceleration of one meter/second². Notice that the magnitude of one newton must be 1/9.8 times the weight of one-kilogram mass, since the weight of a one-kilogram mass acting upon itself, as in free fall, produces an acceleration of 9.8 m/s².

Note that *weight* refers to a *force*, namely, the force that is due to the earth's gravity acting upon an object, and that mass refers to an object's inertia. The terms are often inadvertently interchanged, since they are proportional to one another through the relationship $W = mg$. It certainly is not an obvious relationship; it is a fact we deduce from the common value of acceleration for all objects in free fall.

The English system of units adopts the pound as a unit of *force*, then chooses mass units (slugs) such that a one-pound force acting on a one-slug mass produces an acceleration of one foot/second². The slug is equal to the mass of a 32.2-lb weight. Thus, the mass of an object in slugs is 1/32.2 times its *weight* in pounds. When in doubt, the student should recall that the weight of an object acting unopposed upon the object produces the acceleration of free fall: $1\ g = 32.2\ \text{ft/s}^2$.

Example   A compressed-gas cylinder contains gas at a pressure of 2000 lbs/in.² and weighs 100 lbs. It is accidentally knocked over, breaking off the valve, thus creating a hole with an area of 2 in.². The cylinder is now a rocket! What is its initial acceleration?

Solution   Before the accident, the forces of the confined gas pressing outward on the cylinder walls are balanced in all directions, each square inch of surface supporting an outward force of 2000 lbs. When the valve breaks off, there is a 4000-lb force acting on 2 in.² at one end of the cylinder that is no longer balanced by a compensating force at the valve end. This is shown in Fig. 4-5.

   Thus, a 4000-lb net force acts upon the cylinder (only briefly, since the gas pressure quickly drops as the gas rushes out of the opening). Since the force acting is much larger than the weight of the cylinder, it produces an acceleration much larger than that due to gravity. Let us calculate the acceleration using two systems of units that are familiar in the U.S.

1. $$F = 4000\ \text{lb} \qquad m = \frac{100\ \text{lb}}{32.2\ \text{ft/s}^2} = \frac{100}{32.2}\ \text{slugs}$$

$$a = F/m = \frac{4000\ \text{lb}}{(100/32.2)\ \text{slugs}} = 40 \cdot 32.2\ \text{ft/s}^2 = 1288\ \text{ft/s}^2$$

2. $$a = F/m = \frac{4000\ \text{lb}}{100\ \text{lb}} = 40\ g\text{'s} = 40 \cdot 32.2\ \text{ft/s}^2 = 1288\ \text{ft/s}^2$$

The magnitude of this acceleration makes clear the importance of safety regulations that require protective caps over the valves on high-pressure gas cylinders.

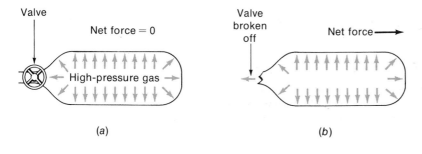

(a)   (b)

Fig. 4-5.   (a) This gas cylinder is in equilibrium. All outward forces are everywhere balanced by outward forces on the opposite wall. (b) Now the gas cylinder is acted upon by an unbalanced force, since the valve is missing.

Newton's second law is not always used in its simplest form. In some fields, such as chemical engineering, it is sometimes convenient to write it in the form

$$F = kma \qquad (4\text{-}6)$$

where k is a constant chosen to maintain the equality when mixed units are used. For example, if we express F in lb, m in lb, and a in ft/s², we would have $k = (32.2 \text{ ft/s}^2)^{-1}$. In this example, no more effort is required to use (4-6) instead of (4-4); since masses in the U.S. are usually given in lb, use of (4-4) requires conversion of lb into slugs in any case.

The strict proportionality between mass and weight given by (4-5) does not imply that the weight of a given body is always the same. The *mass* of a given body is constant and can always be determined by balancing against standard masses with either a gravitational or an inertial balance. But the *weight* of a body is the force exerted by the gravitational attraction that the earth exerts upon it. This force varies slightly at different points on earth. This variation causes the acceleration produced by gravity, g, to vary somewhat over the surface of the earth, as discussed in Chapter 17. For a body of known mass, then, we can determine its weight by use of (4-5) if we know the magnitude of gravitational acceleration at its location. In fact, this value varies so little that we usually use an average value of 9.81 m/s² as if it applies everywhere. On other gravitating bodies, such as our moon or other planets, the gravitational force is quite different in magnitude and can also be found by use of (4-5).

Exercises

6. The moon exerts a weaker gravitational force than the earth. Any object dropped in free fall on the surface of the moon accelerates at 5.32 ft/s². What is the magnitude, in pounds, of the moon's attraction for an object on its surface that weighs 200 lb on the earth's surface?

7. A model rocket has a mass of $1\frac{1}{2}$ kg and produces a thrust (force) of 30 N. What is its initial acceleration if launched in the vertical direction? Do not neglect the force of gravity.

8. The owner of a 3000-lb automobile neglected to set the parking brake. It coasted down his driveway then onto a level street. When the owner looked up it was coasting freely at a steady 6 mi/hr (8.8 ft/s) down the street. He ran after it, grabbed the rear bumper and, dragging his feet, exerted a force of 140 lb

on it. How many feet farther did the car move before it came to rest? How would this number change if it were a 1500-lb auto?

4-8   Inertial Frames

The law of inertia states that the velocity of an object is constant in the absence of a net force. Since the velocity is given by dx/dt, we must specify a coordinate system, or *frame of reference*, that determines x, if this law is to have any meaning. For example, if we stand next to a frozen pond and watch an ice cube coast freely along the surface, we can claim that we are witnessing an illustration of the law of inertia. Yet a passing motorist in an *accelerating* car sees something very different if he measures positions with respect to *his car* rather than with respect to the lake shore. He might at first see the ice cube moving forward with respect to his car, then slow and come to rest *with respect to his car* as he increases speed. Finally, he might well see the ice cube moving backward in his frame of reference as he increases his speed still more. He can conclude that since the ice cube is accelerating, the net force on the cube is not zero. This contradicts the conclusion of the stationary observer.

From the foregoing illustration it is clear that Newton's first law, the law of inertia, is not valid in arbitrary frames of reference. How do we know when it *is* applicable? We proceed as follows. First, we must find *by experiment* a frame of reference in which the law of inertia does hold. Our frozen ice pond is an example. Such a reference frame is called an *inertial frame*. Then any other frame of reference that moves at a *fixed* velocity with respect to this inertial frame must also be an inertial frame. This can be seen by examination of Fig. 4-6, where we consider motion only along the x direction, as we did in the introductory portion of this book. The entire reference frame with coordinates x' and y' is moving at

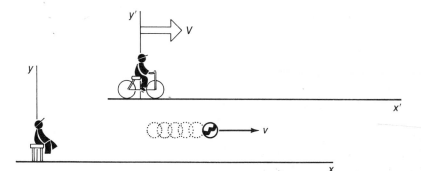

Fig. 4-6. The entire frame of reference composed of x' and y' is moving with the boy on the bike at velocity V to the right with respect to the boy at rest; whose frame of reference contains x and y. Thus a ball moving with velocity v in the lower frame of reference is seen to be moving more slowly by the boy on the bike.

constant velocity V with respect to the frame containing x and y. Thus if a body moves with velocity v in the frame of x and y, it moves at velocity

$$v' = v - V \qquad (4\text{-}7)$$

with respect to the moving frame of x' and y'. Now, if v is constant and V is also a fixed velocity, then (4-7) tells us that v' is also constant. Thus, the object that was moving uniformly in one frame does so also in the other. If the law of inertia holds in one frame, it also holds in the other.

> All inertial frames are related to one another by virtue of the fact that each of them must move only with a constant velocity when seen from any other inertial frame.

Once we have found an inertial frame in which Newton's first law holds, we have also found a frame in which Newton's *second* law holds, for precisely such a frame of reference was implicit in our initial discussion of Newton's second law. This, however, implies that *accelerations* are identical when viewed from any inertial frame of reference. In other words, if we apply a given force F to a given mass m and then view it from various inertial frames of reference, we should find that

$$F = ma \qquad \text{(Newton's second law)}$$

is satisfied in all cases. That the acceleration measured in all inertial frames *is* the same is easily deduced from (4-7), where, we let v (and hence v') vary in accelerated motion but maintain V as constant, as it must be for an inertial frame:

$$v' = v - V \qquad (4\text{-}7)$$

$$a' = \frac{dv'}{dt} = \frac{dv}{dt} - \frac{dV}{dt}$$

But $dV/dt = 0$ since $V = $ constant, and so

$$a' = \frac{dv}{dt} = a$$

Accelerations have the same value for all inertial frames.

A coordinate system attached to an accelerating automobile is an example of a noninertial frame. In such an automobile, an ice cube resting on a horizontal tray appears to accelerate toward the rear of the auto, even though no net force acts upon it. The motion of the ice cube *with respect to the automobile* violates the law of inertia, conservation of momentum, and Newton's second law

($F = ma$). Yet an observer standing still with respect to the road-side outside the automobile would see that the ice cube did not accelerate—the auto simply accelerated out from under it! The frame of reference of the observer who is stationary with respect to the earth is an inertial frame in which Newton's laws of motion are valid. To deduce what happens in the accelerated frame of reference, we apply the laws of motion in a convenient inertial frame, then ask how this motion appears to an accelerated observer. In the present example, the ice cube is acted upon by no net forces, so it moves at constant velocity in an inertial frame. The auto accelerates forward, so we easily deduce that to an observer in the car, the velocity of the ice cube is not constant; it accelerates toward the rear of the car. In this particular case:

a(cube with respect to car) $= -a$(car with respect to earth)

The idea that *any* inertial frame of reference is valid for application of the laws of dynamics is called the *principle of relativity*. It is an observed fact that one of the inertial frames of reference in the universe is the "frame of the fixed stars." This is the frame of reference tied to the average mass distribution of the universe. This observation prompted Mach to attribute the inertia of a body to the presence of the rest of the matter of the universe.* Accelerations with respect to the frame of the fixed stars then determine forces. This view is called *Mach's principle*.

Exercises

9. Two children sit facing one another 3 m apart in the aisle of a passenger train that is moving with a constant velocity of 30 m/s. One of the children holds a toy cart stationary with respect to the train. He then rolls it along the aisle toward the other child, rolling it in the direction of the motion of the train and with uniform velocity with respect to the train. The second child similarly returns it to the first. Each transit takes one second. An observer at a fixed point beside the tracks would measure three different velocities of the cart in his frame of reference during this entire process. Find these three velocities.

---

*Ernst Mach (1838–1916) was an Austrian physicist and philosopher. His philosophical stance was based strictly on phenomena and rejected metaphysical elements. Besides his studies of the foundation of mechanics, he also studied supersonic motion. A velocity scale, based on the speed of sound, is named for him and is expressed in terms of the *Mach number* discussed in Section 22-5.

10. The situation is as in the preceding exercise except (a) the train is accelerating with $a = +\frac{1}{2}$m/s², and (b) the forward most child simply releases the cart, which is assumed to be frictionless. The cart is initially at rest with respect to the train. Find the time required for the cart to reach the other child.

11. Use the conditions of Exercise 10, except assume that one of the children simply holds the cart fixed with respect to the accelerating train. Find the magnitude of the force he must exert on the cart if its mass is 4 kg.

4-9 Summary

Inertial mass, m, is a measure of a body's resistance to changes of velocity.

Momentum is defined by $p = mv$, and the total momentum of a system,

$$p = \sum_{i=1}^{N} m_i v_i$$

is constant if that system is not acted upon by external forces.

Force is defined by Newton's second law: $F = dp/dt$, which becomes $F = ma$ for constant masses. It is important to note that F is the *net* force acting on the mass.

Since all objects in free fall have the same acceleration, g, when acted upon only by their weight, W, we conclude that $W = mg$. If W is given in lb, and m also in lb, then a in the equation $F = ma$ will be in units of g's. More conventional choices of units for acceleration lead to the force and mass units tabulated on p. 80.

An inertial frame is a coordinate system in which the law of inertia, conservation of momentum, and Newton's second law are all valid. Any frame moving with constant velocity with respect to an inertial frame is itself an inertial frame.

Problems

4-1. Two ice skaters, initially stationary on the ice, push away from one another. One of them, with mass of 70 kg, is moving at a speed of 2 m/s. The other is moving in the opposite direction at a speed of 4 m/s. What is the mass of the second skater?

4-2. Two ice skaters are initially stationary on the ice. They push away so that they are separating from one another at 12 ft/s.

If one skater weighs 160 lb and the other 110 lb, what are their separate velocities with respect to the ice?

4-3. A space station that is initially stationary in our chosen frame of reference is struck by a meteorite moving at 28,000 km/hr. The mass of the station is 10^4 kg and that of the meteorite is 23 kg. If the meteorite embeds itself in the space station, what is the velocity of the space station after the impact?

4-4. A train consisting of a few freight cars and an engine is coasting freely on level track toward a loaded freight car, which is stationary with its brakes off. The moving train strikes the single car, the coupler locks them together, and the train + car continues coasting freely down the track. A brakeman noted that the velocity before the collision was 6 ft/s, after the collision it is 5 ft/s, and he knows the weight of the stationary car is 30 tons. Find the weight of the original train.

4-5. A small automobile weighing 1600 lb and moving at 30 mi/hr (44 ft/s) collides inelastically (that is, the cars stick together) with a 6400-lb automobile that is at rest. The brakes of both autos are off, so that they coast freely. (*a*) What is the velocity of the two cars after the collision? (*b*) Assuming the collision lasts 1/10 s, what is the average acceleration of the small car during the collision? (*c*) What average force is supplied by a seat belt to restrain a 160-lb occupant of the small auto during the collision? (*d*) What average force is experienced by a 160-lb occupant fixed in the large auto?

4-6. A 5-g bullet moving horizontally to the right with a velocity of 300 m/s strikes a 600-g steel block that is initially at rest. Following the collision, the bullet moves to the left at 290 m/s. Sketch and quantitatively label a momentum-vs-time graph for the time interval starting slightly before the collision time and ending slightly after. On this graph, plot the momentum of the bullet with a solid line, that of the block with a solid line, and the total momentum with a dotted line.

4-7. A man of mass m stands in the bow of a small boat of mass $M;$ both are at rest with respect to the water. The man then walks at a steady velocity v along the length L of the boat and stops at the rear of the boat. You may assume a lack of friction between the boat and water. (*a*) what is the velocity of the boat with respect to the water if the man's velocity, $v,$ is given with respect to the water? (*b*) Under the conditions of part *a,* how far does the boat move altogether? (*c*) What is the velocity of the boat with respect to the water if the man's velocity, v, indicates his velocity with respect to the *boat?*

4-8. Two boys stand in a boat separated by a distance of 2 m and throw a stone back and forth to one another. The combined mass of the boat and boys is 5 slugs and the mass of the stone is $\frac{1}{4}$ slug. Make this a one-dimensional problem by assuming that the stone moves in a straight, horizontal line between the boys. Further, let it move with a constant velocity of 4 m/s during each throw. Describe the motion of the boat with respect to the water if the boat is initially stationary. Be complete and be quantitative.

4-9. A man is at rest in a chair that is on wheels so that it can roll freely. He throws a rock of mass m with a velocity v_s (the separation velocity) with respect to *himself.* Note that in the inertial frame of the earth, he recoils at a velocity V, the rock moves off at velocity v, and $v_s = v - V$ (recalling that v and V are oppositely directed). The mass of the man and chair together is M. Find V in terms of v_s, m, and M.

4-10.* Repeat Problem 4-9 except assume that the man holds two rocks of mass m. (*a*) He throws the rocks simultaneously at a separation velocity of v_s; find V. (*b*) He throws the rocks sequentially, each with a separation velocity of v_s. Find $V_{final} = V_1 + V_2$, where V_1 is the first velocity gained and V_2 is the second recoil velocity measured with respect to a new inertial frame moving at V_1. (*c*) Which method of throwing the rocks gives the greatest net recoil velocity?

4-11. A 1-kg weight is equivalent to a 2.2-lb weight and 1 m = 3.28 ft. Calculate the conversion factor between pounds of force and newtons.

4-12. A steam locomotive weighs 985,000 lb and can produce an accelerating force of 95,500 lb. (*a*) What is the maximum acceleration of such a locomotive with no load other than itself? (*b*) With this acceleration, how long does it take to reach 60 mi/hr?

4-13. A parachutist with a mass of 80 kg is descending with a constant velocity of 10 m/s. What is the magnitude of the upward force supplied by the parachute?

4-14.* A model rocket producing a thrust of 300 N has an initial mass of $1\frac{1}{2}$ kg. It is launched in the vertical direction, so that the thrust acts upward upon it and its weight acts downward. It burns off all of its fuel at a constant rate in 1.4 seconds. The fuel has a total mass of $\frac{1}{2}$ kg. (*a*) Sketch a graph of *a* vs *t* for the first five seconds of motion of this rocket. (*b*) Find its velocity five seconds after launch. (An integral leading to a natural logarithm is involved in this portion of the problem.)

4-15. A 15-kg mass is accelerated from rest with a force of 100 N. As it moves faster, friction and air resistance create an oppositely directed retarding force given by

$$F_R = A + Bv$$

where $A = 25$ N and $B = 0.5 \dfrac{\text{N}}{\text{m/s}}$.

(*a*) At what velocity does the acceleration equal one-half of the initial acceleration? (*b*) At what velocity does the acceleration vanish?

4-16.* A space capsule with a mass of 500 kg is reentering the earth's atmosphere with an initial velocity of 8×10^3 m/s. Air resistance produces a retarding force of 5×10^4 N. The heat shield is rapidly eroded during reentry, so that the total mass falls from 500 kg to 400 kg in 30 seconds. Assuming a constant retarding force and neglecting the force of gravity, find the velocity at the end of the 30-second interval. You may assume that the eroded material produces no recoil effects; that is, each bit of the eroded material leaves the capsule with zero initial separation velocity. (An integral leading to a natural logarithm occurs in this problem.)

4-17. An object is initially located at $x' = a$ in a coordinate system whose origin is initially at $x = b$ in another coordinate system. Further, the x' coordinate system moves in the positive x direction at a constant velocity V. The object moves in the x' system with a constant velocity v'. (*a*) Write an expression for x' as a function of time for the object. (*b*) Write an expression for x as a function of time for the object in terms of a, b, v', and V.

4-18. A passenger with a mass of 50 kg rides without a seat belt in an auto moving at 20 m/s. The auto crashes into a brick wall, coming to a complete stop almost instantly. The passenger, obeying Newton's first law, continues forward until he is rudely and suddenly stopped by the dashboard of the now stationary auto. He comes to rest in 10^{-2} s. (*a*) Find the passenger's velocity before and after he hits the dashboard in the inertial frame of reference of the brick wall. (*b*) Find the passenger's velocity before and after he hits the dashboard in the inertial frame that always moves with the initial auto velocity of 20 m/s. (*c*) Calculate the average acceleration of the passenger during the 10^{-2} s collision in both frames of reference. (*d*) What is the magnitude of the force exerted upon him by the dashboard? Does this number depend upon which frame of reference you chose?

4-19.   An ice cube with a mass of 50 g rests on a smooth, level tray in an automobile accelerating with respect to the inertial frame of the road. The automobile's acceleration is 100 cm/s². (*a*) Find the force, in dynes, that must be exerted on the ice cube to keep it in place. (*b*) Describe the motion of the ice cube as seen by an observer standing at the roadside if the cube is released. (*c*) Describe the motion of the ice cube as seen by an occupant of the car if the cube is released. (*d*) What sensations other than observation of the ice cube make it clear to the occupant of the car that the car is accelerating?

4-20.   A cart-collision experiment is carried out in a speeding train. In the train, the collision is of the type described in the first line (elastic bumpers) of Table 4-1. The mass of each cart is 1 kg and the initial cart velocity with respect to the train is 2 m/s. (*a*) Find the momentum of each cart before and after the collision in the frame of reference attached to the train. (*b*) Find the momentum of each cart before and after the collision in the frame of reference of the railroad tracks. The train's velocity with respect to these tracks is 10 m/s. (*c*) Is momentum conserved in this second frame of reference?

4-21.   A polar bear with a mass of 350 kg is standing on an initially stationary floating cake of ice that has a mass of 7000 kg. What happens to the cake of ice if the polar bear walks 10 meters (with respect to the ice) in a straight line, then stops?

Chapter 5

Applying the Laws of Dynamics

5-1 Introduction In the preceding chapter we discussed several laws of dynamics: the law of inertia (Newton's first law), conservation of momentum, and Newton's second law. We discussed these laws in the restricted domain of motion along only one dimension. In this chapter we will continue to consider one-dimensional motion, but we will discuss some of the complications that arise when we apply laws of dynamics to various types of systems. We will consider accelerated systems. We will study the way forces are transmitted between parts of a system by struts and cables. We will discuss systems that involve several different masses that exert forces upon one another and are also acted upon by external forces. In order to cope with such systems of interacting masses, we will develop the law that Newton called his third law of motion. Finally, we will discuss the motion of rockets, where the most interesting features are discovered by application of the law of conservation of momentum without explicit mention of forces.

5-2 Passengers in Accelerated Systems

A passenger in an elevator is acted upon by two forces. There is the force N of the elevator floor pressing upward upon the bottom of his feet. He is quite aware of this force through the nerve endings in the soles of his feet. He is also acted upon by a downward directed force, which is his weight, mg. The sum of these two forces is denoted F_{net}, i.e.,

$$F_{net} = N - mg = ma$$

where the minus sign indicates that mg acts downward. Figure 5-1 schematically represents these forces acting upon the inertial mass of the passenger. Figure 5-2 shows the passenger in the elevator under various conditions of acceleration. The upward force on his feet, N, can be conveniently measured by interposing a bathroom scale between his feet and the floor. When the elevator is not accelerating, as when it is at rest or moving with uniform velocity between floors, the net force on the passenger must equal zero, since $F_{net} = ma$ and $a = 0$. Thus the opposing forces N and mg must then be equal in magnitude so that they perfectly cancel, i.e.,

$$N - mg = 0 \qquad (\text{when } a = 0)$$

and hence $N = mg$. The upward force precisely equals the passenger's weight in this case. If the elevator (and hence the passenger) accelerates upward, the force N must exceed the magnitude of mg in order to provide a net upward force. Conversely, a downward acceleration requires $N < mg$, to produce a net force down.

Example   A 120-lb boy in an elevator stands on a scale, which reads 80 lb. Find the acceleration of the elevator.

Solution
$$N - mg = ma$$

$$a = \frac{N - mg}{m} = \frac{80 \ lb - 120 \ lb}{(120/32) \ slug}$$

$$= \frac{-40 \ lb}{3.75 \ slug} = -10.67 \ \text{ft/s}^2$$

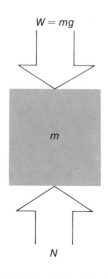

$$W = mg$$

$$m$$

$$N$$

Fig. 5-1.   Schematic representation of the forces acting upon a passenger in an elevator. The upward-directed force N is that of the elevator floor pressing against his feet, while W represents the force of gravity.

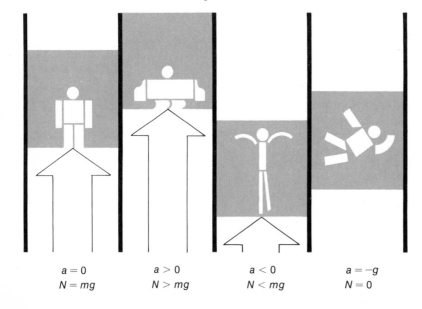

$a = 0$	$a > 0$	$a < 0$	$a = -g$
$N = mg$	$N > mg$	$N < mg$	$N = 0$

Fig. 5-2.   Four cases of motion of the elevator are considered, including the case of free fall $(a = -g)$.

where the negative acceleration indicates that the elevator accelerates downward.

If the cable snaps, so that the elevator accelerates downward at g, the force of gravity on the passenger provides this acceleration with no additional forces between the elevator and the passenger. Accordingly, $N = 0$ in this case. We sometimes say that "the passenger feels weightless" in this situation. Actually, the force of gravity still acts upon him and still is given by $W = mg$. If it were not present he would not be accelerating rapidly down toward an unpleasant fate! However, he is used to living under the *balance* of various forces. The floor presses up on his feet, his body presses down on his legs, and his knee joints are squeezed together by these opposing forces. Myriads of nerve endings there and throughout his body tell him of this normal state of affairs. In free fall, the absence of these sensations is instantly sensed and leads to a panicky sensation that, in some individuals, produces nausea. These fear reflexes have evolved in humans for obvious reasons.

It is interesting to note that the person in a freely falling elevator experiences exactly the same sensations as one who is truly *weightless,* that is, far removed from gravitating objects like the earth. On the other hand, a person in an elevator accelerating upward at $a = g$, far from any gravitating objects, would feel as if he were in the earth's gravitational field, since the floor would press up on him with $F = ma = mg$. In fact, there would be no way for him to distinguish uniform acceleration in the absence of gravity from a state of no acceleration in a gravitational field, unless he looked outside the elevator. Note that if the passenger in the gravity-free, accelerating elevator releases an object, it obeys the law of inertia and continues moving with the velocity it had at the moment of release. The passenger and the elevator continue to accelerate, however, so that the object is left behind — until it is hit by the floor of the elevator. Thus even the phenomenon of free fall is duplicated within the accelerated system without gravity present. This inability to distinguish the effects of gravity from those of an accelerated reference frame is called the *equivalence principle.*

An alternate statement of the equivalence principle is: "The inertia of a body and the gravitational force acting on that body are *exactly* proportional to one another." This exact proportionality leads to the law that all objects accelerate identically in free fall, as we have already seen. Einstein pushed the equivalence principle even further in developing his *general theory of relativity.* Einstein stated that *all* the laws of physics, including those of electricity and magnetism, are such that there is *no* way to distinguish the effects of a gravitational field and those caused by an accelerated frame of reference. This widened form of the equivalence principle is often called the strong form, while the proportionality of inertia and gravitational force is called the weak form of the equivalence principle.

Turning our attention to horizontal conveyances, we can analyze one of the forces that leads to discomfort in buses, trains, or airplanes. Fig. 5-3 shows our passenger accelerating forward. It is apparent that his head will not accelerate along with his body unless some forces are applied to it. Our neck muscles generally provide these forces and can be subjected to excessive strain if the acceleration is too large. However, even for modest values of acceleration, a major cause of discomfort is a *changing* acceleration that causes the passenger continually to change the tension in his muscles to maintain a comfortable position. Accordingly, transportation engineers worry about the rate of change of acceleration and give it a name — the *jerk:*

$$j = da/dt$$

The name is apt, since jerky motion is one with rapidly fluctuating accelerations.

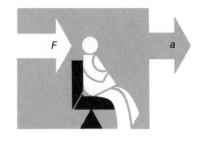

Fig. 5-3. A force *F* must be impressed upon a passenger's head to prevent it from lagging in an accelerated system. In the absence of a headrest, this force is supplied by vertebrae and neck muscles.

Exercises

1. A stationary automobile is struck from behind by a speeding automobile. The initially stationary auto briefly accelerates forward with $a = 40$ g's. How much force must be supplied to an occupant's head in order that it also accelerates forward at 40 g's? The occupant's head weighs 10 lb.

2. Consider an inertial frame of reference in deep interstellar space so that all gravitational forces can be neglected. An elevator is uniformly accelerating with respect to this frame of reference with $a = g$. The direction of this acceleration is always from the floor toward the ceiling of the elevator, so that an occupant believes he is in a "normal" elevator at rest on earth. Consider a ball released three feet from the floor of the elevator at $t = 0$. Its inertia causes it to continue to move at the velocity it possessed at $t = 0$. Calculate the time required for the accelerating floor of the elevator to reach this ball. Compare this time to the time of free fall of a ball falling three feet on earth.

5-3   Newton's Third Law

We have, up to this point, restricted our attention to single objects acted upon by one or more forces in both inertial frames of reference

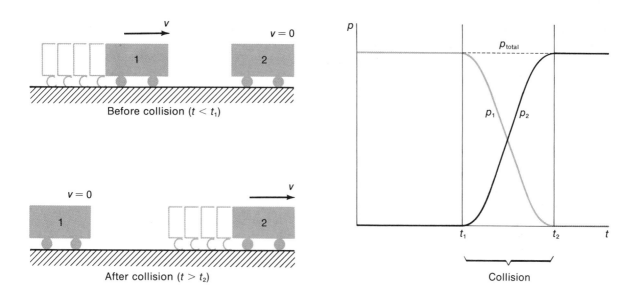

Before collision ($t < t_1$)

After collision ($t > t_2$)

p_{total}

p_1 p_2

t_1 t_2 t

Collision

and accelerating frames of reference. Now we wish to deal with systems containing more than one mass. In order to do so we must consider the effects of *internal* forces, that is, forces that act between elements of a system. For example, a car pulling a trailer constitutes a system that is acted upon by external forces (wind resistance and the forces between the roadway and tires) but also by internal forces (the forces between the car and trailer transmitted via the trailer hitch).

In such a system consisting of two parts, *A* and *B*, one of the first questions that comes to mind is "If *A* exerts a force on *B*, does *B* necessarily exert a force on *A*?" Newton considered this to be such an important question that it became the basis for his third law of motion: "To every action there is always opposed an equal reaction, or: the mutual actions of two bodies upon each other are always equal, and directed to contrary parts." Newton's answer to our question is that if *A* exerts a force on *B*, *B* *necessarily* exerts an equal but oppositely directed force on *A*. If such were not the case, effects contrary to experiment would occur.

For example, consider a cart-collision experiment, such as that sketched in Fig. 5-4. Before the collision, cart 1 has momentum $p_1 = mv_1$ and cart 2 has momentum $p_2 = mv_2 = 0$. After an elastic collision, p_1 is found to be zero while cart 2 has acquired the total momentum originally possessed by cart 1. A graph of these momenta is shown in Fig. 5-4. This graph shows that the total momentum ($p_1 + p_2$) of the system is constant *at all times*. In order for this to be true, every change in the momentum of cart 1 must be mirrored

Fig. 5-4. In an elastic collision between two carts of equal mass, where cart 2 is initially at rest, the momentum of cart 1 drops to zero during the collision, while that of cart 2 quickly rises. Note that the total momentum of the system ($p_1 + p_2$) is constant at all times.

by an equal and opposite change in the momentum of cart 2. But since the *force* of cart 1 on cart 2 is all that can change the momentum of cart 2, and similarly the *force* of cart 2 against cart 1 is all that can change the momentum of cart 1, it must be true that these forces are always equal in magnitude. More formally:

$$p_1 + p_2 = \text{constant}$$

Differentiating:

$$\frac{dp_1}{dt} + \frac{dp_2}{dt} = \frac{d\ (\text{constant})}{dt} = 0$$

Hence

$$\frac{dp_1}{dt} = \frac{-dp_2}{dt}$$

and, since $F = dp/dt$,

$$F_1 = -F_2$$

That is, the force on cart 1 is equal and opposite to that on cart 2. Such a pair of internal forces is called an *action-reaction pair* and their equality is an example of Newton's third law. We see for this simple system that Newton's third law of motion is a necessary consequence of the experimental fact that momentum is always conserved in the absence of *external* forces. Note that in this example, the force of each cart on the other is an internal force of the system consisting of both carts.

Consider now a system containing several masses that can exert forces upon one another *and* upon which external forces act as well. An example is given in Fig. 5-5, where the dotted line con-

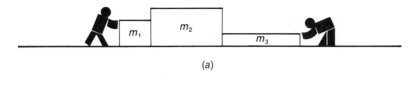

(a)

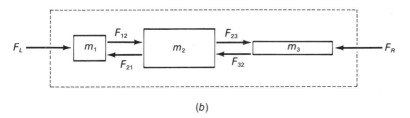

(b)

Fig. 5-5. A system of three bodies is enclosed in a dotted line to indicate that they constitute the system of interest. The forces F_L and F_R act on the system and are called *external forces*. The force pairs between the masses are within the system and are called *internal forces*. (a) Physical example. (b) Force diagram.

tains the three masses we take to be our system. Newton's third law states that any force exerted by mass 1 on mass 2, called F_{12}, is equal and opposite to the force F_{21} exerted by mass 2 on mass 1. Similarly, the forces F_{23} and F_{32} are equal and oppositely directed. Since the total momentum of the entire system is the sum of individual momenta, and since the internal forces cancel and produce no effect upon the total momentum, we conclude that only external forces change the total momentum. The internal forces can change the momentum of individual components of the system but only in such a way that all such changes cancel. For the system of Fig. 5-5, Newton's third law gives us equalities for the magnitudes of each action-reaction pair of internal forces:

$$F_{12} = -F_{21} \qquad (5\text{-}1)$$

and
$$F_{23} = -F_{32} \qquad (5\text{-}2)$$

where the minus sign indicates that these forces act in opposite directions. The net forces acting on *each* mass are

$$\text{mass 1:} \quad F_1 = F_L \pm F_{21} \qquad (5\text{-}3)$$

$$\text{mass 2:} \quad F_2 = F_{12} \pm F_{32} \qquad (5\text{-}4)$$

$$\text{mass 3:} \quad F_3 = F_{23} \pm F_R \qquad (5\text{-}5)$$

Note that F_{21}, F_{32}, and F_R will be negative quantities in the above equations, since they are directed toward the left.

   The change in total momentum for this system is given by the sum of the changes of the three individual momenta. Thus, since

$$p_{\text{total}} = p_1 + p_2 + p_3$$

we have
$$\frac{dp_{\text{total}}}{dt} = \frac{dp_1}{dt} + \frac{dp_2}{dt} + \frac{dp_3}{dt}$$

Applying Newton's second law, $F = dp/dt$, to each mass, this becomes

$$\frac{dp_{\text{total}}}{dt} = F_1 + F_2 + F_3$$

Combining Eq. (5-1) through (5-5), we see that

$$F_1 + F_2 + F_3 = F_L + F_R = \frac{dp_{\text{total}}}{dt}$$

since the action-reaction pairs cancel one another. Thus, the *total* change in momentum for the system is caused only by the *external* forces. Newton's second law therefore generalizes to include many forces acting upon a system of many masses that themselves exert forces upon one another:

$$F_{net} = \frac{dp_{total}}{dt} \tag{5-6}$$

where the words "net" and "total" hide the details discussed above. The result in (5-6) depends upon Newton's third law, "For every action there is an equal and opposite reaction":

$$F_{12} = -F_{21} \tag{5-7}$$

In fact, the success of (5-6) in describing experiments verifies Newton's third law.

Example   A man pushes against his stalled automobile to set it in motion. How many horizontal action-reaction pairs are present?

Solution   First, we note that the force exerted by the man's hands against the back of the auto is mirrored by the force exerted by the auto upon his hands. The latter force is the one he feels against the palms of his hands, the former is the force that acts upon the auto and sets it into motion. The man's feet also exert a force against the ground. If he were standing on a treadmill, this force would start the treadmill moving. Instead, he is standing on the earth, which has such an enormous inertial mass that its response to this force, though very real, is negligible. Because the man exerts a force against the ground, there is an equal reaction force of the ground against his feet. He desperately needs this force if he is going to do much in the way of getting his auto started. Without it he would simply recoil away from the auto when he pushed against it. This is precisely what occurs if he attempts this while standing on ice, where he cannot exert an appreciable horizontal force against the earth.

We might wonder how any motion is ever possible if every force is opposed by a precisely equal reaction force. The important point to recognize is that the action-reaction pairs are *exerted upon different bodies*. For example, if a man pulls a wagon with a 50-lb force, the wagon is acted upon by that force and accelerates. The reaction force of 50 lb in the opposite direction does not act upon the wagon—it acts upon the man.

Consider now a book resting on a table. It exerts a force downward on the table and the table exerts a reaction force back up against the book. The equality of these forces is an example of Newton's third law. If the book is not accelerating, we further conclude that the magnitudes of these book-table forces are equal to *mg*, the force of gravity upon the book. This equality is *not* an example of Newton's third law; we deduce it from Newton's *second* law. For the book:

$$F_{net} = ma = 0$$

and since $F_{net} = F$ (table acting on book) $-mg$ (earth attracting book) we conclude, F (table acting on book) $= mg$ (earth attracting book). But where is the action-reaction twin to the force mg? Since this force is caused by the earth's gravitational attraction upon the book, the book must likewise attract the *earth* with an equal force. Neither book nor earth accelerate in our example because the table between them exerts cancelling forces upon both book and earth.

Let us now remove the table abruptly, so that the book falls toward the earth. Now the book is acted upon only by its own weight, W, and the earth is similarly attracted upward by an equal reaction force. Suppose that the book of mass m_b falls for 0.4 s (about a $\frac{3}{4}$-meter fall); call this time τ. The earth, of mass m_e, is acted upon by the reaction force W, and thus accelerates with a magnitude

$$a_{earth} = \frac{W}{m_e} = \frac{m_b g}{m_e} \qquad \begin{array}{l} F = ma \\ a = \frac{F}{m_e} \end{array}$$

For a $\frac{1}{2}$-kg book we find

$$a_{earth} = \frac{0.5 \text{ kg}}{6 \times 10^{24} \text{ kg}} \; 9.8 \text{ m/s}^2 = 8.2 \times 10^{-25} \text{ m/s}^2$$

In the time τ, the earth moves a distance

$$d = \tfrac{1}{2} a \tau^2 = \tfrac{1}{2}(8.2 \times 10^{-25} \text{ m/s}^2)(0.16 \text{ s}^2) = 6.6 \times 10^{-26} \text{ m}$$

This displacement is 12 orders of magnitude smaller than the diameter of a single atomic nucleus! Small wonder that the earth feels steady underfoot.

Exercises

3. While the book and the earth are falling together, they constitute a system acted upon only by internal forces. Since they were initially at rest, they must move in such a way that their momentum remains zero. Can you use this fact to arrive at the distance moved by the earth?

4. A canoe is propelled by a man pushing a paddle through the water. List some of the action-reaction pairs involved.

5-4   Apparent Failures of Newton's Third Law

It is quite possible to think of interactions between two bodies that violate Newton's third law. If, for example, there is a *time delay* during which the force produced by one body propagates to the

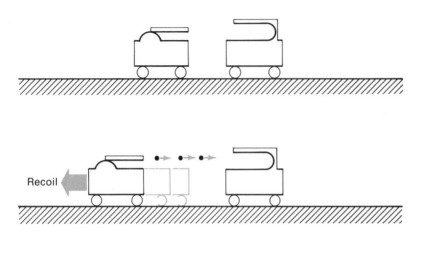

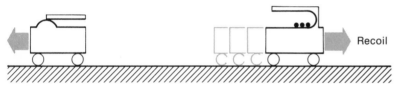

Fig. 5-6. Both carts are initially at rest. The cart that fires the bullets recoils before the cart that absorbs the bullets. An observer unaware of the bullets would conclude that Newton's third law was briefly violated.

other, then Newton's third law might be valid in a time average sense but not always correct at each instant.

Consider the simple mechanism shown in Fig. 5-6. There one cart fires bullets at another. When the bullets are fired, the first cart recoils. When these bullets strike the second cart, they cause it to move. If the bullets were soundlesss and invisible, we would note the first cart moving before the second cart. We would then conclude that Newton's third law was incorrect during this brief time: action was not always equal to reaction. By application of the law of momentum conservation, it should be clear that eventually Newton's third law would appear to be satisfied. After all of the bullets were absorbed, the final momenta of the two carts would be equal and opposite just as if equal and opposite forces had acted.

The two-cart system we have discussed may seem too contrived to be of genuine interest, particularly since action *does* always equal reaction if the observer is aware of each bullet. In fact, however, there is a useful analogy between this bullet-mediated interaction and the electromagnetic force between two electrically charged particles. In both cases the agent that conveys the interaction *itself* possesses momentum. For this reason, Newton's third law does not always appear to hold exactly for electromagnetic interactions if only the charged particles themselves are studied.

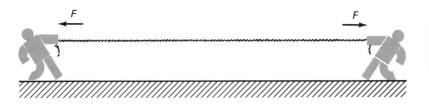

Fig. 5-7. The rope is not accelerating, hence the forces exerted on the rope ends are equal and oppositely directed.

5-5   Tension and Compression

Consider a rope being pulled upon by equal and opposite forces, F, as shown in Fig. 5-7. Since the forces acting *on the rope* are balanced, the rope does not accelerate. At either end of the rope, it is clear that the rope exerts a reaction upon the pulling man with a force of magnitude F. Thus, either man is aware that the rope pulls on him with a force F. The rope thus effectively "transmits" the pulling force, F, from one man to the other. The magnitude of this force is called the *tension* in the rope and is equal to the force exerted on (or by) the rope at either end. This tension exists throughout the rope. If the rope is cut at any point, the force required to hold each of the cut ends in place is equal to the tension at that point.

   Similar arguments can be made for the forces involved in the *compression* of the beam shown in Fig. 5-8. In this case the reaction force exerted by the beam at either end is directed outward and is equal in magnitude to F, the force compressing the beam. Again, the reaction force at either end essentially "transmits" the force exerted on one end of the beam to the object at the other end.

Example   Two horses pull on opposite ends of a rope with equal forces of 200 lb magnitude. Since the rope "pulls back" with a force of 200 lb, the tension in the rope is 200 lb. What is the tension if one end of the rope is tied to a fixed post instead?

Solution   Since the remaining horse still pulls with a force of 200 lb, the fixed post must *also* pull on the other end of the rope with a force of 200 lb. If the two forces did not balance, the rope would accelerate in the direction of the larger force, contrary to our statement of a *fixed* post. As before, Newton's third law shows that the reaction forces at either end of the rope pull (on the horse at one end and the post at the other end) with a magnitude of 200 lb. Thus the tension in the rope is still 200 lb. When we see two horses actively pulling on each end of

Fig. 5-8.   A beam under compression.

the rope, our intuition may tell us that the tension is twice as great as when one end is tied to a post. We see, however, that this is not the case.

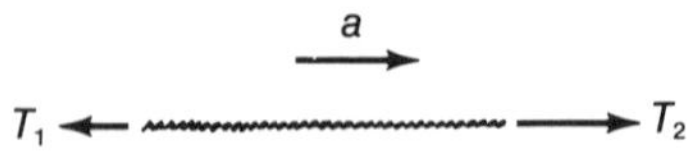

5-6 Forces Transmitted by Accelerating Ropes

If a rope of total mass m is accelerating, it must be acted upon by unbalanced forces. In this case the tensions at the two ends are not equal. Such an accelerating rope is shown in Fig. 5-9, where the tension T_2 must be greater than T_1, since a net force to the right is required.

Example   A force of 20 N is applied at one end of a 10-m long rope, the other end of which is attached to a 4-kg mass, as shown in Fig. 5-10. If the rope has a mass of 1 kg, find the tension in the rope at the end attached to the block. Ignore frictional and gravitational forces.

Solution   Since the difference in tension between the two ends of the rope depends upon the acceleration, we must first find the acceleration of the system. The applied force must accelerate the entire block + rope system, so we may write $a = F/M$, where M is the total mass of block + rope. This value of acceleration can then be used to compute the difference in tensions between the two ends of the rope. Finally, knowing the tension is 20 N where the force is applied, we can find the remaining tension at the block end of the rope. Calling the mass of the rope m and the entire mass M,

$$a = F/M$$

$$F - T_b = ma$$

$$T_b = F - ma = F - m(F/M) = F\left(1 - \frac{m}{M}\right)$$

$$= 20\,\text{N}\,(1 - \tfrac{1}{5}) = 16\,\text{N}$$

The preceding example shows that a rope used to accelerate a block "transmits" a force diminished by the factor $\left(1 - \dfrac{m}{M}\right)$. Since the masses of ropes or cables used to transmit forces between segments of moving systems (such as a wrecker pulling a car) are usually very small compared to the mass of the system, we frequently neglect the mass of the rope entirely and assume that the tension is constant along its entire length. In subsequent problems, the mass of the rope will be stated when it is necessary to take it into account. If the mass of the rope is not stated, assume it is negligible and the tension is constant throughout.

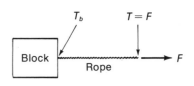

Fig. 5-9.   The tensions at each end of an accelerating rope are unequal. In this case $T_2 > T_1$.

Fig. 5-10.

Exercises

5. A 16-lb rope connects a horse to a cart. The horse pulls with a force of 100 lb and produces an acceleration of $\frac{1}{4}$ g. Neglecting the effects of gravity and friction, (*a*) what is the mass of the cart? (*b*) what is the tension of the rope at the end attached to the cart?

6. In Exercise 5, what is the tension at the center of the rope?

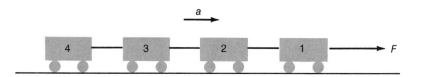

Fig. 5-11. Four trailers acted upon by a force *F*.

5-7. Many-Body Systems

Suppose a system contains several masses that interact with one another, as in Fig. 5-11. The force *F* is an external force that causes the four trailers to accelerate. If no other forces are acting, we can easily find the acceleration of this system:

$$a = F/m_{total} \qquad \text{where} \qquad m_{total} = m_1 + m_2 + m_3 + m_4 \qquad (5\text{-}8)$$

But what about the force transmitted by the couplers between the trailers? Each coupler is under tension and provides a force that is *internal* to the system. By reducing the size of the system, any of these internal forces can be made into an external force, as in Fig. 5-12. It is now clear that the force accelerating trailers 2, 3, and 4 is the tension T_{12} in the coupler between trailers 1 and 2:

$$T_{12} = (m_2 + m_3 + m_4)a \qquad (5\text{-}9)$$

With the aid of (5-8) this becomes

$$T_{12} = \frac{m_2 + m_3 + m_4}{m_1 + m_2 + m_3 + m_4} F$$

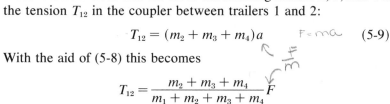

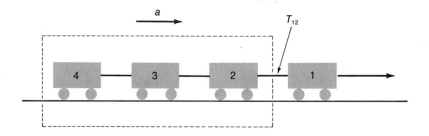

Fig. 5-12. The system that includes only trailers 2, 3, and 4 is accelerated by the tension T_{12} in the coupler between trailers 1 and 2.

We can now proceed to analyze the tensions in the remaining couplers in a similar fashion.

$$T_{23} = \frac{m_3 + m_4}{m_1 + m_2 + m_3 + m_4} F$$

$$T_{34} = \frac{m_4}{m_1 + m_2 + m_3 + m_4} F$$

[handwritten annotations:]
$F =$
$F =$
m
α
m
α
$a = \frac{F}{\text{total } m}$

The key element in finding the internal forces in the above illustrations was a **purposeful restriction in the size of the system** so that the **unknown force would appear in Newton's second law**, as in (5-9), where we could **solve for it.**

[handwritten annotations:]
MUST FIND
a for whole system
$F = ma$

An alternative and somewhat more formal and elegant procedure is to consider each mass of the system separately and write down $F_{net} = ma$ for each. Doing this for trailers 1 through 4 in order we obtain:

$$F - T_{12} = m_1 a$$

$$T_{12} - T_{23} = m_2 a$$

$$T_{23} - T_{34} = m_3 a$$

$$T_{34} = m_4 a$$

(5-10)

If F and the four masses are given, we have four unknowns: the three tensions and the acceleration. Equations (5-10) are thus four simultaneous equations that can be solved for the four unknowns. This procedure leads to the same solutions obtained above. It is worth reemphasizing the point that in order to obtain equations involving the unknowns, it was necessary to consider one mass at a time.

Example A weight of mass m_1 is connected to a cart of mass m_2 by means of a string passed over a pulley, as shown in Fig. 5-13. The system is released so that the weight falls, pulling the cart along. Find the acceleration and tension in the string. Note: (a) The acceleration of the two masses are equal since they are connected by a string of fixed length. (b) The mass of the string and friction are to be neglected. (c) The pulley is to be considered perfect so that it changes only the direction of the string, not its tension.

Solution First, note *well* that the tension in the string does *not* equal the weight of the hanging mass, $m_1 g$. If it did, this mass would be acted upon by two equal and opposite forces, the force of gravity downward and the tension upward. It then would not accelerate, contrary to the

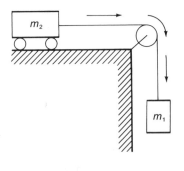

Fig. 5-13.

actual situation. How do we proceed? Isolation of the two masses is the key to a solution. For mass 1 we have the downward force of its weight, m_1g, and the upward force of the tension, T, as yet unknown. Taking the positive direction of acceleration to be down we have:

$$a_1 = \frac{F_{net}}{m_1} = \frac{m_1g - T}{m_1} \qquad (5\text{-}11)$$

The mass m_2 is acted upon only by the tension T:

$$a_2 = T/m_2 \qquad (5\text{-}12)$$

Since the length of the string is fixed, the two masses accelerate together with $a_1 = a_2$, so we may equate (5-11) and (5-12):

$$\frac{m_1g - T}{m_1} = \frac{T}{m_2}$$

Solving for T we find

$$T = \frac{m_1m_2}{m_1 + m_2}g$$

Now we can find the value of acceleration by substituting this value of T into either (5-11) or (5-12). Let us choose the simplest:

$$a = \frac{T}{m_2} = \frac{m_1g}{m_1 + m_2}$$

Note that the dimensions of our result are correct. Further, we can see an intuitively satisfying form in our solution for a. The numerator of our result is the driving force, the weight of m_1. The denominator is the total inertial mass of the system, $m_1 + m_2$. Our result is thus simply Newton's second law for the entire system:

$$a = \frac{F_{net}}{m_{total}} = \frac{m_1g_1}{m_1 + m_2}$$

We arrived at this result, however, not by guessing it would be correct, but by isolating each mass so that (5-11) and (5-12) were obtained without doubt. We then obtained both the tension and the acceleration by combining (5-11) and (5-12).

Exercises

7. In the preceding example substitute the result for T into (5-11) rather than (5-12) to see that the same result is obtained.

8. Three trailers, each with a mass of 2000 kg, are pulled by a tractor. How great is the force exerted by the tractor when the tension in the coupler between the middle and last trailer is 4000 N?

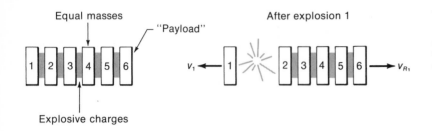

Fig. 5-14. A crude rocket consisting of six equal masses with five explosive charges between them.

5-8 Rocket Motion

We can now use the ideas developed in Chapter 4 to discuss some aspects of rocket motion. In order to more easily see the ideas that are involved, consider the crude device shown in Fig. 5-14. Beginning with the rocket at rest and with no external forces acting, the firing of the first explosive charge will cause mass 1 to move off to the left, while the remainder of the rocket (masses 2–6) recoils off to the right. Since the net momentum of this system was zero before the explosion, it must remain so afterward. Thus, the oppositely directed momenta of mass 1 and the remainder of the rocket must be equal in magnitude:

$$mv_1 + 5mv_{R_1} = 0$$

$$5mv_{R_1} = -mv_1 \qquad (5\text{-}13)$$

where v_1 is the velocity of m_1 and v_{R_1} is the velocity gained by the remainder of the rocket. The minus sign indicates these velocities are in opposite directions. Unfortunately, v_1 is not the quantity that an engineer would measure in determining the effectiveness of the explosive charges. He would clamp the "rocket" down and measure the velocity with which m_1 separated from the rocket. We might call this separation velocity the *exhaust velocity, v_e.* In our example the exhaust velocity is given by

$$v_e = v_{R_1} - v_1 \qquad (5\text{-}14)$$

where v_e is the velocity of exhaust relative to the rocket and v_1 is the velocity of exhaust in the initial frame of reference. Rewriting (5-13) by use of definition (5-14), we find:

$$5mv_{R_1} = -mv_{R_1} + mv_e$$

$$6mv_{R_1} = mv_e$$

$$v_{R_1} = \tfrac{1}{6}v_e$$

The rocket is moving to the right at 1/6 the exhaust velocity.

Now let us view the rocket in a new frame of reference, one that is moving along with it at velocity v_{R_1}. Since this frame of reference is moving at constant velocity with respect to the original frame of reference, it too is an inertial frame of reference in which the conservation of momentum law must be valid. When the second explosive charge is fired, the rocket will gain another increment of velocity, v_{R_2}. A similar derivation to that above now gives:

$$v_{R_2} = \tfrac{1}{5}v_e$$

Similarly,

$$v_{R_3} = \tfrac{1}{4}v_e$$

$$v_{R_4} = \tfrac{1}{3}v_e$$

$$v_{R_5} = \tfrac{1}{2}v_e$$

When the fifth charge is fired, the last mass (which is the "payload") has reached a velocity, with respect to the *initial* frame of reference, that is the sum of the velocity increments $v_{R_1}, v_{R_2}, \ldots$, i.e.,

$$V_{\text{final}} = \tfrac{1}{6}v_e + \tfrac{1}{5}v_e + \tfrac{1}{4}v_e + \tfrac{1}{3}v_e + \tfrac{1}{2}v_e = \tfrac{87}{60}v_e$$

Thus we see that such a rocket can produce a final velocity in excess of its own exhaust velocity. This comes at the expense, however, of a small payload compared to the starting mass of the rocket.

Let us turn our attention to a true rocket, a device that produces a smooth thrust by the continual ejection of mass at its exhaust velocity. The ideas of differential calculus will be useful in this discussion. The continuous flow of mass out of the rocket and into its exhaust may be approximated by the discrete loss of small bits of mass, Δm, at very short time intervals, as in Fig. 5-15. There we see the rocket initially moving with velocity v, gaining an increment of velocity Δv upon ejection of Δm with a separation velocity v_e. The final velocity of Δm is

$$v' = v + \Delta v - v_e \qquad (5\text{-}15)$$

An analysis of the rocket motion in the absence of air resistance and gravity is quite interesting. To do this, all we need is the law of

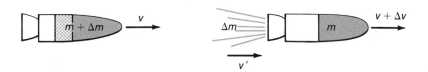

Fig. 5-15. The rocket of total mass $m + \Delta m$ has initial velocity v. It ejects a mass element Δm and gains an increment of velocity Δv. The velocity v' of Δm is actually directed to the left and is thus a negative quantity.

conservation of momentum. Equating the initial momentum of the rocket to the final momentum of rocket plus ejected mass, we have

$$(m + \Delta m)v = m(v + \Delta v) + \Delta m v'$$

By use of (5-15) we find

$$mv + \Delta mv = mv + m\Delta v + \Delta mv + \Delta m \Delta v - \Delta m v_e$$

or
$$m\Delta v = \Delta m v_e - \Delta m \Delta v \qquad (5\text{-}16)$$

If Δm is allowed to become very small, Δv will also become very small. Since we wish to let $\Delta m \to 0$ to represent the continuous flow of exhaust from the rocket, the second term on the right side of (5-16) can be neglected compared to the other terms. We then have

$$\Delta v \approx \frac{\Delta m}{m} v_e$$

This result tells us that the velocity increments, Δv, become larger as m becomes smaller. As more mass elements Δm are ejected, m becomes smaller and smaller. The net change in velocity after many bits of mass are exhausted is the sum of all the velocity increments that have been gained:

$$v_f - v_i = \sum_{j=1}^{N} \Delta v_j \approx \sum_{j=1}^{N} \frac{\Delta m_j}{m_j} v_e \qquad (5\text{-}17)$$

This expression becomes exact as we let each $\Delta m \to 0$, so that the number of elements ejected grows without bound to represent a smooth, continuous mass ejection in the exhaust. The sum in (5-17) then becomes a definite integral:

$$v_f - v_i = v_e \int_{m_i}^{m_f} \frac{1}{m} dm'$$

where dm' is the infinitesimal element of ejected mass. In order to evaluate this integral, we need a relation between the ejected mass, dm', and the mass m. Since each element dm' ejected *decreases m* by an amount dm, we have

$$dm = -dm'$$

so that
$$v_f - v_i = -v_e \int_{m_i}^{m_f} \frac{dm}{m} \qquad (5\text{-}18)$$

The area represented by this definite integral is shown in Fig. 5-16a, where the function to be integrated, $1/m$, is plotted versus the variable m. The result of the integration is found by applying the fundamental theorem of calculus:

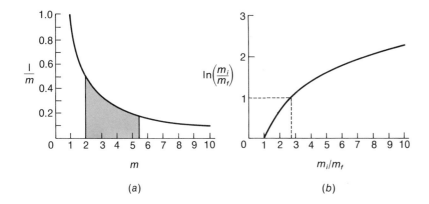

Fig. 5-16. (a) Graph of $1/m$ vs m. The shaded area between $m = 2$ and $m = 5.44$ has an area of $\ln(5.44/2) = \ln 2.72 = 1$. This particular point is indicated on the graph of the ln function in (b).

The value of a definite integral equals the difference between the two indefinite integrals evaluated at the end points.

Thus:

$$\int_{m_i}^{m_f} \frac{1}{m}\, dm = \int \frac{1}{m}\, dm(\text{at } m = m_f) - \int \frac{1}{m}\, dm(\text{at } m = m_i)$$

The name of the function that is the antiderivative of the function $1/m$ is the *natural logarithm of m,* written $\ln(m)$. Thus we have

$$v_f - v_i = -v_e[\ln m_f - \ln m_i] = v_e[\ln m_i - \ln m_f]$$

Utilizing the fact that the difference of the logarithms of two numbers is the logarithm of their ratio, we have

$$v_f - v_i = v_e \ln\left(\frac{m_i}{m_f}\right) \qquad (5\text{-}19)$$

The natural logarithm function is tabulated in most collections of mathematical tables. This function is plotted in Fig. 5-16b for values of its argument between 1 and 10.

Before the rocket is ignited, the mass is constant and $m_i/m_f = 1$. The rocket simply coasts in our inertial frame with uniform velocity in this case, so that $v_f - v_i = 0$. Examination of the graph of the ln function shows that, indeed, (5-19) predicts $v_f - v_i = 0$ for $m_i/m_f = 1$. If 1/4 of the initial mass of the rocket is ejected with exhaust velocity v_e, we find

$$v_f - v_i = v_e \ln \frac{m_i}{m_f} = v_e \ln \frac{1}{3/4} = 0.288 v_e$$

where the value 0.288 was found in a table rather than attempting to read it from the graph in Fig. 5-16b. Note that this gain in velocity is *independent* of whether the mass is ejected quickly, slowly, or even in several different "burns." It also does not matter whether the exhaust is gaseous, liquid, solid, hot, or cold. *All* that matters is: (*a*) what fraction of the initial mass is ejected? and (*b*) what is its exhaust velocity? Of course, exhaust velocity is affected by, for example, the temperature of the exhaust gas. But it is only this exhaust velocity that explicitly appears in the equation of motion of the rocket. The law of momentum conservation is the only dynamical law required to obtain (5-19).

Example   The figures tabulated below are appropriate for the first stage of the Saturn V space launch vehicle.

TABLE 5-1

Total mass at lift-off	3100 tons $= 2.8 \times 10^6$ kg
Rate of exhaust ejection	15 tons/s $= 1.4 \times 10^4$ kg/s
Exhaust velocity	8220 ft/s $= 2500$ m/s
Burn time	2.5 min $= 150$ s
Total mass ejected	2250 tons $= 2.04 \times 10^6$ kg
Final mass remaining	850 tons $= 0.76 \times 10^6$ kg

If this vehicle were launched in interplanetary space so that we could neglect gravity and air resistance, how much velocity would it gain?

Solution   All that we need is the exhaust velocity and the mass ratio:

$$v_f - v_i = -v_e \ln\left(\frac{m_i}{m_f}\right) = 2.5 \times 10^3 \, \frac{m}{s} \cdot \ln\frac{2.8}{0.76}$$

$$= 2.5 \times 10^3 \, \frac{m}{s} \cdot 1.3 = 3.25 \times 10^3 \, \frac{m}{s}$$

$$= 7270 \text{ mi/hr}$$

Since this vehicle is normally launched from earth, the force of gravity reduces the final velocity achieved.

All of our preceding discussion has not required the calculation of a force at any point. Yet it is of interest to talk of the force produced by the rocket that would be measured if we clamped it to a test bed. The design engineer certainly needs the magnitude of this force to design the struts that attach the thrust chamber of the rocket engine to the rest of the vehicle, for example.

We can calculate this force is we return to Eq. (5-16):

$$m\Delta v = \Delta m v_e - \Delta m \Delta v$$

Again, let us neglect the last term since it is the product of two small quantities:

$$m\Delta v \approx \Delta m v_e$$

Now let us divide this approximate equation by the time, Δt, required to eject a quantity of mass Δm.

$$m\frac{\Delta v}{\Delta t} \approx \frac{\Delta m}{\Delta t} v_e$$

Now we will let $\Delta t \to 0$. Of course, as we do so, the quantity of mass ejected in this time interval also approaches zero. The quantity $\Delta v/\Delta t$ will become dv/dt and the quantity $\Delta m/\Delta t$ will become dm/dt. Further, our approximation will become exact in this limit:

$$m\frac{dv}{dt} = \frac{dm}{dt} v_e \qquad (5\text{-}20)$$

The quantity dm/dt is the rate of mass ejection. The left side of (5-20) is the mass of the rocket multiplied by the acceleration of the rocket, which, according to Newton's second law, equals the *force* acting upon the rocket; we call this force the *thrust:*

$$\text{thrust} = \frac{dm}{dt} v_e \qquad (5\text{-}21)$$

Substituting numbers from our Saturn V example, we have

$$\text{thrust} = 1.4 \times 10^4 \, \frac{\text{kg}}{\text{s}} \cdot 2.5 \times 10^3 \, \frac{\text{m}}{\text{s}}$$

$$= 3.5 \times 10^7 \, \text{kg} \, \frac{\text{m}}{\text{s}^2} = 3.5 \times 10^7 \, \text{N}$$

$$= 7.9 \times 10^6 \, \text{lb}$$

At lift-off, the weight of the rocket is 6.2×10^6 lb, so the net upward force is only 1.7×10^6 lb, producing an initial acceleration of

$$a = F/m = \frac{1.7 \times 10^6 \, \text{lb}}{6.2 \times 10^6 \, \text{lb}} = 0.27 \, g\text{'s} = 8.6 \, \text{ft/s}^2$$

This acceleration continually increases as the remaining mass of the rocket decreases.

Example   A jet engine exhausts its combustion products at a rate of 320 lb/s (approximately 5000 ft³/s). Its exhaust velocity is 1000 ft/s. What thrust does it develop?

Solution   Note that mass, in English units, must be given in slugs if we want our answer in units of pounds.

$$\frac{dm}{dt} = \frac{320 \text{ lb/s}}{32 \text{ lb/slug}} = 10 \text{ slug/s}$$

$$\text{thrust} = \frac{dm}{dt} v_e = 10 \frac{\text{slug}}{\text{s}} \cdot 1000 \frac{\text{ft}}{\text{s}}$$

$$= 10^4 \text{ slug ft/s}^2$$

$$= 10^4 \text{ lb}$$

Example   A rocket has a total mass of 5 kg. It ejects mass at a rate of 2 kg/s with an exhaust velocity of 200 m/s. It is fired vertically from the surface of the earth. (*a*) What is its acceleration at lift-off? (*b*) What is its acceleration after it has lost 4/5 its original mass? (Neglect air resistance.)

Solution

(*a*) $a = F/m = \dfrac{\text{thrust} - \text{weight}}{m}$

$$= \frac{\dfrac{dm}{dt} v_e - mg}{m} = \frac{(2 \text{ kg/s}) (200 \text{ m/s}) - 5 \text{ kg} \cdot 9.8 \text{ m/s}^2}{5 \text{ kg}}$$

$$= \frac{400 \text{ N} - 49 \text{ N}}{5 \text{ kg}} = 70.2 \text{ m/s}^2$$

(*b*) $m = \frac{1}{5}(5 \text{ kg}) = 1 \text{ kg}$

$$a = F/m = \frac{400 \text{ N} - 9.8 \text{ N}}{1 \text{ kg}} = 390 \text{ m/s}^2$$

Exercises

9.   Since thrust $= v_e \cdot dm/dt$, the same thrust can be obtained with a large exhaust velocity and small dm/dt, or large dm/dt and small exhaust velocity. Rocket engineers prefer large v_e. Can you see why this is so?

10.   If a rocket continues to accelerate long enough, its velocity with respect to the initial coordinate system becomes larger than the rocket's exhaust velocity. When this occurs, the exhaust is no longer moving backwards in the initial coordinate system, but is actually moving in the direction of the rocket. Is there any contradiction in these statements?

11.   A model solid-fuel rocket engine has a mass of $\frac{1}{10}$ kg. The fuel represents 90% of the mass of the engine, its casing the remaining 10%. It burns its total fuel at a constant rate in $\frac{1}{4}$ s. Its thrust is measured and determined to be 50 N. What is its exhaust velocity?

12. A rocket with a weight of 1000 lb and exhaust velocity of 1000 ft/s is just able to sustain its own weight against the force of gravity at lift-off. If dm/dt is constant, what is the magnitude of its acceleration after 10 s?

13. A payload of 1 metric ton is to be brought to a final speed of 5000 m/s. The rocket has an exhaust velocity of 2500 m/s and is launched in space, so that gravity and air resistance are unimportant. What must the initial mass of the rocket be? The graph of the ln function in Fig. 5-16 is sufficiently accurate for this exercise.

5-9   Summary

Application of Newton's second law to bodies in a frame of reference that is accelerating with respect to an inertial frame is reasonably straightforward. The required net force is determined by the acceleration of the body with respect to the inertial frame. Consideration of the forces acting upon a passenger in an accelerated elevator illustrate the equivalent effects of gravitation and accelerated frames of reference.

Newton's third law, "To every action (force) there is opposed an equal and opposite reaction," is a necessary consequence of the experimental fact of conservation of momentum for isolated systems. When Newton's third law is applied to systems on which external as well as internal forces act, we see that the total momentum of the system is affected only by the external forces. The internal forces cancel in their net effect on the entire system.

Tensions and compressions in ropes and beams effectively transmit forces from place to place without change if (*a*) they are not accelerating or (*b*) their mass is small compared to the remainder of the system.

The accelerations and internal forces of many-body systems acted upon by one or more external forces can be evaluated by carefully isolating one portion of the system at a time and applying Newton's second law to each such portion. This procedure leads to several simultaneous equations that can be solved for the unknown quantities.

We discussed the motion of rockets that are not acted upon by external forces. The law of conservation of momentum allowed us to obtain the final velocity of such a rocket:

$$v_f - v_i = v_e \ln\left(\frac{m_i}{m_f}\right)$$

The only parameters of importance are the exhaust velocity and the ratio of initial mass (fuel plus payload) to final mass (payload).

The force produced by the rocket exhaust is called the thrust and is given by

$$\text{thrust} = v_e \frac{dm}{dt}$$

where dm/dt is the rate of mass ejection.

Problems

5-1.   The occupant of an elevator on earth stands on scales that read 750 N. His mass is 60 kg. What is the direction and magnitude of the elevator's acceleration?

5-2.   In order to experience the sensation of an observer on the surface of the moon, where the acceleration due to gravity is one-sixth g, a physicist stands in an elevator accelerating downward uniformly. (*a*) What value of acceleration should he give the elevator? (*b*) If the elevator shaft is 120 ft long, what is the maximum time he has to conduct experiments?

5-3.   A 2-kg mass hangs from the ceiling of an elevator. The string that suspends it can exert a maximum upward force of 25 N without breaking. If the string is not to break, find the maximum permissible upward acceleration of the elevator.

5-4.   Experiments in the weightless sensations of free fall are to be conducted by dropping an elevator freely for a time then abruptly decelerating it before it hits bottom. If it falls for 140 ft then decelerates uniformly to stop in another 40 ft, (*a*) how long do the occupants feel weightless? (*b*) How much force does the elevator floor exert on a 160-lb occupant during deceleration?

5-5.   A uniform rope with a mass of 4 kg and a length of 16 m hangs from an elevator. Find the tension at its top end, center, and bottom end when the acceleration of the elevator is (*a*) 0, (*b*) 5 m/s² upward, (*c*) 5 m/s² downward, (*d*) 9.8 m/s² downward (free fall).

5-6.   Consider a 160-lb man standing on bathroom scales. He squats down abruptly so that, for a short time, his acceleration is 8 ft/s² downward. (*a*) What do the scales read during that brief time? (*b*) How could he make the reading zero? (*c*) What happens to the scale reading when he suddenly stops accelerating downward?

5-7.   Three trailers, each with a mass of 1 metric ton (10^3 kg), are pulled by a tractor that exerts a force of 2000 N. Find the tension in the coupling between the first and second trailers, and between the second and third trailers. Neglect frictional forces.

5-8. The system shown in Fig. 5-13 consists of equal 2-kg masses. Find the acceleration of the system and the tension in the connecting string. How are these results changed if both masses are doubled?

5-9. A cart is pulled by two masses, as shown in Fig. 5-17. Find T_1, T_2, and a, in terms of m_1, m_2, and m_3.

5-10. A cord with a mass of 1/10 kg per meter of length is used to connect the system of Fig. 5-18. For $m_1 = m_2 = m_3 = \frac{1}{2}$ kg, find the tension acting on cart 1 at the instant shown. Neglect friction. Is this tension constant as the system accelerates? Neglect the sagging of the horizontal parts of the cord.

5-11. The man in Fig. 5-19 weighs 160 lb and the platform weighs 40 lb. Find the tension in the rope when the system is static. Can the man and platform accelerate upward at $a = g/2$?

5-12. A rocket-fuel system of total mass M is at rest on a frictionless horizontal surface. What is the final velocity of the remainder of the rocket with respect to the surface if it (a) ejects a single mass of $\frac{2}{3}M$ at a velocity v_e with respect to the remainder of the rocket; (b) ejects three masses sequentially, each with a mass of $M/5$ and each at velocity v_e with respect to the remainder of the rocket.

5-13. Consider a crude rocket similar to that of Fig. 5-14, which consists of 10 equal masses. The separation velocity for each is v_e. Find the final velocity of the payload (the tenth and last mass) if the rocket begins at rest. Neglect gravitational forces.

5-14. Consider a Saturn V rocket (see Table 5-1). If it were launched in space so that friction and gravity could be neglected, find the gain in velocity at the instant it has burned half its fuel load. The graph of the ln function in Fig. 5-16 is sufficiently accurate for this problem.

5-15. A rocket has an exhaust velocity of 2500 m/s. It is to bring a 1-ton payload initially at rest up to a final velocity of 3750 m/s. How much fuel must it consume? Neglect gravitational forces.

5-16. Find the value to which we must decrease v_e for the Saturn V rocket (see Table 5-1) so that it just balances at launch from the earth, i.e., its initial acceleration is 0. Will the acceleration remain zero for long?

5-17. A rocket with an exhaust velocity of 750 ft/s and initial mass of 20 slugs has an initial acceleration of 75 ft/s². After 5 sec-

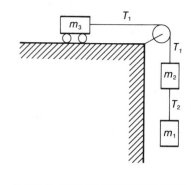

Fig. 5-17.

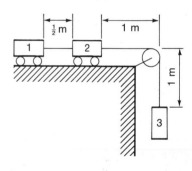

Fig. 5-18.

Fig. 5-19.

onds have elapsed, the acceleration of the rocket has increased to 150 ft/s², though the mass loss rate and exhaust
velocity are constant. What is its mass at this time? Ignore
gravitational forces.

5-18.   A chain is made up of links that weigh 100 g and are spaced
10 cm apart. The chain is lowered onto a scale at a velocity
of 150 cm/s. What is the reading of the scale as a function
of time? (The *average* reading is desired, since the impact
of individual links will cause the scale to jiggle somewhat.)

5-19.   Look in a table of natural logarithms to find the values needed
for calculating the mass ratios required for a rocket to obtain
final velocities of $2v_e$, $4v_e$, $6v_e$, and $8v_e$. For a payload of
fixed mass M, what are the takeoff weights for each case?
Can you see how a multistage rocket could be helpful? Neglect gravity in this problem.

Chapter **6**

Newton's Second Law in Integral Form

6-1 Introduction We have introduced Newton's second law in a form involving differentiation:

$$F = \frac{dp}{dt} \tag{6-1}$$

which becomes

$$F = m\frac{dv}{dt} = ma \tag{6-2}$$

for constant mass. But since differentiation and integration are inverse mathematical operations, we should be able to state the second law in a form involving integration. This is easily done and leads to the concept of *impulse*, which is especially useful in discussing the effects of brief, rapidly varying forces. Newton himself introduced the second law in integral form. He never wrote $F = ma$ in the *Principia* but instead stated that a force F acting for a time Δt causes a change in the momentum of a system:

$$F\Delta t = \Delta p \tag{6-3}$$

which for constant mass becomes

$$F\Delta t = m\Delta v \tag{6-4}$$

If we transpose Δt to the right sides of (6-3) and (6-4) and take the limit as $\Delta t \to 0$, we obtain (6-1) and (6-2). But if we retain the form of (6-3) and sum the effects of the force acting over many time intervals Δt, we obtain

$$\sum_{i=1}^{N} F_i \Delta t_i \approx \Delta p$$

where the approximate equality recognizes the error made if F varies during the time Δt. This equality can be made exact by letting $\Delta t \rightarrow 0$; the sum then becomes an integral:

$$\int_{t_1}^{t_2} F \, dt = \Delta p \qquad (6\text{-}5)$$

where Δp is the change in momentum between times t_1 and t_2. Equation (6-5) is thus called the integral form of Newton's second law.

6-2   The Impulse-Momentum Theorem

The integral form of Newton's second law is usually called the *impulse-momentum theorem*. We can derive it directly from the differential form of the second law, $F = dp/dt$, as follows. Let us take the definite integral of both sides of this equation:

$$\int_{t_1}^{t_2} F \, dt = \int_{t_1}^{t_2} \frac{dp}{dt} \, dt \qquad (6\text{-}6)$$

The right side of (6-6) is simply the sum total of all changes in p that occur between t_1 and t_2 and is thus equal to the net change in p, denoted Δp:

$$\int_{t_1}^{t_2} F \, dt = \Delta p = p_2 - p_1 \qquad (6\text{-}7)$$

Equation (6-7) is the impulse-momentum theorem, where the left-hand side, simply an integral of the force over the given time interval, is called the *impulse*.

We can also arrive at (6-7) by recalling the kinematic effects of a variable acceleration from Chapter 3:

$$\Delta v = v_2 - v_1 = \int_{t_1}^{t_2} a \, dt \qquad (3\text{-}18)$$

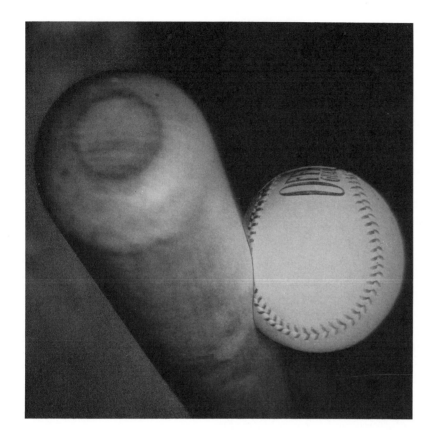

Fig. 6-1. A baseball photographed during the brief time of contact with a bat. The strong deformation of the ball is caused by a peak force near 4000 lb. (Photograph by Dr. Harold E. Edgerton, Massachusetts Institute of Technology, Cambridge, Massachusetts)

From Newton's second law we have $a = F/m$ so that (3-18) becomes

$$v_2 - v_1 = \frac{1}{m} \int_{t_1}^{t_2} F \, dt$$

By transposing m we are again led to (6-7):

$$\int_{t_1}^{t_2} F \, dt = mv_2 - mv_1 = p_2 - p_1 \qquad (6\text{-}7)$$

In Fig. 6-1 we see a baseball photographed during the brief time of its contact with a bat. The strong deformation of the ball indicates that a very large force is exerted upon it. Of course, this force lasts for an extremely short time interval. When such brief, rapidly varying forces act upon a body, it is often not possible or particularly interesting to measure the force as a function of time. However, the net effect of the brief force is measurable through the net change in momentum that it causes. As we have seen, this change is given by the impulse integral:

$$\text{impulse} = \int_{t_1}^{t_2} F\, dt \tag{6-8}$$

For the purposes of illustration, let us first estimate the time-varying force acting upon a struck baseball and then apply the impulse-momentum theorem to see whether it gives a reasonable result. To estimate the time during which the bat and ball are in contact, we can make a simple argument. Suppose the ball is pitched toward the batter at 100 ft/s and that the bat moves toward the ball at 50 ft/s. During the time that the ball and bat are in contact, the ball deforms by approximately $\frac{1}{2}$ in. The time required to cover $\frac{1}{2}$ in. at a relative velocity of 150 ft/s is

$$\Delta t = \frac{s}{v} = \frac{\frac{1}{24}\ \text{ft}}{150\ \text{ft/s}} = 2.8 \times 10^{-4}\ \text{s}$$

However, the relative velocity drops from 150 ft/s to zero during the collision. Let us estimate an average relative velocity of 75 ft/s, which doubles the above time. Also, a similar length of time will be required to set the ball in motion in the opposite direction. Altogether then, the total time of contact between ball and bat is approximately four times the number above. Let us take Δt to be approximately 1.2×10^{-3} s.

The force required to deform the ball as much as that seen in Fig. 6-1 was estimated by placing a baseball and bat in a hydraulic press and measuring the force required to deform the ball. This gave an approximate value of 4000 lb for the magnitude of the force at the moment shown in Fig. 6-1. However, this force rises from zero to the maximum value, then falls back to zero again during the 1.2-ms duration of the bat-ball contact. Thus we are finally led to the crudely estimated force-vs-time graph in Fig. 6-2. The area under this force-vs-time curve gives the value of the definite integral for the impulse over the entire duration of the bat-ball contact. This area is seen to be the area of a triangle with a height of 4000 lb and a base of 1.2 ms.

$$\text{impulse} = \int_{t_1}^{t_2} F\, dt = \tfrac{1}{2}(4 \times 10^3\ \text{lb})(1.2 \times 10^{-3}\ \text{s}) = 2.4\ \text{lb} \cdot \text{s}$$

The change in momentum of the ball must equal the impulse:

$$\Delta p = p_2 - p_1 = \int_{t_1}^{t_2} F\, dt$$

We can calculate p_1 from the velocity of the approaching ball and

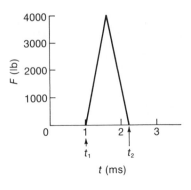

Fig. 6-2. The force the bat exerts upon the ball is approximated by this time behavior.

its mass, which is very nearly 10^{-2} slug. Let us take the initial momentum to be in the negative direction, so that the force exerted on the ball acts in the positive direction and the final momentum of the ball is positive. Solving for this final momentum, we find

$$p_2 = \int_{t_1}^{t_2} F \, dt + p_1 = 2.4 \text{ lb} \cdot \text{s} - (10^{-2} \text{ slug})(10^2 \text{ ft/s})$$

Noting that the dimension lb·s is identical to slug·ft/s, we find

$$p_2 = (2.4 - 1.0) \text{ slug} \cdot \text{ft/s}$$

The final velocity of the ball is thus

$$v_2 = \frac{p_2}{m} = \frac{1.4 \text{ slug} \cdot \text{ft/s}}{10^{-2} \text{ slug}} = 140 \text{ ft/s}$$

We could, of course, turn this procedure around and use known initial and final velocities plus an estimate of the contact time to make an estimate of the peak force.

The numerical values used in this example were realistic. A pitched "fast ball" can reach speeds of 100 mi/hr, so that our value of 100 ft/s is not the fastest possible. By swinging a bat through a photocell timing apparatus, the speed of a swinging bat was found to be between 50 and 80 ft/s for several students. The value of 50 ft/s used in our example is at the lower end of this range. As mentioned previously, the 4000-lb approximate value for the peak force was determined with a hydraulic press. Finally, the weight of a standard baseball is 5 oz, which is very nearly 10^{-2} slug. Our result is also reasonable in that a struck ball typically can travel farther than a pitched ball and must have a larger initial velocity.

Example   A $\frac{1}{2}$-kg ball is dropped from a height of 1.5 m. It rebounds to the same height, where it is caught. Find the impulse produced by the upward-directed force exerted by the floor upon the ball.

Solution   The force of gravity acts continuously on the ball from the moment it is released until it is caught again, as indicated in the force-vs-time curve of Fig. 6-3. The magnitude of this force is $W = mg$ and is directed downward, which is taken to be the negative direction. The force exerted by the floor is impulsive and directed upward, producing the spike in Fig. 6-3. The momentum of the ball at t_1 and at t_2 is zero, since it is released from rest and is caught at the same height at t_2. Thus the change in momentum, Δp, is also zero. We conclude, then, that the total impulse over the entire time is zero:

$$\int_{t_1}^{t_2} F \, dt = \Delta p = 0$$

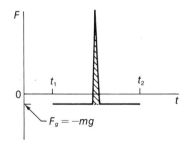

Fig. 6-3.

The total impulse consists of a negative portion produced by W acting for the entire time and a positive portion that acts only during the contact with the floor. The positive portion, shaded in Fig. 6-3, must precisely cancel the negative portion:

$$\int_{t_1}^{t_2} F \, dt = -mg\,(t_1 - t_2) + \text{floor impulse} = 0$$

so floor impulse $= mg\,(t_2 - t_1)$. The time interval is twice that required for a free fall of 1.5 m:

$$h = \tfrac{1}{2}gt^2$$

$$t = \sqrt{2h/g} = \sqrt{\frac{3 \text{ m}}{9.8 \text{ m/s}^2}} = 0.55 \text{ s}$$

Hence $t_2 - t_1 = 1.1$ s and floor impulse $= (\tfrac{1}{2}\text{ kg})\,(9.8 \text{ m/s}^2)\,(1.1 \text{ s}) = 5.4$ kg $\cdot$ m/s $= 5.4$ N $\cdot$ s. An estimate of the time of contact with the floor and the shape of the spike in the F-vs-t curve is required to find the peak force.

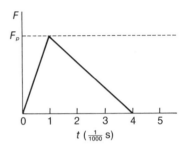

Fig. 6-4.   The force on the pellet in an air gun rises rapidly as air is admitted behind the pellet, then falls as the pellet moves down the barrel.

Exercises

1. A 30,000-lb airplane catapulted from an aircraft carrier reaches a speed of 100 mi/hr at the end of the catapult. Find the net impulse produced by the catapult.

2. A pellet is propelled from an air gun with a force that varies as shown in Fig. 6-4. If the pellet weighs 10 g and leaves the muzzle at 200 m/s, find F_p, the peak force exerted on the pellet.

6-3   Average Force

The *average value* for a set of N different values is defined to be the sum of all N values divided by N. Thus for three values, x_1, x_2, and x_3, we have

$$\bar{x} = \frac{x_1 + x_2 + x_3}{3}$$

where $\bar{x}$ denotes the average value. More generally:

$$\bar{x} = \frac{1}{N} \sum_{j=1}^{N} x_j$$

What if we wish to average a continuously varying quantity over some given span of time, say t_i to t_f? Writing the varying quantity as $x = f(t)$, we can sample the value of x at intervals spaced Δt apart, as in Fig. 6-5. We then have N samples in the larger time interval between t_i and t_f, with the number of samples clearly given by

$$N = \frac{t_f - t_i}{\Delta t}$$

The average value of these N samples is then

$$\bar{x} = \frac{1}{N} \sum_{j=1}^{N} x_j = \frac{1}{N} \sum_{j=1}^{N} f(t_j)$$

Now multiplying and dividing by Δt, we have

$$\bar{x} = \frac{1}{N\Delta t} \sum_{j=1}^{N} f(t_j)\Delta t$$

(a)

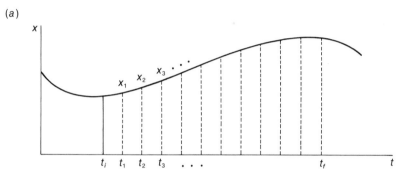

(b)

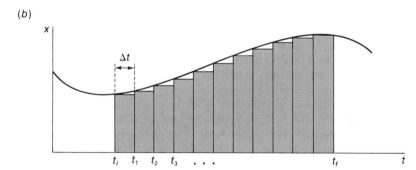

(c)

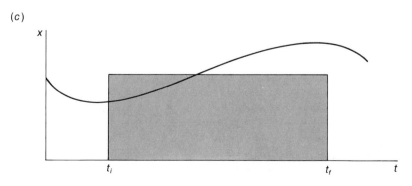

Fig. 6-5. (a) The time interval is broken into N intervals and $\bar{x} = (x_1 + x_2 + x_3 + \cdots + x_n)/N$. (b) Multiplying by the width Δt, we see that $\bar{x} = $ shaded area/($t_f - t_i$). In the limit as $\Delta t \to 0$:

$$x = \int_{t_i}^{t_f} x(t)dt/(t_f - t_i).$$

(c) The area $\bar{x}(t_f - t_i)$ equals that of the definite integral

$$\int_{t_i}^{t_f} f(t)dt.$$

We can obtain an arbitrarily large number of samples and increase the precision of our average by letting $\Delta t \to 0$, so that $N \to \infty$. In the limit our sum becomes the area under the curve of Fig. 6-5, that is, it becomes a definite integral:

$$\bar{x} = \lim_{\Delta t \to 0} \frac{1}{N\Delta t} \sum_{j=1}^{N} f(t_j)\Delta t = \frac{1}{t_f - t_i} \int_{t_i}^{t_f} f(t)\,dt$$

We have used the fact that $N\Delta t$ always equals $t_f - t_i$. As shown in Fig. 6-5, the average value is that *constant* value that produces the same area as the definite integral.

In general, then, we define the average value of a function of time, $f(t)$, over the time interval t_i to t_f by

$$\bar{f} = \frac{\displaystyle\int_{t_i}^{t_f} f(t)\,dt}{t_f - t_i} \tag{6-9}$$

We saw an example of (6-9) in Chapter 3 when we defined $\bar{v} = \Delta x/\Delta t$ and found

$$\Delta x = \int_{t_i}^{t_f} v\,dt$$

so that

$$\bar{v} = \frac{\Delta x}{\Delta t} = \frac{\displaystyle\int_{t_i}^{t_f} v\,dt}{t_f - t_i}$$

For the case of a force that varies with time, we find

$$\bar{F} = \frac{\displaystyle\int_{t_1}^{t_2} F\,dt}{t_2 - t_1} = \frac{\text{impulse}}{\Delta t} = \frac{\Delta p}{\Delta t}$$

That is, the average force is that constant force that produces the same change in momentum, Δp, as the actual varying force over the same time interval. The net change in momentum, then, is given by

$$\Delta p = \bar{F}\Delta t \tag{6-10}$$

This statement, with the definition (6-9), is an alternate form of Newton's second law in integral form.

Exercises

3. Find the average value between $t = 0.1$ s and $t = 0.6$ s for the force graphed in Fig. 6-6.

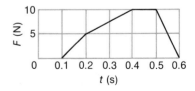

Fig. 6-6.

4. Find the average value between $t = 0.2$ s and $t = 0.5$ s for the force graphed in Fig. 6-6.

5. An average force equal to 30 N acts upon a $\frac{1}{2}$-kg body, which has an initial velocity of 6 m/s, for a time interval of $\frac{1}{2}$ s. Find the body's final momentum.

6. If $x = 4t + t^3$, find $\bar{x}$ between $t = 1$ and $t = 3$. Does $\bar{x}$ lie halfway between x at $t = 1$ and x at $t = 3$? Explain.

6-4 Applications to Collisions

An automobile moving at 30 mi/hr crashes into a brick wall. Can we estimate the forces involved in such a collision? First, we need an estimate of the length of time occupied by the collision. Having seen photographs of wrecked automobiles, it seems reasonable to assume that the front end of the car crumples inward by three feet or so in such a collision. If the car were traveling at a constant 44 ft/s, it would require $\frac{3}{44}$ s to travel that three feet. Since the car is decelerating strongly during that time, however, it is moving at 44 ft/s initially but comes to rest at the end of the collision. Let us assume that the average velocity is 22 ft/s, so that the collision time, Δt, is approximately $\frac{3}{22}$ s.

Since the momentum change of the car is known, we know the impulse that acted upon the car and, hence, the average force. Assuming a 3200-lb car,

$$\Delta p = p_2 - p_1 = 0 - (100 \text{ slug})(44 \text{ ft/s}) = -4400 \text{ lb} \cdot \text{s}$$

Hence
$$\bar{F} = \frac{\Delta p}{\Delta t} = \frac{-4400 \text{ lb} \cdot \text{s}}{\frac{3}{22} \text{ s}} = -32{,}300 \text{ lb}$$

where the minus sign indicates a retarding force. If the deceleration were nonuniform, the peak force would be larger than the average force we have calculated. For example, in the illustration dealing with a struck baseball in Section 6-2, the peak force was 4000 lb but the average force was only 2000 lb. In that example the average value was the mean of the maximum (4000) and minimum (0) values because the force was assumed to rise and fall linearly with time.

It is interesting to write out our result for the auto collision more generally. First,

$$\Delta t = \frac{\Delta x}{\frac{1}{2}v_1}$$

where Δx is the crumpling distance. Then

$$\bar{F} = \frac{\Delta p}{\Delta t} = \frac{-\frac{1}{2}mv_1^2}{\Delta x} \qquad (6\text{-}11)$$

For a given Δx, the average force exerted is proportional to the *square* of the initial velocity—hence the hazards of high velocities multiply more rapidly than our intuition might indicate. However, average force can be reduced if Δx is large. In other words, a car that has a considerable expanse of sheet metal, which crumples readily, (but a rigid passenger compartment to protect the passengers from injury) will minimize the force, as our intuition tells us.* The importance of the quantity $\frac{1}{2}mv^2$ was recognized in early studies of collision and was called the "living force" of an object. We now call this quantity the kinetic energy and will return to it in Chapter 14.

The forces exerted upon passengers in an automobile collision can be estimated in a similar fashion. Using the previous example of a 30-mi/hr collision with a brick wall and a 160-lb occupant, we find

$$F = \frac{\Delta p}{\Delta t} = \frac{-220 \text{ slug} \cdot \text{ft/s}}{\frac{3}{22} \text{ s}} = -1610 \text{ lb}$$

This is the average force required for the passenger to decelerate with the car rather than continue forward into the dashboard. It can be supplied by a conventional seat belt or by other means, such as air bags that automatically inflate between the passenger and dashboard upon impact. Again, we should recall that the peak force may be much larger than the average force.

Exercises

7. How does a 30-mi/hr collision of a car with a brick wall compare with the head-on collision of two cars moving at 30 mi/hr?

8. If two cars of unequal mass make a head-on collision, which car experiences the larger acceleration? How does this affect the forces experienced by the passengers?

9. Making reasonable assumptions, calculate the average force required to restrain a 160-lb occupant of a car that crashes head-on into a similar car. Both cars move at 60 mi/hr initially.

6-5   Newtonian Drag

When a vehicle moves through the atmosphere, it experiences air resistance. This resistance to its motion is caused by *drag forces* acting against the direction of motion. Empirically it is found that

---

*See, for example, "The Crashworthiness of Automobiles" by Patrick M. Miller, *Scientific American,* February 1973.

these drag forces increase linearly with velocity at low speeds but then depend quadratically on velocity at higher speeds. The fact that impulsive collisions produce average forces proportional to the square of the velocity (see Eq. (6-11)) makes us wonder if there is a simple explanation for the v^2-dependence of drag forces.

Since air is composed of many minute molecules, let us consider a flat plate moving with velocity v through a swarm of point masses, as in Fig. 6-7. In a time Δt, the plate strikes the number of molecules contained in the volume swept out by the moving plate, indicated by the dotted lines in Fig. 6-7. This number depends linearly upon the velocity of the plate. If each molecule is brought up to the speed of the plate when it is struck, it acquires a momentum mv, where m is the mass of the molecule. If N is the number of molecules struck in the time Δt, the net momentum gained by these N molecules is

$$\Delta p = Nmv$$

and the average force required to produce this momentum change is is

$$\bar{F} = \frac{\Delta p}{\Delta t} = \frac{Nmv}{\Delta t}$$

Noting that Nm is just the amount of mass, M, swept up in the time Δt, we have

$$\bar{F} = \frac{Mv}{\Delta t} \tag{6-12}$$

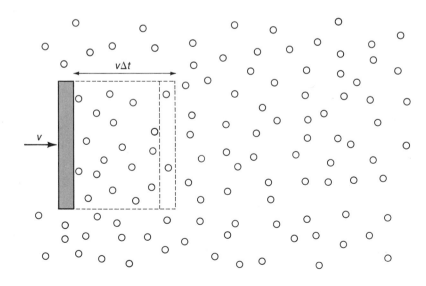

Fig. 6-7. A flat plate sweeps out a distance $v\Delta t$ in a time Δt. The number of molecules struck by the plate increases linearly with v.

But the quantity of mass swept up in a time Δt increases linearly with v and with Δt:

$$M = \text{constant} \cdot v\Delta t \tag{6-13}$$

Combining (6-13) and (6-12) we have

$$\bar{F} = \text{constant} \cdot v^2 \tag{6-14}$$

This is the average force required to speed the molecules up to v as they are struck and, by Newton's third law, it is also the force that opposes the plate moving through the molecules. The quadratic dependence of $\bar{F}$ upon v in (6-14) comes from the v-dependence of the momentum gain multiplied by the v-dependence of the *rate* at which mass is struck.

   The actual situation for a body moving through the air is far more complicated than our simple analysis indicates. At low velocities the swarm of molecules moves in a collective way and flows smoothly around the plate. In this domain of velocities the drag force increases only linearly with velocity. At higher velocities *turbulence* sets in and the flow of air around the plate is not smooth but fluctuates rapidly and chaotically with time. In this domain of velocities the drag depends quadratically upon velocity. The net effect, however, of this turbulence is to cause a portion of the air to momentarily acquire the speed of the plate as it passes. Because of the generality of the impulse-momentum theorem, any mechanism that causes this to occur produces the v^2-dependence of the drag force. Newton first made these general observations, hence the name *Newtonian drag* for this phenomenon.

   We can turn the picture in Fig. 6-7 around and imagine a stream of molecules, representing a wind, impinging upon a wall. Again, we will find a v^2-dependence for the force due to wind pressure. Actually, the molecules of any gas are always in motion with randomly directed velocities. This random motion is called *thermal agitation* and increases with increasing temperature. In Chapter 20 we will see that the average force produced by these random impacts explains the way in which the pressure of an enclosed gas varies with temperature.

Exercises

10. Several mail bags, each with a mass of 25 kg, are spaced one meter from one another in a straight line. They are scooped up by a truck moving at a uniform 10 m/s. Find the magnitude of the average force retarding the motion of the truck.

11. A fire hose squirts 10 kg/s of water against a flat plate. The velocity of the stream is 10 m/s. If the water comes to rest against the plate, find the average force exerted on the plate. Repeat the problem for a stream velocity of 20 m/s.

6-6 Summary

The integral form of Newton's second law is readily obtained by integration of the differential form $F = dp/dt$:

$$\int_{t_1}^{t_2} F \, dt = p_2 - p_1 \qquad (6\text{-}15)$$

The left side of (6-15) is called the impulse and (6-15) is known as the impulse-momentum theorem.

   In cases where a force acts briefly and extremely rapidly, the rapidly varying acceleration from moment to moment may not be known. In such a case, the net result of the force — a change in momentum — is all that we know. This change in momentum is equal to the impulse of the force, however, and if we have a model for the time behavior of the force, we can make estimates of the peak force during the impulse.

   In general, the average value of a function is given by

$$\bar{f} = \frac{\displaystyle\int_{t_1}^{t_2} f(t) \, dt}{t_2 - t_1}$$

In particular, the average force is given by

$$\bar{F} = \frac{\displaystyle\int_{t_1}^{t_2} F \, dt}{t_2 - t_1} = \frac{\Delta p}{\Delta t}$$

   The average force during a collision is found to depend upon the square of the velocity. Similarly, the average force of an object sweeping through a distribution of mass points also depends upon the square of the velocity. The term Newtonian drag describes this situation. The same quadratic velocity dependence is found in some other situations, such as the force exerted by a continuous stream of water against a plate.

Problems

6-1. A railroad boxcar weighing 20 tons is coasting at 4 ft/s. It strikes a stationary train and is coupled to it. If the collision lasts $\frac{1}{30}$ s, estimate the maximum force on the coupler.

6-2.   A $\frac{1}{4}$-lb ball is pitched at 60 ft/s and returns along the same path at 100 ft/s, when a bat produces the force graphed in Fig. 6-8. Find the magnitude of F_p in pounds.

6-3.   A man hops up and down on one foot so that his foot is touching the floor only 1/3 of the time. What is the average force on his foot *during the time of contact?* What is the average force on his foot when the average is carried over a long time?

6-4.   A strain gauge is wired to the rear axle of a test car in order to directly measure the propelling force, with the results sketched in Fig. 6-9. The car is at rest at t_1 and weighs 3200 lb. (*a*) Find the impulse between t_1 and t_2. (*b*) Find the velocity of the car at t_2. (*c*) Find the average force between t_1 and t_2.

6-5.   In Fig. 6-10 we see a golf ball strongly distorted at the moment of impact. The clubhead remains in contact with the ball for a total of 0.5 ms and the initial speed of the ball is 140 mi/hr. The weight of the ball is 1.6 oz. Find the peak force acting on the ball.

6-6.   An alpha particle is a nuclear projectile with a mass of 6.64×10^{-27} kg. If such a projectile moving with a velocity of 1.5×10^7 m/s (1/20 the velocity of light) strikes the nucleus of a gold atom head-on, it will rebound directly back along its original path. The gold nucleus is sufficiently massive that it may be assumed to be stationary. If the collision occurs over a distance of 10^{-13} m, estimate the peak force exerted on the alpha particles. Note: the above conditions are quite realistic for Rutherford's famous alpha-scattering experiment (Section 28-10).

6-7.   Find the average value of a function g between $t = 1$ and $t = 3$, where $g = 10t^2 + 2t - 3$. Compare this to the values of g at the two end points $t = 1$ and $t = 3$.

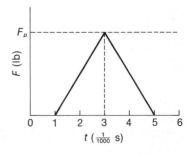

Fig. 6-8.

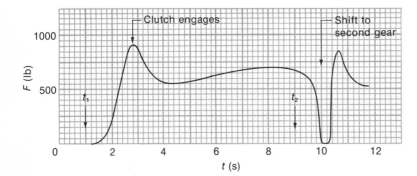

Fig. 6-9.

Fig. 6-10. A flash photograph taken at the moment of impact. (Photograph by Dr. Harold E. Edgerton, Massachusetts Institute of Technology, Cambridge, Massachusetts)

6-8. Find the average value of T between t_1 and t_2, where $T = at^2 + bt + c$.

6-9. A large car moving with a velocity of 20 m/s strikes a small car at rest. The combined crumpling distance of the cars is one meter and they remain locked together after the collision.

	Belch-Fire V8	Tofu 4
Weight	3×10^3 kg	1×10^3 kg
Initial velocity	20 m/s	0
Crumpling distance	0.7 m	0.3 m

Find the average force exerted upon a 70 kg occupant in each car if the occupants rigidly follow the motion of the cars. How is this changed if an air bag allows an additional $\frac{1}{2}$-m motion of the occupant?

6-10. A firehose squirts water at a stationary cup-shaped device that returns the stream back in the opposite direction with an equal velocity. For a flow rate of μ (kg/s) and velocity v (m/s), find the force exerted by the cup on the stream.

6-11. Interstellar space is filled with hydrogen atoms, which have a mass of 1.67×10^{-27} kg. Assume that each cubic meter of space contains 10^6 such atoms. Find the Newtonian drag on a space ship with a frontal area of 1 m² moving at 1/10 the velocity of light, if all the atoms it strikes embed themselves in the front of the ship.

Chapter **7**

Kinematics in Three Dimensions

7-1 **Introduction**   We now wish to generalize mechanics to more than one dimension of motion. As stated in Section 2-1, the complete specification of a moving point generally requires knowledge of three spatial coordinates, all of which are functions of time. If all three coordinates x, y, and z are constants, the point is at rest. If any or all of the coordinates change in time, then a displacement occurs. This displacement may occur entirely in the direction of one of the axes of the coordinate system, but more typically the displacement occurs in some skewed direction that lies between the orthogonal axes of our coordinate system. The fundamental idea of a displacement in a given *direction* constitutes the most primitive form of the concept called a *vector* (from the Latin, meaning a *carrier*). We shall find that the idea of a vector is well suited to the study of kinematics in more than one dimension. In fact, we shall find that many of the basic dynamic quantities, such as force and momentum, are also vector quantities.

7-2 **Scalars and Vectors**

Scalars are quantities that can be represented by a single number, such as temperature, pressure, mass, and loudness. *Vectors* are directed quantities, such as displacement or wind velocity. Consider as an example a displacement. To completely specify a displacement of an object, it is necessary to state the direction as well as the distance over which an object has been moved. Thus, the vector concept generally includes a scalar, which gives the length or *mag-*

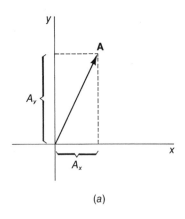

(a)

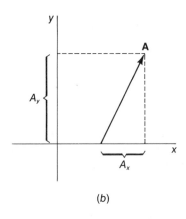

(b)

Fig. 7-1. A vector located in a two-dimensional coordinate system. The magnitudes of its components are not changed by a displacement of the vector, as in (b) above.

nitude of the vector, and a direction. In two dimensions we need two numbers, such as the length and the angle with respect to a chosen axis, to specify a vector; three numbers are required in three dimensions, and, in general, *n* numbers in *n* dimensions.

7-3 Vector Notation

Consider vectors constrained to lie in the plane of this page (which is taken to be flat). A symbol for such a vector is an arrow whose length represents the magnitude of the vector and whose direction gives the direction of the vector. For example,

could stand for a displacement of 2 cm length at an angle of 30° above the horizontal. Another symbol commonly used is a letter with an arrow over it: $\vec{A}$. In print, vectors are usually indicated by bold-faced type: **A**. Since the plane of the page is a two-dimensional space, only two numbers are required to specify such a vector. To illustrate this, let us set up a coordinate system as shown in Figure 7-1. The vertical axis (*ordinate*) is conventionally called the *y* axis, and the horizontal axis (*abscissa*) is called the *x* axis. The vector **A** can be completely specified by its projections, A_x and A_y. To help visualize these projections, imagine a light far above, shining down on the vector **A.** It casts a shadow (hence the word *projection*) of length A_x on the *x* axis. This projection A_x is shown in Fig. 7-1*a*. Similarly, a light shining from the right would cast a shadow of length A_y on the vertical, or *y*, axis. The projections A_x and A_y are often referred to as the *scalar components* of the vector **A**. In Fig. 7-1*b* we see the same vector **A,** located at a different position in the coordinate system. Note that the magnitudes of A_x and A_y are unaffected by the displacement of **A** in the coordinate system. Note also that A_x and A_y are not vectors, but are simply numbers, i.e., scalars.

One may also define the *vector components* of **A**, $\mathbf{A}_x$ and $\mathbf{A}_y$. These are simply vectors with lengths A_x and A_y and pointing in the $+x$ and $+y$ directions, respectively. We will most often utilize the scalar components of vectors in this book.

In Fig. 7-2 we see the same vector **A**, but in this figure we use two different numbers, a length A and an angle θ, to define the vector. The angle θ is usually measured in the counterclockwise direction from the x axis.

If it is really true that the length A and angle θ completely specify a vector and that A_x and A_y are also sufficient for specifying the vector, then we should be able to deduce either pair of numbers from the other. In Fig. 7-3 we see the relation between these number pairs. An important feature of the projection of **A** onto the x and y axes is that the "shadow" is cast at right angles to the axis, as shown in Fig. 7-3. This allows us to use the simple properties of a right triangle to find the projections. We shall need four properties:

1. $\sin \theta = \dfrac{\text{side opposite } \theta}{\text{hypotenuse}}$

2. $\cos \theta = \dfrac{\text{side adjacent to } \theta}{\text{hypotenuse}}$

3. $\tan \theta = \dfrac{\text{side opposite } \theta}{\text{side adjacent to } \theta}$

4. The sum of the squares of the sides equals the square of the hypotenuse. $a^2 + b^2 = c^2$

Given a length A and an angle θ, as shown, we immediately see that

$$A_x = A \cos \theta \qquad \text{and} \qquad A_y = A \sin \theta$$

Thus we derive the x and y components of **A** from known A and θ. Now we want to find A and θ from A_x and A_y. First, A is found by use of the Pythagorean theorem (property 4 above):

$$A^2 = A_x{}^2 + A_y{}^2$$

In order to find θ from A_x and A_y, we simply note that $\tan \theta = A_y/A_x$. Inverting this statement,

$$\theta = \tan^{-1}(A_y/A_x)$$

or θ equals "the angle whose tangent is the ratio A_y/A_x."

Still another method of specifying vectors is to use the English language. For example: "The vector **B** is $4\frac{1}{2}$ meters long and points

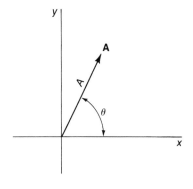

Fig. 7-2. A vector specified by its length, A, and its angle above the horizontal, θ.

$\cos \theta = \dfrac{Ax}{A}$

$Ax = A \cos \theta$

$\sin \theta = \dfrac{Ay}{A}$

$Ay = A \sin \theta$

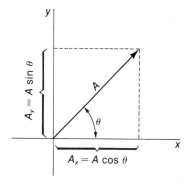

Fig. 7-3. The relationships between the projections of a vector and its length.

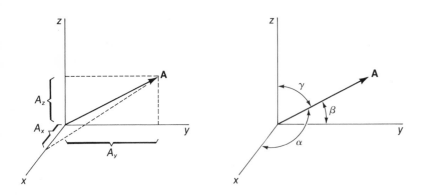

in the direction 12 degrees north of east." Using our formal notation for this case, we would write $B = 4\frac{1}{2}$ m, $\theta = 12°$. Alternatively, we could look up the sine and cosine of 12° and write $B_x = B \cos \theta = 4.4$ m and $B_y = B \sin \theta = 0.94$ m. Either of these pairs of numbers tells us all there is to know about this vector.

In three dimensions we need three numbers to specify a vector, for example, A_x, A_y, and A_z. Using the Pythagorean theorem twice, we see that $A^2 = A_x{}^2 + A_y{}^2 + A_z{}^2$. The angles between the vector and the x, y, and z axes can be labeled, in turn, α, β, γ, as in Fig. 7-4. Then we find $A_x = A \cos \alpha$, $A_y = A \cos \beta$, and $A_z = A \cos \gamma$. These three cosines are often called the *direction cosines* of the vector. The length A and the three direction cosines really constitute only three independent numbers, since the Pythagorean theorem can be used to deduce any one of the cosines once the other two are given. We will seldom need to consider three spatial dimensions in this book and will frequently restrict our consideration to vectors of two dimensions.

Exercises

1. What happens to the components of a vector when we rotate the coordinate system in which the vector is defined, as shown in Fig. 7-5? What happens if, instead, we rotate the vector itself an equal amount in the opposite direction? Is it possible to specify a vector without specifying a coordinate system?

2. A vector has components $A_x = +4$ m, $A_y = +4$ m. What is its length and the angle it makes with the x axis?

3. A vector has components $A_x = -3$ m, $A_y = +4$ m. Find its length and its angle as measured in the counterclockwise direction

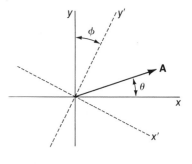

Fig. 7-5. The dotted axes show the result of rotating the x, y coordinate system to a new coordinate system involving x' and y'.

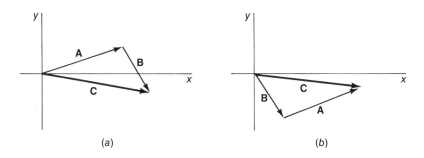

(a)   (b)

Fig. 7-6. (a) The summation of the displacements **A** and **B** is equivalent to that of their resultant, **C.** (b) Interchanging the sequence of displacements does not change the resultant.

from the positive x axis. A rough sketch (on a reduced scale) will help in visualizing the triangle involved here.

4. A vector **A** with a magnitude of 10 units makes an angle of 30° with the x axis, measuring counterclockwise from the x axis. Find the values of its scalar components, A_x and A_y.

7-4 Vector Addition

Graphical methods   Clearly, one of our immediate uses for vectors will be to represent displacements that occur as a result of motion. It will frequently be necessary for us to consider a displacement that is the result of several smaller displacements. The process of summing several displacements to obtain a *resultant* displacement is the prototype for the summation of vectors in general. Fig. 7-6 shows the process for two displacements represented by the vectors **A** and **B,** summed to form their resultant, **C.** The resultant is the single straight line displacement that achieves the same net result as the two consecutive displacements. This resultant is unchanged if we interchange the order of the displacements **A** and **B,** as indicated in Fig. 7-6b. In general, displacement addition consists of finding the single equivalent displacement, the *resultant,* that produces the same effect as all the displacements operating sequentially, as in Fig. 7-7.

   We can generalize the addition process to all types of vectors. Vector summation proceeds by successively placing vectors head to tail, the resultant being that vector that spans the distance from the tail of the first vector to the head of the last. It is important to note that the notion of a vector has nothing to do with its location, only its magnitude and direction. Thus the two vectors in Fig. 7-8 can be summed by moving **B** until its tail coincides with the tip of **A,** as long as we maintain the direction of **B;** that is, **B** in its final location must be parallel to **B** in its original location. The resultant is then easily found in Fig. 7-8c.

(a)

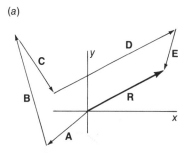

(b)

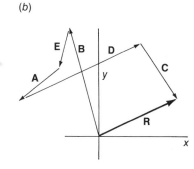

Fig. 7-7. The summation of displacements **A** through **E** forms a resultant, **R.** This resultant is not altered by changing the order of summation to **B, E, A, D, C,** as in (b).

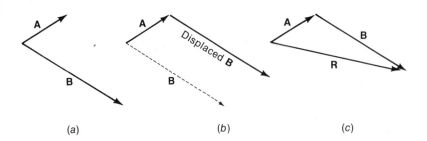

(a)   (b)   (c)

Fig. 7-8. (*a*) Two vectors are positioned inconveniently for summation. (*b*) Repositioning of **B**. (*c*) The sum of **A** and **B** forms their resultant, **R**.

If two vectors are drawn with their tails at a common point, as in Fig. 7-8, a convenient graphical procedure that gives the result of Fig. 7-8*c* is the *parallelogram method* of vector addition. This method is illustrated in Fig. 7-9. A parallelogram with two sides composed of the vectors **A** and **B** is drawn as shown. The resultant, **R** = **A** + **B**, is then given by the diagonal indicated in Fig. 7-9.

Having discussed the sum of vectors, let us now consider the difference of two vectors. Finding the vector **R** = **B** − **A** consists of the addition of **B** and −**A**, where the vector −**A** is found by reversing the sign of each of the components of **A**. The vector −**A** has the same magnitude as **A** but points in the opposite direction, as shown in Fig. 7-10*a*. The parallelogram method can be used to find **B** − **A**, as in Fig. 7-10*b*. A simple method for graphical construction of **B** − **A**, which is just the difference between **B** and **A**, is shown in Fig. 7-10*c*.

Exercises

A ruler, protractor, and graph paper are useful for the following.

5. Vector **A** is 2 units long at an angle of 45° above the horizontal. Vector **B** is 2 units long at an angle of 45° below the horizontal. Find their resultant, **R**.

6. Vectors **A**, **B**, and **C** are given by $A = 6$ cm, $\theta_A = 20°$; $B = 10$ cm, $\theta_B = 80°$, $C = 4$ cm, $\theta_C = 120°$. Find their resultant, **R**.

7. Can the difference of two vectors have a magnitude greater than the magnitude of the sum of the same two vectors? Illustrate with a sketch.

Analytic methods   We need not always use graphical techniques to perform vector sums. The techniques of analytic geometry are available to us and can frequently save us time and effort. Consider the summation of the two vectors **A** and **B**, shown in Fig. 7-11,

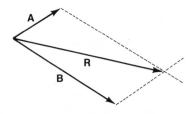

Fig. 7-9. The parallelogram method of vector addition for the pair of vectors shown in Fig. 7-8.

which form an included angle ϕ. The angle ϕ may be given to us or we may have to deduce ϕ from angles θ_A and θ_B given with respect to a coordinate system, as in some of our previous examples. The problem of finding the magnitude of the resultant **R** reduces to finding the third side of a triangle when two sides and their included angle are given. The *law of cosines* is directly applicable: $R^2 = A^2 + B^2 - 2AB \cos \phi$. As an aid in recalling this law, note that it reduces to the Pythagorean theorem when $\theta = 90°$, and that the "correction" to the Pythagorean result makes the square of the resultant smaller than $A^2 + B^2$ when $\phi < 90°$ and greater than $A^2 + B^2$ when $\phi > 90°$, as it should.

Probably the most useful method of vector addition is component addition. This method is illustrated in Fig. 7-12. Looking at the figure, we see that the y component of **R**, the resultant vector, is given by the sum of A_y and B_y. Similarly, note that R_x is given by the sum of A_x and B_x. This method of vector addition reduces the summation of vectors to the summation of its scalar components. Thus the vector equation $\mathbf{A} + \mathbf{B} + \mathbf{C} = \mathbf{R}$ is equivalent to two ordinary equations if we are working in two dimensions:

$$A_x + B_x + C_x = R_x$$

$$A_y + B_y + C_y = R_y$$

In three dimensions, three scalar equations are required:

$$\left. \begin{array}{l} A_x + B_x + C_x = R_x \\ A_y + B_y + C_y = R_y \\ A_z + B_z + C_z = R_z \end{array} \right\} \quad \text{equivalent to } \mathbf{A} + \mathbf{B} + \mathbf{C} = \mathbf{R}$$

We have now reached a rather interesting view of vectors — they are a shorthand notation for an object with several parts, two in two dimensions, three in three dimensions, in general, n in n dimensions. Relativity theory, for example, is most conveniently couched in the language of four-component vectors. Vector equations, such as vector sums, are simply a compact method of stating what would otherwise require n equations for the n components involved.

Example   An airplane flies due north for 110 km, then turns 45° to the right to fly another 50 km, as in Fig. 7-13. Thus the total distance flown is 160 km. If it had instead flown by the shortest possible straight-line route to the same final point, what distance would have been covered?

Solution   (By law of cosines) The straight-line route is simply the *one* vector displacement that produces the same result as the two displacements actually followed, hence it is their vector sum. From the

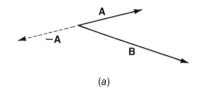

(a)

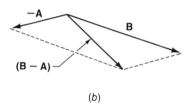

(b)

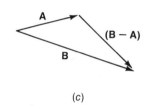

(c)

Fig. 7-10.   The difference between **B** and **A** is found by reversing **A** to form −**A**, as in (a), then adding −**A** to **B**, as in (b). The same result is obtained in one step in (c).

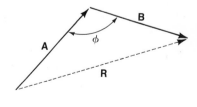

Fig. 7-11.   The law of cosines may be used to solve analytically for the length of **R**.

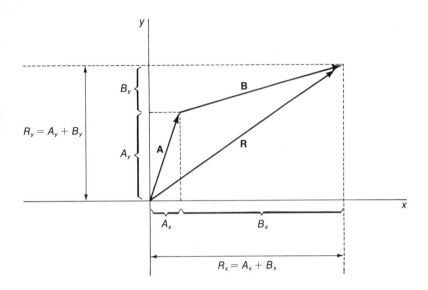

Fig. 7-12. Vector addition by the method of component summation.

law of cosines, $R^2 = A^2 + B^2 - 2AB \cos \phi$. Identifying A, B, and ϕ from Fig. 7-13, we have

$$R^2 = (110 \text{ km})^2 + (50 \text{ km})^2 - 2 \cdot 110 \text{ km} \cdot 50 \text{ km} \cdot (-0.707)$$

$$= 22380 \text{ km}^2$$

$$R = \sqrt{22380 \text{ km}^2} = 150 \text{ km}$$

Example   Find the resultant of three vectors, **A**, **B**, and **C**. They are located in a two-dimensional coordinate system and are specified as follows:

A: $A = 3$ meters $\theta = 40°$ counterclockwise from $+x$ axis

B: $B = 2$ meters $\theta = 90°$ counterclockwise from $+x$ axis

C: $C = 10$ meters $\theta = 200°$ counterclockwise from $+x$ axis

Solution   (By addition of components)

$$A_x = A \cos 40° = 2.3 \text{ meters} \qquad A_y = A \sin 40° = 1.9 \text{ meters}$$

$$B_x = 0 \qquad B_y = B = 2 \text{ meters}$$

$$C_x = C \cos 200° = -9.4 \text{ meters} \qquad C_y = C \sin 200° = -3.4 \text{ meters}$$

The resultant vector, **R**, has components given by:

$$R_x = A_x + B_x + C_x = -7.1 \text{ meters}$$

$$R_y = A_y + B_y + C_y = +0.5 \text{ meters}$$

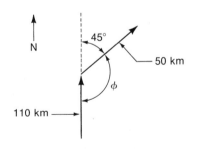

Fig. 7-13.

Thus **R** points to the left (the negative side) of the x axis but lies above the x axis, since it has a positive y component. More specifically, its length is given by

$$R = \sqrt{R_x^2 + R_y^2} = \sqrt{50.4 \text{ m}^2 + 0.25 \text{ m}^2} = 7.12 \text{ m}$$

and its angle above the negative x axis is given by

$$\theta = \tan^{-1}\frac{R_y}{R_x} = \tan^{-1}\frac{0.5}{7.1} = 4°$$

Exercises

8. Two vectors, **C** and **D**, are given by $C = 6$ units, $\theta_C = 80°$; $D = 11$ units, $\theta_D = 10°$. Use the law of cosines to find the magnitude of their resultant vector.

9. Three vectors are given by:

$A_x = 9$ units $\qquad A_y = 5$ units $\qquad A_z = -2$ units

$B_x = -3$ units $\qquad B_y = 5$ units $\qquad B_z = 7$ units

$C_x = 8$ units $\qquad C_y = 10$ units $\qquad C_z = -5$ units

Find the x, y, and z components of the resultant vector.

10. The displacements **A** and **B** are given in component form: $A_x = 8$ m, $A_y = 3$ m; $B_x = -2$ m, $B_y = 5$ m. (a) Find the components of $\mathbf{R} = \mathbf{A} + \mathbf{B}$. ($b$) Find the components of $\mathbf{S} = \mathbf{A} - \mathbf{B}$. ($c$) Find the components of $\mathbf{I} = \mathbf{B} - \mathbf{A}$. ($d$) Find the magnitudes R, S, and I.

7-5   Velocity in Three Dimensions

We have seen that an arbitrary displacement has scalar components that are the sum of component displacements along the x, y, and z axes. When these displacements are divided by the time interval in which they occur, they each define an average velocity along the separate component directions. If, for example, an object is rapidly changing its y coordinate in the positive direction while it is changing its x coordinate quite slowly in a negative direction, its *path* will point primarily along the positive y axis but will be slightly tipped left, toward the negative x direction, as in Fig. 7-14.

In order to sharpen this concept, let us define average velocity components as follows:

$$\bar{v}_x = \frac{\Delta x}{\Delta t} \qquad \bar{v}_y = \frac{\Delta y}{\Delta t} \qquad \bar{v}_z = \frac{\Delta z}{\Delta t}$$

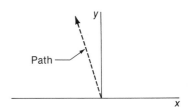

Fig. 7-14.   The path followed by an object with a rapidly increasing y coordinate but slowly *decreasing* x coordinate.

As in Chapters 2 and 3, we will define the instantaneous component velocities to be the appropriate derivatives:

$$v_x = \frac{dx}{dt} \qquad v_y = \frac{dy}{dt} \qquad v_z = \frac{dz}{dt}$$

These component velocities constitute the three components of a
vector **v.** That this vector actually points in the direction of motion
can be seen from the definition of its components. Consider a two-
dimensional illustration. If we wait for an infinitesimal time interval
dt, then infinitesimal displacements will occur given by:

$$dx = v_x \, dt$$
$$dy = v_y \, dt$$

Of course, the direction of the resultant of the displacements dx
and dy is the direction of motion. The angle, θ, above the horizontal
for the vector whose components are dy and dx is

$$\theta = \tan^{-1}\!\left(\frac{dy}{dx}\right)$$

But this is also the angle of the vector **v** formed of v_x and v_y, since

$$\theta = \tan^{-1}\frac{v_y}{v_x} = \tan^{-1}\frac{dy/dt}{dx/dt} = \tan^{-1}\!\left(\frac{dy}{dx}\right).$$

Furthermore, the actual distance, say ds, covered in this time
interval is given by $v \, dt$, where v is the magnitude of the velocity
vector:

$$v = \sqrt{v_x{}^2 + v_y{}^2}$$

This magnitude is frequently called the *speed*, while the word velocity is often reserved for the vector quantity. The velocity vector is
thus seen to be a useful entity. Its direction at any instant of time
points in the direction of the displacement that is occurring at that
time, and its magnitude indicates the rate at which the displacement
occurs. All of the foregoing discussion can be expressed quite compactly using vector notation. Let a position vector, **r,** extend from
the origin of a coordinate system to the point whose motion we are

describing. Note that $r_x = x$, $r_y = y$, and $r_z = z$.

Thus $\mathbf{v} = \dfrac{d\mathbf{r}}{dt}$ is equivalent to $\begin{cases} v_x = \dfrac{dx}{dt} \\[2mm] v_y = \dfrac{dy}{dt} \\[2mm] v_z = \dfrac{dz}{dt} \end{cases}$

Also $d\mathbf{r} = \mathbf{v}\,dt$ is equivalent to $\begin{cases} dx = v_x\,dt \\[2mm] dy = v_y\,dt \\[2mm] dz = v_z\,dt \end{cases}$

If the velocity components are variable we write

$$\left. \begin{aligned} \Delta x &= \int_{t_1}^{t_2} v_x\,dt \\[2mm] \Delta y &= \int_{t_1}^{t_2} v_y\,dt \\[2mm] \Delta z &= \int_{t_1}^{t_2} v_z\,dt \end{aligned} \right\} \quad \text{or} \quad \Delta \mathbf{r} = \int_{t_1}^{t_2} \mathbf{v}\,dt$$

which reduces, *for constant velocity only*, to

$$\left. \begin{aligned} \Delta x &= v_x \Delta t \\ \Delta y &= v_y \Delta t \\ \Delta z &= v_z \Delta t \end{aligned} \right\} \quad \text{or} \quad \Delta \mathbf{r} = \mathbf{v} \Delta t$$

Since velocities are vector quantities, they may be added together as vectors with the methods of Section 7-4.

Example   A velocity vector has constant velocity components $v_y = +8$ m/s, $v_x = +2$ m/s. What is the magnitude and direction of the velocity?

Solution   Applying the Pythagorean theorem,

$$v^2 = v_y{}^2 + v_x{}^2 = 64 \text{ m}^2/\text{s}^2 + 4 \text{ m}^2/\text{s}^2$$

$$= 68 \text{ m}^2/\text{s}^2$$

$$v = 8.25 \text{ m/s}$$

The direction is given by

$$\theta = \tan^{-1}\left(\frac{v_y}{v_x}\right) = \tan^{-1}(4)$$

$$= 76° \text{ above the } x \text{ axis}$$

Example A train moves at 10 ft/s while a passenger moves across the aisle (perpendicular to the direction of the movement of the train) at 3 ft/s. What is the magnitude of his velocity with respect to the frame of reference of the railroad tracks?

Solution The resultant velocity that occurs in the track frame of reference is composed of both the motion of the train along the track and the motion of the passenger in the perpendicular direction. Taking the direction of the tracks as the x axis,

$$v_x = 10 \text{ ft/s}$$

$$v_y = 3 \text{ ft/s}$$

$$v^2 = v_x{}^2 + v_z{}^2 = 109 \text{ ft}^2/\text{s}^2$$

$$v = 10.4 \text{ ft/s}$$

Exercises

11. A golf ball is driven off its tee at an angle of $23°$ above the horizontal at a velocity of 80 ft/s. Find the vertical component of its initial velocity.

12. A boat travels at right angles to the direction of a river's flow with a velocity v_B with respect to the water. The river's current has a velocity v_C with respect to the river bank. Find the total resultant velocity of the boat with respect to the river bank. What is the direction of this resultant velocity?

7-6 Acceleration in Three Dimensions

We have seen that the displacement of an object is a vector quantity, **r**, and that changes in **r** lead to a vector velocity

$$\mathbf{v} = \frac{d\mathbf{r}}{dt} \quad \text{is equivalent to} \quad \begin{cases} v_x = \dfrac{dx}{dt} \\[2mm] v_y = \dfrac{dy}{dt} \\[2mm] v_z = \dfrac{dz}{dt} \end{cases}$$

If the velocity vector is changing, we similarly define a vector acceleration

$$\mathbf{a} = \frac{d\mathbf{v}}{dt} \quad \text{is equivalent to} \quad \begin{cases} a_x = \dfrac{dv_x}{dt} \\[2mm] a_y = \dfrac{dv_y}{dt} \\[2mm] a_z = \dfrac{dv_z}{dt} \end{cases}$$

The inverse statement is

$$\Delta\mathbf{v} = \int_{t_1}^{t_2} \mathbf{a}\, dt \quad \text{or} \quad \begin{aligned} \Delta v_x &= \int_{t_1}^{t_2} a_x\, dt \\[2mm] \Delta v_y &= \int_{t_1}^{t_2} a_y\, dt \\[2mm] \Delta v_z &= \int_{t_1}^{t_2} a_z\, dt \end{aligned}$$

For the case of constant acceleration, these statements reduce to

$$\Delta v_x = a_x \Delta t \quad \Delta v_y = a_y \Delta t \quad \Delta v_z = a_z \Delta t$$

Note that an acceleration does not necessarily change the *magnitude of* **v**. In Fig. 7-15a we see the effect of an acceleration **a** that is not directed along **v**. In a time Δt it produces a change $\Delta\mathbf{v}$ in the velocity vector. The magnitude of **v** is not changed in this example, but the direction of **v** *is* changed. Thus a car that maintains constant speed but changes its direction *has* accelerated. The acceleration does not point in the direction of motion in this case.

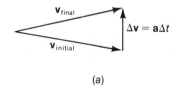

(a)

The equations that were appropriate for one-dimensional motion in Chapter 2 are applicable to each coordinate axis in three dimensions. For the case of *constant* acceleration,

$$x = x_0 + v_{x_0}t + \tfrac{1}{2}a_x t^2$$
$$y = y_0 + v_{y_0}t + \tfrac{1}{2}a_y t^2$$
$$z = z_0 + v_{z_0}t + \tfrac{1}{2}a_z t^2$$

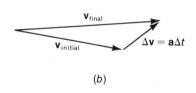

(b)

These three equations can also be written as one vector equation for the position vector:

$$\mathbf{r} = \mathbf{r}_0 + \mathbf{v}_0 t + \tfrac{1}{2}\mathbf{a}t^2$$

Fig. 7-15. An acceleration can act to change the direction of the velocity with no change in the magnitude of the velocity, as in (a). In (b) we see an acceleration that produces a change in both the magnitude and direction of the velocity.

Example An object moves with $v_x = 10$ m/s, $v_y = 0$. However, at $t = 0$, a constant acceleration of $a_x = 0$ and $a_y = 2$ m/s² begins to act. What is the velocity after 4 seconds? In what direction is the object then moving? How far does it move in these 4 seconds?

Solution   The component v_x remains constant at 10 m/s since $a_x = 0$, and v_y increases steadily since $v_y = a_y t$. After 4 seconds, $v_y = (2 \text{ m/s}^2) \cdot 4 \text{ s} = 8$ m/s. At this time $v = \sqrt{v_x^2 + v_y^2} = 12.8$ m/s. Its direction of motion at this time is given by

$$\sqrt{100 + 64} = 12.8 \text{ m/s}$$

$$\theta = \tan^{-1}\left(\frac{v_y}{v_x}\right) = \tan^{-1}\left(\frac{8 \text{ m/s}}{10 \text{ m/s}}\right) = 38.7°$$

above the horizontal. The component displacements at the end of 4 seconds are

$$x = v_x t = 10 \, \frac{m}{s} \cdot 4 \text{ s} = 40 \text{ m}$$

$$y = \tfrac{1}{2} a_y t^2 = \tfrac{1}{2} \cdot 2 \, \frac{m}{s^2} \cdot 16 \text{ s}^2 = 16 \text{ m}$$

$$\tfrac{1}{2} a t^2$$

The total displacement is thus

$$r = \sqrt{x^2 + y^2} = 43 \text{ m}$$

Exercises

13. A body with zero initial velocity accelerates with $a_x = 3$ m/s^2, $a_y = -4$ m/s^2 for a period of 2 seconds. Find the magnitude and direction of **v** at the end of this time.

14. The body in the preceding exercise was initially at a position given by $x = 2$ m, $y = 2$ m. Find its location after 2 seconds.

7-7 Projectile Motion

Consider an object projected from a cliff with an initially horizontal velocity. The force of gravity acting upon the object produces a downward acceleration of $a_y = -g$, where g is the acceleration due to the earth's gravity. The object's velocity in the horizontal direction, to the extent that air resistance can be ignored, is constant. The combination of this constant horizontal velocity component with an ever increasing downward velocity component produces a curved path called a *trajectory*, as shown in Fig. 7-16. Consider the origin of our coordinate system to be located at the base of the cliff. Since v_x is constant, the displacement along the x axis grows linearly, i.e.,

constant horizontal velocity
Increasing downward velocity

trajectory

$$x = v_0 t \qquad\qquad (7\text{-}1)$$

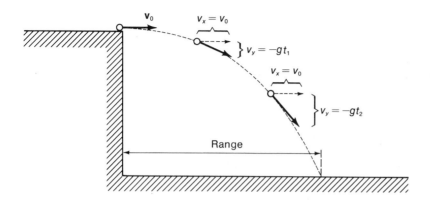

Fig. 7-16. The y component of a projectile's velocity becomes steadily more negative while v_x remains constant.

However, the acceleration along the y axis produces a quadratic dependence of y on time:

$$y = y_0 - \tfrac{1}{2}gt^2 \tag{7-2}$$

where y_0 is the height of the cliff. The dashed line in Fig. 7-16 (the trajectory) is the path followed in x,y space. Equations (7-1) and (7-2), however, give x and y separately as functions of time. In order to obtain the equation of the trajectory, we eliminate t between (7-1) and (7-2). From (7-1), $t = x/v_0$. Substituting into (7-2):

$$y = y_0 - \tfrac{1}{2}g\left(\frac{x}{v_0}\right)^2$$

$$= y_0 - \left(\frac{g}{2v_0^2}\right)x^2 \tag{7-3}$$

Thus the equation for the trajectory (7-3) is in the form $y = y_0 - bx^2$, which is the equation for an inverted parabola centered at $x = 0$ (the cliff face) and with its apex at a distance y_0 above the origin, as shown in Fig. 7-17. The steepness of its sides is given by the coefficient b, which, in the present case, depends upon both the size of g and the initial velocity. Note that the behavior of the y coordinate given by (7-2) is precisely that of an object dropped vertically, while that of the x coordinate given by (7-1) is identical to an object moving along a frictionless, horizontal surface.

The total distance covered (the *range* shown in Fig. 7-16) can be obtained from (7-3) by solving for the value of x that gives $y = 0$. Let us call the range R. From Eq. (7-3):

$$0 = y_0 - \frac{g}{2v_0^2}R^2$$

so that

$$R^2 = \frac{2v_0^2 y_0}{g}$$

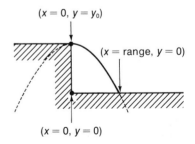

Fig. 7-17. The parabola given by Eq. (7-3) is superimposed upon a sketch of the cliff. The dotted portions are those where Eqs. (7-1) and (7-2) are inapplicable, so that (7-3) is not relevant.

Discarding the negative root for obvious reasons (see Fig. 7-17), we have

$$R = \sqrt{\frac{2y_0}{g}}\, v_0 \qquad (7\text{-}4)$$

This result is quite reasonable, since its dimensions are those of a length and it tells us that a large value of R is obtained by using a tall cliff and/or a large starting velocity. It also tells us precisely how important these factors are, which is more than our intuition can do.

Example   A rock is kicked horizontally off a 100-meter tall vertical cliff at an initial velocity of 7 m/s. How far from the base of the cliff does the rock strike the level ground beyond?

Solution   From (7-4),

$$R = \sqrt{\frac{2y_0}{g}}\, v_0$$

$$= \sqrt{\frac{200 \text{ m}}{9.8 \text{ m/s}^2}} \cdot 7 \text{ m/s}$$

$$= (4.51 \text{ s}) \cdot 7 \text{ m/s} = 31.6 \text{ m}$$

We could also arrive at this result in a slightly different way. The rock falls for the same time, t_0, that a vertically falling rock requires to fall the distance y_0. This time is easily found from (3-13):

$$d = \tfrac{1}{2}gt^2 \qquad (3\text{-}13)$$

which becomes $y_0 = \tfrac{1}{2}gt_0^2$ for our case, so that

$$t_0 = \sqrt{\frac{2y_0}{g}}.$$

During this time the rock moves horizontally with constant velocity v_0. Thus $x = v_0 t$ and, as before,

$$R = v_0 t_0 = \sqrt{\frac{2y_0}{g}}\, v_0$$

Turning now to a more general situation, consider a projectile fired at an angle θ above the horizontal with an initial velocity v_0, as shown in Fig. 7-18. Generally speaking, the x and y motion of a projectile with constant acceleration is given by

$$x = x_0 + v_{0x}t + \tfrac{1}{2}a_x t^2$$

$$y = y_0 + v_{0y}t + \tfrac{1}{2}a_y t^2$$

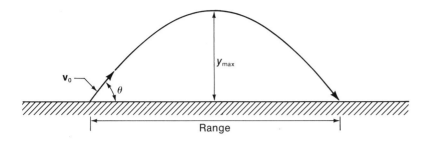

Fig. 7-18.   A projectile fired at an angle θ above the horizontal follows a parabolic trajectory. The apex of the parabola is at a height y_{max} above ground level.

In the present case $a_x = 0$ (neglecting air resistance) and $a_y = -g$. If we choose the origin of the coordinate system at the beginning of the trajectory, we have

$$x = v_{0x}t \tag{7-5}$$

$$y = v_{0y}t - \tfrac{1}{2}gt^2 \tag{7-6}$$

where v_{0x} and v_{0y} are the horizontal and vertical components of the initial velocity:

$$v_{0x} = v_0 \cos \theta \qquad v_{0y} = v_0 \sin \theta$$

We can eliminate t between (7-5) and (7-6) to obtain:

$$y = \left(\frac{v_{0y}}{v_{0x}}\right)x - \left(\frac{g}{2v_{0x}{}^2}\right)x^2 \tag{7-7}$$

which is again the equation of a parabola. The linear term in x translates the apex of this parabola so that it is located above the ground at midrange, as shown in Fig. 7-18.

Exercises

15. If an airplane flying horizontally at constant velocity drops a bomb, it always remains directly beneath the airplane as it falls toward the earth. Explain.

16. A golf ball is driven off its tee with an initial velocity of 100 ft/s at 30° above the horizontal. (*a*) Find v_{0x} and v_{0y}. (*b*) Find the values of x for which y, as given by (7-7), is zero. One of these values of x is the range of the projectile. (*c*) Numerically evaluate your result in part *b* for this particular case to obtain the range of the golf ball.

17. When a projectile reaches its maximum height it is no longer rising nor is it yet falling. In other words, $v_y = 0$ at $y = y_{max}$. (*a*) Use this statement to find the maximum height of the pro-

jectile in Fig. 7-18. (*b*) Now use your result to obtain the maximum height to which the golf ball of Exercise 16 rises.

7-8 Summary

Vectors are directed quantities, the displacement vector being the most fundamental example. A vector in three dimensions is specified by its three components. The components of the sum of several vectors can be obtained by adding the components of the individual vectors, i.e., if $\mathbf{R} = \mathbf{A} + \mathbf{B} + \mathbf{C}$, then

$$R_x = A_x + B_x + C_x$$
$$R_y = A_y + B_y + C_y$$
$$R_z = A_z + B_z + C_z$$

A vector velocity, $\mathbf{v}$, can be defined by $\mathbf{v} = d\mathbf{r}/dt$, where $\mathbf{r}$ is the position vector. Similarly, $\mathbf{a} = d\mathbf{v}/dt$. For constant acceleration we obtain:

$$\left.\begin{array}{l} x = x_0 + v_{0x}t + \tfrac{1}{2}a_xt^2 \\[4pt] y = y_0 + v_{0y}t + \tfrac{1}{2}a_yt^2 \\[4pt] z = z_0 + v_{0z}t + \tfrac{1}{2}a_zt^2 \end{array}\right\} \quad \text{or} \quad \mathbf{r} = \mathbf{r}_0 + \mathbf{v}_0t + \tfrac{1}{2}\mathbf{a}t^2$$

Projectile motion is characterized by a constant negative acceleration in the vertical direction and (in the absence of air resistance) by no acceleration in the horizontal direction. The horizontal (x) and vertical (y) motions are thus given by:

$$x = x_0 + v_{0x}t$$
$$y = y_0 + v_{0y}t - \tfrac{1}{2}gt^2$$

The resultant trajectory is parabolic and has a clearly defined maximum height and range for any given initial velocity and direction.

Problems

7-1. Vector **A** has components $A_x = -3$, $A_y = 3$. Vector **B** has components $B_x = 4$, $B_y = -1$. Find the components of their resultant vector, $\mathbf{R} = \mathbf{A} + \mathbf{B}$. Compare your result to the parallelogram method of addition.

7-2. Two vectors are given by their components: A_x, A_y and B_x, B_y. Find the components of their resultant, $\mathbf{R} = \mathbf{A} + \mathbf{B}$, and show that the resultant is unchanged by the order of summation, i.e., $\mathbf{A} + \mathbf{B} = \mathbf{B} + \mathbf{A}$.

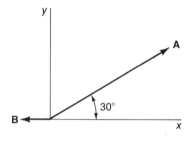

Fig. 7-19.

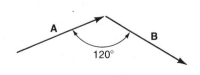

Fig. 7-20.

7-3. Use graphical methods to find $\mathbf{A} + \mathbf{B}$ and $\mathbf{A} - \mathbf{B}$ in Fig. 7-19, where $A = 20$ and $B = 5$.

7-4. Vectors $\mathbf{A}$ and $\mathbf{B}$ are 6 and 4 units long, respectively; they are to be summed as shown in Fig. 7-20. Use the law of cosines to find the magnitude of the resultant vector.

7-5. Find the x and y components of the vectors shown in Fig. 7-21. Find the resultant vector by summing these components and compare your results with the graphical method of addition.

7-6. Two vectors, 3 units and 4 units long, respectively, are at right angles to one another. Find the magnitude and angle (with respect to one of the above vectors) of the resultant vector.

7-7. A three-dimensional vector $\mathbf{C}$ has components $C_x = 11$ units, $C_y = -8$ units, and $C_z = -6$ units. What is its magnitude, C?

7-8. Three vectors are given as shown in Fig. 7-22, with $A = B = C$. Find the magnitude and direction of the resultant of these three vectors.

7-9. The vector velocity $\mathbf{v}_p$ of an airplane with respect to the earth is given by the vector sum of its velocity with respect to the air, $\mathbf{v}_{pa}$, and the velocity of the air with respect to the ground, $\mathbf{v}_w$. If an airplane flies due north with $v_p = 200$ mi/hr, while a wind of 60 mi/hr ($v_w = 60$ mi/hr) blows from west to east, (a) find the magnitude of its airspeed, v_{pa}; (b) find the airplane's heading, that is, find the angle between the long axis of the airplane's body and north.

7-10. A boat has a velocity $\mathbf{v}_B$ with respect to the water. It is travelling in a river that has a uniform velocity $\mathbf{v}_C$, and $v_B > v_C$. (a) In what direction should the boat head to cross from one bank of the river to the other without drifting either up or down stream? (b) If the width of the river is W, compare the time required for the boat to cross the river and return under the conditions of part a with the time required for the boat to go directly upstream a distance W and return. (c) What heading should the boat take to cross from one bank to the other in the minimum possible time without regard for where the boat lands?

7-11. One of the values of x for which $y = 0$ in Eq. (7-7) is the range of the projectile. If the initial velocity is held fixed, the range is a function of θ (from v_{0y} and v_{0x}). (a) Find the two values of θ that give zero range. (b) Find the value of θ that gives the maximum range.

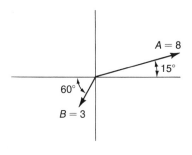

Fig. 7-21.

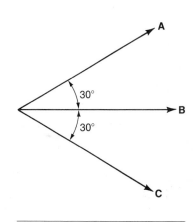

Fig. 7-22.

7-12.* A borate bomber is flying at 300 km/hr at an altitude of 1000 meters. It dives at an angle of 45° below the horizontal when it releases its borate bomb on a small brush fire. (*a*) How far from a point directly beneath the release point will the bomb strike? (*b*) Where will the bomb strike with respect to the point on the ground directly ahead of the airplane (the point the airplane would strike if it continued in a straight line)? (*c*) The point described in part *b* is the point the pilot sees directly ahead of him when he releases the bomb. Can you design a sighting device that is angled down so that the pilot sees the impact point through it? (Design, in this context, means find the angle below the airplane's straight-line trajectory that "looks" at the impact point.)

7-13. Show that the range R of a projectile fired at an angle θ above the horizontal on level ground is given by

$$R = \left(\frac{v_0^2}{g}\right) \sin 2\theta$$

For what angle θ does the range have a maximum value?

Chapter **8**

Dynamics in Three Dimensions

8-1 Introduction We will not introduce new dynamical laws in this chapter. Instead, we will generalize the laws we discussed for one-dimensional motion to include motion in all three spatial dimensions. We will also begin to consider the effects that distinguish an extended body from a point mass. We will thus be led to a definition of the center of mass of extended objects. We will not yet, however, discuss the effects caused by rotations of a body about its center of mass. Translational motion is the name given to motion of a body without rotation. Hence, this chapter introduces the dynamics of translational motion of point masses and extended objects in three dimensions. The vector notation developed in Chapter 7 will be extremely helpful.

8-2 Conservation of Momentum

Just as the momentum of a single mass was defined in one dimension to be

$$p = mv$$

we may define components of momentum for a mass moving in an arbitrary direction:

$$\left.\begin{array}{l} p_x = mv_x \\ p_y = mv_y \\ p_z = mv_z \end{array}\right\} \quad \text{equivalent to} \quad \mathbf{p} = m\mathbf{v} \quad (8\text{-}1)$$

If a system contains N masses, we define the total momentum as the *vector sum* of the individual momenta:

$$\mathbf{p}_{\text{total}} = \sum_{i=1}^{N} \mathbf{p}_i \qquad (8\text{-}2)$$

Experiments with systems containing many bodies show that this total momentum remains constant in both magnitude and direction as long as no external forces act. The law of conservation of momentum thus generalizes to three dimensions:

$$\mathbf{p}_{\text{total}} = \sum_{i=1}^{N} \mathbf{p}_i = \mathbf{constant} \text{ (in the absence of external forces)} \qquad (8\text{-}3)$$

Experiments show that (8-3) holds even if the individual masses collide with one another. Such collisions produce forces that are *internal* to the entire system and do not change the total momentum. In component form (8-3) becomes:

$$\sum_{i=1}^{N} p_{ix} = \text{constant}$$

$$\sum_{i=1}^{N} p_{iy} = \text{constant} \qquad (8\text{-}4)$$

$$\sum_{i=1}^{N} p_{iz} = \text{constant}$$

In fact, we usually use the law of conservation of momentum in the form (8-4) when we come to grips with a numerical problem.

An example that illustrates the conservation of momentum in more than one dimension is shown in Fig. 8-1. The photographs in Fig. 8-1*a* are multiflash pictures looking down upon an air table. The motion of the metal pucks is nearly frictionless, so that the system consisting of the two pucks should maintain constant momentum (Eq. (8-3)) even if they collide with one another. Graphical summation of the final momenta does yield a total momentum that is equal to the original momentum, as shown in part *b* of the figure.

A microscopic example of the law of conservation of momentum is illustrated by the cloud-chamber photograph in Fig. 8-2. The particles involved are moving with 1/20 the velocity of light and have a mass of 6.6×10^{-27} kg. The V-shaped track is produced when one of many alpha particles, moving from left to right, strikes a stationary helium atom's nucleus and causes it to recoil. The alpha particle is itself the nucleus of a helium atom. The alpha particles in this photograph come from the radioactive decay of a heavier

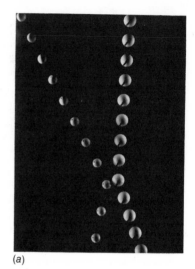

(a)

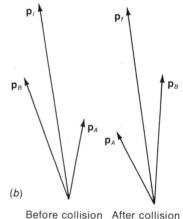

(b)

Before collision   After collision

Fig. 8-1.   (a) A multiflash picture of two metal pucks colliding on a nearly frictionless air table. The individual exposures are taken at uniform intervals, so that the spacing between subsequent images indicates velocity. (From *PSSC Physics*, D. C. Heath and Company, Lexington, Massachusetts, © 1965) (b) Graphical addition of the final momentum vectors shows that $\mathbf{p}_f = \mathbf{p}_i$. The small puck is labeled A, the large one B.

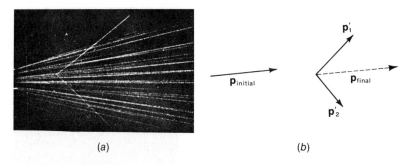

(a)   (b)

Fig. 8-2. (a) The tracks of alpha particles from a radioactive source are visible in this cloud-chamber photograph. Range measurements give their velocity and hence their momentum. (Photograph by P. M. S. Blackett, *Proc. Roy. Soc.*, London (A) 107, 349 (1925). Used by permission of Pergamon Press Limited) (b) The final momentum, $\mathbf{p}_f$, is found to be equal to the initial momentum, $\mathbf{p}_i$.

nucleus. Their tracks are made visible by water droplets that condense in the region of ionization produced by these particles.

In actual practice, views of such collisions from two different directions are used to find the x, y, and z components of each momentum vector. It is then found that, within the experimental accuracy of these measurements, the sum of the components of the momenta of both particles is unchanged, i.e.,

$$p_{1x} + p_{2x} = p'_{1x} + p'_{2x}$$

$$p_{1y} + p_{2y} = p'_{1y} + p'_{2y}$$

$$p_{1z} + p_{2z} = p'_{1z} + p'_{2z}$$

where $\mathbf{p}$ denotes momentum before the collision and $\mathbf{p}'$ denotes momentum after the collision. Together $\mathbf{p}'_1$ and $\mathbf{p}'_2$ define a plane. Since neither of them has a component perpendicular to this plane, $\mathbf{p}_{\text{initial}}$ cannot either. Thus, $\mathbf{p}_{\text{initial}}$ must lie in the same plane.

In some nuclear collisions, particles such as neutrons are produced that have no electric charge. Such uncharged particles leave no tracks in cloud or bubble chambers. In such a situation, research workers frequently apply the laws of momentum conservation to the visible particle tracks and *infer* the presence of an uncharged invisible particle. A striking example, where the invisible particles quickly decay back to two oppositely charged particles, is shown in Fig. 8-3.

Example A 10-kg projectile moving at 140 m/s explodes into three fragments. Choosing the x axis along the initial direction of the projectile, it is found that two of the three fragments have the following masses and velocity components:

$m_1 = 3$ kg: $v_x = 210$ m/s $v_y = -180$ m/s $v_z = 80$ m/s

$m_2 = 4$ kg: $v_x = 105$ m/s $v_y = 40$ m/s $v_z = -60$ m/s

What is the velocity and direction of motion of the other 3-kg fragment? Neglect gravitational forces.

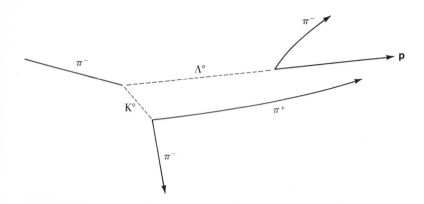

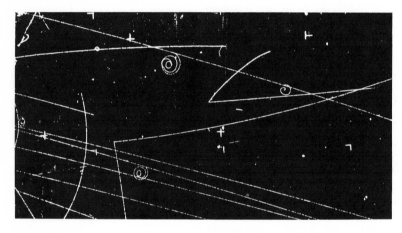

Fig. 8-3. Tracks of subatomic particles in a bubble chamber. The V events result from the decay of invisible neutral particles whose presence and direction are inferred from the momentum of the particles that left visible tracks. (Courtesy University of California, Berkeley, Lawrence Radiation Laboratory)

Solution The three components of the initial momentum of the projectile are

$$p_{0x} = mv_0 = 10 \text{ kg} \cdot 140 \text{ m/s} = 1400 \text{ kg} \cdot \text{m/s}$$

$$p_{0y} = 0 \qquad p_{0z} = 0$$

Since there are no external forces in this example, we have

$$\mathbf{p}_f = \mathbf{p}_i$$

Component by component this becomes

$$p_{fx} = 630 \text{ kg} \cdot \text{m/s} + 420 \text{ kg} \cdot \text{m/s} + p_{3x} = p_{0x} = 1400 \text{ kg} \cdot \text{m/s}$$

$$p_{fy} = -540 \text{ kg} \cdot \text{m/s} + 160 \text{ kg} \cdot \text{m/s} + p_{3y} = p_{0y} = 0$$

$$p_{fz} = 240 \text{ kg} \cdot \text{m/s} - 240 \text{ kg} \cdot \text{m/s} + p_{3z} = p_{0z} = 0$$

Solving for the unknown momentum components of the third fragment, we find:

$$p_{3x} = 350 \text{ kg·m/s}$$

$$p_{3y} = 380 \text{ kg·m/s}$$

$$p_{3z} = 0$$

The velocity components are found by dividing by the mass:

$$V_{3x} = \frac{350 \text{ kg·m/s}}{3 \text{ kg}} = 116.7 \text{ m/s}$$

$$V_{3y} = \frac{380 \text{ kg·m/s}}{3 \text{ kg}} = 126.7 \text{ m/s}$$

$$V_{3z} = 0$$

The magnitude of the velocity of this fragment is thus

$$V_3 = \sqrt{V_{3x}^2 + V_{3y}^2} = 172 \text{ m/s}$$

Since $v_{3z} = 0$, its velocity must lie in the x,y plane at an angle

$$\theta = \tan^{-1}\left(\frac{126.7}{116.7}\right) = 47.3°$$

above the x axis (since $v_y > 0$).

Exercises

1. A billiard ball with a speed of 8 m/s strikes an initially stationary billiard ball, as in Fig. 8-4. Their masses are equal. The initially stationary ball recoils with a speed of 5.66 m/s. Find the final speed of the other ball.

2. An electron moving with a velocity of 10^7 m/s strikes an initially stationary alpha particle. What is the maximum possible velocity of the alpha particle after it is struck, if the electron bounces off with a final velocity still equal in magnitude to 10^7 m/s? An alpha particle's mass is 7,290 times that of an electron.

3. Two billiard balls collide on a pool table. A short time later each, in turn, strikes a side cushion of the table and rebounds from it. What must be included in your "system" in order that its total momentum be conserved, even after the balls strike the cushions?

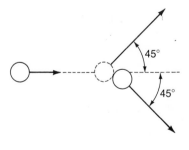

Fig. 8-4.

8-3 Force as a Vector

In one dimension, Newton's second law stated that net force is equal to the time derivative of momentum. We now generalize this law to three dimensions:

$$\mathbf{F}_{net} = \frac{d\mathbf{p}}{dt} \tag{8-5}$$

Component by component, (8-5) states that

$$F_x = \frac{dp_x}{dt} \qquad F_y = \frac{dp_y}{dt} \qquad F_z = \frac{dp_z}{dt}$$

Thus we see that force is also a vector quantity. If we apply (8-5) to a single object of constant mass we obtain

$$\mathbf{F}_{net} = m\mathbf{a} \qquad \text{or} \qquad \begin{cases} F_x = ma_x \\ F_y = ma_y \\ F_z = ma_z \end{cases} \tag{8-6}$$

Implulse also becomes a vector quantity:

$$\int_{t_1}^{t_2} \mathbf{F}\,dt = \mathbf{p}_2 - \mathbf{p}_1 \qquad \text{or} \qquad \begin{cases} \int_{t_1}^{t_2} F_x\,dt = p_{2x} - p_{1x} \\ \int_{t_1}^{t_2} F_y\,dt = p_{2y} - p_{1y} \\ \int_{t_1}^{t_2} F_z\,dt = p_{2z} - p_{1z} \end{cases} \tag{8-7}$$

Example A 3-kg mass moves at 50 m/s in the x direction. A 6-N force acts for 5 s in the y direction. What is the velocity and direction of the mass at the end of the 5-s period?

Solution Consider the velocity in the x direction. Since the component of force in the x direction is zero, this component of the velocity remains unchanged. However, the y component of the velocity, initially zero, is accelerated by the y component of the force:

$$a_y = \frac{F_y}{m} = \frac{6\ \text{N}}{3\ \text{kg}} = 2\ \text{m/s}^2$$

Acting for 5 s, this produces a velocity component of

$$v_y = v_{0y} + a_y t = 0 + 2\ \frac{\text{m}}{\text{s}^2} \cdot 5\ \text{s} = 10\ \frac{\text{m}}{\text{s}}$$

As stated previously, v_x remains constant:

$$v_x = 50\ \text{m/s}$$

So $v = \sqrt{v_x^2 + v_y^2} = \sqrt{2600 \text{ m}^2/\text{s}^2} = 51$ m/s. The mass now moves at an angle θ with respect to the x axis:

$$\theta = \tan^{-1}\left(\frac{v_y}{v_x}\right) = \tan^{-1}\left(\frac{1}{5}\right) = 11.3°$$

Newton's second law in vector form (Eq. (8-5)) is more than just a definition of what is meant by a force vector. Some of its predictions are subject to experimental test. For example, it has been experimentally verified that two forces acting in different directions upon a single body do produce the same net effect as a single force that is the vector sum of the two forces, as required by (8-5). Further, a given force **F** is experimentally found to produce the same vector *change* in momentum regardless of the mass of the body to which it is applied *and* regardless of the initial direction of the body's motion, as required by (8-5).

Exercises

4. A 3200-lb truck is initially moving north at 30 mi/hr. A force of *constant* magnitude and direction acts for 3 s on the truck, causing it finally to move west at 30 mi/hr. Find the magnitude and direction of this force. Begin by finding the north-south and east-west components of the force. These components can be found by application of (8-7).

5. A 1-kg object is initially at rest. A 3-N and 4-N force act on it simultaneously for 1 s and produce a velocity of 5 m/s. What is the angle between these two forces?

8-4   Resolution of Forces

Since forces are vector quantities, they can be resolved into components. We frequently encounter cases in which the response of a system to a force in a *given* direction is quite easily analyzed, yet the applied force is not acting in that particular direction. In many such cases, it is advantageous to resolve the applied force into components that lie along directions for which the resulting motion is easily analyzed.

The motion of the block of mass m on the plane in Fig. 8-5 is an example of such a situation. For simplicity, let us assume that the effects of friction are negligible, as would be the case for an ice cube on a smooth, wet surface. The force of gravity, **W**, is resolved into two components, one along the direction parallel to the plane

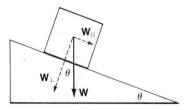

Fig. 8-5.   The force of gravity, **W**, on the block can be resolved into components along the two directions shown: parallel and perpendicular to the inclined plane.

and one along the direction perpendicular to the plane. Calling those components $W_\parallel$ and $W_\perp$, respectively, we have

$$W_\parallel = W \sin \theta$$

$$W_\perp = W \cos \theta$$

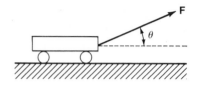

Fig. 8-6.

We know that a force perpendicular to the plane will not cause motion; it simply squeezes the block against the plane. However, a force along the plane will cause an acceleration. In the absence of friction or air resistance, the force $W_\parallel$ is the only force acting in this direction. Denoting the acceleration along this direction by $a_\parallel$, we have:

$$a_\parallel = \frac{W_\parallel}{m} = \frac{W \sin \theta}{m} = \frac{mg \sin \theta}{m} = g \sin \theta \qquad (8\text{-}8)$$

This result is quite reasonable. First the dimensions are correct, noting that $\sin \theta$ is a pure number and hence dimensionless. Second, the behavior of the acceleration is correct at the two extreme angles, $\theta = 0°$ and $\theta = 90°$. At $\theta = 0°$ the plane is horizontal and no acceleration takes place. At $\theta = 90°$, the block is falling vertically and is in free fall, as predicted by (8-8).

Exercises

6. Prove that the angle between $W_\perp$ and W in Fig. 8-5 is equal to θ, the angle of inclination of the inclined plane.

7. A boy pulls with a 50-N force on a 15-kg wagon, as sketched in Fig. 8-6. For $\theta = 30°$, find the acceleration of the wagon. Note that the vertical component of F is insufficient to lift the front wheels of the wagon off the ground, so that its acceleration is in the horizontal direction.

8. A rope supporting a weight W is pulled to one side by a horizontal rope, as shown in Fig. 8-7. If T_1 is the tension in the upper portion of the rope, find T_1 in terms of W and θ.

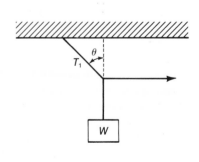

8-5   Frames of Reference

We may now generalize our discussion of inertial frames in Section 4-8 to more than one dimension. Figure 8-8 illustrates the transformation of coordinates involved in viewing a point P from different coordinate systems. The primed coordinate system gives the coor-

Fig. 8-7.

dinates of P as x', y' and z'. Since the origin of this primed coordinate system is itself at a point with coordinates X, Y, and Z in the unprimed coordinate system, we have:

$$x' = x - X$$
$$y' = y - Y \tag{8-9}$$
$$z' = z - Z$$

Now if the point is moving but the systems are stationary with respect to one another, we have:

$$v'_x = \frac{dx'}{dt} = \frac{dx}{dt} - \frac{dX}{dt} = v_x$$

since X is constant. Similarly for the other coordinates,

$$v'_y = v_y \qquad v'_z = v_z$$

However, if the primed system moves at constant velocity $\mathbf{V}$ in the unprimed system, we have

$$v'_x = v_x - V_x \qquad v'_y = v_y - V_y \qquad v'_z = v_z - V_z$$

where $V_x = dX/dt$, $V_y = dY/dt$, and $V_z = dZ/dt$ are the velocity components of the moving frame's origin. Equivalently,

$$\mathbf{v'} = \mathbf{v} - \mathbf{V} \tag{8-10}$$

We thus see that a constant velocity viewed from another coordinate system that *itself* moves at a constant velocity with respect to the first coordinate system still looks like a constant velocity, albeit of different magnitude and direction than the original velocity. We conclude that if the law of inertia is valid in one frame of reference, it is valid in the entire infinity of other reference frames that move at constant velocities with respect to the first. All these frames of reference are called *inertial frames*. As stated in Chapter 4, it is common to think of the frame of reference attached to the bulk of the matter of the universe—often called the "frame of the fixed stars"—as the most primitive or fundamental of these, yet there is no way that mechanics alone can distinguish these various frames of reference. The law of inertia and Newton's second law hold equally well in all of them. The surface of the earth is not a perfect inertial frame, because the velocity of a given point there is always slowly changing due to the earth's rotation on its axis and revolution about the sun. Nonetheless, it is sufficiently close to an inertial frame to be considered one for most purposes.

How can we tell when we are in an inertial frame? How can we write the laws of mechanics when we are not in an inertial frame?

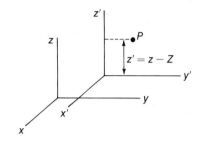

Fig. 8-8. A point P is simultaneously located in two coordinate systems. The primed coordinate system's origin is located at the point X, Y, Z in the unprimed system.

The observer who rides in an accelerating car has no difficulty detecting the noninertial nature of his frame of reference. In order to keep his head from lagging behind, he must tense his neck muscles. He also feels the seat back pressing against his body as he accelerates along with the car. These sensations make it clear to him that he is not in an ordinary, inertial frame of reference. He can continue to write the ordinary laws of mechanics in his accelerated frame of reference only if he introduces *fictitious* forces. He can allege that *something* is pulling back on his head so that he is forced to tense his neck muscles. He may call that "something" a *reaction* to the acceleration. But why not be realistic? With respect to a true inertial frame, his head is accelerating. A force *must* be supplied to produce that acceleration and his neck muscles do just that.

If we adopt the fictitious, reaction-force point of view, how do we compute the force felt by the tensed neck muscles? This "force" has a magnitude

$$F = m_h a_c$$

where m_h is the mass of the head and a_c is the acceleration of the car with respect to an inertial frame. This reaction force, we must remember, is directed opposite to the direction of acceleration. Now we can compute the force supplied by the neck muscles, since the observer's head *in the car frame of reference* is not accelerating, hence no net force acts. Thus the neck muscles provide a force in the direction of acceleration of the car that just offsets the fictional reaction force and produces no net acceleration within the noninertial frame. It seems more direct simply to note that an acceleration *is* occurring and to calculate the single force required to produce the acceleration. Both points of view are effective, but we will emphasize the approach that does not invoke fictitious forces throughout this text. Note that in *either* approach we must know the acceleration with respect to an inertial frame in order to calculate the required force.

Example   A mass hangs from the ceiling of a steadily accelerating car. At what angle does it hang with respect to the vertical?

Solution   The mass is shown in Fig. 8-9. Let us solve this problem in the stationary inertial frame. Let **T** denote the tension in the string. Since the mass must accelerate along with the car, there must be a net force acting on it in the horizontal direction; it is supplied by the horizontal component of the tension in the string, $T \sin \theta$. The force of gravity on the mass is balanced by the vertical component of the string's tension, $T \cos \theta$. These two facts allow us to compute both the magnitude T of the tension and the angle θ. Take x as the direction in

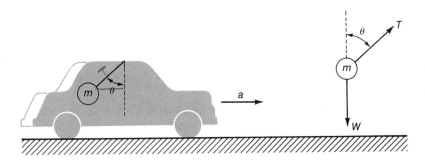

Fig. 8-9. (a) A mass hangs from the ceiling of a steadily accelerating car. (b) The mass is acted upon by the tension in the string and the force of gravity. The tension acts along the direction of the string, gravity acts vertically.

which the car accelerates and y as the vertical. The forces acting on the mass are:

$$F_x = T \sin \theta = ma \qquad F_y = T \cos \theta - mg = 0$$

so that $T \cos \theta = mg$. Hence

$$\frac{T \sin \theta}{T \cos \theta} = \frac{ma}{mg}$$

$$\tan \theta = \frac{a}{g}$$

$$\theta = \tan^{-1}\left(\frac{a}{g}\right)$$

Now squaring and adding the equations for $T \sin \theta$ and $T \cos \theta$,

$$T^2 (\sin^2 \theta + \cos^2 \theta) = m^2 (a^2 + g^2)$$

or

$$T = m (a^2 + g^2)^{1/2}$$

since $\sin^2 \theta + \cos^2 \theta = 1$ always.

The calculation is not greatly changed if we solved the problem in the accelerated frame of reference of the car. Since the ball is not hanging vertically, we say there is a "fictitious force" acting on the mass in the direction pointing to the rear of the car. Then the force supplied by the tension in the string must just offset the net force due to both gravity and this fictitious force. Denoting this net force **F'** we have

$$F_x' = -ma \qquad F_y' = -mg$$

$$F' = (F_x')^2 + (F_y')^2 = m (a^2 + g^2)^{1/2}$$

The direction of **F'** would be to the left and at an angle θ below the horizontal in Fig. 8-9. The angle θ is easily found:

$$\tan \theta = \frac{F_x'}{F_y'} = \frac{a}{g}$$

as was obtained previously. Since this is the direction of the net force, it is also the direction of the hanging string. If a passenger held a helium-filled balloon attached to a string in such a car, how do you suppose it would behave?

Exercises

9. In a given coordinate system, an object's location is given by

$$x = x_0 + v_{0x}t + \tfrac{1}{2}a_xt^2$$

$$y = y_0 + v_{0y}t + \tfrac{1}{2}a_yt^2$$

$$z = z_0 + v_{0z}t + \tfrac{1}{2}a_zt^2$$

Find a displaced coordinate system moving with respect to the initial system in which

$$x' = \tfrac{1}{2}a_xt^2$$

$$y' = \tfrac{1}{2}a_yt^2$$

$$z' = \tfrac{1}{2}a_zt^2$$

10. A hoop is carried 10 ft above a cart moving at a constant velocity of 30 ft/s. A second cart follows with a velocity of 40 ft/s in the same direction as the first (see Fig. 8-10). At what velocity with respect to this second cart should a projectile be fired, as shown in the figure, so that it just reaches the height of the hoop? What should the separation between the carts be so that the ball will pass through the hoop as shown? Note that a reference frame different from the one given simplifies the problem somewhat.

11. A 70-kg man sits in a car accelerating forward at 1.5 m/s². How large is the forward force component exerted on him by the car? How large is the vertical force component exerted on him by the car? What is the total magnitude of the force?

8-6   The Principle of Relativity   NOT ON 2nd HOURLY

The laws of mechanics are valid in *any* inertial frame of reference. What about other areas of physics? Will the laws of electricity and magnetism, for example, also be generally valid in any inertial frame of reference? The *principle of relativity* states that this is so — *all laws of physics are valid in any inertial frame.* The transformations between inertial frames given by Eqs. (8-9) are called *Galilean transformations.* In the beginning of the Twentieth Century, it was noted by many physicists that the laws of electricity and magnetism *were* changed in form if the Galilean transformations were applied

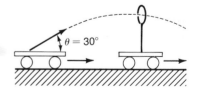

Fig. 8-10.

to them. A rather complicated set of transformation equations were required to preserve the identity of electromagnetic equations when transformed to a new frame of reference. In order to make these transformations simpler, let us rewrite (8-9) for the special case of the primed coordinate system moving along the x axis of the unprimed system with a velocity V:

$$x' = x - Vt$$

$$y' = y$$

$$z' = z \qquad (8\text{-}11)$$

$$t' = t$$

These transformation equations preserve the basic equation of dynamics, $F = ma$, in all inertial frames, since two differentiations of (8-11) leave us with $a'_x = a_x$, $a'_y = a_y$, and $a'_z = a_z$.

The transformation equations that preserve the form of the basic laws of electromagnetic theory in arbitrary inertial frames are:

$$x' = \frac{x - Vt}{\sqrt{1 - V^2/c^2}}$$

$$y' = y$$

$$z' = z \qquad (8\text{-}12)$$

$$t' = \frac{t - xV/c^2}{\sqrt{1 - V^2/c^2}}$$

where c is the velocity of light. Equations (8-12) are called the *Lorentz transformations* after the physicist H. A. Lorentz, who proposed them in 1903. Note that they differ from (8-11) by only a negligible amount if $V \ll c$.

In 1905, Albert Einstein (1879–1955) proposed his *special theory of relativity,* in which the correct transformation equations were taken to be of the Lorentz form, (8-12), rather than Galilean, (8-11). Since it was the Galilean transformations that preserved Newton's first and second laws in all inertial frames, a new formulation of mechanics was required. This new mechanics, called *relativistic mechanics,* made predictions that differed perceptibly from those of classical mechanics only for velocities comparable to that of light. For example, the inertia of a body denoted by its inertial mass m was predicted to vary due to its velocity v according to the formula

$$m = \frac{m_0}{\sqrt{1 - v^2/c^2}} \qquad (8\text{-}13)$$

where m_0 is the mass measured when the body is at rest and c is the velocity of light. Since c is so large, 3×10^8 m/s, this predicted variation of m due to its velocity is very small for ordinary velocities and is usually undetectable. In 1908, however, (8-13) was verified by Bucherer in a study of high-speed electrons.

It is primarily in the realm of atomic and nuclear physics and in some astrophysical processes that speeds occur comparable to that of light. In those cases, the distinction between classical and relativistic mechanics is essential. In this text we are emphasizing classical mechanics, since it applies to so many interesting phenomena and serves as an excellent introduction to a subsequent study of relativistic mechanics and quantum mechanics of this century.

Before leaving the topic of relativity, it should be noted that a form of relativity theory applicable to noninertial frames was developed by Einstein. This theory, called the *general theory of relativity*, gives a profound significance to the equivalence principle* and inherently contains a fundamental theory of gravitation. Einstein attempted to expand this theory so that it would also explain electromagnetism—such a theory would be a *unified field theory*. He did not succeed in this effort, though he sought such a theory until his death in 1955.

Exercise

12.   An auto has a mass of 1500 kg when it is at rest. Use (8-13) to find its mass when it has a velocity of 30 m/s (close to 60 mi/hr).

8-7   Center of Mass

The total momentum of a system is given by

$$\mathbf{p}_{\text{total}} = m_1\mathbf{v}_1 + m_2\mathbf{v}_2 + m_3\mathbf{v}_3 + \cdots + m_N\mathbf{v}_N = \sum_{i=1}^{N} m_i\mathbf{v}_i \quad (8\text{-}14)$$

where i labels each particle in turn; $i = 1, 2, 3, \ldots , N$. If we divide (8-14) by the total mass, M, of the system, we obtain

$$\frac{\mathbf{p}_{\text{total}}}{M} = \frac{\displaystyle\sum_{i=1}^{N} m_i\mathbf{v}_i}{\displaystyle\sum_{i=1}^{N} m_i} = \mathbf{V}_{\text{cm}} \quad (8\text{-}15)$$

*See Section 5-2.

where V_{cm} is called the *center-of-mass velocity*. It represents a sort of average velocity for the entire system. To understand why it is called the center of mass velocity, let us examine an expression that when differentiated, leads us to (8-15):

$$\mathbf{R} = \frac{\sum\limits_{i=1}^{N} m_i \mathbf{r}_i}{\sum\limits_{i=1}^{N} m_i} = \frac{\sum\limits_{i=1}^{N} m_i \mathbf{r}_i}{M} \tag{8-16}$$

The vector $\mathbf{R}$ points to what is called the center of mass of the system. The components of $\mathbf{R}$ are the coordinates of this center of mass:

$$X = \frac{\sum\limits_{i=1}^{N} m_i x_i}{M} \qquad Y = \frac{\sum\limits_{i=1}^{N} m_i y_i}{M} \qquad Z = \frac{\sum\limits_{i=1}^{N} m_i z_i}{M} \tag{8-17}$$

$$X = \frac{\sum\limits_{i=1}^{N} m_i x_i}{M} = \frac{\int x\, dm}{\int dm}$$

Differentiation of (8-16) with respect to time yields (8-15). Thus the velocity V_{cm} is just the velocity of the point called the center of mass. From (8-15) we see that the total momentum of a system is identical to that of a single mass that contains *all* the mass of a system, moving at the velocity of this *center of mass*. That is,

$$\mathbf{p}_{total} = M\mathbf{V}_{cm} \tag{8-18}$$

Let us locate the center of mass of the two bodies of unequal mass shown in Fig. 8-11. In the upper portion of the figure we see m_1 with a mass of 15 g at the origin and m_2 with a mass of 60 g at the point $x = 10$ cm, $y = 0$. The y coordinate of the center of mass is given by (8-17) to be

$$Y = \frac{m_1 y_1 + m_2 y_2}{m_1 + m_2} = \frac{15\text{ g} \cdot 0 + 60\text{ g} \cdot 0}{75\text{ g}} = 0$$

The x coordinate is given by

$$X = \frac{m_1 x_1 + m_2 x_2}{m_1 + m_2} = \frac{15\text{ g} \cdot 0 + 60\text{ g} \cdot 10\text{ cm}}{75\text{ g}} = 8\text{ cm}$$

The point with these coordinates is labeled *c.m.* in the upper portion of Fig. 8-11. In the lower portion of the figure we move the coordinate system, so that the x coordinates are changed. Now

$$X = \frac{m_1 x_1 + m_2 x_2}{m_1 + m_2} = \frac{15\text{ g} \cdot 2\text{ cm} + 60\text{ g} \cdot 12\text{ cm}}{75\text{ g}} = 10\text{ cm}$$

Clearly, the y coordinate of the center of mass is still zero. The important point is that this shift of coordinates in Fig. 8-11 has left

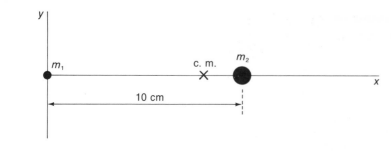

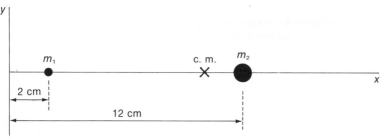

Fig. 8-11. The center of mass (c.m.) is indicated for masses $m_1 = 15$ g, $m_2 = 60$ g. Moving the origin of the coordinate system does not change the position of the c.m. *with respect to the masses.*

the position of the center of mass *unchanged with respect to the position of the masses.*

If two such masses are rigidly linked together, (8-18) tells us that their total momentum is identical to that of a particle whose mass equals their total mass moving with the motion of their center of mass. If no external forces act, conservation of momentum implies that this particular point, the center of mass, will move in a straight line with unchanging velocity regardless of the orientation of m_1 and m_2. This behavior is illustrated in Fig. 8-12, which is a multiflash photograph of two connected masses moving on an air table, which effectively eliminates friction.

Example   Calculate the center of mass of the three masses sketched in Fig. 8-13.

Solution   Any coordinate system may be chosen with no change in our result. Choosing the origin of coordinates at the 5-kg mass, we find

$$X = \frac{3 \text{ kg} \cdot 0 \text{ m} + 5 \text{ kg} \cdot 0 \text{ m} + 10 \text{ kg} \cdot 1 \text{ m}}{18 \text{ kg}} = 0.55 \text{ m}$$

$$Y = \frac{3 \text{ kg} \cdot 2 \text{ m} + 5 \text{ kg} \cdot 0 \text{ m} + 10 \text{ kg} \cdot 0 \text{ m}}{18 \text{ kg}} = 0.33 \text{ m}$$

Now let us consider changes in the motion of the center of mass caused by external forces. Applying Newton's second law to each mass in a system of N masses and summing forces, we have

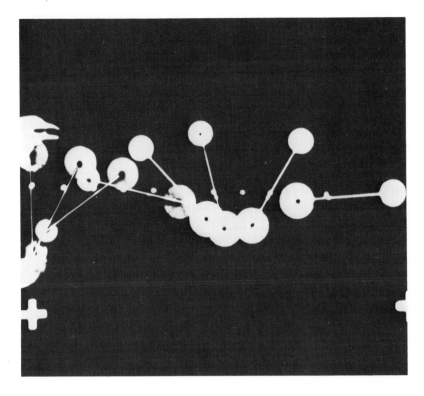

Fig. 8-12. Two pucks are rigidly connected, with one puck having twice the mass of the other. The small white spot marks their center of mass. Note that it follows a straight line even though the two masses are revolving as well as translating. (Adapted from the Ealing filmstrip, "Center of Mass," distributed by BFA Educational Media)

$$\sum_{i=1}^{N} \mathbf{F}_i = \sum_{i=1}^{N} \frac{d\mathbf{p}_i}{dt} \qquad (8\text{-}19)$$

where $\mathbf{p}_i = m_i\mathbf{v}_i$. Some of the forces in the sum on the left side of (8-19) may be internal forces—forces that act between pairs of the masses. However, Newton's third law in vector form is

$$\mathbf{F}_{ij} = -\mathbf{F}_{ji}, \qquad (8\text{-}20)$$

where i and j label *any* pair of the N particles. Since all such internal force pairs cancel in the sum (8-19), we are left with only external forces. Thus if there are, say, J such forces, we have

$$\mathbf{F}_{\text{external}} = \sum_{k=1}^{J} \mathbf{F}_k = \sum_{i=1}^{N} \frac{d\mathbf{p}_i}{dt} \qquad (8\text{-}21)$$

Now let us differentiate (8-18) with respect to time. First we rewrite (8-18):

$$\mathbf{p}_{\text{total}} = M\mathbf{V}_{\text{cm}} \qquad (8\text{-}18)$$

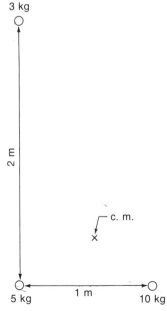

Fig. 8-13.

Then we write out $\mathbf{p}_{\text{total}}$ in more detail:

$$\sum_{i=1}^{N} \mathbf{p}_i = M\mathbf{V}_{\text{cm}}$$

Now we differentiate the above:

$$\sum_{i=1}^{N} \frac{d\mathbf{p}_i}{dt} = M\frac{d\mathbf{V}_{\text{cm}}}{dt}$$

But the left side of this equality can be rewritten using (8-21):

$$\mathbf{F}_{\text{external}} = M\frac{d\mathbf{V}_{\text{cm}}}{dt} = M\mathbf{a}_{\text{cm}} \qquad (8\text{-}22)$$

and we discover that Newton's second law applies in a very general way to systems of many particles. According to (8-22), the effect of the total external force acting on a system is to cause an acceleration of the center of mass just as if the entire mass of the system were concentrated there.

The sketch of Fig. 8-14 illustrates this point for a projectile that explodes in flight. The force of gravity continues to act upon the total mass of the projectile, now scattered into fragments, so that the center of mass of the fragments continues along the parabolic trajectory that was followed by the intact projectile. This idealized illustration would be altered somewhat by the presence of air resistance, which is larger for the irregularly shaped fragments than for the projectile. Inclusion of the forces of air resistance in (8-22), however, would lead to a correct prediction of the path of the center of mass.

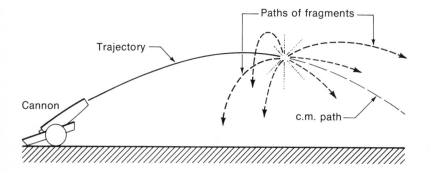

Fig. 8-14. The center of mass of the fragments of a projectile continues along the projectile's original trajectory.

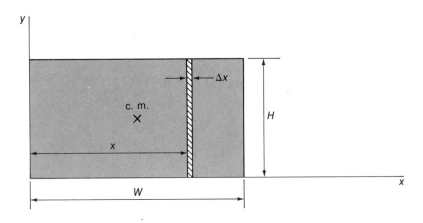

Fig. 8-15. The strip of mass Δm with a width Δx makes a contribution to the center-of-mass sum for this rectangle approximately equal to its mass, Δm, multiplied by its x coordinate, x. The approximate equality becomes exact as $\Delta x \to 0$, so that each portion of Δm is at a distance exactly x from the ordinate.

Exercises

13. Find the center of mass of three bodies whose masses and coordinates are given by:

$$m_1 = 3 \text{ kg} \qquad x_1 = 2 \text{ m} \qquad y_1 = 5 \text{ m} \qquad z_1 = -4 \text{ m}$$

$$m_2 = 6 \text{ kg} \qquad x_2 = 2 \text{ m} \qquad y_2 = 1 \text{ m} \qquad z_2 = 1 \text{ m}$$

$$m_3 = 2 \text{ kg} \qquad x_3 = -5 \text{ m} \qquad y_3 = 2 \text{ m} \qquad z_3 = 1 \text{ m}$$

14. A mass of 2 kg moves along the positive y axis with $v_y = 3$ m/s. A second mass of 3 kg moves along the negative x axis with $v_x = -4$ m/s. Find the velocity and direction of the center of mass for the system consisting of these two masses.

8-8   The Center of Mass of Continuous Bodies   NOT ON 2nd HOURLY

$$\int \rho\, dx = dm$$

Consider the rectangular sheet of uniform thickness shown in Fig. 8-15. It seems reasonable to suppose that its center of mass lies at the symmetry point indicated in the figure, but how can we test this guess? The expressions given in (8-17) are suitable for locating the center of mass of a collection of discrete masses where each mass has well-defined values of its x, y, and z coordinates. Let us, then, mentally divide the rectangular sheet of Fig. 8-15 into extremely narrow mass elements, so that each has a well-defined x coordinate. One such element is shown in Fig. 8-15.

$$X = \frac{\sum_{i=1}^{N} m_i x_i}{M} = \frac{\int x\, dm}{\int dm}$$

The mass of the entire rectangular figure is M. The fraction of M that is contained within the mass element shown is equal to the fraction of the rectangle's area contained in the element, since the

fraction of M (entire mass) contained in element is equal to the area of the element

rectangular sheet is of uniform thickness. Let us call the mass of the element Δm. The area of the element, ΔA, is given by

$$\Delta A = H \Delta x$$

where Δx is the width of the mass element of mass Δm. The area of the entire rectangle is

$$A = WH$$

The fraction of A contained in ΔA is

$$f = \frac{\Delta A}{A} = \frac{\Delta x}{W}$$

so that the mass of this element is

$$\Delta m = fM = \frac{\Delta x}{W} M$$

When the entire rectangle is broken up into N such strips, the x coordinate of the center of mass of the rectangle is then given by (8-17), which becomes

$$X \approx \frac{\sum\limits_{i=1}^{N} m_i x_i}{M} = \frac{\sum\limits_{i=1}^{N} \frac{\Delta x_i}{W} M \cdot x_i}{M} = \frac{1}{W} \sum\limits_{i=1}^{N} \Delta x_i \cdot x_i$$

where the approximate equality is required because not all of the mass of each strip is located *precisely* at x_i. If we let the width of each element tend to zero, however, the expression above becomes exact. In the limit $\Delta x_i \to 0$, the above sum becomes an integral. This integral must cover the entire width of the rectangle:

$$X = \frac{1}{W} \int_0^W x \, dx$$

Since the indefinite integral $\int x \, dx = \frac{1}{2} x^2$, we have

$$X = \frac{1}{W} \left(\tfrac{1}{2} W^2 \right) = \frac{W}{2}$$

so that the location of the center of mass is halfway along the width of the rectangle. A similar procedure, where the rectangle is broken into thin horizontal strips, leads to $Y = H/2$, in agreement with our guess that the center of mass lies at the symmetry point in Fig. 8-15.

Example   The height of the plane figure of uniform thickness shown in Fig. 8-16 is given by

$$h = ax^2 \qquad\qquad (8\text{-}23)$$

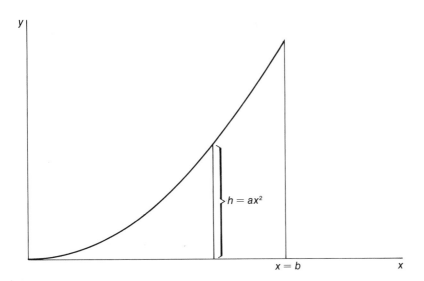

Fig. 8-16.

between $x = 0$ and $x = b$. The total mass of the figure is M. Find the x coordinate, X, of the center of mass of the figure.

Solution   The mass Δm of a strip with a width Δx and height h is proportional to the area of that strip:

$$\Delta m = \sigma h \Delta x$$

where σ is our constant of proportionality. The x coordinate of the center of mass is then

$$X \approx \frac{\sum\limits_{i=1}^{N} \Delta m_i x_i}{M} = \frac{\sigma \sum\limits_{i=1}^{N} h_i \Delta x_i \cdot x_i}{M}$$

where the N strips include the entire plane figure. Using (8-23) this becomes

$$X \approx \frac{\sigma a \sum\limits_{i=1}^{N} x_i^3 \Delta x_i}{M} \tag{8-24}$$

This expression becomes exact when $\Delta x \rightarrow 0$, and (8-24) becomes a definite integral:

$$X = \frac{\sigma a}{M} \int_0^b x^3 \, dx = \frac{\sigma a}{4M} b^4 \tag{8-25}$$

However, σ was not in the given information. We can eliminate it from (8-25) if we realize that

$$M = \sigma A \tag{8-26}$$

where A is the area of the entire figure. This area equals the area under the $h = ax^2$ curve from $x = a$ to $x = b$ and is by definition equal to the definite integral

$$A = \int_0^b h\, dx = a \int_0^b x^2\, dx = \frac{a}{3}b^3 \qquad (8\text{-}27)$$

Thus, from (8-26) and (8-27), we find

$$\sigma = \frac{M}{A} = \frac{3M}{ab^3} \qquad (8\text{-}28)$$

Substituting (8-28) into (8-25), we finally obtain $X = \frac{3}{4}b$, so that the center of mass is close to the right-hand edge of the figure, as we would expect.

In Fig. 8-17 we see an irregular figure for which a calculation similar to that in the preceding example would be quite difficult. In this case the center of mass was found experimentally. The figure was placed on an air table and, by trial and error, a point was found that moved in a straight line, as shown in Fig. 8-17.

Exercises

15. Find the center of mass of the figure of uniform thickness in Fig. 8-18. Note that the figure may be thought of as two rectangles and that we have already located the center of mass for uniform rectangles.

16. What would you suppose is the location of the center of mass of a flat circular figure (such as a coin)? Can you make an argument for your conjecture by drawing some strip-shaped mass elements without going through the entire integration procedure?

17. Find the center of mass of the triangular figure shown in Fig. 8-19. Note that, by symmetry, the center of mass must lie somewhere along the dotted line bisecting the triangle.

8-9   Summary

The concept of momentum can be generalized to three dimensions:

$$\mathbf{p} = m\mathbf{v} \qquad \text{or} \qquad \begin{aligned} p_x &= mv_x \\ p_y &= mv_y \\ p_z &= mv_z \end{aligned}$$

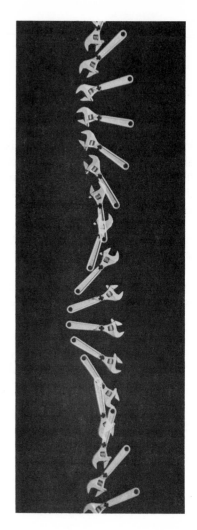

Fig. 8-17. A multiflash photograph of an irregular object coasting on an air table. The black cross was found by trial and error to be the single point that always moves in a straight line and thus marks the center of mass. (From *PSSC Physics*, D. C. Heath and Company, Lexington, Massachusetts, © 1965)

Newton's second law similarly is generalized to three dimensions:

$$\mathbf{F}_{net} = \frac{d\mathbf{p}}{dt}$$

or $\mathbf{F}_{net} = m\mathbf{a}$ (for constant mass).

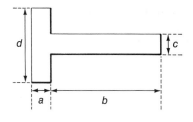

Fig. 8-18.

The acceleration in the above equation must be defined in an inertial reference frame. There is no way, on the basis of mechanics alone, that any one inertial frame can be considered to be the most fundamental frame of reference. The principle of relativity asserts that *no* laws of physics allow us to choose a particular inertial frame as "privileged" in any way.

Einstein applied the principle of relativity to the laws of electromagnetic theory and asserted that the Lorentz transformations give the correct transformation between frames of reference moving with relative constant velocity. Relativity theory requires modifications in the laws of classical mechanics, but these modifications are generally not significant until velocities reach a substantial fraction of c, the velocity of light.

The center of mass of a system of N masses is defined in terms of its three coordinates, X, Y, and Z:

$$
\left.
\begin{array}{l}
X = \dfrac{\sum\limits_{i=1}^{N} m_i x_i}{M} \\[2em]
Y = \dfrac{\sum\limits_{i=1}^{N} m_i y_i}{M} \\[2em]
Z = \dfrac{\sum\limits_{i=1}^{N} m_i z_i}{M}
\end{array}
\right\}
\quad \text{or} \quad
\mathbf{R} = \frac{\sum\limits_{i=1}^{N} m_i \mathbf{r}_i}{M}
\qquad (8\text{-}29)
$$

The total momentum of the N masses is equal to the momentum of a single particle with mass M_{total} and moving at the velocity of the center of mass.

If a system of masses is acted upon by many forces, its center of mass will move exactly as though a single mass, M_{total}, is being acted upon by the net force that is the resultant of all the forces acting on the system.

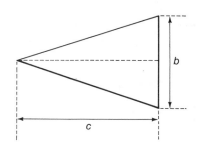

Fig. 8-19.

Problems

8-1. A 150-kg raft is coasting at 2 m/s with respect to the water. A 75-kg occupant dives off at right angles to its motion with a horizontal velocity component of 3 m/s with respect to the water. Find the final magnitude and direction of the raft's velocity. Neglect friction.

8-2. A space capsule with mass m_1 is struck by a meteorite of mass m_2. The meteorite embeds itself in the capsule. Their velocities before the encounter are v_1 and v_2. The velocity of the meteorite makes an angle of θ with the velocity of the capsule. Find the final velocity and direction of the capsule.

8-3. An alpha particle strikes a nucleus in an expansion chamber. Three particles result, the original alpha particle and two fragments of the target nucleus. It is observed that the alpha particle's incident path and the paths of two of the resulting particles all lie in one plane. Can the third resultant particle have a path that does not lie in that plane?

8-4. A sharpshooting neighborhood youth tries to disrupt a tennis game by hitting the tennis ball in flight with a BB from his BB gun. The tennis ball of 80-g mass is deflected 1.5° by the impact and the magnitude of its velocity (15 m/s) is unchanged. If the BB has a mass of $\frac{1}{4}$ g and a velocity of 100 m/s, find the magnitude and direction of its velocity after it strikes the tennis ball. Assume its initially moves toward the tennis ball at right angles to the tennis ball's path.

8-5. An atomic projectile with a mass of 6.6×10^{-27} kg and a velocity 1/20 the velocity of light collides with a stationary target nucleus. The mass of the nucleus is 1.66×10^{-25} kg. After the collision, the projectile is observed moving at an angle of 60° from its initial direction, still at 1/20 the velocity of light. Find the magnitude and direction of the momentum of the recoiling target nucleus. Relativistic effects may be ignored.

8-6. The particle incident from the left in Fig. 8-20 has a momentum of 3×10^{-23} kg·m/s. It causes a nuclear reaction that produces the two detectable particles indicated in the figure. These particles have momenta $p_1 = 1 \times 10^{-23}$ kg·m/s and $p_2 = 2 \times 10^{-23}$ kg·m/s. (*a*) Is there an undetected neutral particle that is also produced in the reaction? (*b*) If so, find the momentum of the neutral particle.

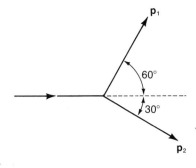

Fig. 8-20.

8-7. An electron has a velocity of $0.98c$. Find its inertial mass if its rest mass, m_0, equals 9.1×10^{-31} kg.

8-8. How fast must a body move before (*a*) its inertial mass is $\sqrt{4/3}$ times its rest mass? (*b*) its inertial mass is twice its rest mass?

8-9. A beam of molecules, each of mass m and moving with velocity v, strikes a wall at a rate of N molecules/s. The molecules strike the wall at an angle θ with respect to the normal to the plane of the wall and rebound at an equal angle θ and with the same velocity, as shown in Fig. 8-21. Find the magnitude of the average force exerted on the wall.

8-10. Two boys pull with 100-N forces on ropes, as in Fig. 8-22. The ropes are attached to a wagon that has a mass of 25 kg. Find the acceleration of the wagon.

8-11. A body with a mass of 20 kg and velocity of 30 m/s moves initially along the x axis. A force with equal x and y components and a magnitude of 40 N acts for 20 s. Find the two possible values of the final magnitude of the body's velocity.

8-12. A 10^4-kg space ship is moving freely at 10^4 m/s due north (Galactic coordinates). A force of the form $F = kt$, where k is equal to 4×10^2 N/s, acts from $t = 0$ to $t = 10^3$ s. This force always points due east. Find the magnitude of the final velocity of the space ship.

8-13.* The air resistance encountered by a bicyclist is given by $F_r = kv^2$, where v is his velocity with respect to the air. This force is in the direction of the air's motion as seen by the cyclist. Show why a cyclist riding at right angles to the wind experiences a greater retarding force component along his line of motion than in the absence of wind. Would this be be true if the air resistance were proportional simply to the velocity?

8-14. A sports car with a mass of 900 kg is moving at 180 km/hr. It rounds a gentle turn and comes out of it moving in a direction at right angles to its original direction with a velocity still equal to 180 km/hr. Choosing the original direction as the x axis and the final direction as the y axis, find the x and y components of the impulse of the applied force required to make this turn.

8-15. A stone is released from the top of the mast of a uniformly moving sailboat. Describe the trajectory of the stone as seen by an observer on the sailboat and by an observer stationary with respect to the water.

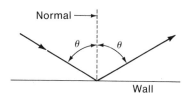

Fig. 8-21.

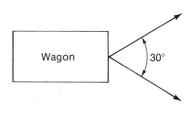

Fig. 8-22.

8-16. A body of mass m approaches another body of mass M with a relative velocity V. Is there a particular reference frame in which the total momentum is zero? How fast does each mass move in this frame of reference? What is the motion of the center of mass in this reference frame?

8-17. A pan of water, shown in Fig. 8-23, is placed in a vehicle. What is the maximum value of uniform acceleration of this vehicle so that the water does not spill?

8-18. A juggler is balancing a stick on his forehead while standing in a train uniformly accelerating with $a = 2$ m/s². Find the angle between the stick and the vertical direction.

8-19. A man of mass m stands in an initially stationary rowboat of mass M. He walks a distance L from the front of the rowboat to the rear, then stops. How far, with respect to the water, does the rowboat move? Does the center of mass of the man + boat system move? Neglect friction between the boat and the water.

8-20. The moon revolves about the center of mass of the earth-moon system approximately once every 27 days. The mass of the earth is 81 times that of the moon and the distance between the center of the earth and the center of the moon is 3.8×10^5 km. Describe the motion of the earth with respect to the center of mass of the earth-moon system.

8-21. The center of mass of some objects is sometimes more easily found by ascribing mass to a part that isn't present, then subtracting its contribution to the center of mass. Find the center of mass of the disk with a hole of radius $R/2$, as shown in Fig. 8-24. The radius of the disk is R.

8-22. An initially stationary 3-kg mass is acted upon by a 3-N force. A second system consists of three masses, each of 1 kg, that are also initially stationary. One of these masses is acted upon by a 5-N force to the right, a second is acted upon by a 2-N force to the left, and the third remains stationary. Note that both systems have the same net mass and net force acting. Show by an explicit calculation that the center of mass of both systems behaves similarly.

8-23. A passenger in a car holds a glass filled to the brim with liquid. The car is accelerating steadily with an acceleration a. In order to keep the liquid from spilling, the passenger must tilt the glass forward. (a) At what angle with respect to the vertical must the glass be held? (b) Now suppose that the same accelerating force is applied to the car, but that it has

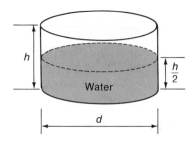

Fig. 8-23.

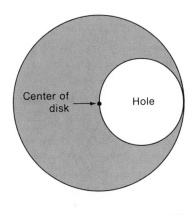

Fig. 8-24.

encountered a hill just steep enough so that it is not accelerating. At what angle with respect to the vertical must the
glass now be held? (*c*) What now is the angle between the
glass and the normal direction with respect to the floor of
the car? (*d*) In general, does it appear that the necessary
angle between the glass and the car's floor is determined by
the inclination of the road or by the position of the accelerator pedal (i.e., the accelerating force supplied to the car)?

8-24.* Find both the *x* and *y* coordinates of the center of mass of
Fig. 8-16.

8-25. Find the center of mass of the plane figure shown in Fig.
8-25.

8-26. Find the center of mass of a solid cube.

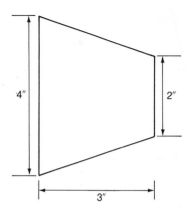

Fig. 8-25.

Chapter **9**

Motion in a Plane

9-1 Introduction In this chapter we will apply our three-dimensional dynamics to some simple but important types of motion. We will restrict our attention to systems in which the trajectories of the masses involved lie in a plane. This type of motion is easy to visualize since it can be sketched on a flat sheet of paper without need for a third dimension. Furthermore, a surprisingly large class of phenomena in nature involves motion in a plane. We will see later that this is due to the *central* character of the classical forces: gravitation and the electrostatic force.* This chapter also provides a convenient point for a discussion of frictional forces that can be easily measured by use of an inclined plane.

9-2 Uniform Circular Motion

In circular motion the *direction* of the velocity vector is continually changing. If the speed along the circular path varies, then the magnitude of the velocity varies as well. For the time being, we will consider only *uniform circular motion,* that is, motion in which the magnitude of the velocity is constant.

When we think of motion in one dimension, it is clear that acceleration cannot occur unless the magnitude of the velocity varies. In more than one dimension, however, we can have a non-zero acceleration even though the magnitude of the velocity is constant. Figure 9-1 illustrates this point. It is clear that a velocity

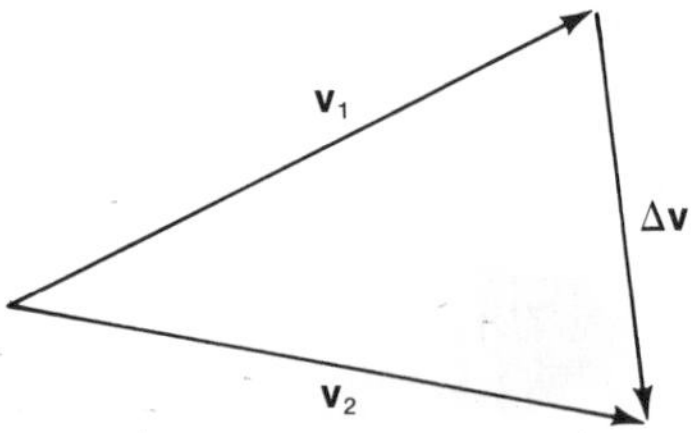

Fig. 9-1. Though $\mathbf{v}_1$ and $\mathbf{v}_2$ have the same magnitude, the *change* in velocity, $\Delta \mathbf{v}$, is non-zero if $\mathbf{v}_1$ and $\mathbf{v}_2$ are not parallel. If $\Delta \mathbf{v}$ occurs in a time Δt, we have $a = \Delta v / \Delta t$.

---

*A *central* force between two objects is a force that acts always along the straight line that joins the objects.

vector of *constant* length cannot change direction without changing its projection (its component) on the *x, y,* or *z* axis. These changes in velocity, divided by the time interval required, are the components of an acceleration vector.

In uniform circular motion, the velocity vector does not change magnitude; it changes only direction. As the velocity vector swings around, the *instantaneous changes* in the velocity vector must always be at right angles to the velocity itself. If this were not the case, the velocity would be increasing or decreasing, in contrast with our assumption of *uniform* circular motion. Since the instantaneous changes in the velocity are at right angles to the velocity, the acceleration is at right angles to the velocity itself.

Referring to Fig. 9-2, note that the vectors v_1 and v_2 are both at right angles to the radius vectors pointing from the center of the circle to the location of the moving point. Thus v_1 and v_2 are separated by the same angle, $\Delta\theta$, as are r_1 and r_2. Fig. 9-3 shows the similar triangles that are produced. From the similarity of the triangles, we have

$$\frac{\Delta r}{r} = \frac{\Delta v}{v} \tag{9-1}$$

where the symbols v and r stand for the constant magnitudes of the vectors. Now we divide (9-1) by the time Δt required for the point to move from position 1 to position 2. Transposing r and v, we obtain

$$\frac{\Delta r}{\Delta t}v = \frac{\Delta v}{\Delta t}r \tag{9-2}$$

If we let $\Delta t \rightarrow 0$ then $\Delta r \rightarrow 0$ also. The limit of $\Delta r/\Delta t$ as Δt vanishes is precisely the definition of the velocity of the point moving along the circumference of the circle. The right side of Eq. (9-2) involves the limit of $\Delta v/\Delta t$ as Δt vanishes, which is the acceleration we wish to find. More formally:

$$\lim_{\Delta t \to 0}\frac{\Delta r}{\Delta t}v = \lim_{\Delta t \to 0}\frac{\Delta v}{\Delta t}r$$

or

$$\frac{dr}{dt}v = \frac{dv}{dt}r$$

Since dr/dt is the velocity and dv/dt the acceleration, we have:

$$v^2 = ar$$

or

$$a = \frac{v^2}{r} \tag{9-3}$$

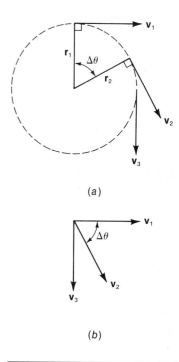

(a)

(b)

Fig. 9-2. (a) The velocity of a point moving along the dotted circular path is continually changing direction. (b) The velocity vectors in (a) are redrawn with their tails at a common origin, so that their changes are more easily visualized.

[handwritten note:] magnitude of acceleration resulting from uniform circular motion with velocity v and radius r

This last equality, (9-3), gives the magnitude of the acceleration that results from uniform circular motion with velocity v and radius r. Note that the dimensions are those of an acceleration:

$$\frac{v^2}{r} = \left[\frac{l^2}{t^2} \cdot \frac{1}{l}\right] = \left[\frac{l}{t^2}\right] = \text{acceleration}$$

The fact that the expression for the acceleration contains the square of v is due to two effects: (1) The larger v is, the larger Δv is for a given angle $\Delta\theta$, so Δv is proportional to v. (2) The time Δt required to cover an angle $\Delta\theta$ diminishes as v increases, so Δt is *inversely* proportional to v. Since the ratio $\Delta v/\Delta t$ is involved in the acceleration, the velocity appears squared in our result. Also, as r is decreased, the time required to sweep out $\Delta\theta$ (at a given velocity) diminishes. Hence the acceleration is proportional to the *inverse* of the radius.

The directions of v and a are illustrated in Fig. 9-4. At any instant the direction of the velocity is that of a tangent to the circle, while the acceleration is directed inward, toward the center of the circle. This acceleration is, accordingly, referred to as a *centripetal* (center-seeking) acceleration.

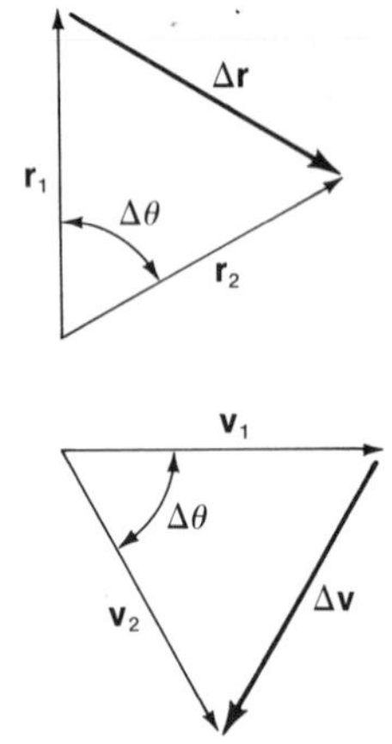

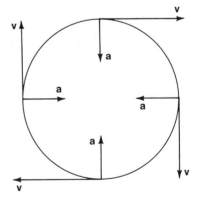

Fig. 9-3. The radius vectors $\mathbf{r_1}$ and $\mathbf{r_2}$ form a triangle together with $\Delta\mathbf{r}$. The triangle formed by $\mathbf{v_1}$, $\mathbf{v_2}$, and $\Delta\mathbf{v}$ is a similar triangle.

Exercises

1. A point on the surface of a spinning automobile wheel is moving with uniform circular motion at 88 ft/s. It is located at a radius of 1.1 ft. Find its centripetal acceleration. Compare it to g.

2. Find the centripetal acceleration of a body fixed on the earth's equator. The velocity due to the earth's rotation is approximately 1000 mi/hr and the earth's radius is approximately 4000 mi.

9-3   Centripetal and Centrifugal Forces

A mass moving in a circular path experiences the centripetal acceleration given by (9-3). According to Newton's second law, it must be acted upon by an inward-directed force:

$$F = ma = m\frac{v^2}{r} \qquad (9-4)$$

If we see an object moving in a circular path, we can infer the presence of an inward-directed centripetal force. Thus when we see a rock whirled around at the end of a string, it is the tension in the string that forces the rock to accelerate continually toward the center of its circular path. Without this force, the rock would not

Fig. 9-4. The velocity is tangent to the circular path, while the acceleration is directed radially inward.

move in a circle at all. In fact, if the string breaks, the rock simply flies away. Similarly, the tread of a spinning automobile tire will not stay in place unless there are sufficient adhesive forces between it and the tire body. Newton used similar reasoning to infer that the earth's gravity caused a force that acts upon the moon and forces it to follow its circular path about the earth. He worked out the idea of centripetal force and applied it to the motion of the moon during the years 1665–1666, when Cambridge University was closed due to an outbreak of the plague. He spent those years at the farm that was his boyhood home and apparently spent most of that time formulating the basic ideas of differential and integral calculus, the laws of mechanics, and the theory of gravitation. He was 25 years old in 1665.*

Example A boy ties a 50-g rock onto a string and spins it about his head at a radius of 79.6 cm at a constant rate of twice per second. What is the tension in the string? (Neglect gravity.)

$T = \dfrac{mv^2}{R}$

Solution The circumference of the circular path is given by $2\pi R$. Since it is covered twice per second, the time of one full revolution is $\Delta t = \frac{1}{2}$ s, and the uniform velocity of the rock is

$$v = \Delta s/\Delta t = 2\pi R/\Delta t = 10 \text{ m/s}$$

The centripetal force is provided by the tension in the string:

$$T = \frac{mv^2}{R} = \frac{0.05 \text{ kg} \cdot 100 \text{ m}^2/\text{s}^2}{0.796 \text{ m}} = 6.28 \text{ N}$$

This is more than 10 times the force of gravity on the stone, hence it and the string whirl in an almost horizontal plane, although the string must "droop" a bit to offset the force of gravity with a vertical component of its tension. (See Section 9-4 and Problem 3.)

In the preceding example, the boy pulls inward on the string to provide the centripetal force that causes the rock to move in a circle. By virtue of Newton's third law, the boy must feel the string pulling *outward* on his hand. This outward-directed reaction force is called a *centrifugal* (center-fleeing) force. The most direct way to calculate its magnitude, however, is to realize that the rock is *accelerating* with respect to an inertial reference frame. That reference frame is the one in which the boy is stationary. This acceleration is, as we previously saw, directed *inward,* hence the force

---

*A brief account of Newton's life and work is given in *Sir Isaac Newton* by E. N. da C. Andrade; New York: The Macmillan Company (1954). Reprinted in paperback by Anchor Books, Doubleday and Company, Inc., Garden City, New York.

acting on the rock is directed inward. That force is supplied by the tension in the string, which pulls equally on the rock and on the boy's hand.

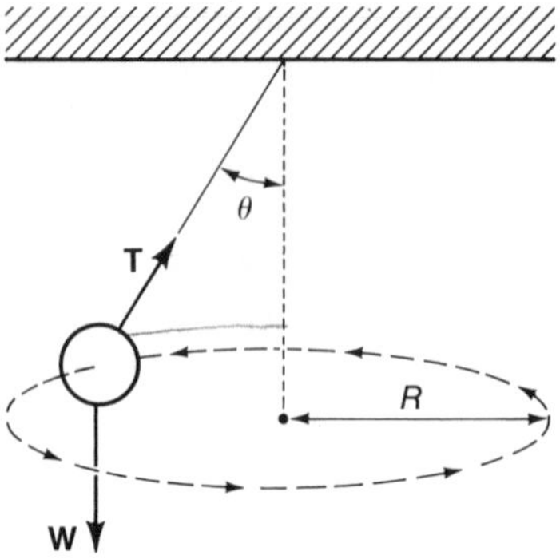

Exercises

3. A race driver usually tries to start "wide" (i.e., on the large-radius side of a turn), cut close to the inside edge of the road in the middle of the turn, and finish "wide" again. Why does he do this?

4. An automobile engine's flywheel spins at 4000 revolutions per minute. It has a radius of 6 in. (*a*) Compare the centripetal acceleration of a point on its rim to *g*. (*b*) What force is required to keep a 1-oz screw in place on its rim?

9-4   Circular Motion with Gravity Present

In the preceding section we discussed centripetal force in the absence of other forces. Let us consider two types of circular motion in which the force of gravity plays a significant role.

Conical pendulum   In Fig. 9-5 we see a mass *m* supported by a string and swinging in a circular path with radius *R*. This path is contained in a horizontal plane. The pendulum bob of mass *m* is acted upon by two forces: its weight **W** and the tension **T** in the string. Since **W** is directed vertically, it has no horizontal component and cannot supply a centripetal force. The tension, however, does have a component directed towards the center of the circular path. The magnitude of the horizontal component of the tension is $T \sin \theta$ and it supplies the needed centripetal force:

$$T \sin \theta = m \frac{v^2}{R}. \tag{9-5}$$

We also know that the *vertical* component of **T** must balance the bob's weight so that it does not accelerate vertically:

$$T \cos \theta = W = mg \tag{9-6}$$

We can find the magnitude of the tension *T* by squaring and adding (9-5) and (9-6):

$$T^2(\sin^2 \theta + \cos^2 \theta) = m^2 \left[g^2 + \left(\frac{v^2}{R} \right)^2 \right]$$

so that
$$T = m \sqrt{g^2 + \frac{v^4}{R^2}}$$

Fig. 9-5. A conical pendulum: the bob swings around in a horizontal circle.

We can find the tangent of the angle θ by dividing (9-5) by (9-6):

$$\frac{T \sin \theta}{T \cos \theta} = \frac{mv^2/R}{mg}$$

$$\tan \theta = \frac{v^2}{gR} \tag{9-7}$$

Finally, we can find the time τ required for one full circle if we first note that

$$\sin \theta = \frac{R}{L} \tag{9-8}$$

where L is the length of the supporting string. We can rewrite (9-7) as

$$\frac{\sin \theta}{\cos \theta} = \frac{v^2}{gR}$$

Then, using (9-8),

$$\cos \theta = \frac{gR}{v^2} \quad \sin \theta = \frac{R^2 g}{v^2 L} \tag{9-9}$$

But v, τ, and R are related by the equation

$$v = \frac{2\pi R}{\tau} \tag{9-10}$$

since $2\pi R$ is the distance covered in the same interval τ. Substituting (9-10) into (9-9), we find

$$\cos \theta = \frac{g\tau^2}{4\pi^2 L}$$

or   PERIOD

$$\tau = 2\pi \sqrt{\frac{L \cos \theta}{g}} \tag{9-11}$$

The time τ required for each full circle of motion is called the *period* of the conical pendulum.

Motion in a vertical circle   Our derivation of the centripetal acceleration given by (9-3) was carried out for uniform circular motion. However, when we took the limit in deriving (9-3), we obtained the *instantaneous* acceleration. Thus, if the speed of a mass varies along its circular path, (9-3) still gives the value of a, the radial acceleration, in terms of v at any given instant.

Consider now a mass attached to a string and swinging in a vertical circle, as in Fig. 9-6. The component of its weight that acts outward along the radial direction is given by

$$W_R = W \cos \theta = mg \cos \theta$$

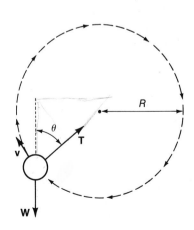

Fig. 9-6.   A mass swinging in a vertical circle is acted upon by the tension T and weight W.

The *net* radial force must be inward and equal to the centripetal force. Thus we have

$$T - W_R = m\frac{v^2}{R}$$

or
$$T = m\left(\frac{v^2}{R} + g\cos\theta\right) \tag{9-12}$$

At the bottom of the circle, $\theta = 0°$ and (9-12) becomes

$$T = m\left(\frac{v^2}{R} + g\right)$$

At the top of the circle, $\theta = 180°$ and (9-12) becomes

$$T = m\left(\frac{v^2}{R} - g\right)$$

$$\frac{v^2}{R} = g$$
$$v = \sqrt{gR}$$
$$V = \sqrt{gR}$$

Exercises

5. A child swings over a creek at the end of a rope, as illustrated in Fig. 9-7. The path followed is a horizontal circle, i.e., in the form of a conical pendulum. If 2π seconds is the time for one full circle, what is the length of the rope?

6. A stone tied to a string of 1-m length is whirled rapidly in a vertical circle. Its speed is steadily slowed until each time the stone is at its maximum height the string goes slack, though it is taut at all other positions. Find the speed of the stone at the top of the circle.

9-5 Sliding Friction

We noted earlier that the difficulty in discovering the law of inertia

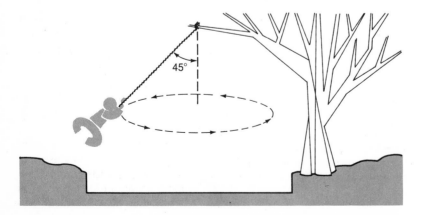

Fig. 9-7. The child swinging in a circle acts as a conical pendulum.

was caused by the fact that most motion we observe is affected by frictional forces. Slide a book on a smooth table top; if it starts with a velocity of one or two feet/second it will come to rest in only a second or so. The situation is depicted in Fig. 9-8.

The force of friction acts in opposition to the motion. It produces a negative acceleration that reduces the velocity of the book. One might suppose that if the two books have equal weight, a book with larger area will experience a larger frictional force. However, each unit of contact area then carries less weight and, thus, does not press as hard against the table. This effect causes the friction between two objects to be *approximately* independent of contact area. A magnified sketch of the contact region is shown in Fig. 9-9. Most surfaces, when viewed on a small enough scale, are quite rough. In order for body 1 to slide over body 2, then, it must alternately lift and drop as the "teeth" of the rough surfaces slide over each other.* Furthermore, the atoms of one body actually form bonds with the atoms in the other body in the isolated points of true contact between the surfaces. The average force required to repeatedly rupture these bonds adds to the frictional forces.** Any force squeezing the two surfaces together causes the force of friction to be larger. The experimental results concerning the frictional force on a solid sliding upon another solid can best be expressed by a simple equation:

$$f_K = \mu_K N \qquad N = mg \qquad (9\text{-}13)$$

where f_K is the force of kinetic friction, μ_K is a number (usually between 0 and 1), called the *coefficient of kinetic friction*, that depends on the surfaces involved, and N is the normal force that acts at right angles to the surfaces. The *normal* force only squeezes the rubbing surfaces together and has no component along the direction of motion. As was stated previously, this equation does not contain the contact area, because an increase of that area is compensated

$$F = ma = \mu_K N$$
$$a = \frac{\mu_K mg}{m}$$
$$a = \mu_K g$$

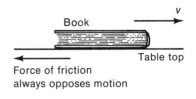

Book v →

Table top

← Force of friction always opposes motion

Fig. 9-8. The force of friction acts in opposition to the motion.

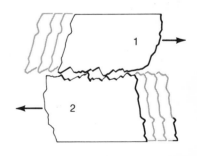

Fig. 9-9. At the microscopic level of observation, most surfaces are rough. Frictional forces are partly due to this roughness.

---

*Later, we will see that conservation of energy seems to imply that the energy lost in lifting over the teeth reappears in the return drop. However, both the energy of lifting and that of deformation of the teeth is lost partly to vibrational energy that propagates through the solid to appear eventually as heat. Some of this energy also excites vibrations in the surrounding air—the hissing or grating noise heard when one solid slides over another.

**For smooth surfaces, such as polished metals, especially in vacuum where the surfaces are free of oxides, this effect is dominant. However, as in the previous footnote, simpleminded application of energy conservation seems to imply that the energy gained during the formation of bonds just cancels the losses due to rupturing the bonds. A more complete discussion is required to understand the irreversibility of the coupling of the interacting atoms with the rest of the thermally fluctuating atoms in the solids. See Section 28-6.

by a decrease in the force per unit area that is squeezing the surfaces together. Furthermore, this frictional force is found to be approximately independent of the velocity of sliding for many materials.

Example   The coefficient of friction, μ_K, between two wooden blocks is 0.2. When the top block, which weighs 15 lb, is pulled steadily over the lower one, how great is the force of friction?

Solution   The normal force N is supplied by the weight of the top block and is equal to 15 lb. Thus

$$f_K = \mu_K N = 0.2(15 \text{ lb}) = 3 \text{ lb}$$

The direction of the force of friction is always so as to oppose the motion. Thus, in order to maintain a constant velocity, an external 3-lb force must constantly act in the direction of motion.

Exercises

7. A book is initially sliding across a table top at 1 m/s. The coefficient of kinetic friction between the book and table top equals 0.4. How long will the book slide before it comes to rest?

    (handwritten notes in margin:) $\mu_K N$ $=ma=\mu_K mg$ $a = \mu_K g$ $.4(16)=4$ $v = v_0 + at$

8. The coefficient of sliding friction between an automobile's tires and a dry concrete road is near unity when the wheels are locked in a "panic stop." For wet tires on a wet road, the coefficient of sliding friction is only half as large. Find the distance a 2000-kg car with locked wheels will slide under both wet and dry conditions with an initial speed of 30 m/s. Are these distances changed for a 1000-kg car?

9-6   Static Friction

If the two surfaces in Fig. 9-9 are at rest with respect to one another, strong bonds may form between the two surfaces at their few points of contact. These bonds must be broken in order to start one surface sliding past the other. For most substances this requires somewhat more force than that required to maintain the motion once it is begun. The initial force required to break these static bonds is again described reasonably well by a coefficient of friction:

$$F_{\text{initial}} = \mu_S N \qquad\qquad (9\text{-}14)$$

where μ_S is the coefficient of static friction. As mentioned above, for most substances

$$\mu_S > \mu_K$$

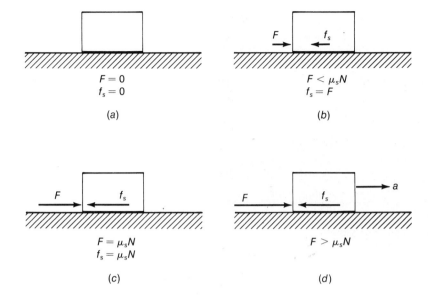

$$F = 0$$
$$f_s = 0$$

(a)

$$F < \mu_s N$$
$$f_s = F$$

(b)

$$F = \mu_s N$$
$$f_s = \mu_s N$$

(c)

$$F > \mu_s N$$

(d)

Fig. 9-10. If no external force is applied parallel to the sliding surface, the frictional force is zero, as in (a). As the external force is increased, as in (b) and (c), the frictional force similarly increases, so that $F_{net} = F - f_s = 0$ and no motion occurs. For external forces larger than $\mu_s N$, as in (d), motion occurs.

If we attempt to set a body into motion by supplying an external force, the force of static friction exactly opposes the external force until the external force finally exceeds the value given by (9-14). Thus the force of static friction can vary between zero and a maximum value depending on whether or not external forces act:

$$f_S \le \mu_S N \qquad (9\text{-}15)$$

The variation of f_S in response to varying external forces is illustrated in Fig. 9-10.

Example   The force that propels an automobile depends upon friction between the driven rubber tires and the roadway. If the rubber simply "lays down" on the pavement and is then "picked up" again as the tire rolls, it is the coefficient of static friction that is involved.* If the wheels are spinning, or if the brakes are locked so the wheels are sliding, it is the coefficient of sliding friction that is involved. The smaller value of the sliding friction coefficient is the reason an experienced driver tries to avoid locking the wheels in an emergency stop.

In the sketch of Fig. 9-11, we see a standard auto for which approximately half the weight is carried by the rear wheels and these are

---

*If the tread of a tire "squirms" as it presses against the road, as it does in standard bias-belted tires, the tread is actually slipping a bit as it rolls. This not only creates heat, it also reduces the coefficient of friction below that of static friction. Radial ply tires are designed to keep the tread fixed with respect to the road while the tread is in contact with the road.

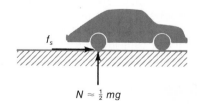

$$N \approx \tfrac{1}{2} mg$$

Fig. 9-11.

Fig. 9-12. In a race car designed for maximum acceleration only, the weight of the car is concentrated over the rear wheels to increase the frictional force between the tires and road. The "wing" provides additional downward pressure at high speed. (Photograph courtesy of *Road and Track Magazine*)

the only wheels which are driven. Thus the maximum frictional force that can be generated between tires and pavement by the engine twisting the drive shaft is

$$f_{max} = \mu_s N \approx \mu_s \frac{mg}{2}$$

The coefficient of friction is approximately unity for rubber tires on dry pavement, so we find that

$$a_{max} = \frac{f_{max}}{m} \approx \frac{g}{2}$$

This is a limit imposed by the nature of the friction between tires and road. This limit can be reached only if the engine is large enough to turn the wheels with this magnitude of force between the rubber tread and road surface. The maximum braking acceleration is roughly twice as great, since brakes act on all four wheels.

In a dragster, a racing car designed for maximum acceleration, nearly all the weight is concentrated over the rear driven wheels by designing it as in Fig. 9-12. The drivers of these cars purposely spin the tires repeatedly as they approach the starting line in order to get the rubber surfaces hot. The rubber then becomes sticky and the coefficient of friction becomes greater than unity. The current records for these cars indicate average accelerations much greater than *g* (see Problem 3-19).

Exercises

9. The normal force in Fig. 9-10 is supplied by the weight W of the block. If $W = 20\ N$, $F = 5\ N$, and $\mu_s = 0.3$, will the block

begin to move? How large is the force of static friction under these conditions?

10. A man, while searching through his pockets for his car keys, places his briefcase on the roof of his car. He then absent-mindedly drives off with the briefcase still on the roof. If the coefficient of static friction between the car roof and briefcase is 0.2, what is the maximum acceleration the car can reach before the briefcase starts to slip?

9-7 Motion on an Inclined Plane

Consider now the motion of objects on an inclined plane with friction acting. In the course of our discussion, we will discover a convenient method for the measurement of coefficients of friction. Figure 9-13 shows a single block of mass m sliding on a plane inclined at an angle θ above the horizontal. The block is acted upon by its weight, $W = mg$, vertically downward. Since the block is free to move only along the inclined plane, only the component of W that is parallel to the plane is effective in producing motion. This component is shown in Fig. 9-13 and is equal to $W \sin \theta$. The force of friction always opposes the motion and is given by $f_K = \mu_K N$, where N is the normal force exerted by the plane on the block at right angles to the sliding surfaces. In the present case, Fig. 9-13 shows that there is a component of the weight that acts at right angles to the plane: $mg \cos \theta$. This force must be exactly offset by N; if this were not the case, the block would accelerate in the nor-

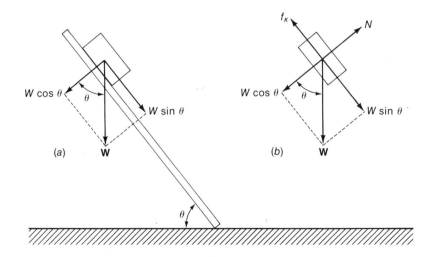

(a)   **W**

(b)   **W**

Fig. 9-13.   (a) The weight **W** of the block can be resolved into two components, one parallel and one perpendicular to the plane. (b) A diagram of all of the forces acting on the block.

mal direction. Since the force of friction acting upon the sliding block opposes the motion,

$$F_{net} = ma$$

becomes $\qquad mg \sin \theta - \mu_K mg \cos \theta = ma$

or $\qquad a = g(\sin \theta - \mu_K \cos \theta)$

If we adjust the angle of the plane so that uniform motion without acceleration occurs, we have a special case:

$$mg \sin \theta_0 - \mu_K mg \cos \theta_0 = 0$$

Solving for the coefficient of friction:

$$\mu_K = \frac{\sin \theta_0}{\cos \theta_0} = \tan \theta_0 \qquad (9\text{-}16)$$

where θ_0 is the particular angle for which $a = 0$.

Thus we have found a quick, convenient method for the measurement of μ_K. A block of material is placed on a plane of the desired composition. The plane is inclined until uniform sliding velocity is observed. The angle of inclination, θ_0, of the plane is noted; then (9-16) gives us μ_K. The coefficient of static friction is similarly found by placing the block at rest upon the plane and then raising the plane until motion just begins. At the angle at which motion begins, $W \sin \theta$ has just exceeded the maximum possible value of the static frictional force. Since

$$f_S \leq \mu_S N$$

a derivation identical to that leading to (9-16) gives

$$\mu_S = \tan \theta_1 \qquad (9\text{-}17)$$

where θ_1 is the critical angle at which motion begins. Representative values for some coefficients of friction measured in this manner are listed in Table 9-1.

TABLE 9-1   Some Coefficients of Friction

Materials	μ_S	μ_K
Wood on wood	0.6	0.4
Brass on steel	0.5	0.45
Greased metals	0.03	0.06
Rubber on concrete	1	0.8
Rubber on wet concrete	0.7	0.5
Steel on wet ice	0.03	0.01

Exercises

11. A 70-kg skier is sliding freely down a slope inclined 10° above the horizontal. For $\mu_K = 0.05$, find the magnitude of the skier's acceleration, neglecting air resistance.

12. A block of wood is placed on a wooden plane. The angle of inclination is increased until the block just begins to slide. Describe in a quantitative way what happens next if the angle is kept fixed at the critical angle θ_1.

9-8   Summary

An example of motion in a plane is that of uniform circular motion. A centripetal acceleration given by

$$a = \frac{v^2}{R} \quad \text{centripetal} \quad (9\text{-}3)$$
acceleration

characterizes such motion. This acceleration is directed toward the center of the circular path. Applying Newton's second law, we see that there must be a centripetal force; this force is given by

$$F = m\frac{v^2}{R} \quad \text{centripetal} \quad (9\text{-}4)$$
force

Such a force may be provided, for example, by the tension in a string.

When the circular motion is in a vertical plane, the external applied force combines with the force of gravity to supply the centripetal force. This external force varies between

$$F_{ext} = m\left(\frac{v^2}{R} + g\right) \quad \text{and} \quad F_{ext} = m\left(\frac{v^2}{R} - g\right)$$

bottom $\theta = 0°$ · · · top $\theta = 180°$

circular motion in a vertical plane

at the bottom and top of the circle, respectively.

In a conical pendulum, the horizontal component of the supporting string's tension supplies the needed centripetal force. The period of such a pendulum is given by

$$\tau = 2\pi\sqrt{\frac{L\cos\theta}{g}} \quad \text{period of} \quad (9\text{-}11)$$
pendulum

$$T = \frac{2\pi R}{V}$$

$$V = \frac{2\pi R}{T}$$

The force of friction between two sliding surfaces is given by

$$f_K = \mu_K N \quad (9\text{-}13)$$

where N is the normal force at right angles to the surfaces in contact. Thus N only squeezes the surfaces together. The force given by (9-13) does not depend upon either the contact area or relative velocity. The frictional forces between many, but not all, materials is adequately represented by (9-13).

The force of static friction varies, depending on how large an external force is applied. It has a maximum value, however, as indicated by

$$f_S \leqslant \mu_S N \tag{9-15}$$

where μ_S is the coefficient of static friction.

A block moving upon an inclined plane is acted upon by its own weight and by frictional forces. The acceleration of a sliding block on an inclined plane is given by

$$a = g(\sin \theta - \mu_K \cos \theta)$$

If the acceleration is zero, the block is sliding uniformly and we find that

$$\mu_K = \tan \theta_0 \tag{9-16}$$

where θ_0 is the angle between the plane and the horizontal. This technique is used to determine coefficients of friction for many materials, some of which are listed in Table 9-1.

Problems

9-1. A centrifuge is a device that produces large accelerations by means of circular motion. If a centrifuge has a radius of 10 cm, how many revolutions per second must it make to produce an acceleration of 50,000 g's?

9-2. A virus particle with a mass of 5×10^{-25} kg is spinning in a centrifuge. Its speed is 343 m/s and it is located at a radius of 0.12 m in the centrifuge. Find the centripetal force acting on the virus particle.

9-3. The boy of the example on page 183 notes that, in fact, the string sweeps out the surface of a shallow cone, as shown in Fig. 9-14. What, in the example given, is the value of ϕ?

9-4. A space station is constructed so that it can continually rotate to provide an artificial gravity for its inhabitants. It is in the form of a cylinder with a radius of 10 m. How many revolutions per minute must it make to provide an acceleration equivalent to normal gravity at its outer surface?

9-5. A ball rolls in a loop-the-loop track, as shown in Fig. 9-15. If the radius of the circular portion of the track is R, what is the minimum speed with which the ball must be moving at the top of the loop in order to maintain contact with the

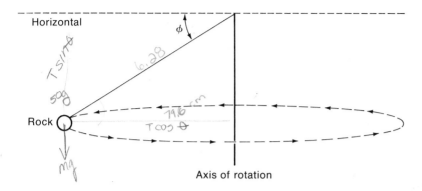

Fig. 9-14.

$t = 2$

Axis of rotation

track? The size of the ball may be considered negligible compared to the radius of the track.

9-6.   An airplane is pulling out of a dive at a speed of 600 mi/hr.
(*a*) If a pilot can endure the forces caused by an acceleration of only 6 *g*'s, what is the minimum radius of curvature
he can follow if his path is a section of a circle in a vertical
plane? Remember that at the bottom of the circle, a force
due to gravity, equivalent to 1 *g* of acceleration, is acting,
so that only 5 *g*'s of centripetal acceleration are allowed.
(*b*) How long, at a constant speed of 600 mi/hr, will it take
for him to change his direction from vertically down to vertically up following such a path? (*c*) What will be the magnitude of the maximum force exerted by the seat on a 160-lb
pilot?

9-7.   The breaking point of a string is 10 lb. A 1-lb weight is
attached to the string and swung in a circle that lies in a horizontal plane. The radius of the circular path is 2 ft and the
height of the circle is 6 ft above the ground. The velocity of
the circling weight is slowly increased until the string snaps.
How far from the center of the circle does the weight strike
the ground?

9-8.   A string is passed through a hole in a table. A weight m_1 is
attached at the bottom of the string, below the table. On the
table top, at the other end of the string, a mass m_2 is set into
rotation in a circular path with a radius and velocity such that
the tension in the string is equal to $m_1 g$ and the system remains in equilibrium. If friction between the weight and the
table and between the cord and the hole is negligible, find
the required velocity for m_2 as a function of the radius of its
circular path. Is this situation stable? That is, if the hanging
weight is slightly raised or lowered, does it tend to return
to its original height?

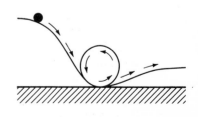

Fig. 9-15.

9-9. Three different station wagons tested recently by a consumer magazine made a full stop from an initial speed of 60 mi/hr in a distance of 140 ft. Find the minimum value of the coefficient of friction between their tires and the road as indicated by this performance.

9-10. A block of wood is placed on an inclined plane of wood. At what angle should the plane be placed so that the block accelerates at $\frac{1}{2}$ m/s?

9-11. A block slides *up* a plane inclined 30° above the horizontal. Its initial velocity is 2 m/s and it slides a distance of 20 cm up the plane before it comes to rest. Find μ_K for the block on this plane.

9-12. A block slides up a plane inclined at an angle θ above the horizontal. Its initial velocity is v_0 and the coefficient of friction is μ_K. Find the distance that the block slides up the plane before it comes to rest.

9-13. A person is in an amusement park device in the form of a cylinder that is spinning about its vertical axis, indicated by the dotted line in Fig. 9-16. The coefficient of static friction between the person's clothing and the cylinder wall is $\mu_S = 0.5$. Find the minimum speed of the outer wall that allows the occupant to remain in place.

9-14. When a car rounds a curve, the force of friction between its tires and the road must supply the needed centripetal force. Find the maximum velocity with which a 4000-lb car can negotiate a circle of 20-ft radius without slipping. Repeat for a 2000-lb car.

9-15. The coefficient of friction of a block sliding as shown in Fig. 9-17 is $\mu_K = 0.3$. The block is pulled by a string passed over a pulley and attached to a weight W. If the mass of the sliding block is 2 kg and the mass of the descending weight is 1 kg, find the acceleration of this system. Remember, as discussed in Section 5-7, that this is a multibody system. In particular, note that the tension in the string cannot be equal to the weight of W if W is to accelerate.

9-16. Find the acceleration of the system shown in Fig. 9-18. Assume the system moves so that m_2 is falling and there is a coefficient of friction μ_K between m_1 and the plane. What is the condition on m_1 and m_2 that produces constant velocity?

9-17. Repeat Problem 9-16 except let m_1 be large enough so that m_2 rises.

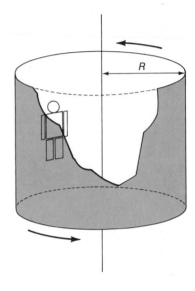

Fig. 9-16.

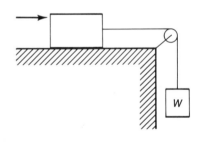

Fig. 9-17.

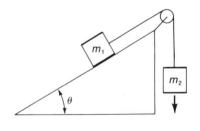

Fig. 9-18.

Chapter 10

Kinematics of Rotations

10-1 Introduction In this chapter we begin to study the motion of extended objects, like wheels or spinning tops, that cannot be treated as point masses. The location and motion of point masses are described precisely by the three-dimensional kinematics of Chapter 7. An extended object, however, can change its orientation as well as its position. In Fig. 10-1*a* we see a body moving without rotation. This type of motion is called *translational motion*. In Fig. 10-1*b* the body is rotating. A combination of rotation and translation is shown in Fig. 10-1*c*. The complete specification of an extended object's motion requires three angles of orientation as well as the three coordinates of its center of mass. At the introductory level with which we are concerned here, however, we will confine our attention to reorientation (or rotation) of an object that is confined to a plane. This will allow us to continue the use of simple vector concepts rather than the somewhat more complex tensor formulation needed for the truly general, three-dimensional dynamics of an extended object. Our more simple results will, nonetheless, apply to a wide variety of interesting physical situations.

10-2 Rotations and Radian Measure

When an extended object rotates, the x, y, and z coordinates of each point in the body continually increase and decrease as the point swings around in a circular path. Use of our familiar x,y,z coordinate system is a complicated way to describe rotations. But a rotation confined to a plane can be described easily by a single

(a)

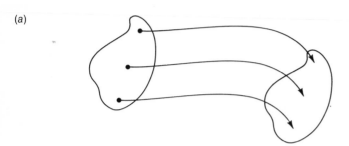

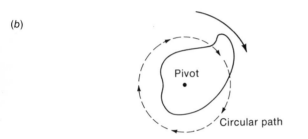

Fig. 10-1. The body in (a) moves with purely translational motion. Each point in the body moves along an identical trajectory. (b) The body is rotating without translational motion. (c) A combination of rotation and translation.

(b)

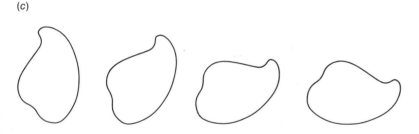

(c)

angle. Most of us are familiar with the *degree* measure of angles. The choice of 360 degrees in one full revolution was made by ancient Babylonians and probably originated with their interests in astronomy and its application to forecasting the seasons, particularly planting times. Since 360 is close to the number of days in the year, the degree is very close to the apparent change in orientation of the constellations from one night to the next. It is thus a unit of convenient size for astronomical purposes.

Consider the length s of the segment of a circle contained in the angle θ, which is indicated in Fig. 10-2. If the circle has a radius r, its entire circumference has a length given by

$$C = 2\pi r$$

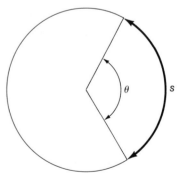

Fig. 10-2. The length s of the segment of the full circle's circumference is included within the angle θ.

The fraction of C contained in θ is equal to the fraction of a full revolution (360°) contained in θ. Thus

$$s = \frac{\theta}{360°}\, 2\pi r \qquad (\theta \text{ measured in } degrees) \qquad (10\text{-}1)$$

We see that s and r are proportional for a given θ. The constant of proportionality is seen to be equal to $2\pi\theta/360°$. Since this proportionality of s and r is frequently utilized in formulating the dynamics of rotating bodies, it is convenient to adopt a new unit of angle so that

$$s = r\theta \qquad (\theta \text{ measured in } radians) \qquad (10\text{-}2)$$

This unit is called the *radian*, abbreviated *rad* and comparison of (10-1) and (10-2) indicates that

$$\theta \text{ (degrees)} = \frac{360}{2\pi}\theta \text{ (radians)}$$

so that $\qquad\qquad$ 1 rad $= 57.3°$ or 2π rad $= 360°. \qquad (10\text{-}3)$

An angle of 1 rad is shown in Fig. 10-3.

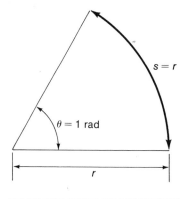

Fig. 10-3. An angle of one radian.

10-3 Angular Velocity and Angular Acceleration

If a rotation through an angle occurs in a given time, we can define an angular velocity ω as follows:

average angular velocity: $\bar{\omega} = \Delta\theta/\Delta t$
instantaneous angular velocity: $\omega = d\theta/dt$

The units of angular velocity are rad/s.

Example   A wheel makes three full revolutions in $\frac{1}{4}$ s. What is its average angular velocity?

Solution   Since there are 2π radians per revolution,

$$\theta = (2\pi \text{ rad/rev}) \cdot 3 \text{ rev} = 6\pi \text{ rad} = 18.8 \text{ rad}$$

so that $\qquad \bar{\omega} = \Delta\theta/\Delta t = 18.8 \text{ rad}/0.25 \text{ s} = 75.2 \text{ rad/s}$

Example   An airplane turns through an angle of 1° in 3 s. Find its average angular velocity in rad/s.

Solution $\qquad\qquad\qquad \Delta\theta = \dfrac{1°}{57.3°/\text{rad}} \qquad \Delta t = 3 \text{ s}$

$$\bar{\omega} = \frac{\Delta\theta}{\Delta t} = \frac{1}{3(57.3)} \text{ rad/s} = 0.0058 \text{ rad/s}$$

Angular acceleration, α, is defined in analogy with linear acceleration, a:

$$\bar{\alpha} = \frac{\Delta\omega}{\Delta t} \qquad \alpha = \frac{d\omega}{dt}$$

Δ angular velocity
Δt

The units of α are rad/s^2.

An interesting point concerning angular quantities is that the unit *rad* has no dimensions. It is a pure number, being the ratio of two lengths. Hence, when an angular acceleration is used in an equation, it carries only the dimensions of $1/t^2$, since radians are dimensionless.

Exercises

1. A point on the rim of a wheel moves 2 m when the wheel turns through an angle of $\frac{1}{4}$ rad. What is the radius of the wheel?

2. If $\theta = 10\ t^2$, find an expression for ω.

3. A wheel is initially turning with $\omega_1 = 10$ rad/s. After an interval of 3 s, it is turning faster: $\omega_2 = 40$ rad/s. Find its average angular acceleration during this time.

10-4 Analogies between Translational and Rotational Kinematics

Since all the foregoing is analogous to our definitions of linear velocity and accelerations, we are led to analogous relationships between one-dimensional kinematics and the kinematics of rotations.

Rotational kinematics   One-dimensional kinematics

$$\omega = d\theta/dt \qquad\qquad v = dx/dt$$

$$\alpha = d\omega/dt \qquad\qquad a = dv/dt$$

$$\Delta\omega = \int_1^2 \alpha\,dt \qquad\qquad \Delta v = \int_1^2 a\,dt \qquad\qquad (10\text{-}4)$$

$$\Delta\theta = \int_1^2 \omega\,dt \qquad\qquad \Delta x = \int_1^2 v\,dt$$

For the special case of constant acceleration:

$$\theta = \theta_0 + \omega_0 t + \tfrac{1}{2}\alpha t^2 \qquad x = x_0 + v_0 t + \tfrac{1}{2}at^2$$

$$\omega = \omega_0 + \alpha t \qquad\qquad v = v_0 + at$$

$$\omega^2 = \omega_0^2 + 2\alpha\theta \qquad\qquad v^2 = v_0^2 + 2ax$$

Example A wheel spins at 2000 rev/min (rpm). It coasts to a stop in 2 min. Assuming constant angular acceleration, find the magnitude of this acceleration and the total angle through which the wheel turns while stopping.

Solution $\omega_0 = \dfrac{2000 \text{ rev/min}}{60 \text{ s/min}} \cdot \dfrac{6.28 \text{ rad}}{\text{rev}} = 209 \text{ rad/s}$

Since α is constant,

$$\alpha = \frac{\Delta \omega}{\Delta t} = \frac{-209 \text{ rad/s}}{120 \text{ s}} = -1.74 \frac{\text{rad}}{\text{s}^2}$$

$$\theta = \theta_0 + \omega_0 t + \tfrac{1}{2}\alpha t^2$$

$$= 0 + 209 \frac{\text{rad}}{\text{s}} \cdot 120 \text{ s} - \frac{1.74}{2}\frac{\text{rad}}{\text{s}^2} \cdot (120 \text{ s})^2$$

$$= 1.26 \times 10^4 \text{ rad} = 2000 \text{ revolutions}$$

Note that we could have used rev as the unit of angle and min as the unit of time throughout this particular example:

$$\alpha = \frac{-2000 \text{ rev/min}}{2 \text{ min}} = -1000 \frac{\text{rev}}{\text{min}^2}$$

and $\theta = 2000 \dfrac{\text{rev}}{\text{min}} \cdot 2 \text{ min} - \tfrac{1}{2} \cdot 1000 \dfrac{\text{rev}}{\text{min}^2} \cdot (2 \text{ min})^2$

$$= 2000 \text{ rev}$$

Exercises

4. An electric motor starts from rest with a constant angular acceleration of 250 rad/s². (a) How long does it require to complete three full revolutions? (b) What is its angular velocity after three full revolutions? (c) How long does it require to make the fourth full revolution?

5. The angular velocity of a wheel is given by

$$\omega = a + bt^2$$

Find an expression for (a) the angular acceleration α, and (b) the net angle θ through which the wheel turns between t_1 and t_2.

10-5 Tangential Velocity and Acceleration

A point moving along a circular path has a velocity that is always tangent to the circle and perpendicular to the radius, as shown in Fig. 10-4. Since the distance through which such a particle moves is given by $s = r\theta$, its tangential velocity can be found by differentiation:

$$v_T = \frac{ds}{dt} = r\frac{d\theta}{dt} \quad \text{(for } r = \text{constant)}$$

$$= r\omega \qquad \text{tangential velocity}$$

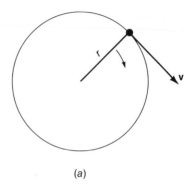

(a)

If, on the other hand, r varies so that the point moves along a spiral path, then we define the radial velocity

$$v_R = dr/dt$$

Since the radial velocity is perpendicular to the tangential velocity, the total velocity is given by

$$v = \sqrt{v_R^2 + v_T^2} \tag{10-5}$$

A point moving along a circular path experiences a radial acceleration given by

$$a_R = v^2/r \tag{9-3}$$

as discussed in the preceding chapter. Since the radius is constant, v is simply the tangential velocity and is given by $v = r\omega$ so that

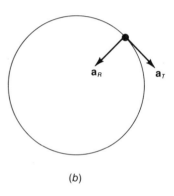

(b)

$$a_R = \frac{v^2}{r} = \frac{(r^2\omega^2)}{r} = r\omega^2 \tag{10-6}$$

If ω is not constant, then v_T is not constant, implying a *tangential* acceleration

$$a_T = \frac{dv_T}{dt} = \frac{d(r\omega)}{dt}$$

$$= \frac{r\,d\omega}{dt} \quad \text{(for } r = \text{constant)}$$

$$= r\alpha$$

Fig. 10-4. (a) The velocity of a point following a circular path is tangent to that path. (b) This point experiences a radial acceleration a_R and, if ω is not constant, a tangential acceleration a_T as well.

For *circular* motion ($r = $ constant) with ω variable or constant, we can collect all this information together:

$$v_T = \omega r \qquad \text{tangential velocity} \tag{10-7}$$

$$a_T = \alpha r \qquad \text{tangential acceleration} \tag{10-8}$$

$$a_R = v^2/r = \omega^2 r \tag{10-9}$$

$$\text{radial acceleration}$$

Example If $r = 10$ cm, $\omega = 5$ rad/s, and $\alpha = 10$ rad/s² find the tangential velocity, and the magnitude and direction of the acceleration.

Solution The tangential velocity is given by

$$v = \omega r = 5\ \frac{rad}{s} \cdot 10\ cm$$

$$= 50\ \frac{cm}{s}$$

The radial acceleration is directed inward and given by

$$a_R = r\omega^2 = 10\ cm \cdot 25\ \frac{rad^2}{s^2} = 250\ \frac{cm}{s^2}$$

(since the radian is dimensionless). The tangential acceleration is perpendicular to the radius:

$$a_T = \alpha r = 10\ \frac{rad}{s^2} \cdot 10\ cm = 100\ \frac{cm}{s^2}$$

Thus, the total acceleration is directed both inward, toward the center of the circle, and forward, in the direction of motion, as shown in Fig. 10-5. The magnitude of the total acceleration is

$$a = \sqrt{a_R^2 + a_T^2} = 269\ cm/s^2$$

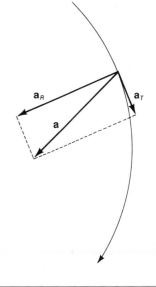

Fig. 10-5.

Exercises

6. A 128-lb person stands on a rotating platform at a radius of 10 ft. If at a given instant the platform has an angular velocity of 4 rad/s and an angular acceleration of -4 rad/s², what is the magnitude of the instantaneous force required to keep him in position?

7. A mass of 3 kg is located on the rim of a wheel of $\frac{1}{2}$ m radius. At $t = 0$, $\omega = 0$ and $\alpha = 15$ rad/s². Assuming constant α, find the direction and magnitude of the force required to keep the mass in place at the following times: (a) at $t = 0$; (b) at $t = \frac{1}{3}$ s; (c) after one full revolution.

10-6 The Rolling Wheel

When a wheel of radius R rolls along a surface without slipping, it must rotate with an angular velocity that produces exactly the tangential velocity needed to cause the lower point of the wheel to be always at rest with respect to the road. The velocity of any point on the wheel is a combination of the tangential velocity of that point with respect to the axle and the forward velocity V of the axle itself.

combination of V_T and forward V

Velocity at any point on the wheel — vector sum of forward V and tangential velocity

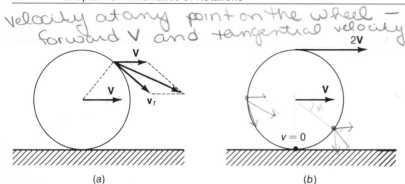

(a)   (b)

2V
V
V
v_T
V
v = 0

Fig. 10-6. (*a*) A rolling wheel. The velocity at any point is the vector sum of the forward velocity of the axle *V* and the tangential velocity with respect to the axle. (*b*) At the bottom of the wheel these two velocities must cancel, so $v_T = V$. At the top they are parallel and so add to 2*V*.

Since $v_T = \omega R$ and points oppositely to V at the point of contact with the ground, that point will be at rest with respect to the ground if ω is such that $V = v_T = \omega R$. Thus the point momentarily touching the ground is at rest, the axle moves forward at V, and a point at the top of the wheel moves forward at $2V$, as in Fig. 10-6. The wheel rotates with $\omega = V/R$.

Bottom $v_T = V$ v = 0
TOP of wheel 2V

At points other than the top or bottom of the wheel, v_T and V are not parallel. The resulting motion of a point on the rim of a rolling wheel is called a *cycloid* and is shown in Fig. 10-7. This is the trajectory of such a point as seen by an observer stationary with respect to the road. If we view the spinning wheel from a frame of reference attached to its axle, we simply see a stationary axle with points on the rim of the wheel following a simple, circular path.

Though these views appear different, they should not lead to differing physical consequences, since they merely represent different views of the same phenomenon. Consider a rock embedded in an automobile tire of one foot radius rolling at 60 mi/hr. In Fig. 10-8 we see what happens to a following car if the rock is cast loose just after the stone leaves the ground. This view is in the frame of reference of the cars as seen by a passenger of a car moving at the same speed in a parallel lane. The moment the rock leaves the tire tread, it follows the dictates of the law of inertia. That is, it moves

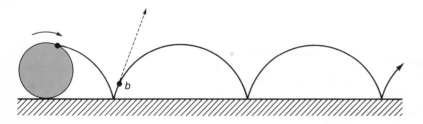

b

Fig. 10-7. The path traced by a point on the rim of a rolling wheel. The dotted line indicates the path a stone follows if cast loose from the tire at *b*.

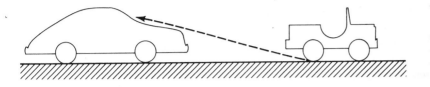

Fig. 10-8. Seen from a frame of reference moving with the cars, the stone flies tangentially off the spinning wheel and strikes the second car's windshield.

in a straight line (except for the effect of gravity, which gradually curves its trajectory) with a velocity of

$$V_T = \omega R = V_{car} = 60 \text{ mi/hr}$$

The luckless motorist following receives a cracked windshield.

But now let us look at the event as seen by an observer standing at the side of the road. To him the rock follows the cycloidal path shown in Fig. 10-7. The rock flies loose at a point such as that labeled *b* in Fig. 10-7. Following the law of inertia, the rock moves in the direction shown by the dotted line. Thus, to the roadside observer, the rock appears to be rising just after the first car passes. However, an instant later, a second car comes along at 60 mi/hr and strikes the rock with its windshield, thereby cracking the windshield.

Our two different views of this event agree in the end, though they do present apparently differing pictures. Their differences, however, are only the simple differences caused by motion of the observer.

Exercises

8. In Fig. 10-7 we see that a point on the wheel in contact with the ground is *momentarily* at rest. Is it accelerating?

9. Since the views of a rolling wheel by a stationary observer and an observer moving with the wheel differ only by a constant velocity, will they agree on the acceleration of points on the wheel? What if the car accelerates?

10. Find the magnitude and direction of the velocity of a point on the rim of a wheel rolling at velocity *V* when that point is halfway between the road and the top of the wheel.

10-7   The Angular Velocity Vector

It will prove helpful to associate a definite *direction* with a rotating system—a direction that gives us the orientation of the system. A moment's reflection convinces us that if an object is rotating about

$\vec{\omega}$ · angular velocity vector
magnitude – how rapidly a system rotates
direction – axis of rotation

some axis, as in Fig. 10-9, the only fixed direction is that of the axis of rotation itself. Accordingly, this direction is taken for a vector **ω,** whose magnitude tells us how rapidly a system rotates and whose direction gives the axis of rotation. It is called the *angular velocity vector.* The ambiguity as to which way along the axis the vector should point is resolved by a *right-hand rule,* illustrated in Fig. 10-9. To apply the rule to an object rotating about an axis, one imagines the plane of rotation to be the flat head of a screw turning in the same direction. The shank has a right-handed thread and its direction of advance is the direction we choose for our vector **ω.** An alternative method of finding its direction is to curl the fingers of the right hand so that they wrap around the axis of rotation with the finger tips pointing in the direction of tangential velocity. The thumb then rests on the axis and points in the direction of **ω.**

Exercise

11. A bicycle with wheels of 66-cm diameter rolls at 10 m/s. Find the direction and magnitude of the angular velocity vector of its front wheel.

10-8 The Vector Product

Frequently in our analysis of rotational motion, we will find ourselves calculating a product that involves two vectors. Calling the two vectors **A** and **B,** the product of interest has a magnitude equal to the product of B and the component of **A** perpendicular to **B.** If we use $A_\perp$ to denote the component of **A** that is perpendicular to **B,** then the product of **A** and **B** has a magnitude C given by

$$C = A_\perp B$$

For two vectors making an angle θ with respect to one another, as in Fig. 10-10, this becomes

$$C = AB \sin \theta \qquad (10\text{-}10)$$

since $A_\perp = A \sin \theta$

The product we have described is called the *vector product* or the *cross product* and is written

$$\mathbf{A} \times \mathbf{B} = \mathbf{C} \qquad (10\text{-}11)$$

The magnitude of **C** is given by (10-10) and the *direction* of **C** is perpendicular to the plane containing **A** and **B.** Since there are two

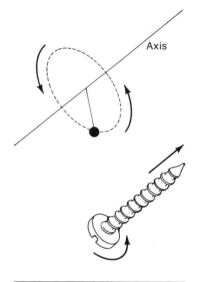

Axis

Fig. 10-9. The axis of rotation defines two possible directions. The right-hand rule selects the direction of advance of a right-hand screw as the direction of the angular velocity.

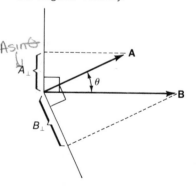

$A \sin \theta$

Fig. 10-10. The magnitude of the vector product of **A** and **B** is given by $A_\perp B$ or $B_\perp A$. The direction of this product vector is perpendicular to the plane containing **A** and **B** and into the page for **A** × **B,** out of the page for **B** × **A.**

possible directions perpendicular to that plane, we use a right-hand rule to define the direction of **C**:

> Place the tails of **A** and **B** together. Then rotate the tip of **A** toward **B**. The direction a right-hand screw would advance is the direction of **C**, where **C** = **A** × **B**.

For the vectors shown in Fig. 10-10, **A** × **B** points into the page. Reversing the order of the cross product does not change its magnitude but does reverse its direction:

$$\mathbf{B} \times \mathbf{A} = -\mathbf{A} \times \mathbf{B}$$

Thus, **B** × **A** points out of the page in Fig. 10-10.

An example of the use of the cross product is indicated in Fig. 10-11. There we see a point located in a rotating rigid body. The vector **r** points from an arbitrary point on the axis of rotation to the point P. The point P follows the circular path shown in Fig. 10-11 as the body rotates. The radius $r_\perp$ of that circular path is given by

$$r_\perp = r \sin \theta$$

and the tangential velocity of P along its circular path is

$$v_T = \omega r_\perp$$

We recognize in this result the magnitude of a cross product,

$$\mathbf{v} = \boldsymbol{\omega} \times \mathbf{r} \tag{10-12}$$

Furthermore, the right-hand rule definition of the direction of a cross product ensures that (10-12) correctly gives the direction of the velocity. Thus (10-12) gives both the magnitude and direction of the velocity of P. In similar fashion it can be shown that when **r** and **v** are not perpendicular, $\boldsymbol{\omega}$ is given by

$$\boldsymbol{\omega} = \frac{\mathbf{r} \times \mathbf{v}}{r^2} \tag{10-13}$$

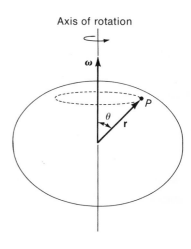

Axis of rotation

Fig. 10-11.   The point P is fixed in a rotating rigid body. It follows the dotted circular path with a velocity given by **v** = ω×**r**.

CROSS PRODUCT OF IDENTICAL VECTORS EQUALS ZERO

Exercises

12. Show that **A** × **A** = 0.

13. Let **A** be a vector running from left to right along the bottom edge of this page. Let **B** be a vector running from bottom to top along the left edge of this page. Find **C** in convenient units, where **C** = **A** × **B**.

14. Show that both (10-12) and (10-13) agree with (10-7) when **v** is perpendicular to **r** and **r** is perpendicular to ω.

10-9   Summary

When we use radian measure for angles, define the angular velocity $\omega = d\theta/dt$, and the angular acceleration $\alpha = d\omega/dt$, we obtain the following connections between rotational and translational quantities:

$$s = r\theta$$

$$v = r\omega$$

$$a = r\alpha$$

Rotational kinematic quantities obey relations analogous to those of translational kinematics, e.g.:

$$\theta = \theta_0 + \omega_0 t + \tfrac{1}{2}\alpha t^2 \qquad \text{for } \alpha = \text{constant}$$

An object on the rim of an angularly accelerating wheel is subject to both a tangential and a radial acceleration:

$$a_T = \alpha r \qquad \text{\textit{tangential acceleration}}$$

$$a_R = v^2/r = \omega^2 r \qquad \text{\textit{radial acceleration}}$$

The angular velocity may be represented by a vector $\boldsymbol{\omega}$ whose direction lies along the axis of rotation and is given by a right-hand rule. This vector contains the only stationary direction associated with a rotation and has a length given by the magnitude of the angular velocity. The vector product or cross product is written

$$\mathbf{C} = \mathbf{A} \times \mathbf{B} \qquad \text{where} \qquad C = AB \sin\theta$$

and the direction of $\mathbf{C}$ is given by a right-hand rule. The tangential velocity of any point P fixed in a rotating rigid body is conveniently written

$$\mathbf{v} = \boldsymbol{\omega} \times \mathbf{r} \qquad \text{\textit{tangential velocity}}$$

where $\boldsymbol{\omega}$ is the angular velocity of the rotating body and $\mathbf{r}$ extends to P from any point on the axis of rotation.

Problems

10-1.   The angular acceleration of a wheel is 12 rad/s². Find the time required for its angular velocity to change from 100 rad/s to 340 rad/s.

10-2.   A wheel with an initial angular velocity ω_0 reaches an angular velocity of $5\omega_0$ while it turns through an angle of 6 rad. Find its uniform angular acceleration α.

10-3. A grindstone initially spinning at $t = 0$ with $\omega = 3600$ rpm coasts to a stop in 2 min. Assume constant angular acceleration and find (a) the angular acceleration, (b) the total number of revolutions, (c) the tangential velocity of a point at 10 cm radius at $t = 1$ minute, (d) the radial acceleration of a point at 10 cm radius at $t = 1$ minute.

10-4. The length of the day increases by 1 ms per century. Find the angular acceleration of the earth in rad/s².

10-5. Two wheels are constructed, as in Fig. 10-12, with four spokes. The wheels are mounted one behind the other so that an observer normally sees a total of eight spokes but sees only four spokes when they happen to align with one another. If one wheel spins at 6 rev/min while the other spins at 8 rev/min, how often does the observer see only four spokes?

Fig. 10-12.

10-6. Two wheels are mounted side by side and each is marked with a spot on its rim. The two spots are aligned with the wheels at rest, then one wheel is given a constant angular acceleration of 0.1 rad/s² and the other, 0.05 rad/s². (a) When will the two marks become aligned again for the first time? (b) When will they become aligned for the second time?

10-7. A point follows a spiral path given by

$$r = r_0(1 + kt^2) \qquad \text{and} \qquad \theta = Ct$$

Find an expression for the magnitude of the velocity of this point at any time t.

10-8. A man on a merry-go-round stands so that his center of mass is 4 m from its center. He finds he must lean in toward the center at an angle of 15° from the vertical in order to maintain his balance. Find the constant angular velocity of the merry-go-round.

10-9. Suppose the man of Problem 10-8 walks steadily outward along a radius of the (rotating) merry-go-round. Does he have to provide a force component on himself that is perpendicular to the radius? Explain.

10-10. A car with wheels of 0.3-m radius is moving at 10 m/s and, at that instant, accelerating in the direction of motion with an acceleration of 2 m/s². Find the instantaneous total acceleration of a point on the rim of the wheel if it is rising and is halfway between the road and the top of the wheel.

Find this acceleration first in the car's frame of reference, then in the frame of the road. Which value should be used in calculating the force on a mass at that point?

10-11. Define a vector **C** by the equation

$$\mathbf{A} \times \mathbf{B} = \mathbf{C}$$

If **A** and **B** lie in the plane of this page with their tails at a common point and are separated by an angle θ, find the magnitude and direction of **C** when $\theta = 0°$, $90°$, $180°$, $270°$, and $360°$. If $\mathbf{D} = \mathbf{B} \times \mathbf{A}$, how are **C** and **D** related?

10-12. Use Eq. (10-12) to find the tangential velocity of a point on the earth's surface halfway between the north pole and the equator. The earth's radius is 6.4×10^3 km.

Chapter *11*

Rotational Dynamics

11-1 Introduction In the last chapter we described the rotation of extended bodies in terms of their angular velocity, ω, and their angular acceleration, α. In this chapter we describe the ways that forces can change the angular velocity of a system. We will discuss the *torque* or twisting ability of a force. Just as we found that there are analogies between velocity and angular velocity, and between acceleration and angular acceleration, we will find rotational analogs to the dynamical quantities of mechanics. Thus we will define the *moment of inertia* of a rotating body, which is analogous to the mass of a translating body, and *angular momentum,* which is analogous to ordinary momentum.

Before we proceed to a study of torque and angular momentum, it will be helpful to discuss differentiation of products of scalar functions and of vectors.

11-2 Differentiation of Vector Products

Let us begin by finding the derivative of the product of two scalar functions. Consider two different functions of time, for example, $u(t)$ and $v(t)$. The derivatives of these functions with respect to time are written

$$\frac{du}{dt} \quad \text{and} \quad \frac{dv}{dt}$$

or $u'(t)$ and $v'(t)$. These derivatives have a fundamental definition, i.e.,

$$\frac{du}{dt} = \lim_{\Delta t \to 0} \frac{\Delta u}{\Delta t}$$

which can be written out more completely:

$$\frac{du}{dt} = \lim_{\Delta t \to 0} \frac{u(t + \Delta t) - u(t)}{\Delta t}$$

Now, turning to the differentiation of a product, let $f = uv$, that is,

$$f(t) = u(t) \cdot v(t)$$

Then $$f + \Delta f = (u + \Delta u) \cdot (v + \Delta v)$$

$$= uv + u\Delta v + v\Delta u + \Delta u \Delta v$$

Subtracting $f = uv$ from the above, we obtain

$$\Delta f = u\Delta v + v\Delta u + \Delta u \Delta v$$

Next we divide by Δt:

$$\frac{\Delta f}{\Delta t} = u\frac{\Delta v}{\Delta t} + v\frac{\Delta u}{\Delta t} + \Delta u\frac{\Delta v}{\Delta t}$$

Taking the limit as $\Delta t \to 0$,

$$\frac{df}{dt} = \lim_{\Delta t \to 0} \frac{\Delta f}{\Delta t} = u\frac{dv}{dt} + v\frac{du}{dt} + 0 \cdot \frac{dv}{dt}$$

since $\Delta u \to 0$ as $\Delta t \to 0$. Finally,

$$\frac{df}{dt} = u\frac{dv}{dt} + v\frac{du}{dt} \tag{11-1}$$

An alternative notation is $f' = uv' + vu'$.

Example   The expression $y = at^5$ can be written $y = uv$, where $u = at^2$ and $v = t^3$. Show that application of (11-1) to this product agrees with the result of directly differentiating the original function using the rule (2-8):

$$\frac{d}{dt}(Bt^n) = nBt^{n-1}$$

Solution   Direct differentiation by application of (2-8) gives

$$\frac{dy}{dt} = 5at^4$$

Separating the function y into a product,

$$y = at^2 \cdot t^3$$

and applying (11-1), we obtain

$$\frac{dy}{dt} = at^2 \cdot \frac{d(t^3)}{dt} + t^3 \cdot \frac{d(at^2)}{dt}$$

$$= at^2 \cdot 3t^2 + t^3 \cdot 2at$$

$$= 5at^4$$

in agreement with our first result. In actuality, the rule (2-8) is obtained by repeated application of (11-1), starting with the easily derived results for $n = 1$ and $n = 2$.

Now let us consider the vector product

$$\mathbf{C} = \mathbf{A} \times \mathbf{B}$$

If $\mathbf{A}$ and $\mathbf{B}$ vary with time, how does $\mathbf{C}$ vary? We can find $d\mathbf{C}/dt$ as follows. Adding arbitrary vector increments to $\mathbf{A}$ and $\mathbf{B}$, we find

$$\mathbf{C} + \Delta\mathbf{C} = (\mathbf{A} + \Delta\mathbf{A}) \times (\mathbf{B} + \Delta\mathbf{B})$$

$$= \mathbf{A} \times \mathbf{B} + \mathbf{A} \times \Delta\mathbf{B} + \Delta\mathbf{A} \times (\mathbf{B} + \Delta\mathbf{B})$$

By subtraction we obtain

$$\Delta\mathbf{C} = \mathbf{A} \times \Delta\mathbf{B} + \Delta\mathbf{A} \times (\mathbf{B} + \Delta\mathbf{B})$$

Dividing by the time Δt required for these changes to occur, we find

$$\frac{\Delta\mathbf{C}}{\Delta t} = \mathbf{A} \times \frac{\Delta\mathbf{B}}{\Delta t} + \frac{\Delta\mathbf{A}}{\Delta t} \times (\mathbf{B} + \Delta\mathbf{B}) = \mathbf{A} \times \frac{\Delta\mathbf{B}}{\Delta t} + \frac{\Delta\mathbf{A}}{\Delta t} \times \mathbf{B} + \frac{\Delta\mathbf{A}}{\Delta t} \times \frac{\Delta\mathbf{B}}{\Delta t}\Delta t$$

and in the limit as $\Delta t \to 0$, we find

$$\frac{d\mathbf{C}}{dt} = \mathbf{A} \times \frac{d\mathbf{B}}{dt} + \frac{d\mathbf{A}}{dt} \times \mathbf{B} \qquad (11\text{-}2)$$

The order is important in this result. The second term is not equivalent to

$$\mathbf{B} \times \frac{d\mathbf{A}}{dt}$$

since the vector cross product changes sign upon reversal of order.

Exercises

1. Let $u = at^2$ and $v = bt + ct^3$. Define $f = uv$ and find f' by application of (11-1). Compare your result to that obtained by first multiplying u and v together and then differentiating by use of the rules (2-8) and (2-9).

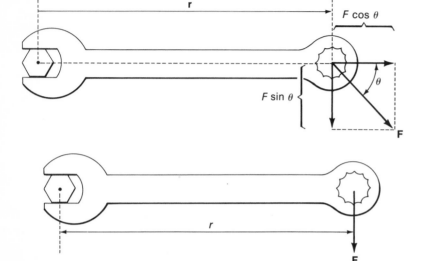

Fig. 11-1. The torque due to the force F is $\Gamma = rF$.

Fig. 11-2. Only the component of **F** perpendicular to the lever arm produces a twisting effect.

2. Let **r** be a position vector pointing to the location of a point mass. Then the velocity of that point mass is given by $\mathbf{v} = d\mathbf{r}/dt$. Differentiate the expression $\mathbf{r} \times \mathbf{v}$, utilizing the rule (11-2).

11-3   Torque and Angular Momentum

The force in Fig. 11-1 causes a rotation to occur. The tendency to turn depends upon both the magnitude of the force and the length of its lever arm, r. The same turning ability can be produced by a small force and large lever arm or a large force and small lever arm. Thus if we try to open a door pushing close to the hinge, it requires a much larger force than if we push far from the hinge. The twisting ability of a force is called the *torque* Γ. We might try defining

$$\Gamma = rF$$

since this would at least qualitatively agree with our observations. However, if the force does not act perpendicular to the lever arm, as in Fig. 11-2, only the perpendicular component, $F \sin \theta$, acts to produce a twist. The other component, $F \cos \theta$, simply pulls or pushes sideways on the nut without producing a rotation. If we define **r** as a vector pointing from the axis of rotation to the point of application of the force, we could write

$$\Gamma = \mathbf{r} \times \mathbf{F} \qquad (11\text{-}3)$$

The magnitude of Γ is then given by

$$\Gamma = rF \sin \theta$$

handwritten notes:
tendency to turn
depends on
① magnitude of force
② length of lever arm, r

$\Pi = rF$

⊥ component $F \sin \theta$

product of r and $F \sin \theta$
(perpendicular component of F)

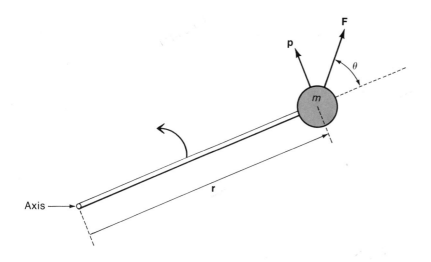

Fig. 11-3. A mass m attached to a rigid rod of length r pivots about an axis.

which is the desired product of r and the perpendicular component of **F**. Furthermore, the vector direction of **Γ** is the same as the direction of the vector **ω**, which represents the rotational velocity produced by the torque. We will adopt (11-3) as our definition of torque and we will soon see that it is a good definition.

Now consider the extremely simple rotating system of Fig. 11-3. We define the *angular momentum* of this system by

$$\mathbf{L} = \mathbf{r} \times \mathbf{p} \qquad (11\text{-}4)$$

where **p** is the momentum of the mass. For this particular case **p** and **r** are perpendicular so that

$$L = rp = rmv$$

Since $v = r\omega$, we find

$$L = (mr^2)\omega$$

That is, the angular momentum equals the product of mr^2 and the angular velocity, ω. This is analogous to

$$p = mv$$

where the momentum is equal to the product of mass and velocity. In the rotational case, the quantity mr^2 plays the role of the inertial mass. The quantity mr^2 is called the *moment of inertia* of the revolving mass.

Now let us take the time derivative of (11-4), utilizing the rule (11-2):

$$\frac{d\mathbf{L}}{dt} = \mathbf{r} \times \frac{d\mathbf{p}}{dt} + \frac{d\mathbf{r}}{dt} \times \mathbf{p} \qquad (11\text{-}5)$$

We can see that the second term on the right side of (11-5) is always zero if we write it out in more detail:

$$\frac{d\mathbf{r}}{dt} \times \mathbf{p} = \frac{d\mathbf{r}}{dt} \times m\mathbf{v} = \frac{d\mathbf{r}}{dt} \times m\frac{d\mathbf{r}}{dt}$$

We see now that the terms of this cross product are parallel, so the product vanishes. We are thus left with only the first term of (11-5):

$$\frac{d\mathbf{L}}{dt} = \mathbf{r} \times \frac{d\mathbf{p}}{dt}$$

and, since Newton's second law gives us

$$\mathbf{F} = \frac{d\mathbf{p}}{dt}$$

we have

$$\frac{d\mathbf{L}}{dt} = \mathbf{r} \times \mathbf{F} = \mathbf{\Gamma} \qquad (11\text{-}6)$$

using the definition contained in (11-3). This result is the rotational analog of Newton's second law:

$$\mathbf{\Gamma} = \frac{d\mathbf{L}}{dt} \qquad \text{is analogous to} \qquad \mathbf{F} = \frac{d\mathbf{p}}{dt}$$

We can thus deduce the following:

1. *The angular momentum* **L** *remains constant unless a torque* **Γ** *acts.*
2. *When a torque is present it causes a change in the angular momentum at a rate given by* (11-6).

If several forces act to twist a system about a fixed axis, we can sum their torques to find a net torque; it is the net torque that is equal to the rate of change of the angular momentum of the system. Further, if several masses rotate about a common origin, we can sum their angular momenta to obtain the total angular momentum. Thus for *m* forces lying in a plane acting on *n rigidly connected* masses rotating in that plane, we have

$$\mathbf{\Gamma}_{\text{total}} = \sum_{i=1}^{m} \mathbf{r}_i \times \mathbf{F}_i = \sum_{j=1}^{n} \mathbf{r}_j \times \frac{d\mathbf{p}_j}{dt}$$

$$= \frac{d}{dt} \sum_{j=1}^{n} \mathbf{r}_j \times \mathbf{p}_j = \frac{d\mathbf{L}_{\text{total}}}{dt} \qquad (11\text{-}7)$$

as long as the $\mathbf{r}_j$ vectors are constant. One example of this situation is shown in Fig. 11-4.

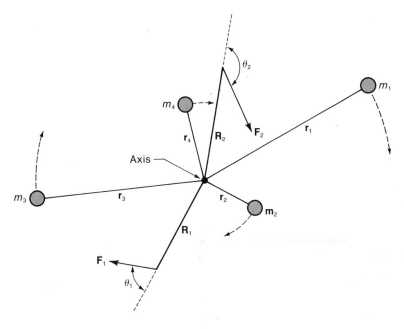

Fig. 11-4. Clockwise rotation of several masses connected by rigid rods to an axis of rotation. The torque produced by $\mathbf{F_1}$ and $\mathbf{F_2}$ acts to increase the angular momentum of this system.

For n masses that remain at fixed distances from an axis of rotation, the total angular momentum about any point on that axis can be written in terms of $\boldsymbol{\omega}$ as follows:

$$\mathbf{L}_{\text{total}} = \sum_{j=1}^{n} \mathbf{r}_j \times \mathbf{p}_j = \sum_{j=1}^{n} m_j \mathbf{r}_j \times \mathbf{v}_j$$

Since each mass follows a circular path as it rotates with angular velocity $\boldsymbol{\omega}$, we have

$$\mathbf{v}_j = \boldsymbol{\omega} \times \mathbf{r}_j$$

where $\boldsymbol{\omega}$ is the same for each of the masses, since they are all rigidly connected. Thus $\mathbf{r}_j$ and $\mathbf{v}_j$ are perpendicular. If the masses all lie in a plane with $\boldsymbol{\omega}$ perpendicular to the plane, we find

$$\mathbf{L}_{\text{total}} = \left(\sum_{j=1}^{n} m_j r_j^2 \right) \boldsymbol{\omega} \qquad (11\text{-}8)$$

$L = mr^2 w$

where the quantity in parentheses is the *total moment of inertia I* about the axis of rotation:

$$I = \sum_{j=1}^{n} m_j r_j^2 \qquad (11\text{-}9)$$

$I = mr^2$

$L = Iw$

Hence

$$\mathbf{L}_{\text{total}} = I\boldsymbol{\omega} \qquad (11\text{-}10)$$

total angular momentum = total moment of Inertia × angular velocity

Differentiating (11-10) with respect to time,

$$\frac{d\mathbf{L}_{total}}{dt} = I\frac{d\omega}{dt} = I\alpha \qquad (11\text{-}11)$$

for I constant, i.e., for a rigid rotating assembly. Substituting (11-11) into (11-7), we finally obtain

$$\Gamma = m\alpha \qquad \Gamma = I\alpha \qquad (11\text{-}12)$$

where α is the angular acceleration. Equation (11-12) is analogous to Newton's second law in the form

$$\mathbf{F} = m\mathbf{a}$$

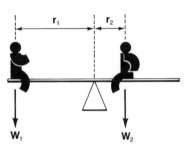

Fig. 11-5.

moment of Inertia is analogous to mass

Again we see that the role of the moment of inertia is analogous to that of mass. There is an important difference, however. Because the coordinates r_j enter into (11-9), the moment of inertia depends upon the location of the point about which we calculate the moment of inertia. This should not surprise us, however, for it can indeed be more difficult to set a system of masses into rotation about one point than another. For example, if we hold a meter stick at its center and twist it into rotation, less torque is required (for a given α) than if we twist it about one end. We will investigate this point in more detail in Section 11-4.

Example   (The law of levers)   Consider the situation shown in Fig. 11-5. If the system is balanced, we have

$$\frac{d\mathbf{L}}{dt} = 0$$

Using (11-7) to calculate Γ about the pivot point, we see that

$$\Gamma_{total} = \mathbf{0}$$

Taking the positive direction for Γ to be out of the page, we can write

$$\Gamma_{total} = r_1 W_1 - r_2 W_2$$

since $\mathbf{r}$ and $\mathbf{W}$ are perpendicular. We have seen that Γ_{total} must be zero, so we find

$$r_1 W_1 = r_2 W_2$$

which is the law of levers. The twisting ability of a force acting at right angles to a lever is given by the product of the force and the distance between its point of application and the pivot.

Example   Four 2-kg masses are connected by $\frac{1}{4}$-m spokes to an axle, as in Fig. 11-6. A force of 24 N acts on a lever $\frac{1}{2}$ m long to produce an angular acceleration α. Find the magnitude of α.

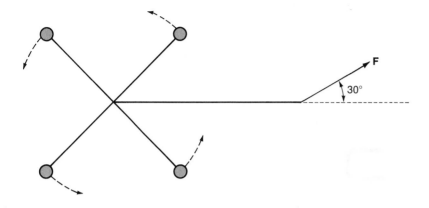

Fig. 11-6.

$$\Gamma = I\alpha$$
$$L = I\omega$$
$$\Gamma = rF\sin\omega$$
$$I = mr^2$$

$$\Gamma = I\alpha$$

Solution $$\alpha = \frac{\Gamma}{I} = \frac{\overset{rF\sin\theta}{\frac{1}{2}\,\text{m}\cdot 24\,\text{N}\,(\sin 30°)}}{\underset{mr^2}{4\cdot 2\,\text{kg}\cdot(\frac{1}{4}\,\text{m})^2}} = \frac{6\,\text{N}\cdot\text{m}}{\frac{1}{2}\,\text{kg}\cdot\text{m}^2}$$

Noting that $N = \text{kg}\cdot\text{m/s}^2$, we have $\alpha = 12\,\text{s}^{-2}$. The dimensions of α are correctly those of s^{-2}, since radians are themselves dimensionless. Our four-mass system undergoes an angular acceleration of 12 rad/s².

Exercise

3. Two equal masses m are connected to an axle and pulley, as in Fig. 11-7. A string is wound around the pulley and a force **F** pulls on the string. Note that the friction between the string and pulley ensures that **F** acts always tangentially to the pulley rim. The radius of the pulley is r and the radius of the circular path followed by the masses is R. (*a*) Find the magnitude and direction of the torque created by the force **F** about the axle. (*b*) Find this system's moment of inertia about the axle, neglecting the mass of all parts except the two masses m. (*c*) Find the angular acceleration of this system. (*d*) If the system begins at rest, find its angular velocity at a time interval τ after the constant force **F** was applied.

11-4   Moment of Inertia for Extended Mass Distributions NOT ON 2nd HOURLY

Frequently we want to treat systems in which the mass is not concentrated into points but extended continuously, as in a disk. First, consider the simplest case, a wheel with its mass concentrated in a rim and a negligible mass in its spokes, as in Fig. 11-8. If we break

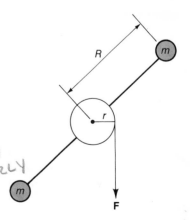

Fig. 11-7.

this continuous mass distribution into n mass elements Δm, we can calculate the moment of inertia:

$$I = \sum_{i=1}^{n} \Delta m_i r_i^2$$

But since $r_i = r$ is the same for each element Δm_i, it can be factored out of the sum. Thus the moment of inertia for the entire wheel is simply

$$I = r^2 \sum_{i=1}^{n} \Delta m_i = mr^2$$

$I = mr^2$

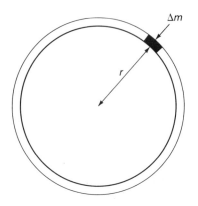

Fig. 11-8. A wheel with its entire mass concentrated at the rim. Each element of mass, Δm, is at the same distance, r, from the axis of rotation.

Now consider a system whose mass elements lie at varying radii from the axis, such as Fig. 11-9. Our first guess at the rod's moment of inertia might be to consider all the mass to be located at its center, $r = L/2$, which is the location of its center of mass. Recall, however, that the center of mass was found by averaging the contribution of each mass element multiplied by r, not by r^2. The center of mass of an extended mass distribution is accordingly called the *first moment* of the mass, while the moment of inertia is called the *second moment*.

A quick numerical calculation shows us that the assumption of an "effective radius" at $L/2$ is incorrect. First, compute the moment of inertia with this assumption and obtain $I = m(L/2)^2 = \frac{1}{4}mL^2$. Next, consider the mass of the left half of the rod to lie at its center of mass, that is, put $m/2$ at $r = L/4$; and the mass of the right half to lie at *its* center of mass, $m/2$ at $3L/4$. The sum of the moments of inertia of the two masses gives the total moment of inertia:

$$I_{\text{total}} = \frac{m}{2}\left(\frac{L}{4}\right)^2 + \frac{m}{2}\left(\frac{3L}{4}\right)^2 = \frac{10}{32}mL^2$$

This result is nearly $\frac{1}{3}mL^2$, yet our first assumption gave $\frac{1}{4}mL^4$. In fact, if we subdivide the mass into four parts, we obtain $\frac{21}{64}mL^2$, and each finer subdivision approaches still closer to $\frac{1}{3}mL^2$. The important point here is that a *linear* average, like that for the center of mass, is inappropriate for finding the moment of inertia, in which each mass contributes in proportion to the *square* of its radius. Integral calculus permits us to choose mass elements so small, however, that each element lies at a unique radius and hence has an exactly defined moment of inertia. The sum of all these contributions to the moment of inertia will be an integral.

As in Fig. 11-9, consider an element of mass Δm that lies a distance r from the axis. Its moment of inertia is only a portion of the entire moment of inertia, so we may write

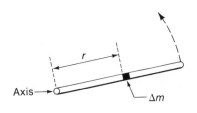

Fig. 11-9. A rod of length L rotating about an axis at one end.

$$\Delta I \approx \Delta m \cdot r^2$$

where ΔI is the moment of inertia of the mass Δm. The approximate equality is required since not all parts of Δm are at precisely the same radius r. The summation of all n of these infinitesimal contributions gives us the total moment of inertia:

$$I_{total} \approx \sum_{i=1}^{n} \Delta m_i r_i^2$$

The mass contained in Δm is the product of the mass per unit length of the rod multiplied by the length of the element Δm, i.e.,

$$\Delta m = \lambda \Delta r$$

so that
$$I_{total} \approx \lambda \sum_{i=1}^{n} r_i^2 \Delta r$$

If we take the limit as $\Delta r \to 0$, the approximate equality becomes exact and the sum becomes an integral:

$$I_{total} = \lambda \int_{r=0}^{r=L} r^2 \, dr \qquad (11\text{-}13)$$

Carrying out the integration and substituting upper and lower limits, we obtain

$$I_{total} = \tfrac{1}{3}\lambda [L^3 - 0^3] = \tfrac{1}{3}\lambda L^3$$

This result can be put into far more useful form by noting that λ, the mass per unit length, equals the total mass of the rod divided by its length: $\lambda = m/L$. Thus

$$I_{total} = \tfrac{1}{3} m L^2 \qquad \text{ROD} \qquad (11\text{-}14)$$

We can obtain the moment of inertia of a rod of length *L pivoted at its center* by noting that it is twice that of a rod with a length of $L/2$ pivoted at one end. We can thus utilize the result (11-14):

$$I = 2\left(\frac{1}{3}\right)\left(\frac{m}{2}\right)\left(\frac{L}{2}\right)^2 = \frac{1}{12} m L^2$$

Thus a rod pivoted at its center has only one-fourth the moment of inertia of the rod pivoted at one end.

The above result can also be obtained directly from (11-13) simply by changing the limits of the integral to $-L/2$ and $+L/2$, as required to include the entirety of the centrally pivoted rod:

$$I = \lambda \int_{-L/2}^{+L/2} r^2 \, dr = \frac{\lambda}{3}\left[\frac{L^3}{8} - \left(\frac{-L^3}{8}\right)\right] = \frac{\lambda}{12} L^3$$

Substituting $\lambda = m/L$, we have

$$I = \frac{1}{12}mL^2$$

As we indicated at the end of Section 11-3, the moment of inertia *does* depend upon the point about which it is calculated. This simply reflects the difference in the torque required to produce a given angular acceleration when we twist the rod about different points on the rod. The greatest torque is required (hence *I* is largest) when we twist the rod about one end, for we then place the various elements of mass of the rod as far from the pivot point as possible.

A slightly more complicated case for calculation of the moment of inertia is shown in Fig. 11-10. There we consider the moment of inertia of a solid cylinder rotating about its axis. The moment of inertia of one thin cylindrical shell centered about the axis of the solid cylinder is given by $\Delta I = \Delta m r^2$.

In this case the quantity Δm depends upon r, since the larger shells near the outside of the cylinder contain more mass than those near the center. Thus we must express Δm in terms of r. The mass *density* ρ of the solid comprising the cylinder is defined by the mass per unit volume:

$$\rho = m/V$$

Then the mass contained in any volume element, such as the volume ΔV of the shell in Fig. 11-10, is given by

$$\Delta m = \rho \Delta V$$

If the height of the cylinder is L and the shell has radius r and thickness Δr, we have:

$$\Delta V = 2\pi r L \Delta r$$

so that $$\Delta m = 2\pi L \rho r \Delta r$$

The moment of inertia of this shell is thus

$$\Delta I = \Delta m \cdot r^2 = 2\pi L \rho r^3 \Delta r$$

Summing all the ΔI's and taking the limit as $\Delta r \to 0$, we are led to an integral:

$$I = 2\pi L \rho \int_{r=0}^{r=R} r^3 \, dr = 2\pi L \rho \frac{R^4}{4} = \frac{\pi}{2} L \rho R^4 \qquad (11\text{-}15)$$

Since $\rho = M/V = M/\pi R^2 L$, we can rewrite (11-15) to read

$$I = \tfrac{1}{2}MR^2 \qquad (11\text{-}16)$$

SOLID CYLINDER.

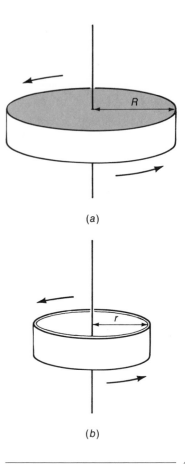

(a)

(b)

Fig. 11-10. (a) The total moment of inertia of a solid cylinder of radius *R* is obtained by integrating the infinitesimal contributions of thin, cylindrical shells, shown in (b), which have radii lying between 0 and *R*.

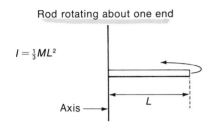

Rod rotating about one end

$I = \frac{1}{3}ML^2$

Axis —→

L

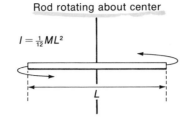

Rod rotating about center

$I = \frac{1}{12}ML^2$

L

Fig. 11-11.   Moments of inertia for several solids.

UNIFORM DISC

½mr²

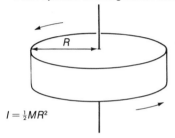

Solid cylinder rotating about axis

R

$I = \frac{1}{2}MR^2$

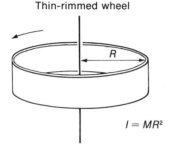

Thin-rimmed wheel

R

$I = MR^2$

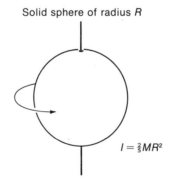

Hollow cylinder

R_2 R_1

$I = \frac{1}{2}M(R_1^2 + R_2^2)$

Solid sphere of radius R

$I = \frac{2}{5}MR^2$

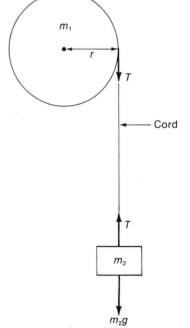

m_1

r

T

Cord

T

m_2

m_2g

Similar procedures may be followed to obtain the moments of inertia of various solids with simple geometrical shapes, with the results given in Fig. 11-11. It is probably worth mentioning that the result for the hollow cylinder does not contain a sign error – the *sum* of the radii is correct.

Example   A solid cylindrical flywheel has a cord wrapped around its circumference, from which is suspended a weight, as shown in Fig. 11-12. Find the angular acceleration of the flywheel if it has a radius of 10 cm, a mass of 2 kg, and the weight has a mass of 3 kg.

Fig. 11-12.

Solution   The weight, W, produces a tension T in the cord. This tension multiplied by its lever arm (r in this case) gives the torque acting on the flywheel. We are immediately tempted to set $T = m_2g$. However, if $T = m_2g$, it pulls *up* on the weight with a force that just cancels the downward pull of gravity on the weight. Thus *no* acceleration occurs, which is not the case. Let us solve for the acceleration of the weight and the angular acceleration of the wheel separately, then combine our results. Newton's second law for translational motion of the weight is

$$m_2g - T = m_2a \tag{1}$$

while Newton's second law for rotational motion of the wheel is

$$\Gamma = I\alpha$$

This can be rewritten

$$rT = \tfrac{1}{2}m_1r^2\alpha$$

As long as the rope does not slip on the rim of the wheel, we have

$$a(\text{weight}) = a(\text{rope}) = a(\text{rim of wheel}) = \alpha r$$

so that $\alpha = a/r$ and we find

$$rT = \tfrac{1}{2}m_1ra$$

or

$$T = \frac{m_1}{2}a \tag{2}$$

Substitute (2) into (1): $$m_2g - \frac{m_1a}{2} = m_2a$$

$$\left(m_2 + \frac{m_1}{2}\right)a = m_2g$$

$$a = \frac{m_2}{m_2 + \dfrac{m_1}{2}}g$$

For our case, $a = \dfrac{3}{3+1}g = \dfrac{3}{4}g = 7.3 \text{ m/s}^2$, so

$$\alpha = a/r = \frac{7.3 \text{ m/s}^2}{0.1 \text{ m}} = 73 \text{ rad/s}^2$$

Exercises

4. A solid sphere, as in Fig. 11-11, is mounted on an axle. A string is wound around its equator and the free end pulled with a force of 6 N. The mass of the sphere is 2 kg and its radius is 5 cm. Find the angular acceleration.

5.   A revolving door has a moment of inertia of 20 kg·m². A man pushes on it 30 cm from its axis of rotation with a force of 60 N. In what direction should he push to achieve the maximum angular acceleration? Find the maximum angular acceleration he can produce.

11-5 Conservation of Angular Momentum

Equation (11-7) tells us that if no torque acts upon a system, its angular momentum is constant. A gyroscope is a device with a rapidly rotating wheel mounted on three bearings so that no external torques can be applied to it. Fig. 11-13 shows such a device. A navigational gyro in an aircraft relies on the fact that the direction of the angular momentum of the spinning wheel stays fixed in space. The wheel is kept spinning by an airstream directed at its periphery and no matter how the airplane is turned or banked, no external torques can be transmitted to the wheel. Since the angular momentum is given by $\mathbf{L} = I\boldsymbol{\omega}$, and since $\boldsymbol{\omega}$ points along the axis of rotation of the wheel, this implies that the direction of the wheel's axle remains fixed in space while the airplane turns, banks, and rolls about it. Thus, regardless of external visibility, the pilot always has available a fixed reference direction. Scales are fixed to the frames attached to the bearings and windows are provided in an enclosed housing so that the pilot can refer to a convenient numerical scale calibrated in 360°, as in a conventional compass.

   Figure 11-14 illustrates a case in which only one component of an angular momentum vector is conserved. A man is standing on a turntable so that no torques can be transmitted to him about a vertical axis. That is, no vertical torques can be applied to him, though the ability of the bearing in the turntable to support a sideways force *does* allow horizontal torques to be applied to him. If the man at rest holds a spinning wheel with its angular momentum pointing vertically, the man + wheel system has a total vertical

Fig. 11-13. A gyro consists of a rapidly spinning wheel mounted upon three bearings, so that no external torques can be applied to it. (Courtesy Fisher Scientific Co., Educational Materials Division, Chicago, Illinois)

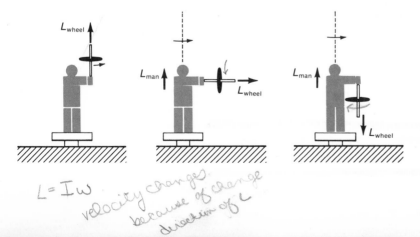

Fig. 11-14. A man initially at rest stands on a turntable and holds a spinning wheel with its angular momentum vector pointing up. As he changes the direction of the wheel's angular momentum, he begins to rotate.

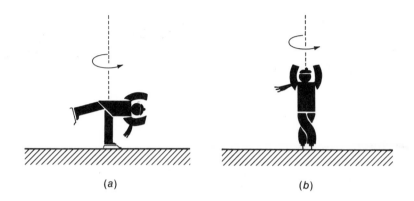

Fig. 11-15. (a) An ice skater begins a spin with large moment of inertia. (b) He finishes with a small moment of inertia.

(a)   (b)

angular momentum equal to L_{wheel}. If he turns the axis of the wheel away from the vertical, the vertical component of the wheel's angular momentum diminishes. Taking z as the vertical direction:

$$L_z(\text{wheel}) = L_{wheel} \cos \theta$$

where θ is the angle of the wheel's axis from the vertical. Since the *total* value of L_z must remain constant (no torques in the z direction), the man and turntable must now begin to rotate, so that their angular momentum makes up for the reduced vertical component of the wheel's angular momentum. When the wheel is returned to its original orientation, the rotation of the man and turntable ceases.

The results of this experiment are quite independent of the method by which the z component of the wheel's angular momentum is changed—as long as no external torques act. For example, the man on the initially stationary turntable can simply stop the spinning wheel (with its axis held vertically all the while) with his hand. He and the turntable will then turn, so that the total value of L_z will be unchanged. If, still holding the wheel's axis vertically, he uses his free hand to start spinning the wheel, he will discover that he and the turntable come to rest just when he gets the wheel spinning at its original speed.

A dramatic illustration of angular momentum conservation is provided by an ice skater's spin. Initially, an ice skater swings into a turn, then stands on the toe of one skate, rotating slowly about that pivot point. The ice skater constitutes a torque-free system since the contact between the skate toe and the ice is essentially frictionless. Typically the skater begins the spin with arms and one leg extended and the torso horizontal, as shown in Fig. 11-15. This places most of the mass of the skater's body as far as possible from the axis of rotation and thus maximizes the moment of inertia. The skater then slowly changes position to that shown in part *b* of the figure. Since this puts the bulk of the skater's mass close to the axis

moment of inertia maximized by putting mass as far away as possible from the axis of rotation

of rotation, the moment of inertia is greatly reduced. However, conservation of angular momentum tells us that the product, $I\omega$, remains constant. Since I decreases, ω must increase. In this manner, skilled skaters can achieve rotational velocities so large that they appear blurred to the observer.*

A similar phenomenon appears to take place in the evolution of stars. Stars apparently are formed by the self-gravitation of clouds of dust and gas in interstellar space. Any small angular velocity the dust cloud may have possessed (such as that due to the slow rotation of the entire galaxy) becomes tremendously increased as the star shrinks from a diffuse, extended cloud to a relatively small, dense object. Incidentally, the process of gravitational contraction heats the star to the point of luminescence and beyond, eventually raising the internal temperature to the point where self-sustaining nuclear reactions begin. Some large stars apparently become violently unstable near the end of their evolution, when their nuclear energy sources are expended. It is thought that these stars explode violently, producing a supernova explosion. The outer layers of the star are blown into space, leaving behind a core of extremely dense matter in which the atomic nuclei are actually in contact. In this process, a mass comparable to our own sun shrinks to an object only a few kilometers in radius, thus greatly reducing its moment of inertia. Though our own sun rotates once every 25 days, a supernova remnant in the nearby Crab Nebula** rotates 30 times per second. Such stars frequently produce short pulses of radio and light emissions and are called *pulsars*.

Exercises

6. An ice skater, as in Fig. 11-15*a,* has a moment of inertia of 8 slug·ft² and is rotating at 1 rev/s. She diminishes her moment of inertia to 1 slug·ft², as in Fig. 11-15*b*. What is her angular velocity?

7. At various seasons of the year the earth's atmosphere has a net average motion with or against the direction of the earth's rotation. Explain how this affects the earth's rotation period.

---

*Since a retinal image in the human eye persists for approximately 1/20 of a second, we could crudely estimate the angular velocity by the extent of blurring.

**The present location of this object, called the Crab Pulsar, coincides with the position of a brilliant supernova reported by Chinese astronomers in 1054 A.D. The Crab Nebula itself is an expanding cloud of dust and gases that is the present remnant of the outer layers of the star. Much more information on supernova remnants can be found in the article by Jeremiah P. Ostriker in the January 1971 issue of *Scientific American*.

8. Two flywheels are arranged so that they can be coupled together by means of a clutch, but no external torques act on this two-flywheel system. One of the flywheels ($I_1 = 8 \text{ kg} \cdot \text{m}^2$) is rotating with an angular velocity of 50 rad/s, while the other flywheel ($I_2 = 5 \text{ kg} \cdot \text{m}^2$) is at rest. They are then coupled together with the clutch so that they must rotate with a common angular velocity. What is that final angular velocity?

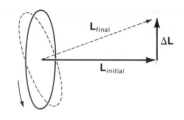

Fig. 11-16. A change in **L**, Δ**L**, at right angles to the angular momentum produces a final angular momentum in a new direction.

11-6 Rotational Motion with Torque Perpendicular to the Angular Momentum

Equation (11-7) indicates that changes in angular momentum are vector quantities that point in the direction of the applied torque. If we apply a torque at right angles to the angular momentum, the angular momentum vector will turn to a new orientation. This is illustrated in Fig. 11-16. A case in which the torque is at right angles to the angular momentum and hence produces a change Δ**L** at right angles to **L** is shown in Fig. 11-17. The wheel is supported by the tension **T** in a string attached to one end of the axle. Clearly, if the wheel were not spinning and the other end of the axle were unsupported, the wheel would fall and swing back and forth like a pendulum at the end of the string. However, if the wheel *is* spinning, it has angular momentum that, in Fig. 11-17, points to the right. The weight of the wheel does not generate a torque about the center of the wheel, since it acts effectively at the wheel's center. Consider the torque about the center of the wheel that is caused by the tension pulling upward on the end of the axle. This torque points directly into the page, as shown by applying the right-hand rule for $\mathbf{\Gamma} = \mathbf{R} \times \mathbf{T}$. Hence Δ**L** must also point into the page, and the new angular momentum vector is *not* below the old one but in the same horizontal plane, pointing back into the page.

Viewed from overhead, the angular momentum vector moves counterclockwise. Since this vector points along the axle of the

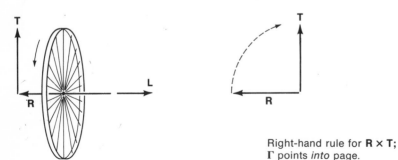

Right-hand rule for **R** × **T**; $\mathbf{\Gamma}$ points *into* page.

Fig. 11-17. A rapidly spinning bicycle wheel is suspended by a string at one end of an axle at a distance R from the center of the wheel.

spinning wheel, the wheel itself must be changing its orientation. This phenomenon is called *precession* and is easily demonstrated with a toy gyroscope. Contrary to our intuitive notions, a spinning wheel supported only at one end of its axle will not fall, but instead slowly precesses about that end. The rate at which the axle rotates in space is called the precessional angular velocity, Ω_p. Since there is no net acceleration of the wheel's mass up or down, the vertical forces must balance. Gravity pulls down on the wheel with a force mg and the support must supply an upward force that equals mg.

Let us examine the precession phenomenon in more detail by use of an example.

Example A small $\frac{1}{2}$-kg gyroscope wheel has a moment of inertia of 0.01 kg $\cdot$ m² and spins at 1000 rev/min. It is suspended by a string at a point on its axle 5 cm from its center of mass. Find its rate of precession, Ω_p.

Solution Sketching the situation in Fig. 11-18, we see that the torque due to the upward force mg is $\Gamma = mgR$ and points into the page. This produces a change in the direction of **L** given by $\Gamma = dL/dt$. Viewed from overhead, this torque will cause **L** to precess counterclockwise, as seen in the sketch of Fig. 11-19. In the sketch, $\Delta\theta$ is the change in direction of **L** caused by the applied torque. We have from the geometry of Fig. 11-19,

$$\Delta L \approx L\Delta\theta$$

Dividing by Δt, the time required for the change ΔL, and taking the limit as $\Delta t \rightarrow 0$, we find

$$\frac{dL}{dt} = L\frac{d\theta}{dt} = L\Omega_p$$

The left side of this equation is the applied torque (since $\Gamma = dL/dt$) and the right side contains the desired rate of precession, Ω_p. Hence

$$\Gamma = L\Omega_p$$

or

$$\Omega_p = \Gamma/L$$

In our example, $\Gamma = mgR = 0.25$ N $\cdot$ m and the angular momentum is given by

$$L = I\omega = \left(\frac{1}{100}\text{ kg} \cdot \text{m}^2\right)(1000\text{ rev/min})\left(0.105\ \frac{\text{rad/s}}{\text{rev/min}}\right)$$

$$= 1.05\text{ kg} \cdot \text{m}^2 \cdot \text{rad/s}$$

Thus

$$\Omega_p = \frac{\Gamma}{L} = \frac{0.25\text{ N} \cdot \text{m}}{1.05\text{ kg} \cdot \text{m}^2 \cdot \text{rad/s}} = 0.24\text{ rad/s}$$

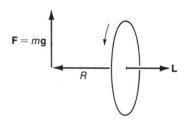

F = mg   R   L

Fig. 11-18.

$$\Omega_p = \frac{\Gamma}{L}$$

T points into the page

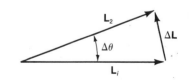

Fig. 11-19.

The gyroscope precesses slowly, since it takes

$$\frac{6.28 \text{ rad}}{0.24 \text{ rad/s}} = 26 \text{ s}$$

for one full turn of the gyroscope axle's orientation.

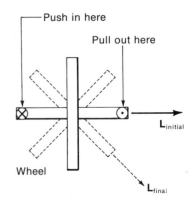

Fig. 11-20. A downward-directed torque must be applied to change $\mathbf{L}_i$ to $\mathbf{L}_f$. It is produced by pulling out of the page on the right handle, pushing in on the left.

In the preceding example we examined the motion produced by a given torque. We can invert the problem and ask, "What torque is required to change the orientation of the axle of a spinning wheel?" In Figure 11-20 we see a spinning wheel fitted with two handles. In order to cause the right handle to drop while the left handle rises, a torque must be applied that points down. The right-hand rule tells us that we can supply such a torque by pulling the right handle toward us and/or pushing the left handle away. Paradoxically, if we wish the right handle to move lower we pull on it! Spinning wheels evidently respond at right angles to the direction our intuition expects.

A well-known trick of instrument technicians in the Navy in the days of World War II was to place a rapidly spinning (but quiet) gyro on a fixed axis inside a black box and then ask a sailor to deliver the box to a location that required him to make a turn enroute. The results were usually mildly catastrophic.

Exercises

9. Consider the situation shown in Fig. 11-17. (*a*) What happens if the wheel is spinning in the opposite direction? (*b*) What happens if the wheel is spinning as shown but the point of support is at the right end of the axle rather than the left?

10. A bicycle wheel is observed to be precessing as in Fig. 11-17. Its weight is 5 lb, its moment of inertia is 1/3 slug · ft², and it spins at 30 rad/s. The point of support is located 4 in. from the center of gravity of the wheel. Find the precessional angular velocity, Ω_p.

11. A bicycle rider suddenly leans to the right, thus applying a torque to the front wheel. (*a*) What is the direction of that torque? (*b*) How does that torque affect the direction of the angular momentum of the rotating front wheel? (*c*) If the rider has his hands off the handle bars, so that no other torques are exerted on the front wheel, does the effect that you found in part *b* make it more or less difficult for the rider to maintain control of the bicycle? See, for example, "The Stability of the Bicycle" by David E. H. Jones in the April 1970 issue of *Physics Today*.

11-7 An Alternative Treatment of Gyroscopic Phenomena

Most of us agree that the behavior of spinning wheels when we push or pull their axes is a bit puzzling. They seem to respond to our efforts with a motion at right angles to that expected. Suppose we review the ideas of Section 11-3, beginning with Newton's second law for translational motion:

$$\mathbf{F} = \frac{d\mathbf{p}}{dt}$$

and then performing a purely mathematical operation:

$$\mathbf{r} \times \mathbf{F} = \mathbf{r} \times \frac{d\mathbf{p}}{dt}$$

With appropriate definitions, this results in

$$\boldsymbol{\Gamma} = \frac{d\mathbf{L}}{dt}$$

A straightforward application of this last equation "explains" gyroscopic phenomena. Since that equation is derived directly from Newton's second law, we should be content with our explanation. Yet, for most of us, this procedure is sufficiently abstract to leave us feeling a bit shortchanged. Torques are simply pushes or pulls on a lever, so we would like to be able to use Newton's second law directly on the masses in a rotating wheel and "see" how one applies these forces to achieve a change in orientation. Such a procedure is quite cumbersome for a truly extended continuous object, but it can be carried out for a small number of point masses.

Consider a rotating wheel made up of four masses on four spokes, as shown in Fig. 11-21. An observer wishes to change the orientation of the axle, as shown in parts b and c of the figure. In Fig. 11-21c we see that the top mass, m_1, and bottom mass, m_3, have had to change the direction of their momentum. Accordingly, because of Newton's second law, a force to the *right* was exerted on the top mass, a force to the *left* on the bottom mass. Masses 2 and 4 were bodily moved but their momenta were not changed appreciably in the process (if the wheel is rotating rapidly in comparison to the desired precession). Thus, in order to swivel the axle as shown in part b of the figure, a force had to be exerted on the top of the wheel toward the right, and on the bottom of the wheel to the left. These forces constitute a torque that could as well have been provided by pulling *up* on the left handle (out of the page in Fig. 11-21b) while pushing down on the right.

Paradoxically, pulling up on one handle and down on the other does not cause the left handle to rise while the right drops; instead,

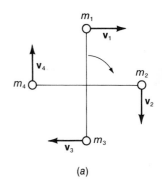

(a)

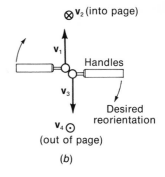

(b)

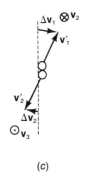

(c)

Fig. 11-21. Four masses rotating about an axle. In (b) an overhead view shows the top mass moving forward, the bottom mass backward, while m_2 is dropping and m_4 is rising. If the axle is swiveled to a new position, as shown in (c), m_1 and m_3 experience a large change in momentum, m_2 and m_4 do not.

it causes the left handle to swivel forward (clockwise) and the right to swivel backwards. Application of Newton's second law, $\mathbf{F} = d\mathbf{p}/dt$, directly to these rotating masses *has* shown us that they respond in an unexpected fashion. Our results are in complete accord with a conventional treatment based on the rotational form of Newton's second law, $\mathbf{\Gamma} = d\mathbf{L}/dt$, which is far more easily applied to extended objects than a point-by-point application of the linear form of Newton's second law.

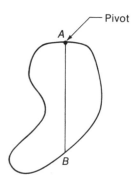

11-8 Center of Gravity

In order to calculate the torque exerted by a force, we must know the length of the lever arm through which it acts. We must know, then, the point of application of the force. In the case of the force of gravity acting upon an *extended* body, there is not a single point at which the weight of the body acts. Instead, each portion of the body is attracted toward the earth, so that there are many parallel force vectors distributed over the entire mass. The effect of these distributed forces is identical to the effect of a single force, *mg*, acting at a point called the *center of gravity* of the object.

The center of gravity of an object with arbitrary shape can be found by the procedure shown in Fig. 11-22. By its definition, the center of gravity must lie directly below the pivot point and hence somewhere along *AB*. Suspending the body from a different point locates the center of gravity somewhere along *CD*. The intersection of *AB* and *CD* is thus the location of the center of gravity.

As indicated in Fig. 11-23, a body suspended at its center of gravity experiences no gravitational torque about this point and will be stable in any orientation.

For *n* masses distributed at various positions on the *x* axis, we can equate the individual torques about the origin to that produced by the entire weight acting at the center of gravity:

$$m_1 g x_1 + m_2 g x_2 + m_3 g x_3 + \cdots + m_n g x_n = M g X$$

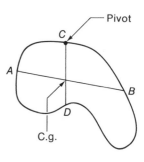

Fig. 11-22. The center of gravity hangs directly below the pivot so it is somewhere along the line *AB*. Hanging it from a different pivot point indicates that the center of gravity lies somewhere on *CD*. The intersection of these lines locates the center of gravity.

Pivot at c.g.

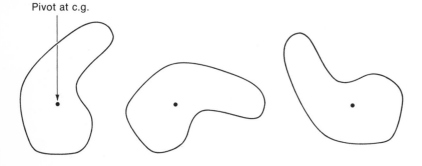

Fig. 11-23. The body shown in Fig. 11-22 has a small hole drilled at the center of gravity. This hole is slipped over a smooth pin so that the body can pivot clockwise and counterclockwise freely. It remains balanced in any orientation.

where M is the total mass of the n masses and X the position of the center of gravity. Factoring out g and transposing M, we find

$$X = \frac{m_1x_1 + m_2x_2 + m_3x_3 + \cdots + m_nx_n}{M}$$

But this is precisely the definition of the location of the center of *mass* of the n objects given by Eq. (8-17). The same argument shows that the location of the center of gravity along the y and z axes is also identical to the center of mass locations given by (8-17).*

For continuous mass distributions, the sums of (8-17) become integrals, as discussed in Section 8-8.

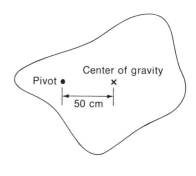

Fig. 11-24.

Exercises

12. A body with a mass of 4 kg is shown in Fig. 11-24. Find the magnitude of the torque its weight exerts about the pivot point shown.

13. A rectangular plate of length l, uniform thickness, and mass m is free to pivot, as shown in Fig. 11-25. Find the torque its weight exerts about the pivot.

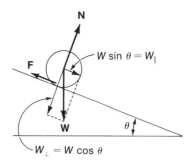

Fig. 11-25.

11-9   Rolling on an Inclined Plane

In Fig. 11-26 we see a cylinder on an inclined plane. As discussed in Section 11-8, the center of gravity of a system gives the effective location of the force of gravity acting upon that system. Thus, the weight of the cylinder effectively acts at the center of the cylinder, as shown in Fig. 11-26. The force of friction acts to oppose the motion of the cylinder down the plane. Note that it is the torque produced by the force of friction that causes the cylinder to rotate about its c.m. Since the force of friction cannot exceed $\mu_S N$, where μ_S is the coefficient of static friction and N is the normal force, a cylinder placed on a slippery surface may not experience sufficient torque to give it sufficient angular acceleration to roll without slipping. Rolling without slipping implies that $\omega = v/r$, where ω is the angular velocity of the cylinder and v its translational velocity. Differentiating, we find that $\alpha = a/r$. Applying Newton's law in its translational form, $\mathbf{F} = m\mathbf{a}$, and its rotational form, $\Gamma = I\alpha$, allows us to deduce the acceleration of the cylinder and the value of μ_S required to ensure rolling without slipping.

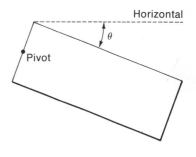

Fig. 11-26.   A cylinder of radius r placed upon an inclined plane is acted upon by the force of gravity, **W**, the normal force, **N**, and the force of friction, **F**.

*The identity of c.m. and c.g. occurs only if g has the same value over the entire mass distribution. In a nonuniform gravitational field, the c.m. and c.g. no longer coincide exactly.

The component of the force of gravity acting parallel to the inclined plane, $W \sin \theta$, acts to accelerate the cylinder. The force opposing the acceleration is **F**, the frictional force. So we have:

$$W \sin \theta - F = ma \qquad (11\text{-}17)$$

Since the force of gravity and the normal force both pass through the center of the cylinder, they exert no torque. Only the frictional force produces a torque *about the center of the cylinder*. Thus if Γ is the torque acting about the center and I is the moment of inertia of the cylinder about its center, Newton's second law in rotational form, $\Gamma = I\alpha$, becomes

$$Fr = \tfrac{1}{2}mr^2\alpha$$

Then, since $\alpha = a/r$, we obtain

$$F = \tfrac{1}{2}ma \qquad (11\text{-}18)$$

Note that (11-18) was obtained by *assuming* rolling without slipping ($\alpha = a/r$). If frictional forces are not as large as given by (11-18), it simply means that slipping must occur. Substituting (11-18) into (11-17):

$$W \sin \theta = ma + \tfrac{1}{2}ma$$

$$mg \sin \theta = \tfrac{3}{2}ma$$

$$a = \tfrac{2}{3}g \sin \theta$$

Comparing this result with (8-8), we see that a cylinder that rolls without slipping has only 2/3 the acceleration of an object sliding without friction. The frictional force, however, has a maximum value given by

$$F \leqslant \mu N = \mu W \cos \theta$$

The maximum possible value of this force defines the boundary between rolling with and without slipping:

$$F_{\max} = \mu mg \cos \theta$$

But $F = \tfrac{1}{2}ma = \tfrac{1}{3}mg \sin \theta$. Hence

$$\mu mg \cos \theta_{\max} = \tfrac{1}{3}mg \sin \theta_{\max}$$

$$\tan \theta_{\max} = 3\mu$$

For example, a coefficient of friction appropriate for wood on wood ($\mu \approx \tfrac{1}{2}$) predicts:

$$\theta_{\max} = \tan^{-1}(3 \cdot \tfrac{1}{2}) = \tan^{-1}(1.5) = 56.3°$$

For $\theta < 56.3°$, a wooden cylinder will roll without slipping, while for $\theta > 56.3°$, the cylinder will slip down the plane.

Exercises

14. When a can of baked beans is rolled down a plane inclined at 45°, the entire contents of the can rotate with the can. However, when a can of fruit juice is rolled down the plane, the can rotates around the juice, so that the juice has only translational motion. Estimate the acceleration of the can of beans and the can of juice, neglecting the mass of the can. (The results are useful if the labels of several cans are lost.)

15. Discuss the motion of a solid cylinder, initially at rest, that starts sliding down a smooth inclined plane ($\mu \approx 0$), then reaches a portion of the inclined plane where $\mu = \frac{1}{3} \tan \theta$. The angle of inclination of the plane is θ and μ is the coefficient of sliding friction.

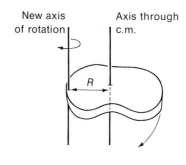

Fig. 11-27. An object rotating about an axis that does not pass through the center of mass.

11-10   The Parallel Axis Theorem   NOT ON and HOURLY

In some of the applications that appear in this chapter, we have evaluated torques with respect to the center of gravity of rotating objects. Since a torque is a product of both the force and its lever arm, the magnitude of a torque generated by a force must change when we evaluate that torque with respect to a different axis. Nonetheless, it would seem that the motion predicted by rotational dynamics should not depend upon our whims in the choice of axis. Indeed, it does not. If we evaluate the moment of inertia of our system with respect to a shifted axis, it, too, is modified in such a way that the new values of torque and moment of inertia predict the same result when used in the equation $\Gamma = I\alpha$.

The calculation of the moment of inertia of an object with respect to an axis that does not pass through its center of mass is accomplished by application of the *parallel axis theorem*. Figure 11-27 illustrates an object of mass M rotating about such an axis. This axis is a distance R away from a parallel axis that *does* pass through the center of mass. The parallel axis theorem states that

$$I = I_{\text{cm}} + MR^2$$

That is, the total moment of inertia about an arbitrary axis is that of a point mass located at the center of mass plus the moment of inertia about that center of mass.

Proof   Consider the several rigidly connected mass points sketched in Fig. 11-28. Each of the masses m_i are a distance r_i from the c.m. They are each a distance r_i' from a new pivot that is a distance R from the c.m. From the law of cosines,

$$(r_i')^2 = r_i^2 + R^2 - 2r_i R \cos \theta_i$$

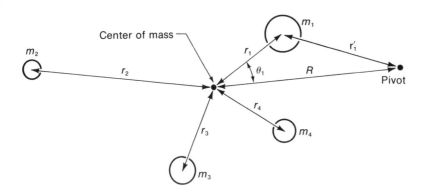

Fig. 11-28. Four rigidly connected masses lie at distances r_1, r_2, r_3, and r_4 from their center of mass. We wish to find their moment of inertia with respect to the pivot shown. The distance r_1' between m_1 and the pivot is shown.

The moment of inertia of each mass m_i about the new axis is

$$I_i' = m_i(r_i')^2 = m_i r_i^2 + m_i R^2 - 2m_i r_i R \cos \theta_i$$

The total moment of inertia is thus

$$I' = \Sigma_i m_i r_i^2 + \Sigma_i m_i R^2 - 2(\Sigma_i m_i r_i \cos \theta_i) R$$

The first term defines the moment of inertia with respect to the center of mass. The second term is just MR^2, where M is the total mass, and the parentheses in the third term enclose M times the location of the center of mass. However, since the quantity in parentheses gives M times the distance of the center of mass *from* the center of mass (our starting point), it must equal zero. Thus we have

$$I' = I_{cm} + MR^2$$

as was to be proved.

Example   Given the moment of inertia of a rod pivoted at its center of mass, use the parallel axis theorem to deduce its moment of inertia when it is pivoted at one end.

Solution   In Section 11-4 we saw that the moment of inertia of a rod pivoted at its center of mass is $\frac{1}{12}ML^2$. Pivoting the rod at one end moves the axis to a distance $L/2$ away from the center of mass. The parallel axis theorem then gives us

$$I = \tfrac{1}{12}ML^2 + M(L/2)^2 = \tfrac{1}{3}ML^2$$

which agrees with the result obtained in Section 11-4.

In the preceding example, the moment of inertia about one end could just as easily have been calculated directly without use of the parallel axis theorem. Such is not the case, however, for a rolling wheel. At any instant it is pivoting about its point of contact, as sketched in Fig. 11-29. We know its moment of inertia about its center (for a thin-rimmed wheel it is MR^2, for a solid wheel, $\frac{1}{2}MR^2$),

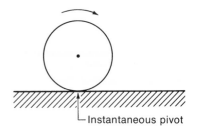

Fig. 11-29.

but the moment-of-inertia integral is not easily evaluated for a pivot point on the rim. The parallel axis theorem immediately gives the result:

$$I = I_{cm} + MR^2$$

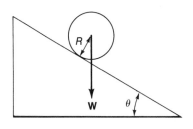

Fig. 11-30.

Example   A solid wheel rolls without slipping down an inclined plane. Find its acceleration.

Solution   As sketched in Fig. 11-30, take the point of contact as the pivot. Then the force of friction contributes nothing to the torque, since it has no lever arm about that point. The weight of the wheel does produce a torque, however:

$$\Gamma = WR \sin \theta$$

The angular acceleration is $\alpha = \Gamma/I$, where I is the moment of inertia with respect to our chosen pivot point. Thus

$$\alpha = \frac{WR \sin \theta}{\frac{1}{2}MR^2 + MR^2} = \frac{2}{3}\frac{W \sin \theta}{MR} = \frac{2}{3}\frac{g \sin \theta}{R}$$

The acceleration is given by

$$a = \alpha R = \frac{2}{3}g \sin \theta$$

as was concluded in the preceding section, where we evaluated torques about the center of mass and used the moment of inertia about the center of mass.

11-11   Summary

In this chapter we found that we could modify Newton's second law into a rotational form by use of the vector cross product:

$$\mathbf{F} = \frac{d\mathbf{p}}{dt}$$

$$\mathbf{r} \times \mathbf{F} = \mathbf{r} \times \frac{d\mathbf{p}}{dt}$$

which can be written as

$$\Gamma = \frac{d\mathbf{L}}{dt}$$

when we define $\Gamma = \mathbf{r} \times \mathbf{F}$ and $\mathbf{L} = \mathbf{r} \times \mathbf{p}$. For a rigid plane figure with an axis of rotation perpendicular to the plane, we found

$$\mathbf{L} = I\omega$$

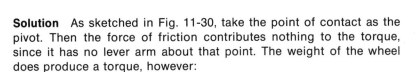

where the moment of inertia I is found from

$$I = \sum_{i=1}^{n} m_i r_i^2$$

for a discrete collection of masses. For a continuous mass distribution, this sum becomes an integral. Results for several simple shapes are given in Fig. 11-11.

Systems with no external torques acting upon them conserve angular momentum. Examples are given by a navigational gyro, a spinning ice skater, and a shrinking star.

When a torque acts in a direction parallel to the angular momentum, it merely speeds or slows the angular velocity of a system. However, when a torque acts at right angles to the angular momentum of a system, it changes the direction of the angular momentum, that is, it changes the direction of the system's axis of rotation. The motion of the system's axis in this case is clearly defined by the equation $\Gamma = dL/dt$, even though the result is not always in agreement with our intuition. The phenomenon of gyroscopic precession is an example of such a case.

The force of gravity produces parallel forces on all parts of an extended body. The center of gravity is that point at which a single force, equal to the weight of the entire body, produces the same result as the actual, distributed weight of the body. This point coincides with the center of mass.

A wheel rolling on an inclined plane is an example of a system that both rotates and translates. Two different analyses, one considering torques about the center of mass, the other, torques about the point of contact, give identical results for the acceleration of the wheel.

Problems

11-1. An electric drill motor produces a 15-N·m torque. It is attached to a 30-cm diameter sanding disc with a moment of inertia of 0.01-kg·m². Find the tangential velocity of a point on the rim of the initially stationary disc 0.01 s after the motor is turned on.

11-2. A flywheel in the form of a solid cylinder with a radius of 1 m and a mass of 10^3 kg is initially spinning with an angular velocity of 1000 rev/min. Bearing friction causes a constant torque of 30 N · m, which slows the rotation. How much time is required to bring the flywheel to rest?

11-3. Two masses are connected by a string passed over a pulley. The pulley is in the form of a solid cylinder of mass m_1 and radius r. The two hanging masses are m_2 and m_3. Find the magnitude of the acceleration of m_3 in terms of m_1, m_2, m_3, r, and g. Neglect the weight of the string and friction in the bearing.

11-4. As in Problem 9-8, a string is passed through a hole in the center of a smooth table and a mass attached to the string swings continuously in a circular path centered on the hole. (a) If the mass is given an initial tangential velocity v_0 at an initial radius r_0, find the force a person must exert on the end of the string below the table to supply the tension needed to maintain the mass in its circular path. (b) If the string is pulled downward, the radius of the path decreases. Use conservation of angular momentum to find the angular velocity as a function of radius, given the initial conditions of part a. (c) Find the required tension in the string as a function of radius for the initial conditions given in part a. (d) Does your result change the conclusion reached for the last portion of Problem 9-8?

11-5. A 160-lb man stands on a turntable with his arms outstretched. In each hand he holds a 40-lb weight, and he is initially turning at $\frac{1}{2}$ rev/s. Estimate his angular velocity after he pulls the weights close to his body.

Do this problem
See other Physics book

11-6. A particle of mass m moves in a straight line with constant velocity. Show that its angular momentum with respect to any fixed point is a constant.

11-7. A rifle bullet with a mass of 20 g is fired at 600 m/s so that it strikes the rim of a wheel tangentially. The wheel is initially at rest, has a radius of 10 cm, a mass of $1\frac{1}{2}$ kg, and is in the form of a solid cylinder. If the bullet embeds itself in the rim of the wheel, find the angular velocity of the wheel after the impact. The results of the preceding problem should be useful.

11-8. Two ice skaters of equal mass approach one another on parallel trajectories that are separated by 60 cm. They each have a velocity of 3 m/s with respect to the ice. They link arms as they meet, keeping themselves always separated by 60 cm. Is angular momentum conserved? How fast are they now rotating about their mutual center of mass?

EXTRA

11-9. Presently, the moon orbits about the earth once every 27 days, always keeping one face toward the earth. The earth turns on its axis once per day. Eventually, tidal friction will cause the earth to turn so that it keeps one face always toward the moon. Because the total angular momentum of the earth-moon system must be constant, the moon must recede to a greater distance as this occurs. (*a*) Prove the latter statement. (*b*) Taking the orbit of the moon to be centered on the center of the earth (not quite true but close), calculate the final radius of the moon's orbit at the end of this process.

11-10. If the polar ice caps of the earth were to melt so that all their water returned to the oceans, the oceans would be 100 ft deeper. How would the rotational velocity of the earth be changed? Can you make a rough estimate of the change produced in the length of a day?

11-11. A satellite carrying an ultraviolet telescope is placed in orbit about the earth so that astronomical observations will not be hampered by the earth's atmosphere. The satellite's designers have fixed a rapidly spinning wheel inside the satellite. Due to conservation of angular momentum, the satellite always maintains a fixed orientation with respect to the stars, even though the satellite's center of mass moves in a circular orbit. Describe a system of rocket thrusters that can change the orientation of this satellite by any desired amount. Pay particular attention to the thrust direction required to change the direction of the axis of the spinning wheel.

11-12. A ship in rough water rolls severely about its long axis. It also pitches (motion in which the bow rises while the stern falls and vice versa), but this motion is much less severe than the rolling motion in a very large ship because the ship is long enough to "average out" the effects of all but the largest waves. Consider a large, rapidly rotating flywheel placed in the hold of such a ship, with its axis fixed to the ship and pointing vertically. How would this device affect the motion of the ship?

11-13. A spinning bicycle wheel is suspended as in Fig. 11-17, but not with its axis horizontal. If the angle of its axis below the vertical is ϕ, find the angular precession rate, Ω_p. Show that it agrees with the equation $\Omega_p = \Gamma/L$ for all values of ϕ.

11-14. A thin rod of mass m_1 is freely pivoted at its center. A mass m_2 is attached halfway between the pivot and one end of the rod. Find the angular acceleration when the rod is horizontal.

11-15. A wooden meter stick is pivoted at one end and has a small cup containing a ball fastened to its upper surface. If this cup is at the extreme free end of the meter stick, which is released from a horizontal position, the end of the meter stick accelerates so rapidly that the ball is left behind and falls out of the cup. Where along the meter stick can the cup be fastened so that the ball just stays in the cup as the stick is released?

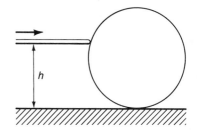

Fig. 11-31.

11-16.* A rapidly spinning disk is dropped vertically onto a smooth, horizontal surface. Its axis of rotation is horizontal, so that after it hits the surface it begins to roll. If it has a mass of 2 kg, a radius of 10 cm, and if μ_K between the disk and surface equals 0.1, what is the value of its initial acceleration? Assume no bouncing.

11-17.* A billiard ball is struck at a height h by a horizontally directed force exerted by a cue stick, as in Fig. 11-3. Find the value of h required so that the ball begins rolling without slipping for (a) μ between ball and table = 0; (b) μ between ball and table = 0.1.

11-18. A solid sphere is placed on a plane inclined 20° above the horizontal. What is the minimum value of the coefficient of friction between the ball and the plane to permit rolling without slipping?

Chapter **12**

Statics

12-1 Equilibrium Statics concerns itself with the interplay of forces in a structure at rest. It may seem a bit strange to have postponed a discussion of stationary systems until after we have discussed dynamics. Historically, statics evolved a bit ahead of dynamics in the time of Leonardo da Vinci (1452–1519) and Galileo (1564–1642). Nonetheless, it is quite difficult to truly understand why a system does *not* move until we understand what *does* cause motion. Since a body at rest has constant velocity ($v = 0$ always), its acceleration also vanishes. Accordingly, since the acceleration of a body is produced by the sum of all forces acting upon it, the *net* force acting upon it must also vanish. Thus:

$$\Sigma_i \mathbf{F}_i = m\mathbf{a} = \mathbf{0} \qquad (12\text{-}1)$$

for a stationary body. When the forces upon a body balance, we say it is in *equilibrium*. Equation (12-1) is called the *first condition for equilibrium*.

It is possible, however, for a body to have zero net force acting upon it and yet be subject to a nonzero torque. An example is shown in Fig. 12-1. Though the center of mass of the body does not accelerate, the body experiences an angular acceleration. Thus, a *second condition for equilibrium* is that all torques acting upon a body must balance:

$$\Sigma_i \mathbf{\Gamma}_i = \mathbf{0} \qquad (12\text{-}2)$$

If both conditions (12-1) and (12-2) are satisfied, we can be sure the body is in equilibrium. Conversely, if we *observe* a body at rest, the conditions (12-1) and (12-2) *must* apply.

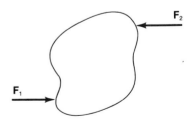

Fig. 12-1. If $\mathbf{F}_1 = \mathbf{F}_2$ the first condition for equilibrium is met, but a net torque acts.

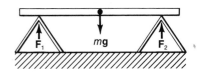

12-2   Application of the Equilibrium Conditions

If conditions (12-1) and (12-2) are satisfied, the object is in equilibrium and, once at rest, will remain at rest. As an example, consider a plank supported on two sawhorses, as in Fig. 12-2. Three forces act upon the plank: its weight, $m\mathbf{g}$, and the forces supplied by the supports, $\mathbf{F}_1$ and $\mathbf{F}_2$. Friction between the supports and the plank could provide horizontal components to $\mathbf{F}_1$ and $\mathbf{F}_2$, but we will consider the forces to be vertical. Intuitively, it seems that for the symmetric situation shown, equilibrium occurs when the load is split equally between the supports, i.e., $F_1 = F_2 = mg/2$. Let us apply the two conditions for equilibrium to check this guess. First, the sum of the two upward-directed forces $\mathbf{F}_1$ and $\mathbf{F}_2$ must equal the downward force of gravity, as can be seen by application of Eq. (12-1). Since $\mathbf{F}_1$ and $\mathbf{F}_2$ are directed oppositely to $m\mathbf{g}$,

Fig. 12-2. A uniform plank supported by two symmetrically placed sawhorses.

$$F_1 + F_2 - mg = 0$$

Hence
$$F_1 + F_2 = mg \qquad (12\text{-}3)$$

This last result, (12-3), gives us only the *sum* of F_1 and F_2. We have to apply Eq. (12-2) in order to obtain values for F_1 and F_2 separately. The *line of action* of the weight of the plank may be taken through its center of gravity, as discussed in the preceding chapter. Let each support be a distance l from the center of mass of the plank. Then the torques due to the three forces can be evaluated as follows. First, choose the left-hand support or any convenient point to be the origin of a coordinate system, as shown in Fig. 12-3. Let us take torques into the page as positive, those out of the page as negative. Then the torques due to each of the three forces are:

$$\mathbf{\Gamma}_1 = \mathbf{r}_1 \times \mathbf{F}_1 \qquad \Gamma_1 = r_1 F_1 \sin\theta = x_1 F_1 = 0 \qquad (\text{since } x_1 = 0)$$

$$\mathbf{\Gamma}_2 = \mathbf{r}_2 \times \mathbf{F}_2 \qquad \Gamma_2 = r_2 F_2 \sin\theta = x_2 F_2 = -2lF_2$$

$$\mathbf{\Gamma}_3 = \mathbf{r}_{\text{cm}} \times m\mathbf{g} \qquad \Gamma_3 = lmg$$

So the torque due to $\mathbf{F}_1$ is zero (it has a lever arm of zero length) and the torques due to the weight of the plank, $m\mathbf{g}$, and the force $\mathbf{F}_2$ are oppositely directed. Substituting our results into equation (12-2):

$$\Sigma_i \mathbf{\Gamma}_i = 0$$

$$2lF_2 - lmg = 0$$

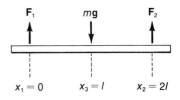

hence
$$F_2 = \frac{mg}{2} \qquad (12\text{-}4)$$

Fig. 12-3. Coordinates used in evaluating the torques acting in Fig. 12-2.

Now substituting (12-4) back into (12-3) gives

$$F_1 + \frac{mg}{2} = mg$$

hence
$$F_1 = \frac{mg}{2}$$

Our original guess that F_1 and F_2 are each equal to one-half the total weight is thus confirmed. This would not have been the case if the supports were not equally spaced from the c.g.

Application of (12-2) does not, at first glance, seem as straightforward as applying (12-1). The difficulty is that any one of the torques Γ_i that enters into (12-2) depends upon the particular coordinate system chosen, since the distance from the point of application of a force to the origin of the coordinates enters directly into the definition of a torque:

$$\mathbf{\Gamma} = \mathbf{r} \times \mathbf{F}$$

However, the *results* obtained from (12-2) are *independent* of our choice of coordinates. As an example, measure the distances from the center of gravity of the plank rather than from the left end in the previous example. Now the torque due to the weight of the board vanishes, since its lever arm is zero, and the torques due to $\mathbf{F}_1$ and $\mathbf{F}_2$ are equal but oppositely directed:

$$\Gamma_1 = +F_1 l \quad \text{(directed into the page)}$$
$$\Gamma_2 = -F_2 l \quad \text{(directed out of the page)}$$
$$\Gamma_3 = 0$$

Applying (12-2), $F_1 l - F_2 l = 0$, i.e.,

$$F_1 = F_2$$

Since we have $F_1 + F_2 = mg$ from (12-3), we obtain:

$$F_1 = F_2 = \frac{mg}{2}$$

which agrees with our previous result. In general, the point chosen for an origin in the evaluation of torques may be anywhere inside or outside of the system without changing the result obtained by application of (12-1) and (12-2). Hence we usually pick the origin at the point of application of one of the forces to eliminate one term in the evaluation of (12-2).

Example A diving board is supported by two pipe supports, as sketched in Fig. 12-4. These pipes are clamped to the board so that

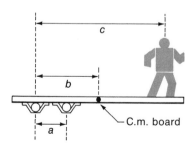

Fig. 12-4.

they can supply either an upward- or a downward-directed force. The supports are separated by a distance a, the center of mass of the board is a distance b from the rear support, and a boy is a distance c from the rear support. If the weight of the board is **W** and that of the boy is **B**, find the forces supplied by the two supports at equilibrium.

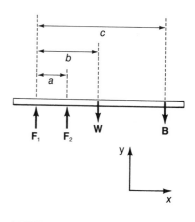

Fig. 12-5.

Solution   Let us label the various forces acting on the board as in Fig. 12-5. Applying (12-1):

$$F_{1x} + F_{2x} = 0$$

$$F_{1y} + F_{2y} - W - B = 0 \qquad (12\text{-}5)$$

The first equation for the horizontal components of the support forces simply states that they must be equal and opposite. Such force components would exist if, for example, either clamp connecting the board to the pipe supports were not centered properly over the pipe. Usually these horizontal components are quite small.

The second equation (12-5), for the y components of the forces contains the more interesting information concerning the relation of the vertical support-force components to the weight of the board and boy. However, it can only give us the *sum* of F_{1y} and F_{2y}, not the value of these two forces separately. When we measure lever arms from the left-hand pipe support, application of (12-2) gives us:

$$-F_{2y}a + Wb + Bc = 0$$

where torques directed into the page are chosen to be positive. Solving for F_{2y}:

$$F_{2y} = (b/a)W + (c/a)B \qquad (12\text{-}6)$$

Our result in (12-5) can now be used to find F_{1y}:

$$F_{1y} = W + B - F_{2y}$$

$$= \left(1 - \frac{b}{a}\right)W + \left(1 - \frac{c}{a}\right)B$$

Since both b and c are larger than a, we have found that F_{1y} is negative, that is, it points in the opposite direction from that assumed in the above sketch. This makes good sense, since the weight of the board + boy is beyond the right-hand pipe support, so that a downward force is required at the left-hand pipe support. In the absence of such a downward-directed force, the board would simply pivot about the right-hand pipe support with the right end of the board falling, the left end rising. Take some representative figures for the weights and dimensions in this example; let us say $W = 100$ lb, $B = 120$ lb, $a = 4$ ft, $b = 6$ ft, and $c = 10$ ft. Then we have:

$$F_{2y} = (b/a)W + (c/a)B = \tfrac{3}{2}W + \tfrac{10}{4}B$$

$$= 150 \text{ lb} + 300 \text{ lb} = 450 \text{ lb} \qquad \text{(upward)}$$

$$F_{1y} = \left(1 - \frac{b}{a}\right)W + \left(1 - \frac{c}{a}\right)B = -\tfrac{1}{2}W - \tfrac{3}{2}B$$

$$= -50 \text{ lb} - 180 \text{ lb} = -230 \text{ lb} \qquad \text{(downward)}$$

Example   A signboard of length l is held in place by a cable and a support, as shown in Fig. 12-6. The center of mass of the board is at its center. If the board weighs 80 lb and $\theta = 30°$, find the tension **T** in the cable.

Solution   The support must supply an upward force to help hold up the sign, but it also supplies an outward-directed force, since the tension in the cable tends to pull the sign into the wall. We label the vertical component of this force F_y and the horizontal component, F_x. A diagram showing all forces acting upon the sign is sketched in Fig. 12-7. Such a sketch is helpful in solving a statics problem since it helps us to see clearly which forces act upon the object and in which direction they act. Applying Eq. (12-1), we obtain

$$F_y + T \sin \theta - W = 0 \qquad (12\text{-}7)$$

$$F_x - T \cos \theta = 0 \qquad (12\text{-}8)$$

Applying (12-2) and using the wall support as a pivot point, we find

$$W(l/2) - lT \sin \theta = 0 \qquad (12\text{-}9)$$

Solving (12-9) for $T \sin \theta$ we obtain

$$T \sin \theta = W/2$$

Substituting this result into (12-7):

$$F_y = W - T \sin \theta = W - W/2 = W/2 = 40 \text{ lb}$$

Thus we have found that half the weight of the sign is carried by the vertical component of the wall force, the other half by the vertical component of the cable tension. The tension is found easily from $T \sin \theta = W/2$ or

$$T = \frac{W}{2 \sin \theta} = 80 \text{ lb}$$

and F_x is found from (12-8):

$$F_x = T \cos \theta = \frac{W}{2} \cdot \frac{\cos \theta}{\sin \theta} = \frac{W}{2} \cot \theta = 69.3 \text{ lb}$$

Reviewing the procedures followed in the preceding examples:

1.   A force diagram is drawn.
2.   Both conditions for equilibrium are applied. Incorrect guesses as to the direction of applied forces are unimportant, since the solution will give a minus sign for such wrong guesses. How-

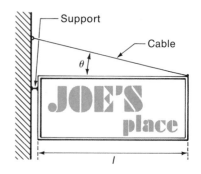

Fig. 12-6.

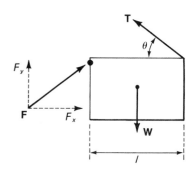

Fig. 12-7.

ever, it is important that all possible forces acting upon the object be taken into account, and that signs be treated with care. In applying (12-2), any convenient point may be taken as the origin of the coordinates.

3. Since the equilibrium-condition equations are vector equations, several scalar equations usually result. Generally, no one of these equations provides the solution desired, but solving them as a system of simultaneous equations for several unknowns provides the desired results.

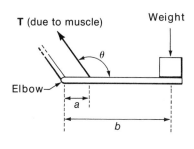

Fig. 12-8.

Exercises

1. Consider the boy on the diving board in the first example of this chapter. Using the numerical values suggested at the end of that example, find where the boy should stand in order to allow F_1 to vanish.

2. When a weight is held in the hand, the muscles of the upper arm provide the necessary tension so that the forearm is static and does not pivot about the elbow joint. The sketch in Fig. 12-8 illustrates the essence of the situation when a weight is held with the forearm horizontal. Ignoring the weight of the forearm, find the required muscular tension **T** in terms of three variables: the angle θ, the distance a from the elbow to the point of attachment of the muscles' tendons, and the distance b between the elbow and the weight. How can you take the weight of the forearm itself into account? Making reasonable assumptions, find **T** when the weight weighs 20 lb.

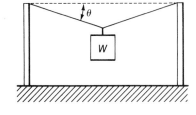

Fig. 12-9.

3. A clothesline rope is stretched between two posts, as shown in Fig. 12-9. If the load W weighs 20 lb and the angle θ made by both sides of the rope with respect to the horizontal is 10°, find the tension **T** in the rope. Ignore the weight of the clothesline.

12-3   Frictional Forces

In many static systems, some of the forces that act are frictional forces. An obvious example is that of a ladder leaning at an angle against a wall. We know that the ladder will slip and fall flat if the surfaces involved are too slippery. There is nothing inherently new about the effects of frictional forces when they are applied in the equilibrium conditions (12-1) and (12-2), but since the maximum magnitude of a frictional force depends upon the normal force, there is an extra degree of complexity in the analysis of such systems.

For simplicity consider only one frictional force to act in the situation shown in Fig. 12-10. Such is the case if the upper contact

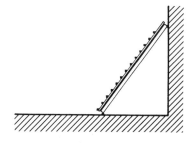

Fig. 12-10. Equilibrium of a ladder propped against a wall requires a frictional force at one or both of its points of contact.

of the ladder with the wall is frictionless ($\mu = 0$) while the coefficient of friction between the ladder and the floor is nonzero. A force diagram is sketched in Fig. 12-11. The force of the floor on the ladder is labeled $\mathbf{F}_f$ and that of the wall on the ladder is $\mathbf{F}_w$. The force $\mathbf{F}_f$ has a horizontal component $\mathbf{H}$ due to the frictional force. However, this frictional force cannot exceed μN, where N is the vertical component of $\mathbf{F}_f$. The force $\mathbf{F}_w$, on the other hand, has no frictional component and is thus normal to the wall. Condition (12-1) for the vertical force components becomes $N - mg = 0$, where m is the mass of ladder, or

$$N = mg$$

This result states that the normal force exerted by the floor must equal the weight of the ladder, as indeed it must since no other upward-directed forces are available to support the ladder's weight. For the horizontal components:

$$H - F_w = 0$$

and since $H \leq \mu N$, we have

$$\mu N \geq F_w$$

Substituting $N = mg$, this becomes

$$\mu mg \geq F_w \qquad (12\text{-}10)$$

This last inequality simply indicates that μ *cannot* be arbitrarily small. As our own observations indicated earlier, if the surface at the base of the ladder is too slippery, it will not remain static.

To apply condition (12-2) to the torques about the bottom end of the ladder, which is of length l, we need the torque due to weight of ladder, Γ_L, and that due to the force of the wall, Γ_w:

$$\Gamma_L = \mathbf{r} \times m\mathbf{g}$$

$$= \tfrac{1}{2}mgl \sin \theta$$

where θ is the angle between $\mathbf{g}$ and the lever arm. The angle θ is complementary to the angle ϕ shown in Fig. 12-11, so we may write:

$$\Gamma_L = \tfrac{1}{2}mgl \cos \phi$$

Similarly evaluating Γ_w, we have:

$$\Gamma_w = lF_w \sin \phi$$

These torques act in opposition to one another (Γ_L points into the page, Γ_w out), so condition (12-2) becomes:

$$\tfrac{1}{2}mgl \cos \phi - lF_w \sin \phi = 0$$

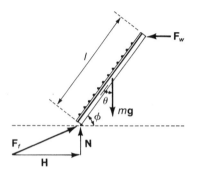

Fig. 12-11. Three forces act upon the ladder of Fig. 12-10. The force $\mathbf{F}_f$ exerted by the floor is shown resolved into horizontal and normal components.

Solving for F_w we obtain

$$F_w \sin \phi = \tfrac{1}{2}mg \cos \phi$$

$$F_w = \tfrac{1}{2}mg\frac{\cos \phi}{\sin \phi} = \tfrac{1}{2}mg \cot \phi$$

Substituting this result into (12-10), we find

$$\mu mg \geqslant \tfrac{1}{2}mg \cot \phi$$

or
$$\mu \geqslant \tfrac{1}{2}\cot \phi$$

We have thus obtained a condition on the coefficient of friction, μ, that must be satisfied if the system is to be in equilibrium. The result is intuitively satisfying since it states that (a) μ can be arbitrarily large but must not be less than a certain quantity; (b) as ϕ becomes smaller (the ladder approaches the horizontal), this minimum value of μ must get larger (since $\cot \phi \rightarrow \infty$ as $\phi \rightarrow 0$), or else the ladder will fall. Since most surfaces have $\mu < 1$, the ladder will generally have to be placed at an angle steeper than 27° for stability, unless a cleat is provided on the floor at the foot of the ladder or, equivalently, if the foot of the ladder is dug into soft earth.

All of the foregoing has assumed a vanishing coefficient of friction at the upper end of the ladder, so we expect that in most actual uses of ladders, the angle can be somewhat smaller than these results would dictate.

Exercises

4. Would the above results for the leaning ladder be changed if a man stood halfway up the ladder?

5. In the leaning ladder example, which equilibrium equations would be changed by the addition of friction at the upper end of the ladder?

6. A rope is attached to a 20-kg plank, as shown in Fig. 12-12. The rope is horizontal and attached to the plank 3/4 of the way up from the bottom. If $\phi = 75°$, (a) find the minimum value of the coefficient of friction at the base of the plank for equilibrium; (b) find the tension in the rope at equilibrium.

12-4   Summary

Two conditions must be met before an object can be said to be in equilibrium.

1. The forces must balance: $\Sigma_i \mathbf{F}_i = \mathbf{0}.$ (12-1)

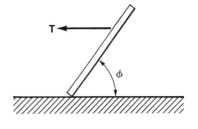

Fig. 12-12.

2.   The torques must balance: $\Sigma_i \Gamma_i = 0.$ (12-2)

When a rigid body is observed to be at rest, conditions (12-1) and (12-2) must hold. Though the magnitude of each torque in (12-2) depends on the chosen coordinate origin (or pivot point), the *results* of application of (12-2) and (12-1) are unchanged by this choice.
    An effective procedure for solving statics problems is:

1.   Sketch the system.
2.   Draw a force diagram including all possible forces. Label all relevant forces and lengths. Do not worry about your choice of direction for a force in your diagram: if your choice was incorrect, a minus sign in your solution will indicate this.
3.   Apply both conditions for equilibrium. This will generally lead to *several* equations involving *several* unknowns. Solving this system as a set of simultaneous equations will give the unknowns in terms of the knowns, if you began with sufficient information.

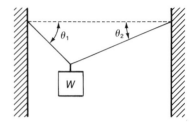

Fig. 12-13.

Problems

12-1.   A weight is suspended off-center on a clothesline, as shown in Fig. 12-13. The tension in the left side of the line, T_1, is not equal to that in the right side of the line, T_2. For $\theta_1 = 20°$, $\theta_2 = 10°$, find T_1 and T_2.

12-2.   A shelf of negligible mass projects 2 m out from a wall. It is firmly fastened to the wall. A strut, as sketched in Fig. 12-14, is attached to the shelf at an angle $\theta = 30°$. A 500-kg mass is located with its center of mass a distance x from the wall. Find the force of compression on the strut as a function of x.

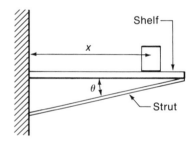

Fig. 12-14.

12-3.   A wire supports a uniform beam, as sketched in Fig. 12-15. The mass of the beam is 130 kg. How much mass can be hung at the end of the beam without exceeding the strength of the wire if this strength is 2800 N?

12-4.   A 15-ft ladder with a weight of 50 lb is propped against a wall. The ladder is at an angle of 60° above the horizontal. If the coefficient of friction between the ladder and wall is negligible, and that at the foot of the ladder equals 0.3, how far up the ladder can a 160-lb man climb before the ladder slips?

12-5.* A man pushes horizontally against the side of a refrigerator at a distance x above the floor in order to slide it across the floor. The coefficient of friction is $\mu = \frac{1}{2}$; the refrigerator has

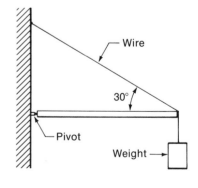

Fig. 12-15.

height h and width w. Assuming that the refrigerator's center of gravity is at its geometrical center, find the maximum value of x so that the refrigerator does not tip as it is pushed.

12-6.* A horizontal rod is held against a wall by a string, as shown in Fig. 12-16. Friction between the rod and wall prevents the rod from falling. If the coefficient of friction, μ, is just large enough to provide equilibrium, find μ in terms of θ.

12-7. Two arms are pivoted at their base, as shown in Fig. 12-17. A block of 50-kg mass is pinched between them, as shown. Find the minimum possible value of the coefficient of friction between the arms and the block that will keep the block from slipping for the following two cases: (a) the arms have negligible weight compared to the block; (b) each arm has 20-kg mass.

12-8.* A series of bricks are stacked as sketched in Fig. 12-18. Each brick has a length l. (a) What arrangement of bricks causes the top brick to project as far as possible beyond the table edge? (b) Show that with a sufficient number of bricks, it is possible to get the center of gravity of the top brick beyond the edge of the table. (c) What is the minimum number of bricks required?

12-9. Two painters, each weighing 160 lb, stand on an 80-lb plank supported by two ropes, as in Fig. 12-19. One of them is $2\frac{1}{2}$ ft from the left end, the other is in the center. Find the magnitude of the tension in each rope and the angle θ_2.

12-10. A box with a height equal to its width is placed on a flatbed truck. Small cleats are nailed to the truck bed around the edge of the box so that it cannot slide. It can, however, tip. Find the value of acceleration of the truck for which the box just begins to tip. Once it begins to tip, will it continue

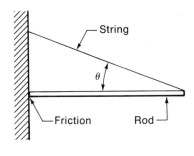

Fig. 12-16.

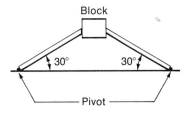

Fig. 12-17.

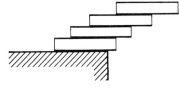

Fig. 12-18.

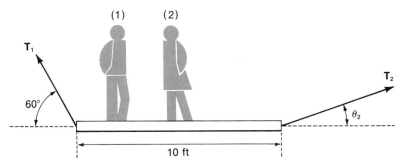

Fig. 12-19.

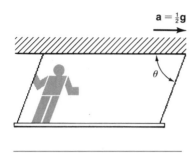

$a = \frac{1}{2}g$

θ

Fig. 12-20.

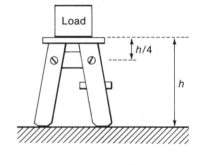

Load

h/4

h

Fig. 12-21.

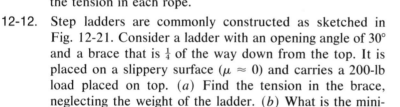

45°   45°

Rope

Fig. 12-22.

over onto its side or will it come to a new equilibrium position if the acceleration is maintained at the critical value that just causes tipping?

12-11.* A 90-kg man stands 1 m from the left end of a 4-m, 45-kg plank that is suspended as in Fig. 12-20. The support is accelerating toward the right with a uniform acceleration of $\frac{1}{2}$ **g**, and the system is in equilibrium (θ = constant). Find the tension in each rope.

12-12. Step ladders are commonly constructed as sketched in Fig. 12-21. Consider a ladder with an opening angle of 30° and a brace that is $\frac{1}{4}$ of the way down from the top. It is placed on a slippery surface ($\mu \approx 0$) and carries a 200-lb load placed on top. (*a*) Find the tension in the brace, neglecting the weight of the ladder. (*b*) What is the minimum value of μ so that this brace is not needed?

12-13. A 20-lb rope hangs from a ceiling, as in Fig. 12-22. (*a*) Find the tension in the rope at each of its points of support. (*b*) Find the tension in the rope at its lowest point.

12-14. A rectangular door of uniform thickness is supported by two symmetrically placed hinges that are separated by a distance equal to the width of the door, as in Fig. 12-23. Find the magnitude and direction of the horizontal components of the forces exerted on the door by the hinges.

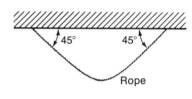

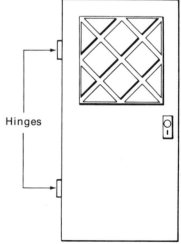

Hinges

Fig. 12-23.

Chapter 13

Properties of Materials

13-1 Introduction In the preceding chapter we developed techniques to evaluate the forces that a static structure must endure. In this chapter we will describe some properties of materials that are utilized in the fabrication of structures. The ultimate strength, weight, flexibility, and reliability of structures depend on these properties. In the course of our discussion, we will search for properties that depend only upon the material itself, not the particular shape or size in which it is fabricated. We will discover that some properties that are shared by all materials have a common underlying explanation in terms of the atomic structure of these materials. Thus, we will see an example of a situation that occurs frequently in the natural sciences. Our search for knowledge of a practical sort needed in the design of structures leads us to a deeper understanding of the nature of matter itself. It should be pointed out that the inverse process — in which an initially basic, nonapplied line of research leads to knowledge of an immediately useful form — is a frequent occurrence as well.

13-2 Density

One of the properties that characterize any given material is its *density, ρ*. Density is defined as the *mass per unit volume* of a substance:

$$\rho = m/V$$

In the metric system of units, the mass scale was originally chosen so that one cubic centimeter of water has a mass of one gram. Thus in the cgs system of units, the density of water is 1 g/cm³. In mks units, the density of water is

$$\rho_w = 1 \ \frac{g}{cm^3} \cdot \frac{10^{-3} \ kg}{g} \cdot \frac{10^6 \ cm^3}{m^3} = 10^3 \ \frac{kg}{m^3}$$

Oftentimes the density of a substance is put in ratio to that of water. This defines the *specific gravity:*

$$\text{specific gravity} = \frac{\rho}{\rho_w}$$

The specific gravity is thus a dimensionless number whose numerical value equals that of the density in cgs units.

The densities of various solids and liquids vary over only a limited range, as seen in Table 13-1. On an atomic scale, these substances are characterized by the fact that their atoms are nearly in contact. For this reason, both solids and liquids are referred to as *condensed matter.*

TABLE 13-1   Densities of Various Liquids, Solids, and Gases

Substance	Density g/cm³	Density kg/m³	lb/ft³
Water	1.00	1.00 × 10³	62.4
Kerosene	0.82	8.2 × 10²	51.2
Glycerine	1.26	1.26 × 10³	78.7
Aluminum	2.7	2.7 × 10³	169
Copper	8.8	8.8 × 10³	549
Iron	7.85	7.85 × 10³	490
Gold	19.3	1.93 × 10⁴	1205
Air*	1.29 × 10⁻³	1.29	0.0805
Argon*	1.78 × 10⁻³	1.78	0.111
Hydrogen*	8.99 × 10⁻⁵	8.99 × 10⁻²	0.00561
Xenon*	5.85 × 10⁻³	5.85	0.365

*At standard atmospheric pressure and at the temperature of the freezing point of water.

The densities of gases listed in Table 13-1 are seen to be roughly 1/1000 of those for condensed matter. At the atomic level this implies a larger average spacing (roughly 10 times larger) between atoms in a gas. Because the density of gases depends strongly upon pressure and temperature, the table values are given at a specified temperature and pressure. The temperature and pressure indicated in the table is called *standard temperature and pressure,* abbreviated **STP**. We will discuss the behavior of gases under changing conditions of temperature and pressure more fully in Chapter 20.

Exercises

1. A solid spherical iron pellet has a diameter of 2 mm. Find its mass in grams.
2. Estimate the weight, in lb, of the air contained in an average living room.

13-3 Atomic Structure of Matter

The quantity of mass contained in individual atoms varies in discrete steps from that of the atoms of the lightest element, hydrogen, to that of the atoms of the heaviest element, uranium.* Accordingly, an atomic mass scale can be devised in which the mass of a single hydrogen atom is taken to be unity and successively heavier elements have masses that are nearly integer multiples of this basic element, e.g., helium-4, lithium-6, carbon-12, etc. For reasons involving experimental convenience, the presently accepted SI atomic mass scale defines the atomic mass unit as follows:

> An atom of the most common isotope of carbon, carbon-12 (^{12}C), is considered to have a mass of exactly 12 atomic mass units (amu). One amu thus equals 1/12 the mass of a single carbon-12 atom.

The connection between this microscopic mass scale, the amu, and our macroscopic scale involving g or kg is obtained through *Avogadro's number, N_0*:

$$N_0 = 6.02 \times 10^{23}$$

This is the number of carbon-12 atoms contained in 12 grams of carbon-12. It is also, then, approximately equal to the number of atoms in one gram of hydrogen-1, 6 grams of lithium-6, etc. The masses of these atoms are not *precisely* divisors of the mass of carbon-12 because of effects on the order of 1% caused by strong nuclear binding energies and the Einstein mass-energy principle.

---

*Uranium is the heaviest naturally occurring element found on earth. Heavier elements can be produced in the laboratory by nuclear reactions and may also be produced in stellar explosions. These heavier elements are radioactive and have limited lifetimes. See, for example, "The Synthetic Elements: IV" by Glenn T. Seaborg and Justin L. Bloom *Scientific American,* April 1969.

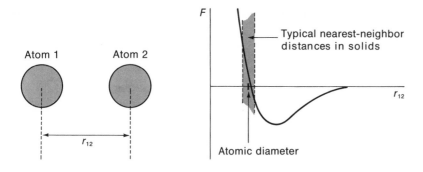

Fig. 13-1. Typically, inter-atomic forces are repulsive for small separation distances, attractive for distances comparable to atomic size, and negligible for large distances.

Vanadium, for example, has an atomic mass equal to 50.94 amu. Thus 50.94 g of vanadium contains N_0 atoms. In the SI system of units, Avogadro's number has been taken as the unit of *amount of substance,* the *mole:*

> The mole is the amount of substance of a system that contains as many elementary entities as there are atoms in 0.012 kg of carbon-12.

Though atoms do not have sharp edges, the bulk of their matter is contained within a reasonably well defined volume. It is thus possible to speak approximately of the radius of an individual atom. These radii depend somewhat upon the specific chemical element to which the atom belongs, with a typical radius being approximately equal to 10^{-10} m.

Atoms interact with other atoms via electromagnetic forces, discussed in Chapter 28. All atoms attract one another at separation distances of a few atomic radii. At small separations, when the atoms begin to interpenetrate one another, the interatomic force becomes repulsive. In Fig. 13-1 we see a representative graph of the magnitude of the interatomic force as a function of separation distance, where a negative force indicates attraction, a positive force repulsion. By Newton's third law, the forces felt by each of the interacting pair of atoms are equal and oppositely directed.

When a group of atoms that strongly attract one another are placed together, they move under the influence of their mutual interatomic forces until no *net* forces act upon any of them. Because the forces from *nearest-neighbor* atoms are strongest, the typical equilibrium distances between nearest neighbors are such as to minimize the nearest-neighbor forces, as indicated in Fig. 13-1.

As we will see in more detail in Chapter 20, there is a chaotic jiggling motion of the atoms in a substance, a motion that increases with increasing temperature. This *thermal agitation* tends to disrupt the tendency of atoms to occupy precise equilibrium positions with respect to nearest neighbors. For sufficiently low temperatures, however, individual atoms can permanently remain in equilibrium position with only a slight jiggling motion about their average location due to thermal agitation. In this situation, the entire group of atoms remains rigidly locked in position and we call the substance a *solid*. Raising the temperature can increase thermal agitation to the point that atoms no longer remain in a rigid array but still remain within the range of interatomic forces. This state of matter is called the *liquid state*. A further increase in thermal agitation can cause atoms to move beyond the range of interatomic forces and into the *gaseous state*.

It seems that atoms in a solid will seek locations at a distance from each nearest-neighbor atom such that all interatomic forces are zero. This, however, is impossible, since some of the atoms, such as next-nearest neighbors, are bound to be far enough away to produce attraction if others are close enough to produce no net force. Instead, the atoms move about until they find a happy medium between all the competing forces where there is no *net* force on any given atom. This usually results in a crystalline structure, a regular, periodic arrangement of the atoms that is most stable for the particular collection of atoms. An example is the cubic lattice structure of ordinary salt shown in Fig. 13-2. Some solids form large crystals that reveal the basic symmetry of their lattice structure, but most form a jumbled array of microcrystals that are not easily observable. Such is the case with most metals, for example. An example is shown in Fig. 13-3.

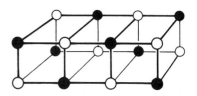

Fig. 13-2. The crystalline structure of sodium chloride (NaCl). The sodium atoms are shown as open circles, the chlorine atoms as solid circles. This particular structure is called a *cubic lattice*.

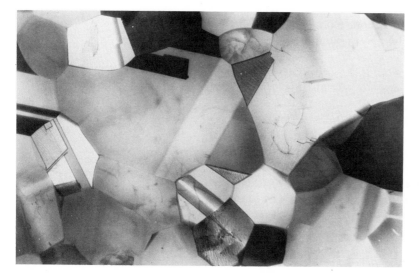

Fig. 13-3. An electron micrograph of fine-grain stainless steel at a magnification of 10,000× shows a jumbled array of microcrystals. The irregular lines in some of the crystals are imperfections called *dislocations*. (Courtesy U.S. Steel)

Some amorphous materials such as tar, plastics, and glass more closely resemble an extremely viscous liquid than a true solid. These substances tend to have a poorly defined melting point, simply softening more and more as they are heated, while crystalline solids maintain a rigid structure until thermal agitation breaks their atomic bonds. The "true" solids thus melt abruptly at a well-defined temperature. As rigid as glass appears to us at room temperature, it nonetheless flows or "creeps" appreciably so that a pane of glass that stands vertically becomes measurably thicker at the bottom over a period of years. A crystalline solid, such as a metal, does not exhibit large "creeping" behavior until it is subjected to rather strong forces.

When we stretch a macroscopic object, we cause each pair of atoms along the direction of stretching to separate beyond their normal separation distance. The total of their interatomic forces is felt as a force that pulls back in the direction that will restore the solid to its original dimensions. Compression similarly displaces atoms from their resting position and causes a force that acts to restore the original shape of the solid.

As can be seen in Fig. 13-1, the approximate range at which atoms bind into solids is one where the interatomic force between nearest neighbors is very nearly a linear function of distance. Accordingly, the force exerted by a deformed solid obeys the linear relationship shown in Fig. 13-4. There we consider the force exerted by a sample of material that is stretched or compressed. The variable x measures the extent of departure from its equilibrium length, with positive x representing stretching and negative x representing compression. The negative force for positive x indicates that the force exerted by the bar acts in a direction to reduce x back to zero (equilibrium). This behavior can be represented by a linear equation known as *Hooke's law:*[*]

$$F = -kx \qquad (13\text{-}1)$$

where k is the *force constant* or *stiffness* of the bar. For sufficiently large forces, the interatomic forces depart from linearity; and for even larger forces, the microcrystals slip past one another, producing permanent deformation. Still greater forces rupture the material. A large value of k indicates that large forces are required to produce even a small deformation. A small value of k indicates an easily deformed substance.

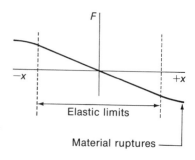

Fig. 13-4. The magnitude of a force **F** exerted by a deformed solid is linearly proportional to the extent of deformation up to its elastic limit.

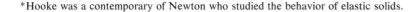

[*]Hooke was a contemporary of Newton who studied the behavior of elastic solids.

Exercises

3. The atomic mass of cobalt is 58.93 amu. How many cobalt atoms are in a 4-g sample of cobalt?

4. Find the mass, in grams, of a single atom of carbon-12 from the definitions of the amu and the mole.

13-4   Stress, Strain, and Elastic Moduli

The deformation produced by a force acting on a sample of material depends upon the size and shape of the sample. Thus a slender iron wire is more readily stretched, and has a lower force constant, k, than a large iron bar. If we wish to find a quantity that describes the rigidity of a particular material, independent of the size of the sample, we must take into account the expected variation of k with sample size.

Examination of Fig. 13-2 provides a clue as to the way in which k varies with sample size. If the sample has a large cross section, the number of atomic bonds that share the load is large, so that each atomic bond carries a smaller share of the load. Since the number of atomic bonds involved is proportional to the area of the sample and each bond exhibits the nearly linear behavior indicated in the shaded region of Fig. 13-1, we expect the deformation produced to depend on the applied force *per unit area*.

A solid object may have forces acting on it in such a way as to produce stretching, bending, or a uniform decrease in volume. In general, the force per unit area that produces the deformation is called the *stress*, while the fractional change in shape that occurs is called the *strain*. The ratio of stress to strain for a given material is called its *elastic modulus* and is a characteristic of the *material*, not its shape.

Young's modulus, Y, is defined to be the value of the force per unit area that produces a stretching or compression of a rod of the material divided by the fractional change produced:

$$Y = \frac{F/A}{\Delta L/L} \qquad N/m^2 \qquad (13\text{-}2)$$

Example   A 1-m-long wire of 1 mm diameter is observed to stretch 0.5 mm when a 2-kg weight is hung on it. Find Young's modulus for the material composing the wire.

Solution   The cross-sectional area of the wire is $A = (\pi/4)d^2 = 7.85 \times 10^{-7}$ m², while the force is $W = mg = 19.6$ N. The fractional change in length is

$$\Delta L/L = \frac{0.5 \text{ mm}}{1000 \text{ mm}} = 5 \times 10^{-4}$$

Thus $Y = \dfrac{F/A}{\Delta L/L} = \dfrac{19.6}{(5 \times 10^{-4})(7.9 \times 10^{-7})} \text{ N/m}^2 = 5 \times 10^{10} \text{ N/m}^2$

In the above example, note that the units of Young's modulus are those of a force divided by an area, since the quantity $\Delta L/L$ in the denominator of the definition is dimensionless. In general, the various elastic moduli we define will have these dimensions.

When a force per unit area acts over the entire surface of a block of material, the volume of the block is reduced. The value of the force per unit area is called the *pressure*. The change in pressure divided by the accompanying fractional change in volume, $\Delta V/V$, defines the *bulk modulus, B*:

$$B = \frac{-\Delta p}{\Delta V/V} \qquad \text{N/m}^2 \quad \text{force/area} \qquad (13\text{-}3)$$

where the minus sign indicates that an increase in pressure produces a decrease in volume.

Fig. 13-5 illustrates a *shear* deformation produced by equal and oppositely directed forces applied to the top and bottom faces of a block. The top and bottom of the block each has an area A. The deformation produced is measured by $\Delta L/L$. For small angles the deformation $\Delta L/L$ equals the *shear angle, ϕ*. The *shear modulus* is defined by

$$S = \frac{F/A}{\Delta L/L} \qquad \text{N/m}^2 \quad \text{force/area}$$

In each of the three preceding definitions, the numerator contains a force divided by an area while the denominator contains a dimensionless ratio. Thus, the elastic moduli all have the dimensions of force/area. Table 13-2 gives representative elastic moduli for several different materials.

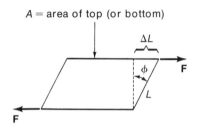

$A = $ area of top (or bottom)

Fig. 13-5. Shear deformation produced by shearing forces.

TABLE 13-2   Elastic Moduli for Several Materials*

	Y (N/m²)	S (N/m²)	B (N/m²)
Aluminum	7×10^{10}	2.4×10^{10}	7.0×10^{10}
Brass	9.0×10^{10}	3.5×10^{10}	6.1×10^{10}
Steel	20×10^{10}	8.1×10^{10}	16×10^{10}
Glass	7×10^{10}	3×10^{10}	5×10^{10}

*See, for example, *Machinery's Handbook* by Erik Oberg and F. D. Jones: New York, The Industrial Press (1962).

(a)

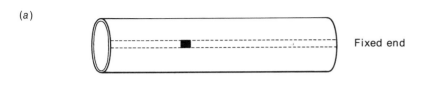

Fixed end

Fig. 13-6. A thin-walled tube with radius R is twisted by an applied torque Γ. The square element shaded in (a) is deformed into a parallelogram in (b) by the shear stress. (c) The shearing angle, ϕ, is related to the twisting angle, θ, by $\phi = \dfrac{R}{L}\theta$.

(b)

$\Gamma \longrightarrow$

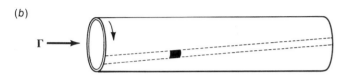

(c)

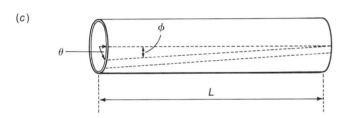

When a tube is twisted, as in Fig. 13-6, the material of its walls is subjected to a shearing stress. The force **F** producing the shearing stress acts at a right angle to the radius R to produce a torque

$$\Gamma = RF$$

The force acts on the cross-sectional area of the tube wall,

$$A = 2\pi Rt$$

where t is the wall thickness. The stress is thus given by

$$\text{stress} = \frac{F}{2\pi Rt} = \frac{\Gamma}{2\pi R^2 t}$$

The shear angle ϕ is related to the twisting angle θ by

$$\phi = \frac{R}{L}\theta$$

where both ϕ and θ are given in radians. Dividing stress by strain to obtain shear modulus, we find

$$S = \frac{\Gamma/2\pi R^2 t}{\theta R/L} = \frac{\Gamma L}{2\pi R^3 t\theta}$$

The torque Γ that produces a twist θ in a tube is thus given by

$$\Gamma = \left(\frac{2\pi R^3 t S}{L}\right)\theta$$

Exercises

5. A solid steel post 15 cm in diameter and 3 m long stands on end and supports a 2000-kg weight. By how much does it compress?

6. A 1-cm³ block of brass is subject to shearing forces of 500 N applied to opposite faces. (*a*) Find the shearing strain, $\Delta L/L$. (*b*) Find the shearing angle in radians.

7. A piece of glass with a volume of 25 cm³ sinks to the ocean bottom, where it is subjected to a pressure of 5×10^6 N/m². Find the change in its volume.

8. A cylindrical tube of brass 1 cm long with a 1-cm diameter and a wall thickness of $\frac{1}{2}$ mm is fixed at one end. A torque Γ acting about the cylinder's axis of symmetry is applied at the other end, thus producing shear in the walls of the cylinder. Find the value of K in the equation

$$\Gamma = K\theta$$

    for this particular brass cylinder.

13-5   Strength of Materials

As shown in Fig. 13-4, most materials under tension or compression eventually exceed their elastic limit, the point beyond which force is no longer simply proportional to displacement. Between the elastic limit and the point of failure, many materials take on a permanent deformation so that, if the stress is removed, the material does not return to its original equilibrium shape. The point at which this occurs is called the *yield point;* Table 13-3 gives representative yield points (under tension) for several materials. The values depend upon alloying and annealing.

Fig. 13-7. Photograph of iron whiskers (magnified 25 times), which are nearly perfect crystals. Their strength is 10^6 lb/in.². (Courtesy U.S. Steel)

TABLE 13-3   Yield Points under Tension

Material	Yield Point (N/m²)	Yield Point (lb/in.²)
Aluminum	6.8×10^7	10×10^3
Brass	1.1×10^8	13×10^3
Steel	6.0×10^8	90×10^3

The fact that these yield points are much smaller than the elastic moduli in Table 13-2 simply indicates that most materials can stretch only a small fraction of their length ($\Delta L/L \approx 1/1000$ for aluminum) before taking on a permanent deformation. The *ultimate strength* is that stress that leads to failure of the material. The ultimate strength of most of the materials in Table 13-3 is roughly twice the yield point. The behavior of a given material depends upon alloying and annealing, since both of these processes can change the basic crystalline structure of a substance, as well as the size of the microcrystals that combine to form a macroscopic sample.

If a substance were composed of a perfect single crystal, we would expect it to have the strength of the atomic bonds themselves. The ultimate strength of a perfect crystal of iron, for example, would be 1.6×10^6 lb/in.² on this basis.* Performance near this ultimate limit is observed for the whisker-like crystals shown in Fig. 13-7. The many imperfections (called *dislocations*) of large metal samples have a tendency to slip past one another under stress and cause the much lower observed strengths of metals.

*See "The Strength of Steel" by Victor F. Zackey, *Scientific American,* August 1963; and "Observing Dislocations in Crystals" by W. C. Dash and A. G. Tweet, *Scientific American,* October 1961.

Causing a metal to flow by stretching, bending, or hammering can also cause changes in its elastic moduli and yield point. A piece of copper, for example, that has been heated to a dull red heat and then cooled is quite soft (it has a very low yield point). If it is stretched, or flexed back and forth a number of times, it abruptly becomes quite rigid. This effect is called *work-hardening*. When the material is in the soft state, the various crystal imperfections slip past one another readily. As they slip, however, more and more of them "catch" upon irregularities until the mass of crystals is locked into a more rigid structure. When enough imperfections combine at the surface to produce a crack, the opening stages of failure begin. Details of this process are still poorly understood but obviously of vital importance to the design of reliable structures.*

13-6 Summary

The density of a substance is defined to be its mass per unit volume and is thus a characteristic of the material itself rather than of a specific sample of the material. The densities of most solids and liquids lie in the range of 1–10 g/cm³. The density of gases varies with temperature and pressure far more than that of solids and liquids. At standard temperature and pressure (STP), gases have approximately 1/1000 the density of solids or liquids.

The underlying microscopic structure of matter involves atoms that are nearly in contact for solids and liquids. In solids the effects of thermal agitation are small enough so that the atoms stay ordered in a rigid array. Increasing the temperature of many solids causes this rigidity to break down but leaves the atoms with an average spacing roughly equal to that of the solid. This is called the liquid state of the substance, as it can flow readily. Still higher temperatures can increase thermal agitation to the point that atoms are completely separated from one another — the gaseous state.

Forces applied to solid bodies can cause tension, compression, or shear deformation. At the microscopic level these deformations are understood to be due to the cooperative motion of the solid's constituent atoms. The force per unit area that produces a deformation is called the stress, while the *fractional* change in shape produced by the stress is called the strain. The ratio of stress to strain thus has the units of force per unit area.

The three elastic moduli are simply the ratio of stress to strain for stretching and compressing a rod (Young's modulus), bulk com-

*See "Fracture in Solids" by John J. Gilman, *Scientific American,* February 1960.

pression (bulk modulus), and shear deformation (shear modulus). The defining equations are:

$$Y = \frac{F/A}{\Delta L/L}$$

$$B = \frac{F/A}{\Delta V/V}$$

$$S = \frac{F/A}{\Delta L/L}$$

These elastic moduli are characteristics of the substance, independent of the size or shape of a given specimen.

The yield point of a material is that stress at which the material deforms so much that it does not completely return to its original shape after the stress is removed. The ultimate strength of many materials is about twice their yield point. The ultimate strengths of most materials are several orders of magnitude smaller than their elastic moduli. This indicates that these materials can stretch by only a very small fraction of their length before failure.

Problems

13-1.   A copper tube with an outer diameter of 2 cm and an inner diameter of 1.5 cm has a length of 2.5 m. Find its mass.

13-2.   A truck has a load limit of 20,000 lb. How many pieces of iron pipe 20 ft in length, $1\frac{1}{2}$ in. in outside diameter, and with a wall thickness of $\frac{1}{8}$ in. can the truck carry?

13-3.   A typical deep ocean depth is 10,000 ft. Consider each square foot of ocean bottom to support a column of water with a 1-ft² cross section, extending from the ocean bottom to the surface. What is the net force, due to the weight of water, acting down on each square foot of ocean bottom?

13-4.   The molecular weight of water is 18 amu. Find the number of water molecules contained in one liter ($= 10^3$ cm³) of water.

13-5.   Consider a substance with an atomic weight of 50 amu, a density of 6 g/cm³, and a simple cubic crystal structure such as that shown in Fig. 13-2. Calculate the length of one of the cube edges between two adjacent atoms.

13-6.   A hollow glass tube with 1-mm-thick walls and a 1-cm outside diameter stands on end and supports a 50-N weight. Find its fractional change in length, $\Delta l/l$.

13-7. A rod that stretches and shortens under tension and compression obeys Hooke's law, $F = -kx$. Using Young's modulus in Table 13-2, find the force constant, k, in N/m for a 1-cm-diam aluminum rod that is 2 m long. How does this result change if the rod is $\frac{1}{2}$ cm in diameter? How does this result change if the rod is 1 m long and 1 cm in diameter?

13-8. An automobile jack, sketched in Fig. 13-8, pushes up on the bottom of a hollow aluminum rod. Two steel wires supply the tension to hold the top plate down on the rod; the wires are thus under tension. Assume that the table top, jack, and top plate are essentially rigid, so that the rod compresses and the wires stretch. The rod has an inner diameter of 0.9 cm, an outer diameter of 1.0 cm, and a length of 40 cm; while the two wires each have a diameter of 0.5 mm and a length of 70 cm. How far does the piston move when the force it exerts is raised from 50 N to 250 N?

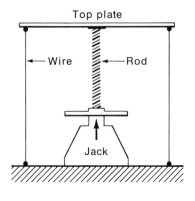

Fig. 13-8.

13-9.* A 1-mm-diam aluminum wire is lightly stretched horizontally between two walls 10 m apart. A load of 50 kg is hung at its center. Both halves of the wire now make equal angles, θ, below the horizontal. Calculate this angle.

13-10. An aluminum tube with a 4-cm diameter, a 1-mm wall thickness, and a length of 10 m is clamped at one end and a torque is applied to the other end. Find the twisting angle θ for $\Gamma = 50$ N · m. Repeat for a similar steel tube.

13-11. The steel drive shaft of a bus is 10 m in length with a diameter of 10 cm and a wall thickness of 2 mm. Find the torque exerted at one end of this shaft that will produce a twisting angle of 0.2 rad.

13-12. A piece of aluminum sinks to the ocean bottom, where it is subjected to a pressure of 5×10^6 N/m². What is its density there if its density was originally 2.7 g/cm³?

13-13. Rocks at the base of continents at a depth of 30 km are subjected to a pressure of 8×10^8 N/m². Find their fractional change in volume, $\Delta V/V$, if their bulk modulus equals that of steel.

13-14. Aluminum has a specific gravity of 2.7 and a bulk modulus of 7×10^{10} N/m². Find the pressure required to increase its specific gravity to 2.71.

13-15. An elevator weighs 1100 lb and has a maximum load limit of 1200 lb. (Thus, its maximum total weight is 2300 pounds.) The elevator is suspended by three steel cables, each of which can operate the loaded elevator alone and still be stressed at only 70% of its yield point. If the maximum acceleration of the elevator is $\frac{1}{5}g$, what must the diameter of the cables be? You may consider the cables to be solid cylinders whose weight is small compared to their load and use the value for steel in Table 13-3.

13-16.* An automobile wheel has an outer diameter of 32 in. and has four spokes connected to an inner hub of 8-in. diameter, as shown in Fig. 13-9. If a 3200-lb car is driven by two such wheels (with a torque supplied to the inner hub by a driven axle) with a maximum acceleration of $\frac{1}{2}g$, find the minimum diameter of the aluminum spokes that produces a shear strain half of that of the yield point of aluminum (a safety factor of two). Take the yield point for aluminum in shear to be 3/4 of that for tension.

13-17. An $1\frac{1}{2}$-m-long automobile drive shaft is fabricated of steel tubing with a radius of 4 cm and wall thickness of 2 mm. Find the maximum permissible torque that can be sustained by this shaft if it is not to exceed its yield point. Take the yield point for steel in shear to be 3/4 that under tension.

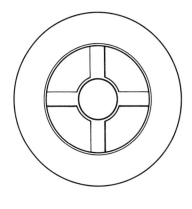

Fig. 13-9.

Chapter *14*

Kinetic Energy

14-1 Introduction The physical concept of energy may be the most important single idea in an introduction to physics. The principle of conservation of energy begins innocently enough in mechanics as the work–kinetic energy theorem, is brought into a widely useful form with the introduction of potential energy, and finally appears in electromagnetic phenomena, heat, wave motion—in short, it appears in nearly all areas of physics. This universality of the energy concept makes it possible, for example, to follow the flow of nuclear energy from the interior of the sun into radiant energy through space, into chemical energy in green plants, and finally into mechanical energy in the muscles of an animal that feeds on those plants. The law of conservation of energy tells us that though energy may change its form, its total quantity must remain constant. In effect, this conservation law tells us what can and cannot occur in nature—what is and is not energetically possible—regardless of details of the process under study.

14-2 The Work–Kinetic Energy Theorem

The work–kinetic energy theorem considers the effect of a force acting on a body for a given *distance*. (You may recall that the impulse-momentum theorem dealt with the effect of a force acting for a given *time*.) If a *constant* force F acts on a mass m to move it a distance Δx in the direction of the force, it produces an acceleration $a = F/m$ that changes the velocity of the mass. Hence if v_i is the initial velocity of the mass and v_f the final velocity, then by Eq. (3-17),

$$v_f^2 = v_i^2 + 2a \cdot \Delta x$$

so that

$$v_f^2 = v_i^2 + 2\frac{F \cdot \Delta x}{m}$$

Rearranging terms,

$$F \cdot \Delta x = \tfrac{1}{2}mv_f^2 - \tfrac{1}{2}mv_i^2 \qquad (14\text{-}1)$$

[handwritten annotations: "work done by a force" pointing to left side; "kinetic energy" pointing to right side]

The left side of (14-1) is defined to be the *work* done by a force, while the right side represents the change in the quantity $\tfrac{1}{2}mv^2$. This latter quantity is called the *kinetic energy* of the mass. Equation (14-1) could be written

$$W = \Delta K \qquad (14\text{-}2)$$

where

$$K = \tfrac{1}{2}mv^2 \qquad (14\text{-}3)$$

and

$$W = F \cdot \Delta x \qquad (14\text{-}4)$$

Equation (14-1) is the work–kinetic energy theorem for the case of a constant force. It states that the work done on a mass equals the change in kinetic energy. Note that a force acting in the same direction as the displacement produces an *increase* in kinetic energy, since the product $F \cdot \Delta x$ has a positive sign; a force acting against a displacement gives negative work and a *decrease* in kinetic energy. In the SI system of units, the dimensions of work are newton · meters. A newton · meter is called a *Joule,* abbreviated J. In English units the dimensions of work are ft · lb. In cgs units the unit of work, the dyne · centimeter, is called the *erg.* Since 1 cm $= 10^{-2}$ m and 1 N $= 10^{-5}$ dyne, 1 erg $= 10^{-7}$ J.

The dimensions of work and of torque both happen to be a length multiplied by a force. They are different entities, however, as torque is a vector and work is a scalar. To avoid confusion, torque is sometimes written m · N, or cm · dyne, or lb · ft, in opposite order from work units.

The units of kinetic energy are kg · m²/s² in SI units and slug · ft²/s² in the English system of units. Since the newton equals one kg · m/s² and the lb force equals one slug · ft/s², the units of work and of kinetic energy agree, as they must.

Example   A cart is rolling steadily with a velocity of 15 m/s. A braking force of 5 N is exerted while the cart moves a distance of 15 m. What is the final velocity? Note that the force acts for *longer* than one second, even though the initial velocity is 15 m/s and the force acts over a 15-m distance. The action of the force continually decreases the velocity, so the time interval over which it acts is not immediately known, though it can be calculated.

Solution The initial kinetic energy of the cart is given by $\frac{1}{2}mv_i^2$ and its final kinetic energy by $\frac{1}{2}mv_f^2$. Applying the work–kinetic energy theorem:

$$F \cdot \Delta x = \tfrac{1}{2}mv_f^2 - \tfrac{1}{2}mv_i^2$$

Solving for v_f:

$$v_f^2 = \frac{2F \cdot \Delta x}{m} + v_i^2 \tag{14-5}$$

Since F and Δx are oppositely directed, the quantity $F \cdot \Delta x$ will be negative:

$$v_f^2 = (15 \text{ m/s})^2 - 2\,\frac{(5 \text{ N} \cdot 15 \text{ m})}{2 \text{ kg}} = 225 \text{ m}^2/\text{s}^2 - 75 \text{ N} \cdot \text{m/kg}$$

Since $N = kg \cdot m/s^2$, note that the dimensions $N \cdot m/kg$ become m^2/s^2, so that

$$v_f^2 = (225 - 75) \text{ m}^2/\text{s}^2 = 150 \text{ m}^2/\text{s}^2$$

and
$$v_f = 12.2 \text{ m/s}$$

In the preceding example, note that (14-5) is simply a restatement of (3-17), since F/m is the acceleration. The work–kinetic energy theorem for constant forces and one dimension of motion provides nothing inherently new. Its usefulness becomes apparent only when it is generalized to include variable forces and more than one dimension.

14-3 Variable Forces in One Dimension

If a force acts on a moving body but is *not* constant, we must use the methods of integral calculus to compute the net work done by the force. Consider the force that is a function of x graphed in Fig. 14-1a. In Fig. 14-1b, the work done by the variable force is approximated by a sum of terms. Each term in the sum represents the work done by a *constant* force whose value is approximately equal to the minimum value of the true force on a small interval Δx. Each shaded rectangle in Fig. 14-1b has an area given by $F_i \cdot \Delta x_i$, where the index i takes on values $i = 1, 2, 3, 4, \ldots$ corresponding to each of the rectangles shown. The total work done between c and d is approximately the sum of the areas of all the inscribed rectangular strips, i.e.,

$$W_{\text{total}} \approx F_1 \cdot \Delta x_1 + F_2 \cdot \Delta x_2 + F_3 \cdot \Delta x_3 + \cdots + F_n \cdot \Delta x_n$$

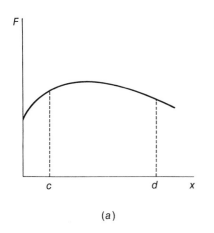

(a)

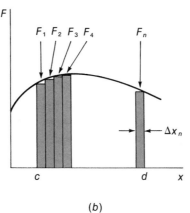

(b)

Fig. 14-1. (a) A force that varies with position acts upon a body that moves along the x axis. (b) The total work done by the force between c and d is approximated by breaking the interval into small segments of width Δx_1, Δx_2, Δx_3, . . . , Δx_n. The sum of the areas of these strips closely approximates the total work done by the force as it moves from c to d.

More briefly,
$$W_{\text{total}} \approx \sum_{i=1}^{n} F_i \cdot \Delta x_i$$

As the width Δx of each rectangular strip is made smaller, the net work obtained by the above prescription more and more closely approximates the exact value of the work performed by the continuously variable force. Thus we are led to the *work integral:*

$$W_{\text{exact}} = \lim_{\Delta x_i \to 0} \sum_{i=1}^{n} F_i \cdot \Delta x_i = \int_{c}^{d} F \, dx \qquad (14\text{-}6)$$

We now take this integral as our definition of work. The value of this work integral depends on the relationship between F and x. However, it is easy to show that this integral *always* equals the change in the kinetic energy, thus establishing the work–kinetic energy theorem for variable forces in one dimension.

To show this we rewrite (14-6) utilizing Newton's second law, so that F can be written

$$F = m \frac{dv}{dt} \qquad \text{or} \qquad F = m \lim_{\Delta t \to 0} \frac{\Delta v}{\Delta t}$$

Hence (14-6) becomes

$$W = \lim_{\Delta x_i \to 0} m \sum_{i=1}^{n} \frac{\Delta v_i}{\Delta t_i} \Delta x_i \qquad (14\text{-}7)$$

As $\Delta x_i \to 0$, so do the quantities Δt_i and Δv_i since they are, respectively, the time in which the mass moves a distance Δx_i and the velocity gained over that distance. We can thus rewrite (14-7) as follows:

$$W = \lim_{\Delta t_i \to 0} m \sum_{i=1}^{n} \frac{\Delta x_i}{\Delta t_i} \Delta v_i = m \int_{v_1}^{v_2} v \, dv$$

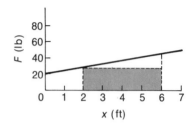

where v_1 is the velocity of the mass m at position c in Fig. 14-1 and v_2 is the velocity at position d. This integral is easily evaluated since the antiderivative of v is $v^2/2$:

$$W = \tfrac{1}{2}mv_2{}^2 - \tfrac{1}{2}mv_1{}^2 \tag{14-8}$$

which is the work–kinetic energy theorem for variable forces. Using the definitions (14-3) and (14-6), we can again write this theorem in the form $W = \Delta K$.

Fig. 14-2.

Example The linearly increasing force graphed in Fig. 14-2 acts upon a 4-slug body. (a) Find the change in the body's kinetic energy between $x_1 = 2$ ft and $x_2 = 6$ ft. (b) If it initially moves with $v_1 = 8$ ft/s, find v_2.

Solution $W = \displaystyle\int_{x_1}^{x_2} F \, dx =$ the area under the F-vs-x curve in Fig. 14-2 between $x = 2$ ft and $x = 6$ ft. This area is the sum of the shaded rectangle and the triangular portion above the rectangle. From the geometry of the graph, we see that

$$W = \int_{x_1}^{x_2} F \, dx = 30 \text{ lb} \cdot 4 \text{ ft} + \tfrac{1}{2}(20 \text{ lb} \cdot 4 \text{ ft}) = 160 \text{ ft} \cdot \text{lb}$$

and this work is equal to the change in kinetic energy. Alternatively, we could substitute the equation of the line that represents F in the graph and evaluate the integral. The equation of the line is:

$$F = F_0 + kx$$

where $F_0 = 20$ lb and $k = 5$ lb/ft in the above example.

$$\int_{x_1}^{x_2} F \, dx = \int_{x_1}^{x_2} F_0 \, dx + k \int_{x_1}^{x_2} x \, dx$$

$$= F_0(x_2 - x_1) + \tfrac{1}{2}k(x_2)^2 - \tfrac{1}{2}k(x_1)^2$$

$$= 20 \text{ lb} \cdot 4 \text{ ft} + \tfrac{1}{2}\left(5\,\frac{\text{lb}}{\text{ft}} \cdot 36 \text{ ft}^2\right) - \tfrac{1}{2}\left(5\,\frac{\text{lb}}{\text{ft}} \cdot 4 \text{ ft}^2\right)$$

$$= 160 \text{ ft} \cdot \text{lb}$$

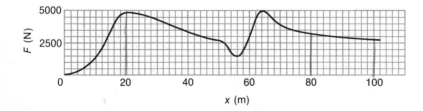

Fig. 14-3.

This answer was obtained previously using the graphical method of evaluation. Now to find the final velocity:

$$W = \Delta K = \tfrac{1}{2}mv_2{}^2 - \tfrac{1}{2}mv_1{}^2$$

$$v_2{}^2 = 2W/m + v_1{}^2$$

$$= 2 \cdot 160 \text{ ft} \cdot \text{lb}/4 \text{ slug} + 64 \text{ ft}^2/\text{s}^2$$

$$= 144 \text{ ft}^2/\text{s}^2$$

$$v_2 = 12 \text{ ft/s}$$

Exercises

1. A 5-kg cart is initially moving with a velocity of 30 m/s. A force of 120 N, which acts in the same direction as the cart's motion, is applied for a distance of 100 m. Neglecting friction, (*a*) find the initial kinetic energy; (*b*) find the final kinetic energy; (*c*) find the final velocity.

2. A 160-lb skier encounters a level area at a velocity of 30 ft/s. If the coefficient of friction between his skis and the snow is given by $\mu = 0.05$, find the distance he coasts before coming to rest.

3. A test car with a mass of 1000 kg is accelerated by the force F represented in the graph of Fig. 14-3. If the car starts from rest, find its velocity at $x = 20$ m, 80 m, and 100 m.

14-4 Work Done by a Spring

A spring obeys Hooke's law: $F = -kx$, where x is measured from the equilibrium position of the spring.* The constant k is called the *spring constant* and has the dimensions of force/length. This simple force law closely approximates many forces found in nature. The work W_{12} done by such a force between positions x_1 and x_2 is easily evaluated:

$$W_{12} = \int_{x_1}^{x_2} F\, dx = \int_{x_1}^{x_2} -kx\, dx = -k \int_{x_1}^{x_2} x\, dx = +\tfrac{1}{2}k(x_1{}^2 - x_2{}^2) \quad (14\text{-}9)$$

Example A spring stretches 10 cm when a 2-N force is applied to it. The spring is then attached to the mass shown in Fig. 14-4. If $x_1 = 20$ cm and $x_2 = 10$ cm, find the work done by the spring from x_1 to x_2.

---

*The minus sign indicates that when the spring is stretched $(x > 0)$, the force exerted by the spring acts in the direction of decreasing x: the negative x direction. Similarly, a compressed spring $(x < 0)$ exerts a force in the positive x direction.

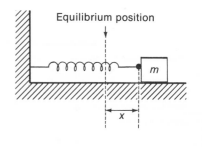

Equilibrium position

Fig. 14-4.

Solution   The spring constant for this spring is found from the information given above:

$$k = -F/x = -(-2 \text{ N})/0.1 \text{ m} = 20 \text{ N/m}$$

The minus signs cancel because the force exerted *by the spring* is in the opposite direction to the displacement. Now that we have a numerical value for k, we can compute the work done by the spring between $x = 20$ cm and $x = 10$ cm. We expect this work to be positive, since the spring force now acts in the direction of the displacement. Applying equation (14-9),

$$W_{12} = \tfrac{1}{2}(20 \text{ N/m})\,[(0.2 \text{ m})^2 - (0.1 \text{ m})^2] = 0.3 \text{ N} \cdot \text{m}$$

We could, of course, now find the change in kinetic energy of the mass by application of the work–kinetic energy theorem.

Exercises

4. An experimenter stretches a spring from x_1 to x_2, with $x_2 > x_1$. Is the work done by the experimenter positive or negative? Is the work done by the force of the stretched spring positive or negative?

5. A spring is compressed from $x_1 = -5$ cm to $x_2 = -15$ cm. The amount of work done by the spring is -1 N·m. Find the value of the spring constant, k.

6. Two straps of rubber are arranged as a slingshot. Their combined spring constant is $k = 50$ lb/ft. A 2-oz stone is loaded into the slingshot, which is then stretched 8 in. beyond its equilibrium length and released. The mass of the rubber straps is negligible. (*a*) Find the kinetic energy of the stone as it leaves the slingshot. (*b*) Find the stone's velocity.

14-5   Work in Three Dimensions

There are many situations in which the force acting upon an object is not parallel to the displacement of the object. In order to generalize the work–kinetic energy theorem to include such situations, it is helpful to resolve the force into components perpendicular to the displacement and parallel to the displacement. We will call these components $F_\perp$ and $F_\parallel$, respectively. The component of force perpendicular to the motion of a body merely serves to change the direction of motion. Thus, $F_\perp$ does not change the magnitude of the velocity and hence does not change the kinetic energy. The component of **F** parallel to the displacement is indicated in Fig. 14-5 and *does* change the kinetic energy of the body.

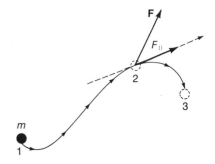

Fig. 14-5.   A body moves from position 1 to 2 to 3 along a curved path. The instantaneous direction of motion of the body at position 2 is the tangent to the path, indicated by the dotted line. Only the component of the force **F** that is parallel to this direction can change the *magnitude* of the body's velocity. This component is called $F_\parallel$.

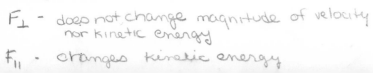

$F_\perp$ – does not change magnitude of velocity nor kinetic energy

$F_\parallel$ – changes kinetic energy

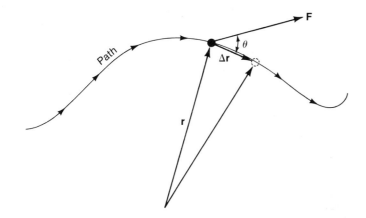

Fig. 14-6. The position of the body is specified by **r,** and the magnitude of its displacement is Δr. The angle between **F** and $\Delta \mathbf{r}$ is θ and $F_{\parallel} = F \cos \theta$.

We can specify the position of the body moving along an arbitrary path with a position vector **r,** as in Fig. 14-6. The displacement of the body is given by $\Delta \mathbf{r}$. The increment of work ΔW done by the force as the body is displaced by $\Delta \mathbf{r}$ is given by

$$\Delta W \approx F_{\parallel} \cdot \Delta r$$

where the approximate equality becomes exact as $\Delta r \rightarrow 0$. As in Section 14-3, the sum of all the increments between r_1 and r_2 gives the net work done between r_1 and r_2. In the limit as each $\Delta r \rightarrow 0$, the work is exactly given by

$$W = \int_{r_1}^{r_2} F_{\parallel} \, dr \qquad (14\text{-}10)$$

Note that $F_{\parallel}$ depends not only upon the direction of the force **F** but upon the direction of the path. In fact, $F_{\parallel}$ at any given position along the path depends upon the direction of the path in *that* position.

Since $F_{\parallel}$ equals $F \cos \theta$, we may also write

$$W = \int_{r_1}^{r_2} F \cos \theta \, dr \qquad (14\text{-}11)$$

At first, application of (14-11) seems a hopeless task since we must know θ and **F** all along the path and they are both, in general, varying. Fortunately, there are many forces in nature that have such simple properties that the work they perform is easily evaluated. One of these is the force of gravity near the surface of the earth; the next section illustrates the simplicity of application of (14-11) to this force.

Example A constant force **F** lies in the x,y plane and is always parallel to the x axis. This force acts continually upon a mass m as it moves from the point $x=0, y=0$ to the point $x=C, y=C$, as shown in Fig. 14-7. Find the work done by the force.

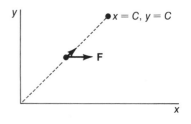

Solution The work done is given by

$$W = \int_{\substack{x=0 \\ y=0}}^{\substack{x=C \\ y=C}} F(\cos 45°) \, dr$$

and since F is constant, it may be factored out of the integral:

$$W = F \cos 45° \int_{\substack{x=0 \\ y=0}}^{\substack{x=C \\ y=C}} dr = \frac{\sqrt{2}}{2} F \int_{\substack{x=0 \\ y=0}}^{\substack{x=C \\ y=C}} dr$$

Fig. 14-7.

The remaining integral may be thought of as the sum of all the infinitesimal steps dr from the point $x=0, y=0$ to the point $x=C, y=C$ along the indicated path. That is, it is simply equal to the length l of the path:

$$l = \sqrt{C^2 + C^2} = \sqrt{2} \cdot C$$

so that we find

$$W = \frac{\sqrt{2}}{2} F\sqrt{2} \cdot C = FC$$

In the preceding example, and in many other cases, it is convenient to think of dr as an actual increment of the path length; since we are led to the exact integral from the inexact sum in the limit as $\Delta r \to 0$. When we think of dr as an actual quantity it is called an *infinitesimal*. Thinking of dr in this way often leads to correct conclusions. The history of the development of ideas regarding infinitesimals and some recent developments in mathematical thinking on this topic are contained in "Nonstandard Analysis" by Martin Davis and Reuben Hersh, in the June 1972 issue of *Scientific American*.

Exercises

7. Repeat the preceding example except assume that the path followed is first along the x axis to the point $x = C, y = 0$, then vertically to the same final point.

8. A mass m is subject to a force that points always in the positive y direction with a magnitude given by $F = Ky^2$. Find the work done by this force on the mass m if it starts from the origin of coordinates and moves a distance h along (a) the x axis; (b) the y axis; (c) a line between the x and y axes inclined at 45° with respect to the x axis.

14-6 Work Done by Gravity

Suppose a bead is threaded over a wire that is bent in the fashion shown in Fig. 14-8. Consider the bead and wire to be well polished, so that frictional forces may be neglected. If the bead is released at y_1 it starts sliding down the wire, following each turn and twist as it goes. It is acted upon by the force of gravity but also by the forces due to the constraint of the wire. The constraint forces are those that cause the bead to swing its velocity vector back and forth as it follows the twisting path. The *net* force on the bead is thus complicated: it varies both in magnitude and direction along the path. However, the absence of frictional forces assures us that the wire cannot provide a force on the bead with a component parallel to the path—it is too slippery ($\mu = 0$). We are assured, then, that the constraint forces are perpendicular to the velocity and merely change its direction, not its magnitude. Since the constraint force has no component parallel to the path, it does no work.

The force of gravity can have a projection parallel to the path, but it is a varying projection. Consider the arbitrary segment of the path in Fig. 14-9. The quantity that appears in the work integral is $F\,dr\cos\theta$. But note that $dr\cos\theta$ is simply dy in our case, since $dr\cos\theta$ is a projection of dr onto the y direction. As the bead moves a distance dr along the path, it falls a distance dy, where $dy = dr\cos\theta$.

The question of sign arises here. As long as the bead moves downhill, the quantity $F\,dr\cos\theta$ is positive, since the force has a component in the same direction as the displacement. If portions of the wire path carry the bead uphill, the quantity $F\,dr\cos\theta$ will be negative, since $\cos\theta$ is negative for θ between 90° and 180°, which corresponds to uphill motion. Thus positive work is done by gravity in downhill portions of the path, negative work in uphill portions. If we take the positive y direction as upward, then $F = -mg$ since gravity acts downward. In this case dy is a negative quantity in downhill portions of the path, so the product $F\,dy$ is positive there. For the net work done by gravity as the bead slides from a height y_1 to a new height y_2, we have:

$$W = \int_{y_1}^{y_2} F\,dr\cos\theta = \int_{y_1}^{y_2}(-mg)\,dy = -mg\int_{y_1}^{y_2}dy = mg(y_1 - y_2)$$

which is a simple result that is *independent* of the path from y_1 to y_2. In Fig. 14-8, $y_1 > y_2$ so that the net work done by gravity is positive. This work causes the kinetic energy of the bead to increase as it slides down the wire. Applying the work–kinetic energy theorem:

$$W = mg(y_1 - y_2) = \tfrac{1}{2}mv_2^2 - \tfrac{1}{2}mv_1^2 \tag{14-12}$$

Equation (14-12) tells us that $y_1 > y_2$ (starting point above finish)

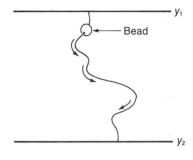

Fig. 14-8. A bead threaded over a slippery wire.

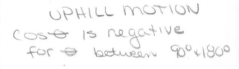

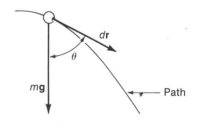

Fig. 14-9. A small segment of the path followed by the bead in Fig. 14-8. Note that the angle θ is simply the angle between the wire and the vertical, since the force of gravity acts vertically.

implies $v_2 > v_1$, that is, the velocity increases. When $y_1 < y_2$ the object is moving uphill and, of course, loses velocity.

Looking back over the calculations that led to Eq. (14-12), it is apparent that the same result obtains for a car rolling down a hill, or for a roller coaster, or for a projectile near the earth's surface — as long as frictional forces are negligible.

Example   A stone is thrown horizontally from a cliff at $v_0 = 10$ m/s. What is the magnitude of its velocity when it strikes the ground 20 m below?

Solution   The initial kinetic energy is $\frac{1}{2}mv_0{}^2$; the work done by gravity is $mg(y_0 - y_f)$, where $y_0 - y_f$ is the height of the cliff; and the final velocity, v_f, may be obtained from the work–kinetic energy theorem. For convenience, let us set $y_0 - y_f = h$. Then

$$mgh = \tfrac{1}{2}mv_f{}^2 - \tfrac{1}{2}mv_0{}^2$$

Solving for v_f:

$$v_f{}^2 = v_0{}^2 + 2gh = 100 \text{ m}^2/\text{s}^2 + 2(9.8 \text{ m/s}^2)(20 \text{ m}) = 492 \text{ m}^2/\text{s}^2$$

$$v_f = 22.2 \text{ m/s}$$

Note that all we obtained was the magnitude of the final velocity, not its direction. This illustrates the scalar nature of kinetic energy (and of work)—it is simply a number without vector properties. In contrast, the impulse-momentum theorem for three dimensions involves vector quantities and can give vector information.

Exercises

9.   Looking at the equation $\frac{1}{2}mv_f{}^2 = mg(y_0 - y_f) + \frac{1}{2}mv_0{}^2$ and solving for v_f, it is clear that if $y_0 - y_f$ is sufficiently negative (uphill motion), we get an imaginary value for v_f (i.e., $v_f{}^2 < 0$). What does this mean?

10.   In the preceding example, the final velocity of the stone was found by the work–kinetic energy theorem. Find this velocity by the methods of Section 7-7. Is there any basic difference in the two methods of solution?

11.   A roller coaster starts at position 1 in Fig. 14-10 with a velocity of 5 ft/s. Find its velocity at points 2 and 3. Assume negligible friction.

12.   Use the work–kinetic energy theorem to find the maximum height a projectile can reach when it is fired upward at v_0.

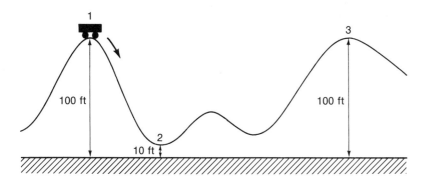

100 ft

2

10 ft

100 ft

3

Fig. 14-10.

14-7 Kinetic Energy of Rotation

When an extended object rotates, the parts of it at large distances from the axis of rotation move rapidly, while those close to the axis of rotation move slowly. In order to obtain its total kinetic energy of rotation, we must sum the contributions from all the various portions of the object. An individual point mass m, moving about a pivot point as in Fig. 14-11, has a velocity $v = r\omega$. Its kinetic energy is thus

$$K = \tfrac{1}{2}mv^2 = \tfrac{1}{2}mr^2\omega^2 = \tfrac{1}{2}I\omega^2$$

where I is the moment of inertia of this single mass. If we consider n point masses at various fixed radii from a fixed axis of rotation, all rotating with an angular velocity ω, we obtain:

$$K = \sum_{i=1}^{n} \tfrac{1}{2}m_i v_i^2 = \tfrac{1}{2}\left[\sum_{i=1}^{n} m_i r_i^2\right]\omega^2 \qquad (14\text{-}13)$$

But we recognize the quantity in brackets as the moment of inertia of this collection of masses (see Section 11-2), so we may write

$$K = \tfrac{1}{2}I\omega^2 \qquad (14\text{-}14)$$

where I is the moment of inertia of the system.

If the rotating system is a continuous extended object, we break it up into infinitesimal mass elements dm, and the sum in (14-13) becomes the moment of inertia integral. Once again, the quantity in brackets is simply the moment of inertia of the rotating object, so (14-14) is generally valid for any mass distribution rotating as a rigid body about a fixed axis. Note that (14-14) is quite analogous to (14-3), with I replacing m and ω replacing v:

$$K = \tfrac{1}{2}I\omega^2 \qquad \text{analogous to} \qquad K = \tfrac{1}{2}mv^2$$

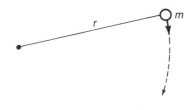

Fig. 14-11. A point mass m moves about a pivot point with a velocity $v = \omega r$.

$$W = \Delta K = \tfrac{1}{2}I\omega_2^2 - \tfrac{1}{2}I\omega_1^2$$

The work done by a constant torque Γ twisting through an angle θ is easily shown to be

$$W = \Gamma\theta$$

which is analogous to

$$W = F \cdot \Delta x$$

Generalizing to a torque that depends upon θ, we obtain

$$W = \int_{\theta_1}^{\theta_2} \Gamma \, d\theta \qquad\qquad (14\text{-}15)$$

which is analogous to

$$W = \int_{x_1}^{x_2} F \, dx$$

Example   A flywheel with a radius of 20 cm and a moment of inertia of 0.5 kg · m² has a cord wrapped around its perimeter. This cord is pulled with a constant tension of 50 N while the wheel turns through an angle of 4 radians. If the initial angular velocity is 3 rad/s, find the final angular velocity.

Solution   The torque exerted by the force of 50 N at the perimeter is given by the product of the radius of the wheel and the force, since this force is at right angles to the radius, which goes between the center of rotation and the point of application of the force. This torque acts through an angle $\theta = 4$ radians.

$$\Gamma = rF$$

$$W = \Gamma\theta = rF\theta = 0.2 \text{ m} \cdot 50 \text{ N} \cdot 4 \text{ rad} = 40 \text{ J}$$

Applying the work–kinetic energy theorem:

$$W = \Delta K = \tfrac{1}{2}I\omega_2^2 - \tfrac{1}{2}I\omega_1^2$$

$$\omega_2^2 = \frac{2W}{I} + \omega_1^2$$

$$= \frac{80 \text{ J}}{0.5 \text{ kg} \cdot \text{m}^2} + 9 \text{ rad}^2/\text{s}^2 = 169 \text{ rad}^2/\text{s}^2$$

$$\omega_2 = 13 \text{ rad/s}$$

Example   In Section 11-5 we saw that when an ice skater changes her moment of inertia from I_1 to I_2, her rotational velocity changes as predicted by conservation of angular momentum:

$$I_1\omega_1 = I_2\omega_2$$

so that

$$\omega_2 = \frac{I_1}{I_2}\omega_1 \qquad\qquad (14\text{-}16)$$

If $I_1 > I_2$ then $\omega_2 > \omega_1$. But what about her kinetic energy of rotation? Does it remain constant? Let us see. The initial kinetic energy is

$$K_1 = \tfrac{1}{2}I_1\omega_1{}^2$$

The final kinetic energy is

$$K_2 = \tfrac{1}{2}I_2\omega_2{}^2$$

Using (14-16), we find

$$K_2 = \tfrac{1}{2}I_2\left(\frac{I_1}{I_2}\omega_1\right)^2 = \frac{I_1}{I_2}\left(\tfrac{1}{2}I_1\omega_1{}^2\right) = \frac{I_1}{I_2}K_1 \tag{14-17}$$

and since $I_1 > I_2$, the kinetic energy has increased! How is it that the angular momentum remains constant while the kinetic energy increases?

First, the nearly frictionless contact of the ice skates assures that no external torques act, so that angular momentum is conserved. Similarly, no work is done by external torques to increase the kinetic energy. But some work must have been done to cause this increase—where and how? The work that was done was that of the ice skater in the process of moving elements of mass from a large radius to a smaller radius against the centrifugal force. In fact, a straightforward integration of the work done by the ice skater in the process of changing from a large-I configuration to a low-I configuration yields exactly the amount of work needed to agree with (14-17). See Problem 14-14.

Exercises

13. A solid, disc-shaped flywheel with a mass of 4 metric tons and a radius of 1.5 m is rotating at 80 rad/s. It is losing kinetic energy at a rate of 100 J/s. How long will it take to come to rest?

14. A drill motor applies a torque of 15 N · m to a sanding disc with a moment of inertia of 0.02 kg · m². If the sanding disc is initially at rest, how much kinetic energy does the disc have after four full revolutions?

14-8   Combined Translation and Rotation

When a mass translates, its kinetic energy is given by $K = \tfrac{1}{2}mv^2$. When it rotates, $K = \tfrac{1}{2}I\omega^2$. What is the kinetic energy of a rigid body, such as a rolling wheel, that does both? As is shown in the Appendix at the end of this chapter, the total kinetic energy is simply the sum of two terms:

$$K = \tfrac{1}{2}mv_{cm}{}^2 + \tfrac{1}{2}I\omega^2 \tag{14-18}$$

where m is the total mass, v_{cm} is the velocity of the center of mass, and I is the moment of inertia about the center of mass. The first term in (14-18) is the translational kinetic energy due to motion of the object's center of mass, while the second term is the rotational kinetic energy due to rotation about the center of mass.

Example A wheel with mass m, radius r, and moment of inertia I rolls downhill. If it starts from rest and changes its altitude by an amount h, find its final velocity.

Solution We have seen in Section 14-6 that gravity does work on a mass that moves to a lower height. The work is given by

$$W = mgh$$

By the work–kinetic energy theorem,

$$W = \Delta K$$

So we have

$$mgh = \tfrac{1}{2}mv^2 + \tfrac{1}{2}I\omega^2$$

for the rolling wheel that starts from rest and loses height by an amount h. Recalling that rolling implies $v = \omega r$ (see Section 10-7),

$$mgh = \tfrac{1}{2}mv^2 + \tfrac{1}{2}I\frac{v^2}{r^2}$$

If the wheel is a uniform disc with $I = \tfrac{1}{2}mr^2$, we obtain

$$v^2 = \tfrac{4}{3}gh$$

This result is compatible with our earlier result (see Section 11-9), that the acceleration of a cylinder placed on an inclined plane is given by

$$a = \tfrac{2}{3}g\sin\theta$$

Problem 14-12 illustrates this point. Note that our present result, however, is more general since it applies when the wheel follows any path downhill (as long as it maintains contact with the surface and rolls without slipping).

Exercises

15. A solid spherical ball is placed at rest on a smooth surface and allowed to roll downhill, so that its elevation decreases by h. Find its final velocity.

16. A cart of 2-kg mass is fitted with four wheels, each of $\tfrac{1}{2}$-kg mass and each with a radius of 4 cm. The wheels are uniform discs. The cart rolls downhill so that its elevation decreases by 3 m. Find its final velocity if it begins at rest.

14-9   Power

The *rate* at which work is done by a force is called *power:*

$$P = \frac{dW}{dt} \qquad (14\text{-}19)$$

In MKS units the dimensions of power are Joules/second and are given the name watt; the kilowatt $= 10^3$ watts is a commonly used unit of power. In the English system of units, a unit of power is a ft·lb/s. A more commonly used unit of power is the *horsepower,* which is defined to be 550 ft·lb/s. The conversion between kilowatts and horsepower (usually abbreviated kW and hp) is easily found:

$$1 \text{ kW} = 10^3 \text{ N} \cdot \text{m/s}$$

$$= (10^3 \text{ N} \cdot \text{m/s}) (0.225 \text{ lb/N}) (3.28 \text{ ft/m})$$

$$= 738 \text{ ft} \cdot \text{lb/s} = 1.34 \text{ hp}$$

or $\qquad\qquad\qquad$ 1 hp $= 746$ W

The kW is a unit of convenient size for expressing the mechanical power output of electric motors and automobile engines. Though the kW is commonly used to measure electrical power, as discussed in Section 33-7, it is not widely used for mechanical devices, even in countries utilizing the metric system of units.

Example A 160-lb man can run steadily upstairs at a rate that increases his height above ground by 3 ft/s. The power he generates is given by the work done (*mgh*) per unit time:

$$P = 160 \text{ lb} \cdot 3 \text{ ft/s} = 480 \text{ ft} \cdot \text{lb/s} = 0.87 \text{ hp}$$

Actually, the power generated by his muscles exceeds this by some amount, since some of the work done by those muscles is used simply to overcome frictional forces. The overall efficiency of conversion of chemical energy into mechanical energy by a man running uphill is about 25%.* In fact, he must dissipate some power while at "rest" simply to sustain basic metabolic processes. A resting adult dissipates energy at a rate of approximately 100 W ($= 0.12$ hp).

Multiplication of power by a time interval leads us back to a quantity of energy. A bizarre energy unit that uses mixed time units has come into common usage in the electrical industry and is called the kilowatt · hour:

$$1 \text{ kW} \cdot \text{hr} = 10^3 \text{ W} \cdot 3600 \text{ s} = 3.6 \times 10^6 \text{ J} = 3.6 \text{ megajoules}$$

*See "Muscle Engines" by Rodolfo Margaria, *Scientific American,* March, 1972.

A typical mechanical source of power, such as an automobile engine, generates a torque Γ that, turning through angle θ, produces work. The spinning shaft of the motor turns at a rate $d\theta/dt = \omega$, so that the rate at which work is done is found by differentiation of $W = \Gamma\theta$:

$$p = dW/dt = \Gamma d\theta/dt = \Gamma\omega$$

$p = \Gamma\omega$

Consider an automobile engine that generates 150 lb·ft of torque while turning at 3600 rev/min. Converting rev/min to rad/s, we obtain

$3600 \text{ rev}/\text{min} \left(2\pi \text{ rad}/\text{rev} \right) \left(\frac{1}{60} \text{ s} \right)$

$$p = 150 \text{ lb} \cdot \text{ft} \cdot 377 \text{ rad/s} = 5.66 \times 10^4 \text{ ft} \cdot \text{lb/s}$$

$$= 103 \text{ hp}$$

It is typical of internal combustion engines that they generate maximum power only when turning rapidly. It is for this reason that gears are used to couple the rapidly rotating engine drive shaft to the more slowly turning wheels. The reduction in speed is accompanied by an increase in torque but their product—the power—remains constant. The knack of choosing the best gear ratio to obtain maximum straightaway speed in a racecar, for example, consists of finding a gear ratio such that the engine turns at the number of rev/min that produces peak horsepower exactly when the car has reached maximum speed. A choice of too large a gear ratio, for example, results in the engine's turning faster than when it generates maximum power at full speed. Since the engine is not at full power, then, it is clear that a larger top speed can be obtained by decreasing the gear ratio in this instance. A wide variety of devices —gears, pulleys, torque converters—are used in machinery to match the optimum-performance specifications of a power source to the speed and torque demands of the intended service.

Example   Two engines are possibilities for the propulsion of a truck. One of them generates a maximum of 120 hp at 1200 rev/min, while the other generates a maximum of 170 hp at 4800 rev/min. In the intended service, maximum torque is needed at 200 rev/min. Which engine is preferable (all other things, such as reliability and cost, being equal)?

Solution   The higher-horsepower engine will be preferable, for it is always possible to convert the angular velocity of a given engine to that desired by means of gears or a torque converter. Since the power is given by the product of torque and angular velocity, and the magnitude of power is unchanged (aside from frictional losses) by gear multiplication, it is clear that the peak torque at 200 rev/min will be given by the 170-hp engine. Its torque at 4800 rev/min is

$$\Gamma = P/\omega = \frac{170 \text{ hp} \cdot 550 \text{ ft} \cdot \text{lb/s}}{4800 \dfrac{\text{rev}}{\text{min}} \cdot 0.105 \dfrac{\text{rad/sec}}{\text{rev/min}}} \longleftarrow \left(2\pi \text{ }^{rad}\!/_{rev}\right)\left(\tfrac{1}{60}\text{ s}\right)$$

$$= 1.85 \times 10^2 \text{ lb} \cdot \text{ft}$$

After conversion to 200 rev/min by a gear train, the torque is

$$\Gamma = \frac{170 \text{ hp} \cdot 550 \text{ ft} \cdot \text{lb/s}}{200 \dfrac{\text{rev}}{\text{min}} \cdot 0.105 \dfrac{\text{rad/sec}}{\text{rev/min}}} = 4.45 \times 10^3 \text{ lb} \cdot \text{ft}$$

Example In the article "Flywheels" by Richard F. Post and Stephen E. Post in the December 1973 issue of *Scientific American*, it is proposed that a flywheel be used to store sufficient mechanical energy to drive a small car for a range of 200 miles. They indicate that 30 kW · hr of stored energy suffices. For a 100-kg flywheel with a radius of 50 cm, find the required angular velocity of the flywheel.

Solution The kinetic energy stored in the spinning flywheel is given by

$$K = \tfrac{1}{2}I\omega^2$$

so that the angular velocity required is

$$\omega = \sqrt{\frac{2K}{I}}$$

The 30 kW · hr is equivalent to 108 megajoules and the moment of inertia is given by

$$I = \tfrac{1}{2}mr^2 = 12.5 \text{ kg} \cdot \text{m}^2$$

Thus $\omega = \sqrt{\dfrac{216 \times 10^6 \text{ J}}{12.5 \text{ kg} \cdot \text{m}^2}} = 4.16 \times 10^3 \text{ rad/s} = 39{,}700 \text{ rpm}$

Exercises

17. A man can use his arm muscles, pulling on a rope, to generate $\tfrac{1}{2}$ hp over a span of a minute or so. How long does it take him to pull a 50-lb load of bricks to the top of a 100-ft-high building?

18. An engine developing 100 hp is just able to keep a certain car moving on a level surface at a constant 94 mi/hr. Find the magnitude of the sum of all the retarding forces on the car at that speed. These are forces due to rolling friction, air resistance, and bearing friction.

19. An electric motor consumes energy at a 2-kW rate. If it converts 90% of this power into mechanical power, find its torque at 1650 rev/min.

14-10 Kinetic Energy in Relativity   NOT ON and HOURLY

Einstein's special theory of relativity treats the dynamics of objects that move with velocities comparable to that of light $(3 \times 10^8 \text{ m/s})$. It is found that the laws of classical mechanics, which we have studied in this text, are modified in this high-velocity regime. For example, in relativistic dynamics the kinetic energy of a mass m_0 moving with velocity v is given by

$$K = \frac{m_0 c^2}{\sqrt{1 - \dfrac{v^2}{c^2}}} - m_0 c^2 \qquad (14\text{-}20)$$

where c is the velocity of light and $m_0 c^2$ is called the particle's "rest energy." In classical mechanics we have

$$K = \tfrac{1}{2} m v^2 \qquad (14\text{-}21)$$

Equation (14-20) is now considered to be the correct statement. Since classical mechanics existed for 200 years before relativity theory, one wonders how such a glaring discrepancy as that between Eqs. (14-21) and (14-20) could have been overlooked. The explanation is simple enough: at everyday velocities, where classical mechanics is applied, there is no appreciable discrepancy.

To see this point, we will utilize the binomial series:

$$(1+x)^n = 1 + nx + \frac{n(n-1)}{2!}x^2 + \frac{n(n-1)(n-2)}{3!}x^3$$

$$+ \frac{n(n-1)(n-2)(n-3)}{4!}x^4 + \cdots$$

where $\cdots$ indicates that, in general, an infinite number of terms follows. Applying this series to $(1+x)^3$, we obtain

$$(1+x)^3 = 1 + 3x + \frac{3 \cdot 2}{1 \cdot 2}x^2 + \frac{3 \cdot 2 \cdot 1}{1 \cdot 2 \cdot 3}x^3 + \frac{3 \cdot 2 \cdot 1 \cdot 0}{1 \cdot 2 \cdot 3 \cdot 4}x^4 + 0$$

Note that the fifth term is zero and that all following terms are zero. Hence

$$(1+x)^3 = 1 + 3x + 3x^2 + x^3$$

However, for any power that is not a positive integer, this series does not terminate but is an infinite series. We will see, however, that in our application, all but the first few terms are negligible. The first term in Eq. (14-20) can be written

$$m_0 c^2 \left(1 - \frac{v^2}{c^2}\right)^{-1/2} = m_0 c^2 \left[1 + \tfrac{1}{2}\frac{v^2}{c^2} + \tfrac{3}{8}\frac{v^4}{c^4} \cdots \right] \qquad (14\text{-}22)$$

by applying the binomial series. Now note that for ordinary velocities, the ratio v/c is quite small, so that v^2/c^2 and v^4/c^4 are extremely small fractions. For example, a rocket leaving the earth
typically moves at about 10^4 m/s. Since the speed of light is 3×10^8
m/s, we have

$$v/c = \tfrac{1}{3} \times 10^{-4}$$

$$v^2/c^2 \approx 10^{-9}$$

$$v^4/c^4 \approx 10^{-18}$$

Thus, even for this large velocity, the third term in (14-22) is nine
orders of magnitude smaller than the second term and can be comfortably ignored (higher terms are still smaller). Accordingly, we
obtain for (14-20),

$$K = m_0 c^2 \left(1 - \frac{v^2}{c^2}\right)^{-1/2} - m_0 c^2 \approx m_0 c^2 + \tfrac{1}{2} m_0 v^2 - m_0 c^2 = \tfrac{1}{2} m_0 v^2$$

where the approximate equality is extremely accurate for ordinary
velocities. Thus it is that classical mechanics was not discovered
to be in error until particles (such as high-energy electrons) that
moved with velocities close to that of light were studied early in
the twentieth century.

Examining (14-20), we see that the kinetic energy of a moving
body approaches infinity as $v \rightarrow c$. Since this energy must be supplied by the work done on the body, it is clear that c is a "speed
limit" for any material body. There is a particle that moves precisely at c, called the *photon*. In modern quantum theory, a photon
is the carrier of any kind of electromagnetic radiation, including
light. There is no discrepancy with (14-20) for the case of photons,
however, as the rest mass of the photon is identically zero.

There have been many experimental attempts to discover
particles that move faster than light, but, thus far, they have been
unsuccessful. The name *Tachyon* has been given to these hypothetical particles. One of the motivations for supposing the possible
existence of such particles is the observation that (14-20) is finite
for speeds greater than c if the mass of the particle is given by an
imaginary number. Interested readers will find a discussion in "Particles that Go Faster than Light" by Gerald Feinberg in the February 1968 issue of *Scientific American*.

14-11   Summary

The work–kinetic energy theorem for variable forces with arbitrary direction can be written

$$W = \int_{r_1}^{r_2} F \cos \theta \, dr = \tfrac{1}{2}mv_2{}^2 - \tfrac{1}{2}mv_1{}^2$$

Both sides of the last equality are scalars. Thus, the theorem cannot give us information concerning the direction of the velocity but only its magnitude. The theorem is useful when we know the path followed by the mass or when the force does not vary in direction.

The work done by a spring with spring constant k is given by

$$W = \tfrac{1}{2}k(x_1{}^2 - x_2{}^2)$$

where x is the displacement of the spring from its equilibrium length.

For the constant force of gravity near the earth's surface, we find that

$$W = mg(y_1 - y_2) \qquad W = mgh$$

WORK DONE BY Gravity

regardless of the path followed. Accordingly, frictionless motion along any path leads to a velocity variation that depends only upon height. One note of caution: this does not imply that the same *time* is required for an object to fall vertically through a height h and to slide down an inclined plane with the same height, h (see Problem 14-17).

For rotational motion of a rigid body, the kinetic energy is given by

$$K = \tfrac{1}{2}I\omega^2$$

and work by

$$W = \int_{\theta_1}^{\theta_2} \Gamma \, d\theta$$

For combined rotation of a rigid body about its center of mass and translation of the center of mass, we have

$$K_{\text{total}} = \tfrac{1}{2}mv_{\text{cm}}^2 + \tfrac{1}{2}I\omega^2$$

where I is the moment of inertia about the center of mass.

Power is the rate of doing work: $P = dW/dt$. An engine that delivers its maximum power at a given rotational velocity can be coupled to a load that requires a different angular velocity by means of gears or torque converters without a change in power (aside from friction).

In relativistic mechanics it is found that kinetic energy is not given simply by $\tfrac{1}{2}mv^2$ but rather by

$$K = \frac{m_0 c^2}{\sqrt{1 - \dfrac{v^2}{c^2}}} - m_0 c^2$$

However, for ordinary velocities this expression reduces to the classical expression $K = \frac{1}{2}mv^2$ very accurately. The kinetic energy in relativistic mechanics approaches infinity as $v \to c$. The infinite amount of work thus required to bring a particle up to $v = c$ implies that c is an absolute speed limit in the universe.

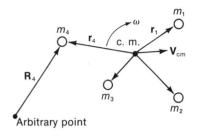

Fig. 14-12. Several masses rotating about their center of mass, which is also translating.

APPENDIX: Kinetic Energy of a Translating, Rotating, Rigid Body

The position of n particles labeled $i = 1, 2, 3, \ldots, n$ is given with respect to an arbitrary point by the position vectors $\mathbf{R}_i$. Figure 14-12 illustrates this for four masses and shows $\mathbf{R}_4$ explicitly. Any of the $\mathbf{R}_i$ can be written as a sum of $\mathbf{R}_{cm}$, the position vector that locates the center of mass, and $\mathbf{r}_i$, the position vector of the ith mass with respect to the center of mass.

$$\mathbf{R}_i = \mathbf{R}_{cm} + \mathbf{r}_i$$

Differentiating, we obtain

$$\mathbf{V}_i = \mathbf{v}_{cm} + \mathbf{v}_i$$

where $\mathbf{V}_i$ is the velocity of the ith particle relative to the origin and $\mathbf{v}_i$ its velocity relative to the center of mass.

The kinetic energy is the sum of all the individual kinetic energies:

$$K_{total} = \frac{1}{2}\Sigma_i m_i V_i^2$$

Using the law of cosines to add $\mathbf{v}_{cm}$ and $\mathbf{v}_i$ we find

$$K_{total} = \frac{1}{2}\Sigma_i m_i (v_{cm}^2 + 2v_{cm}v_i \cos\theta_i + v_i^2)$$

where θ_i is the angle between $\mathbf{v}_i$ and $\mathbf{v}_{cm}$.

$$K_{total} = \frac{1}{2}\Sigma_i m_i v_{cm}^2 + (\Sigma_i m_i v_i \cos\theta_i) v_{cm} + \frac{1}{2}\Sigma_i m_i v_i^2 \quad (14\text{-}23)$$

The first term of (14-23) is one-half the total mass of the system multiplied by the center-of-mass velocity squared. The second term contains a sum in parentheses that we recognize as the component of the total momentum of the center of mass that is parallel to $\mathbf{v}_{cm}$. However, it represents the momentum of the center of mass with respect to the center-of-mass frame of reference and hence vanishes. Each term within the last sum in (14-23) can be written $\frac{1}{2}m_i r_i^2 \omega^2$ only if the masses rotate as a rigid body about their center of mass:

$$K_{total} = \frac{1}{2}M_{total} v_{cm}^2 + \frac{1}{2}\Sigma_i m_i r_i^2 \omega^2$$

$$= \frac{1}{2}M_{total} v_{cm}^2 + \frac{1}{2}I\omega^2 \quad (14\text{-}24)$$

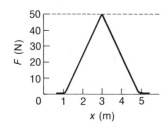

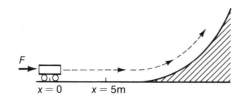

Fig. 14-13.

Equation (14-24) is the result we wished to show: the kinetic energy of a rigid collection of objects is given by the sum of the translational kinetic energy of the center of mass and the rotational kinetic energy about the center of mass.

Problems

14-1. An automobile has a coefficient of rolling friction of 0.04. A driving force equal to 1/10 of its weight is applied to it. If the car starts from rest, how far must it move before its velocity equals 20 ft/s?

14-2. The force acting on a mass is given by $F = C_1 + C_2 x$, where C_1 and C_2 are constants and x is the position. Find the work done by the force while the mass moves from the position x_1 to x_2 if the force is always parallel to the displacement.

14-3. The force graphed in Fig. 14-13 is applied horizontally to a 5-kg cart, which then coasts up a ramp, as shown. Find y_{max}, the maximum height reached by the cart.

14-4. A block starts from rest and slips down a frictionless surface, as in Fig. 14-14, then reaches a portion of the path where the coefficient of friction is $\mu = 0.5$. This portion is an inclined plane tipped 10° below the horizontal. Find the distance, L, that is covered by the block on the inclined plane before it comes to rest.

14-5. Find the power generated by a jet engine with 10,000-lb thrust when it is moving at (a) 250 mi/hr; (b) 500 mi/hr.

14-6. The drag forces on a 2800-lb car total 80 lb at 45 mi/hr. Find the power required to drive this car 45 mi/hr up a hill that is 10° above the horizontal.

14-7. The drag forces on an airplane are given by $D = C_1 v + C_2 v^2$, where $C_1 = 12$ N/m/s and $C_2 = 5 \times 10^{-3}$ N/m²/s². Find the power (in kW) required to move this airplane at a steady 600 km/hr.

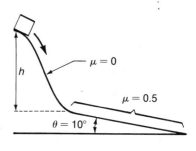

Fig. 14-14.

14-8.   Compare the amount of work required to move an automobile a given distance at two different constant speeds,
v and $2v$, when the drag forces on the auto are (*a*) constant, (*b*) proportional to v, and (*c*) proportional to v^2.

14-9.   It is said that stop-and-go driving consumes more gasoline
than driving at a constant speed. Compare the amount of
work required to bring a 3200-lb car up to 30 mi/hr to that
required to move the car at a constant 30 mi/hr for one
block (250 ft), if the total drag forces at 30 mi/hr are 100 lb.

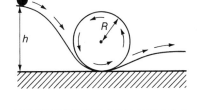

Fig. 14-15.

14-10. Compare the kinetic energy of the earth's rotation on its
axis to its translational kinetic energy due to its orbital
motion about the sun.

14-11. The day is increasing in length by 1/1000 of a second per
century due to a loss in the earth's rotational kinetic energy
caused by tidal friction. Calculate the average power that
the tides are dissipating.

14-12. In Section 11-3 it was shown that a cylinder rolling on an
inclined plane experiences an acceleration $a = \frac{2}{3}g \sin \theta$. In
Section 14-8 (Eq. (14-18)) we showed that such a wheel
starting from rest obtains a velocity given by $v^2 = \frac{4}{3}gh$.
Show that these results are compatible.

14-13. A solid spherical ball is placed at rest on a loop-the-loop
track, as in Fig. 14-15. It rolls without slipping down the
ramp and, if it has sufficient velocity, goes up and around
the circle without losing contact with the track. Find the
minimum height h that allows this. Your answer will be
in terms of R, the radius of the loop. See Prob. 9-5.

14-14.* As discussed in an example in Section 14-7, an ice skater
must do work to decrease her moment of inertia from I_1 to
I_2 so that conservation of angular momentum can increase
ω by the formula

$$\omega_2 = \frac{I_1}{I_2}\omega_1$$

Use a simple model to show that this work equals the gain
in kinetic energy of rotation, which is given by (14-17).
Assume that a single mass m rotates with speed ω_1 at a
radius r_1, so that its initial moment of inertia is mr_1^2. The
mass is slowly pulled into a smaller radius, so that $I_2 = mr_2^2$
is smaller than I_1. The inward centripetal force is given by

$$F = m\omega^2 r$$

Integrate this force from r_1 to r_2 to find the net work done in pulling m from r_1 to r_2. Does it account for the gain in kinetic energy?

14-15.* The drag forces on a racing car are given by $F = a + bv$, where a is 100 lb, b is 4.5 lb/ft/s, and v is the velocity of the race car in ft/s. If the radius of the driving wheels is 14 in., find the gear ratio that couples the engine to the rear wheels so that maximum velocity can be obtained. The power versus rotational velocity for the engine is given in Fig. 14-16.

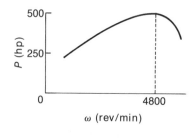

Fig. 14-16.

14-16. Suppose that the drag forces on a car are given by a constant term that represents bearing friction, rolling friction, etc., added to a term that varies with the square of the velocity to represent air resistance. Consider a medium-sized American car. If you have ever pushed a stalled car on a level road, you can estimate the constant term. Let us use a value of 100-lb. From your rough knowledge of the horsepower of such a car and its top speed, you can also estimate the coefficient of the v^2 term. For this problem take the power available at the rear wheels to be 100 hp and the top speed to be 100 mi/hr. Now that you have values for a and b in

$$F_{\text{drag}} = a + bv^2$$

compare the energy required to move the car over a distance of 10 mi at a constant 55 mi/hr to that required at a constant 70 mi/hr.

14-17. A mass starts from rest and slides down a frictionless surface from point a to a point c, as shown in Fig. 14-17. By conservation of energy we know the final velocity is $v = \sqrt{2gh}$ for all three paths shown: ac, abc, and $ab'c$. Find the time required to traverse each path.

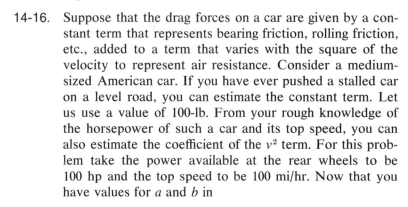

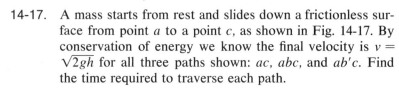

Fig. 14-17. A mass slides from a to c along any of three paths: ac, abc or $ab'c$. Consider the corner at b' to be sufficiently smooth so that the mass coasts from b' to c with the full velocity it gained in falling to b'.

Chapter *15*

Potential Energy and Conservation of Energy

15-1 Introduction In this chapter we will continue and extend our discussion of energy to include *potential energy*. When a projectile is thrown vertically upward, the force of gravity does negative work on it so that its kinetic energy diminishes until at its peak it has no kinetic energy ($v = 0$). After the projectile reaches its maximum altitude and begins to fall, the work done by gravity increases the kinetic energy until it strikes the ground, arriving at the ground with the same velocity, and thus kinetic energy, as it started (if air resistance is negligible). The negative work done by gravity during the ascent is precisely mirrored by the positive work on the descent. The force of gravity apparently "borrows" kinetic energy from the projectile and then returns it on the descent. Hence the name *potential energy* is given to the energy that is expanded against the earth's gravity. Once a body has been elevated, the potential for "repayment" is there.

In a similar fashion, a moving mass can be brought to rest by compression of a spring. The lost kinetic energy is returned to the mass when the spring bounces back. Forces that repay the energy given to them are called *conservative* forces, since they conserve that energy for later return. Some forces, such as that due to friction, never return lost energy in the form of kinetic energy. Such forces are called *nonconservative* or *dissipative*. We will see in subsequent chapters that this dissipated kinetic energy is not truly lost. Most typically it has been transformed into heat, another form of energy.

15-2   Gravitational Potential Energy

The work done by gravity when a mass changes height from y_1 to y_2 is, as we have seen,

$$W_{12} = mg(y_1 - y_2)$$

The negative of this quantity is the work we do against the force of gravity when we raise an object of mass m. We will see that this work reappears as kinetic energy when the mass falls back to y_1. Accordingly, let us define the potential energy U_g for the earth's gravity by

$$\Delta U_g = -mg(y_1 - y_2) = mgy_2 - mgy_1$$

In other words,

$$U_g = mgy \tag{15-1}$$

The change in kinetic energy of a mass acted upon only by gravity is given by the work done by gravity:

$$\tfrac{1}{2}mv_2{}^2 - \tfrac{1}{2}mv_1{}^2 = mgy_1 - mgy_2$$

and we can rearrange terms to obtain

$$\tfrac{1}{2}mv_1{}^2 + mgy_1 = \tfrac{1}{2}mv_2{}^2 + mgy_2$$

This is equivalent to

$$K_1 + U_1 = K_2 + U_2 \tag{15-2}$$

where K is kinetic energy and U is gravitational potential energy. If we define the *total mechanical energy* by $E = K + U$, then (15-2) becomes

$$E_1 = E_2$$

Equation (15-2) is a statement of the *law of conservation of energy*. Looking at (15-2) we see that it means simply that any increase (or decrease) in kinetic energy must be exactly offset by a decrease (or increase) in potential energy. Of course, it was for this reason that we defined potential energy as we did. Thus far we have developed the law for masses moving under the influence of gravity alone.

Example   A 2-kg mass is dropped from a height of 100 m. Demonstrate the constancy of its total energy.

Solution   Taking ground level to be the origin of our y axis (this choice will be shown to be unimportant), the initial energy as the ball is released is given by

$$E_1 = U_1 + K_1 = mgy_1 + \tfrac{1}{2}mv_1{}^2$$

$$= 2 \text{ kg} \cdot 9.8 \, \frac{m}{s^2} \cdot 100 \text{ m} + 0$$

$$= 1960 \text{ J}$$

After the ball has fallen $s = 100$ m we have

$v_2{}^2 = 2gs$ *(handwritten)*

$$v_2{}^2 = 2gs = 2 \cdot 9.8 \cdot 100 \text{ m}^2/\text{s}^2 = 1960 \text{ m}^2/\text{s}^2$$

$$E_2 = U_2 + K_2 = mgy_2 + \tfrac{1}{2}mv_2{}^2$$

(handwritten: no potential energy)

$$= 0 + 1960 \text{ kg} \cdot \text{m}^2/\text{s}^2 = 1960 \text{ J}$$

If we had chosen $y = 0$ as the starting point, we would have written:

$$E_1 = U_1 + K_1 = 0 + 0 = 0 \text{ J}$$

$$E_2 = U_2 + K_2 = 2 \text{ kg} \cdot 9.8 \, \frac{m}{s^2} \cdot (-100 \text{ m}) + \tfrac{1}{2} \cdot 2 \text{ kg} \cdot v_2{}^2$$

$$= -1960 \text{ J} + 1960 \text{ J} = 0 \text{ J}$$

In the above example the value of the potential energy depends upon the arbitrarily chosen location of the origin of coordinates. Nonetheless, the energy conservation law is satisfied, since both sides of (15-2) are affected equally by this choice.

Exercises

1. A car with an initial velocity of 10 m/s rolls down a 30° slope for a distance of 80 m. Find its final velocity by use of Eq. (15-2).
2. A stone is thrown upward with an initial velocity of 20 ft/s. Use (15-2) to find its velocity after it has risen 4 ft.

15-3   Elastic Forces

The elastic force due to a spring closely obeys Hooke's law:

$$F = -kx$$

In Chapter 14 we saw that the work done by a spring when the end of the spring moves from x_1 to x_2 (measured from the equilibrium position) is

$$W = \tfrac{1}{2}kx_1{}^2 - \tfrac{1}{2}kx_2{}^2 \qquad (14\text{-}9)$$

Thus we define its elastic potential energy to be given by

$$U = \tfrac{1}{2}kx^2 \qquad (15\text{-}3)$$

(handwritten: elastic potential energy)

With the potential energy defined by (15-3), the work–kinetic energy theorem once again leads to a conservation of total energy for a mass acted upon only by an elastic force. The total energy is now given by:

$$E = U + K = \tfrac{1}{2}kx^2 + \tfrac{1}{2}mv^2 \qquad (15\text{-}4)$$

and by application of the work–kinetic energy theorem, with the work given by (14-9), we see that

$$U_1 + K_1 = U_2 + K_2 \qquad \text{or} \qquad E_1 = E_2$$

That is, E remains constant. Whether the spring is compressed $(x < 0)$ or stretched $(x > 0)$, its potential energy increases. If a mass is connected to a stretched spring and released, the spring will pull the mass toward the equilibrium point $(x = 0)$. As it does, the spring loses potential energy and the mass gains kinetic energy. Thus the total energy, $U + K$, is constant.

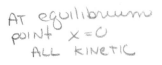

AT equilibrium point $x = 0$ ALL KINETIC

Once $x = 0$ is reached, the mass will overshoot and change its kinetic energy back into stored potential energy of the spring. Thus, in the absence of friction, a mass attached to a spring will oscillate back and forth, continually changing the energy of the system from kinetic to potential and back again.

Example A spring is attached to a mass, as shown in Fig. 15-1, and frictional forces between the block and surface are negligible. The spring is stretched to $+x_0$ then released from rest. Find the square of its velocity at (a) $x = x_0/2$; (b) $x = 0$; (c) $x = -x_0$.

Solution The initial energy is given by

$$E_1 = U_1 + K_1 = \tfrac{1}{2}kx_0^2 + 0$$

(since the initial velocity is zero). Since the total energy remains constant, the final energy must equal the initial energy.

(a)
$$U_2 + K_2 = U_1 + K_1$$

$$\tfrac{1}{2}k\left(\frac{x_0}{2}\right)^2 + \tfrac{1}{2}mv^2 = \tfrac{1}{2}kx_0^2$$

$$\tfrac{1}{2}mv^2 = \tfrac{1}{2}kx_0^2 - \tfrac{1}{8}kx_0^2$$

$$v^2 = \tfrac{3}{4}\frac{k}{m}x_0^2$$

(b)
$$0 + \tfrac{1}{2}mv^2 = \tfrac{1}{2}kx_0^2$$

so that
$$v^2 = \frac{k}{m}x_0^2$$

(c)
$$\tfrac{1}{2}k(-x_0)^2 + \tfrac{1}{2}mv^2 = \tfrac{1}{2}kx_0^2$$

$$v^2 = 0$$

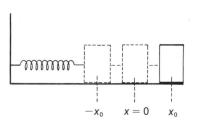

Fig. 15-1.

At $x = -x_0$ the mass has compressed the spring and has come to rest at a distance beyond the equilibrium point that is equal in magnitude to its original distance from the equilibrium point. The mass will now accelerate back to the right, repeating the entire sequence of events.

When both elastic and gravitational forces act on a body, the work done on the body consists of two terms. If a spring is oriented vertically, so that the force on the body due to the spring is

$$F_s = -ky$$

and the gravitational force is

$$F_g = -mg$$

we find the work done by both the elastic and gravitational forces to be

$$W = \tfrac{1}{2}k(y_1^2 - y_2^2) + mg(y_1 - y_2) = -\Delta U_s - \Delta U_g$$

where U_s and U_g are given by (15-3) and (15-1), respectively. Applying the work–kinetic energy theorem, we find

$$\Delta K = \tfrac{1}{2}mv_2^2 - \tfrac{1}{2}mv_1^2 = \tfrac{1}{2}k(y_1^2 - y_2^2) + mg(y_1 - y_2)$$

That is, any decrease (increase) in potential energy is balanced by an increase (decrease) in kinetic energy. We thus define the total energy to be the sum of three terms: the kinetic energy, the elastic potential energy, and the gravitational potential energy. The sum of these three energies is the total energy, and the work–kinetic energy theorem assures us that the total energy will be constant:

$$K_1 + U_{s_1} + U_{g_1} = K_2 + U_{s_2} + U_{g_2}$$

Exercises

3. It is found that a given spring is compressed by 10 cm when a 10-N force is applied to it. This same spring is connected to a 4-kg mass, as in Fig. 15-1, and compressed so that $x_1 = -50$ cm. The mass is then released. Find the velocity of the 5-kg mass at $x_2 = +40$ cm.

4. A vertically oriented spring attached to the floor has a force constant $k = 5 \times 10^3$ N/m. It is compressed by 10 cm and clamped there. A 5-kg mass is placed on top of the spring, which is then released to project the mass vertically upward. How high above the equilibrium position of the unloaded spring will the mass rise? The mass is not attached to the spring, so the spring simply acts as a vertical catapult.

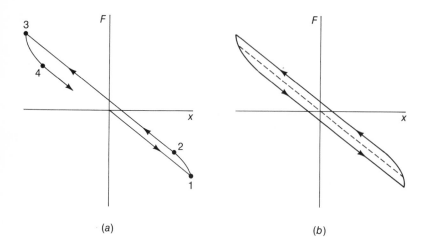

(a)   (b)

Fig. 15-2. An exaggerated illustration of the behavior of a spring. The dotted line shows Hooke's law, the solid line the true behavior as it is first stretched, then compressed.

15-4   Conservation of Energy and Dissipative Forces

NOT ON
2ND HOURLY

We have seen that elastic and gravitational forces both lead to a useful definition of potential energy. We have defined potential energy so that the law of conservation of mechanical energy holds:

$$\Delta U + \Delta K = 0$$
or
$$U + K = \text{constant}$$
or
$$U_1 + K_1 = U_2 + K_2$$

(15-5)

where U is the potential energy, which typically depends on one or more coordinates, and K is the kinetic energy, which depends upon the square of velocity. Forces for which (15-5) can be written, like gravitation, are called *conservative* forces since the conservation of mechanical energy is applicable. The force of friction, however, *always* opposes the motion of an object. It always leads to a decrease in kinetic energy unless an external force supplies the energy lost to friction. Actually, this energy is not *lost;* it is transformed into heat, which is just another form of energy. If we take a sufficiently microscopic view, we discover that the "lost" kinetic energy appears in the form of the increased kinetic energy of the motion of the atoms that constitute the friction-warmed objects. The discovery that heat energy is directly related to the energy of atomic motion was a high point of nineteenth century physics and will be discussed in Chapter 19.

An interesting example of a dissipative force is given by the mechanical *hysteresis* of a spring. In actual practice springs do not *exactly* obey Hooke's law, as we usually assume. Instead, the displacement of the spring as it is elongated and compressed lags slightly behind the value predicted by $F = -kx$. This situation is shown in Fig. 15-2. The material of the spring yields slightly as it is stretched and causes the lagging phenomenon called hysteresis. This flow involves the slippage of some atoms past others, which causes internal friction within the material of the spring itself. This internal friction causes heating of the spring, which can become quite pronounced if it is repeatedly stretched and compressed. In Fig. 15-2*a* we see the lag of the spring from 1 to 2 when the motion first reverses and the opposite lag, from 3 to 4, when the direction of motion is again reversed. Figure 15-2*b* shows the *hysteresis loop* traced out by the force as the spring is repeatedly stretched and compressed.

Fig. 15-3 shows the work done by the spring when it obeys Hooke's law. The work done between x_1 and x_2 is given by

$$W = \int_{x_1}^{x_2} F\, dx = -k \int_{x_1}^{x_2} x\, dx$$

This work is proportional to the shaded area but changes sign depending on whether the spring is elongating (x increasing) or compressing (x decreasing). Thus the negative work done by the spring from x_1 to x_2 is cancelled by the positive work done by the spring returning from x_2 to x_1, i.e., total energy is conserved.

However, when hysteresis is present, as in Fig. 15-4, the work done from x_1 to x_2 is *not* cancelled by the work done from x_2 back to x_1. The difference in work for the two directions of motion is precisely the area contained within the hysteresis loop, which lies between x_1 and x_2. This area represents work (or energy) lost by

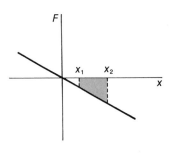

Fig. 15-3. The work done by the spring between x_1 and x_2 is equal to the shaded area.

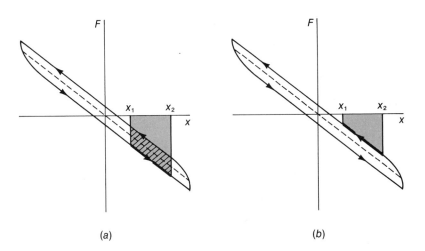

(a)   (b)

Fig. 15-4. The work done by the spring between x_1 and x_2 is not cancelled on the return when hysteresis is present. The difference in work between these two directions of motion equals the shaded area *within* the hysteresis loop between x_1 and x_2.

the spring. The net energy lost during one full cycle around the hysteresis loop is just the total area encompassed by the loop, as shown in Fig. 15-5. This area has the dimensions of work, and hence of energy. Somewhat more formally, we may say:

$$\oint F\,dx = \int_A^B F\,dx + \int_B^A F\,dx = \text{hysteresis loss}$$

where the circle on the integral sign is a shorthand notation used to indicate one complete cycle of motion.

From the preceding consideration it is clear that the repetitive flexing of any object that does not exhibit perfect Hooke's-law behavior will lead to a loss of energy. This is true whether the object is a spring, the main spar of an aircraft wing, or the cone support of a loudspeaker. This energy generally appears as heat. The hysteresis loop followed by the steel in a ball bearing can partly explain the small loss of energy that occurs. Though this loss is small, the fact that different segments of a steel ball undergo compression and expansion as the ball rolls under a load implies an energy loss, so that the coefficient of rolling friction is not zero. The design of a high-quality ball bearing is thus related to a choice of materials with suitable hysteresis properties. The hysteresis exhibited by most materials is much less than that indicated in Figures 15-2, 15-4, and 15-5 and is not visible on graphs of that scale.

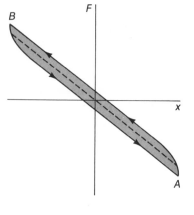

Fig. 15-5. The total loss of energy for one full cycle of motion, say from A to B and back to A, is just equal to the area enclosed by the hysteresis loop.

Exercise

5. A steel beam flexed by an amount x exhibits the hysteresis behavior shown in Fig. 15-6. Find the net energy loss for one cycle of flexing.

15-5   The General Relation between Force and Potential Energy

If a force is not dissipative, we have seen that a potential energy can be defined for that force. We have already considered the force of gravity and forces exerted by ideal springs as examples. The general approach we used was to calculate the work "stored" in the system, work that could be returned and hence represented *potential* energy. When we stretch a spring, for example, we do positive work, since the force we exert is in the same direction as the displacement. The spring force itself does negative work, since the force *it* exerts opposes the displacement. It is the work *we do* that is stored by the spring and that is called potential energy. But since Newton's third law tells us that the spring's force is exactly

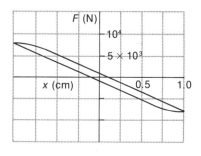

Fig. 15-6.

equal and opposite to that that we exert, we may equally well write the potential energy change in terms of the *spring* force if we remember the minus sign:

$$U_2 - U_1 = - \int_{x_1}^{x_2} F_{\text{spring}} \, dx = - \int_{x_1}^{x_2} -kx \, dx$$

Thus, as we saw in Section 15-3, we choose $U = \frac{1}{2}kx^2$ as the potential energy of a spring, so that $K_1 + U_1 = K_2 + U_2$.

A general definition of potential energy for nondissipative forces in one dimension is given by

$$U(x_2) - U(x_1) = - \int_{x_1}^{x_2} F \, dx \qquad (15\text{-}6)$$

or more briefly,

$$U(x) = - \int F \, dx \qquad (15\text{-}7)$$

Notice that this definition of potential energy defines only the *change* in potential energy. This is as it should be, for in any application of (15-2), it is only changes of potential energy that are involved.

Whenever an integral equation such as (15-7) is given, we can write an equivalent differential equation. This is because the derivative and integral are inverse operations:

$$U(x) = - \int F \, dx \qquad (15\text{-}8)$$

Differentiating both sides:

$$\frac{d}{dx} U(x) = - \frac{d}{dx} \int F \, dx = -F$$

so that

$$F = -\frac{dU}{dx} \qquad (15\text{-}9)$$

Equation (15-9) gives us the differential equation that is equivalent to the integral equation (15-8). One can obtain (15-8) simply by integrating both sides of (15-9). The differential form, $F = -dU/dx$, points out a useful feature of a graph of potential energy for a system. The force acting at any value of the position x is given by the negative of the slope of $U(x)$ at that point. In other words, the force points in the "downhill" direction on a plot of U vs x.

A plot of U vs x has another valuable feature in that any bounds on the possible motion of a system due to energy conservation are easily seen on such a plot. Consider, for example, a mass attached

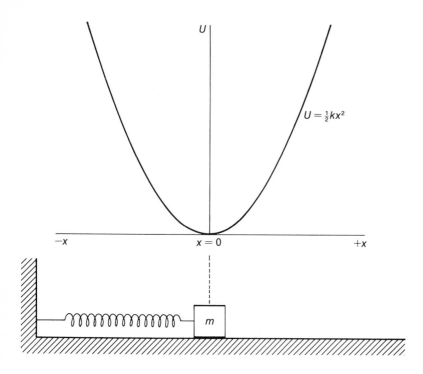

to a spring that exerts a force $F = -kx$ upon the mass. Suppose that no other significant forces act upon the mass. Intuitively, we know that if we pull on the mass, stretching the spring, and then release it, the mass will accelerate back toward the equilibrium point. However, when it reaches the equilibrium point, where $F = 0$, it will have acquired a large velocity and so will overshoot. The inertia of the mass will carry it on, causing the spring to compress until the mass finally comes to rest for an instant. But now the compressed spring will reaccelerate the mass in the opposite direction, once again causing an overshoot. The mass thus oscillates back and forth. The key features of this sort of motion can be shown clearly on a plot of $U(x)$, as graphed in Fig. 15-7. The direction of the force is in the "downhill" direction of such a graph, which is to the right on the left half ($x < 0$) of the potential energy curve and to the left on the right half. This agrees with what we already know about Hooke's law. We can, of course, obtain Hooke's law from the potential energy by application of (15-9):

$$F = -\frac{dU}{dx} = -\frac{d}{dx}\left(\tfrac{1}{2}kx^2\right) = -kx$$

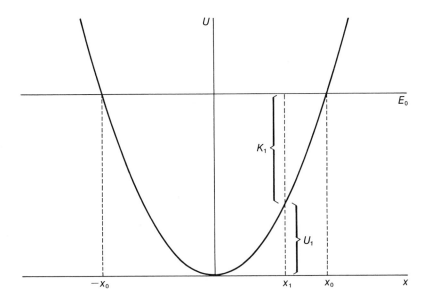

Fig. 15-8. The initial energy of the mass released from rest is given by E_0. The value of $K = \frac{1}{2}mv^2$ can be found *anywhere* by finding the difference between E_0 and U at that point, as illustrated at x_1.

Suppose now that x_0 is the point from which the mass shown in Fig. 15-7 is released. Since $K_0 = 0$ (if v is initially zero), the total energy is just the initial potential energy:

$$E_0 = U_0 + K_0 = \tfrac{1}{2}kx_0^2 + 0$$

This initial value is indicated by the horizontal line in Fig. 15-8. From the law of conservation of energy (Eq. 15-5) the total energy of this system must always equal E_0. When the mass is released at x_0, it moves toward smaller x. As it does, it gains kinetic energy. The value of the kinetic energy *at any point* is simply the difference between $U(x)$ at that point and E_0. An example is shown at the point x_1 in Fig. 15-8.

Now, following the motion of the mass toward the negative x direction, we see that it has maximum kinetic energy at $x = 0$ and then, as the compression of the spring slows the mass, the kinetic energy diminishes as x approaches $-x_0$. At $x = -x_0$, the potential energy once again equals E_0, thus leaving zero kinetic energy. This point is appropriately called a *turning point,* for now the force acting to the right reaccelerates m back toward $x = 0$.

The regions to the left of $x = -x_0$ and to the right of $+x_0$ are "forbidden." We see this because at either of those points, $K = 0$ and hence $v = 0$. Further, the force is pushing back toward $x = 0$. Thus, the mass cannot penetrate the forbidden region. Since $U > E_0$ in the forbidden region, K would have to be negative there. But $K = \frac{1}{2}mv^2$ is always positive.

It is interesting to note, at this point, that the area of mechanics that is capable of dealing with atomic phenomena, quantum mechanics, does not lead to a *strictly* forbidden region. Quantum mechanics instead states that the *probability* of finding a system in this energetically forbidden region is small and quickly approaches zero if the system strays very far into the classically forbidden region. Although the ideas of quantum mechanics are novel and of the utmost importance in understanding atomic phenomena, we are getting somewhat ahead of ourselves. A reasonable degree of understanding of classical mechanics and of wave motion is an excellent preparation for an introduction to quantum mechanics; so let us return to the classical realm.

Example   A 3-kg mass is attached to a spring with a force constant of 75 N/m. The mass is set into motion at $x = 0$ (the equilibrium position of the spring) with an initial velocity of 5 m/s. Find the turning points of the subsequent motion.

Solution   The mass moves in the positive x direction until all of its kinetic energy is absorbed into potential energy. It then moves back toward the origin, overshoots, and moves toward the negative x direction until, again, it has changed all of its kinetic energy into potential energy. Initially,

$$E_0 = U_0 + K_0 = \tfrac{1}{2}k(0)^2 + \tfrac{1}{2}mv_0^2$$

The turning points are those points at which the mass is momentarily at rest, i.e., $K = 0$. Setting $K = 0$ in the relation $E = U + K$, we obtain $U = E_0$ or

$$\tfrac{1}{2}kx_t^2 = E_0 = \tfrac{1}{2}mv_0^2$$

where x_t is the value of x at the turning point:

$$x_t = \pm v_0 \sqrt{m/k} = \pm \sqrt{\frac{3 \text{ kg}}{75 \text{ N/m}}} \cdot 5 \text{ m/s} = \pm 1 \text{ m}$$

$$\tfrac{1}{2}x^2 = v_0 (m/k)$$

A useful property of a graph of U vs x is that it quickly allows us to distinguish between stable and unstable equilibrium. Wherever the $U(x)$-vs-x graph has zero slope, the force is zero and, hence, the system can be in equilibrium there. If the curvature is positive (a positive second derivative, which means the curve "holds water") the equilibrium is stable, for a small displacement leads to a restoring force. The equilibrium point of a spring is an example. If, on the other hand, the graph of $U(x)$ has negative curvature at an equilibrium point, any small displacement leads to a force that acts to increase the displacement. Such a situation is called *unstable equilibrium* and is exemplified by a roller skate balanced on a hilltop that falls off to either side.

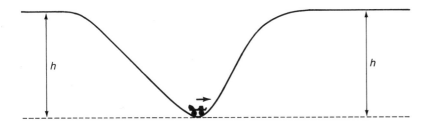

Fig. 15-9.

Exercises

6. The potential energy of a body is given by $U = C_1 x^3 + C_2$. Find an expression for the force that acts upon the body.

7. A $\frac{1}{2}$-slug mass is attached to a spring with a force constant of 250 lb/ft. The spring is stretched 6 in., then released (so that $v_0 = 0$). Find the velocity of the mass at $x = \pm 3$ in.

8. A roller skate is caught in a trough of depth h, as in Fig. 15-9. It has a velocity v_0 when it is crossing the bottom of the trough. (a) In the absence of friction, find the minimum value of v_0 required for the skate to get out of the trough. (b) If v_0 is only half the escape velocity found in part a, sketch the turning points of this system.

15-6 Interatomic Forces

The precise nature of the force between two atoms varies with the atomic species involved and with the nature of the bond, if any, that is formed, i.e., covalent or ionic. A reasonably typical behavior is shown in Fig. 15-10. There, the interatomic force is given in terms of r, the separation distance between two atoms. Positive values of F in Fig. 15-10 indicate repulsion (the forces act to increase r), while negative values indicate attraction. If we now integrate the force over distance to obtain the potential energy and assume that $U \to 0$ as $r \to \infty$, we obtain $U(r)$ as given in Fig. 15-11. The units of U are electron volts (eV) with $1 \text{ eV} = 1.602 \times 10^{-12} \text{ erg} = 1.602 \times 10^{-19} \text{ J}$. The reader should check, at least qualitatively, that application of $F_r = -du/dr$ to the graph of Fig. 15-11 *does* lead to the graph in Fig. 15-10. The reader may well wonder how forces as small as 10^{-5} dyne (one newton $= 10^5$ dyne) are measured. In fact, both the theoretical calculation and experimental measurement of interatomic interactions yield direct information on the atomic kinetic and potential *energies* rather than the forces involved. Thus, the atomic physicist works directly with information of the sort

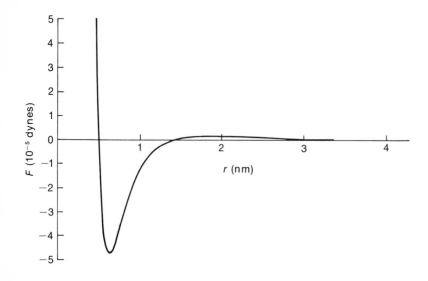

Fig. 15-10. The force between two atoms as a function of r, their center-to-center spacing. Units of F are dynes and units of r are nanometers (10^{-9} m).

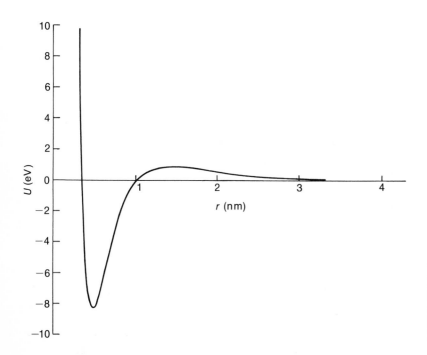

Fig. 15-11. Potential energy between two atoms. Units of U are electron volts ($= 1.6 \times 10^{-19}$ joule) and units of r are nanometers.

shown in Fig. 15-11. Figure 15-10 was, in fact, derived from Fig. 15-11 to show how the force itself must behave.

The potential-energy diagram of Fig. 15-11 allows us to understand some of the behavior exhibited by two interacting atoms. If the atoms are initially far apart, a potential energy barrier prevents their close approach unless they have sufficient kinetic energy. If their initial energy (given by their initial kinetic energy at large r, where $U \approx 0$) is less than the height of the barrier near $r = 1.5$ nm, they will rebound at r_1 (distance of closest approach), as shown in Fig. 15-12. If they are given sufficient kinetic energy to surmount this barrier, as in Fig. 15-13, they come very close together until the forces of repulsion caused by their interpenetration bring them to rest at r_2. This is a turning point, since the force now accelerates the two atoms away from one another again.

How then, do two atoms ever bind together to form a molecule? If their energy is too low, a barrier prevents their close approach, but if their energy is high enough to permit close approach, it is also high enough for the atoms to escape back over the barrier. If the available energy of the system can be decreased while the atoms are in the region $r_A > r > r_B$, shown in Fig. 15-14, they will be trapped. The value of r then oscillates between the limits $r_{\min}$ and $r_{\max}$ for a pair of atoms bound together. The pair of atoms thus

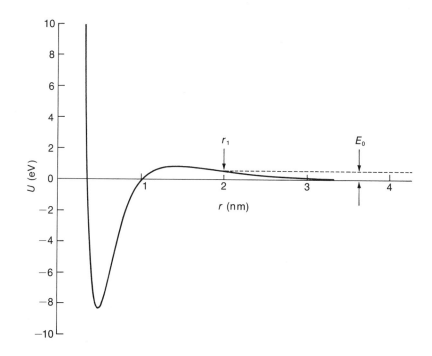

Fig. 15-12. Two atoms approach one another but have insufficient total energy to surmount the barrier near $r = 1.5$ nm.

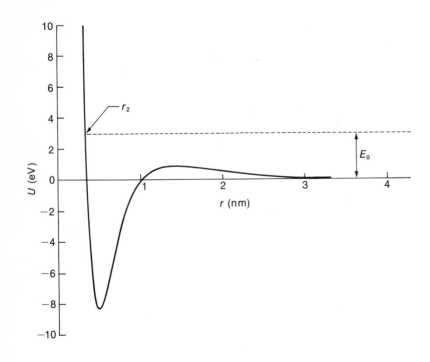

Fig. 15-13. Two atoms approach one another with initial energy large enough to overcome the barrier.

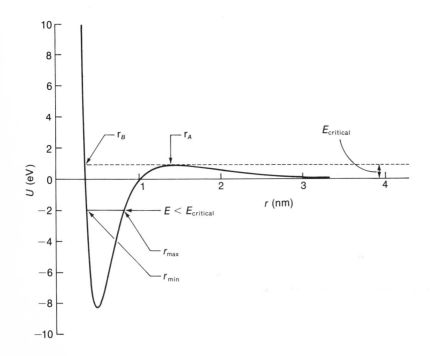

Fig. 15-14. When $E < E_{critical}$ the two atoms can be bound to one another with r oscillating between the values r_{min} and r_{max}, as illustrated for one particular value of E.

bound constitute a molecule. The energy of vibrational motion is referred to as *molecular excitation energy*. The mechanism whereby colliding atoms lose energy at just the right range of r to form a bound state necessarily involves a third body. This third body can be another atom with which one of the reacting atoms collides, so that it loses energy and becomes bound. The third "body" can also be electromagnetic radiation given off, which takes away sufficient energy to allow capture. This latter process is called *radiative capture*.*

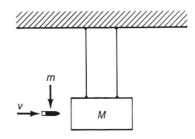

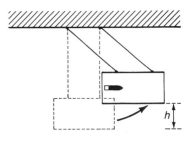

Fig. 15-15. A rifle bullet of mass m embeds itself in a block of wood, causing the block to rise to a maximum height h as it recoils from the impact.

Exercises

9. For simplicity, assume that one of the atoms involved in the situation of Fig. 15-14 is so massive that it is essentially stationary as the other atom approaches. The kinetic energy of the system is then entirely given by the kinetic energy of the moving atom. Taking that atom to be a chlorine atom ($m = 6 \times 10^{-26}$ kg) find its velocity at large r (where $U \approx 0$) and at $r = 2$ nm if $E_0 = 1$ eV. (1 eV $= 1.6 \times 10^{-12}$ erg $= 1.6 \times 10^{-19}$ J)

10. Take the situation to be as in Exercise 9 except that the chlorine atom drops down to a bound state at $E = -2$ eV while it is in the well (where the potential energy is negative), as in Fig. 15-14. A molecule has thus been formed. (*a*) How much energy had to be given off so that this could occur? (*b*) If this occurred to a macroscopic quantity of atoms, say N atoms where $N = 6.02 \times 10^{23}$ (Avogadro's number), how much energy, in joules, would be released?

15-7   Ballistic Pendulum

A ballistic pendulum is a device for accurately measuring the velocity of a projectile, such as a rifle bullet. An analysis of its operation involves several of the dynamical principles we have discussed. The operation is quite simple. A large block of wood is suspended as shown in Fig. 15-15. The bullet with velocity v embeds itself in the block of wood, causing it to swing back some distance. As it does, its center of gravity rises by the amount labeled h in the figure.

The experimental data obtained is h; the question is, "What is v?" The answer requires some care in the application of the laws of conservation of momentum and conservation of energy. At the

---

*Radiative capture can occur directly or involve the excitation of internal energy in one of the atoms, which then subsequently emits radiation.

moment of collision, no *external* forces act along the x axis on the bullet-block system, so we know that the x component of momentum will be conserved. The energy of the system, however, is *not* conserved since strong forces slow the bullet. (Though these are *internal* forces and do not change the net momentum, the same cannot be said for kinetic energy.) Thus the bullet loses far more kinetic energy than that gained by the block. Such a collision is called *inelastic*. Some of the "lost" energy is dissipated as sound waves from the loud noise of impact, but most of the lost energy appears as heat. In fact, the lead bullet frequently is partially melted by this heat.

Once the block is recoiling, it rises along the arc of its swing so that gravity does negative work upon it. The kinetic energy of the block decreases to zero at the peak of its swing, but the sum of its gravitational potential energy and its kinetic energy must remain constant during this phase of the motion. This conclusion depends upon the lack of work done by any but gravitational forces during the upward swing.

Calling the mass of the bullet m and that of the block M, we have, by conservation of momentum,

$$mv = (m + M)V$$

where v is the velocity of the bullet before it strikes the block and V is the recoil velocity of the block with the bullet embedded in it. Equating the initial kinetic energy of the recoiling block and bullet to the potential energy at the top of its swing, we have, by conservation of energy,

$$\tfrac{1}{2}(m + M)V^2 = (m + M)gh$$

or
$$V = \sqrt{2gh}$$

Combining this result with the result obtained above we have

$$v = \frac{m + M}{m}\sqrt{2gh}$$

In order to make the measurement more accurate, it is customary to measure x, the horizontal recoil distance, and then to use the geometry of the circular arc followed by the block to compute h. Accuracy is increased simply because x is generally much larger than h so that small errors in its measurement are relatively less important than similar small errors in a measurement of h.

Exercises

11. A ballistic pendulum measurement yields the data tabulated below. Find the rifle bullet's velocity.

Rifle: elephant stopper special

Bullet: 30 g $\quad = .03 \, kg$

Block: 40 kg

h: 3 cm

$$v = \frac{70}{30}\left(\sqrt{2(9.8)(.3)}\right)$$

$$\frac{40.03}{.03}$$

$$\frac{76}{30}\left(\frac{\sqrt{2(9.8)(.3)}}{2.4}\right)$$

12. The velocity of a rifle bullet decreases steadily due to air resistance, so that the velocity measured by a ballistic pendulum is not the muzzle velocity of the rifle. If the rifle is placed too close to the block, however, the explosive force of the gases produced by the gunpowder influence the block. Can you devise a *series* of ballistic pendulum measurements that will give the true muzzle velocity of a rifle?

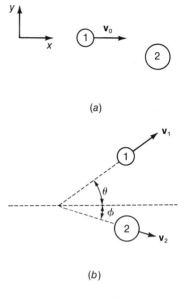

(a)

(b)

Fig. 15-16. Collision of two objects seen (a) before the collision and (b) after the collision.

15-8   Collisions

The law of conservation of mechanical energy, combined with the conservation of momentum, allows a precise and general description of collisions. Those collisions in which no kinetic energy is lost are called *elastic* collisions, while those in which some kinetic energy is lost are called *inelastic* collisions. A macroscopic example of a nearly elastic collision is that of two billiard balls colliding. The collision of two protons is a microscopic example of elastic scattering. In an inelastic collision, such as that of a bullet and a ballistic pendulum, kinetic energy is lost but does reappear in another form: heat energy in that example. At the microscopic level the collision of two atoms can be inelastic if one or both of the atoms is raised to an excited state. The energy lost from the motion of the projectiles in this case appears as internal excitation energy of one or both atoms.

elastic — no kinetic energy lost

inelastic — some kinetic energy is lost

In Fig. 15-16 we see two masses before and after an elastic collision. Conservation of momentum and energy allows us to write several equations for this elastic collision.

Conservation of p_x: $\quad m_1 v_0 = m_1 v_1 \cos \theta + m_2 v_2 \cos \phi \quad$ (15-10)

Conservation of p_y: $\quad 0 = m_1 v_1 \sin \theta + m_2 v_2 \sin \phi \quad\quad$ (15-11)

Conservation of energy: $\frac{1}{2} m_1 v_0^2 = \frac{1}{2} m_1 v_1^2 + \frac{1}{2} m_2 v_2^2 \quad\quad$ (15-12)

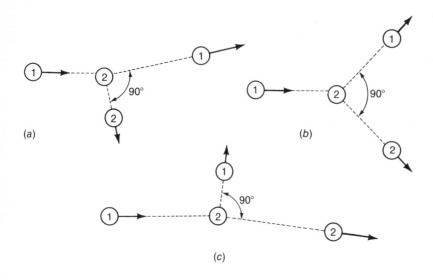

Fig. 15-17. Elastic collisions between equal masses: (a) grazing collision, (b) glancing collision, (c) nearly head-on collision.

If the collision is inelastic with the loss of an amount of kinetic energy E_i, we must modify the last equation:

$$\tfrac{1}{2}m_1v_0^2 = \tfrac{1}{2}m_1v_1^2 + \tfrac{1}{2}m_2v_2^2 + E_i \qquad (15\text{-}13)$$

If a quantity Q of energy is given off in the collision, as in an exothermic chemical reaction, the released energy can be added to the left side of (15-12):

$$\tfrac{1}{2}m_1v_0^2 + Q = \tfrac{1}{2}m_1v_1^2 + \tfrac{1}{2}m_2v_2^2$$

For the case of equal masses m_1 and m_2, we see that (15-12) is particularly simple for an elastic collision:

$$v_0^2 = v_1^2 + v_2^2 \qquad (15\text{-}14)$$

Further, the law of momentum conservation in this case becomes simply

$$\mathbf{v}_0 = \mathbf{v}_1 + \mathbf{v}_2$$

That is, $\mathbf{v}_0$, $\mathbf{v}_1$, and $\mathbf{v}_2$ must make a closed vector triangle. The result (15-14) makes it clear that this triangle is a *right* triangle, so we have the general result that $\mathbf{v}_1$ and $\mathbf{v}_2$ are always at right angles to one another for the elastic scattering of equal masses. Figure 15-17 illustrates a few examples. Pool players will quickly recognize some of the collisions depicted and their use in placing a ball in the desired pocket.* Nuclear physicists use this result to help de-

---

*In that game the 90° angle between the recoiling balls can be slightly altered by purposely imparting spin to the incident ball.

termine which of the events they observe are due to the collisions of protons on protons when they use a beam of protons to bombard a target that contains protons and other, heavier nuclei.

In most cases where the masses are not equal all of equations (15-10) through (15-12) may have to be applied to obtain a desired result. If the collision is not elastic, (15-13) replaces (15-12).

Example   A nuclear physicist bombards a target containing ^{12}C nuclei with a beam of protons, each of which has a kinetic energy of 15 MeV (1 MeV $= 10^6$ eV). He detects scattered protons at $\theta = 90°$ from the incident direction. What is the kinetic energy of such protons if each has been elastically scattered by a single ^{12}C nucleus?

Solution   Let us denote the mass of the proton by m, that of a ^{12}C nucleus (twelve times as massive as the proton) by M. The incident proton has velocity v_0, the outgoing proton v_1, and the recoiling nucleus v_2. Applying Eqs. (15-10)–(15-12) for the case of $\theta = 90°$, we obtain:

$$mv_0 = Mv_2 \cos\phi \qquad \text{(conservation of } p_x)$$

$$mv_1 = Mv_2 \sin\phi \qquad \text{(conservation of } p_y)$$

$$\tfrac{1}{2}mv_0^2 = \tfrac{1}{2}mv_1^2 + \tfrac{1}{2}Mv_2^2 \qquad \text{(conservation of energy)}$$

Squaring the first two equations, adding, and recalling that $\sin^2\phi + \cos^2\phi = 1$:

$$m^2(v_0^2 + v_1^2) = M^2 v_2^2$$

Hence
$$\tfrac{1}{2}Mv_2^2 = \frac{m}{M}(\tfrac{1}{2}mv_0^2 + \tfrac{1}{2}mv_1^2)$$

Substituting this last result into the third equation:

$$\tfrac{1}{2}mv_0^2 = \tfrac{1}{2}mv_1^2 + \frac{m}{2M}mv_0^2 + \frac{m}{2M}mv_1^2$$

Denoting the incident kinetic energy K_0 and the scattered proton's energy by K_1, we have:

$$\left(1 - \frac{m}{M}\right)K_0 = \left(1 + \frac{m}{M}\right)K_1$$

$$K_1 = \frac{M - m}{M + m}K_0$$

In our example,

$$K_1 = \frac{12 - 1}{12 + 1}K_0 = \frac{11}{13}(15 \text{ MeV}) = 12.69 \text{ MeV}$$

Thus the nuclear physicist knows that the protons he observes at 90° with a kinetic energy of 12.69 MeV are those that scattered elastically from ^{12}C nuclei within the target. The remainder of the initial energy, 2.31 MeV, is not lost: this is the kinetic energy of a recoiling ^{12}C nucleus.

Exercises

13. Two automobiles of equal mass moving at right angles to one another collide inelastically. They stick together and continue on together. If their initial velocities were equal, find their final velocity and direction immediately after the collision. What fraction of their original kinetic energy was lost in the collision?

14. A proton strikes an initially stationary helium nucleus, which is four times as massive as the proton. After the collision the proton is moving at 10° with respect to its initial direction, while the helium nucleus recoils at 30° on the other side of this initial direction. Find the ratio of the final velocities of the proton and the helium nucleus.

15-9 Summary

A conservative force is one that "repays" work done against it. Defining the change in potential energy as the negative of work done by the force, we are led to *conservation of total mechanical energy:*

$$\Delta U + \Delta K = 0$$

where

$$\Delta U = -\int_1^2 F \cos \theta \, dr = -W$$

is the work done against the force and

$$\Delta K = \tfrac{1}{2}mv_2{}^2 - \tfrac{1}{2}mv_1{}^2$$

The potential energy of a mass in the earth's gravitational field (close to the earth, so that g can be considered constant) is given by $U = mgh$. The potential energy of a spring is given by

$$U = \tfrac{1}{2}kx^2$$

where k is the spring constant and x is the displacement of the spring from its equilibrium length.

A dissipative force in one dimension is one for which

$$\oint F \, dx \neq 0$$

For a conservative force we may write

$$U(x) = -\int F \, dx$$

and $F = -dU/dx$. A graph of U vs x then gives many important features of a system: its turning points, the magnitude and direction of the force, and it equilibrium points (both stable and unstable).

Conservation of energy coupled with conservation of momentum can be applied to the analysis of collisions. Such an analysis can be used to predict the behavior of the remainder of a system when only a portion of the system's motion has been measured. Examples are given by the use of a ballistic pendulum to measure muzzle velocities and the analysis of atomic collisions.

Problems

15-1.   A 7-kg mass is raised vertically 5 m by a force F of 150 N. (*a*) How much work is done by F? (*b*) How much work is done by the force of gravity, mg? (*c*) What is the net work done on the mass? (*d*) How much kinetic energy is gained by the mass?

15-2.   A projectile is fired with velocity v_0 at an angle of θ above the horizontal. As it rises, its kinetic energy decreases but never falls to zero (unless $\theta = 90°$). Find the minimum possible value of kinetic energy for a given v_0 and θ.

15-3.   A 32-lb projectile is projected horizontally from a 50-ft cliff at 50 ft/s. Using the methods of Section 7-7, compute the magnitude of its velocity when it reaches the ground. Using this velocity as velocity v_2 and the initial 50 ft/s as velocity v_1, show that $E_1 = E_2$.

15-4.   A ball of mass m is attached to a pivot by a light, rigid wire of length l. When released from a horizontal position, it follows the circular arc, as shown in Fig. 15-18. (*a*) How fast is the ball going when it reaches the lowest point on its swing? (*b*) Find the tension in the wire as a function of the angle θ from the horizontal. Make a rough sketch of the tension versus θ.

15-5.   A mass m falls a distance h and strikes a spring. It compresses the spring by an amount d before it comes to rest. If energy is conserved in this process, find d in terms of m, h, and the spring constant k.

15-6.* A spring obeys Hooke's law:

$$F = -ky \qquad \text{and} \qquad U = \tfrac{1}{2}ky^2$$

A mass m is attached to the spring, which is suspended vertically so that the weight of the mass stretches the spring

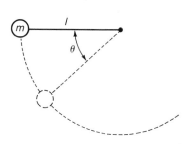

Fig. 15-18.

to a new equilibrium position. Now $U = \frac{1}{2}ky^2 + mgy$, where y is measured with respect to the original equilibrium position of the unloaded spring. Show that if we choose a new coordinate, y', that is measured with respect to this new equilibrium position, we obtain

$$U = \frac{1}{2}k(y')^2 + \text{constant}$$

so that an object hanging on a spring moves about its new equilibrium point just as it would move about the original equilibrium point of the spring in the absence of gravity.

15-7. When one mole of carbon atoms combine with one mole of oxygen atoms to form one mole of carbon monoxide, 1.1×10^5 J of energy are produced. What is the energy release (in eV) of each pair of carbon and oxygen atoms as they combine to produce one carbon monoxide molecule?

15-8. A block with mass m slides on a horizontal surface with a coefficient of friction μ. (a) How much work is done by the force of friction when the block slides from x_1 to x_2? (b) How much work is done by friction when the block is returned from x_2 to x_1? (c) Can you define a potential energy $U(x)$ for this frictional force?

15-9. A body slides down a frictionless plane of length l that is inclined at an angle θ above the horizontal. At the bottom of the inclined plane, the body slides onto a level plane with coefficient of friction μ. How far does it slide on this level portion?

15-10. A 500-MW electrical power plant stores its off-peak surplus of electrical energy by pumping water uphill into a reservoir. The water can then drive an electrical generator during times of high demand. If we wish to store 50% of its output for 8 hr in a reservoir 100 m above the power plant, what is the required capacity of the reservoir?

15-11. A body free to move along the x axis is acted upon by a force given by

$$F = ax + bx^2$$

(a) Find the change in its potential energy U between x_1 and x_2. (b) Write an expression for $U(x)$ such that $U(0) = 0$.

15-12. Compare the following two systems for storing energy. One is an automobile spring with a force constant $k = 1.5 \times 10^4$ N/m. This spring can flex a maximum distance of

30 cm. The other is the 1800-kg automobile itself. It can be elevated to store energy in gravitational potential energy.

15-13. A 20-g bullet with a velocity of 500 m/s strikes a 5-kg ballistic pendulum. Find the kinetic energy of the bullet-block system just before and just after impact.

15-14.* A proton elastically strikes an initially stationary electron head-on. Find the recoil velocity of the electron as a fraction of the proton's initial velocity. The proton is 1836 times as massive as the electron.

15-15.* A proton with a speed of $0.1c$ collides with another proton, which is initially at rest. After the collision, one of the protons is observed moving at an angle of 30° away from its initial direction. Find the angle of the other proton with respect to the initial direction and find the velocity of both protons, assuming an elastic collision. Neglect relativistic effects.

15-16. Two football players, one with a weight of 180 lb, the other with a weight of 220 lb, are running downfield at 30 ft/s to catch a pass. They are converging so that their paths intersect at an angle of 30°. They make an inelastic collision—that is, they stick together and move in a straight line for an instant after the collision until they fall to the ground. How much energy is lost in the inelastic collision?

15-17. As in Problem 9-8, a string is passed through a hole in the center of a smooth table and a mass attached to the string swings continuously in a circular path centered on the hole. The mass has an initial angular velocity ω_0. Conservation of angular momentum allows us to find the way in which ω varies as the string is pulled shorter from below the table (see Problem 11-4). (a) Find the increase in kinetic energy of the rotating mass as the radius is shortened from r_1 to r_2. (b) Find the work done by someone pulling on the string under the table to shorten the radius of the rotating mass from r_1 to r_2.

Chapter *16*

Simple Harmonic Motion

16-1 Introduction A mass attached to a spring will, once displaced, oscillate back and forth about its equilibrium position. If the spring obeys Hooke's law and if friction is negligible, the resulting motion is an example of simple harmonic motion. This type of motion is encountered frequently in nature and in various engineering applications. The ideas we encounter in this chapter will reappear frequently — for example, in the discussion of wave motion, of alternating-current electrical circuits, and of atomic vibrations.

The analysis we carry out in this chapter for a mass attached to a spring does more than introduce mathematical concepts for subsequent use: it uses several dynamical principles we have studied and thus develops these concepts more fully. Though the motion of an isolated mass acted upon by a spring may not appear to be a particularly exciting topic, it will seem of more general interest if we give a few examples of systems that closely resemble the mass-spring system: an atom held in its crystalline lattice position by interatomic forces, the cone of a loudspeaker held in position by a springy support, an automobile wheel acted upon by the combined forces of the automobile suspension and the flexed rubber tire, a bridge displaced by wind gusts and supported by steel girders or cables, or a steel-frame skyscraper whose base is abruptly displaced by an earthquake shock. All of these systems can oscillate with a characteristic frequency once they are displaced. Their motion may be more complicated than that of simple harmonic motion but, as we will see in Chapter 23, their motion can still be described in terms of a sum of simple harmonic oscillations of different frequencies.

16-2 The Equation of Motion

The force acting on a mass attached to a spring is simply given by Hooke's law:

$$F = -kx$$

as we have seen earlier. But Newton's second law also tell us that

$$F = ma = m\frac{d^2x}{dt^2}$$

Combining Hooke's law with Newton's second law gives us

$$-kx = m\frac{d^2x}{dt^2}$$

or

$$\frac{d^2x}{dt^2} = -\frac{k}{m}x \qquad (16\text{-}1)$$

Equation (16-1) is called the *equation of motion* for the mass-spring system, since solutions of (16-1) describe the way in which the system moves. Equations that contain derivatives, such as (16-1), are called *differential equations*. In particular, (16-1) is an example of a *second-order* differential equation, since it contains a second derivative. A general theorem about second-order equations such as (16-1) is that there are two independent solutions to the equation and that any other possible solution to the equation can be constructed as a sum of these two independent solutions.* We do not have the space to be more specific here but we do not need to be.

Even a qualitative study of (16-1) tells us a good deal about the way in which x must depend upon t to satisfy (16-1). As was mentioned in Chapter 3, the second derivative of a function is simply related to the curvature of its graph. A positive value for d^2x/dt^2 indicates that a graph of x vs t is cup-shaped in a way that "holds water." A negative value for d^2x/dt^2 indicates the curve has the shape of an inverted cup. We can use these qualitative ideas to make a rough sketch of x vs t that will satisfy (16-1).

Let us begin by assuming that x is positive; that is, the mass has stretched the spring beyond its equilibrium point. A graph of x vs t must then be curved downward, since

$$\frac{d^2x}{dt^2} = -\left(\frac{k}{m}\right)x \qquad (16\text{-}1)$$

where k and m are fixed and positive for a given mass and spring and we are assuming x is positive. As seen in Fig. 16-1, this down-

*See, for example, *Calculus with Analytic Geometry*, 2nd Ed., by Edwin J. Purcell; New York: Meredith Corporation (1972), pp. 880–883.

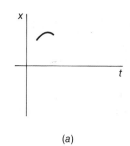

(a)

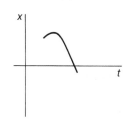

(b)

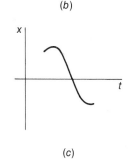

(c)

Fig. 16-1. Assume that x(t) satisfying Eq. (16-1) has an initially positive value (stretch the spring); the graph must then curve downward, as in (a). As x approaches zero, the curvature approaches zero also, so that the graph is nearly linear there, as indicated in (b). When x becomes negative, the curvature becomes positive, as in (c) so that, eventually, x will recross the abscissa.

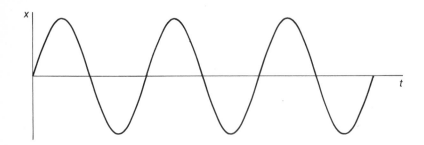

Fig. 16-2. A more precise graph of a function $x(t)$ that satisfies Eq. (16-1). Only a segment of the t axis is shown; the function is periodic and continues indefinitely. Note that the curvature (acceleration) is maximum for large values of $|x|$ and that $x(t)$ is a straight line (constant velocity) when $x \approx 0$.

ward curvature means that x must eventually begin to decrease. As x decreases, (16-1) tells us that d^2x/dt^2 becomes smaller also; that is, $x(t)$ becomes a more nearly linear function. Finally, as in Fig. 16-1c, when x becomes negative, (16-1) tells us that the curve becomes cup-shaped upward. Eventually, then, x becomes again positive and the whole chain of events reoccurs.

Physically speaking, when $x > 0$, the mass accelerates toward $x = 0$. When it arrives at $x = 0$ there are *no* forces acting upon it, so it continues at constant velocity until $x < 0$. Now the spring begins to compress so that acceleration back toward $x = 0$ begins. This process repeats itself indefinitely and at regular time intervals, so the motion of the mass is called *periodic*.

A function of time, $x = f(t)$, that exactly satisfies (16-1) is graphed with reasonable precision in Fig. 16-2. We will discuss the nature of this function in more detail in the next section and then return to a quantitative treatment of the harmonic oscillator in the following section.

Exercises

1. The sketch of Fig. 16-3a shows a small portion of $x = f(t)$. Qualitatively extend this function to larger values of t under the condition that

$$\frac{dx}{dt} = +x$$

2. Qualitatively extend x to larger values of t in Fig. 16-3b for

$$\frac{dx}{dt} = -x$$

16-3   Sinusoidal Functions of Time

The function graphed in Fig. 16-2 is a *sinusoidal* function of time. It is of the form

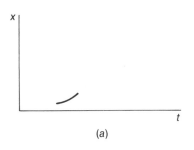

(a)

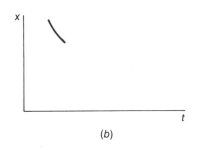

(b)

Fig. 16-3.

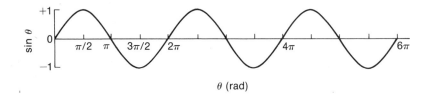

$$x = A \sin(\omega t) \qquad (16\text{-}2)$$

where A and ω are constants. In (16-2) and in subsequent discussions, the *argument* of the sine function – the quantity in parentheses – is expressed in *radians*. The dimensions of ω must therefore be t^{-1}, as they were for the angular velocity, ω. The close connection that exists between simple harmonic motion and circular motion will become more apparent as we go along.

When its argument is expressed in radian measure, the sine function begins at zero for zero argument, becomes unity when its argument reaches $\pi/2$ (90°), falls back to zero for an argument equal to π (180°), and reaches -1 at an argument of $3\pi/2$, as shown in Fig. 16-4 for a plot of $\sin \theta$ vs θ.

If we now make the argument of the sine function increase steadily *with time*, as occurs in (16-2), the sine function will oscillate continually between the values ±1 as time evolves. In Fig. 16-5 we see a plot of $\sin(\omega t)$ versus time for three different values of ω. One complete oscillation of the sine function, as between $t = 0$ s and $t = 8$ s at the top of Fig. 16-5, is called one *cycle* of oscillation. The time required for one full cycle is called the *period,* often denoted T. In the middle curve of Fig. 16-5, the period is $T = 4$ s; in the lower curve, the period is $T = 2$ s.

The number of complete cycles that occur every second is called the *frequency* of oscillation denoted f. The frequency is simply related to the period:

$$f = \frac{1}{T} \qquad (16\text{-}3)$$

The units of frequency, cycles per second (cps), are called *hertz,* abbreviated Hz. Finally, the *angular frequency* ω is related to the frequency by

$$\omega = 2\pi f \qquad (16\text{-}4)$$

where the units of ω are rad/s.

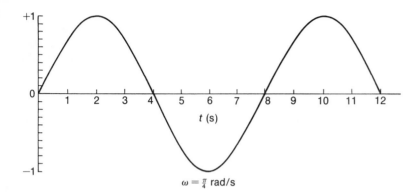

Fig. 16-5. The function sin(ωt) is plotted against time for three values of ω. The time scale is the same for all three graphs. As ω is increased, the frequency of oscillations increases.

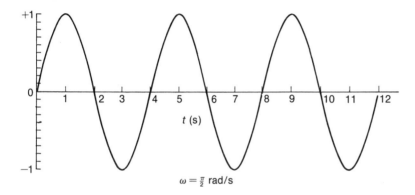

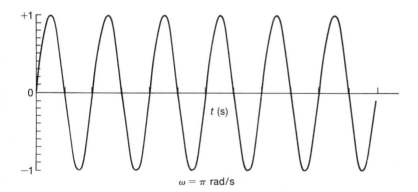

The coefficient A that multiplies the sine function in (16-2) is called the *amplitude*. Since the sine function itself oscillates between ±1, the amplitude is simply a vertical scale factor, as indicated in

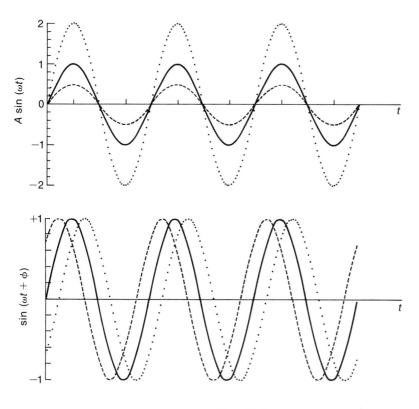

Fig. 16-6.   A plot of $A \sin(\omega t)$ for three values of A: $A = 2$ (dotted curve), $A = 1$ (solid curve), and $A = \frac{1}{2}$ (dashed curve).

Fig. 16-7.   A plot of $\sin(\omega t + \phi)$ for $\phi = \pi/4$ (or $\phi = 45°$) given by the dashed line, for $\phi = 0$ (the solid line), and for $\phi = -\pi/4$ (the dotted line).

Fig. 16-6. If we wish to slide the sine function to the left or right along the t axis, we can modify (16-2) by adding a *phase angle* ϕ to its argument:

$$x = A \sin(\omega t + \phi) \qquad (16\text{-}5)$$

The effect of adding a positive value of ϕ is to start the sine function farther along in its argument at $t = 0$ than if ϕ were zero. This is illustrated in Fig. 16-7. Of course, if $\phi = 2\pi$ we are back to the original function, since the sine function repeats every 2π.

If we graphically differentiate

$$x = \sin(\omega t)$$

we are led to the derived function shown in the center of Fig. 16-8. This derived function looks suspiciously like the sine function itself, shifted in phase. In fact it *is* just that, apart from a scale factor:

$$f(t) = \sin(\omega t) \qquad \text{implies} \qquad f'(t) = \omega \sin(\omega t + \tfrac{1}{2}\pi)$$

and because the sine and cosine are cofunctions, i.e.,

$$\sin(\theta + \tfrac{1}{2}\pi) = \cos\theta$$

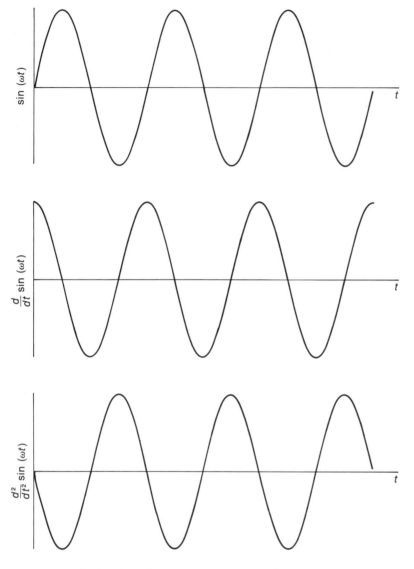

Fig. 16-8. Differentiation of $\sin(\omega t)$. Near $t = 0$ its slope is maximum and positive, so its derivative is largest and positive there. At its first peak the slope of $\sin(\omega t)$ is zero, so the value of its first derivative is zero there. Just beyond the first peak of $\sin(\omega t)$ the slope becomes negative, as indicated by its negative first derivative. A second differentiation leads to the curve at the bottom.

we can equivalently write:

$$f'(t) = \omega \cos(\omega t)$$

The scale factor ω occurs because the larger ω is, the more rapidly $\sin(\omega t)$ varies and, hence, the steeper its slope and the larger its derivative.*

---

*If we had chosen other than radian measure for the argument of the sine function, we would have an additional constant scale factor multiplying the derivative.

We will not prove our statements concerning differentiation of the sine and cosine functions in this chapter. They are fully discussed in any introductory calculus text.* For our purposes the salient points are

$$\left.\begin{array}{l} \dfrac{d}{dt} \sin(\omega t) = \omega \cos(\omega t) \\[2mm] \dfrac{d}{dt} \cos(\omega t) = -\omega \sin(\omega t) \end{array}\right\} \qquad (16\text{-}6)$$

Thus the *second* differentiation in Fig. 16-8 leads us from the cosine function back to the sine function at the bottom of the figure, but multiplied by a minus sign and the factor ω^2:

$$\frac{d^2}{dt^2} \sin(\omega t) = -\omega^2 \sin(\omega t) \qquad (16\text{-}7)$$

If we had started with the cosine function we would have obtained a similar result:

$$\frac{d^2}{dt^2} \cos(\omega t) = -\omega^2 \cos(\omega t) \qquad (16\text{-}8)$$

Multiplication of both sides of (16-7) or (16-8) by an amplitude, of course, changes nothing.

The important point to be made here is that the sine and cosine function, $f(t) = A \sin(\omega t)$ and $g(t) = B \cos(\omega t)$, are the two *unique* functions that have the property that two differentiations lead us back to the starting function (multiplied by a minus sign and a scale factor). We are thus assured that they solve (16-1), the equation of motion for a mass-spring system, since that is all that is required.

We could reverse our discussion and "discover" the sine and cosine functions through the study of a specific differential equation, (16-1). However, most of us first encounter the sine and cosine functions in a study of trigonometry, where they are related to properties of a right triangle.

We can now pull the ideas of this and the preceding section together and quantitatively solve the equation of motion (16-1) for a harmonic oscillator. We claim that either

$$x = A \sin(\omega t) \qquad (16\text{-}9)$$

or

$$x = B \cos(\omega t) \qquad (16\text{-}10)$$

---

*See, for example, *Calculus and Analytic Geometry*, 4th Ed., by George B. Thomas, Jr.; Reading, Mass.: Addison-Wesley Publishing Co. (1968), pp. 155–158.

will satisfy (16-1),

$$\frac{d^2x}{dt^2} = -\frac{k}{m}x \qquad (16\text{-}1)$$

Let us demonstrate this for the cosine solution and leave the sine solution for the reader in Exercise 4. Substitute for x from (16-10) into Eq. (16-1):

$$\frac{d^2}{dt^2} A \cos(\omega t) = -\frac{k}{m} A \cos(\omega t)$$

Applying (16-8):

$$-\omega^2 A \cos(\omega t) = -\frac{k}{m} A \cos(\omega t)$$

Thus the equation of motion is satisfied *if*

$$\omega^2 = \frac{k}{m} \qquad \omega = \sqrt{\frac{k}{m}}$$

or, by application of (16-4),

$$f = \frac{1}{2\pi} \sqrt{\frac{k}{m}} \qquad (16\text{-}11)$$

Thus we have shown that sinusoidal motion satisfies (16-1) provided that the frequency of that motion satisfies (16-11), which gives f in terms of the stiffness of the spring and the mass of the object attached to it. The stiffer the spring, or the lighter the mass, the higher the frequency.

Exercises

3. Use the rule for differentiation of products discussed in Section 11-2 to find dx/dt when

$$x = At^2 \cdot \sin(\omega t)$$

4. Substitute (16-9) directly into (16-1) and show that (16-1) is satisfied if $\omega^2 = k/m$.

5. Since the dimensions of frequency are t^{-1}, the dimensions of $\sqrt{k/m}$ should also be t^{-1} according to (16-11). Show that this is so.

6. A $\frac{1}{2}$-slug mass is attached to a spring with a force constant of 32 lb/ft. Find the number of full cycles of oscillation made each second.

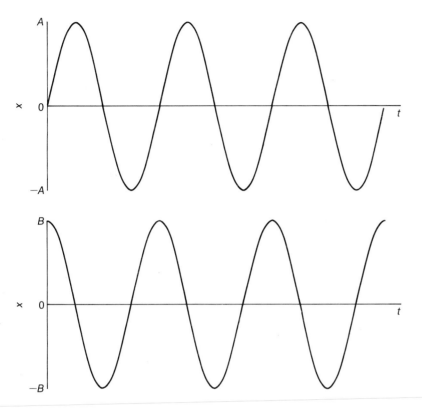

Fig. 16-9. Either of the solutions (16-9) or (16-10) satisfies the equation of motion, (16-1). In the lower graph, (16-10) corresponds to pulling the mass aside to $x = +B$ and releasing it at precisely $t = 0$. In the upper graph, (16-9) corresponds to giving the mass an initial velocity at $t = 0$, so it coasts to A before beginning its return.

16-4 Characteristics of Simple Harmonic Motion

The solution given by (16-9) is somewhat distressing in that A is *any* constant, although it is apparent that it must have the dimensions of length. Actually, this arbitrariness is just as it should be, for the magnitude of A simply gives the maximum excursions of the mass away from $x = 0$, i.e., $-A \leqslant x \leqslant +A$. Since we nowhere *specified* the magnitude of these excursions in arriving at (16-1) or its solution, (16-9), it is not surprising that our solution is not specific on this point. The same discussion also applies to (16-10).

As shown in Fig. 16-9, the oscillatory motion of the solution (16-9) begins with $x = 0$ at $t = 0$ and is limited by the values $\pm A$. The motion given by (16-10) begins with $x = B$ at $t = 0$. The choice between the sine or cosine solution simply amounts to a choice of where we will pick the zero on our time scale. The difference between the sine and cosine solutions is seen to be $\frac{1}{4}$ cycle by comparison in Fig. 16-9.

Example   A $\frac{1}{4}$-slug mass is attached to a spring with $k = 25$ lb/ft. It is stretched to $x_0 = \frac{1}{2}$ ft then released with $v_0 = 0$. Describe its subsequent motion.

Solution   It seems natural, although certainly not essential, to choose $t = 0$ at the moment of release of the stretched spring. In that case, the cosine solution is tailored to our needs, since it has maximum x at $t = 0$. So for this example,

$$x = B \cos(\omega t)$$

At $t = 0$, $\cos(\omega t) = 1$, hence we find B easily:

$$x_0 = B \cos(0) = B$$

Thus, $B = x_0 = \frac{1}{2}$ ft. The constant ω is given by $\sqrt{k/m}$:

$$\omega = \sqrt{k/m} = \sqrt{\frac{25 \text{ lb/ft}}{\frac{1}{4} \text{ slug}}} = 10 \text{ s}^{-1}$$

The motion then is given by $x = B \cos(\omega t)$, where $B = \frac{1}{2}$ ft and $\omega = 10 \text{ s}^{-1}$. Thus, for example, at $t = 0.15$ s the x coordinate is given by

$$x = \tfrac{1}{2} \text{ ft} \cdot \cos(10 \text{ s}^{-1} \cdot 0.15 \text{ s}) = \tfrac{1}{2} \text{ ft} \cdot \cos(1.5 \text{ rad})$$

$$= 0.035 \text{ ft}$$

Interestingly enough, the frequency of a mass-spring system is unchanged if the mass hangs from the spring under the influence of gravity. The effect of gravity is simply to offset the location of the equilibrium point, as was indicated in Problem 15-6 of the preceding chapter.

As we saw in Section 16-3, we can adjust the time at which x crosses the t axis by adding a phase angle ϕ to the argument of (16-9):

$$x = C \sin(\omega t + \phi) \tag{16-12}$$

where C is the amplitude of this sinusoidal motion. The expression (16-12) also satisfies (16-1), since

$$\frac{d}{dt} C \sin(\omega t + \phi) = \omega C \cos(\omega t + \phi)$$

and

$$\frac{d}{dt} C \cos(\omega t + \phi) = -\omega C \sin(\omega t + \phi)$$

As we noticed earlier, setting $\phi = \pi/2$ (90°) makes (16-12) identical with the cosine solution, (16-9). Thus (16-12) can describe any sort of simple harmonic motion we might encounter.

The velocity and acceleration of the mass m with displacement given by (16-12) are easily found by differentiation:

$$x = C \sin(\omega t + \phi)$$

$$v = dx/dt = \omega C \cos(\omega t + \phi) \qquad (16\text{-}13)$$

$$a = dv/dt = \omega C \frac{d}{dt} \cos(\omega t + \phi)$$

$$= -\omega^2 C \sin(\omega t + \phi) = -\omega^2 x \qquad (16\text{-}14)$$

$$x = C \sin(\omega t + \phi)$$
$$v = \omega C \cos(\omega t + \phi)$$
$$a = -\omega^2 x$$

Example A "shaker" is a device that tests the ability of instruments to withstand violent acceleration by rapidly oscillating a test platform to which they are firmly attached. Consider such a shaker with an amplitude of 6 in. and a frequency of 100 Hz. What is the maximum value of the acceleration?

Solution By Eq. (16-14), $a = -\omega^2 C \sin(\omega t + \phi)$, so the acceleration oscillates between the limits of $\pm\omega^2 C$ as the sine function oscillates between ± 1. In our example, $\omega = 2\pi f = 6.28 \times 10^2 \text{ s}^{-1}$ and $C = \frac{1}{2}$ ft. Thus

$$a_{max} = \omega^2 C = (6.28 \times 10^2)^2 \text{ s}^{-2} \cdot \tfrac{1}{2} \text{ ft}$$

$$= 1.97 \times 10^5 \text{ ft/s}^2 = 6.1 \times 10^3 g$$

Exercises

7. Give the value of ϕ to be used in (16-12) if we wish to describe simple harmonic motion that begins at $t = 0$ with (a) the spring compressed to $-C$, then released; (b) the mass at $x = 0$ but with a velocity $+v_0$; (c) the mass at $x = 0$ but with a velocity $-v_0$; (d) the spring stretched to $+C$, then released.

8. An automobile engine has pistons that weigh 2 lb each. Each revolution of the crankshaft produces one full cycle of oscillation of the piston. When the engine is running at 3600 rev/min, find the maximum force exerted on a piston, assuming simple harmonic motion with an amplitude of 2 in. The actual motion of the piston is not exactly simple harmonic but close enough so that your result will give a good estimate of the forces required.

16-5 Energetics of Simple Harmonic Motion

Our previous discussion of the energy of a mass-spring system in Section 15-5 showed that the total energy of the system is constant.

If our solution for the motion of the mass-spring system is correct, it too must give constant energy. We can explicitly calculate the value of the potential energy and kinetic energy and, indeed, show that their sum is a constant:

$$U = \tfrac{1}{2}kx^2 = \tfrac{1}{2}kC^2 \sin^2(\omega t + \phi) \qquad (16\text{-}15)$$

$$K = \tfrac{1}{2}mv^2 = \tfrac{1}{2}m\omega^2 C^2 \cos^2(\omega t + \phi)$$

since $\qquad\qquad v = dx/dt = \omega C \cos(\omega t + \phi)$

Note that K can be rewritten using the fact that $\omega^2 = k/m$:

$$K = \tfrac{1}{2}m\left(\frac{k}{m}\right)C^2 \cos^2(\omega t + \phi) = \tfrac{1}{2}kC^2 \cos^2(\omega t + \phi) \quad (16\text{-}16)$$

The total energy is the sum of the potential and kinetic energies:

$$E = U + K = \tfrac{1}{2}kC^2 \sin^2(\omega t + \phi) + \tfrac{1}{2}kC^2 \cos^2(\omega t + \phi)$$

Since the sine and cosine are out of phase with one another, the energy shifts back and forth between kinetic and potential energy. It is easy to show their sum is constant, as indicated in Fig. 16-10:

$$E = \tfrac{1}{2}kC^2[\sin^2(\omega t + \phi) + \cos^2(\omega t + \phi)]$$

$$= \tfrac{1}{2}kC^2 = \text{constant} \qquad\qquad (16\text{-}17)$$

since $\sin^2 \theta + \cos^2 \theta = 1$ always.

When a spring with an attached mass is stretched to some position x_0, the energy stored in it is $U_0 = \tfrac{1}{2}kx_0^2$; when released it will oscillate with an amplitude $C = x_0$. Equation (16-17) shows that the total energy, though it continually changes from potential to kinetic and back again, remains equal to this initial value.

Exercise

9. When the work done by the spring on an oscillating mass is positive, the kinetic energy of the mass is increasing. When the work done is negative, the kinetic energy is decreasing. Break up a cycle of motion of the mass into four parts: (a) x goes from 0 to $+A$; (b) x returns from $+A$ to 0; (c) x goes from 0 to $-A$ (d) x returns from $-A$ to 0. Show that the work done in these quadrants has the appropriate sign to cause the changes in kinetic energy predicted by (16-16).

16-6   Relations between Simple Harmonic Motion and Uniform Circular Motion

The position of a point moving along a circular path with constant

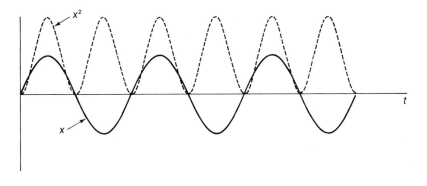

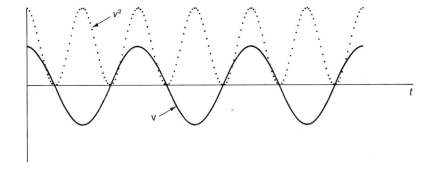

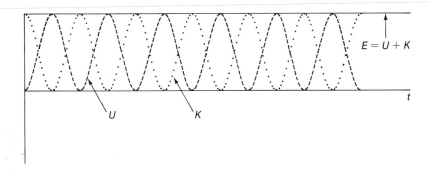

Fig. 16-10. The upper graph shows x and x^2 while the middle graph shows v and v^2. The lower graph indicates that the sum of $K = \frac{1}{2}mv^2$ and $U = \frac{1}{2}kx^2$ is a constant. This constant is the total energy, E.

velocity can be described by using coordinates suited to this purpose, i.e.,

$$R = \text{constant} \quad \text{and} \quad \theta = \omega t$$

where ω is a constant. The resulting tangential velocity is

$$v_T = \omega R$$

However, uniform circular motion can also be described in rectangular Cartesian coordinates, as shown in Fig. 16-11. We can expect this description to be a bit more complicated since rectangular (Cartesian) coordinates are not as well suited to circular motion as the radius-and-angle (polar) coordinates we used previously.

The x and y coordinates of a point on the circle are given by the x and y components of a radius vector **R** pointing from the center of the circle to the point on the circular path. From Fig. 16-11 we see:

$$x = R \cos \theta \qquad (16\text{-}18)$$

$$y = R \sin \theta \qquad (16\text{-}19)$$

If the point is moving with uniform circular motion we have $\theta = \omega t$, so that (16-18) and (16-19) become

$$x = R \cos(\omega t) \qquad (16\text{-}20)$$

$$y = R \sin(\omega t) \qquad (16\text{-}21)$$

Thus we see that the motion of the x or y component of uniform circular motion is identical with that of an object undergoing simple harmonic motion of amplitude R. A simple demonstration of this correspondence is illustrated in Fig. 16-12.

The result we obtained earlier for the radial acceleration,

$$a_R = -\omega^2 R$$

can also be obtained from a Cartesian description of uniform circular motion. We differentiate (16-20) and (16-21) twice to obtain the x and y components of the acceleration. We then sum these components, using the Pythagorean theorem:

$$a_x = -\omega^2 R \cos(\omega t)$$

$$a_y = -\omega^2 R \sin(\omega t)$$

$$a_{\text{total}} = -\omega^2 R \sqrt{\cos^2(\omega t) + \sin^2(\omega t)} = -\omega^2 R$$

where the minus sign indicates that the acceleration is directed inward along **R**.

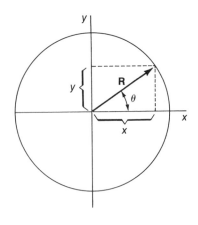

Fig. 16-11. The x and y coordinates of a point moving along a circular path.

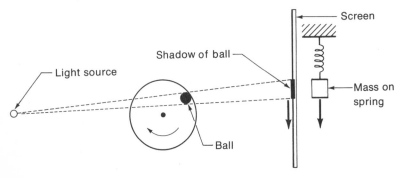

Fig. 16-12. A demonstration that illustrates the simple harmonic motion of the projection of a point moving uniformly around a circle. When ω for the rotating wheel is chosen to match that for the mass on a spring, the shadow and mass remain perfectly synchronized until friction reduces the amplitude of the mass and spring.

Exercises

10. Differentiate (16-20) and (16-21) to obtain v_x and v_y. Show that the velocity *vector* is pointing in the $+y$ direction at $t = 0$, and that it rotates counterclockwise thereafter.

11. A point undergoes simple harmonic motion in both the x and y directions simultaneously. If $y = A \sin(\omega t)$ and $x = A \sin(\omega t + \phi)$, find ϕ so that the particle moves in a circle.

16-7 Pendulum Motion

Whenever a mass is acted upon by a restoring force linearly dependent on displacement, simple harmonic motion results. An approximate example that finds application in timekeeping is the simple pendulum. The restoring force is provided by the component of gravity that acts along the path followed by the mass (the bob) at the end of the pendulum. This force is indicated in Fig. 16-13. Distance along the arc followed by the bob is denoted s and given by

$$s = l\theta$$

where l is the length of the supporting wire or string. Considering the component of force along the path and the acceleration it produces, we have:

$$F = ma$$

$$-mg \sin\theta = m(d^2s/dt^2)$$

$$= ml(d^2\theta/dt^2)$$

so that

$$d^2\theta/dt^2 = -\frac{g}{l}\sin\theta \qquad \text{acceleration} \qquad (16\text{-}22)$$

where the minus sign indicates a *restoring* force.

Functions that have the property that their second derivative is proportional to the sine function exist and are hence solutions of (16-22), but they are somewhat complicated.*

The solution of (16-22) is greatly simplified if we take advantage of the series expansion of the sine function (see Appendix):

$$\sin\theta = \theta - \frac{\theta^3}{3!} + \frac{\theta^5}{5!} - \frac{\theta^7}{7!} + \cdots$$

where θ is expressed in radians. For small values of θ, all terms beyond the first term are very small and can be ignored. Making

*They are called *elliptic integrals of the second kind.*

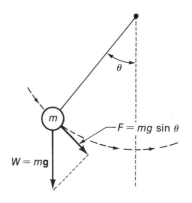

Fig. 16-13. The component of weight that accelerates the pendulum bob is shown. The other component, *mg* cos θ, contributes to the tension in the supporting wire or string.

this approximation we can rewrite (16-22):

$$d^2\theta/dt^2 = -(g/l)\theta \qquad (16\text{-}23)$$

assuming $\qquad\qquad \sin\theta = \theta \qquad\qquad (16\text{-}24)$

This is the same form of equation that we previously saw lead to simple harmonic motion. A solution is

$$\theta = C\sin(\omega t)$$

where C is the amplitude of the swing and

$$\omega = \sqrt{g/l}$$

The frequency of the motion is given by

$$f = \omega/2\pi = \frac{1}{2\pi}\sqrt{g/l} \qquad (16\text{-}25)$$

The period of the pendulum is given by

$$T = 1/f = 2\pi\sqrt{l/g}$$

This time interval is independent of the amplitude of the swing *as long as Eq. (16-24) is satisfied.* This occurs only for small angles. The appendix at the end of this chapter gives some numerical illustrations of the variation of T with amplitude.

The simple pendulum is an idealized system—a point mass supported by a weightless string. A more realistic description of any pendulum takes into account the effect of the moment of inertia of its bob as it rocks back and forth. Such a pendulum consisting of an extended mass distribution is called a *physical pendulum.* A torque about the pivot point is generated by the force of gravity, which effectively acts at the center of gravity, as shown in Fig. 16-14. The angular acceleration produced by this torque depends on the moment of inertia of the physical pendulum about its pivot point:

$$\Gamma = I\alpha$$

or $\qquad\qquad -mgd\sin\theta = I\,\dfrac{d^2\theta}{dt^2}$

where the minus sign indicates that the torque acts to *reduce* θ. If $\sin\theta \approx \theta$, we obtain

$$\frac{d^2\theta}{dt^2} = -\left(\frac{mgd}{I}\right)\cdot\theta$$

Once again we have a differential equation that states that the second derivative of a function, $\theta(t)$, is proportional to the function itself. As we saw earlier, this function is a sinusoidal function:

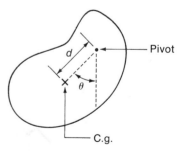

Fig. 16-14.   An arbitrarily shaped mass hung on a pivot forms a physical pendulum.

$$\theta = C \sin(\omega t)$$

where $\qquad\qquad \omega = \sqrt{mgd/I} \qquad\qquad$ (16-26)

Finally, consider the *torsional pendulum* in Fig. 16-15. The restoring torque due to the twisting support rod accelerates the disc with a moment of inertia I:

$$\Gamma = -K\theta = I\alpha$$

so that $\qquad\qquad \dfrac{d^2\theta}{dt^2} = \left(\dfrac{-K}{I}\right)\theta$

Note that no small-angle approximation is involved here. A solution is given by $\theta = \theta_0 \sin(\omega t)$, where θ_0 is the amplitude and

$$\omega = \sqrt{K/I} \qquad\qquad \text{(16-27)}$$

The balance wheel in a watch is a torsional pendulum. It oscillates back and forth under the influence of the restoring force of a small spring called a hair spring. Loss in amplitude due to friction is offset by the energy supplied by the escapement mechanism driven by the main spring. As this main spring runs down, the amplitude of the balance wheel's oscillation decreases somewhat. However, due to the fact that the period of simple harmonic motion is *independent* of its amplitude, the watch maintains accurate time as long as any energy remains in the main spring.

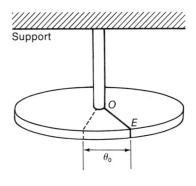

Fig. 16-15. A torsional pendulum is set into oscillation by twisting the line *OE* to the position of the dotted line, then releasing it.

$$\omega = \sqrt{g/\ell} \qquad \text{simple pendulum}$$

$$\omega = \sqrt{mgd/I} \qquad \text{physical pendulum}$$

$$\omega = \sqrt{K/I} \qquad \text{torsional pendulum}$$

Exercises

12. A mass is to be hung on a string to make a simple pendulum. How long should the string be if we want the pendulum to oscillate once every four seconds? Does it matter how large the mass is?

13. A physical pendulum is made by suspending a rod of uniform thickness from one end and letting it swing back and forth. Find the frequency of this pendulum as a function of its length.

14. Show that the dimensions of (16-25), (16-26), and (16-27) are all t^{-1}.

16-8 Summary

A mass acted upon only by a linear restoring force, i.e., $F = -kx$, moves in simple harmonic motion. Newton's second law for such a system gives us

$$\frac{d^2x}{dt^2} = -\left(\frac{k}{m}\right)x$$

This differential equation is solved by either a sine function or a cosine function:

$$x = A \sin(\omega t)$$

or

$$x = A \cos(\omega t)$$

where ω is given by

$$\omega = \sqrt{k/m}$$

[handwritten: $\omega = \sqrt{\frac{k}{m}}$]

and A is the amplitude of the motion. The choice of the sine or cosine solution (or a sum of both) amounts to a choice of the phase of the motion. The acceleration is easily found to be

[handwritten: $x = C \sin(\omega t + \phi)$]
[handwritten: $v = \omega C \cos(\omega t + \phi)$]
[handwritten: $a = -\omega^2 x$]

$$a = -\omega^2 A \sin(\omega t)$$

when $x = A \sin(\omega t)$

The frequency, f, and period, T, are related to ω by the equations

[handwritten: $T = \frac{2\pi}{\omega}$]

[handwritten: $f = \frac{1}{2\pi}\sqrt{\frac{k}{m}}$] $f = \omega/2\pi$ $T = 2\pi/\omega$ *[handwritten: $T = 2\pi\sqrt{\frac{m}{k}}$]*   *[handwritten: $f = \frac{\omega}{2\pi}$]*

This frequency is unchanged if the mass hangs from the spring in a uniform gravitational field (see Problem 15-6).

The energy of a harmonic oscillator oscillating with amplitude A is constant and given by

$$E_{\text{total}} = \tfrac{1}{2}kA^2$$

This total energy is the sum of the kinetic energy, $\tfrac{1}{2}mv^2$, and the potential energy, $\tfrac{1}{2}kx^2$.

The motion of a harmonic oscillator is identical to the projection of uniform circular motion with the same value of ω and with a radius equal to the amplitude of the harmonic oscillator.

A simple pendulum and a physical pendulum undergo simple harmonic motion (frequency independent of amplitude) if the angle of swing is small, so that the approximation $\sin \theta = \theta$ is valid. Under these conditions,

[handwritten: $T = 2\pi\sqrt{\frac{\ell}{g}}$ $f = \frac{1}{2\pi}\sqrt{\frac{g/\ell}{}}$]

$$\omega = \sqrt{g/l} \quad \text{(simple pendulum)}$$

$$\omega = \sqrt{mgd/I} \quad \text{(physical pendulum)}$$

[handwritten: $d = $ distance from pivot to center of mass]

where d is the distance from the pivot to the center of mass and I is the moment of inertia about that pivot point.

A torsional pendulum acted upon by a linear restoring torque $\Gamma = -K\theta$ undergoes simple harmonic motion of the angle θ:

$$\theta = \theta_0 \sin(\omega t)$$

where $\omega = \sqrt{K/I}$.

**APPENDIX: Expansion of the Sine Function
and Small-Angle Approximation**

Though we tend to think of the sine function as a ratio of sides of
a triangle, it is just as precisely defined by the fact that it is a solution of Eq. (16-1); that is, it is the negative of its second derivative.
We can construct a polynomial function with this property, but it
has an infinite number of terms. Consider the following infinite
series:

$$f(x) = x - \frac{x^3}{3!} + \frac{x^5}{5!} - \frac{x^7}{7!} + \cdots$$

where 3! means $1 \cdot 2 \cdot 3 = 6$. In general, $n! = 1 \cdot 2 \cdot 3 \cdots n$. Differentiating, we obtain

$$f'(x) = 1 - \frac{3x^2}{3!} + \frac{5x^4}{5!} - \frac{7x^6}{7!} + \cdots$$

$$= 1 - \frac{x^2}{2!} + \frac{x^4}{4!} - \frac{x^6}{6!} + \cdots$$

$$f''(x) = 0 - x + \frac{x^3}{3!} - \frac{x^5}{5!} + \frac{x^7}{7!} - \cdots$$

But this is just the negative of $f(x)$. So we have found a function
such that $f''(x) = -f(x)$. This function is the sine function:

$$\sin x = x - \frac{x^3}{3!} + \frac{x^5}{5!} - \frac{x^7}{7!} + \cdots$$

The tables of sine functions in mathematical handbooks are calculated from this series, where enough terms are used to obtain
the desired degree of accuracy. The cosine function is a similar
series, involving even powers of x:

$$\cos x = 1 - \frac{x^2}{2!} + \frac{x^4}{4!} - \frac{x^6}{6!} + \frac{x^8}{8!} - \cdots$$

The fact that it is possible to differentiate these infinite series term
by term and the uniqueness of the sine and cosine functions as
solutions of (16-1) are topics dealt with in a calculus course.

In the approximation we made to solve the simple pendulum
problem, we assumed that

$$\sin \theta = \theta$$

The error involved for various values is tabulated below.

θ (degrees and minutes)	θ (radians)	sin θ	% error
2°52′	0.05	0.04998	0.04
5°44′	0.10	0.09983	0.17
8°36′	0.15	0.14944	0.38
11°28′	0.20	0.19867	0.66
17°11′	0.30	0.29552	1.5
22°55′	0.40	0.38942	2.7
28°39′	0.50	0.47943	4.3

The effect of using our simple approximation to solve (16-22) is surprisingly small and is indicated below.

Maximum Amplitude (degrees)	True Period / Approximate Period
2°	1.00006
4°	1.0003
6°	1.0005
8°	1.0008
10°	1.0019
20°	1.0076
30°	1.017
40°	1.03
50°	1.05
70°	1.10
90°	1.18

Exercises

15. Carefully plot $f(\theta) = \theta$ and $f(\theta) = \sin \theta$ on the same graph for $0 < \theta < 90°$. At what value of θ can you distinguish them in your graph?

16. Suppose a clock is carefully calibrated while its pendulum swings with an amplitude of 10°, then the swing amplitude decreases to 8°. How much time will the clock gain per day?

Problems

16-1. Sketch a rough graph of $f(\theta)$ vs θ for its first few cycles, where $f(\theta) = \sin(\theta + \phi)$, and where ϕ is a constant equal to (a) $-\pi/2$; (b) 0; (c) $\pi/2$; (d) π; (e) $3\pi/2$; (f) 2π.

$T = 2\pi/\omega$

16-2. Sketch a rough graph of $f(t)$ vs t for its first few cycles, where $f(t) = \sin(\omega t)$ and ω is a constant with value (a) π s^{-1}; (b) 2π s^{-1}; (c) 4π s^{-1}; (d) 10 s^{-1}.

16-3. A general solution (both amplitude and phase are adjustable) to (16-1) is given by

$$x = C \sin(\omega t + \phi)$$

Show that an equivalent solution is given by

$$x = A \sin(\omega t) + B \cos(\omega t)$$

In particular, find the relationships between the parameters C and ϕ of the first solution to the parameters A and B of the second. It will be helpful to recall the following trigonometric identities:

$$\sin^2(\theta) + \cos^2(\theta) = 1$$

$$\sin(x + y) = \sin x \cos y + \cos x \sin y$$

16-4. In Fig. 16-9 the upper graph corresponds to the solution when the mass is given an initial velocity v_0 at $t = 0$ with zero initial displacement. Find the magnitude of that initial velocity in terms of A and ω.

16-5. A certain spring stretches by 8 cm while a 1-kg weight is hung on it. The weight is then removed and a 2-kg weight is attached to the spring. What is the period of the resulting oscillation?

16-6. A harmonic oscillator has an amplitude of $\frac{1}{2}$ m and a frequency of 100 Hz. Find the velocity for $x = 0$ and for $x = \frac{1}{4}$ m. Compare these velocities to the average velocity found by dividing the amplitude by the time required to cover that distance (the time of $\frac{1}{4}$ cycle).

16-7. An atom is oscillating about its equilibrium position in a crystal lattice with a frequency of 10^{11} Hz and an amplitude of 0.1 nm. Find (a) its maximum velocity, and (b) its maximum acceleration.

16-8. A penny is placed on a piston that oscillates up and down with simple harmonic motion and an amplitude of 15 cm. The frequency of oscillation is slowly increased until a faint "clink" can be heard once each cycle. This is caused by the penny momentarily losing contact at the top of each stroke. At what frequency does this first occur?

$f = \dfrac{\omega}{2\pi}$

16-9. While the door of an elevator is open, an occupant notes that if he flexes his knees once quickly, he sets the elevator

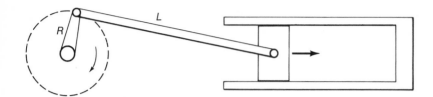

Fig. 16-16.

cage oscillating up and down with an amplitude of 2 cm and a frequency of 3 Hz. The elevator cable is made of steel with a Young's modulus as tabulated in Chapter 13. The length of the cable is 20 m and the total mass of the elevator cage is 800 kg. Find the diameter of the steel cable. What is the frequency at a lower floor, when the cable is 30 m long? Neglect the mass of the cable.

16-10. Integrate the work done by the spring in a simple harmonic oscillator over $\frac{1}{4}$ cycle of motion from $x = 0$ to $x = A$. Show that this work equals the change in kinetic energy over that $\frac{1}{4}$ cycle.

16-11. During one full cycle of motion of a mass attached to a spring, the energy changes from potential to kinetic and back again (twice) but its magnitude is unchanged. It must by true, then, that the total work done by the force of the spring is zero over one full cycle. Prove this directly by evaluating the work integral for one cycle.

16-12* Look at the interatomic potential graphed in Fig. 15-11. Assume that this represents the interaction between a fluorine atom and a much heavier atom that may be considered to be stationary. The interatomic spacing is near 0.5 nm and the energy is about −4 eV, so that a bound system is formed. There is a small amount of kinetic energy, so that r oscillates back and forth about its equilibrium value. Assuming that the shape of the well is parabolic near the equilibrium minimum, estimate the "spring constant" k from the information given in the graph. Find the frequency of vibration of the fluorine atom, assuming simple harmonic motion.

16-13. A piston is driven by a crank rotating with constant angular velocity, as sketched in Fig. 16-16. Show that the motion of the piston closely approximates simple harmonic motion if $L \gg R$.

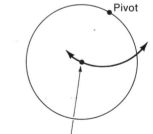

Center oscillates along path shown

Fig. 16-17.

16-14. A wheel rotates with a uniform angular velocity of 150 rad/s. Find the y component velocity at $t = 0.1$ s for a point that was at $x = 6$ cm, $y = 6$ cm at $t = 0$.

16-15. Find the length of a simple pendulum that beats seconds (one full cycle every two seconds), as in a grandfather clock.

16-16. A meter stick is suspended from one end and swung as a pendulum. Find its period. Now consider the meter stick suspended at the 49-cm point and find its period.

16-17. A solid disc with a diameter of 20 cm and a mass of $\frac{1}{2}$ kg is hung by a pivot at its edge, as sketched in Fig. 16-17. Find its period as a physical pendulum assuming small angles of oscillation.

16-18. Estimate the moment of inertia and frequency of the balance wheel in a watch. Find the torsional spring constant for the hair spring.

16-19. A bullet of mass m is fired with velocity v at a block of mass M, as in Fig. 16-18a. The block is initially in equilibrium, attached to a spring with spring constant k. This constant is determined by the static measurement shown in Fig. 16-18b, where the spring is stretched by an amount d when the weight is hung over a pulley. Find the position, velocity, and acceleration of M six-tenths of a second after the bullet strikes the block and buries itself in M if $d = 3$ cm, $W = 6$ kg, $v = 250$ m/s, $m = 20$ g, and $M = 3980$ g.

16-20* Consider a steel rod that is pivoted at an arbitrary point along its length and allowed to swing freely as a physical pendulum. The plane in which it swings is vertical. Find an expression for the frequency of oscillation in terms of the length l of the rod and the distance d of the pivot from the end. How is this expression modified if the plane containing the pendulum swing is tipped away from vertical by an angle Φ?

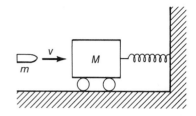

(a)

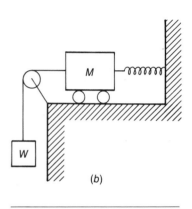

(b)

Fig. 16-18.

Chapter *17*

Gravitation

17-1 Introduction Gravitation has played a key role in the development of man's understanding of motion. The early natural philosophers frequently explained free fall by asserting that the "natural place" of any object was in contact with the earth. This did not lead to new knowledge, but it was hard to argue with such an explanation. Galileo dispelled some erroneous concepts concerning free fall and studied motion on inclined planes to deduce some kinematic and dynamic ideas. These included the result that the distance traversed by a uniformly accelerating object is in proportion to the square of the elapsed time and also the statement of the law of inertia. Galileo is frequently credited with a free fall experiment from the leaning tower of Pisa which showed that, except for air resistance, different masses fall with the same acceleration. Actually, it is unknown whether or not he performed such studies, but he did firmly state the above conclusion. He based his conclusion in part upon the following clever argument, partly because scholars of his day were more interested in logic than in experiment.

Suppose a mass, m_1, is falling toward the earth and a larger mass, m_2, falls more rapidly because of its greater mass. The situation is sketched in Fig. 17-1a. What happens if we break m_2 into two halves? We might guess that these two halves, having less mass, would each fall more slowly (as in Fig. 17-1b). Now tie the two halves together with a chain. They are once again the heavy object and fall rapidly, as in Fig. 17-1c. The only difference between the situation sketched in b and c is the presence of the chain. If the chain pulls on half (1) to make it move faster, it must *retard* (2).

If it pulls on (2) to make *it* move faster, it must retard (1). Hence it cannot be that a heavier object accelerates more rapidly than a light one, for the necessary additional force of one portion of itself on another would cancel its effect by the reaction back on the first part. In this argument Galileo was close to the notion of action-reaction, although its precise form, $\mathbf{F}_{12} = -\mathbf{F}_{21}$, was not necessary for his argument.

The final discovery of the law of universal gravitation awaited developments in astronomy. The key data that was needed was provided by fantastically precise measurements of planetary motion, which were made in the late sixteenth century by Tycho Brahe. The telescope was not yet invented by Galileo and all of Tycho's measurements were made by simple, yet accurate, sighting instruments. With tremendous dedication Tycho accumulated 20 years of planetary position data with an accuracy of 1/60 degree. At his death, Tycho entrusted Johannes Kepler, one of his students, with the task of editing and publishing his planetary tables. Kepler was fascinated with the possibility of deducing a precise mathematical form for the orbits of the planets. The data for the planet Mars finally provided the answer. The investigation was complicated by the fact that the earth itself moves about the sun, as does Mars. Kepler used a clever technique to "unscramble" the separate motions of the earth and Mars from the data, which consisted only of sightings of Mars from the moving earth; his technique is related by most introductory astronomy textbooks.

Having obtained a set of forty points defining Mars' orbit, Kepler could see that the orbit was somewhat oval in shape but searched in vain for a mathematical description. He tried equations of eccentric circles and ellipses with no success.* Kepler finally found that an ellipse, with the sun at one of its foci, gave a beautiful description of the data. Further work led him to announce three general laws of planetary motion, still known as *Kepler's laws*. These laws are:

1. Each planet moves in an ellipse with the sun at one focus.

2. The radius vector from the sun to a planet sweeps out equal areas in equal times.

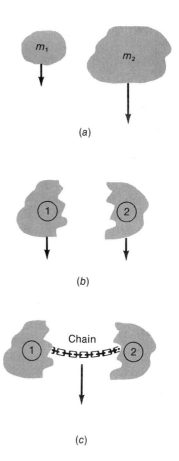

Fig. 17-1.

*His frustration can be heard in the following passage of a letter written by Kepler after years of calculations: "While triumphing over Mars, and preparing for him, as for one already vanquished, tabular prisons and equated eccentric fetters, it is buzzed here and there that my victory is vain, and that the war is raging anew. For the enemy left at home a despised captive has burst all the chains of the equations, and broken forth from the prisons of the tables." The story of Kepler's efforts is told in "How Did Kepler Discover His First Two Laws?" by Curtis Wilson in the March 1972 issue of *Scientific American*.

3. The squares of the times of revolution (or years) of the planets
    are proportional to the cubes of their average distances from
    the sun.

These laws provided the basis by which Newton (born in 1642,
twelve years after Kepler's death) was finally to discover the law
of universal gravitation. Part of the difficulty of understanding
planetary motion had been that earlier workers could not imagine
that the same physics that worked here on earth could be involved
in the motion of the heavens. It is difficult to speak too highly of
Newton's brilliance and effort. To put it briefly: in order to solve
the problem of planetary motion, Newton enunciated his three laws
of motion (i.e., he invented mechanics), he invented the differential
and integral calculus, and he proposed a law of gravity; he then
applied all these ideas to the motion of planets and solved the
differential equation of their orbits to show that they are ellipses
with the sun at one focus.* Though he made many other important
contributions in optics and astronomy, it is primarily for his construction of mechanics, the calculus, and the law of gravity that
he is honored.

17-2   The Inverse Square Law of Gravity

Many people in Newton's days conjectured on the possibility that
a force of gravity extended from one object to another and provided the necessary force to keep the planets in orbit. Some even
conjectured that this force might fall off as the inverse square of
the separation of two gravitating objects. But no one had any idea
of how to prove or disprove such a hypothesis. Newton first considered the difference between the pull of the earth upon an object
at its surface and upon the distant moon. He calculated the centripetal acceleration of the moon in its orbit about the earth and
found it to be 2.6×10^{-3} m/s². This value of acceleration was in the
same proportion to g, the gravitational acceleration at the earth's
surface, as the ratio of the square of the earth's radius to the square
of the radius of the moon's orbit:

$$\frac{a_{\text{moon}}}{g} = \frac{r_{\text{earth}}^2}{R_{\text{moon}}^2}$$

The situation is shown in Fig. 17-2. The result suggests that the
force of gravity decreases as $1/r^2$, where r is the distance from the

---

*More generally, he showed that any orbit for an inverse-square force law is a conic
section: a circle, ellipse, parabola, or hyperbola, depending on the energy and angular
momentum of the orbit.

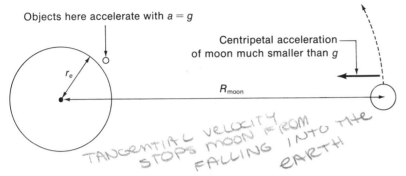

Objects here accelerate with $a = g$

Centripetal acceleration of moon much smaller than g

r_e

R_{moon}

TANGENTIAL VELOCITY STOPS MOON FROM FALLING INTO THE EARTH

Fig. 17-2. Any object in free fall at the earth's surface accelerates towards the center of the earth with $a = g$. The moon is also in free fall and continually accelerates toward the center of the earth with $a = (2.65 \times 10^{-4})\, g$. Because of the moon's large tangential velocity, it never reaches the earth but follows an orbit around it.

center of the earth to the object acted upon by the force. Newton further surmised that the strength of one object's gravitational attraction on a second object depends linearly on its mass, m_1. But, since action equals reaction (Newton's third law), the second object in Fig. 17-2 must exert a gravitational force back on the first and this force, presumably, would depend on the mass m_2 of the second object. A gravitational force that satisfies all of the above hypotheses is given by

$$F_{12} = G\,\frac{m_1 m_2}{r_{12}^2} \qquad (17\text{-}1)$$

where F_{12} stands for the force of gravity exerted by m_1 on m_2; r_{12} is the center-to-center distance between m_1 and m_2; and G is a constant. The reaction force, F_{21}, given by a similar prescription certainly satisfies Newton's third law. However, the center-to-center definition of r_{12}, though it fits the earth-moon data, leaves much to be desired. What, for example, do we take to be the "center" of an irregularly shaped object? It is tempting to guess that this "center" should be the center of mass, but that cannot be. The simple situation shown in Fig. 17-3 illustrates the difficulty. Suppose that $F_{12} = G\, m_1 m_2 / r_{cm}^2$ in that situation. It seems unreasonable to suppose the gravitational force changes if we snap m_1 into two masses, a and b, but leave them in place. Yet, summing the forces upon m_2 of those two halves of m_1, now using the c.m. distances r_a and r_b in (17-1), leads to a different result than when we used the original center of mass. Exercise 17-2 illustrates this point.

Newton hypothesized instead that two *point* masses interact gravitationally according to (17-1). For point masses there can be no ambiguity in the meaning of r_{12}. For extended masses we must consider the contribution of each infinitesimal bit of mass to the total gravitational force. Since each bit of mass lies, in general, at a different distance from the object being attracted, each will con-

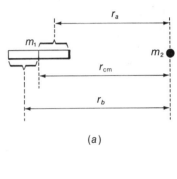

(a)

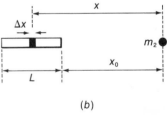

(b)

Fig. 17-3. (a) The distance from the center of mass m_1 to a point mass m_2 is labeled r_{cm}. The distance from the centers of mass of the two halves of m_1 to the point mass m_2 are labeled r_a and r_b. (b) The rod is subdivided into elements Δx in length in order to calculate the net gravitational force between m_1 and m_2.

tribute according to the inverse of *its* distance squared. For an extended object m_1 acting upon a point mass m_2 we write:

$$F_{12} = G \lim_{\Delta m \to 0} \sum_{i=1}^{n} \frac{\Delta m_i \cdot m_2}{r_{i2}^2}$$

where m_1 is broken into n pieces. As Δm_i approaches zero, n must approach infinity. This limit we now recognize as an integral, but Newton was held back for several years by this difficulty. He resolved it eventually by inventing the integral calculus. As we shall show in a later section, integration over a *spherically symmetric* mass distribution does yield a gravitational force on a second body equal to that obtained by assuming the spherical object acts from its center.

For the situation shown in Fig. 17-3, the required integral is deduced by a method quite similar to that used in obtaining the moment of inertia of a rod in Section 11-4. The rod of length L is broken up into mass elements of length Δx, so that if λ is the mass per unit length of the rod, we find that

$$F = G \lim_{\Delta x \to 0} \sum_{i=1}^{n} \frac{\lambda \Delta x_i \cdot m_2}{x_i^2} = G \lambda m_2 \int_{x_0}^{x_0 + L} \frac{dx}{x^2}$$

where x is measured from m_2 to the mass element and where x_0 is the fixed distance between the end of the rod and the point mass m_2.

For two point masses, the direction of the force of gravity is along the line joining them. It can be written in vector form:

$$\mathbf{F}_{12} = -G \frac{m_1 m_2}{r_{12}^3} \mathbf{r}_{12}$$

where $\mathbf{r}_{12}$ points from m_1 to m_2 and the minus sign indicates that m_1 *attracts* m_2.

Exercises

1. In Newton's era, scientists were actively considering the possibility that the force of gravity might vary as the inverse square of the separation distance. See if you can arrive at this guess as they did. Imagine that any massive object continually sends out some sort of "influence" *in all directions* into the space surrounding it. (These influences were called *vortices* by some.) As one recedes farther and farther from the source of these influences, they are spread more and more thinly, in the sense

of their number per unit area. Does this idea lead to an inverse square law? Explain.

2. Consider the situation shown in Fig. 17-3, where m_1 is a rod of length l with its midpoint a distance $2l$ away from a point mass m_2. Find, according to Eq. (17-1), the force of gravity between m_1 and m_2, (a) assuming that m_1 has its mass concentrated at its center of mass; (b) assuming that the mass of the rod is concentrated at the two centers of mass of the two halves of the rod. (c) Finally, find the exact force between the rod and point mass by evaluating the integral

$$F = G\lambda m_2 \int_{3l/2}^{5l/2} \frac{dx}{x^2}$$

where $\lambda = m_1/l$.

17-3   The Cavendish Experiment

The gravitational force is extremely weak compared to other forces in nature. It is only because we are close to an extremely large, massive object—the earth—that we are so aware of a gravitational force. The gravitational attraction between two large autos parked side by side is approximately 10^{-4} N, which is close to 3×10^{-5} oz. In Newton's time no one was able to observe effects of this force between laboratory-sized objects. The theory of gravitation was instead arrived at by considering the motion of objects of planetary size, where this force is quite large. Newton developed his theory in the interval from 1666 to 1687* but the first precision laboratory observation of this force was made by Lord Cavendish in 1798. The experimental configuration he used is shown in Fig. 17-4. He carefully calibrated the torsional constant, κ, of the thin fiber. Then, since $\Gamma = \kappa\theta$ for such a fiber, a measurement of the deflection angle θ produced by placing the masses M close by gave the value of Γ, the torque produced by the gravitational forces. He varied the spacing and masses to verify the proportionality of the force to the product of the masses and the inverse square of their separation distance.** He also obtained a value for the constant G that appears

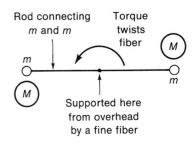

Fig. 17-4. The configuration used in the Cavendish experiment. The small twist produced by the weak gravitational forces between m and M was measured by observing the deflection of a beam of light reflected from a mirror attached to the rod connecting m and m.

---

*The year 1687 marks the publication date of Newton's *Principia Mathematica*. He did not publish his work as a brief paper written in terse scientific jargon, as we so often do today. His was such a major contribution that he wrote a textbook (the *Principia*). It was a tremendous effort for him, as most physics students can well believe, and he would never have accomplished it without the continual urging and support of Halley, the astronomer for whom Halley's comet is named.

**The proportionality to the inverse square of separation is and was much more accurately obtained by extensive planetary orbit data. Nonetheless, this behavior had not previously been demonstrated at small separation distances.

in (17-1). The presently accepted value for G is

$$G = 6.6732 \times 10^{-11} \text{ N} \cdot \text{m}^2/\text{kg}^2$$

$G = 6.67 \times 10^{-11} \text{ N} \cdot \text{m}^2/\text{kg}^2$

A few moments reflection leads us to the startling observation that the Cavendish measurement of G also measured the mass of the earth. The gravitational force on an object of mass m at the earth's surface is experimentally known to be mg. But, we also know it is given by (17-1), with r_{12} equal to R_e, the distance between the center of the earth and its surface, where the mass m is located. Then we have

$$mg = G\frac{m_e m}{R_e^2} \qquad (17\text{-}2)$$

After Cavendish determined G, the only unknown in (17-2) was m_e, the mass of the earth. Solving for m_e we find:

$$m_e = \frac{R_e^2 g}{G} = \frac{(40.6 \times 10^{12} \text{ m}^2)(9.8 \text{ m/s}^2)}{6.673 \times 10^{-11} \text{ N} \cdot \text{m}^2/\text{kg}^2}$$

$m_e = 5.98 \times 10^{24} \text{ kg}$

$$= 5.98 \times 10^{24} \text{ kg}$$

The volume of the earth is given by

Volume of Earth

$$V = \tfrac{4}{3}\pi R_e^3 = 1.08 \times 10^{21} \text{ m}^3$$

$V = \tfrac{4}{3}\pi R_e^3$

A sphere of water of this size would have a mass of 1.08×10^{24} kg, since a cubic meter of water has a mass of 10^3 kg. Thus, the specific gravity of the earth is

$$\frac{\rho_{\text{earth}}}{\rho_{\text{water}}} = \frac{5.98}{1.08} = 5.5$$

The average density of the earth is $5\frac{1}{2}$ times that of water, even though the surface rocks of the earth have a density of only about half this amount.

So it was that Cavendish, observing the interaction of small lead masses in his laboratory, had in effect weighed the earth. Furthermore, his measurement provided important geological information at the same time, since the average density of the earth is higher than its surface density. The earth's inner regions must be much more dense than the outer regions, a fact of profound significance to one attempting to understand the structure of the earth.* Presently accepted theories describe the earth's core as liquid

---

*Since the volume occupied by a spherical object is proportional to the cube of its radius, the "inner regions" of the earth have to be quite dense to substantially affect the average density of the earth.

(since it will not propagate transverse seismic waves) with a high density (as discussed above) and at a high temperature.*

Exercises

3. The fiber in a Cavendish experiment has a torsional constant $\kappa = 5 \times 10^{-8}$ N · m. Both small masses have mass $m = 10$ g and are separated by a 40-cm rod (neglect the rod's mass). The large masses M each has a mass of 5 kg and is placed to produce maximum torque on the system while the center-to-center spacing of both pairs of m and M is 10 cm. Find the total deflection angle, θ, produced by this torque.

4. A truck with a mass of 5×10^3 kg is parked with its "center" 3 m from a 50-kg boy. Estimate the gravitational force on the boy produced by the nearby truck's mass.

17-4 Gravitational and Inertial Mass

The appearance of the mass m in the law of gravity is a rather remarkable coincidence of nature. The product of masses in (17-1) determines the strength of a *force*, the force of gravity. Mass as we use it in Newton's second law describes the *inertia* of a body — it describes the way a body will respond to an applied force. The mass that appears in Newton's second law is called *inertial* mass, while that that appears in the law of gravity is called *gravitational* mass.

The strict proportionality of these two aspects of mass leads to identical values of g for any free-falling object. This is seen by writing Newton's second law for a free-falling object of mass m near the surface of the earth:

$$F = ma$$

or

$$G\frac{m_e m}{R_e^2} = ma$$

where the m on the left side is a gravitational mass and that on the right an inertial mass. If gravitational and inertial mass are exactly proportional, we can always cancel the quantity m.**

---

*See "The Fine Structure of the Earth's Interior" by Bruce A. Bolt, *Scientific American*, March 1973.

**It is not really meaningful to speak of more than *proportionality* between the two aspects of mass since the empirically determined gravitational constant G will absorb any such proportionality constant.

$$G \frac{m_e}{R_e^2} = a$$

so that all free-falling objects at the earth's surface fall with an identical acceleration called g:

$$g = G \frac{m_e}{R_e^2} = 6.67 \times 10^{-11} \frac{N \cdot m^2}{kg^2} \cdot \frac{5.98 \times 10^{24} \ kg}{(6.37 \times 10^6 \ m)^2}$$

$$= 9.83 \ m/s^2$$

Actually, on the earth's surface, the observed value of g varies between 9.78 and 9.83 m/s² with latitude, as discussed in Section 17-8. At any given location, however, it is identical for all objects, within the precision of the measurements.

Measuring g for a variety of objects is not the only way to show the identity of gravitational and inertial mass. One method was devised and carried out by Newton. He constructed a simple pendulum with a hollow bob into which he could put various substances. To obtain the period of a pendulum as given by Eq. (16-25):

$$T = 2\pi \sqrt{l/g} \qquad (16\text{-}25)$$

we must be able to cancel the gravitational mass and inertial mass to obtain (16-22):

$$\frac{d^2\theta}{dt^2} = -\frac{g}{l} \sin \theta \qquad (16\text{-}22)$$

The timing of a pendulum swing can be carried out to great precision by timing the total duration of many swings and dividing by the number of swings. Newton did this and found that T was always given by (16-25), thus verifying that m always cancelled in (16-22).

In a rather clever experiment that involves the combination of centrifugal forces due to the earth's rotation and the force of gravity, Eötvös was able to check proportionality of the two types of mass with great accuracy. In experiments carried out in 1909, he found that the proportionality was exact to one part in 10^8 for many different substances.

17-5 Circular Orbits

Consider a projectile fired from a cannon, as in Fig. 17-5. At low initial velocities the projectile falls quickly to the earth. However, if the initial velocity were high enough and air resistance negligible, might it not be possible that the earth's surface "falls away" due to its curvature fast enough so that the projectile will never intersect it? A more productive way of phrasing this question is to ask

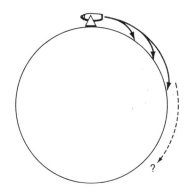

Fig. 17-5. A projectile fired with ever increasing velocities has greater and greater range.

whether the force of gravity on the projectile can just exactly supply the centripetal force required to cause the projectile to move in a circular orbit just above the earth's surface. The force on such a "satellite" will be directed toward the center of the earth, as shown in Fig. 17-6. In order for it to follow a circular path, there must be a centripetal force provided by gravity. The centripetal force F_c is given by

$$F_c = \frac{mv^2}{r} = F_g$$

so that

$$\frac{mv^2}{r} = G\frac{m_e m}{r^2}$$

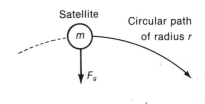

Fig. 17-6.   The force on a satellite is directed toward the center of the earth.

Solving for the velocity that satisfies the above condition,

$$v^2 = G\frac{m_e}{r} \qquad (17\text{-}3)$$

Just above the surface of the earth, we set r equal to the radius of the earth:

$$v^2 = G\frac{m_e}{R_e} = \frac{(6.67 \times 10^{-11}\ \text{N} \cdot \text{m}^2 \cdot \text{kg}^{-2})\,(5.98 \times 10^{24}\ \text{kg})}{6.37 \times 10^6\ \text{m}}$$

$$= 62.5 \times 10^6\ \text{m}^2/\text{s}^2$$

so that $\qquad v = 7.9 \times 10^3\ \text{m/s} = 17,700\ \text{mi/hr}$

Since the earth is 24,900 mi in circumference, it would take 1.4 hr for each full revolution by such a satellite. Most of us are already aware, from news broadcasts, that an artificial satellite in near earth orbit (so that r is very close to the R_e) takes about an hour and a half for each orbital period.

When we derived (17-3), the mass of the satellite cancelled out of the result. This is because of the strict proportionality of gravitational and inertial mass. If an astronaut is a passenger in a satellite, he follows precisely the same trajectory as does the satellite itself. The earth's gravity acts upon him to keep him in orbit just as it acts upon the satellite to keep it in orbit. Thus, no force between the satellite and its occupant is required. This rather unnatural state of affairs is called "weightlessness," though the force of gravity on the astronaut is by no means zero! If it were, he would follow a straight line into outer space—until stopped by the wall of the satellite.

We can also find how the orbital period depends upon the distance between the earth's center and the satellite. Since the time required for each orbital revolution is proportional to the radius and inversely proportional to the velocity, we can perhaps already see

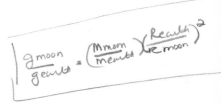

from (17-3) that the square of the orbital period is proportional to the cube of the radius. Let the time of one revolution be T. Then from Eq. (17-3),

$$v = \sqrt{G\frac{m_e}{r}}$$

$$T = \frac{2\pi r}{v} = \frac{2\pi r}{\sqrt{G\dfrac{m_e}{r}}}$$

$$T^2 = \left(\frac{4\pi^2}{Gm_e}\right)r^3$$

[handwritten: orbital period depends on the distance between the earth's center and the satellite]

Since the expression in parentheses is constant (for any given central body), we have deduced **Kepler's third law** from dynamics combined with the law of gravity.

Note also that observation of a satellite's orbital period about any object, if we know its orbital radius, allows us to determine the mass of the central body. In this way the masses of various constituents of the solar system have been measured. In fact, nature has been kind enough to make a substantial fraction of all stars in our galaxy members of double star systems. In some cases astronomers have been able to measure both the orbital period and separation of double stars that orbit one another and to determine their masses.* The relatively small number of stars whose mass has been determined in this fashion has, nonetheless, led to the mass-luminosity relation that has allowed astronomers to infer the masses of many other stars.**

[handwritten: escape velocity $v = \sqrt{2g\,R_e}$]

Exercises

5. An artificial satellite is to be placed in orbit about the earth, in the plane of its equator, so that it appears stationary to an observer on the earth, that is, its orbital period is to be 24 hr. Find the required radius of its orbit.

---

*It is particularly evident for double-star systems, where the two stars have comparable masses, that one star cannot be fixed while the other revolves about it (apply the conservation of momentum law to the system). Instead, *both* stars orbit about their common center of mass. The moon-earth system also rotates about its "balance point," but since the mass of the earth is 81 times that of the moon, this point is actually inside the earth, the moon being only 60 earth radii distant. It is thus reasonably accurate to treat the earth as fixed with the moon rotating about it.

**See, for example, George Abell, *Exploration of the Universe*, p. 442 and ff, 2nd Edition, Holt, Rinehart and Winston, New York, 1969.

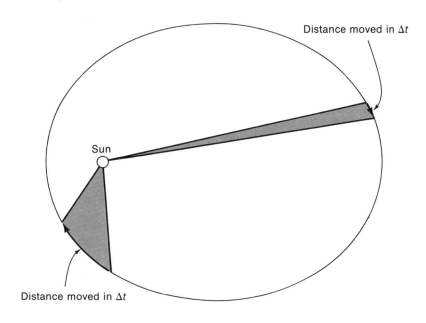

Distance moved in Δt

Sun

Distance moved in Δt

Fig. 17-7. The path of a planet is an ellipse with the sun at one focus. When it is closer to the sun, the planet moves more rapidly, in such a way that equal areas are swept out in equal times by the radius vector pointing from the sun to the planet.

6. The earth's mass is 81 times that of the moon, while its radius is 3.7 times larger than that of the moon. Find the gravitational acceleration (in units of g) of an object in free fall at the moon's surface.

17-6 Kepler's Second Law

Kepler's first law states that the orbits of planets are ellipses, but its proof is beyond our scope. Kepler's second law states that the radius vector from the sun to a planet sweeps out equal areas in equal times (see Fig. 17-7). As the planet moves along its elliptical path, it moves most rapidly in those portions where it is close to the sun. The reader who does not recall the properties of an ellipse can construct one with the aid of two fixed pegs, a loop of string, and a pencil. Figure 17-8 shows the method used to draw the ellipse. The closer the two pegs are to one another, the more nearly the ellipse resembles a circle. The more widely spaced the pegs (which locate the foci), the more *eccentric* the ellipse.

Now consider the area swept out by a planet's radius vector, as shown in Fig. 17-9. The arc length S swept out by the tip of the radius vector is equal to $v\Delta t$. The area swept out is equal to the area of a triangle with two sides equal to R_{av}, the average length of R, and with the third side equal to the component of S perpen-

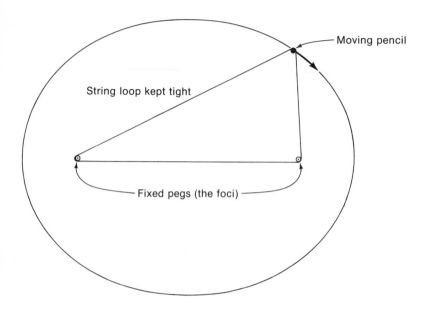

Moving pencil

String loop kept tight

Fixed pegs (the foci)

Fig. 17-8. A tightly stretched string on two pegs guides a pencil around an ellipse.

dicular to R. Consider now the area swept out in the same time, Δt, at two different points on the orbit:

$$\text{area}_1 = R_1 S_1 \sin \theta_1 = R_1 v_1 \Delta t \sin \theta_1$$

$$\text{area}_2 = R_2 S_2 \sin \theta_2 = R_2 v_2 \Delta t \sin \theta_2$$

From Kepler's second law:

$$R_1 v_1 \Delta t \sin \theta_1 = R_2 v_2 \Delta t \sin \theta_2$$

Cancelling Δt and multiplying by m, the planet's mass:

$$m v_1 R_1 \sin \theta_1 = m v_2 R_2 \sin \theta_2$$

Recognizing these as vector cross products, we find:

$$\mathbf{R}_1 \times m\mathbf{v}_1 = \mathbf{R}_2 \times m\mathbf{v}_2$$

or

$$\mathbf{R}_1 \times \mathbf{p}_1 = \mathbf{R}_2 \times \mathbf{p}_2$$

Recalling from Chapter 11 that the angular momentum of a single mass is given by $\mathbf{L} = \mathbf{r} \times \mathbf{p}$, we finally obtain

$$\mathbf{L}_1 = \mathbf{L}_2$$

Thus, Kepler's second law is simply a statement of the conservation of angular momentum. It applies to a planet because the force of gravity on the planet acts directly along the sun-planet radius vector and thus cannot produce a *torque* on the planet. Kepler's

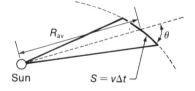

R_{av}

Sun $S = v\Delta t$ θ

Fig. 17-9. The area swept out by the planet's radius vector in a time Δt is proportional to $R_{av} S \sin \theta$.

second law, then, does not depend upon the inverse square nature of the force of gravity but only upon the fact that its direction lies along the line joining the two masses. Any force that has this property is called a *central force*.

17-7 Spherical Mass Distributions

By comparing the force of gravity on the earth's surface to that at the orbit of the moon, Newton was led to the inverse square law of gravity. Newton's "demonstration" of this law depended on the assumption that the earth's gravity behaves as though all the mass of the earth were concentrated at its center. It certainly isn't obvious that this assumption is correct. The gravitational effect of the rod in Fig. 17-3 is *not* what results if all its mass is concentrated at its center. For the case of the earth, those parts of it that are close underfoot have an enormously larger effect on us than those parts far away. On the other hand, the distant portions of the earth contain more mass than the comparatively small volume close underfoot. Do these effects just cancel for the case of a spherical earth?

Newton developed integral calculus to answer this question. Consider the hollow, spherical shell of mass in Fig. 17-10. A ring-shaped element of the shell's mass is shaded in the figure. All of this ring-shaped region is at the *same* distance, *x*, from the mass *m*. Only one distance, *x*, enters into the denominator of (17-1) for the force of gravity exerted by the ring on the mass. However, there is a slight geometrical complication in that the force vectors from different portions of the ring do not simply add together. Two such force vectors due to opposite segments of the ring are shown in Fig. 17-11. All of the force vectors involved lie on a cone of half-angle α. By symmetry, their resultant lies on the axis of the cone and has magnitude $F \cos \alpha$, where **F** is the force exerted on *m* due to the gravitational attraction of the mass contained in the ring, if that mass were at one point on the ring.

Since this ring of mass is an infinitesimal element of all the mass in the spherical shell, we will call the mass of the ring dM. Note that the radius of the ring is $R \sin \theta$. Its circumference is thus $2\pi R \sin \theta$. Its width is $R\, d\theta$ and its thickness is that of the spherical shell, ΔR. If the density of the mass in the shell is ρ, we have:

$$dM = \rho\, dV = \rho \cdot 2\pi R \sin \theta \cdot R\, d\theta \cdot \Delta R$$

$$= 2\pi \rho R^2 \sin \theta\, d\theta \cdot \Delta R$$

The force on *m* due to dM is

$$F = G\frac{m\, dM}{x^2} \cos \alpha \qquad (17\text{-}4)$$

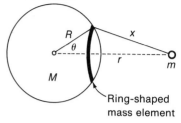

Hollow sphere

Ring-shaped mass element

Fig. 17-10. A hollow shell of mass *M* with radius *R* attracts a mass *m* that is a distance *r* from the center of the sphere. The distance between *m* and all points of the ring-shaped mass element shown is *x*.

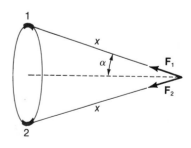

Fig. 17-11. The force vectors caused by opposing sides of the ring cancel in the vertical direction but project onto the *r* direction as $F \cos \alpha$.

But x is related to r, R, and θ by the law of cosines:

$$x^2 = r^2 + R^2 - 2rR \cos \theta \qquad (17\text{-}5)$$

and $\cos \alpha$ is given by

$$\cos \alpha = \frac{r - R \cos \theta}{x} \qquad (17\text{-}6)$$

Differentiation of Eq. (17-5) with respect to θ gives

$$2x \, dx = 2rR \sin \theta \, d\theta \qquad (17\text{-}7)$$

since r and R are constant in the situation of Fig. 17-10. Combining Eq. (17-4), (17-5), (17-6), and (17-7), we obtain:

$$F = G \frac{\Delta R \pi \rho m R}{r^2} \left(\frac{r^2 - R^2}{x^2} + 1 \right) dx$$

This is the force due to one ring of the spherical shell at a certain value of x. The total force will be the integral of this force over all possible values of x:

$$F_{\text{total}} = G \frac{\Delta R \pi \rho m R}{r^2} \int_{r-R}^{r+R} \left(\frac{r^2 - R^2}{x^2} + 1 \right) dx \qquad (17\text{-}8)$$

where, again, r and R are constant. The result is

$$F_{\text{total}} = G \frac{(4\pi R^2 \rho \Delta R) m}{r^2}$$

The quantity in brackets is the total mass M of the shell,

$$F_{\text{total}} = G \frac{Mm}{r^2} \qquad (17\text{-}9)$$

just as if two point masses were involved. Thus, for a mass located *outside* the spherical shell, the gravitational force exerted on a mass point it identical with that for the extended mass concentrated at its center. Since a solid sphere can be built up from many hollow, spherical shells, the same result holds for the solid earth.

When the point mass is *inside* the hollow shell, a rather remarkable cancellation occurs. As shown in Fig. 17-12, the gravitational attraction of one side of the spherical shell on an off-center point mass is cancelled by the opposing pull of the other side of the shell. Though one side is closer to the point, that side contains less mass, as indicated in Fig. 17-12.

More formally, this can be seen by making the limits of the integral in (17-8) suitable for an interior point. These limits then become $R + r$ and $R - r$ and the integral becomes identically zero.

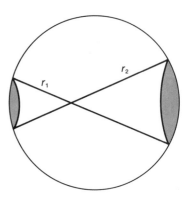

Fig. 17-12. The ratio of the two areas of the spherical shell contained within the double cone is proportional to $(r_2/r_1)^2$. But the gravitational force is inversely proportional to the distance squared. The net effect is a perfect cancellation of gravitational forces anywhere within a hollow sphere.

Example   A hole is drilled through the earth along a diameter, as in Fig. 17-13. A small mass m is released at one end of the hole. What happens?

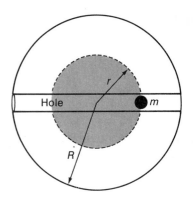

Solution   Assuming that the earth is of uniform density, a very simple result occurs. At any given distance r from the center of the earth, only the spherical mass with radius smaller than r attracts the mass. A shell of earth lying at larger radius produces no net force. The inverse square of the distance to the center of the earth increases rapidly as we approach the center. However, the remaining mass *inside* a given radius is proportional to the cube of that radius. The net result is that the force of gravity falls linearly as we approach the center, being exactly zero at the center. From Fig. 17-13 we see that the mass M' within radius r is given by

$$M' = \tfrac{4}{3}\pi r^3 \rho$$

Fig. 17-13.   The mass outside the dotted circle does not affect m.

The magnitude of the force of gravity on the mass m is:

$$F = G\frac{M'm}{r^2} = G \cdot \frac{4}{3}\pi\rho m\,\frac{r^3}{r^2}$$

$$= \tfrac{4}{3}\pi\rho\, Gmr \qquad (17\text{-}10)$$

Note that the force of gravity at the earth's surface is given by

$$F_g = G\frac{Mm}{R^2} = \tfrac{4}{3}\pi\rho\, GRm = mg$$

where R is the earth's radius. Hence we can rewrite (17-10):

$$F = -\left(\frac{mg}{R}\right)r \qquad (17\text{-}11)$$

where the minus sign indicates that the force is directed toward smaller r. This is a simple linear restoring force, so we expect simple harmonic motion for the mass m. The period is given by

$$T = 2\pi/\omega$$

where $\omega = \sqrt{k/m}$, with k being the quantity in parentheses in Eq. (17-11). Thus

$$T = 2\pi\sqrt{R/g}$$

$$= 2\pi\,(6.37 \times 10^6 \text{ m})^{1/2}\,(9.8 \text{ m/s}^2)^{-1/2}$$

$$= 5.09 \times 10^3 \text{ s}$$

This is the period of time required for a mass dropped into a hole through the earth to reappear at the starting point. This period is equal to 1.4 hr—precisely the time required for one full revolution *about* the earth for a satellite in a circular orbit just above the earth's surface.

Exercises

7. Find the force exerted by the earth on a 160-lb astronaut 1000 mi above the surface of the earth.

8. A mass dropped into a diametrical hole through the earth moves in simple harmonic motion. What would the period of this motion be if the density of the earth were 3.0 g/cm³ rather than the actual 5.5 g/cm³?

17-8 Variation of g

The value of g, the acceleration produced by gravity at the surface of the earth, is not quite constant over the entire surface of the earth. There is a smooth change in the sea-level value of g from 9.832 m/s² at either pole of the earth to 9.780 m/s² at the equator. The difference (0.052 m/s²) amounts to a $\frac{1}{2}\%$ change. The mean value of 9.806 m/s² occurs in the vicinity of 45° latitude. This general variation of g is due to (*a*) the departure of the earth's shape from a perfect sphere and (*b*) the centrifugal forces due to the earth's rotation.

The shape of the earth is quite close to an ellipsoid that bulges slightly at the equator. The distance from the center of the earth to sea level is 21 km (13 mi) larger at the equator than at the poles. Thus, the value of g is slightly smaller at the equator due to the increased distance from the center.

Furthermore, the acceleration caused by the earth's rotation reduces the *net* inward force in the frame of reference of the rotating earth at the equator. This rotational acceleration is given by

$$a_R = \omega^2 R_e$$

where
$$\omega = \frac{1 \text{ rev}}{24 \text{ hr}} \cdot \frac{1 \text{ hr}}{3600 \text{ s}} \cdot \frac{2\pi \text{rad}}{1 \text{ rev}} = 7.26 \times 10^{-5} \text{ rad/s}$$

so
$$a_R = 0.0336 \text{ m/s}^2$$

which is over one-half of the observed difference between g at the pole (where no centrifugal effects occur) and g at the equator. The remainder of the difference is due to the previously discussed departure from a spherical shape.

There are also small-scale variations in g caused by local variations in the density and shape of the earth's crust. In fact, one of the first crude estimates of the magnitude of G, before Cavendish's experiment, was made by determining the gravitational effect of a mountain, which caused the direction of plumb lines on

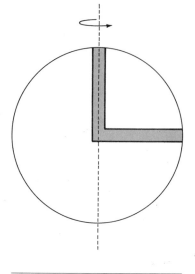

Fig. 17-14. The spinning of the earth on its axis causes sea level to be farther from the center of the earth at the equator than at the pole. Newton calculated the equilibrium depths of water in an L-shaped pipe to deduce the difference.

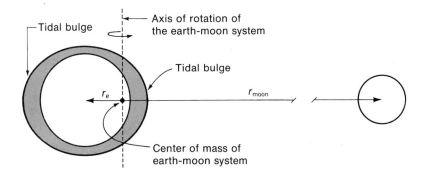

opposite sides of the mountain to depart from that expected for a uniform earth.

Newton first explained the departure of the earth's shape from spherical as due to the centrifugal effects of its rotation. Assuming that the earth's crust is plastic enough to allow it to flow to an equilibrium shape, it is clear that the spinning of the earth will cause an equatorial bulge. Newton used the clever model shown in Fig. 17-14 to calculate the difference between sea level at the pole and at the equator. By calculating the difference in depths of water required for equilibrium in the L-shaped pipe, he correctly explained the nonspherical shape of the earth.

Just as the earth attracts the moon, the moon attracts the earth. Its influence is most easily detected in the effects it produces on the earth's oceans. Figure 17-15 indicates the tidal effect of the moon's gravitation. If the earth did not have continents and did not rotate on its own axis, there would be one bulge in the ocean level directly under the moon and another one on the opposite side: one caused by the increased attraction of the moon compared to its average effect on the earth, and the other by the decreased attraction on the far side. In the actual situation, the earth rotates and repeatedly rams the margins of the continents into the tidal bulge. The depth of the tidal bulge is only 21 in., but since the earth's equatorial velocity is near 1000 mi/hr, the tides slop up to considerably greater depths at the shorelines. A *tidal wave* is the name given to a wave generated by seismic disturbances on the ocean floor. Their close kinship to tides is apparent when one realizes that they are typically of small amplitude (one foot), long wavelength (hundreds of miles), and high velocities (hundreds of mi/hr) in deep ocean water. It is usually only when they strike coastlines that they rear up to damaging heights.

The loss of energy caused by the tides impinging on the continents steadily slows the earth's rotation rate. Since the total angular momentum of the earth-moon system remains constant (except for the smaller effects of solar tides), the moon is receding steadily to larger radii. Tidal effects are thus quite important in the evolution of the earth-moon system. They are also important to other planets and must be considered in any successful model of the solar system.*

17-9   Gravitational Potential Energy

In Chapter 15 we calculated the potential energy of a mass near the surface of the earth and found it to be

$$U = mgy$$

so that changes in elevation by an amount h produce changes in U:

$$\Delta U = mgh$$

This result, however, depends upon the assumed constancy of the earth's gravity, so that $W = mg$ regardless of the elevation y. Clearly this cannot be precisely true, for we have seen that the force of the earth's gravity falls inversely with the square of the distance from its center. However, since the earth's surface is already some 4000 mi from its center, most changes in elevation we deal with have negligible effect. We can use the binomial series discussed in Section 14-10 to see better how the acceleration due to gravity varies for small height changes.

For points outside the earth we have

$$g' = G\frac{M_e}{r^2}$$

where g' indicates that this value will depend upon r, equaling g only for $r = R_e$. Writing r as $R_e + h$, where h is the height above the earth, we find

$$g' = G\frac{M_e}{(R_e + h)^2} = G\frac{M_e}{R_e^2}\left(1 + \frac{h}{R_e}\right)^{-2} \tag{17-12}$$

Since h/R_e is generally a very small number, we can use a binomial expansion, dropping the higher-order terms. Recalling that

$$(1 + x)^n = 1 + nx + \frac{n(n - 1)}{2!}x^2 + \frac{n(n - 1)(n - 2)}{3!}x^3 + \cdots$$

---

*See "Tides and the Earth-Moon System" by Peter Goldreich, *Scientific American,* April 1972.

and considering h/R_e as our variable x, we obtain

$$\left(1 + \frac{h}{R_e}\right)^{-2} = 1 - 2\frac{h}{R_e} + 3\left(\frac{h}{R_e}\right)^2 - \cdots \qquad (17\text{-}13)$$

Since $h/R_e = 2.5 \times 10^{-4}$ for $h = 1$ mi, we can comfortably ignore the term containing the square of this ratio. The higher-order terms are even less significant.

Rewriting (17-12) with the aid of (17-13), we obtain:

$$g' \approx G\frac{M_e}{R_e^2}\left(1 - 2\frac{h}{R_e}\right) = g\left(1 - 2\frac{h}{R_e}\right)$$

Thus for a height of $h = 100$ m above the earth's surface, the ratio of g' to g is given by

$$g' = \left(1 - 2 \cdot \frac{10^2 \text{ m}}{6.37 \times 10^6 \text{ m}}\right)g = 0.9999686g$$

The fractional departure of g' from g is only

$$\frac{\Delta g}{g} = \frac{2 \cdot 10^2}{6.37 \times 10^6} = 3.14 \times 10^{-5}$$

However, if we consider a rocket or satellite which moves some distance from the earth, the $1/r^2$ variation becomes important. In order to find the variation of gravitational potential energy with distance from the earth's center, we must now integrate the force given by (17-1). Since the force of gravity is radially directed toward the center of the earth, it is only the component of displacement parallel to the radius that appears in the work integral. Just as an arbitrary $dr\cos\theta$ became dy in Section 14-6, we need to consider only changes in r, the distance from the center of the earth, in this calculation. Thus changes in the gravitational potential energy are given by:

$$\Delta U_g = -\int_{r_1}^{r_2} F_g \, dr$$

Writing $F_g = -GM_em/r^2$, where the minus sign indicates an attraction, we have

$$\Delta U_g = GM_em \int_{r_1}^{r_2} \frac{dr}{r^2} = -GM_em\left(\frac{1}{r_2} - \frac{1}{r_1}\right) \qquad (17\text{-}14)$$

Thus we may take

$$U_g = -G\frac{M_em}{r} \qquad (17\text{-}15)$$

Example   A rocket of mass m is slowly lifted from the surface of the earth to an altitude of $R_e = 6.37 \times 10^6$ m above its surface. Find the net work done; that is, find its increase in gravitational potential energy.

Solution   Directly from (17-14) we find

$$\Delta U_g = GM_e m\left(\frac{1}{R_e} - \frac{1}{2R_e}\right)$$

Since we know that

$$g = G\frac{M_e}{R_e{}^2}$$

we can rewrite our result in a more convenient form:

$$\Delta U_g = mgR_e{}^2\left(\frac{1}{R_e} - \frac{1}{2R_e}\right) = mg\left(\tfrac{1}{2}R_e\right)$$

The mass m was lifted a total distance of R_e and the work done was only half that required if the force of gravity had not diminished with distance.

escape velocity

$V = \sqrt{2g R_e}$

Exercises

9. A 1-kg mass falls from rest from a distance above the earth's surface equal to the earth's own radius, i.e., $r_{\text{initial}} = 2R_e$ with respect to the center of the earth. Find the kinetic energy of the mass when it strikes the surface of the earth.

10. A mass falls from an infinite distance to the surface of the earth. Find its velocity when it reaches the earth's surface.

17-10 Summary

The law of gravity between point masses m_1 and m_2 is given by

$$F = G\frac{m_1 m_2}{r_{12}{}^2}$$

$\dfrac{m_s}{(d_s)^2} = \dfrac{m_e}{(d_e)^2}$

where G is the gravitational constant with a value of 6.67×10^{-11} N $\cdot$ m²/kg². In order to evaluate the gravitational effects of an extended mass distribution, the above inverse square law must be integrated over the extended object. For a sperical mass distribution (hollow or solid), the result for an object outside the sphere is as if all the mass of the sphere were concentrated at its center.

A hollow spherical mass produces no gravitational force on an object inside of it. A solid spherical mass distribution of uniform density produces a gravitational force on a mass within it: the force is directly proportional to the distance from the center of the sphere.

A satellite in circular orbit about a central body is continually accelerating toward the center of its orbit. This inward acceleration is caused by the inward-directed force of gravity. Solving Newton's second law $(F = ma)$ for such a satellite gives Kepler's third law:

$$T^2 = \left(\frac{4\pi^2}{Gm}\right)r^3 \qquad \longleftarrow$$

where T is the satellite's period, r its orbital radius and m the mass of the central body. Kepler's second law — that equal areas of a planet's orbit are swept out by the sun-planet vector in equal times — is a direct consequence of the conservation of angular momentum.

The variation of g with latitude on the earth's surface is due in part to the earth's departure from a spherical shape and in part to the earth's rotation. The tides are a direct consequence of the gravitational forces that the moon exerts on the earth's oceans.

For small variations in height above the earth's surface, the force of gravity is nearly constant, with

$$\Delta U = mgh$$

For large variations, however, we must integrate the varying force given by (17-1), in which case

$$\Delta U = -GM_e m\left(\frac{1}{r_2} - \frac{1}{r_1}\right)$$

[handwritten annotations in margin:]

From rest

$0 - \dfrac{Gm_em}{r} = \frac{1}{2}mv^2 - \dfrac{Gm_em}{r}$

$\Delta G = \dfrac{Gm_e}{r_e} = gR_e$

Problems

17-1. Calculate the magnitude of the force of gravity between the earth and moon. Compare this with the magnitude of the sun's attraction for the moon.

17-2. How far from the earth must one go before the gravitational attraction of the sun and of the earth are comparable?

17-3. Find the magnitude of gravitational acceleration at the circular orbit of a satellite that orbits the earth once every 24 hr. What is the orbital period of an earth satellite that is at a radius such that the acceleration due to the earth's gravity is $g/2$?

17-4. A satellite orbits the earth once every 4 hr. Another identical satellite orbits the moon every 4 hr. Compare the force of the earth's gravity upon the earth satellite to that of the moon's gravity upon the moon satellite.

17-5. A satellite is placed in orbit about the moon at an altitude of 100 mi above the moon's surface. Find its orbital period. (Some of the necessary data are contained in Exercise 6 on page 353.)

17-6. A satellite is placed into orbit just above the surface of planet X. The radius of the planet is 2×10^6 m and the observed period of satellite revolution is 84.8 min. Find the density of planet X.

17-7. If a hole is drilled through the earth along a chord instead of a diameter, a mass within it exhibits simple harmonic motion, just as in the example on page 357. Show that this is the case and that the period is the same as that for the motion along a diameter. Assume no friction between the mass and the tunnel wall and assume constant density for the earth.

17-8. Use the binomial expansion to show that (17-14) is consistent with $\Delta U = mgh$ near the earth's surface if $h \ll R_e$.

17-9. Find the force of gravity acting upon an astronaut in orbit 100 mi above the surface of the earth.

17-10. A mass m is released from rest (with respect to the earth) at a distance of 3 earth radii from the center of the earth. Find its velocity when it strikes the earth.

17-11. Find the work required to lift slowly a 1-kg body against the earth's gravity from the surface of the earth to an infinite distance from the earth.

$$\Delta U_g = G \frac{m_e}{R_e} = g R_e.$$

17-12. Find the initial velocity of a vertically fired projectile that will just coast from the earth's surface to an infinite distance. This velocity is the *escape velocity* for the earth. Compare the earth's escape velocity to that for the moon.

17-13. Compare the earth's escape velocity (see Problem 17-12) to the velocity of a satellite in orbit near the surface of the earth.

17-14. A rocket loaded with radioactive wastes is to be fired into the sun for permanent disposal. The rocket is fired directly opposite the earth's orbital velocity, so that when the rocket is well clear of the earth, it is stationary in the inertial frame tied to our solar system. The rocket then falls radially in toward the sun. Calculate the work required to do this for a 5×10^6-kg rocket, (*a*) ignoring the presence of the earth's gravity; (*b*) including the effect of the earth's gravity.

Chapter **18**

Behavior of Fluids

18-1 Introduction A *fluid* is a substance that is capable of flow. This flow may be rapid or slow, smooth or turbulent. A comparison between smooth and turbulent flow is shown in Fig. 18-1. The lines along which the fluid moves in smooth flow are called *streamlines*. *Liquids* are fluids that have densities generally similar to those of solids and that are not easily compressible, that is, their bulk moduli are similar to those of the solids tabulated in Chapter 13. *Gases* are fluids of low viscosity that have low densities and are typified by the ease with which they may be compressed. Sea-level air at room temperature, for example, has a bulk modulus 10^7 times smaller than that of water.

Because a fluid can flow, it is also typified by its inability to support shear forces. It is for this reason that oil is introduced between metal surfaces, such as engine bearings, to reduce friction between those surfaces. The layer of oil close to the stationary surface in Fig. 18-2 is nearly stationary, while successive layers of oil, closer and closer to the moving surface, have progressively larger velocities that finally equal that of the top plate. Fluid motion in layers as depicted in Fig. 18-2 is called *laminar flow*.

Though a fluid cannot support shear forces for any length of time, there are small frictional forces between layers of the fluid. Thus moving liquids generate small shear forces, called *viscous forces*. Viscosity can be measured by timing the fall of a steel ball between two fixed depths in a fluid sample. The ball will fall slowly in substances like honey or heavy oil; these substances are said to have *high viscosity*. The ball will fall rapidly in fluids of low viscosity, such as water or gasoline.

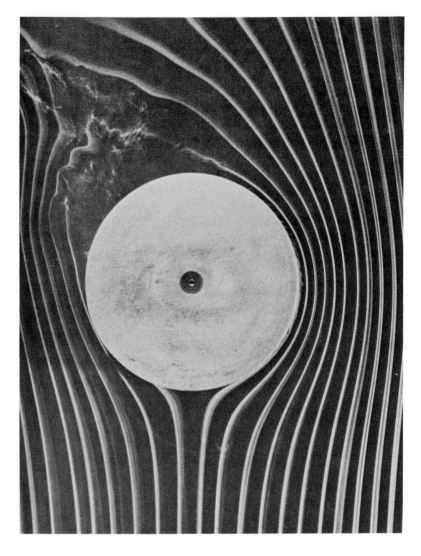

Fig. 18-1. Streamline flow of air moving past a rotating cylinder is revealed by smooth smoke streaks in the moving air. Behind the cylinder there is a region of turbulence. (Courtesy Dr. F. N. M. Brown)

18-2 Pressure in Static Liquids

When we defined the bulk modulus in Chapter 13, we introduced the concept of *pressure*, the force per unit area:

$$p = F/A$$

In Fig. 18-3 we see a cylindrical vessel containing a liquid with a depth h. The force of gravity on the liquid produces a downward force acting on the bottom of the vessel. The force on the bottom is

$$F_b = m_l g$$

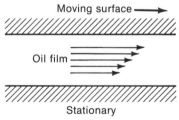

Fig. 18-2. A film of oil introduced between two metal surfaces acts as a lubricant because it is unable to support shear forces.

where m_l is the mass of the liquid. Since m_l equals the density ρ of the liquid multiplied by the volume of liquid and this volume, in turn, equals the product of the area A and depth h, we can write:

$$F_b = \rho A h g$$

The pressure on the bottom is then:

$$p = \frac{F_b}{A} = \rho g h$$

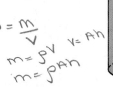

$\rho = \dfrac{m}{V}$ $V = Ah$

$m = \rho V$

$m = \rho Ah$

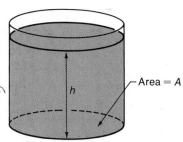

The pressure is seen to be *independent* of the area of the vessel. It depends only upon the density of the liquid and its depth.

Fig. 18-3.   A cylindrical vessel of cross-sectional area A filled to a depth h.

Example   A beaker with a cross-sectional area of 80 cm² and a mass of 20 g is placed on a scale, as in Fig. 18-4. It is filled with 800 g of water, which occupies a volume of 800 cm³ and thus fills the beaker to a depth of 10 cm. The net force on the scale is equal to the weight of the 20-g beaker added to the force on the bottom of the beaker caused by the pressure of the water. This pressure is

$$p = \rho g h$$

and the force is

$$F = pA = \rho g h A = Mg = 800 \text{ g} \cdot 980 \text{ cm/s}^2 = 7.84 \times 10^5 \text{ dynes}$$

where M is the mass of water. Added to this force is the weight of the beaker:

$$W = mg = 20 \text{ g} \cdot 980 \text{ cm/s}^2 = 1.96 \times 10^4 \text{ dynes}$$

for a total force of 8.036×10^5 dynes. Since the scale is calibrated in mass units, its reading is $F/g = 820$ g, as it should be, for this is the total mass placed upon it. Now a finger is thrust into the water, displacing 8 cm³ of water, so the depth becomes $h' = 10.1$ cm. The force on the bottom is now

$$F' = \rho g h' A = 7.9184 \times 10^5 \text{ dynes}$$

so that the total force is

$$F_{total} = F' + W = (7.9184 \times 10^5 + 1.96 \times 10^4) \text{ dynes}$$
$$= 8.1144 \times 10^5 \text{ dynes}$$

Hence the scale reading is

$$\frac{F_{total}}{g} = 828 \text{ g}$$

Though only 820 g is placed on the scale, the scale reads 828 g, as if another 8 cm³ of water had been added to the beaker. *Indeed*, the scale has no way of distinguishing between an increased depth due to the true addition of 8 cm³ of water or a displacement of 8 cm³ of water by inserting an object into the water.

pressure only depends on h

Fig. 18-4.

If we now consider a vessel with the same depth but a different shape, as in Fig. 18-5, it seems that the pressure on the bottom of this vessel would differ from that of the cylinder, but this is not the case. To see this, we must consider a more general question: How does the pressure vary with depth *within a liquid?* In Fig. 18-6 we have isolated one element of that liquid. It has a thickness Δh and the area of its top (or bottom) is A_1. The area of any of its sides we denote A_2. Since liquids in static equilibrium do not swirl about, we conclude that each element of the liquid must be at rest and, hence, in equilibrium. The pressure exerted by the remainder of the liquid on this element must produce forces F_3 and F_4 that cancel, F_5 and F_6 that cancel, and F_2 must be just enough larger than F_1 to support the weight of this liquid element:

$$F_3 = p_3 \cdot A_2$$

$$F_4 = p_4 \cdot A_2$$

Hence $\qquad\qquad\qquad p_3 = p_4 \qquad\qquad\qquad$ (18-1)

Similarly, $\qquad\qquad\quad p_5 = p_6 \qquad\qquad\qquad$ (18-2)

Now we must have $F_2 > F_1$, so that:

$$p_2 A_1 - p_1 A_1 = \rho g A_1 \cdot \Delta h$$

$$p_2 - p_1 = \rho g \Delta h$$

or $\qquad\qquad\boxed{\Delta p = \rho g \Delta h} \qquad\qquad$ (18-3)

Thus if the density is constant, the change in pressure between two points at different depths within a liquid is simply linearly dependent on the magnitude of this difference. Equations (18-1) and (18-2) tell us that there is no change in pressure associated with horizontal changes in position; only depth changes produce changes in pressure. If the variation in pressure produces an appreciable variation in density, as in the earth's atmosphere, the situation is more complicated (see Appendix).

The pressure at a given depth is independent of the direction in which it acts: left, right, up, or down. This can be seen by the following simple argument. Since fluids cannot support shear, the force they exert upon a surface must be perpendicular to that surface. Thus, if the element of liquid in Fig. 18-6 were modified into a prism with its top face inclined as shown in Fig. 18-7, the net force down would be given by the vertical projection of F_1. The area of the top is increased, however, so that the net downward force is unchanged:

$$F_{\text{down}} = F_1 \cos \theta = p_1 A_1' \cos \theta$$

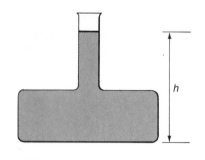

Fig. 18-5.

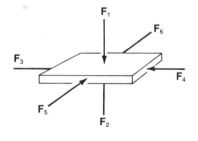

Δh produces ΔP

Fig. 18-6.   A small rectangular element within the liquid in Fig. 18-3.

where A_1' indicates that the top area is larger than it was in Fig. 18-6 due to the fact that it is tipped. In terms of A_1 we find that

$$A_1' = \frac{A_1}{\cos \theta}$$

so that
$$F_{\text{down}} = p_1 \cdot \frac{A_1}{\cos \theta} \cdot \cos \theta = p_1 A_1$$

exactly as it did previously. Thus the prism is also in vertical equilibrium. We have *assumed* that $F = pA$, independent of the orientation of the area and found equilibrium. Since we *observe* that all elements of a static fluid are in equilibrium, we invert our argument to conclude that:

> At any given depth, the pressure exerted by a fluid is independent of the orientation of the surface against which it is exerted.

Since variations of pressure within the static fluid depend only upon depth, any additional pressure applied to the surface of the fluid, such as atmospheric pressure or pressure applied by a piston, must also be added to the pressure found at every other point in the fluid. This result was first stated in 1653 by Blaise Pascal (1623–1662) and is called *Pascal's law:*

> *Pressure applied to an enclosed fluid is transmitted undiminished to every portion of the fluid and the walls of the containing vessel.*

The hydraulic press is a device that capitalizes upon Pascal's law to multiply a force. Its operation is indicated in Fig. 18-8. The force F_1 acts on the piston with area A_1 to produce a pressure given by

$$p = F_1/A_1$$

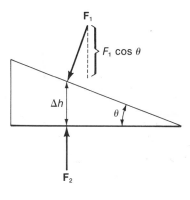

Fig. 18-7. Inclining the top of the fluid element increases the top area but decreases the vertical projection of $\mathbf{F}_1$. These two effects cancel exactly.

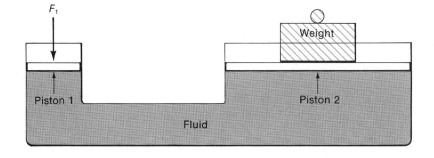

Fig. 18-8. The principle of the hydraulic press. A small force F_1 applied to piston 1 balances a much larger force F_2 due to the weight placed on piston 2. The force F_1 is smaller by the ratio A_1/A_2 than the force F_2.

This same pressure acts upon the larger piston with area A_2 to produce a force

$$F_2 = pA_2 = F_1 \frac{A_2}{A_1} \qquad (18\text{-}4)$$

Thus the original force F_1 is multiplied by the ratio A_2/A_1 to produce a larger F_2. This device does not, however, give something for nothing, since the piston A_1 must move much farther than piston A_2. In fact, if A_1 moves a distance d_1 displacing a volume $A_1 d_1$, then

$$d_2 = \frac{A_1}{A_2} d_1 \qquad (18\text{-}5)$$

in order that the same volume be displaced by A_2. The work done in pushing A_1 through a distance d_1 is

$$W_1 = F_1 d_1$$

and the work produced by A_2 is $W_2 = F_2 d_2$, which upon substitution from (18-4) and (18-5) becomes

$$W_2 = F_1 \frac{A_2}{A_1} \frac{A_1}{A_2} d_1 = W_1$$

Thus the *net work* produced by the press precisely equals that done upon it.

Exercises

1.   The vessel in Fig. 18-5 has a cross-sectional area of 150 cm² in the lower portion, while the cross-sectional area of the upper narrow portion is only 10 cm². It is filled with water to a depth of $h = 100$ cm. Find the pressure exerted by the water at the bottom of the vessel.

2.   Explain why the liquid stands at the same depth in all the vessels of Fig. 18-9.

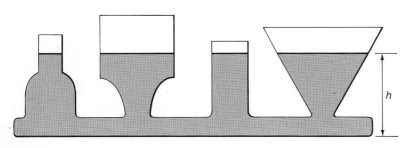

equal pressure at the base must give equal height above the base

Fig. 18-9. Liquid stands at the same depth, h, in connected vessels of a variety of shapes.

3. A customer examines two bicycle pumps, one with a 1-in.-diam piston that sells for $3.95 and another with $1\frac{1}{2}$-in.-diam piston that sells for $4.95. The salesman asks if the customer wishes to pump low-pressure (30–40 lb/in.²) tires or high-pressure (60–80 lb/in.²) tires. When he replies that he wants to pump high-pressure tires, the salesman immediately says that the larger pump is required. What is your conclusion?

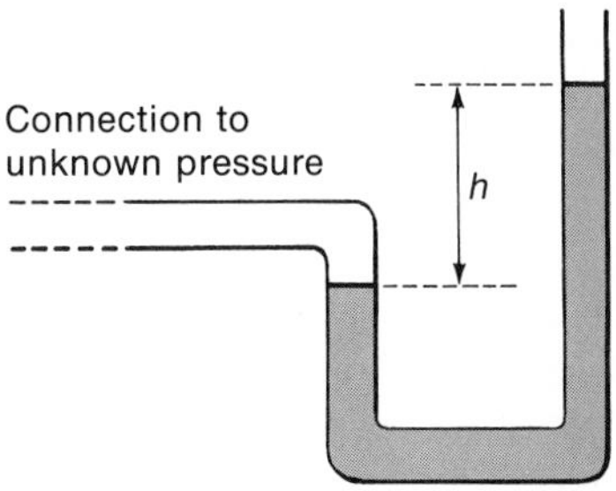

18-3 Pressure Measurements and Atmospheric Pressure

One type of pressure gauge, a *manometer,* balances an unknown pressure against the pressure due to a column of liquid, as in Fig. 18-10. Another type of pressure gauge, shown in Fig. 18-11, contains a thin metal diaphragm that is flexed by applied pressure. The movement of the diaphragm is indicated by a pointer.

If one side of a pressure gauge is connected to a good vacuum and the other side left open to the atmosphere, it will read the pressure exerted by the earth's atmosphere. This atmospheric pressure is 1.01×10^5 N/m² (14.7 lb/in.²) at sea level, decreasing exponentially with altitude, as indicated in the Appendix to this chapter.

Evangelista Torricelli (1608–1647) experimented with columns of liquid supported by the pressure of the earth's atmosphere. The arrangement he used is shown in Fig. 18-12. Since there is a vacuum above the liquid, the pressure the liquid produces at the base of the liquid column is just $\rho g h$, where ρ and h refer to the liquid. At equilibrium, this pressure must be offset by atmospheric pressure. Torricelli originally worked with columns of water, but the height of such a column of water is inconvenient for study (slightly over 33 ft.) He then invented the mercury barometer, in which mercury (density 13.6 times that of water) replaced the water, so that the instrument was less than 1 m high. In units of the height of such a mercury column, the pressure of the earth's atmosphere at sea level is about 76 cm of mercury (usually written 760 mm Hg). Most

Connection to
unknown pressure

Fig. 18-10. A U-tube manometer in which a difference in height, *h,* is produced by an applied pressure. This pressure is given by $\rho g h$, where ρ is the density of the filling liquid.

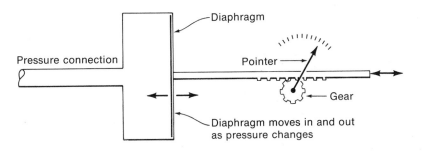

Diaphragm

Pressure connection

Pointer

Gear

Diaphragm moves in and out as pressure changes

Fig. 18-11. A pressure gauge that utilizes a metal diaphragm linked to a pointer.

American barometers in use for detecting slight changes in atmospheric pressure due to weather phenomena are calibrated in inches of mercury—thirty inches being close to the average pressure.

Water is often pumped from a well by a vacuum pump at the top of a pipe that dips into the water. The maximum distance that water can be lifted in such an arrangement is limited to the height that can be supported by atmospheric pressure: about 33 ft.*

When the right side of the manometer in Fig. 18-10 is open to the atmosphere, its reading gives the *difference* between the pressure applied to it and atmospheric pressure. This type of pressure reading is called *gauge pressure*. If the right side of the manometer is connected to a vacuum, it reads *absolute* pressure.

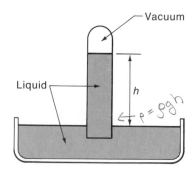

Fig. 18-12. A column of liquid held up by atmospheric pressure.

Exercises

4. How can a pressure gauge like that shown in Fig. 18-11 be arranged to read absolute pressure?

5. Find the atmospheric pressure in dynes/cm² when a mercury barometer reads 76 cm. The density of mercury is 13.6 g/cm³.

6. The pressure exerted by the earth's atmosphere at sea level is 14.7 lb/in.². A cylindrical vessel with a diameter of 2 ft has the air pumped out of it, so that atmospheric pressure on the outside is no longer balanced by pressure on the inside. Find the net force on one end of the cylindrical container.

18-4 Buoyancy

When a body is immersed into a fluid, as in Fig. 18-13, a net upward force called the *buoyant* force is exerted upon the body. The cause of this force is the difference in pressure on the lower and upper portions of the body. Since this pressure is due to a difference in depth, it is easy to calculate directly the buoyant force for a simple case such as a rectangular solid oriented horizontally within the fluid. For an irregular solid, such a calculation is difficult due to the variation of pressure at various points of the solid's surface. Obviously such a problem can be attacked with integral calculus, but a simple and ingenious solution for the magnitude and direction of the buoyant force was found long before calculus was invented. We will describe this simple solution, which leads to Archimedes' principle.

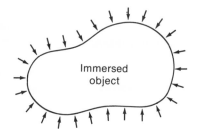

Fig. 18-13. A fluid comes into intimate contact with an immersed object and can exert only normal forces upon it. The forces on the lower surface are greater than those on the upper surface due to the increased depth.

---

*For an apparent exception to this rule, see "How Sap Moves in Trees" by Martin H. Zimmerman, *Scientific American,* March 1963.

Let us find the net upward force exerted on the object in Fig. 18-13. Imagine that the object is removed and replaced by fluid identical to that surrounding it. In order to fix our attention on the particular volume of liquid that now replaces the immersed object, imagine a surface enclosing this volume that has precisely the shape of the immersed object's surface. If the enclosed volume is filled with fluid, the fact that this volume is in equilibrium tells us that the *net* force exerted by the remainder of the fluid on this fluid element is just exactly that required to keep the enclosed fluid in equilibrium. Thus the net force is upward-directed and equal to $m_f g$, where m_f is the mass of the fluid contained in the volume, as shown in Fig. 18-14.

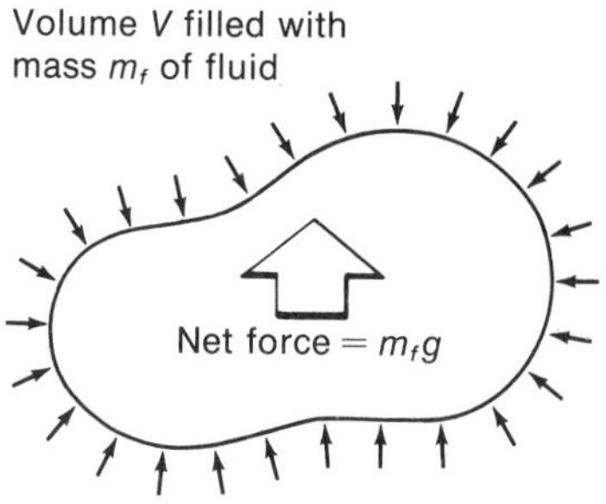

Now replace the imaginary volume of the previous paragraph with the actual object of mass m_0. The fluid must still come into contact with every portion of the surface as before and exerts the same forces upon it everywhere. The net force on the immersed object must simply be the force required to support an equivalent volume of fluid. This force is the buoyant force and is given by

$$F_b = m_f g = \rho_f V g \qquad (18\text{-}6)$$

where m_f is the mass of fluid displaced by the immersed object. This mass is given in the last portion of (18-6) by the density of the displaced liquid multiplied by the volume of the immersed object. This buoyant force acts vertically upward.

Equation (18-6) is called *Archimedes' principle* and was deduced, in a manner similar to the above, by Archimedes c. 250 B.C. Archimedes' principle states that the magnitude of the buoyant force is equal to the weight of displaced fluid. If the buoyant force is less than the weight of the immersed object, the net force on the object will be directed downward and it will sink. If the weight of the object is less than the buoyant force, the net force will be directed upward and the object will float. Since the mass of the immersed object is given by its density, ρ_0, multiplied by its volume, its weight is

$$W = m_0 g = \rho_0 V g \qquad (18\text{-}7)$$

The buoyant force is given by (18-6):

$$F_b = \rho_f V g$$

The net upward force on a *completely immersed* object is thus

$$F_{\text{net}} = F_b - W = (\rho_f - \rho_0) V g \qquad (18\text{-}8)$$

Thus bodies with a density greater than that of the fluid sink, while those with lower density float.

Volume V filled with mass m_f of fluid

Net force = $m_f g$

Fig. 18-14. The net force on the volume V of the fluid is just equal to the weight of the fluid it contains.

Example   A balloon is fabricated of 2 g of rubber and inflated with $\frac{3}{4}$ g of helium. The balloon has a volume of 4 liters. Find the net upward force acting on the balloon.

Solution   The balloon displaces 4 liters of air. At 1.29 g/liter, the mass of displaced air is 5.1 g. Let us compute the forces involved in the cgs system of units. The weight of displaced air gives the buoyant force:

$$F_b = W_a = m_a g = 5.1 \text{ g} \cdot 980 \text{ cm/s}^2$$

$$= 5 \times 10^3 \text{ dynes}   \text{(upward)}$$

$$W_{balloon} = 2.75 \text{ g} \cdot 980 \text{ cm/s}^2 = 2.7 \times 10^3 \text{ dynes}   \text{(downward)}$$

The net upward force is

$$F_{net} = (5 - 2.7) \times 10^3 \text{ dynes} = 2.3 \times 10^3 \text{ dynes}$$

(1 ounce $= 2.9 \times 10^4$ dynes).

Example   A block of wood with a density of 0.6 g/cm³ is in the shape of a cube, 10 cm on each edge. It floats as sketched in Fig. 18-15. How much of the block projects above the water?

Solution   The block must displace enough water to equal its own weight. Since the block has a density 6/10 that of water, 6/10 of it must be below the surface. We prove this as follows:

$$W_{block} = \rho_{wood} V_{block} g   \text{(downward)}$$

$$\rho_{wood} = 0.6 \rho_{water}$$

$$F_b = W_{water} = \rho_{water} (f V_{block}) g   \text{(upward)}$$

where f is the fraction below the surface. For equilibrium, the net force on the block must vanish:

$$\rho_{wood} V_{block} g = \rho_{water} (f V_{block}) g$$

$$\rho_{wood} = \rho_{water} f$$

$$f = \rho_{wood} / \rho_{water} = \frac{0.6 \rho_{water}}{\rho_{water}} = 0.6$$

Thus 0.6 of the block is below the surface, 0.4 above: $h = 4$ cm in our example.

Buoyant forces sometimes operate in situations where the fluid is so viscous that we do not expect buoyant effects. In rainy areas, for example, it is unwise to empty a swimming pool in the winter. If the ground surrounding the pool becomes sufficiently saturated with water it can force the pool up out of the ground or perhaps crack the floor or sides of the pool. Similarly, underground oil-storage tanks that are nearly empty have been known to erupt (usually with awesome slowness) from water-laden lawns during heavy rains.

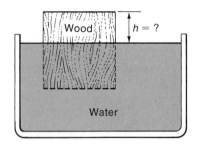

Fig. 18-15.

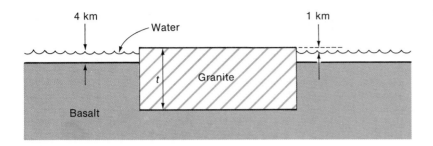

Fig. 18-16. A model for calculation of the buoyant forces on a continent.

Though we normally think of the earth's crust as being quite rigid, it is capable of flow over long spans of time in the geological sense. Thus the current geological view of a mountain range is somewhat like that of an iceberg. Material that is less dense than surrounding crustal material floats with a portion protruding above the surrounding plains—much like the wood block in the last example. When we see a mountain range, then, we only see a portion of the entire structure; a great deal of the low-density material is lying below the surface.

Similarly, the continents themselves appear to be floating in the earth's mantle.* This mantle, which underlies the oceans, is a dense, black, basaltic rock, while the continents appear to be made of a less dense mixture of granite and sedimentary rock. In Fig. 18-16 a crude model of the situation is indicated. Using a specific gravity of 3.0 for the basalt mantle, 2.65 for the granite, and 1.0 for the oceans, we find that the continent must be 31 km thick if it is simply floating in equilibrium. Indeed, seismic information does yield a value near 30 km for the thickness of continents.

Since the thickness of the crust is greater where the less dense material is located, a cancellation occurs so that the force of the earth's gravity changes very little from one region to another (apart from the general behavior discussed in the preceding chapter), unless that region is *not* in equilibrium. Thus, geologists can find regions that are out of equilibrium, such as a rising mountain range, by looking for gravitational anomalies. These are regions where g differs from the value expected at that latitude.

Exercises

7. A diver underwater lifts a stone with a volume of 0.1 m³ and finds he must exert a force of 760 N to lift it. Find the density of the stone.

*See "Plate Tectonics" by John F. Dewey, *Scientific American*, May 1972; or "The Evolution of the Andes" by David E. James, *Scientific American*, August 1973.

8. A solid cube 1 cm on each edge is oriented with its top and bottom faces horizontal below the surface of a vessel of water. The top face is at a depth of 1 cm, so that the bottom face is at a depth of 2 cm. Calculate the force down on the top surface and the force up on the bottom surface. Compare the net upward force given by the difference of these forces to that predicted by Archimedes' principle.

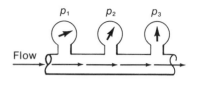

Fig. 18-17. Frictional forces between a moving liquid and the pipe walls cause a steady decrease in pressure, i.e., $p_1 > p_2 > p_3$.

18-5 Pressure in Moving Fluids

When a fluid flows through a pipe, a pressure drop exists along its length (see Fig. 18-17). This drop in pressure is caused by friction between the moving fluid and the walls of the pipe. An isolated element of the fluid moving with constant velocity v is shown in Fig. 18-18. The pressure difference between its two ends is just that required to offset the forces of friction. Since the pressure changes uniformly, the *pressure gradient, dp/ds,* where s is along the direction of flow, is constant and negative. For fluids with low viscosity this pressure gradient will be small. In the absence of friction the pressure will be constant throughout the pipe.

Now consider a region where the pipe is constricted, as in Fig. 18-19. A hasty use of our intuition tells us that, as long as friction is small, the liquid is under greater pressure in the constricted area since it seems the liquid is "squeezed" into the smaller pipe. Actually, the reverse is true. Qualitatively, we can see this must be so. The fluid moves slowly in the pipe of large diameter but must move rapidly in the small diameter pipe, since the same net flow moves throughout the system. Thus any element of liquid making the transition from the wide pipe to the narrow pipe must accelerate. The only way this can occur is if a net force acts to the right on such an element as it makes the transition in Fig. 18-19. This net force is provided by a pressure difference, with $p_1 > p_2$.

In order to make this qualitative argument precise, we have to find the change in flow velocity produced by a pipe of changing cross section. We have seen, qualitatively, that there will be a pressure change, so fluids that are easily compressible (gases) will also change density. Most liquids, however, have large enough bulk moduli that they will not appreciably change density. Since a varying density adds another degree of complication to the discussion, let us derive results applicable to fluids with constant density. In a fixed time interval Δt, a volume of liquid $V_1 = A_1 v_1 \Delta t$ must pass point 1 in Fig. 18-20. In the same time interval, a volume $V_2 = A_2 v_2 \Delta t$ must pass point 2. Now *if the liquid is incompressible, V_1 must equal V_2,*

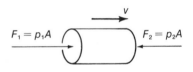

Fig. 18-18. An element of fluid of length l and cross-sectional area A moving with velocity v. $F_1 - F_2 > 0$ and balances the force of friction.

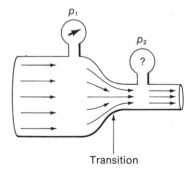

Fig. 18-19. Liquid flowing along a pipe that becomes constricted.

for the fluid cannot simply disappear between points 1 and 2 (unless we have a leaky pipe). Hence we have:

$$V_1 = V_2 \qquad (18\text{-}9)$$

so that

$$A_1 v_1 \Delta t = A_2 v_2 \Delta t$$

and

$$A_1/A_2 = v_2/v_1$$

or

$$v_2 = v_1 A_1/A_2 \qquad (18\text{-}10)$$

Equation (18-10) is called the equation of continuity since it stems from the fact that liquid cannot appear or disappear between points 1 and 2 but must continue from one point to another.

We know what change in velocity is produced by a cross-sectional area change. Now we must find the change in pressure that occurs. The simplest procedure is to calculate the work done on an element of fluid as it moves through the transition region and to equate this work to the increase in kinetic energy that occurs. The work done by the force F_1 shown in Fig. 18-21 is given by $F_1 \cdot l_1$ and that by F_2 is $F_2 \cdot l_2$. Since $F_1 = p_1 A_1$ and $F_2 = p_2 A_2$, we have:

$$\text{net work} = W = p_1 A_1 l_1 - p_2 A_2 l_2$$

$$= p_1 V_1 - p_2 V_2 = (p_1 - p_2)V$$

where we have dropped the subscript on V because the equal volumes flow in equal time by Eq. (18-9).

The increase in kinetic energy is given by:

$$K = \tfrac{1}{2}mv_2^2 - \tfrac{1}{2}mv_1^2 = \tfrac{1}{2}\rho V v_2^2 - \tfrac{1}{2}\rho V v_1^2$$

Hence, since W must equal K,

$$(p_1 - p_2)V = \tfrac{1}{2}\rho V(v_2^2 - v_1^2)$$

$$p_1 - p_2 = \tfrac{1}{2}\rho(v_2^2 - v_1^2) \qquad (18\text{-}11)$$

Equation (18-11) is *Bernoulli's equation;* the pressure difference produced is often called the *Bernoulli effect*. Note that, as anticipated, if $v_2 > v_1$ then $p_1 > p_2$.

Example   Water at a pressure of 5×10^4 N/m² flows at 3 m/s through a pipe of $\frac{1}{50}$-m² cross-sectional area, which then reduces to $\frac{1}{100}$-m² cross section. What is the velocity and pressure in the smaller cross-section pipe?

Solution   The final velocity is given by (18-10):

$$v_2 = v_1 A_1/A_2 = 2v_1 = 6 \text{ m/s}$$

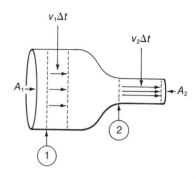

Fig. 18-20.   As the streamlines crowd together into the smaller cross section, A_2, the velocity increases.

$$v_D = \sqrt{2gh}$$

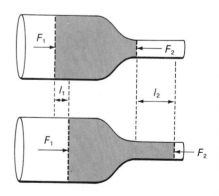

Fig. 18-21.   The work due to F_1 is $F_1 \cdot l_1$ while, at the same time, that due to F_2 is $F_2 \cdot l_2$.

The pressure change is

$$p_1 - p_2 = \tfrac{1}{2}\rho\,(v_2{}^2 - v_1{}^2)$$

$$= \tfrac{1}{2}10^3\,\frac{kg}{m^3}\cdot\left(\frac{36\ m^2}{s^2} - \frac{9\ m^2}{s^2}\right)$$

$$= 1.35 \times 10^4\ kg\cdot m/s^2/m^2$$

$$= 1.35 \times 10^4\ N/m^2$$

Hence

$$p_2 = p_1 - (p_1 - p_2) = 5 \times 10^4\ N/m^2 - 1.35 \times 10^4\ N/m^2$$

$$= 3.65\ N/m^2$$

The above calculations neglect the small additional pressure drop due to friction.

If the two points at which we wish to compare pressures are not at the same height, there will be an additional pressure difference, ρgh. A complete statement of Bernoulli's theorem is thus:

$$p + \tfrac{1}{2}\rho v^2 + \rho gh = \text{constant} \tag{18-12}$$

if friction is negligible.

Bernoulli's principle can also be applied to fluids that are not confined to pipes, as long as they flow smoothly so that their motion is described by streamlines, as defined in the introduction to this chapter. Since each element of fluid follows a streamline, fluid elements cannot cross streamlines. Thus a bundle of streamlines can replace our pipe in deriving Bernoulli's equation with no change in the result. Equation (18-11) can therefore be applied to a smoothly flowing fluid that is not confined to a pipe, as long as we follow a streamline from point 1 to point 2.

Exercises

9. Water flows through a pipe of 10-cm diameter that narrows to 4-cm diameter. The pressure in the wider pipe is 5×10^5 dynes/cm². The pressure in the narrow pipe is 3×10^5 dynes/cm². What is the flow velocity in the large-diameter segment? Neglect frictional forces.

10. A rectangular house 10 m × 15 m faces into the wind. The front door is open, but all the other doors and windows are closed. The stationary air within the house has a higher pressure than that flowing rapidly over the roof. Find the net force lifting the roof if the wind velocity is a uniform 100 km/hr.

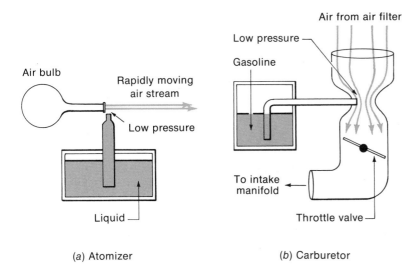

Fig. 18-22. (*a*) The normal atmospheric pressure on the liquid raises it up the tube since there is a low pressure at the top of the tube. The liquid is dispersed as small droplets as it enters the air stream. (*b*) Rapidly moving air produces a low pressure region in a gasoline engine's carburetor. Atmospheric pressure then forces gasoline into the intake air, where it is vaporized and drawn into the engine's cylinder.

(*a*) Atomizer

(*b*) Carburetor

18-6   Applications of the Bernoulli Effect

Applications of the pressure change due to motion of fluids are so numerous that only a few can be mentioned here. In many devices, the pressure drop in a swiftly moving air stream is used to elevate a liquid into the air stream to produce a fine spray of droplets, as illustrated in Fig. 18-22.

In Fig. 18-23 are shown two demonstrations of the Bernoulli effect. The elevated ping-pong ball in part *a* is held there by the pressure differential between slowly and rapidly moving air. In *b* it appears that air rushing down through the spool would simply blow the card off the bottom. But, paradoxically, the rapidly moving air between the spool and card produces low pressure, so that the card is supported against its own weight. Air flow over an airplane wing produces "lift" in a similar manner, since the upper surface of the wing is curved slightly so that air must move more rapidly over its upper surface than over its lower surface. The wing may also be placed at a small angle (*angle of attack*) to the air flow so as to deflect the air downward. This produces a positive pressure on the lower surface as well as the Bernoulli effect. An airplane flying upside down is totally dependent upon a large angle of attack.

The curve of a thrown, spinning ball is best understood in two steps. First, consider a stationary ball spinning rapidly about an axis. As is shown in Fig. 18-24*a,* air is dragged around the ball by frictional forces. Now if the ball is thrown from right to left, we can view it in an inertial frame of reference moving with the ball. In that

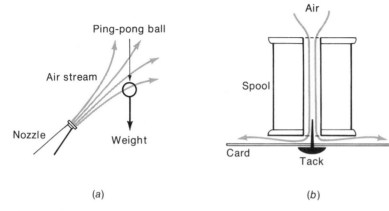

Ping-pong ball

Air stream

Nozzle

Weight

Air

Spool

Card

Tack

(a)   (b)

Fig. 18-23. (a) The greater pressure of stationary air keeps a ping-pong ball elevated in an air stream against the force of gravity. (b) A card + thumbtack are held up by an air flow that would appear to blow the card away from the spool. (The thumbtack merely serves to keep the card centered).

frame of reference air is streaming past the ball from left to right. Note that the air above the ball in Fig. 18-24b is moving more rapidly than that below the ball. The pressure above the ball is thus lower than that below. In Fig. 18-25 an overhead view of a spinning ball shows how a simple curve develops.

The airspeed indicator on an aircraft uses Bernoulli's principle quantitatively. In Fig. 18-26 we see that a tube (called the *Pitot tube*) extends forward and is connected to a gauge that records the pressure difference between the Pitot tube and an opening on the side of the aircraft. Air cannot enter the Pitot tube since the pressure gauge does not allow air to pass through. Thus the air just at the tip of the Pitot tube is stationary, while that at the opening on the side of the aircraft is moving at the airspeed of the aircraft. The pressure difference read by the gauge is just:

$$p_1 - p_2 = \tfrac{1}{2}\rho(v_2{}^2 - v_1{}^2)$$
$$= \tfrac{1}{2}\rho(v_2{}^2 - 0)$$

where ρ is the density of air and v_2 is the airspeed. In practice, the gauge is calibrated directly in units of mi/hr or km/hr, assuming a density ρ for sea-level air. The pilot then corrects the indicated airspeed to true airspeed by using a table that gives the effect of decreasing air density at increasing altitudes. If you had difficulty with Exercise 10 previously, you might try it again.

Exercises

11. The waist gunner on a World War II "flying fortress" bomber fires a gun in a direction perpendicular to the bomber's motion. The bullets fired by the gun are given a spin about an axis along their direction of motion to stabilize them gyroscopically.

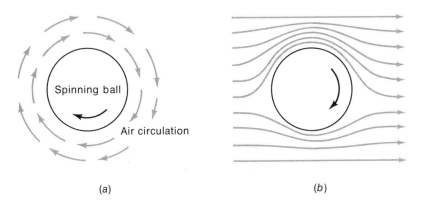

(a)   (b)

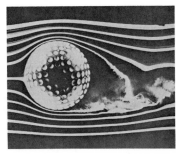

Fig. 18-24. (a) A spinning ball drags air around with its spinning motion. (b) The combination of air moving past the ball plus the rotating air pattern of (a) produces an asymmetry. The air above the ball has a higher velocity than that below. (c) A smoke-streak photograph of air-flow past a golf ball rotating clockwise. (Courtesy Dr. F. N. M. Brown)

Describe the way in which the Bernoulli effect influences the bullets, depending upon whether they are fired out of the left or the right side of the bomber. Assume the bullets spin clockwise as seen by the gunner.

12. The Pitot tube shown in Figure 18-27 is connected to a U-tube manometer filled with a liquid with a density of 1.5 g/cm³. Assume a density of 10^{-3} g/cm³ for air. Plot a calibration curve that gives v, the velocity of air flow, as a function of h, the manometer reading, for h between 0 and 5 cm.

18-7   Summary

A fluid is typified by its inability to support shear forces when in static equilibrium. Both gases and liquids are fluids. The pressure in a static fluid varies only with depth, with the change in pressure given by

$$\Delta p = \rho g h$$

smaller area — higher pressure

Any pressure applied to a confined fluid is transmitted undiminished to all portions of the fluid and the walls of its container.

The fact that pressure increases with depth leads to a buoyant force on objects immersed in fluids. The magnitude of this force is given by Archimedes' law:

$$F_b = \rho_f V g$$

where ρ_f is the density of the fluid and V is the volume displaced by the immersed object.

When fluids move steadily, there are two additional sources of pressure variation. First, frictional forces produce a pressure drop along a pipe carrying a moving fluid. Second, if we follow a

Low pressure

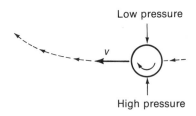

High pressure

Fig. 18-25. The curved path followed by a rapidly rotating ball.

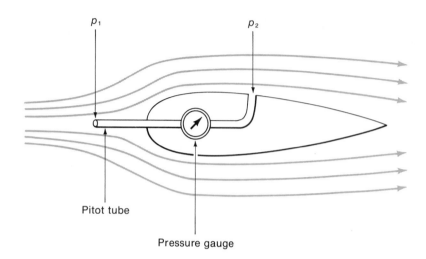

p_1 p_2

Pitot tube

Pressure gauge

Fig. 18-26. The Pitot tube of an airspeed indicator is connected to a pressure gauge. The other side of the pressure gauge connects to an opening on the side of the aircraft past which air moves at the air speed of the aircraft.

fluid along either a pipe or a bundle of streamlines, the pressure is found to decrease in regions of high velocity. This is called the Bernoulli effect and is described by the equation

$$p + \tfrac{1}{2}\rho v^2 + \rho g h = \text{constant}$$

along the pipe (neglecting frictional forces) or along a streamline.

APPENDIX: Pressure Variation in Compressible Fluids

Equation (18-3) applied to finite differences in depth, Δh. Considering infinitesimal differences, we obtain

$$dp/dh = \rho g$$

If ρ is a constant, as it very nearly is in water and other nearly incompressible fluids, this differential equation is solved by

$$p = \rho g h$$

If h is measured from the bottom of the fluid (as is convenient in our intended application) instead of from the surface, we obtain

$$dp/dh = -\rho g$$

But now if the density *depends upon* the local value of the pressure, as in the earth's atmosphere, this differential equation is not as easily solved. However, the solution is quite straightforward if the function $\rho = f(p)$ is simple enough. In the earth's atmosphere, the

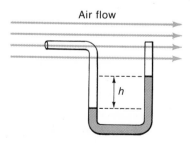

Air flow

h

Fig. 18-27.

density of air is very nearly proportional to the pressure. Writing this proportionality as $\rho = Bp$ we have:

$$dp/dh = -Bpg$$

or

$$dp/p = -Bg\, dh$$

Integrating both sides:

$$\int_{p_0}^{p} \frac{dp}{p} = -\int_{0}^{h} Bg\, dh$$

where p_0 is the pressure at zero height. Taking antiderivatives,

$$\ln p - \ln p_0 = -Bgh$$

so that $\ln(p/p_0) = -Bgh$ or

$$p = p_0 e^{-Bgh}$$

Thus we see that in the earth's atmosphere the pressure falls exponentially with increasing altitude. This will be true as long as the density is simply proportional to the pressure. In Section 20-3 we will see that this is very nearly true only for constant temperature. The pressure in the earth's atmosphere varies roughly in an exponential manner as given above but with some variations caused by changes in temperature with altitude.

Problems

18-1. A diver at a depth of 10 ft tries to breathe through a length of hose that extends above the surface of the water. Estimate the net force on his chest caused by the difference between the water's pressure and the atmospheric pressure in his lungs.

18-2. Find the gauge pressure and absolute pressure in N/m² at a depth of 10 m in water.

18-3. A hydraulic automobile jack has a lifting piston with a 4-cm diameter. Find the required diameter of the small piston if it is to be driven by a force of 250 N while the jack supports a load of 500 kg.

18-4. A rectangular dam has a width of 20 ft and backs up water to a depth of 15 ft. The pressure near the top of the dam is small and rises to a maximum value at the full depth of 15 ft. Find the net force of the water on the dam. (The density of water is 62.5 lb/ft³.)

18-5. The density of air at sea level is 1.29 kg/m³. Find the weight of air contained in an otherwise empty truck with dimensions 3 m × 3 m × 6 m. When the truck is driven onto a set of scales to determine its empty weight, is the weight of the air included in the reading? Explain.

18-6. A rowboat is floating next to a dock. A 160-lb man steps into the boat and it sinks 3 in. deeper into the water. What is the cross-sectional area of the boat?

18-7.* The occupant of a boat in a small pond begins to throw stones overboard. Does the level of the boat with respect to the water sink or rise? Does the level of the lake surface sink or rise? What data do you need to determine the net rise of the boat? Calculate the rise in terms of these data.

18-8. You are faced with the original problem that baffled Archimedes for a time. A crown is alleged to be made of gold, but how can one be sure? When immersed in water it displaces 98 cm³ of water. Its weight in air is 1470 g. (*a*) Is it made of pure gold? (*b*) How much does it weigh when suspended in water? (The density of gold is 19.3 g/cm³.)

18-9. A piece of gold-silver alloy weighs 1200 g in air. When immersed in water it weighs 1100 g. Find the ratio of gold to silver in the alloy. The specific gravities of gold and silver are 19.3 and 10.5, respectively.

18-10. A body with a volume of 100 cm³ is suspended in a beaker of water, as in Fig. 18-28. The lower scale reads 1150 g while the upper scale reads 800 g. The mass of the beaker is 50 g. (*a*) Find the quantity of water in the beaker. (*b*) Find the density of the suspended body.

18-11. A cork of volume V_c and density ρ_c is tied to the bottom of a cylindrical bucket with a cord. The bucket has area A and is filled to a depth h, as shown in Fig. 18-29. (*a*) Find the tension T in the cord. (*b*) Show that the combination of the pressure on the bottom of the bucket and the tension in the cord gives a *net* downward force on the bucket equal to the weight of the bucket's contents.

18-12. A cube of ice (specific gravity 0.92) floats in a glass of water filled to the brim. The ice melts. Does the water level drop, stay the same, or does it overflow? Explain.

18-13. Ocean waves are generated by winds blowing over the water surface. Explain how the Bernoulli effect is involved in wave formation.

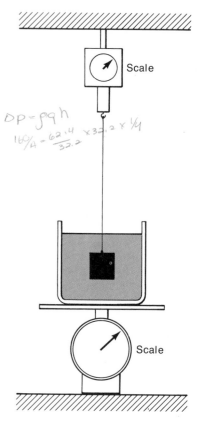

Fig. 18-28.

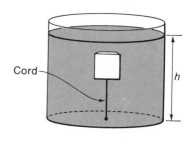

Fig. 18-29.

$$DP = \tfrac{1}{2}\rho\,(v_2{}^2 - v_1{}^2)$$

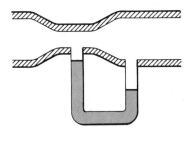

18-14. The airflow velocity under an aircraft wing is 100 m/s. If the velocity over the top of the wing can be made 20% larger than that over the bottom, how many square meters of wing area are needed to support an airplane with a mass of 2500 kg?

18-15. A house has an open window facing into the wind and a closed window on the side of the house. Sketch some streamlines for the air flow. (*a*) Find the difference in pressure between the inside and the outside of the closed window. (*b*) If the wind blows past the side of the house at 120 mi/hr and air has a density of 2.5×10^{-3} slugs/ft³. Find the total force on a 4 × 5-ft window.

18-16. A U-tube manometer filled with water is connected as in Fig. 18-30. The wide and narrow portions of the pipe have diameters of 5 cm and 2.5 cm, respectively. The difference in height of the water columns is observed to be 6 cm. Find the velocity of a gas with $\rho = 1.2$ kg/m³ flowing through the pipe.

Fig. 18-30.

Chapter *19*

Temperature and Heat

19-1 Introduction In the next three chapters we turn our attention to phenomena involving heat. Our work in mechanics has required only three fundamental quantities: length, mass, and time. We now add a fourth fundamental quantity: temperature. We will find that heat phenomena and mechanical phenomena are fundamentally related. The study of this interrelationship is a major portion of the science of *thermodynamics*.

The notion of "hot" and "cold" is primitively tied to various sensory organs of the human body. Yet this sense of hot and cold is more than subjective, for we observe many physical processes that are changed in reproducible ways that correlate with our own sense impressions: water evaporates more slowly on cold days than on warm days and even freezes when it is cold enough. Many substances expand slightly when heated and contract when cooled. This phenomenon is put to use in many types of thermometers.

In Chapter 13 we briefly mentioned thermal agitation, the random jiggling motion that atoms execute about their equilibrium locations in a solid. In a macroscopic quantity of matter, there is a great deal of atomic kinetic energy and interatomic potential energy tied up in this motion. Furthermore, the amplitude of this atomic motion, and the energy associated with it, increases with increasing temperature. We give the name *heat energy* or *internal energy* to this energy associated with thermal atomic motion within a substance. In this chapter we will make the concept of temperature a precise one and disentangle two different concepts: *temperature* and *quantity of heat*.

19-2 Thermometry

Temperature is the name given to a precise quantitative measure of "hotness"; a device which measures temperature is called a *thermometer*. The advent of quantitative measurements accompanied precise studies of thermal expansion. Gases, as well as many liquids and solids, exhibit thermal expansion, and one of the first thermometers, due to Galileo, used the expansion and contraction of air to measure temperature. The apparatus he used is shown in Fig. 19-1. A partial vacuum in the bulb allows the water column to stand at a convenient height. When the air remaining in the bulb becomes warmer, it expands so that *h* decreases. As it cools, the liquid level rises again.

The thermal expansion of a liquid is utilized in the common type of thermometer shown in Fig. 19-2. The liquids most frequently used are alcohol (dyed red) and mercury. Fahrenheit was a scientific-instrument maker in Western Europe and used devices like that pictured in Fig. 19-2. In order to provide a numerical scale, he defined the lowest temperature he could obtain with an ice-salt mixture to be the zero of his scale and the temperature of the human body to be 96 on his scale, with 96 equal intervals ruled between these points on his thermometer.* These divisions were called degrees of temperature. Extrapolating his scale to higher temperatures led to an assignment of 212 degrees to the boiling point of water. This *Fahrenheit* temperature scale is not particularly satisfactory since its reference points are not easily reproducible by other workers.

The Celsius temperature scale rectified this difficulty by assigning two fixed points to a temperature scale: 0°C is the temperature of a mixture of ice and water at normal atmospheric pressure (the *ice point*) and 100°C is the boiling point of water at atmospheric

*The present assignment of 98.6°F to body temperature reflects experimental inaccuracies of the original measurements. The assignment of 96° to body temperature is a bit curious. Since 100°F is about the highest temperature encountered on a summer day in Western Europe, one wonders if the 100 mark was perhaps arrived at on this basis.

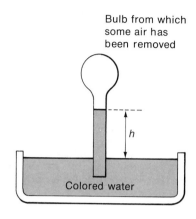

Fig. 19-1. An air thermometer of the type built by Galileo. As the air remaining in the bulb expands and contracts, the height *h* of the water column decreases and increases.

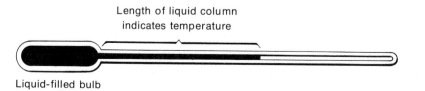

Length of liquid column indicates temperature

Liquid-filled bulb

Fig. 19-2. A liquid-in-glass thermometer that depends on the thermal expansion of the liquid it contains. The relatively small expansion of the liquid in the bulb is made easily visible by allowing it to expand into a very fine bore hole. Thus a small increase in the net volume of the liquid produces a large change in the length of the liquid column in the bore hole. The expansion of the glass itself is small.

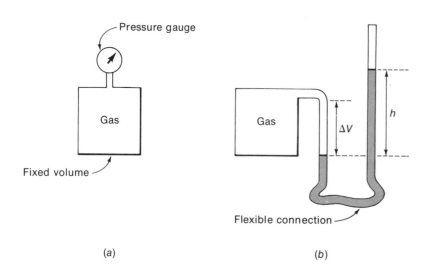

Fig. 19-3. (*a*) A constant-volume gas thermometer. (*b*) A constant-pressure gas thermometer. The U-tube is always readjusted so that *h* is constant and the gas is under constant pressure. As the temperature varies, the gas will occupy more or less volume as indicated by the reading, Δ*V*.

pressure.* Since there are 180 Fahrenheit degrees between the ice point at 32°F and the boiling point at 212°F, a Fahrenheit degree is 5/9 as large as a Celsius degree. Thus since 68°F is 36 Fahrenheit degrees above the ice point, it is 20 Celsius degrees above the ice point, which is 0°C. Hence we see that 68°F corresponds to 20°C. A simple equation for conversion between these scales is

$$°F = \tfrac{9}{5}(°C) + 32$$

Not all materials expand and contract in a similar manner when heated and cooled. A temperature scale could be based, for example, on the thermal expansion properties of mercury, but then it would be found that the expansion of alcohol is not quite a linear function of this temperature. If we had chosen the expansion properties of alcohol for our temperature scale, we would notice that other materials expand in a way that is not exactly linearly dependent upon the temperature. This is an unsatisfactory state of affairs, for our temperature scale then depends upon the particular substance we use to define the scale. Fortunately, this difference from one substance to another does not occur for the thermal expansion properties of gases (as long as the gases are at low pressures and densities).

Two types of gas thermometers are shown in Fig. 19-3. The important point is that temperature scales set up with different

---

*This scale was originally called the centigrade scale. Anders Celsius (1701–1744) was a Swedish astronomer who urged acceptance of this scale.

gases and with *either* a constant-volume or constant-pressure thermometer agree. Accordingly, a temperature T is defined in terms of gas thermometry. Experimentally, the readings of all gas thermometers decrease linearly with decreasing temperature at such a rate that an extrapolation to a zero reading corresponds to a Celsius temperature of $-273.15°C$. This temperature is used to define the zero (sometimes called *absolute* zero) of the *Kelvin* temperature scale. The units of the Kelvin scale are chosen to be equal to those of the Celsius scale. The location of this "absolute zero" point involves an extrapolation, since all gases liquefy (at which point a gas thermometer ceases to function) at temperatures above absolute zero. Helium remains gaseous to temperatures below those for any other gas. It too, however, liquefies at about 4 degrees above absolute zero.

Gas thermometers can be used to find an unknown temperature by comparing it to a standard temperature. Originally, the freezing and boiling points of water were taken as standard temperatures. However, when extremely accurate temperature measurements were made, it was discovered that the freezing and boiling points of water were sensitive to impurities dissolved in the water. The temperature at which the solid, liquid, and gaseous states of extremely pure, air-free water coexist, the *triple point,* was found to be a more convenient standard. This point is approximately $0.01°C$ above the freezing point and the currently accepted temperature scale in SI units defines one fixed point, the triple point of water. This scale is called the *thermodynamic temperature scale* and its unit is the *kelvin*, abbreviated K.

The kelvin, unit of thermodynamic temperature, is the fraction 1/273.16 of the thermodynamic temperature of the triple point of water. The kelvin is also the SI unit for temperature intervals.

When a constant-volume gas thermometer is used for precise measurements, its gas volume is brought to the temperature of the triple point of water. The pressure of the gas is then p_0 at the temperature T_0, 273.16K in this case. The gas is then brought to the new temperature to be measured: this temperature is given by

$$T_x = \frac{p_x}{p_0} T_0$$

where p_x is the pressure of the gas thermometer at the unknown temperature. The same discussion applies to a constant-pressure gas thermometer, with volume readings substituting for pressure readings.

Thermometry involving other methods—the variation of electrical resistance with temperature, for example—can be tied to the gas-thermometer scale by calibrating where the gas scale is useful and then using these other methods to carry the definition of temperature into the extremely low temperature region near absolute zero. By international agreement, several calibration points have been assigned fixed temperatures on the basis of careful measurements made by the National Standard Laboratories. The techniques for interpolating between these points are also specified: gas thermometry, electrical resistance variation, optical pyrometry, etc.* The Comité International des Poids et Mesures adopted the "International Practical Temperature Scale" in 1968 with the assigned temperatures shown in Table 19-1. Thermodynamic temperatures (SI units) are found by adding 273.15 to the Celsius temperatures.

TABLE 19-1

	Assigned Temperature
Freezing point of gold	1064.43°C
Freezing point of silver	961.93°C
Freezing point of zinc	419.58°C
Boiling point of water	100.00°C
Triple point of water	0.01°C
Boiling point of oxygen	−182.962°C
Triple point of oxygen	−218.789°C
Boiling point of neon	−246.048°C
Boiling point of hydrogen	−252.87°C
Equilibrium between the liquid and vapor phases of equilibrium hydrogen at 33330.6 N/m² pressure	−256.108°C
Triple point of hydrogen	−259.34°C

A comparison of temperature scales is given in Table 19-2.

TABLE 19-2   Comparison of Various Temperature Scales

Scale	Abbreviation	Absolute Zero	Ice Point	Boiling Point
Thermodynamic (SI)	K	0K	273.15K	373.15K
Celsius	C	−273.15°C	0°C	100°C
Fahrenheit	F	−459.67°F	32°F	212°F
Rankine	R	0°R	491.67°R	671.67°R

*A complete discussion is given in *Physics Today,* published by the American Institute of Physics (Sept. 1971). p. 32.

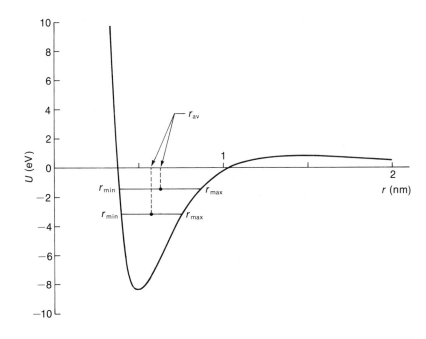

Fig. 19-4. Typical interatomic potential for a pair of atoms. For a given value of total energy the atoms vibrate toward and away from each other between r_{min} and r_{max}. When their energy is increased their average spacing, r_{av}, increases.

Exercises

1. What is the Celsius temperature that corresponds to the Fahrenheit average body temperature (98.6°F)?

2. The gas in a constant-volume gas thermometer is brought to the temperature of the triple point of water. Its absolute pressure is recorded as 82 mm Hg. The thermometer is then brought to the temperature of boiling sulfur and the pressure recorded is 215.5 mm Hg. Find the temperature of boiling sulfur.

19-3   Thermal Expansion: A Microscopic View

Having established a consistent temperature scale in Section 19-2, we can examine the phenomenon of thermal expansion in a quantitative manner. As we will soon see, temperature is closely related to the kinetic energy of thermal agitation of the atoms of a substance. As temperature increases, the atoms "jiggle" or vibrate more and more violently. At sufficiently high temperatures, this thermal kinetic energy becomes large enough to break bonding between atoms of a solid, so that melting occurs. Looking at Fig. 19-4, we see that two atoms bound together by typical interatomic forces can vibrate back and forth between the limits r_{min} and r_{max}. As kinetic energy of vibration is increased (higher temperatures), it can be seen from Fig. 19-4 that r_{min} becomes smaller and r_{max} larger. But because the slope of the curve $U(r)$ is much more gentle in the r_{max} region, r_{max} increases much more than r_{min} decreases. Thus the *average* spacing between the two atoms increases with increasing temperature. Similarly, the average spacing between each pair of nearest-neighbor atoms in a solid increases with temperature.

A solid object made up of many atoms thus expands as its temperature is increased, and the average atom-to-atom spacing increases. The precise details of how this increase is related to temperature depends on the relative slopes of the interatomic potentials in the regions near r_{min} and r_{max}. Most materials expand in a way that depends nonlinearly on temperature, as illustrated in Fig. 19-5. Some materials, in fact, exhibit a negative coefficient of thermal expansion over some temperature ranges. Evidently, the interatomic potential energy for these materials has a steeper slope near r_{max} than near r_{min}.

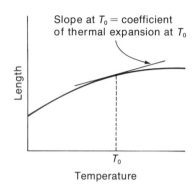

Fig. 19-5. An exaggerated graph of length change of a sample material as temperature changes. The slope of this curve at any particular temperature is called the *coefficient of thermal expansion* at that temperature.

19-4   Thermal Expansion: A Macroscopic View

When a solid object with a positive thermal-expansion coefficient is heated, it expands uniformly in all dimensions, as indicated in Fig. 19-6. The change in volume is conveniently described by the *volume-expansion coefficient:*

$$\beta = \frac{1}{V}\frac{dV}{dT} \quad \text{volume-expansion} \tag{19-1}$$

Note that β is just the slope of the volume-vs-temperature curve normalized to a unit volume. Thus, if β is constant over a temperature interval ΔT, we may write:

$$\Delta V = \beta V \Delta T \tag{19-2}$$

If β varies, (19-2) will still be a good approximation for a small temperature change ΔT. If an object is in the form of a long rod, we may be more interested in its length change than in its volume change. We define the *length-expansion coefficient:*

$$\alpha = \frac{1}{L}\frac{dK}{dT} \quad \text{length-expansion} \tag{19-3}$$

If the object expands and contracts isotropically (independent of direction), the length-expansion coefficient is simply related to the volume-expansion coefficient (see Problem 19-1):

$$\alpha = \beta/3$$

Values of β for several substances are given in Table 19-3.

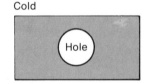

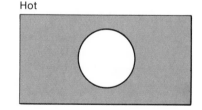

Fig. 19-6. When an object expands thermally, all dimensions increase, including those of any holes. The expansion is much exaggerated above.

TABLE 19-3

Substance	Thermal-Expansion Coefficients near 20°C (1/°C)
Alcohol, ethyl	75×10^{-5}
Aluminum	7.2×10^{-5}
Brass	6.0×10^{-5}
Carbon disulfide	115×10^{-5}
Copper	4.2×10^{-5}
Glass (Pyrex)	0.9×10^{-5}
Glass (soda)	5×10^{-5}
Glycerin	50×10^{-5}
Invar	0.27×10^{-5}
Mercury	18×10^{-5}
Steel	3.6×10^{-5}

Example   A steel beaker with a volume of 100 cm³ is filled to the brim at 20°C with ethyl alcohol. How much alcohol will overflow the beaker if the temperature of the entire system is raised from 20°C to 30°C?

Solution   The volume of the beaker will change by an amount

$$\Delta V_b = \beta V \Delta T$$

$$= 3.6 \times 10^{-5} (°C)^{-1} \cdot 10^2 \text{ cm}^3 \cdot 10°C$$

$$= 3.6 \times 10^{-3} \text{ cm}^3$$

For the alcohol we have

$$\Delta V_a = 75 \times 10^{-5} (°C)^{-1} \cdot 10^2 \text{ cm}^3 \cdot 10°C$$

$$= 75 \times 10^{-3} \text{ cm}^3$$

Since the alcohol expands more than the beaker, some will overflow:

$$\text{overflow} = V_a - V_b = 7.14 \times 10^{-2} \text{ cm}^3$$

Exercises

3. A soda-glass flask with a volume of 250 cm³ is filled to the brim with glycerin at 60°C. The flask and glycerin are then cooled to 0°C. How much glycerin (at 0°C) must be added to keep the flask brimful?

4. Railroad tracks are generally laid in segments. Consider steel rails of 60-ft lengths. What is the change in length of each segment between a cold winter day (−10°C) and a hot summer day (40°C)? Don't forget $\alpha = \beta/3$.

19-5   Heat

When two isolated objects of unequal temperature are placed in contact, they tend to reach a common, intermediate temperature. This process can be slowed considerably by placing materials like asbestos, glass wool, or crumpled paper between the objects. Such materials are called *thermal insulators*. If we enclose several objects at different initial temperatures in an enclosure made of thermal insulator, it is an experimental fact that the objects all will come to a common temperature. When this has occurred, the system is said to be in *thermal equilibrium*. This fact tells us quite a bit about the nature of temperature and is often called the *zeroeth law of thermodynamics*.

It seems natural enough to assume that *something* flows from a hot object to a cold object (or vice versa) as they come to a common temperature. Early investigators called this hypothetical substance *caloric* and assumed it was a weightless, invisible fluid that was present in large quantity in a hot body, less in a cold body. The study of what we now call heat capacity of objects was called calorimetry originally and it still retains that name. We now know that, at the microscopic level, some of the kinetic energy of the thermally agitated atoms in the hotter object is transferred to the atoms of the cooler object. The average value of the interatomic *potential* energy is somewhat dependent upon temperature also. The total loss of both forms of atomic energy by a cooling object is gained by the warming object as long as they are isolated from other objects.*

Heat is simply a form of energy. The fact that it is hidden in the invisible domain of atomic motion obscured this fact until the early part of the nineteenth century. Careful observations made by Benjamin Thompson (1753–1814) laid the foundations for the discovery. Thompson was an American who, during a multifaceted career, was for a time Minister of War for Bavaria (he was later made Count Rumford of Bavaria). As part of his duties he supervised cannon construction. He observed that the water placed in the cannon bore to keep it cool while drilling boiled away continually as long as the drill was in operation. At the time it was thought that the fine subdivision of the metal by the action of the drill caused the metal to release its caloric. But when the drill was allowed to become so dull that it no longer cut at all, the water still boiled as

---

*By atomic energy we mean potential energy and kinetic energy of atoms. This term is occasionally used incorrectly to describe energy liberated in nuclear reactions, as in a nuclear reactor. The term "nuclear energy" or "nuclear power" is more apppropriate and coming into wider usage.

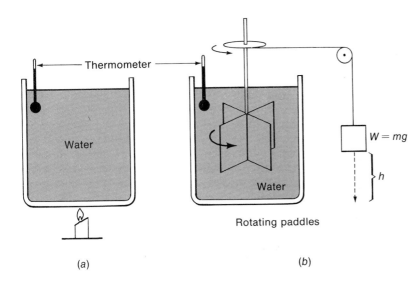

long as mechanical energy was supplied. Thompson correctly guessed that heat was a form of energy that could be supplied by mechanical means. He did not, however, make quantitative studies of this process.

James Joule (1818–1889) in England carried out quantitative measurements that clearly showed the connection between mechanical energy and heat energy. The essence of his procedure is illustrated in Fig. 19-7. The quantity of heat required to raise the temperature of one gram of water by one degree Celsius is defined to be one *calorie*. In Fig. 19-7a we see heat added to a container of water to raise its temperature. In Fig. 19-7b the same change in temperature is produced by doing mechanical work. The precise amount of work is given by mgh, where h is the distance through which the mass m falls while causing the paddles to rotate. Joule discovered that a fixed quantity of work done on the water (or any other substance he studied) was always equivalent to one calorie of heat. The presently accepted value (sometimes called the *mechanical equivalent of heat*) is 4.186 joules/calorie. This result is independent of whether the energy is supplied in the form of mechanical work through stirring of a liquid or rubbing of solids, or whether the energy is supplied by electrical means.

4.186 Joules/calorie

19-6 Calorimetry

Calorimetry is the science of measuring quantities of heat. We have already defined a unit of heat—the calorie—which is the amount of heat required to raise the temperature of one gram of water by one degree Celsius. The experiments of Joule (and many others since) have shown that the calorie is simply a unit of energy, so we could just as well measure heat in joules. However, calorimetry began before this was understood. Further, the calorie is tailored nicely to heat measurements, for it is defined in terms of practical units of mass and temperature measurement, so we continue to use the calorie as an energy unit where thermal energies are involved.

Heat capacity is the name given to the ability of an object to take up heat. A material that can absorb a large amount of heat yet changes its temperature only slightly has a large heat capacity. One that experiences a large temperature change with the addition of a small quantity of heat is said to have a small heat capacity. The heat capacity per unit mass of a substance is called its *specific heat*. If a quantity of heat, ΔQ, added to a mass m of material produces a change in its temperature ΔT, the specific heat, c, is defined by

$$c = \frac{\Delta Q}{m \Delta T}$$

When heat is measured in calories, m in grams, and T in °C, the specific heat of water has a value of unity, for the calorie is *defined* in terms of the temperature rise of one gram of water. The specific heats of some substances are given in Table 19-4 below. As can be seen there, most materials have specific heats considerably below that of water.

TABLE 19-4   Specific Heats of Some Substances

Substance	Specific Heat (cal/g · °C)
Aluminum	0.212
Concrete	0.16
Copper	0.094
Ethyl alcohol	0.58
Glass	0.12–0.20
Ice	0.48
Lead	0.031
Mercury	0.033
Water	1.00

Specific heats are determined by exchanging heat between a substance whose specific heat is unknown and another substance whose specific heat is known. The primary definition of the calorie relates all specific heats to that of water.

Example A block of aluminum at $T = 80°C$ and weighing 20 g is dropped into 100 g of water at a temperature of 20°C. If no external heat flows into or out of this system, find the final temperature of the aluminum + water when they reach thermal equilibrium.

Solution The aluminum will cool by an amount ΔT_a while the water will warm by an amount ΔT_w, so that they reach a common temperature T_f. Since no heat flows in or out of the system, any heat lost by the aluminum will be gained by the water.

$$\Delta T_a = (T_2 - T_1)_a = T_f - 80°C$$

$$\Delta T_w = (T_2 - T_1)_w = T_f - 20°C$$

$$\Delta Q_a = c_a m_a \Delta T_a$$

$$\Delta Q_w = c_w m_w \Delta T_w$$

$$\Delta Q_a + \Delta Q_w = 0$$

$$c_a m_a \Delta T_a = -c_w m_w \Delta T_w$$

$$c_a m_a (T_f - 80°C) = -c_w m_w (T_f - 20°C)$$

$$(c_a m_a + c_w m_w) T_f = c_w m_w (20°C) + c_a m_a (80°C)$$

$$T_f = \frac{c_w m_w (20°C) + c_a m_a (80°C)}{c_a m_a + c_w m_w}$$

$$= \frac{(1 \text{ cal/g} \cdot °C) 100 \text{ g} \cdot 20°C + (0.212 \text{ cal/g} \cdot °C) 20 \text{ g} \cdot 80°C}{(0.212 \text{ cal/g} \cdot °C) 20 \text{ g} + (1.0 \text{ cal/g} \cdot °C) 100 \text{ g}}$$

$$= \frac{2339.2 \text{ cal}}{104.14 \text{ cal/°C}} = 22.46°C$$

A larger unit of heat is the kilocalorie, the amount of heat that causes a 1°C rise in temperature of 1 kg of water. It is also called the large calorie or the Calorie (with a capital C). The units used to measure the energy content of foods are kilocalories. The muscles of the human body during heavy exertion are able to utilize (by oxidation reactions) 220 cal/minute/kg body weight and convert about 25% of this energy into useful mechanical work.*

Exercises

5. A slice of bread is placed inside a steel container and surrounded by enough oxygen to burn it entirely. This container is immersed in 2 kg of water. The bread is burned and it is found

*See "The Sources of Muscle Energy" by Rodolfo Margaria, *Scientific American,* March 1972.

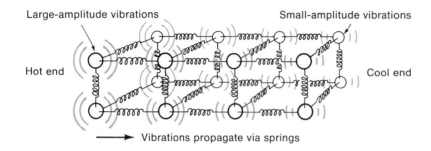

Large-amplitude vibrations   Small-amplitude vibrations

Hot end   Cool end

Vibrations propagate via springs

Fig. 19-8. A solid conducts heat by propagating thermal vibration from atom to atom. Its behavior is similar to that of a collection of masses connected by springs, where the springs represent the interatomic forces.

that the temperature of the water rises by 30 K. If the heat capacity of the steel container is negligible compared to that of the water, find the number of calories of heat released by the oxidation of the slice of bread. Nutritionists usually use the kilocalorie (10^3 calories) to measure the energy release of foods.

6. An electrical heater embedded in a 1-kg block of copper consumes electrical power at a 1-kW rate for 60 s. If all this energy is retained as heat in the copper block, what is the temperature change of the copper block? Assume that the heat capacity of the electrical heater itself may be neglected compared to that of the copper block.

19-7   Conduction of Heat

Heat moves from one object to another or from one portion of an object to another by a variety of mechanisms. *Conduction* is the flow of heat directly through a material. The underlying microscopic process is illustrated in Fig. 19-8. A high-temperature region, characterized by large-amplitude thermal vibrations, sets nearby atoms into increased-amplitude vibrations at the expense of its own thermal energy. Heat is thus conducted throughout a solid body.

On a macroscopic scale we describe the heat-conducting properties of a substance by its coefficient of thermal conductivity, which is defined in terms of the rate at which heat flows through a slab of material. Fig. 19-9 illustrates the definition of k, the coefficient of thermal conductivity. If T_1 is the temperature of the left side and T_2 is the temperature of the right, and $T_1 > T_2$, then heat flows from left to right in the figure. The rate of flow is given by dQ/dt, the number of calories per second that flow from left to right. Experimentally, it is found that dQ/dt is proportional to the temperature difference $\Delta T = T_1 - T_2$ and to the cross-sectional area of the bar, and inversely proportional to its length. Accordingly, we can write

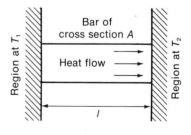

Region at T_1

Bar of cross section A

Heat flow

Region at T_2

l

Fig. 19-9. Heat flows through a bar of cross section A and length l.

$$\frac{dQ}{dt} = kA\frac{\Delta T}{l} \qquad\qquad (19\text{-}4)$$

which defines k, the coefficient of thermal conductivity. Table 19-5 exhibits some thermal conductivities.

TABLE 19-5   Thermal Conductivities for Some Materials

Substance	k (kcal/s $\cdot$ m $\cdot$ °C)
Air	5.6×10^{-6}
Aluminum	0.0048
Brick	1.7×10^{-4}
Copper	0.0092
Ethyl alcohol	4.2×10^{-5}
Glass	1.4×10^{-4}
Steel	0.0011

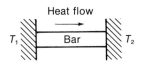

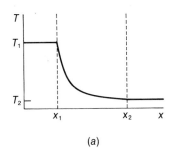

(a)

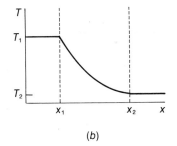

(b)

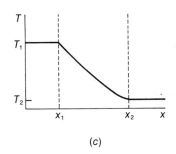

(c)

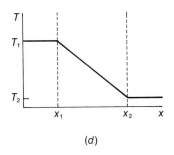

(d)

Fig. 19-10. A bar of material at temperature T_2 is clamped between two surfaces at temperatures T_1 and T_2. At first, (a) only the end near the hot surface warms, but eventually, (d) a steady state is reached when equal quantities of heat per unit time enter the left end of the bar, move past any point x in the bar, and flow out of the right end of the bar.

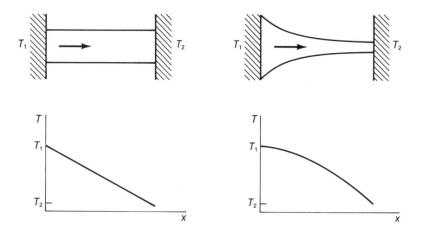

Fig. 19-11.   If the cross-sectional area is constant, dT/dx is constant for a steady state. A varying cross section, as in part (b), implies a varying slope of the T-vs-x curve in order that the same quantity of heat per unit time passes each value of x in the conductor.

Now considering a slab of material of infinitesimal thickness, dx; we write

$$\frac{dQ}{dt} = -kA\frac{dT}{dx}\tag{19-5}$$

where the quantity dT/dx is called the *temperature gradient*. Because heat flows from high temperatures to low temperatures, dT/dx is negative when dQ/dt is positive, so we introduce a minus sign. In Fig. 19-10 plots of T vs x are shown for various times after a uniform conducting bar is clamped between two surfaces that differ in temperature by a constant amount. At first the region near the hot surface is warmed, but the high temperature gradient there quickly moves heat to regions farther away. The curve of T vs x gradually approaches a straight line. This condition (Fig. 19-10d) is called a *steady state*, for now every calorie of heat flowing into the left end of the bar flows out of its right end. The T-vs-x curve is now necessarily linear since (19-5) tells us that a uniform temperature gradient dT/dx is required for uniform heat flow at all values of x.

Consider now the bar of nonuniform cross section in Fig. 19-11. When equilibrium is reached, dT/dx must vary so that the heat flow is the same at each value of x in spite of the varying cross section. Where the cross-sectional area is small, the temperature gradient is large. An example of an application of this result is given in the Appendix to this chapter.

Example   A steel bar 10 cm long with 2-cm² cross section makes firm end-to-end contact with a copper bar of the same dimensions. The two together make a bar 20 cm long and are clamped between two surfaces, as shown in Fig. 19-12. If T_1 is 100°C, T_2 is 0°C, and a steady state has been reached, find T_3, the temperature of the junction be-tween the copper and steel bars.

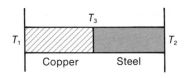

Fig. 19-12.

Solution For a steady state the temperature gradients must be uni-form along the lengths of the copper bar and the steel bar, since both have constant cross-sectional area. However, the lower heat conduc-tivity of steel, k_s, requires a larger temperature drop across the steel segment than across the copper segment. The difference in tempera-ture drops can be found by imposing the condition of steady state: the same amount of heat flows through each bar.

$$k_c A \frac{T_1 - T_3}{l} = \frac{dQ}{dt} = k_s A \frac{T_3 - T_2}{l}$$

$$\frac{T_1 - T_3}{T_3 - T_2} = \frac{k_s}{k_c}$$

and since $k_s < k_c$, the temperature drop across the copper $(T_1 - T_3)$ is less than that across the steel. More specifically:

$$\Delta T_c = (T_1 - T_3) = \frac{0.0011 \text{ kcal/s} \cdot \text{m} \cdot °\text{C}}{0.0092 \text{ kcal/s} \cdot \text{m} \cdot °\text{C}} (T_3 - T_2)$$

or
$$\Delta T_c = 0.12 \Delta T_s$$

Now let us solve for T_3:

$$T_3 = \frac{T_1 + \frac{k_s}{k_c} \cdot T_2}{1 + \frac{k_s}{k_c}} = \frac{100°\text{C} + 0}{1.12} = 83.6°\text{C}$$

Exercises

7. A pane of glass 50 cm × 100 cm and 2 mm thick has a tempera-ture on its inner face of 5.3°C and a temperature on its outer face of 5°C. Calculate the steady-state rate of heat flow through the pane.

8. A bar of aluminum 10 cm long is firmly butted against a bar of steel 5 cm long. Both bars have identical and uniform cross-sectional areas and are together clamped between two surfaces 15 cm apart. The surface against the aluminum bar is at 10°C while that against the steel bar is at 80°C. Find the temperature at their junction.

19-8 Radiation

Another mechanism for the transfer of heat is *radiation*. Any accel-erating electric charge emits electromagnetic radiation and this radiation carries energy. Since atoms are constructed of electrically charged particles, the jiggling thermal motion of the atoms produces electromagnetic radiation. This *radiant* energy produces the sensation of warmth when we stand in front of a fireplace. A "red hot" iron bar radiates energy rapidly. As it cools, it continues to radiate but at an ever slower rate. The transport of heat energy by radiation does not require a medium. Radiant energy from the sun reaches us through the vacuum of interplanetary space. The rate of loss of thermal energy by this mechanism obeys *Stefan's law:*

> The amount of energy radiated from each unit area of a warm body's surface is proportional to the fourth power of the *absolute* temperature:

$$\frac{dQ/dt}{A} = \epsilon \sigma T^4 \qquad (19\text{-}6) \qquad \sigma = 5.67 \times 10^{-8}\, \text{W}/\text{m}^2 \cdot \text{K}^4$$

where dQ/dt is the rate of energy loss and A the area of the radiating surface.

The constant of proportionality, σ, has the value 5.67×10^{-8} W/ m$^2 \cdot$ K^4. The *emissivity, ϵ,* varies between 0 and 1. A black body is one that completely absorbs radiation and is also the best possible radiator. It has emissivity equal to unity. A surface that reflects well, such as polished metal, also radiates poorly and has a small emissivity. When an object with area A at a given temperature T_1 is surrounded by an enclosure at a different temperature T_2, it *gives off* heat at a rate

$$\left(\frac{dQ}{dt}\right)_1 = \epsilon \sigma A T_1^4$$

and *absorbs* heat at the rate

$$\left(\frac{dQ}{dt}\right)_2 = \epsilon \sigma A T_2^4$$

The net loss (or gain) of heat is

$$\left(\frac{dQ}{dt}\right)_{net} = \epsilon \sigma A (T_1^4 - T_2^4)$$

Objects near room temperature (300K) emit radiant energy at rather long wavelengths (infrared radiation), while objects heated to higher temperatures emit more of their energy at shorter wavelengths. At the temperature of a tungsten incandescent lamp filament (3300K), much of the radiation is in the very short wavelengths, near 10^{-6} m. This short-wavelength radiation is visible light.

Exercises

9. Find the rate (in watts) at which radiation is given off by a tungsten surface with $A = \frac{1}{10}$ cm², $\epsilon = 0.35$, and $T = 3300$K.

10. Estimate the radiative heat loss rate of a person in an enclosure at a temperature of $-20°C$. Assume that the person's radiating surface is at a temperature of $30°C$. Compare this rate to that when the walls of the enclosure are at room temperature, $20°C$. Don't forget to convert these temperatures to the Kelvin scale.

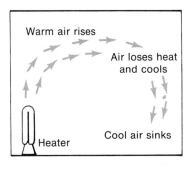

Fig. 19-13. The lower density of warm air, caused by its thermal expansion, provides the buoyant force that drives convection currents.

19-9   Convection

Because fluids change density when heated, an interesting heat-transport mechanism called *convection* can occur. It involves mass transport and requires the presence of a gravitational field. In Fig. 19-13 we see a typical situation in which convection occurs. The warm air, buoyed up by the more dense cool air in the room, moves away from the heater. However, once this warm air reaches a cooler portion of the room, it cools (and in turn warms that portion of the room), becomes more dense, and sinks to form the ever circulating convection current.

Besides their technical applications in home heating, for example, convection currents occur frequently in nature. The large amount of solar energy delivered to the earth's equatorial regions is redistributed over the globe by a series of convection currents, i.e., winds in the atmosphere. Within the sun itself, convection currents aid in transporting the heat generated by nuclear reactions in its interior out to its surface where it can be radiated into space. Only in recent years has it become apparent that some very slow convection processes operate in the seemingly solid crust of the earth.* These processes reveal themselves in the slow movement ($\approx$ cm/yr) of continents with respect to one another.

19-10   Summary

Quantity of heat and *temperature* are two related but distinct concepts. The thermal expansion of gases defines a uniform, reproducible temperature scale that can be extrapolated to very high or

---

*See, for example, "Plate Tectonics" by John F. Dewey, *Scientific American*, May 1972.

very low temperatures by other techniques for measuring temperature.

Most materials expand upon heating and contract when cooled. We define the coefficient of volume expansion,

$$\beta = \frac{1}{V}\frac{dV}{dT}$$

and the coefficient of linear expansion,

$$\alpha = \frac{1}{L}\frac{dL}{dT}$$

Quantity of heat is conventionally measured in calories or kilocalories. Since a quantity of heat is a form of energy, we can also measure it in joules or ft·lb. The conversion between joules and calories is known as the mechanical equivalent of heat.

4.186 J/cal

The specific heat c of a substance is defined as the number of calories required to raise the temperature of one gram of that substance by one Celsius degree:

$$c = \frac{\Delta Q}{m\Delta T}$$

ΔQ = cmΔT

Heat can be transported by three mechanisms: conduction, radiation, and convection. Conduction involves propagation of the thermal motion of atoms from one region to another via interatomic forces. We define a coefficient of thermal conductivity via the equation governing heat flow by conduction:

$$\frac{dQ}{dt} = -kA\frac{dT}{dx}$$

Radiation is the transport of energy via electromagnetic waves, such as radio waves, infrared waves, light, and x rays. At low temperatures, all materials give off long-wavelength radiation. As the temperature is raised, the rate of emission of energy increases rapidly and the wavelength of that radiation becomes shorter. The total rate of radiation is given by Stefan's law:

$$dQ/dt = \epsilon\sigma AT^4$$

Convection transports heat by transporting masses of warmed material. The mechanism driving the mass transport is buoyancy due to the reduced density of the warmer material.

APPENDIX: Conduction of Heat through a Cylindrical Shell

If the cross-sectional area available for heat flow is nonuniform, the temperature gradient will vary. If the inner surface of a cylindrical

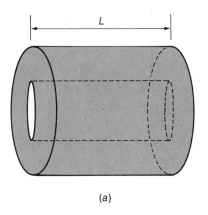

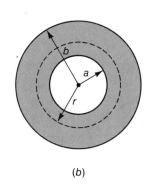

Fig. 19-14.

(a)   (b)

shell (such as the insulating jacket around a steam pipe) is at a high temperature, heat will flow toward the outer, cool surface. However, then the cross-sectional area changes steadily from a small area near the inner surface to a large area near the outer surface. This situation is sketched in Fig. 19-14. In part *b* of the figure we see that the area of the inner surface is

$$A_1 = 2\pi aL$$

and that of the outer surface is

$$A_2 = 2\pi bL$$

In general, the area through which heat can flow while passing from the inside to the outside of the cylinder is given by

$$A = 2\pi rL$$

Applying (19-5),

$$\frac{dQ}{dt} = -kA\frac{dT}{dr}$$

$$= -k2\pi rL\frac{dT}{dr}$$

Rearranging terms:

$$\frac{dr}{r} = \frac{-2\pi Lk}{(dQ/dt)}dT$$

Integrating both sides of the last equation,

$$\int \frac{dr}{r} = \frac{-2\pi Lk}{(dQ/dt)}\int dT$$

Note that the quantity in the denominator, dQ/dt, is simply a constant, being equal to the rate at which heat flows out of the cylindrical surface. To obtain the net temperature difference between the inside and outside of the cylindrical shell, we integrate the right side of the last equation from T_1 to T_2, the left side is correspondingly integrated from the inner radius, a, to the outer radius, b:

$$\int_a^b \frac{dr}{r} = \frac{-2\pi Lk}{(dQ/dt)} \int_{T_1}^{T_2} dT$$

Carrying out the integrations, we find

$$\ln(b/a) = \frac{2\pi Lk}{(dQ/dt)}(T_1 - T_2)$$

or

$$\frac{dQ}{dt} = 2\pi Lk\frac{T_1 - T_2}{\ln(b/a)}$$

Thus the rate of heat flow depends linearly on the temperature difference $(T_2 - T_1)$ and logarithmically on the ratio of the inner radius to outer radius of the cylindrical shell.

Example   The insulation around a 14-cm-diam steam pipe is 10 cm thick and has a conductivity $k = 0.00035$ cal/s $\cdot$ cm $\cdot$ °C. If the inside temperature is 100°C and the outside temperature is 50°C, find the rate at which heat is lost for each meter of pipe length.

Solution   The inner radius is 7 cm while the outer radius is 12 cm. Hence

$$\frac{dQ}{dt} = 2\pi Lk\frac{T_1 - T_2}{\ln(b/a)}$$

$$= (6.28)(100 \text{ cm})\left(0.00035 \frac{\text{cal}}{\text{s} \cdot \text{cm} \cdot \text{°C}}\right) \cdot \frac{100\text{°C} - 50\text{°C}}{\ln(12/7)}$$

$$= 0.22 \frac{\text{cal}}{\text{s}} \cdot \frac{50}{0.539}$$

$$= 20.4 \text{ cal/s}$$

Problems

19-1.   Since any arbitrarily shaped body can be approximated by an assembly of many rectangular solid elements, consider one such element with edges L_1, L_2, and L_3. Show for this element, and hence for the entire body, that

$$\beta = 3\alpha$$

(see Equations (19-1), (19-2), and (19-3)).

19-2. A steel cable 100 m long and 1 cm in diameter is rigidly clamped between two points with a tension of 10^3 N. Find the tension in the cable when it is cooled by 20°C.

19-3. A thermometer of the type shown in Fig. 19-2 is made of pyrex with mercury as the fluid. The volume of the bulb is $\frac{1}{2}$ cm³. Find the diameter of the bore hole that will cause the mercury column to move 1 mm/°C.

19-4. Rivets are sometimes cooled before insertion into rivet holes so that they make an extremely tight fit when they warm to normal temperature. Consider an aluminum rivet that is cooled by dry ice to −78°C before insertion into a hole with 0.1 in diameter. What is the rivet's room-temperature (20°C) diameter if it just fits in the hole when cooled? The material in which the hole is drilled is at room temperature.

19-5. A block of material of unknown specific heat, weighing 25 g with a temperature of 90°C, is dropped into 150 g of water at 10°C. They reach a final common temperature of 24°C. Find the unknown specific heat.

19-6. An automobile with a mass of 2000 kg moving at 20 m/s is brought to a stop in 10 s by applying the brakes. The kinetic energy of the car is converted to thermal energy in the brakes by the friction between the brake shoes and the brake drum. (*a*) Find the average rate, in cal/s, at which heat is produced in the brakes. (*b*) If all of this heat is used to warm an 8-liter bucket of water, find the final temperature of the water if it starts at room temperature (20°C).

19-7. In the preceding problem, find the temperature rise of the brake drums if the heat goes nowhere else. The specific heat of the drum material is 0.2 and the total mass is 25 kg.

19-8. A 40-kg block of wood with an initial velocity of 20 m/s slides along a horizontal surface with a coefficient of friction $\mu = 0.3$ and eventually comes to rest. How much heat is produced? How is this result changed if μ is doubled to 0.6?

19-9. A block of copper at a temperature of 800°C is dropped into 120 g of water at 20°C. Find the mass of the copper block if the final temperature of the water + copper system is 80°C.

19-10. A slice of bread yields 75 Calories upon oxidation. If 25% of this energy can be converted to mechanical work by the human body, find the distance h that a 75-kg man can climb with the expenditure of this much energy.

19-11. A soldering iron is placed against the end of a copper bar with 4-cm² cross-sectional area and 10-cm length. The full 100 W of the soldering iron's heat flows down the copper bar. Find the temperature difference between the two ends of the bar.

19-12. A recent magazine article stated that energy can not be saved by shutting off heat to a room and closing its door, because the thermal insulation of the outer walls is better than that of the interior walls. The same article said that energy is not saved by turning down the thermostat at night because the furnace has to produce extra heat in the morning to rewarm the house. Discuss both of these claims.

19-13. Radiant energy from the sun strikes the earth at a rate of 1.36×10^3 watts/m² where the absorbing surface is at right angles to the sun's rays. This number must vary as $1/r^2$, where r is the distance from the sun, since the total radiation of the sun spreads over all directions in space and a sphere surrounding the sun at a given distance has an area of $4\pi r^2$. If the sun radiates as a blackbody ($\epsilon = 1$), find the surface temperature of the sun. The sun's radius is 6.95×10^{10} cm. The radius of the earth's orbit is 1.49×10^{13} cm.

Chapter *20*

States of Matter

20-1 Introduction We traditionally think of three states of matter: solid, liquid, and gaseous. Some substances, such as water, are commonly observed in all three states. At low temperatures, water is in its solid form, ice, and as the temperature is increased, it becomes liquid and then a gas (water vapor). Some substances do not exhibit all of these states. Many compounds, for example, dissociate before they reach a temperature sufficient to vaporize them.

At the microscopic level, increasing temperature produces greater thermal agitation. This first breaks the rigidity of interatomic bonds, causing a solid to become a liquid. When these bonds are completely disrupted, the liquid becomes a gas. If the temperature is raised far beyond that required to produce a gas (to about 10,000K), the atoms themselves are partially dissociated. Each atom loses one or more electrons, leaving behind a positive ion. The cloud of positive ions and electrons is, overall, electrically neutral but, unlike an ordinary gas, each element of this "gas" is not electrically neutral. Such a substance is called a *plasma* and can be considered the "fourth state of matter." The glowing substance in a neon sign is an example of a plasma. Because its constituents are electrically charged, it conducts electricity readily and behaves in many ways differently from ordinary gases. This fourth state of matter seems unusual to us but, in fact, a large fraction of all the matter in the entire universe is in the form of a plasma.

For our present purposes, we may restrict our attention to the first three states of matter, since they will suffice to introduce us to the concepts of thermodynamics and to explain the fundamental

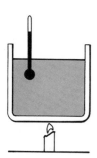

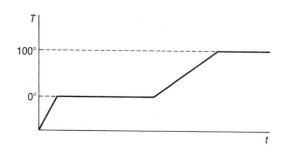

Fig. 20-1. Heat is applied to an ice sample at a rate dQ/dt. The temperature increases steadily until the melting point, then remains constant at 0°C until all the ice is melted.

ideas that underly the operation of heat engines. Since most of our present methods of utilizing energy involve heat engines at some point, their study is quite important. We will concentrate our attention on gases, since they are most often the working fluids of contemporary heat engines.*

20-2 Phase Changes

The transition of a substance from one state to another is called a *phase change*. The melting of ice is an example of such a phase change. Fig. 20-1 illustrates the variation of temperature observed as a sample of ice is heated until it melts and then further heated until it boils. At the melting point, the temperature of the sample remains constant until all of the ice is melted, then rises steadily to the boiling point. Once again, the temperature remains constant at 100°C until all of the water has been converted to steam. If the steam is confined to the vessel, it will then proceed to increase further in temperature as long as heat is supplied.

The stability of temperature during a phase change is precisely why such points were chosen for calibration points in the temperature scale described in the preceding chapter. During a phase change, any energy supplied is used to break interatomic bonds, so the temperature cannot increase until all bonds are broken and the change is complete. In the reverse process—steam condensing into water, for example—the condensation occurs at a precise temperature and heat is given off. This heat must be carried away in order for condensation to continue. At the microscopic level, when two molecules come close enough to "fall into" one another's deep potential energy wells, they acquire large kinetic energy. If this molecular kinetic energy is carried away from the substance as heat, the molecules can drop in energy, forming a liquid from the gas phase. Thus, heat is given off during the condensation.

---

*One of the possible energy sources of the future, the nuclear fusion reactor, may very well utilize a high-temperature plasma as its working fluid.

The heat required to cause melting of one unit mass is called the *heat of fusion*. An equal quantity of heat is given off when the liquid returns to the solid phase. Similarly, the *heat of vaporization* is absorbed when a liquid boils and reappears when the vapor condenses into a liquid again. The quantity of heat, Q, required for a phase change is given by $Q = mL$, where m is the mass of material involved and L is the appropriate heat of fusion or vaporization. The quantity L of heat for a phase change is called the *latent heat*. The temperature at which a substance melts or boils depends upon applied pressure, as do its latent heats. The values given in the table occur at normal atmospheric pressure.

phase change
$Q = mL$
↑
heat of fusion
or
vaporization

TABLE 20-1   Latent Heats for Several Substances
at Atmospheric Pressure

Substance	Melting Point (°C)	Heat of Fusion (cal/g)	Boiling Point (°C)	Heat of Vaporization (cal/g)
Water	0	79.7	100	539.2
Ethyl alcohol	−115	24.9	79	204.3
Nitrogen	−210	6.2	−196	37.8
Gold	1063	16.1	2966	446.0

Many solids can exist in more than one form of crystalline structure. The particular structure in which such a solid exists depends upon temperature and pressure. These changes of crystal symmetry are also phase changes and are often accompanied by their individual latent heats.

Example   How much ice at 0°C must be dropped into 100 g of water at 80°C in order to cool the mixture to 20°C? Ignore the heat capacity of the container.

Latent heat
ice melts   80 cal/g

Solution   Each gram of water must lose 60 calories to cool to 20°C. The ice must absorb 80 cal/g just to melt into water at 0°C, then another 20 cal/g to warm to 20°C. If all this occurs in an insulated vessel, the heat loss equals the heat gain. Letting the unknown mass of ice be m_i and the mass of water, m_w:

$$m_i L + m_i c \Delta T_i = m_w c \Delta T_w$$

$$m_i(80 \text{ cal/g}) + m_i(1 \text{ cal/g} \cdot °C)20°C = 100g(1 \text{ cal/g} \cdot °C)60°C$$

$$m_i = \frac{6000 \text{ cal}}{100 \text{ cal/g}} = 60 \text{ g}$$

Exercises

1. An ice plant produces 3 tons of ice per hour. The water supplied
    to the plant is at 16°C and the ice is stored at −10°C. Find the
    minimum rate at which heat must be extracted from the water
    in cal/s. Express this number in horsepower as well. The spe-
    cific heat of ice is 0.48.

2. A pan containing 1 kg of water at 20°C is placed on a 1-kW
    electric heater. Assuming that all this energy goes into heating
    the water, find (*a*) the time required to reach the boiling point,
    and (*b*) the rate at which water is converted to steam, once the
    boiling point is reached. (*c*) What is the total time required for
    the pan to completely run dry?

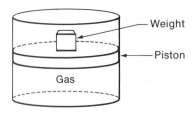

Fig. 20-2. A gas is confined
to a cylinder closed with a mov-
able piston. Under constant
pressure (weight/area of piston)
the volume occupied by the gas
varies linearly with *T*.

20-3 Behavior of Gases

At low densities, gases behave in a simple manner. If the temperature of a gas in a container of fixed volume is varied, the pressure
varies linearly with the thermodynamic temperature:

$$p = \text{constant} \cdot T \qquad (20\text{-}1)$$

If the pressure is fixed, as in Fig. 20-2, and the thermodynamic
temperature is varied, the volume occupied by the gas will vary
linearly with the temperature:

$$V = \text{constant} \cdot T \qquad (20\text{-}2)$$

We now view (20-1) and (20-2) as consequences of our definition
of the thermodynamic temperature scale that was discussed in
Chapter 19. The law given by (20-2) was originally discovered by
use of mercury thermometers over a limited temperature range by
Joseph Louis Gay-Lussac in 1802.

   If the temperature of the gas is fixed but the volume is varied,
as in Fig. 20-3, the pressure of the gas will vary inversely with the
volume:

$$p = \text{constant}/V$$

or
$$pV = \text{constant} \qquad (20\text{-}3)$$

This latter result, (20-3), was discovered by Robert Boyle in 1660
and is known as *Boyle's law*. Finally, if the quantity of substance
within the container in Fig. 20-3 is varied while the volume and
temperature remain fixed, it is found that the pressure is proportional to the quantity of substance

$$p = \text{constant} \cdot n \qquad (T \text{ and } V \text{ fixed}) \qquad (20\text{-}4)$$

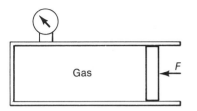 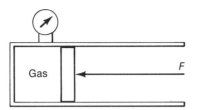

Fig. 20-3. The pressure of a gas is inversely proportional to the available volume when its temperature is held fixed.

where n is the number of moles (defined in Section 13-3) of gas within the volume.

All of these laws can be combined into one statement, known as the *ideal gas law:*

$$pV = nRT \qquad (20\text{-}5)$$

where n is the number of gram moles of gas within the volume and R is an experimentally determined constant that has the same value for all gases. In order for (20-5) to be valid, T must be measured from absolute zero. We will use the thermodynamic temperature expressed in Kelvins. If p is measured in N/m^2, V in m^3, and T in K, we may determine R from the fact that one mole of a gas at standard temperature (273.15K) and standard pressure (one atmosphere $= 1.013 \times 10^5$ N/m^2) occupies a volume of 22.4 liters (one liter $= 10^{-3}$ m^3).

$$R = \frac{pV}{nT} = (1.013 \times 10^5 \text{ N/m}^2)(22.4 \times 10^{-3} \text{ m}^3)(\text{mole})^{-1}(273.15\text{K})^{-1}$$

$$= 8.314 \text{ J/mole} \cdot \text{K}$$

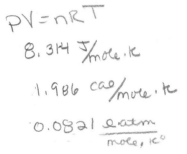

For energy measurements in calories we obtain

$$R = 1.986 \text{ cal/mole} \cdot \text{K}$$

and for pressure in atmospheres, volume in liters we obtain

$$R = 0.0821 \text{ liter} \cdot \text{atm/mole} \cdot \text{K}$$

The ideal gas law is illustrated in Fig. 20-4, in which p is plotted against V for various fixed values of T. Since $p \propto a/V$ when T is fixed, each of the curves shown there has the form of a *hyperbola* (one of the conic sections). Curves of the sort shown in Fig. 20-4 with the temperature held constant are called *isotherms*. Any physical process that causes a gas to vary its pressure and volume in such a way that the temperature remains fixed is called an *isothermal process*.

Our discussion up to this point has revolved around the ideal gas law. What about the "nonideal" gases? Just what is it that makes a gas "ideal"? We say a gas is an ideal gas when it behaves ac-

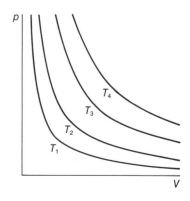

Fig. 20-4.   A family of isothermal curves for an ideal gas: $T_1 < T_2 < T_3 < T_4$.

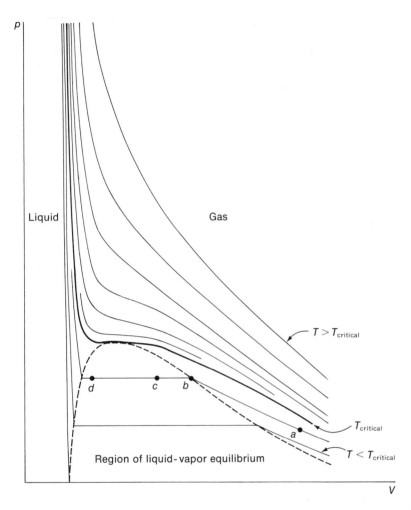

p

Liquid

Gas

$T > T_{\text{critical}}$

d c b

T_{critical}

a

$T < T_{\text{critical}}$

Region of liquid-vapor equilibrium

V

Fig. 20-5. At sufficiently high temperature and low pressure, a gas obeys the ideal gas law. As the temperature is lowered, the actual isothermals depart more and more noticeably from ideal gas behavior. Eventually a change to the liquid phase occurs. For $T > T_{\text{critical}}$ only the gaseous phase occurs.

cording to Eq. (20-5), as graphed in Fig. 20-4. If a real gas is made more dense by increasing pressure and/or cooling, it begins to depart from ideal behavior. If carried far enough, the increased density can finally lead to liquefaction—a phase change. The entire gamut of this behavior is shown in Fig. 20-5. There it can be seen that for sufficiently high temperatures, the gas behaves as an ideal gas. At lower temperatures the dependence of p upon V is no longer exactly as given by (20-5), and at still lower temperatures, a phase change occurs. This is illustrated in Fig. 20-6, where the sketches labeled a through d correspond to the points labeled on one of the isothermal curves of Fig. 20-5.

There is a critical temperature above which no phase change occurs and below which the phase change does occur. Substances below their critical temperature, when they are at sufficiently low

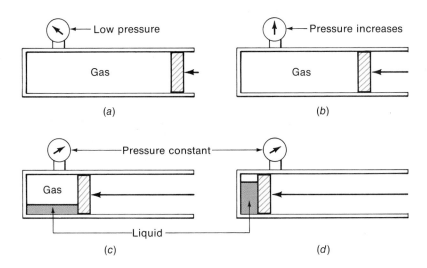

Fig. 20-6. A gas sample is compressed, at constant temperature, in a piston-cylinder apparatus. When the gas begins a phase change to the liquid state, the pressure ceases to rise until this phase change is complete.

pressure so that they are in the gaseous phase, are called *vapors*. Vapors will condense to the liquid phase if sufficiently compressed. Gases above the critical temperature simply become more and more dense as they are compressed. They never make a discontinuous phase change: they simply approach the density and compressibility of a liquid smoothly as compression increases.

Exercises

3. Two moles of oxygen gas are placed in a 5-liter container at a temperature of 20°C. What is the pressure?

4. Three liters of a gas initially at 300K are heated to 500K. The volume of the container is allowed to expand so that the pressure of the gas does not increase. What is the final volume of the container?

20-4   Kinetic Theory of Gases

We now know that substances, including gases, are composed of molecules. One of the early triumphs of the molecular theory of matter was an explanation of the properties of gases based on just a few simple assumptions involving a molecular model of gases. This theory is called the *kinetic theory of gases*. The few assumptions involved are that (*a*) gases are composed of large numbers of molecules,* (*b*) these molecules make elastic collisions with

---

*This condition is well satisfied. At standard temperature and pressure (273K and 760 mm of mercury), one liter of gas contains 2.7×10^{22} molecules.

each other and with the walls of a container, (*c*) the molecules themselves occupy a negligible portion of the volume of the container, (*d*) these molecules are in continual random motion, and (*e*) the molecules move freely between collisions.

According to kinetic theory, the pressure exerted by a gas is due simply to the continual impacts of its molecules upon the container walls. Consider a single, elastic impact of a molecule against a wall. The component of the molecule's velocity that is perpendicular to the wall simply changes sign during the collision, since the mass of the molecule is negligible compared to that of the wall. The parallel component is unaffected, as indicated in Fig. 20-7. The net change in momentum is thus perpendicular to the wall. Now if molecules of mass *m* are striking the wall at a rate of M collisions per second, we can compute the average rate of momentum change for the molecules:

$$(\Delta p/\Delta t)_{\text{wall}} = Mm(2v_x)$$

where $2v_x$ is the net change of the x component of each molecule's velocity. The average force exerted on the wall by the molecular impacts is equal to the average rate of their momentum change:

$$F_{\text{av}} = \Delta p/\Delta t = 2Mmv_x$$

If one molecule is rebounding back and forth between two parallel walls separated by a distance L, the time Δt required for each round trip is $\Delta t = 2L/v_x$ so that

$$M = \frac{1}{\Delta t} = \frac{v_x}{2L}$$

For N molecules,

$$M = \frac{Nv_x}{2L}$$

The net force on the wall is thus

$$F = 2Mmv_x = \frac{Nmv_x^2}{L}$$

Since the molecules do not all have identical velocities, the average force exerted on the wall depends upon $\overline{v_x^2}$, the average value of v_x^2:

$$F_{\text{av}} = \frac{Nm\overline{v_x^2}}{L} \qquad (20\text{-}6)$$

For any given molecule,

$$v^2 = v_x^2 + v_y^2 + v_z^2$$

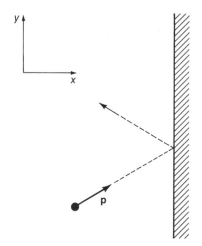

Fig. 20-7. A molecule of mass *m* rebounds elastically from a wall. The component of its momentum perpendicular to the wall, p_x, changes sign while p_y is unaffected.

DERIVATION OF
PV=nRT

so that averaging over each term gives us

$$\overline{v^2} = \overline{v_x^2} + \overline{v_y^2} + \overline{v_z^2}$$

Since the motion of the molecules is assumed to be random, there cannot be any difference between $\overline{v_x^2}$, $\overline{v_y^2}$, and $\overline{v_z^2}$, so we can say

$$\overline{v_x^2} = \tfrac{1}{3}\overline{v^2} \tag{20-7}$$

Substituting (20-7) into (20-6), we obtain

$$F_{av} = \frac{Nm\overline{v^2}}{3L}$$

The pressure on the wall is the force divided by the area of the wall, which is L^2 for a cubical container:

$$p = \frac{F_{av}}{L^2} = \frac{Nm\overline{v^2}}{3L^3}$$

Noting that L^3 is the volume V of the box,

$$p = \frac{Nm\overline{v^2}}{3V} \quad \text{or} \quad pV = \tfrac{1}{3}Nm\overline{v^2}$$

The total number of molecules, N, is just Avogadro's number, N_0, times the number of moles, n, in the container.

$$pV = n\frac{N_0 m\overline{v^2}}{3}$$

Factoring out the average kinetic energy per molecule, $\bar{K} = \tfrac{1}{2}m\overline{v^2}$, we have

$$pV = n \cdot \tfrac{2}{3}N_0\bar{K} \tag{20-8}$$

Equation (20-8) now looks like the ideal gas equation:

$$pV = nRT$$

if the product RT is identified with $\tfrac{2}{3}N_0\bar{K}$. By use of kinetic theory we have been able to deduce the ideal gas law, which is an empirical law, from a few simple hypotheses. Furthermore, for the first time we see in the result (20-8) a microscopic view of the meaning of temperature. High temperatures correspond to large kinetic energies of gas molecules, low temperatures to small kinetic energies. At the absolute zero of temperature, all of the kinetic energy of the gas molecules is lost—they are motionless.* All real gases

---

*This result is correct in the view of classical mechanics. Quantum mechanics informs us, however, that even at absolute zero, a small "zero point motion" persists. This zero point energy is in no way available for conversion to external energy and does not alter our general conclusions.

liquefy before this point, but if they did not, the product of pressure and volume would be zero at absolute zero. The effects of finite molecular size and intermolecular forces are discussed further in the Appendix at the end of this chapter.

20-5 Work Done in Isothermal Expansion

The expansion of a heated gas is often used in heat engines to provide useful work from thermal energy. An example is shown in Fig. 20-8. There, the temperature of the gas is held constant so that an isothermal expansion takes place. The work done by the gas when the piston moves outward a distance ds is given by

$$dW = F\,ds = pA\,ds$$

where A is the area of the piston. Noting that $A\,ds$ is an element of *volume, dV*, swept out by the piston when it moves a distance ds, we may write:

$$dW = p\,dV$$

The work done by the gas when the volume it occupies changes from V_1 to V_2 is

$$W_{12} = \int_{V_1}^{V_2} p\,dV \qquad (20\text{-}9)$$

Figure 20-9 shows this integral graphically. Using the ideal gas law, (20-5), we may write

$$p = \frac{nRT}{V}$$

and substituting in (20-9), we obtain

$$W_{12} = nRT \int_{V_1}^{V_2} \frac{dV}{V} = nRT[\ln V_2 - \ln V_1]$$

$$= nRT \ln\!\left(\frac{V_2}{V_1}\right) \qquad (20\text{-}10)$$

Equation (20-10) is the work done by n moles of gas expanding (or contracting) from volume V_1 to V_2 at a fixed temperature T.

Example One tenth mole of gas at 400K is contained in a cylinder and allowed to expand from a volume of 1 liter to a volume of 2 liters with the temperature held constant. It is then cooled to 200K and compressed, with constant temperature, to 1-liter volume again. Find the net work done by the gas.

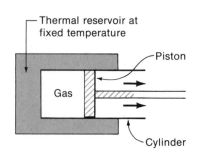

Fig. 20-8.   A gas, maintained at a fixed temperature by a thermal reservoir, expands against a moving piston to do useful work.

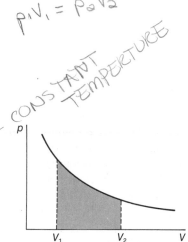

Fig. 20-9. The integral in Eq. (20-9) corresponds to the shaded area above.

Solution   The work done by the expanding gas at 400K is given by (20-10):

$$W_{12} = nRT \ln\left(\frac{V_2}{V_1}\right) = \frac{1}{10} \text{ mole} \cdot \left(8.314 \frac{\text{joules}}{\text{mole} \cdot \text{K}}\right) \cdot 200\text{K} \cdot \ln 2 = 230 \text{ J}$$

For the return stroke,

$$W_{12} = nRT \ln\left(\frac{V_1}{V_2}\right) = \frac{1}{10} \text{ mole} \cdot \left(8.314 \frac{\text{joules}}{\text{mole} \cdot \text{K}}\right) \cdot 200\text{K} \cdot \ln\left(\tfrac{1}{2}\right)$$

$$= (166.3)\,(-0.693) \text{ J} = -115 \text{ J}$$

The minus sign indicates that work is done upon the gas on the return stroke. The net work done is

$$W_{\text{net}} = W_{12} + W_{21} = (230 - 115) \text{ J} = 115 \text{ J}$$

We see, then, that taking a fixed quantity of gas through a thermal cycle involving first a high-temperature expansion and then a lower-temperature compression has produced useful mechanical work. As we will see in the next chapter, this work has come from the thermal energy required to heat the high-temperature gas and to maintain its temperature during the expansion.

Exercise

5. One-half mole of gas is at a pressure of 5 atm and a temperature of 500K. It expands isothermally until its pressure drops to 2 atm. Find the work done by the gas.

20-6   Work Done in Adiabatic Expansion

In most heat engines, the working gas is rapidly heated, then allowed to expand behind a moving piston. This process is typically so rapid that there is insufficient time for transfer of heat between the cylinder walls and the gas. This type of expansion, where no heat exchange occurs, is called *adiabatic* expansion. In an isothermal expansion, the energy expended by the gas in moving the piston is supplied in the form of heat from the thermal reservoir, holding the temperature of the gas constant while the gas does the external work. The work done by an adiabatically expanding gas, on the other hand, is supplied at the expense of the kinetic energy of the gas molecules. Thus, an adiabatically expanding gas decreases in temperature as it expands, so that its pressure drops more rapidly than that of an expanding gas held at constant temperature. The situation is depicted in Fig. 20-10. For an isothermal expansion the gas obeys the equation

$$pV = \text{constant}$$

$$p_1 V_1 = p_2 V_2$$

or

$$p = \text{constant}/V$$

But for an adiabatic expansion, it is found that

$$pV^\gamma = \text{constant} \tag{20-11}$$

or

$$p = \text{constant}/V^\gamma \qquad \boxed{p_1 V_1^\gamma = p_2 V_2^\gamma}$$

where γ is a certain constant that is always greater than 1 (typically, $\gamma = \frac{5}{3}$ for monatomic gases). It follows that in an adiabatic process, p falls more steeply with increasing volume than in an isothermal process.

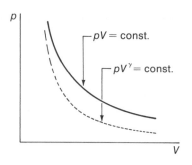

Fig. 20-10. The pressure of an adiabatically expanding gas decreases more rapidly than for isothermal expansion.

Exercise

6. A quantity of gas initially at 50 atm pressure and occupying $\frac{1}{10}$ m³ volume expands adiabatically to twice its initial volume (with $\gamma = \frac{5}{3}$). Find the final pressure of the gas.

20-7 Summary

Matter can exist in various states, e.g., solid, liquid, or gas. We call the various possible states of a given substance *phases* and the transformation from one phase to another a *phase change*. The heat required to produce a phase change in a unit of mass is called the *latent heat, L*. The equation $Q = mL$ gives the net heat absorbed or liberated by a mass m in a phase change. During the absorption or liberation of latent heat, the temperature is constant: only the phase is changing.

phase change
$Q = mL$

CONSTANT TEMP.

Gases at either a reasonably low pressure or high temperature obey a simple law called the *ideal gas law:*

$$pV = nRT$$

where p is the pressure, V is the volume, n the number of moles of gas, and T the absolute temperature. The ideal gas constant, R, is empirically determined and identical for all gases.

This same gas law can be derived from the kinetic theory of gases, in which it is assumed that a gas consists of randomly moving molecules that occupy a negligible volume and that make elastic collisions with the walls and each other. The kinetic theory result is

$$pV = n \cdot \tfrac{2}{3} N_0 \bar{K}$$

where N_0 is Avogadro's number and $\bar{K}$ is the average kinetic energy per molecule. Comparing the ideal gas law and the kinetic theory result, we see that the temperature for an ideal gas is proportional to the average molecular kinetic energy.

When a gas expands from volume V_1 to volume V_2, it does work. For an ideal gas at constant temperature, the work done is:

$$W_{12} = nRT \ln\left(\frac{V_2}{V_1}\right) \qquad \text{ISOTHERMAL}$$

Somewhat less work is done in an adiabatic expansion, where the gas is not maintained at constant temperature but allowed to cool as it expands.

APPENDIX: Van der Waals' Equation

In a gas at standard temperature and pressure, there are 2.7×10^{19} molecules/cm³. Each molecule (for a monatomic gas, such as argon) has a diameter of about 2×10^{-10} m and the average intermolecular spacing is 15 diameters. The molecules have an average velocity of about 10^3 m/s and collide with another molecule on the average of once for every 500 diameters traveled. The volume occupied by the molecules can be compared to the total volume available. Since each molecule takes up only $\frac{1}{15}$ of the intermolecular spacing, and since the volume occupied is given by a length cubed, we find the ratio of the volume occupied by a molecule, V_m, to the available volume, V_a, to be:

$$\frac{V_m}{V_a} \approx \left(\frac{1}{15}\right)^3 = \frac{1}{3375}$$

Thus, at standard temperature and pressure, the molecules themselves occupy less than one-thousandth of the available volume. Such gases behave very nearly like ideal gases.

Two major corrections to the simple ideal gas law can be made on the basis of our understanding of the microscopic, kinetic theory of gases. First, a fraction of the available volume is occupied by the molecules themselves; accordingly, we should subtract a small constant term from the true volume. Second, the molecules attract one another with a force that, for large separation distances, is inversely proportional to the seventh power of their separation distance. Such a force is called a *van der Waals force*. The pressure exerted by the molecules on a wall is thus smaller than expected due to the average force of the remaining molecules pulling back on the molecules that approach the wall. Since the pressure in the gas law is the true pressure on the wall, we should add a small term to the

$$W = \int_{V_1}^{V_2} p \, dV$$

$$P_1 V_1 = P_2 V_2$$

$$P_1 V_1^{\gamma} = P_2 V_2^{\gamma}$$

$$\frac{P_1 V_1}{T_1} = \frac{P_2 V_2}{T_2}$$

gauge pressure, which compensates for the decreased reading. This modified pressure is the "true" pressure inside the gas.

The retarding force exerted by gas molecules on those molecules near the wall is proportional to the number density of the molecules, as long as the force range is short compared to average intermolecular spacing. Further, the number of molecules near the wall that will be affected is also proportional to the number density, N/V. Thus, the pressure must be modified by a term proportional to $1/V^2$. Making this correction to the gauge pressure p, as well as the correction for finite molecular volume, we obtain the van der Waals' equation:

$$\left(p + \frac{a}{V^2} \right)(V - b) = nRT$$

where the constant a will depend on both the strength and the range of the intermolecular force. The constant b is, in principle, calculable from knowledge of the molecular volume. When the constants a and b are adjusted to match data for real gases, a very nice agreement is obtained in the region of the p-V diagram above the critical temperature. Below the critical temperature, liquefaction occurs and cannot be described by van der Waals' equation, which applies only to the gaseous phase.

The pressure correction, a/V^2, represents two-body attractions between molecules, taken pair by pair, the idea being that only two molecules at a time are within each other's range of force.

If the force range is long enough, or equivalently if the density of the gas is high enough, three-body interactions become important and lead to a small $1/V^3$ correction. At successively higher densities we can expect successive terms containing higher and higher inverse powers of V to appear in the equation of state.

Problems

20-1. A 1500-kg auto is moving at $v = 30$ m/s. It is braked to a stop rapidly, so that its kinetic energy is transformed into heat in the brake mechanism. If this heat is utilized to melt ice that is initially at 0°C, what quantity of ice will be melted into water at 0°C?

20-2. Use the conditions of Problem 20-1, except assume that the heat will be used to transform water at 100°C to steam at 100°C. What quantity of water will be vaporized?

20-3. Farmers once placed large pans of water in fruit cellars to protect the fruit from freezing. Fruit typically freezes at a temperature slightly lower than the freezing point of pure water. Explain how the pans of water protected the fruit. How many calories of heat are given off by the freezing of a 30-gal washtub of water?

20-4. A vessel of negligible heat capacity contains 800 gm of water at 80°C. How many grams of ice at 0°C must be dropped into the water to cool it to 50°C?

20-5. In a hot-water heating system, hot water is delivered to radiators, where it cools by giving off heat to the room. In a steam heating system, steam condenses in the radiator to give off its heat of vaporization to the room. If a quantity m of water in a hot-water system cools by 20°C while passing through the radiator, what quantity of condensing steam will provide the same amount of heat?

20-6. Compare the volume required for heat storage in two solar-heated houses. One house utilizes a tank of water that is heated to 50°C during sun lit periods, then cools to 25°C to give off heat during darkness and cloudy weather. The other house utilizes the solid-liquid phase change that occurs at 31°C in Glauber's salt, with a latent heat of 51.3 cal/g. The Glauber's salt is cycled over the same range of temperature, 50°C–25°C. It has a specific gravity of 1.6 and a specific heat of 0.38 cal/g in the liquid phase, 0.25 cal/g in the solid phase.

20-7. An ice cube at −15°C and weighing 50 g is dropped into 200 g of water at 80°C. The water is in an aluminum container (also at 80°C) that has a mass of 50 g. If the container and its contents are placed in an insulated enclosure, so that no other heat flows in or out of this system, find the final temperature of the system.

20-8. In order to achieve a partial vacuum, a technician heats a closed vessel to red heat (800°C), opens the valve to let the heated air inside drop to atmospheric pressure, then closes the valve. He next cools the vessel in an ice bath. What now is the pressure in the vessel?

$$\frac{P_1 V_1}{T_1} = \frac{P_2 V_2}{T_2}$$

20-9. A bubble of air is released from a scuba diver's mask at a depth of 100 m. The water temperature there is 13°C. The bubble rises to the surface, where the temperature is 21°C. If the initial volume of the bubble was 1 cm³, find its volume just as it reaches the surface.

20-10. Find the density of an ideal gas with atomic mass A in terms of arbitrary pressure and temperature. Evaluate your expression for air ($A = 29$) at standard temperature and pressure.

20-11. At the beginning of the compression stroke in a diesel engine, the cylinder contains 1 liter of air at atmospheric pressure. At the end of the stroke, the air has been reduced in volume to 60 cm³. Find the final temperature and pressure of the air if the initial temperature is 20°C and the compression is adiabatic ($\gamma = \frac{7}{5}$).

20-12. A beam of hydrogen atoms with a kinetic energy of 250 eV per atom strikes a surface at normal incidence and is completely absorbed. Find the average force exerted on the surface if the beam delivers 10^{15} atoms/s.

20-13. Utilize the comparison between the ideal gas law and the kinetic theory result, (20-8), to calculate the average speed of oxygen molecules at standard temperature and pressure. The mass of an oxygen molecule is 5.3×10^{-23} g. Find the average speed of a hydrogen molecule ($m = 1.7 \times 10^{-24}$ g) under the same conditions.

20-14. A cylinder with one end closed and a diameter of 10 cm is fitted with a piston that moves back and forth so that the space between the piston and the closed end of the cylinder varies between 1 cm and 10 cm. Oxygen at STP is admitted into the cylinder with the piston in the 10-cm position. The oxygen is then compressed to the 1-cm position while the temperature is kept at 273K. The oxygen is next heated to 473K and allowed to drive the piston back to the 10-cm position while being maintained at 473K. Find the net work done by the gas.

20-15. One mole of oxygen gas at STP is confined in a piston-cylinder arrangement with a weight placed on the piston so that the oxygen is always under an absolute pressure of 2 atm. The oxygen is then heated to 373K. How much work is done by the expanding gas?

20-16. Find the work done by a gas that obeys van der Waals' equation,

$$\left(p + \frac{a}{V^2}\right)(V - b) = nRT$$

as it expands from V_1 to V_2 at constant T.

20-17. Find the work done by a gas expanding adiabatically, so that

$$pV^\gamma = K$$

as it expands from V_1 to V_2.

Chapter 21

Thermodynamic Laws and Heat Engines

21-1 Introduction Many heat engines use a piston-cylinder arrangement that contains a gas. The gas is heated, thus increasing the average molecular velocity. The molecules bombard the piston and, as it moves outward, the force due to the pressure of the gas does work on the piston. The energy given to the moving piston is taken from molecular kinetic energy. Each time a molecule rebounds from an outward moving piston, it rebounds with a slightly smaller velocity than it would if the piston were stationary. This is the mechanism whereby the random, kinetic motion of the gas molecules is transferred to orderly, useful motion of a piston. It is clear that we can never extract *all* of the thermal energy of the gas unless we carry the piston expansion so far that all of the kinetic energy of the gas molecules is spent. However, before this can occur, the temperature of the gas will fall to absolute zero. There are two limitations that prevent this occurrence.

First, all substances liquefy before they reach absolute zero. Some however, such as helium gas, nearly reach absolute zero before liquefaction and it seems that we could extract a very large fraction of this gas's thermal energy without difficulty. The second, more fundamental, limitation is due to the thermal motion of the molecules that constitute *any* container in which the gas can be placed. In order to allow the gas to cool nearly to absolute zero (to extract nearly all its thermal energy) the container itself must also be at a temperature near absolute zero. But where are we to find a container near absolute zero? If we attempt to provide one here on earth, we must expend more energy in refrigeration than we can gain

by the operation of a heat engine into such a low temperature reservoir. It is true that we can let a heated gas surrounded by a thermal
insulator expand *once* to temperatures below that of its surroundings, but we cannot do this repeatedly in a cyclic process, as required for a useful engine.

In this chapter we will examine limits that nature imposes on
the operation of heat engines. The first limit is simply a statement of
conservation of energy and is called the first law of thermodynamics.
The second states that no cyclic process can transform all of the
thermal energy of a substance into mechanical energy for the reasons discussed above; it is called the second law of thermodynamics.

you can't win
you can't even
break even

21-2 The First Law of Thermodynamics

The first law of thermodynamics is a statement of conservation of
energy. Since we have seen that heat is simply a form of energy, the
total energy of a system is now recognized to include heat energy as
well as macroscopic kinetic and potential energy.* The heat added
to a system must equal the work done by the system plus any change
in its *internal energy*. Internal energy is the name we give to energy
at the microscopic level within the system. In general, this energy
can be both potential and kinetic in form, but for an ideal gas the
internal energy is simply the sum of all the molecular kinetic
energies:

internal energy
$$U_{\text{ideal gas}} = nN_0\bar{K} \tag{21-1}$$

where U is the internal energy and $\bar{K}$ is the average kinetic energy
per molecule. Interatomic potential energy is unimportant for an
ideal gas because its molecules are so widely spaced that their interactions are negligible. In the preceding chapter we saw that

$$pV = n \cdot \tfrac{2}{3} N_0 \bar{K} \qquad \text{(kinetic theory)}$$

and
$$pV = nRT \qquad \text{(observation)}$$

Hence for an ideal gas,

$$\tfrac{2}{3} N_0 \bar{K} = RT$$

and
$$\bar{K} = \frac{3}{2} \frac{R}{N_0} T$$

---

*We have already seen that heat energy is simply *atomic* kinetic and potential
energy.

The ratio R/N_0 is defined to be *Boltzmann's constant*, denoted k. Thus the average molecular kinetic energy for an ideal gas is given by

$$\bar{K} = \tfrac{3}{2}kT \tag{21-2}$$

Combining (21-1) and (21-2), we see that

$$U_{\text{ideal gas}} = \tfrac{3}{2}nRT \tag{21-3}$$

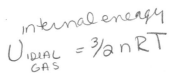

internal energy
$U_{\text{IDEAL GAS}} = \tfrac{3}{2} n RT$

In general, U need not be a function of temperature alone. For example, for a nonideal gas in which intermolecular forces are non-negligible, compression and expansion varies the intermolecular spacing and hence the net molecular potential energy. In this case, the internal energy depends upon pressure as well as temperature.

Calling the heat added to a system ΔQ, the work done by the system ΔW, and the change in its internal energy ΔU, the first law of thermodynamics can be written:

$$\Delta Q = \Delta W + \Delta U \tag{21-4}$$

Equation (21-4) simply states that any energy flowing into the system is balanced by energy flowing out of the system or by a gain in internal energy. The sign convention generally chosen is that that is natural to an engineer interested in producing an efficient heat engine. Such an engine takes in heat, ΔQ, gives off useful work, ΔW, and may use some of the absorbed heat to change its internal energy, ΔU. Thus heat *added to* the system is positive and work done *by* the system is positive. In the situation shown in Fig. 21-1, the sign of the work is negative, since it is work done *on* the system. In this case, the change in internal energy, ΔU, is equal to the sum of the work done and heat added.

Now let us apply (21-4) to the isothermal expansion of an ideal gas. Recalling that for an *ideal* gas the internal energy depends only on temperature, we see that $\Delta U = 0$ for an isothermal process. Hence we have simply $\Delta Q = \Delta W$. Thus, upon isothermal expansion from V_1 to V_2, an ideal gas must absorb heat:

$$\Delta Q = \Delta W = nRT \ln\left(\frac{V_2}{V_1}\right) \tag{21-5}$$

where the last equality comes from Eq. (20-10).

Exercises

1. Find the internal energy in 2 moles of an ideal gas at 400K.

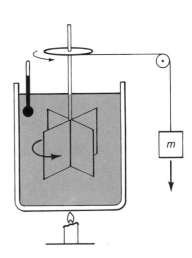

Fig. 21-1. A bucket of water is the system. Heat, Q, is being added to it, as well as work, W.

2. One of your fellow students shows you a novel design for a heat engine. It absorbs 500 cal in each cycle, exhausts 100 cal and produces 2500 J of work in each cycle. Should he recheck his figures? Explain.

21-3 Specific Heats of an Ideal Gas

If heat is added to a gas at a fixed volume, the added heat simply goes into increased internal energy, hence into increased temperature. But if instead the gas is to be maintained at constant pressure, it must expand as heat is added. As it expands, it does external work and hence any added heat must divide itself between the work done by the gas as it expands and an increase in internal energy of the gas. It is convenient to define a specific heat for each of these cases. It is also convenient to define the specific heat of a gas in terms of the number of *moles* of the gas present, rather than its mass. The molar specific heat at constant pressure is denoted C_p, while that at constant volume is denoted by C_V. Thus by definition, at constant volume:

$$\Delta Q = C_V n \Delta T \tag{21-6}$$

For an ideal gas, the added energy, ΔQ, must go into molecular kinetic energy:

$$\Delta Q = n N_0 \Delta \bar{K} = \Delta U \tag{21-7}$$

where $\Delta \bar{K}$ is the average increase in molecular kinetic energy. From Eqs. (21-3) and (21-7), we see that

$$\Delta Q = \tfrac{3}{2} n R \Delta T \tag{21-8}$$

and substituting into (21-6), we obtain

$$\tfrac{3}{2} n R \Delta T = C_V n \Delta T$$

or

$$C_V = \tfrac{3}{2} R$$

The value of R is 1.986 cal/mole $\cdot$ K, so we expect that $C_V = \tfrac{3}{2} R$ = 2.979 cal/mole $\cdot$ K. Table 21-1 lists experimentally determined values for C_V for various gases. The monatomic gases, helium and argon, have values of C_V that fit the prediction nicely, while the diatomic and triatomic gases display larger values. This points out a limitation of our simple kinetic theory, where we assumed the molecules were like billiard balls, with no internal structure. This assumption is nearly justified for gases with monatomic molecules, where nearly all of the thermal energy goes into translational kinetic energy. But complex molecules can have kinetic energy of vibration and rotation as well. This possibility was neglected in writing (21-7) and so our conclusion, $C_V = \frac{3}{2}R$, is valid only for monatomic gases.

$C_V = \frac{3}{2}R$
only for
monoatomic gases

TABLE 21-1

Gas	C_V	C_p	$C_p - C_V(\text{cal/mol} \cdot \text{K})$
Helium (He)	2.98	4.97	1.99
Argon (A)	2.98	4.97	1.99
Hydrogen (H_2)	4.88	6.87	1.99
Nitrogen (N_2)	4.96	6.95	1.99
Oxygen (O_2)	5.03	7.03	2.00
Carbon dioxide (CO_2)	6.80	8.83	2.03
Sulfur dioxide (SO_2)	7.50	9.65	2.15

$C_V = 2.98 \text{ cal/mole K}$

In order to calculate C_p, the molar specific heat for a gas at constant pressure, we must take into account the fact that some of the absorbed heat goes into external work when the temperature is increased, for the pressure cannot remain constant unless the gas is allowed to expand. Consider the addition of an infinitesimal quantity of heat, dQ, to a gas. Its internal energy increases by dU and it does a quantity of work $p\,dV$:

$$dQ = dU + p\,dV \qquad (21\text{-}9)$$

The quantity dU can be found from (21-7) and (21-8):

$$dU = \tfrac{3}{2}nR\,dT$$

In order to express $p\,dV$ in terms of dT, we differentiate the ideal gas law; $pV = nRT$. Recalling that p, n, and R are constant, we obtain

$$p\,dV = nR\,dT$$

Now rewriting (21-9),

$$dQ = \tfrac{3}{2}nR\,dT + nR\,dT$$

$$= C_p n\,dT$$

where the last equality is the *definition* of molar specific heat. Solving for C_p:

$$C_p = \tfrac{3}{2}R + R = \tfrac{5}{2}R \qquad (21\text{-}10)$$

This result applies to an ideal, monatomic gas. The first two entries in Table 21-1 indicate that the prediction is borne out:

$$C_p = \tfrac{5}{2}R = 4.965 \text{ cal/mole} \cdot \text{K}$$

$C_p = \frac{5}{2}R$
$C_p = 4.97 \text{ cal/mole} \cdot \text{K}$

Note that the derivation that led to (21-10) can give

$$C_p = C_V + R \qquad (21\text{-}11)$$

where all that we require is that $pV = nRT$. Thus for *any* ideal gas, we expect the difference $C_p - C_V$ to equal a constant with a value of 1.986 cal/mole · K. Table 21-1 shows that this conclusion is very closely realized for many gases.

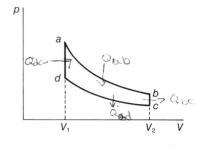

Fig. 21-2.

Example   One-fifth mole of helium (considered an ideal gas) is enclosed in a piston-cylinder arrangement, then taken through the cycle shown in Fig. 21-2: isothermal expansion at 400°C from a to b, cooling at constant volume to 100°C from b to c, isothermal compression at 100°C from c to d, heating at constant volume from d back to a. Let $V_2 = 3V_1$. (1) Compute the heat absorbed in the steps d–a and a–b. (2) Compute the heat given off in the steps b–c and c–d. (3) Compute the net work done over the complete cycle.

heat absorbed
a-b expansion
d-a heating at constant volume
heat given off
cooling
compression

Solution   Let the high temperature be T_1 and the low temperature T_2. (1) In the process d–a, heat is absorbed at constant volume:·

$$Q_{da} = nC_V(T_1 - T_2) \qquad \text{heat at constant volume}$$

In the process a–b, heat must be absorbed to replace exactly the work done in this process, since it is isothermal and the internal energy of an ideal gas is constant at constant temperatures. Applying (20-10):

heat = work done
$\int p\,dV$

$$Q_{ab} = nRT_1 \ln\left(\frac{V_2}{V_1}\right) \qquad \text{isothermal expansion}$$

(2) The heat given off in going from b to c and from c to d is similarly calculated:

$$Q_{bc} = nC_V(T_1 - T_2) \qquad \text{Temperature change at constant volume}$$
$$\text{cooling at constant volume}$$
$$Q_{cd} = nRT_2 \ln\left(\frac{V_2}{V_1}\right) \qquad \text{isothermal compression}$$

The net heat absorbed (heat in minus heat out) is:

$$Q_{in} - Q_{out} = nR(T_1 - T_2) \ln\left(\frac{V_2}{V_1}\right)$$

(3) The net work done in this cycle is the difference between the work done *by* the gas in process a–b and the work done *on* the gas in process c–d:

$$W_{out} = nR(T_1 - T_2) \ln\left(\frac{V_2}{V_1}\right)$$

and is precisely equal to the net heat absorbed, as energy conservation demands. Substituting numerical values:

$$W_{out} = (\tfrac{1}{5} \text{ mole}) (8.314 \text{ J/mole} \cdot \text{K}) (673K - 373K) \ln (\tfrac{3}{7})$$

$$= 548 \text{ J}$$

In the process of this cycle, the amount of heat absorbed is

$$Q_{in} = nC_V(T_1 - T_2) + nRT_1 \ln \left(\frac{V_2}{V_1}\right)$$

$$= 623.5 \text{ J} + 1229.4 \text{ J}$$

$$= 443.3 \text{ cal}$$

The heat given off is

$$Q_{out} = nC_V(T_1 - T_2) + nRT_2 \ln \left(\frac{V_2}{V_1}\right)$$

$$= 623.5 \text{ J} + 681.4 \text{ J}$$

$$= 312.2 \text{ cal}$$

Thus our heat engine absorbs 443.3 cal/cycle, gives off 312.2 cal/cycle as exhaust heat, and the difference, 131.1 cal = 548 J, is useful work. The inevitability of exhaust heat is at the heart of the second law of thermodynamics, which we will discuss in Section 21-7.

handwritten annotations:

$eff = \dfrac{Wout}{Q_{IN}} =$

$\dfrac{W}{Q_{ab} + Q_{da}}$

$Q_{ab} + Q_{da} - Q_{bc} - Q_{cd} = W$

Exercises

3. Two moles of oxygen gas are heated from room temperature (20°C) to the boiling point of water (*a*) at constant pressure; (*b*) at constant volume. How many calories of energy are required in each case?

4. A heat engine operates in the cycle given in the preceding example. Its working gas is 0.1 mole of nitrogen, its high temperature 800°C, and its low temperature 300°C. The ratio V_2/V_1 equals 10. Find the net work done per cycle.

21-4 Efficiency of Heat Engines

As pointed out in the introduction to this chapter, no heat engine can convert 100% of the supplied heat into useful mechanical work. We define the *efficiency* of a heat engine to be the ratio of work produced per cycle to the heat supplied per cycle. The same units of energy, of course, must be used in computing this ratio:

$$\text{eff} = \frac{W_{out}}{Q_{in}} \qquad = \quad 1 - \frac{Q_{OUT}}{Q_{IN}}$$

All of the heat engines we will consider are cyclic devices. They carry a working fluid through a cycle of operation that repeats over and over. There can be no accumulation, or loss, of internal energy in the working fluid since it continually returns to its initial state. Thus, when we apply the first law of thermodynamics to a cyclic heat engine, we need to consider only heat supplied to or exhausted by the engine and work done by the engine. By the first law of thermodynamics, we know that W_{out} must equal the difference between the supplied heat, Q_{in}, and the exhaust heat, Q_{out}.

$$\text{eff} = \frac{W_{out}}{Q_{in}} = \frac{Q_{in} - Q_{out}}{Q_{in}} = 1 - \frac{Q_{out}}{Q_{in}} \qquad (21\text{-}12)$$

Fig. 21-3 illustrates this schematically.

For the engine in the example of Section 21-3, the efficiency is

$$\text{eff} = \frac{W_{out}}{Q_{in}} = \frac{131.1 \text{ cal}}{443.3 \text{ cal}} = 0.296 = 29.6\%$$

A gasoline engine is a more realistic example of a heat engine. It operates by compression of a gasoline vapor + air mixture that is burned quickly to raise the temperature of the gases for the power stroke. The expansion is essentially adiabatic and the opening of the exhaust value, the exhaust stroke, and the intake stroke basically amount to cooling the working fluid back to room temperature.

The essential features of the internal combustion gasoline engine's operation are shown in Fig. 21-4. This cycle of operation is called the *Otto cycle*, after the inventor of the internal combustion engine. Recalling that $pV^{\gamma} = \text{constant}$ for the adiabatic compression and expansion processes, it is possible to compute the net work done in the Otto cycle (see Problem 21-7). Computing the efficiency of the Otto cycle, we find

$$\text{eff} = 1 - \frac{1}{(V_2/V_1)^{\gamma-1}}$$

where the ratio V_2/V_1 is called the compression ratio. For a value of γ applicable to air and a compression ratio of 7, we obtain an efficiency of approximately 50%. In actual practice, frictional forces and heat losses reduce the overall efficiency below this value.

21-5   The Carnot Cycle

Sadi Carnot (1796–1832) was a French engineer interested in obtaining the maximum possible efficiency from heat engines. Considering the crude state of development of steam engines in his

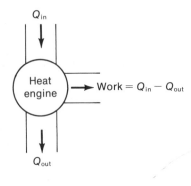

Fig. 21-3. A heat engine gives off a quantity of work equal to the difference between the heat input and the exhaust heat.

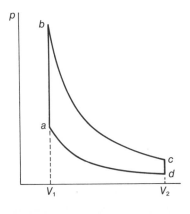

Fig. 21-4. The Otto cycle. Combustion heats the gas from *a* to *b*. Adiabatic expansion along *b–c* provides the power stroke. Exhaust and intake accomplish *c–d*, while the adiabatic compression stroke is along *d–a*.

time and the fact that the first law of thermodynamics was scarcely understood or accepted, his work was brilliant. He adopted a theoretical approach that was instrumental in arriving at the second law of thermodynamics. Since greater thermal efficiencies are obtained by increasing the temperature difference between input and output, he reasoned that *maximum* efficiency could be obtained by extracting heat at a *constant* temperature from a high-temperature reservoir via an isothermal expansion followed by an adiabatic expansion, then reversing these cycles to exhaust heat at a *constant* lower temperature to a low-temperature reservoir.

This cycle of operation, illustrated in Fig. 21-5, is called a *Carnot cycle*. No engine actually operates in a Carnot cycle: its importance is due to the fact that its efficiency is the maximum possible between a high-temperature source at T_1 and a low-temperature exhaust reservoir at T_2. This Carnot efficiency is independent of the working fluid used in the cycle. Further, the efficiency is very simply related to the temperatures T_1 and T_2:

$$\text{eff(Carnot)} = 1 - \frac{T_2}{T_1} \qquad (21\text{-}13)$$

where the temperatures are measured on the Kelvin scale.

To calculate the efficiency of a Carnot cycle, we proceed as follows. As in the example in Section 21-3, we find the heat absorbed in the isothermal expansion to be

$$Q_{\text{in}} = nRT_1 \ln\left(\frac{V_b}{V_a}\right)$$

and the heat given off in the isothermal compression to be

$$Q_{\text{out}} = nRT_2 \ln\left(\frac{V_c}{V_d}\right)$$

To find the efficiency from (21-12), we need the ratio

$$\frac{Q_{\text{out}}}{Q_{\text{in}}} = \frac{T_2 \ln(V_c/V_d)}{T_1 \ln(V_b/V_a)} \qquad (21\text{-}14)$$

We can find the ratios V_c/V_d and V_b/V_a by using the fact that $pV =$ constant along the isothermal portions of the cycle and $pV^\gamma =$ constant along the adiabatic portions:

$$p_a V_a = p_b V_b$$
$$p_c V_c = p_d V_d$$
$$p_b V_b{}^\gamma = p_c V_c{}^\gamma$$
$$p_d V_d{}^\gamma = p_a V_a{}^\gamma$$

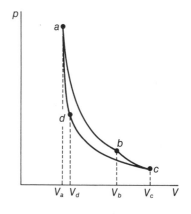

Fig. 21-5. A Carnot cycle. An isothermal expansion absorbs heat at T_1 along *a–b*, followed by an adiabatic expansion along *b–c* until temperature T_2 is reached. The isothermal compression along *c–d* gives off exhaust heat at T_2, then the adiabatic compression from *d* to *a* heats the working fluid back to T_1.

Multiplying all four of these equations together and cancelling common factors, we find

$$(V_b V_a)^{\gamma-1} = (V_c V_a)^{\gamma-1}$$

or

$$\frac{V_b}{V_a} = \frac{V_c}{V_d}$$

Substituting this result into (21-14), we find

$$\frac{Q_{\text{out}}}{Q_{\text{in}}} = \frac{T_2}{T_1}$$

so that the efficiency is

$$\text{eff} = 1 - \frac{Q_{\text{out}}}{Q_{\text{in}}} = 1 - \frac{T_2}{T_1}$$

It can be shown that no other cycle is more efficient than the Carnot cycle and its efficiency is thus an upper limit for realizable engines.*

All reversible cycles that operate between fixed temperatures T_1 and T_2 have the Carnot efficiency, $1 - (T_2/T_1)$. Irreversible cycles, such as those involving explosions of gasoline-air mixtures, have lower efficiencies. There are *reversible* cycles that have lower efficiencies than the Carnot efficiency, namely, those that take in heat, not at a fixed high temperature but over a span of temperatures ranging downward from the value of the high-temperature reservoir. The following example illustrates this point.

Example An engine operates on the cycle used in the example of Section 21–3: isothermal expansion at T_1 followed by cooling to T_2, then isothermal compression at T_2 and reheating to T_1. Suppose the working fluid is an ideal gas with $C_V = 3R$. Find the engine's efficiency and compare it to the Carnot engine's efficiency.

Solution Proceed as in the example of Section 21–3. The heat taken in at V_1 in the process d–a is

$$Q_{da} = nC_V\Delta T = nC_V(T_1 - T_2)$$

Note that the temperature of the gas varies from T_2 up to T_1 in this portion of the cycle. The heat absorbed along the isothermal a–b is calculated at the end of Section 21-2:

---

*Rudolf Diesel began the development of the engine that bears his name in an attempt to achieve the efficiency of the Carnot cycle. The series of compromises with practical limitations that led to the modern diesel engine are described in "Rudolf Diesel and his Rational Engine" by Lynwood Bryant, *Scientific American*, August 1969.

$$Q_{ab} = nRT_1 \ln\left(\frac{V_2}{V_1}\right)$$

The net heat absorbed is $Q_{in} = Q_{da} + Q_{ab}$. The heat rejected is

$$Q_{out} = Q_{bc} + Q_{cd} = nC_V(T_2 - T_1) + nRT_2 \ln\left(\frac{V_1}{V_2}\right)$$

$$\text{eff} = \frac{Q_{in} - Q_{out}}{Q_{in}} = \frac{(T_1 - T_2)\ln(V_2/V_1)}{3(T_1 - T_2) + T_1\ln(V_2/V_1)}$$

For simplicity in this example, let us choose (V_2/V_1) to be 2.78, since $\ln(2.78) = 1$. For $T_1 = 600K$, $T_2 = 300K$, we obtain

$$\text{eff} = \frac{300}{900 + 600} = 20\%$$

For a Carnot cycle working between the same two temperatures,

$$\text{eff} = 1 - \frac{T_2}{T_1} = 1 - \frac{300}{600} = 50\%$$

In general, any heat engine will have an efficiency less than that for a Carnot cycle operating between the same upper and lower temperatures.

Exercises

5. A Carnot engine exhausts heat at $T_2 = 300K$, which is close to ordinary room temperature. What upper temperature T_1 is quired to give 90% efficiency? Compare this temperature to the melting point of gold (1337.6K).

6. In a modern steam-turbine power plant, the inlet temperature is typically 810K and the exhaust temperature 310K. Find the efficiency of a Carnot cycle operating between these two temperatures. The actual efficiency achieved in these steam turbine power plants is 41%.*

21-6   Refrigerators

Using the same schematic representation we used earlier for a heat engine, the action of a refrigerator is illustrated in Fig. 21-6. The first law of thermodynamics tells us that Q_{out} will be the sum of Q_{in} and W. An extremely efficient refrigerator is one that transfers large quantities of heat from the cold source to the hot reservoir

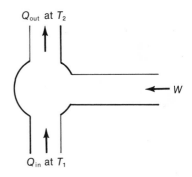

Fig. 21-6.   A refrigerator takes in heat from a low-temperature source (T_1) and mechanical work from an energy source, and gives off exhaust heat to its higher-temperature surroundings (T_2).

---

*See "The Conversion of Energy" by Claude M. Summers, *Scientific American*, September 1971.

with only a small expenditure of work. There is nothing in the first law of thermodynamics that prevents an arbitrarily efficient refrigerator. In actuality, though, the performance of a refrigerator is quite limited.

Let us define a coefficient of performance, P:

$$P = \frac{Q_{\text{in}}}{W} = \frac{Q_{\text{in}}}{Q_{\text{out}} - Q_{\text{in}}}$$

where W and Q are considered positive quantities. The coefficient of performance can vary between zero and infinity depending on the amount of work required in the refrigeration cycle.

The cycle used in the example at the end of the preceding section can be run backwards to make a refrigerator.

$$Q_{\text{in}} = Q_{dc} - Q_{ad}$$

$$Q_{\text{out}} = Q_{ba} - Q_{cb}$$

$$P = \frac{Q_{\text{in}}}{Q_{\text{out}} - Q_{\text{in}}} = \frac{T_1 \ln(V_2/V_1) - 3(T_2 - T_1)}{T_2 \ln(V_2/V_1) - 3(T_2 - T_1) - T_1 \ln(V_2/V_1) + 3(T_2 - T_1)}$$

Again, let us choose $(V_2/V_1) = 2.78$ for convenience. For $T_1 = 270\text{K}$ and $T_2 = 300\text{K}$, we obtain

$$P = \frac{270 - 90}{300 - 90 - 270 + 90} = 6$$

For a Carnot refrigerator, which utilizes the Carnot cycle of Fig. 21-5, we have

$$P = \frac{Q_{\text{in}}}{Q_{\text{out}} - Q_{\text{in}}} = \frac{T_1}{T_2 - T_1}$$

Using the above temperatures,

$$P = \frac{270}{300 - 270} = 9$$

In general, no refrigeration cycle has a higher coefficient of performance than a Carnot cycle.

Exercises

7. How much of the mechanical energy supplied to a refrigerator's mechanism appears in its exhaust heat? What is wrong with leaving the refrigerator door open to cool off the kitchen?

8.   A refrigerator operates between $T_1 = 270K$ and $T_2 = 310K$. Find its ideal (Carnot) coefficient of performance. How many calories per second can it remove from the cold enclosure if power is supplied to it at a 250-W rate ($\approx \frac{1}{3}$ hp)?

21-7   The Second Law of Thermodynamics

A Carnot cycle is the most efficient possible, yet its efficiency approaches 100% only when its exhaust temperature is lowered to absolute zero or its inlet temperature is raised far beyond what ordinary materials can survive. The impossibility of converting all of a quantity of thermal energy into useful mechanical work is the essence of the *second* law of thermodynamics. There are several ways to state the second law, but a simple statement is:

> No process is possible such that the sole result is the absorption of heat from a reservoir, where all the heat is converted into mechanical work.

While the first law imposes overall conservation of energy upon thermal processes, the second law states that there are additional restrictions when it comes to transforming heat into useful mechanical work. The first law says, "You can't win"; the second law says, "You can't break even."

The underlying reason for the second law is the random, chaotic nature of thermal energy coupled with its universal presence. The random nature of the motion of molecules in a gas, for example, forces us to cool it completely to absolute zero to utilize all this molecular energy. Yet, unless the surroundings of the gas are themselves at absolute zero, we cannot so cool it in a useful, cyclic device without expending an excessive amount of energy in a refrigerator.

Thermal energy *can* be utilized as long as it is "running downhill." When heat flows from a high temperature to a lower temperature, we can utilize a portion of its energy. The maximum possible fraction we can use is given by the efficiency of the Carnot cycle, with most practical heat engines being even less efficient. Presently, the universe contains high-temperature energy sources (such as stars) and low-temperature regions. Eventually, when all energy production is exhausted, all of the universe will reach thermal equilibrium. At that time, the energy content of the universe will still be what it is today. This energy, however, will be unavailable, since all parts of the universe will then be at the same temperature.

This dismal scene is usually referred to as the "heat death" of the universe.*

Exercise

9. The oceans of the earth contain approximately 1.5×10^{24} g of water. How much heat can we obtain by cooling the oceans by 10°C? How does this compare with 10^{22} J, which is the estimated needs** of the entire earth's population between the years 1960 and 2000? What prevents our use of this thermal energy?

21-8 Entropy

A high-temperature gas is characterized by a great deal of disorder. Because the random thermal motion of gas molecules diminishes as the temperature is lowered, a low-temperature gas is somewhat more orderly. When the gas condenses into a liquid, the degree of disorder diminishes abruptly as the molecules come nearly into contact with one another and the latent heat of vaporization is given off. Finally, when the liquid solidifies, the disorder of the substance decreases again as the molecules fall into an ordered array and the latent heat of fusion is given off.

The thermodynamic *entropy*, *S*, of a system is a measure of its disorder and is defined by:

$$\Delta S = \int_{T_1}^{T_2} dQ/T \qquad (21\text{-}15)$$

where dQ represents heat added to a system in a reversible process and T is the absolute temperature. Only *changes* in entropy are defined in classical thermodynamics but it can be shown that the entropy is zero at absolute zero. Since entropy increases as heat is added to a system, large entropy corresponds to a high degree of disorder.

This somewhat vague assertion is made more precise and justified in a branch of physics called *statistical mechanics*. There, it is asserted on the basis of probability theory that any system will

---

*This time lies tens of billions of years in the future. Some cosmological models that involve a pulsating universe terminate each cycle of pulsation with a "rebirth" of the expanding universe. The time scales involved, if those models are correct, will probably never allow a complete "heat death."

**See "Energy and Power" by Chauncey Starr, *Scientific American*, September 1971.

tend toward that state that has the greatest possible number of ways of being formed. Such a state is, in a very real sense, the most disordered state possible. Thus the universe evolves in such a way that entropy increases with the ultimate "heat death" corresponding to maximum possible entropy. In such a universe, no order remains. There are no longer hot places and cool places, no longer high-density regions and low-density regions: only a uniformly chaotic assembly of molecules in continual thermal motion.

Example When a piece of ice melts, it is clear that molecular disorder has increased. Show that the thermodynamic definition of entropy, (21-15), leads to an increase in entropy for this process.

Solution Since the melting of ice takes place at constant temperature, the integral in (21-15) is particularly easy to evaluate:

$$\Delta S = \frac{1}{T} \int dQ = \frac{Q}{T}$$

where Q is the heat added to the ice to cause it to melt. Thus Q is given by the latent heat of fusion for water:

$$\Delta S = \frac{mL}{T}$$

For 100 g of ice we have an increase of entropy given by

$$\Delta S = \frac{100 \text{ g} \cdot 80 \text{ cal/g}}{273 \text{K}} = 29.3 \text{ cal/K}$$

Exercises

10. Find the increase in entropy when 100 g of water are heated from 273K to 323K.

11. When a quantity of water is cooled, the entropy of the water decreases. How is this compatible with the statement that the entropy of the universe always *increases*?

21-9 Thermal Consequences of Power Production

When power is produced by a heat engine for consumption by society, a certain amount of waste heat is produced by the power source. This waste heat is the exhaust heat, which is quite substantial for typical efficiencies of heat engines. From the definition of efficiency:

$$\text{eff} = \frac{W}{Q_{in}} = \frac{Q_{in} - Q_{out}}{Q_{in}}$$

we see that

$$Q_{\text{out}} = Q_{\text{in}} (1 - \text{eff}) = W \left(\frac{1}{\text{eff}} - 1 \right)$$

In Table 21-2, the average efficiencies of several different power sources are listed along with the ratios of their exhaust heat to their useful power output.

TABLE 21-2

Power Source	Use	Efficiency(%)	Exhaust Heat/ Useful Output
Steam turbine	Electrical power	40	1.5
Diesel engine	Mechanical power	38	1.63
Automobile engine	Mechanical power	25	3
Steam locomotive	Mechanical power	8	11.5

The effects of exhaust heat on the environment are frequently undesirable. A power plant that discharges its exhaust heat into a river, for example, can upset the sometimes delicate ecological balances and result in the loss of fish species. On a large enough scale, weather phenomena can even be affected by man-made heat sources. As energy production around the globe continues to increase, the possibilities for serious detrimental effects, called *thermal pollution,* also rise.

It is clear that the most efficient possible power sources are called for and thermodynamics teaches us that high efficiency requires high input temperatures. This is one of the reasons for current interest in the possibility of nuclear fusion reactions as a power source. These reactions occur at temperatures of millions of degrees Kelvin, so that theoretical efficiencies could approach 100%. However, even with 100% efficiency at the power source, we must not forget that the most likely end product of "useful" power output is heat. The mechanical energy expended by an automobile is dissipated in friction and appears as heat. The light produced by an electric light is eventually absorbed on some surface and becomes heat, except for the small fraction that escapes into outer space. The energy given to the beam of electrons within a television picture tube becomes heat when the electrons strike the screen. Of course, some of the electron beam's energy goes into "useful" light production but that, too, is absorbed eventually to become heat.

Except for hydroelectric power, which cannot supply all of the needed power, our traditional power sources are "brute force" sources. We expend chemical or nuclear energy, which not only consumes a limited natural resource but upsets the natural energy balance of the earth. Long-range research, accordingly, leans towards more subtle methods of power utilization. The use of solar power is an example. The net flux of solar energy striking the earth is a staggering 5.4×10^{24} J *per year*. This number can be put into perspective when we compare it to the estimate of 10^{22} J for the total power requirements of the earth's population in the 40-year span from 1960 to 2000. The diversion of a fraction of the incident solar flux into mechanical or electrical energy production does not change the overall energy balance of the earth, since this energy eventually appears as heat in either case.

The many schemes currently under consideration to make ecologically balanced uses of natural energy sources cannot be covered here. The interested reader can see, for example, an article titled, "The Flow of Energy in an Industrial Society" by Earl Cook in the September 1971 issue of *Scientific American*.

21-10   Summary

The first law of thermodynamics is a generalization of the law of conservation of energy with the recognition that heat is a form of energy. On a microscopic scale we recognize that this energy is simply the sum of the kinetic and potential energies of atoms in the substance. In thermodynamics we take a large-scale view and call this energy *internal energy*.

$\Delta Q = \Delta W + \Delta U$

An ideal monatomic gas has an internal energy U that is simply the sum of molecular kinetic energies and is equal to $U = \frac{3}{2}nRT$.

$U = \frac{3}{2} nRT$

When a gas expands isothermally from V_1 to V_2, heat is converted to mechanical work in the amount

$$W = nRT \ln \left(\frac{V_2}{V_1}\right)$$

Two molar specific heats, C_p and C_V, are defined for a gas. The specific heat at constant pressure, C_p, is larger because of the work done by the expanding gas. For a monatomic ideal gas,

$$C_V = \tfrac{3}{2}R \qquad C_p = \tfrac{5}{2}R$$

while for any ideal gas,

$$C_p = C_V + R$$

The efficiency of a heat engine is defined to be

$$\text{eff} = \frac{W_{\text{out}}}{Q_{\text{in}}}$$

For a Carnot cycle the efficiency is given by

$$\text{eff} = 1 - \frac{T_2}{T_1}$$

where T_2 is the low (exhaust) temperature and T_1 the high (input) temperature. Any non–Carnot cycle heat engine has an efficiency that is less than that of the Carnot cycle.

For a refrigerator a coefficient of performance, P, can be defined, where

$$P = \frac{Q_{\text{in}}}{W}$$

and for a Carnot refrigerator this is

$$P = \frac{T_1}{T_2 - T_1}$$

No refrigerator can have a larger coefficient of performance than a Carnot refrigerator operating between the same two temperatures.

The second law of thermodynamics concerns the *availability* of thermal energy. In general, no process is possible that produces pure mechanical energy from heat.

Entropy is a measure of the disorder of a system and always tends to increase. The ultimate fate of the universe may be a "heat death," in which entropy is maximized.

Problems

21-1. Two moles of helium gas are heated from 300K to 400K at constant volume. (*a*) Find the quantity of heat required. (*b*) Find the final pressure if the initial absolute pressure is 2 atm. (*c*) What is the volume of the container?

21-2. Two moles of helium gas are heated from 300K to 400K at constant pressure. (*a*) Find the quantity of heat required. (*b*) Find the final volume if the initial volume is 22.4 liters. (*c*) Find the pressure. (*d*) How much work is done by the expanding gas?

21-3. An ideal gas expands isothermally, with $T = 900K$, from V_1 to V_2. To what temperature should the gas be cooled before isothermal compression so that on the return stroke from V_2 to V_1, only half as much work will be done as that produced by the expansion?

21-4. An ideal gas at a temperature of 800K expands isothermally from 0.2 m³ to 20 m³ and does a quantity of work W_1. It is then cooled to 400K and compressed isothermally to 10 m³. This compression requires a quantity of work W_2. Find the ratio W_2/W_1.

21-5. A heat engine operates on the cycle graphed in Fig. 21-2. Its working fluid is 0.05 mole of a monatomic ideal gas. It cycles between $T_1 = 800K$ and $T_2 = 400K$ at 50 cps. The ratio V_2/V_1 equals 7.73. Find its power output in watts and in horsepower.

21-6. A heat engine operating on the cycle of Fig. 21-2 has an exhaust temperature $T_2 = 300K$. The ratio V_2/V_1 equals 5. Find the temperature T_1 required to achieve an efficiency of 60%. How does this compare with the temperature T_1 required for a Carnot engine to achieve 60% efficiency with $T_2 = 300K$?

TRY eff of 40%

21-7.* A heat engine operates in the following cycle with a monatomic ideal gas as its working fluid. (*a*) Heat the gas from T_2 to T_1 at constant volume, V_1. (*b*) Let the gas expand adiabatically to V_2. (*c*) Cool the gas at constant volume, V_2. (*d*) Compress the gas adiabatically to V_1 at T_2. Calculate its efficiency in terms of the given variables.

21-8. A power plant burns coal to produce a high temperature of 1000K. It exhausts heat into a stream that is at 300K. If it produces power at a rate of 1000 MW (10^9 J/s): (*a*) find the minimum possible rate of exhaust-heat rejection, i.e., assume Carnot efficiency. (*b*) How many m³/s of cooling water are then required so that the stream will not be heated by more than 2°C? How does this compare with your estimate of the flow in a small river?

21-9. Suppose the cycle discussed in the example of Section 21-3 is operated in reverse as a refrigerator between the temperatures 270K and 300K. If it operates at 4 cps, find (*a*) the rate of removal of heat from the low-temperature reservoir (in cal/s): (*b*) the power required to operate it (in W).

21-10. Consider the refrigerator discussed in Problem 21-9. If it operates to keep a low-temperature reservoir cool (270K) and a high-temperature reservoir warm (300K), a heat engine can be operated between these high- and low-temperature reservoirs. Consider the heat engine to be a Carnot engine. Compare the power produced by the Carnot engine to that required by the refrigerator in order to keep the Carnot engine running.

21-11. An ordinary kitchen refrigerator operates with a low temperature near 0°C and an exhaust temperature that is that of its cooling coils, which are necessarily above room temperature. Assuming this exhaust temperature to be 40°C, find the ideal (Carnot) coefficient of performance. At what rate must power be supplied to such a refrigerator to remove 400 cal/s from the cold enclosure?

21-12. An apartment dweller decides to turn his window-mounted air conditioner around backwards so that it cools the outside air and warms his apartment. If the outside air is at a temperature of 10°F and the inside air is at 68°F, what is the *minimum* quantity of work required to bring each calorie of heat inside? If this work is supplied at 100% efficiency by electrical energy, what has been gained over converting the electrical energy directly into heat with a 100% efficiency? That is, find the ratio of electrical energy required per calorie for the two methods of heating.

21-13. When 100 g of water at 0°C is mixed with 100 g of water at 100°C, equilibrium is reached when the 200 g of water is at 50°C. Show that the net entropy of the water has increased in this process.

21-14. Calculate the change in entropy of n moles of an ideal (monatomic) gas isothermally expanding at T_0 from V_1 to V_2. Repeat the calculation for the reverse process. What can you say about the overall entropy change of the universe in these processes?

Chapter **22**

Wave Motion

22-1 Introduction Wave motion occurs in a wide variety of physical settings. Examples are sound waves, waves on a stretched violin string, and light waves. Both sound waves and waves on a stretched string are mechanical waves. We can understand their properties on the basis of Newtonian mechanics. Light waves belong to a class of waves called *electromagnetic waves*. In order to describe and understand them fully, we require some background in electromagnetic theory. Still, a great deal can be understood about light once we know it is a wave, regardless of what type of wave it is.

In this and the next chapter we will focus our attention on mechanical waves. The language and techniques we develop in our study of mechanical waves will find application to other types of waves in subsequent chapters. As we will see in Chapter 40, even matter exhibits wavelike properties when it is viewed at the atomic level. In this chapter we will concentrate on describing waves. The next chapter will emphasize an explanation of mechanical wave motion in terms of the fundamental laws of mechanics and thermodynamics.

22-2 Description of Waves

When holding one end of a tightly stretched spring, it is easy to produce a wave by abruptly jerking the end up and then down, as in Fig. 22-1. This wave appears as a "kink" that runs along the length of the spring. Such a traveling disturbance is a wave in the form of a single pulse. If you watch water waves near a coast, you

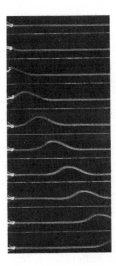

Fig. 22-1. When the end of a stretched spring is jerked abruptly, the disturbance takes time to propagate along the spring. The disturbance propagates at a constant velocity, u, called the *wave velocity*. (From *PSSC Physics*, D. C. Heath and Company, Lexington, Massachusetts, © 1965)

may see a procession of continually undulating waves so that a cork fixed at one location bobs up and down endlessly as the waves pass under it. Such a disturbance is a wave train and can be arbitrarily long. The unifying feature that makes us describe both the single pulse and the periodic undulations as wave phenomena is that they are both disturbances that travel with a well-defined velocity. Mechanical waves typically involve the disturbance of some *medium*—such as a spring or the water surface. The traveling disturbance produces some motion of the medium but the medium itself does not move along at the wave velocity: only the disturbance moves at that velocity. It is for this reason that mechanical waves can transport energy without a net mass transport.

If the motion caused by the passage of a wave is perpendicular to the direction of propagation of the wave itself, the wave is called *transverse*. An example is given by the wave in Fig. 22-2a. In a *longitudinal* wave, the motion caused by the passage of the wave is along the direction of propagation. Such a longitudinal wave in a stretched coil spring is shown in Fig. 22-2b. In two or three dimensions, the directions in which the waves move are called *rays*, while the surfaces of like displacement (such as wave crests) are called *wave fronts*, as shown in Fig. 22-3.

A mathematical description of a traveling wave is rather straightforward. Consider the triangle-shaped pulse shown in Fig. 22-4. This "triangle function," $y = T(\xi)$, is defined to have a value of unity at $\xi = 0$ then falling linearly to zero at both $\xi = +1$ and $\xi = -1$. The letter ξ was chosen as the argument of this function in order to avoid confusion with spatial dimensions. If we wish to describe a *traveling* triangular pulse we can replace the argument ξ of the function by a new argument, $x - ut$, where u is a constant equal to the wave velocity. The peak of the triangle function occurs when its argument is zero, by definition. When this argument is $x - ut$, the peak occurs at larger and larger values of x as t increases, because then $x - ut = 0$ or $x = ut$. Thus, simply replacing ξ in $T(\xi)$ by $x - ut$ generates a triangle wave function that propagates toward positive x, as shown in Fig. 22-5. Similarly, replacement of ξ by $x + ut$ causes this wave to propagate toward negative x.

A function that describes a moving wave will depend upon *both* space and time. At any given instant of time, a wave varies with distance, while at any given point in space, the passage of a wave causes variations with time. Generally speaking, any function $f(\xi)$ can have its argument replaced by $ax - bt$ to produce a motion to the right along the x axis, or by $ax + bt$ to correspond to motion to the left. Similarly, motion along the y or z axes can be described by arguments in the form $(ay \pm bt)$ or by $(az \pm bt)$. To find the velocity u_w of the wave motion along the x axis, we need merely to find the

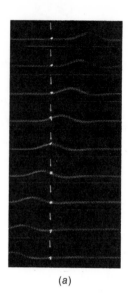

(a)

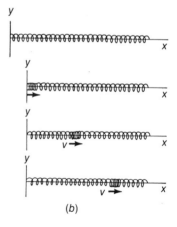

(b)

Fig. 22-2. (a) A transverse wave moving from right to left along a stretched spring with a ribbon tied at one point. The ribbon moves up and down as the wave passes but does not move along with the pulse. (From *PSSC Physics*, D. C. Heath and Company, Lexington, Massachusetts, © 1965) (b) Longitudinal waves moving along a stretched spring.

derivative of x with respect to t when $ax - bt$ is constant, since a fixed value of the argument corresponds to a fixed point on the wave. Thus $ax - bt =$ constant implies that

$$x = \left(\frac{b}{a}\right) t + \frac{\text{constant}}{a}$$

and we find $u_w = dx/dt = b/a$. For the example shown in Fig. 22-5, the value of a is unity and the value of b is 3 m/s.

Exercises

1. Write an argument for a wave function $f(\xi)$ so that it propagates in the negative y direction at 5 m/s.

2. Can you write an expression for the triangle function $T(\xi)$? (You may have to consider the range of ξ in parts.)

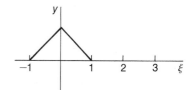

Fig. 22-3. Circular wave fronts spreading out from a point where a stone was dropped into a pond. The rays indicate the direction of motion of the wave fronts.

22-3   Sinusoidal Waves

We now are able to write a sinusoidal function of space and time that will describe a sinusoidal wave train moving with a wave velocity u. The waves we describe might be due to variations in the height of a water surface, the displacement of a stretched string, or the pressure in a sound wave. Whatever the physical quantity of interest is, let us call it $\psi(x,t)$. A sine wave form is given by:

$$\psi(x,t) = A \sin(ax - bt)$$

with $u = b/a$. A more convenient form is given by

$$\psi(x,t) = A \sin 2\pi\left(\frac{x}{\lambda} - \frac{t}{\tau}\right) \tag{22-1}$$

where A is the amplitude of the wave, λ is the wavelength, and τ is the period. The velocity of this wave is found by differentiating x with respect to t when x and t are chosen to maintain a constant point on the wave, that is, a constant value of the argument. Thus, when

$$\frac{t}{\tau} - \frac{x}{\lambda} = \text{constant}$$

we have

$$x = \frac{\lambda t}{\tau} - \lambda \cdot \text{constant}$$

so that

$$u = \frac{dx}{dt} = \frac{\lambda}{\tau} \quad \text{distance} \tag{22-2}$$

$$\text{time}$$

$$\boxed{u = \lambda f}$$

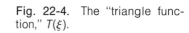

Fig. 22-4. The "triangle function," $T(\xi)$.

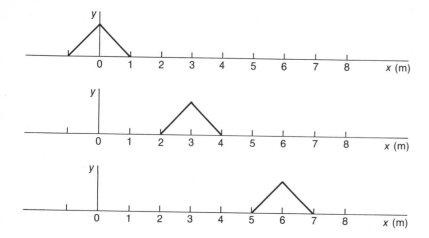

Fig. 22-5. Three successive plots of $T(x - ut)$ vs x are shown for $t = 0$, $t = 1$ s, and $t = 2$ s. The value of u is 3 m/s.

Since λ has the dimension of a length and τ the dimension of time, (22-2) checks dimensionally.

The meanings of the wavelength, λ, and the period, τ, are illustrated in Fig. 22-6. There $\psi(x,t)$ is plotted against x at five different value of t, so that the manner in which ψ depends on time can be visualized. Each of these plots is thus a "snapshot" of the moving wave. Note that the wavelength, λ, is the *distance* over which the wave goes through one full cycle of oscillation in any one of the plots. Note also that the *time* for the wave to cover one wavelength is τ, the period.

An equivalent expression is frequently used in which the factor 2π is absorbed into two constants in the numerator rather than in the denominator:

$$\psi(x,t) = A \sin(kx - \omega t)$$

where k is called the *wave number* and ω the *angular frequency*. Let us now summarize the relationships among the parameters describing sinusoidal waves:

Period: τ

Frequency: $f = 1/\tau$

Angular frequency: $\omega = 2\pi f$

Wavelength: λ

Wave number: $k = 2\pi/\lambda$

Wave velocity: $u = \lambda/\tau = \lambda f = \omega/k$

If we take a snapshot of a sinusoidal wave at a fixed time, we obtain

$$\psi(x) = A \sin(kx - \text{constant})$$

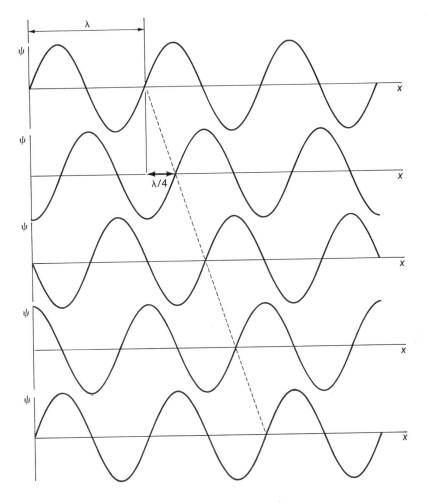

Fig. 22-6. Illustration of the meaning of λ and τ. The successive plots of $\psi(x,t)$ vs x are made at five different values of t, each one $\tau/4$ larger than the previous. A given point on the wave, such as that connected from one graph to the next by a dotted line, advances $\lambda/4$ each time.

If, instead, we observe the wave at a fixed point on the x axis, we obtain

$$\psi(t) = A \sin(\text{constant} - \omega t)$$

which is equivalent to (16-2), the equation for simple harmonic motion. Thus a transverse sinusoidal wave propagating along a stretched string causes any *fixed* point to move perpendicularly to the string with simple harmonic motion.

Exercises

3. Show that $u = \omega/k$.

4. Given that the expression

$$y(x,t) = 2 \text{ m} \cdot \sin\left[2\pi\left(\frac{x}{10 \text{ m}} - \frac{t}{2 \text{ s}}\right)\right]$$

describes some surface water waves moving along the x axis, find (a) their amplitude; (b) their frequency; (c) their wavelength; (d) their wave number; (e) their wave velocity.

22-4 Sound

Sound is a wave that is longitudinal in a gas or liquid and either longitudinal, transverse, or a mixture in a solid. In air, the speed of sound is roughly independent of pressure but it is dependent on temperature. At standard temperature and pressure, the velocity of sound in air is 331 m/s. Table 22-1 gives the velocity of sound in various substances. The increased velocity of sound in air at increased temperature is related to the increased thermal velocity of the molecules. Similarly, the large value for the speed of sound in hydrogen results from the large value of thermal velocities for these light molecules.

SOUND - LONGITUDINAL
GAS

EITHER TRANSVERSE
OR LONGITUDINAL
OR MIXTURE
IN A SOLID

DEPENDENT ON TEMP.

TABLE 22-1   Velocity of Sound

Substance	Temperature (°C)	m/s	ft/s
Air (sea level, STP)	0	331.3	1,087
Air (50 atm pressure)	0	334.7	1,098
Air (sea level)	20	344.0	1,129
Hydrogen (STP)	0	1,269.5	4,165
Water	15	1,450.0	4,760
Copper	20	3,560.0	11,700
Aluminum	20	5,100.0	16,700
Iron	20	5,130.0	16,820

There are three properties that a listener readily attributes to a reasonably steady sound, such as a note produced by a musical instrument. First, the *pitch* of the sound is associated with the frequency of the sound. A low pitch corresponds to a low frequency, a high pitch to a high frequency. It is now cutomary to assign a frequency of 440 Hz to the note A above middle C. Most sounds contain more than one frequency. The *quality* of a tone is related to the particular mixture of frequencies that are present. This mixture varies from one musical instrument to another and allows us to distinguish tones produced by different instruments even when they play the same note. The lowest frequency present in the mixture of frequencies usually determines the pitch that a listener will ascribe to that sound. Finally, the *intensity* or *loudness* of a sound is related to the amplitude of the sound waves.

At the maximum sound intensity that the human ear can tolerate, the pressure swings above and below normal atmospheric pressure by about only 0.03%. At a frequency of 1000 Hz this corresponds to a longitudinal displacement amplitude of the average position of the air molecules of about 0.001 cm. For the faintest sound that can be heard at 1000 Hz frequency, the displacement amplitude is one million times smaller, only 10^{-9} cm. Measurements made with many subjects indicate that over this enormous range of sensitivity of the ear, the sensation of loudness is closely proportional to the logarithm of the amplitude. Thus, each increase by a factor of 10 in the amplitude produces a similar increase in the sensation of sound, with six such steps taking us from the faintest to the loudest sounds.

The *intensity* of sound is conventionally defined by the rate at which it transports energy. The work required to produce one oscillation of a sound wave is proportional to the *pressure excess* of the wave (the excess pressure above the equilibrium value) multiplied by the displacement amplitude of the air (since work = force × distance). Since the displacement and pressure excess are proportional to each another in a sound wave, the energy content of a sound wave is thus proportional to the square of the amplitude. Since this energy is carried along with the wave, the intensity of the wave can be defined in terms of the rate at which it carries energy for each unit area of wavefront. These intensity units are thus power per unit area. In open surroundings, sound waves propagate spherically outward from a spherically symmetric source. Thus the power emitted by the source is spread over larger and larger areas as the wavefront progresses outward. Since the increase in area of spherical wavefronts is proportional to the square of their radius, the intensity of sound decreases as the inverse square of the distance from the source.*

The standard definition of the intensity level, β, in decibels is given by:

$$\beta = 10 \log (I/I_0) \qquad (22\text{-}3)$$

---

*If the sound waves are confined by enclosures, the rate of intensity decrease cannot be so simply described.

where I_0 is a reference intensity taken at 10^{-16} W/cm², which is the lower limit of auditory response. This form of intensity definition is chosen because of the logarithmic response of the ear. The units of β are called *decibels*, abbreviated dB. The *bel* (10 decibels), an inconveniently large unit, was originally chosen in honor of the inventor of the telephone, Alexander Graham Bell (1847–1922). The loudest sounds the ear can tolerate are about 120 dB, which corresponds to 10^{-4} W/cm². At intensities of 80 dB, the ear can hear tones between 20 and 20,000 Hz, while at 20 dB intensity, the response is from 200 to about 15,000 Hz.*

Exercises

5. Sound traveling through an iron rail propagates much faster than in air (Table 22-1). How different are the transit times of sound from a train one mile distant as heard through the air and through the rails?

6. Find the ratio of the power carried by sound waves at the threshold of pain to that carried by the sound of a quiet whisper. (See Table 22-2 and Eq. (22-3).)

22-5 The Doppler Effect

When a source of sound is stationary in air and emitting a pure sinusoidal tone of 1000 Hz, it produces spherical sound waves with a wavelength of just under a foot. Thus, the wave fronts are a concentric series of ever expanding spheres, each differing from the next by one foot in radius. A stationary listener, some distance away, hears a 1000-Hz tone, since 1000 wave fronts impinge upon his ear each second. However, if the listener is moving toward the source, he intercepts more waves per second than are being radiated, that is, he hears a higher pitch. Alternatively, if the source is moved toward a stationary listener, the emitted waves are crowded together in front of the moving source to produce a shorter wavelength. When these shortened waves pass by a stationary observer, he too hears a higher pitch. These changes in pitch brought about by relative movement of the source and observer are called the

TABLE 22-2 Sound Intensities

Sound	Intensity (dB)
Loudest rock "music"	135
Threshold of pain	120
Elevated train	90
Busy street traffic	70
Noisy office	60
Speaking voice at 3 ft	40
Whisper	20
Rustle of leaves	10
Threshold of hearing	0

---

*This reduction of audible frequency range for softer sounds is partially compensated in many high-fidelity sound systems by a "loudness" control, which emphasizes low and high frequencies as the volume is reduced.

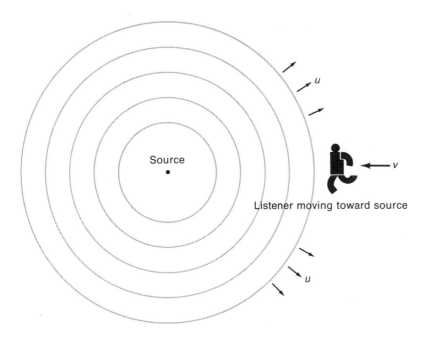

Listener moving toward source

Fig. 22-7. The Doppler effect, in this case due to motion of the listener, is caused by his ear "sweeping up" more cycles per second as he moves toward the source.

*Doppler effect.** In Fig. 22-7 and Fig. 22-8, the causes of the Doppler effect are schematically illustrated. If we restrict our attention to source and observer motion along a line joining them, we can easily calculate the shift in frequency due to the Doppler effect. When the source is stationary and the observer is stationary, he receives $(u/\lambda)t$ waves per time interval t. If he moves toward the source at a velocity v_o, the relative speed of the waves and the observer is increased to $u + v_o$ so that he receives waves at a higher rate. The modified frequency he hears, f', is given by the net number of waves (cycles) received per time interval t:

$$f' = \frac{u + v_o}{\lambda}$$

In terms of the old frequency,

$$f' = \frac{u + v_o}{u/f} = f\left(1 + \frac{v_o}{u}\right)$$

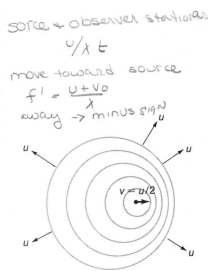

sorce + observer stationary

$u/\lambda t$

move toward source

$f' = \dfrac{u + v_o}{\lambda}$

away → minus sign

$v = u/2$

---

*Christian Johann Doppler (1803–1853) pointed out that the color of a luminous body is changed by relative motion of the source and observer, since light is a wave phenomenon with the color determined by the frequency of the observed light. The name *Doppler effect* is now used for this effect in any type of wave motion.

Fig. 22-8. When the source of sound moves, the sound waves are crowded together on one side, spread out on the other.

If the observer is moving away from the source, there is a minus sign in the quantity in parentheses. Thus, in either case, we find

$$f' = f\left(1 \pm \frac{v_o}{u}\right) \qquad (22\text{-}4)$$

where the plus sign is for motion toward the source and the minus sign for motion away.

When the source is in motion, the sound waves produced in front of it are shortened. The undisturbed wavelength is $\lambda = u\tau = u/f$. But the source moving at velocity v_s "catches up" with the wavefront by an amount $\Delta\lambda = v_s\tau = v_s/f$ in the time between successive cycles. The true wavelength in front of the source is thus

$$\lambda' = \lambda - \Delta\lambda = (u/f) - (v_s/f) = (u - v_s)/f$$

For the change in wavelength behind the source, we get

$$\lambda' = \lambda + \Delta\lambda = (u/f) + (v_s/f) = (u + v_s)/f$$

Putting these results in terms of a modified frequency, we have

$$f' = \frac{u}{\lambda'} = f\left(\frac{u}{u \pm v_s}\right) \qquad (22\text{-}5)$$

where the minus sign is for the source moving toward the observer, the plus sign for motion away from the observer. Combining (22-4) and (22-5) to take into account both source and observer motion with respect to the air we finally obtain:

$$f' = f\left(\frac{u \pm v_o}{u \mp v_s}\right) \qquad (22\text{-}6)$$

There is no need to memorize a sign convention since common sense will determine the correct choice of sign as long as the reader recalls that the pitch is always raised when the observer and the source approach one another.

Example A stationary source emits a 1000-Hz tone and a stationary observer hears that tone. However, a large sheet of plywood is held behind the source so that sound waves originally moving away from the observer are reflected back toward him. If the sheet of plywood is moved toward him at 20 m/s, what frequency does he attribute to the reflected tone?

Solution The reflected wave first strikes the plywood moving toward it, so that the frequency variation of sound waves at the plywood surface is given by (22-4) for a moving observer (the plywood is the "observer"). The plywood is now a moving source of sound waves, so

SOURCE IN MOTION

MINUS → source moving toward observer

plus → source moving away from

that (22-5) is applicable. Both effects raise the frequency when motion of the reflector is toward the observer. The net result in this case is

$$f' = f \left(\frac{u + v_p}{u - v_p} \right)$$

where v_p is the velocity of the plywood. Thus

$$f' = 1000 \text{ Hz} \cdot \left(\frac{344 \text{ m/s} + 20 \text{ m/s}}{344 \text{ m/s} - 20 \text{ m/s}} \right)$$

$$= 1123 \text{ Hz}$$

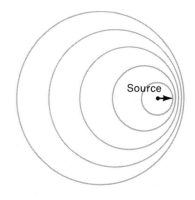

Fig. 22-9. A source moving near the speed of sound.

This effect is utilized in radar measurements of automobile velocities. The speed of the radar waves is so large (3×10^8 m/s) that the fractional shift of frequency caused by the moving car is extremely small. However, the initial frequency of the radar waves is great enough (several thousand megahertz) that the *change* in frequency is several hundred Hz, which is rather easily measured. It does call for a frequency *stability* of the radar transmitter on the order of one part in a billion but only for the brief time interval between emission and reception of the radar waves.

The expression for the observed frequency in (22-6) diverges (tends toward infinity) as the motion of the source towards the observer approaches u, the speed of sound. This is due to the extreme crowding of the wavefronts, shown in Fig. 22-9, as the source velocity approaches the speed of sound. For velocities at and beyond the speed of sound, (22-6) is no longer applicable, as the assumptions made in its derivation are not fulfilled.

Consider the situation shown in Fig. 22-10, when the source velocity exceeds the speed of sound. Since the source — which is at

SONIC

source velocity exceeds the speed of sound

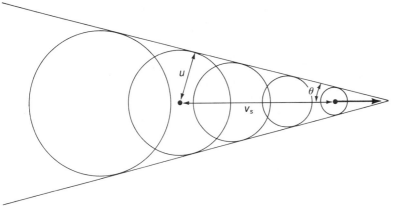

Fig. 22-10. The wave fronts from a source exceeding the speed of sound form a cone. The half-angle of the cone is given by $\sin \theta = u/v_s$.

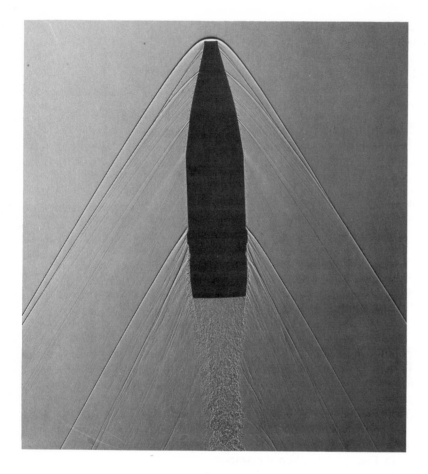

Fig. 22-11. The shock waves from a supersonic projectile are shown in this spark photograph. (Courtesy, U.S. Army Ballistic Research Laboratories, Aberdeen Proving Ground, Maryland 21005)

the center of each successive spherical wavelet—moves faster than the wavefront from the preceding cycle, the successive wavefronts all touch a conical surface that has a half-angle given by

$$\sin \theta = \frac{u}{v_s}$$

When any projectile moves with a velocity exceeding that of sound, it generates a *shock wave* with the conical shape shown in Fig. 22-10. A spark photograph of such a shock wave due to a bullet moving faster than sound is shown in Fig. 22-11. Velocities greater than that of sound are called *supersonic*. Aircraft in supersonic flight generate shock waves that not only affect the aircraft but also produce effects ranging from a mild sound to actual damage as they pass over the ground, depending in intensity upon the size and altitude of the generating aircraft. These effects collectively are called "sonic boom" since the shock wave formed is similar to that formed by an explosion.* Since it is the speed of sound that determines the speed with which an aircraft will enter the supersonic domain, a velocity scale that sets the speed of sound equal to unity is often used by aircraft engineers. Velocities expressed in these units are called *Mach numbers.***

Exercises

7. The fundamental frequency of a train's whistle is 400 Hz. What frequency is heard by a stationary listener if the train approaches him at 60 mi/hr?

8. A source emitting sound with a frequency *f* and an observer are stationary, but a wind blows from the source toward the observer with velocity *w*. Is there a Doppler shift? What frequency does the observer hear?

22-6   Interference

When two waves simultaneously propagate through the same region, they combine to form a *composite wave*. This process is called *constructive interference* when the two waves reinforce each other and *destructive interference* if they cancel one another. For many types of waves, the way in which several waves add together is particularly simple—the net displacement at each point (and at each time) is the algebraic sum of the displacements of all waves present. When this is the case, we say the waves obey a *superposition principle*. The superposition principle holds when the restoring forces in the medium are linearly dependent on displacement, as is the case for most wave phenomena we shall discuss.

Consider two sinusoidal waves of the same frequency traveling together in the same direction. If their crests coincide, we say they are *in phase*. As long as the superposition principle holds, the amplitude produced by the waves in phase is just the sum of the two amplitudes. Waves that are exactly out of phase, with the crest of one falling in the trough of the other, produce a resultant wave with an amplitude equal to the difference of the two wave amplitudes.

---

*See "Sonic Boom" by Herbert A. Wilson, Jr., *Scientific American,* January 1962.

**Named for Ernst Mach (1838–1916), an Austrian physicist and philosopher. See also *Mach's principle* in Section 4-8.

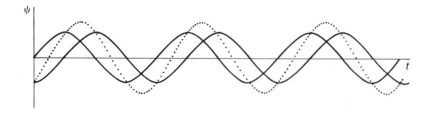

Fig. 22-12. Two sinusoidal waves one-quarter cycle (90°) out of phase are summed to produce the wave shown as the dotted line. Note that this resultant wave has the same frequency as either of the two interfering waves.

When two sinusoidal waves of the same frequency are not exactly in phase or out of phase, the result is not as simple, but the resultant wave is sinusoidal and maintains the original frequency of the two interfering waves. This is shown graphically in Fig. 22-12 for two waves of equal amplitude and frequency that are one-quarter cycle out of phase. The general result for arbitrary phase differences can be obtained analytically as well as graphically.

Consider an observer at a fixed point so that two waves produce time dependences given by:

$$\text{wave 1:} \qquad \psi_1 = A \sin(\omega t)$$

$$\text{wave 2:} \qquad \psi_2 = A \sin(\omega t + \phi)$$

where ϕ is the phase difference. The sum of these waves can be found by use of a trigonometric identity:

$$\sin a + \sin b = 2 \cos\left(\frac{a-b}{2}\right) \sin\left(\frac{a+b}{2}\right) \qquad (22\text{-}7)$$

In our application this identity becomes:

$$\psi_{\text{total}} = \psi_1 + \psi_2 = 2A \cos\left(\frac{\phi}{2}\right) \sin\left(\omega t + \frac{\phi}{2}\right)$$

Thus the resultant wave has a phase halfway between ψ_1 and ψ_2 and an amplitude of $2A \cos(\phi/2)$, which lies between twice the individual amplitudes (when $\phi = 0°$) and zero (when $\phi = 180°$).

When two waves of slightly different frequencies combine, the phenomenon of *beats* occurs. Beats are easily heard when two tones are heard that differ by only a few Hz. The resultant tone is heard to fluctuate in loudness several times per second. This rate of fluctuation is called the *beat frequency*. The cause of its occurrence is easily seen in Fig. 22-13. The two waves are in phase for a time, then out of phase for a time, so their resultant varies in amplitude.

Intuitively we might guess that the rate at which the amplitude fluctuates is related to the difference in frequency between the

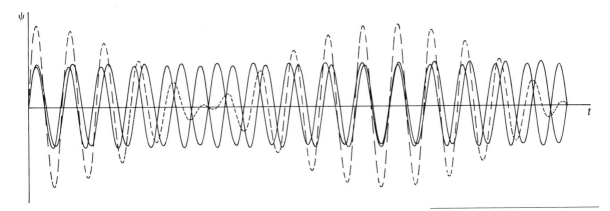

waves — the more similar the frequencies, the longer it takes to vary from in-phase to out-of-phase. We can demonstrate this directly. Consider an observer at a fixed point so that the passage of the two wave trains produces two sinusoidal oscillations with time:

$$\text{wave 1:}\qquad \psi_1(t) = A\sin(\omega_1 t)$$

$$\text{wave 2:}\qquad \psi_2(t) = A\sin(\omega_2 t)$$

where we recall that $\omega = 2\pi f$. Summing these two waves and again using (22-7):

$$\psi_{\text{total}} = A[\sin(\omega_1 t) + \sin(\omega_2 t)]$$

$$= 2A\cos\left[\left(\frac{\omega_1 - \omega_2}{2}\right)t\right]\sin\left[\left(\frac{\omega_1 + \omega_2}{2}\right)t\right]$$

The sine term is a sinusoidal oscillation with a frequency that is the *average* value of the original frequencies. The cosine term is a sinusoidal oscillation at one-half the *difference frequency* of the two waves. If this difference is small, the wave amplitude will be slowly modulated by the cosine term, so that the amplitude continually varies from zero to twice the amplitude of the original waves. Since the cosine term passes through zero twice per complete cycle, the frequency of beats is just equal to the difference frequency, $f_1 - f_2$. Thus, the number of beats per second is just equal to the difference in frequencies, expressed in Hz, of the two waves. The beat phenomenon is used extensively in electronics to produce a more easily amplified low-frequency signal from very high frequency signals. This technique in electronics is called *heterodyning*.

Fig. 22-13. Two waves with slightly differing frequencies (solid curves) interfere constructively at some times, destructively at others. The result (dashed curve) is equivalent to a wave with a frequency intermediate between the initial frequencies and with an *envelope* that varies periodically.

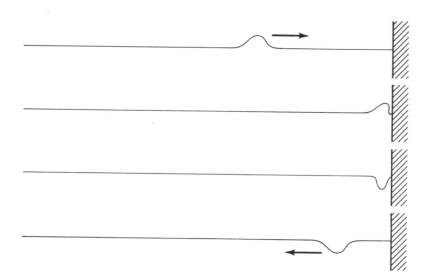

Fig. 22-14. A pulse on a stretched string is inverted upon reflection from a rigid support.

Exercises

9. Two tuning forks are accurately adjusted so that they each produce a tone with a frequency of 440 Hz. A lump of wax is then attached to the tip of one of the forks. It is now observed that three beats per second are heard when the two forks are sounded simultaneously. What are the two possible frequencies for the fork with the wax attached? Which of these two do you suppose is the correct choice?

10. At the end of the example in Section 22-5, it was stated that a small frequency shift in reflected radar waves caused by the Doppler shift can be easily detected. Can you see how the beat phenomenon is used in this application?

22-7 Standing Waves

When a transverse wave on a stretched string encounters a rigid support at the end of the string, it is reflected back along the string as an inverted wave. This inversion and reflection is illustrated in Fig. 22-14 for a single pulse. The physical reasons for this behavior are qualitatively discussed in Section 23-4, but for our present purposes we need only the experimental fact of inversion upon reflection from a fixed point.

If we imagine a long wave train incident upon the clamped end of a stretched string, it is clear that the inverted reflected wave must interfere in some way with the incident wave. The sum of the

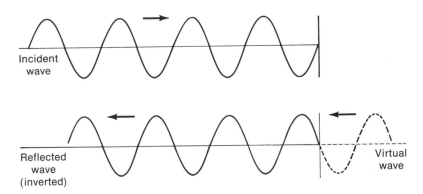

Incident wave

Reflected wave (inverted)

Virtual wave

Fig. 22-15. The relationship of the reflected wave to the incident wave is more readily seen by imagining a "virtual wave" incident upon the end of the string from the other side of the support. The incident wave and reflected wave are shown separately here. In fact, they coexist and interfere with one another, as shown in the next two figures.

incident and reflected wave produces a resultant wave. Since the reflected wave is simply inverted, it is as if an inverted wave appeared out of the wall (or clamp) at the end of the string, as illustrated in Fig. 22-15. The combination of the incident and reflected waves for a long sinusoidal wave train is thus seen to be identical to the sum of two sinusoidal waves propagating in opposite directions. Their resultant sum is indicated at several different times in Fig. 22-16. Note that a stationary point, called a *node*, is produced at the wall, as indeed must be the case since this point is rigidly clamped by the support.

As the incident and reflected waves slide past one another, they sometimes cancel and sometimes add. A succession of nodes spaced every one-half wavelength exists along the string at fixed

Fig. 22-16. The incident wave, ψ_1, and the reflected wave, ψ_2, are shown separately at four successive times, each $\frac{1}{4}$ cycle apart. The resultant wave, $\psi_1 + \psi_2$, is shown below ψ_1 and ψ_2 in each case.

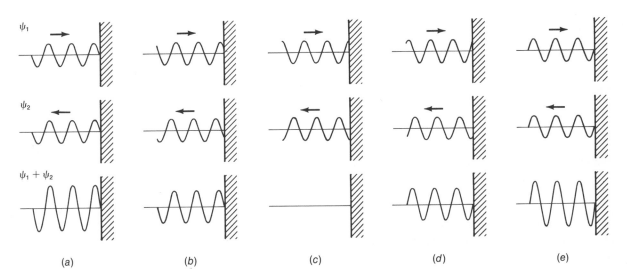

ψ_1

ψ_2

$\psi_1 + \psi_2$

(a)   (b)   (c)   (d)   (e)

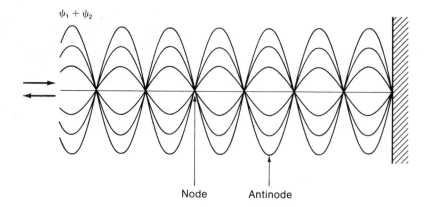

$\psi_1 + \psi_2$

Node   Antinode

Fig. 22-17. A succession of nodes and antinodes can be seen in this time exposure of the combined incident and reflected waves.

points. The time-exposure envelope of the total wave shown in Fig. 22-17 illustrates this point. Since the nodes stay fixed, such resultant waves are called *standing waves*. Each point of the string moves up and down with simple harmonic motion but with an amplitude that depends upon its location along the string.

When the string is clamped at both ends, a new phenomenon occurs: only waves whose half-wavelength is an integral factor of the total string length can exist as standing waves. Any other wavelengths cannot possibly produce nodes at *both* ends of the string simultaneously. But since both ends of the string are, indeed, forced to be stationary, such noninteger wavelengths are not allowed.

CLOSED AT BOTH ENDS

The standing waves produced by the first few allowed wavelengths are shown in Fig. 22-18. Since an integral number of half-wavelengths must fit into the length L, it is easy to deduce the allowed wavelengths λ:

$$n(\lambda/2) = L \qquad (n = 1, 2, 3, \ldots)$$

Thus

$$\lambda = \frac{2L}{n} \qquad (n = 1, 2, 3, \ldots) \qquad (22\text{-}8)$$

The allowed frequencies f can be computed from the allowed wavelengths if we know the wave velocity u in the string:

$$f = \frac{u}{\lambda} = \frac{nu}{2L} \qquad (n = 1, 2, 3, \ldots) \qquad (22\text{-}9)$$

The lowest allowed frequency is given by $n = 1$ in (22-9). This frequency is called the *fundamental frequency*. The various higher

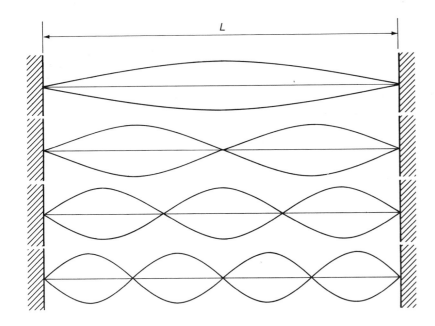

L

Fig. 22-18. When a string is clamped at points a distance L apart, only certain wavelengths can exist as standing waves on the string. The allowed wavelengths are given by $L = n(\lambda/2)$ where $n = 1, 2, 3, \ldots$.

allowed frequencies are called *harmonics*, the second harmonic corresponding to $n = 2$, the third harmonic to $n = 3$, and so on.*

Strings used in musical instruments can have their fundamental frequencies adjusted in several ways:

1. by changing their length, as when a guitarist presses against a string at various distances from a clamped end of the string;

2. by changing their tension—the initial tuning of a guitar or violin is accomplished by adjusting the string tension; this changes u and hence f;

3. by adjusting the mass per unit length of the string. The strings for the lowest notes on a piano are wound with wire to give them a large mass per unit length. This also changes u and hence f.

When two strings are adjusted so that one has a fundamental frequency exactly twice that of the other, the higher-frequency string has *all* of its harmonics coinciding with harmonics of the lower-frequency string. The resultant sound is pleasant (harmonious) and

---

*The "first harmonic" is the fundamental frequency. The frequencies above the fundamental one are also called *overtones*, with the second harmonic then called the *first* overtone, the third harmonic the second overtone, and so on.

the two strings are said to be one *octave* apart. Certain other ratios of fundamental frequencies produce a large number of coinciding harmonics. An example is the frequency ratio 6:5:4, known by musicians as the "perfect major triad chord."

OCT AVes
$f_2 = 2f_1$

Exercises

11. A string is stretched between two supports that are $\frac{1}{2}$ m apart. What is the wavelength of sinusoidal waves that produce the fourth harmonic of this string?

12. A string with a wave velocity of 40 m/s is stretched between supports $\frac{1}{2}$ m apart. Find the string's fundamental frequency.

13. When the string of Fig. 22-16 is in configuration *c*, where is all the energy? Put in another way, how does the string once again become displaced, as in Fig. 22-16*d*?

22-8 Summary

Mechanical waves consist of either longitudinal or transverse disturbances propagating with a well-defined velocity in a medium. A convenient mathematical description of a disturbance that moves with a fixed velocity is given by making $(ax \pm bt)$ the argument of the function that describes the shape of the disturbance. The plus sign in the above expression causes the disturbance to move to the left, while the minus sign corresponds to motion to the right.

Sinusoidal waves describe an endless wave train that, at any fixed location, produces simple harmonic motion of the medium. A convenient description of sinusoidal waves is given by

$$\psi = A \sin 2\pi\left(\frac{x}{\lambda} - \frac{t}{\tau}\right)$$

or by

$$\psi = A \sin(kx - \omega t)$$

where τ is the period, λ the wave length, ω the angular frequency, and k the wave number. The wave velocity is found to be

$$u = \frac{\lambda}{\tau} = \frac{\omega}{k}$$

The frequency is given by

$$f = \frac{1}{\tau} = \frac{\omega}{2\pi}$$

Sound in air is a longitudinal wave disturbance. Though we generally think of sound propagating through air, similar compressional waves can propagate through liquids and solids as well. The intensity of a sound wave is defined in terms of the rate at which it carries energy per unit area of its wavefront. Spherical sound waves diminish in intensity as the inverse square of the distance from their source. The conventional scale of sound intensity is logarithmic because the response of the human ear is very nearly logarithmic. The units of this scale, decibels, are defined by:

$$\beta = 10 \log (I/I_0)$$

where I_0 is a reference intensity of 10^{-16} W/cm^2, which is just perceptible to the average listener.

The Doppler shift is a change in frequency caused by motion with respect to the medium of either the source of sound or the listener. When the source moves, the effect is caused by a crowding of wavefronts in front of the source and a spreading out behind the source. When the listener moves, the effect is caused by his intercepting more waves per second if he moves toward the source, or less if he moves away. The frequency shift produced with either or both types of motion is given by

$$f' = f\left(\frac{u \pm v_o}{u \mp v_s}\right)$$

where u is the velocity of sound, v_o the velocity of the observer, and v_s the velocity of the source; all three with respect to the medium.

A conical shock wave is produced by any object moving through air with a velocity greater than that of sound. Such a shock wave is produced by supersonic aircraft and is called sonic boom. The half-angle, θ, of the conical shock wave is given by $\sin \theta = u/v$, where u is the velocity of sound and v the velocity of the moving object.

Interference refers to the effects caused by the simultaneous presence of two or more waves. For waves traveling together with the same frequency, interference can be constructive, destructive, or anywhere in between, depending on their phase difference. In any case, the resultant wave has the same frequency as the interfering waves. When two waves differ in frequency, beats are produced with a frequency equal to the difference in frequency of the two interfering waves.

When a wave on a stretched string is reflected by a rigid support, the wave is inverted. The sum of a sinusoidal wave with its own reflection produces a standing wave pattern with nodes spaced every half-wavelength from the fixed support. If both ends of a string are clamped, only certain distinct wavelengths (and frequencies) can exist as standing waves. The allowed wavelengths and frequencies on a string of length L are:

$$\lambda = 2L/n \qquad (n = 1, 2, 3, \ldots)$$
$$f = nu/2L \qquad (n = 1, 2, 3, \ldots)$$

$u = \lambda f$

$u/\lambda = f$

$\dfrac{u}{f} = \lambda$

Problems

22-1. Graph the function $\psi = f(\xi)$ between $\xi = -1$ and $\xi = +2$ where $f = 0$ for $\xi < 0$ and for $\xi > 1$, $f = 1$ for $0 \leqslant \xi \leqslant 1$. Replace ξ by $5t - 2x$ and make graphs of ψ vs x for $t = 0$, 1, 2.

22-2. A sinusoidal sound wave has a frequency of 1000 Hz in air at STP. Find its wavelength (in feet), wave number, and angular frequency.

22-3. Find the fractional change in wavelength, $\Delta\lambda/\lambda$, when a 1000-Hz sound wave passes from air at STP into air at 20°C.

22-4. Consider a loudspeaker radiating sound energy in all directions (spherical wavefronts) at a rate of 100 W. At what distance from the loudspeaker is the sound intensity 120 dB?

22-5. A sound source produces spherical wavefronts. At a distance of 10 m from the source, the sound intensity is 10 dB. At what distance is the sound just barely audible?

22-6. A jet engine makes a whine of 12,000 Hz fundamental frequency. It approaches a stationary observer at 400 mi/hr, passes close by him, and continues away from him, always moving along a straight line. (*a*) Qualitatively describe the way the observed pitch varies as the jet passes. (*b*) What are the fundamental frequencies heard by the observer when the engine is far away and approaching him and when it is far away and receding?

22-7. A 10,000-Hz sound source is driven back and forth along a straight line that points at a stationary observer. The source motion along this line is simple harmonic with an amplitude of 80 cm and a frequency of 10 cps. Sketch a graph of the frequencies heard by the observer over the span of 0.5 s. Assume $u = 340$ m/s.

22-8. A jet aircraft at an altitude of 30,000 ft flies straight and level over an observer with a speed of Mach 1.5. Assuming a constant value of 10^3 ft/s for the speed of sound, (*a*) where is the aircraft when the sonic boom strikes the observer? (*b*) How long after the aircraft is directly overhead does this occur?

22-9. One of the reasons for the swept-back wings of a jet aircraft is to prevent the shock wave formed by supersonic flight from crossing over the wing surface. The apex of the V formed by the swept wings of a given supersonic aircraft forms an angle of 120° (half-angle = 60°). What do you suppose is the maximum velocity for which it is designed at an altitude where the speed of sound is 640 mi/hr?

22-10. A wave is produced by the sum of two waves of the same frequency that have amplitudes A and $2A$. (*a*) What phase difference between them produces the maximum amplitude? (*b*) What phase difference produces the minimum amplitude? (*c*) What are these maximum and minimum amplitudes? (*d*) At a phase difference of 60°, what is the resultant amplitude?

22-11. A string 30 cm long has a wave propagation velocity of 264 m/s. Find its fundamental frequency.

22-12. Two strings are tuned one octave apart in fundamental frequency. Which harmonics of the lower-frequency string coincide with the harmonics (and fundamental frequency) of the high-frequency string?

22-13. An automobile is fitted with an oscillator producing a standard frequency f_0. A sending station is located on a highway and has a whistle that also produces this same sound frequency, f_0. By mixing the Doppler-shifted frequency received by the approaching car with the standard frequency within the car, beats are produced. (*a*) Find the relation between the number of beats per second and the automobile's velocity. (*b*) Show how the *distance* covered by the automobile is related to the total number of beats counted. Can you see how this method can be applied to aircraft navigational control by replacing the sound waves with radio waves?

22-14. A small radio loudspeaker is placed over an air column of adjustable length L, as shown in Fig. 22-19. The sound waves reflect from the water surface and return to the loudspeaker in phase, out of phase, or at some intermediate phase, depending upon the value of L. When the loudspeaker is driven at a steady frequency of 1000 Hz, an

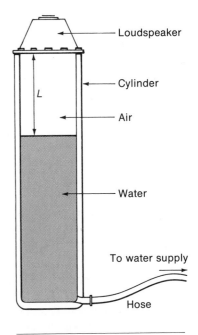

Fig. 22-19. The length L of the air column is varied by changing the depth of water in the cylinder.

observer hears a minimum intensity of sound for $L = 23$, 39, 56, 73, 89.5, and 107 cm. These values are not all separated by exactly the same interval because of the difficulty of distinguishing the exact position of each minimum. Using these data determine an average experimental value for the wavelength of the sound waves in the air column. Calculate the velocity of sound from this result.

Chapter **23**

Wave Dynamics

SECTIONS 1-4

23-1 Introduction In the preceding chapter we found ways to describe waves and discussed the behavior of sound waves and waves on stretched strings. In the first portion of this chapter we will show how the velocity of wave propagation on a taut string can be predicted from first principles (Newton's laws). The additional insight gained in this manner is not only intellectually appealing, it allows us to understand which variables are important in determining the behavior of waves. Such knowledge is useful both in our understanding of the wave phenomena that occur in nature and in the design of instruments that utilize wave phenomena.

23-2 Wave Velocity on a Stretched String

It is an experimental fact that if a tightly stretched string is disturbed in a direction perpendicular to its length this disturbance will propagate rapidly along the string as a transverse wave. The initial disturbance can be created by abruptly displacing one end of the string, as is done when we jerk one end of a stretched garden hose to produce a wave on the hose. Alternatively, a segment of the string can be set into transverse motion abruptly with little or no initial displacement. Such is the case when a piano string is struck by the hammer that is driven from the keyboard of the instrument. Still other forms of excitation of transverse waves on a string are plucking and bowing the string of a guitar or violin. Whatever the method of excitation, these transverse waves move away from the point of excitation at a well-defined velocity. What determines

TRANSVERSE WAVE
DISTURBANCE IN A
PERPENDICULAR DIRECTION
TO IT'S LENGTH

this velocity? We will begin to answer this question by qualitatively examining the interplay between the forces of tension within the string and the string's inertia. We will see that this interplay determines the wave velocity.

In Fig. 23-1 a stretched string is sketched at successive times after it is struck at a point. The distortion of the string causes the tension forces that act along the direction of the string to exert vertically directed forces. The forces produce accelerations that then lead to displacements in the vertical direction. As can be

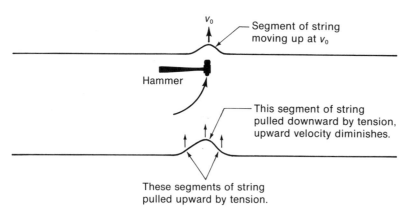

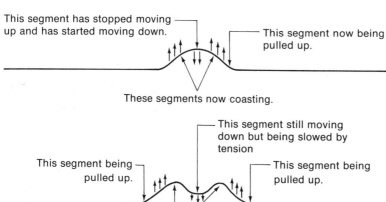

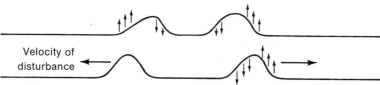

Fig. 23-1. A hammer blow produces an upward velocity of a segment of a tightly stretched string. The curvature produced near this segment of string causes the tension forces to slow the upward motion of the central segment and start adjacent portions of the string into motion. The arrows indicate directions and magnitudes of the velocity of various portions of the string. The final result is a pair of disturbances propagating away from the impact point.

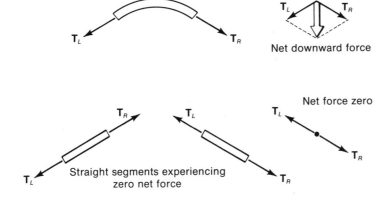

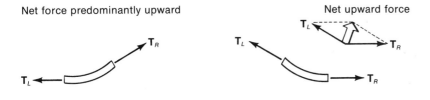

Net downward force

Net force zero

Straight segments experiencing
zero net force

Net force predominantly upward

Net upward force

Fig. 23-2. Three regions of interest are shown corresponding to the situation shown third from the top in Fig. 23-1. The effect of tension in the curved string segments is to produce a net force toward the center of curvature of that segment. Straight segments, even if inclined from their normal orientation, experience zero net force.

seen in the lower portion of Fig. 23-1, these transverse motions lead to longitudinal propagation of the disturbance, that is, propagation along the length of the string. Note, however, that the matter composing the string does not move longitudinally; only the disturbance does. Segments of the string itself move only in the transverse direction.

It is the tension within the string, coupled with distortion of the string's shape, that causes the transverse accelerations. The way the forces producing those accelerations come about is indicated in Fig. 23-2. If the string is not too severely distorted, the tension remains constant in value throughout the string. Thus, if a string segment is not curved but merely tipped from its normal orientation, the forces due to tension simply cancel. On the other hand, the curved segments of string experience a net transverse force. The more sharply curved a segment is, the greater the net force upon it.

A mathematically simple calculation of the wave velocity in a stretched string can be carried out by viewing the traveling pulse from the particular inertial frame of reference that is moving along with a single pulse. In this frame the rope appears to stream past the observer at the wave velocity, u. We do not know the magnitude of u, but hope to deduce it. Since the mass of the rope follows a curved path as it streams through the region of the pulse, centripetal forces must be supplied by components of the tension. One segment of the rope, shown shaded, is momentarily isolated in Fig. 23-3. This small segment has a radius of curvature R and contains a quantity of mass m given by:

$$m = \mu s = \mu R \theta \qquad (23\text{-}1)$$

where μ is the mass per unit length of the rope and s is the length of the shaded segment. The tensional forces at the two ends of the segment cancel in the horizontal direction but add vertically to give a downward directed force

$$F = 2T \sin \frac{\theta}{2}$$

For small angles we may make the approximation

$$\sin \frac{\theta}{2} \approx \frac{\theta}{2}$$

In this approximation F becomes

$$F = T\theta \qquad (23\text{-}2)$$

This transverse force, **F**, acts in the direction required to produce the centripetal acceleration that the string experiences as it moves through the region of interest. The required centripetal acceleration a_c for the radius of curvature R is:

$$a_c = u^2/R$$

Applying Newton's second law to the accelerated mass m, we obtain:

$$F = m\frac{u^2}{R}$$

and with m given by (23-1) and F by (23-2), we have

$$T\theta = \mu R \theta \frac{u^2}{R}$$

or

$$T = \mu u^2$$

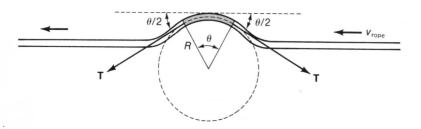

Fig. 23-3. In the frame of reference of the travelling pulse, the rope streams by with $v_{rope} = u$, the wave velocity. The required centripetal force is supplied by components of the tension. In this illustration gravitational forces are small compared to tension.

The cancellation of both R and θ in the above expression indicates that all segments of the string with any radius of curvature whatsoever simultaneously satisfy Newton's second law at this particular velocity, u. Solving for u:

$$u = \sqrt{T/\mu} \qquad (23\text{-}3)$$

Thus, in a stationary frame of reference, we see all portions of the pulse propagating with a single velocity that is determined only by the tension in the string and its mass per unit length. The dimensions of the right side of (23-3) are easily shown to be those of a velocity:

$$\sqrt{T/\mu} = (\text{force} \cdot \text{mass}^{-1} \cdot \text{length})^{1/2}$$

and since force equals mass $\times$ acceleration:

$$\sqrt{T/\mu} = (\text{mass} \cdot \text{length} \cdot \text{time}^{-2} \cdot \text{mass}^{-1} \cdot \text{length})^{1/2}$$

$$= \frac{\text{length}}{\text{time}}$$

A complicated pulse shape will have regions with widely differing radii of curvature, but the fact that u given by (23-3) is independent of curvature indicates that all regions of the pulse will propagate with the same velocity, so that the pulse retains its shape as it moves along the string.

Exercises

1. In Fig. 23-3 the segment of string shown shaded has a net downward force on it. There are two regions where the net force on a string segment caused by tension forces has an upward component. Can you see that the required centripetal acceleration in those two regions is also in the upward direction?

2. If μ is given in kg/m and T in N, what will be the units of the velocity given by (23-3)? What choice of units for μ and T leads to a velocity in ft/s?

23-3   Velocity of Sound Waves – compressional waves

Sound waves are compressional waves. Consider, for example, the sound waves made by the vibrating cone of a loudspeaker. When the loudspeaker cone moves outward briefly and rapidly, it compresses the air immediately in front of it; but that compressed air rapidly expands into the next adjacent region farther away from the loudspeaker. In the process, the air in this adjacent region becomes compressed. As shown in the middle portion of Fig. 23-4,

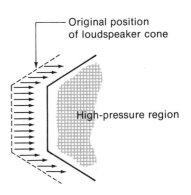

Original position of loudspeaker cone

High-pressure region

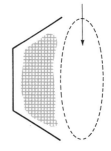

Adjacent air layer now accelerates due to excess pressure acting on its left side

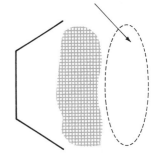

Next air layer now accelerates, and so on

Fig. 23-4.   A brief outward movement of a loudspeaker cone causes a compressional wave to move outward. The speed with which this disturbance can move is limited by the inertia of the air itself.

this compressed air does two things. First, the pressure it exerts on the air ahead of it (to the right in Fig. 23-4) sets that air into motion away from the loudspeaker. Second, the pressure the compressed air exerts on the air behind it (between the speaker cone and the compressed region) brings that moving air to a stop. The process continues, so that a compressional wave is rapidly communicated from one region to the next and moves ever farther away from the source.

Two factors are involved in determining the speed with which this compressional wave will move: the compressibility of the medium and the inertia of the medium. If the medium has low compressibility (corresponding to a large bulk modulus), then a small motion of the medium produces a large pressure excess and, hence, a large value of acceleration of the adjacent layers. Thus we expect a large bulk modulus to lead to high propagation velocities. Secondly, if the density of the medium is low, the inertia of any given layer is small and the wave disturbance propagates rapidly. Conversely, high density leads to low propagation velocities.

To make our discussion more precise, consider the arrangement shown in Fig. 23-5. When the piston is displaced to the right, a compressional wave runs along the tube. The motion of this pulse lends itself to simple analysis because the motion is confined to one dimension. In Fig. 23-6 we view the moving compressional pulse

large BULK MODULUS →
high propagation velocities

high density →
low propagation velocities

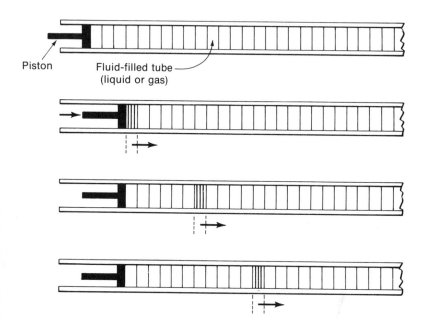

Piston

Fluid-filled tube
(liquid or gas)

Fig. 23-5. A brief displacement of the piston causes a compressional wave to move along the tube. The uniformly spaced lines indicate regions of normal density, while the crowded lines indicate the region of compression.

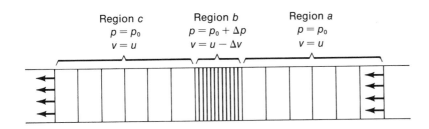

Region c
$p = p_0$
$v = u$

Region b
$p = p_0 + \Delta p$
$v = u - \Delta v$

Region a
$p = p_0$
$v = u$

Fig. 23-6. In the frame of reference of the moving pulse, fluid is streaming from right to left. The excess pressure within the pulse slows fluid entering from the right, but the fluid accelerates as it leaves the left side of the pulse.

from a frame of reference that moves along with the pulse. The fluid moves along at normal pressure and density in region a. It then enters the pulse (region b), which is at higher pressure. The pressure difference between a and b produces a net retarding force on the entering fluid so that it slows down upon entering this region. The compression so produced is indicated by the crowding of lines in region b. Finally, the fluid is reaccelerated to its original velocity as it passes back to normal pressure in region c.

Consider the fluid divided into segments each Δx long. The mass contained in each segment is

$$m = \rho A \Delta x \tag{23-4}$$

where ρ is the density of the undisturbed fluid and A the cross-sectional area of the pipe. As each segment enters region b at velocity u, the fluid slows an amount Δv to $u - \Delta v$ in the time required for the next slab to take its place, namely

$$\Delta t = \frac{\Delta x}{u} \tag{23-5}$$

The net force acting on the slab entering the pulse is produced by the pressure difference Δp between regions a and b:

$$F = \Delta p \cdot A$$

The impulse-momentum theorem tells us that this force acting for the time Δt must produce a change in momentum:

$$\Delta p \cdot A \cdot \Delta t = -m \Delta v$$

where the minus sign indicates that the increased pressure (positive Δp) produces a decrease in velocity. Utilizing (23-4) for m and (23-5) for Δt, we have:

$$\Delta p \cdot A \cdot \frac{\Delta x}{u} = -\rho A \Delta x \Delta v$$

or

$$\Delta p = -\rho u \Delta v \tag{23-6}$$

The change in velocity, Δv, of one of the elements entering region b is simply related to the change in volume, ΔV, caused by the crowding of the fluid in this region:

$$\frac{\Delta v}{u} = \frac{\Delta V}{V} \tag{23-7}$$

Recalling the definition of bulk modulus:

$$B = \frac{-\Delta p}{\Delta V/V} \tag{13-3}$$

we can rewrite (23-6) with the aid of (23-7) and (13-3):

$$\Delta p = \rho u^2 \frac{\Delta p}{B}$$

or

$$u = \sqrt{B/\rho} \tag{23-8}$$

We already guessed that a large value of the bulk modulus leads to high propagation velocities, while a large value of density leads to low propagation velocities. We now see precisely how these factors influence wave velocity.

For sound waves in a gas, the bulk modulus is conveniently expressed in terms of the undisturbed gas pressure p_0. The passage of a sound wave is usually so rapid that the compression is adiabatic, making (20-11) applicable

$$pV^\gamma = \text{constant} \tag{20-11}$$

where γ is the ratio of specific heats C_p/C_V. Differentiating:

$$\frac{dp}{dV} = \text{constant} \cdot (-\gamma)V^{-\gamma-1}$$

A small change Δp thus is related to ΔV by:

$$\Delta p \approx \text{constant} \cdot (-\gamma)[V^{-\gamma-1}]\Delta V \tag{23-9}$$

Dividing (23-9) by (20-11), we obtain:

$$\frac{\Delta p}{p} \approx -\gamma\frac{\Delta V}{V} \tag{23-10}$$

From the definition of the bulk modulus,

$$B = \frac{-\Delta p}{\Delta V/V} \tag{13-3}$$

and (23-10) we see that

$$B \approx \gamma p$$

where p is the undisturbed pressure.

Equation (23-8) for the speed of sound can now be rewritten for a gas:

$$u \approx \sqrt{\gamma p/\rho}$$ (23-11)

$\gamma = c_P/c_V$

Substituting values appropriate to air at STP, we obtain

$$u \approx [(1.4)(1.01 \times 10^5 \text{ N/m}^2)(1.29 \text{ kg/m}^3)^{-1}]^{1/2} = 331 \text{ m/s}$$

which agrees with the measured value of the speed of sound in air at STP.

When the undisturbed pressure of a gas is increased, its density also increases. These two effects offset one another in (23-11), so that the speed of sound is not very strongly dependent upon pressure. In fact, for an ideal gas, the speed of sound is completely independent of pressure. Changes in the temperature, on the other hand, lead to changes in the ratio p/ρ and hence affect the speed of sound. For an ideal gas we have

$$pV = nRT$$

and since the density, ρ, is proportional to $1/V$, we may write

$$p/\rho \propto T$$

which upon substitution into (23-11) gives

$$u \propto \sqrt{T}$$

Example A commercial jet airliner can operate up to Mach 0.9 (9/10 the speed of sound). In winter, at a typical cruising altitude of 35,000 ft, the pressure is 172 mm Hg, the density is 3.92×10^{-1} kg/m³, and the temperature is $-56.3°C$. Find, in mi/hr, the cruising speed that corresponds to Mach 0.9 under these conditions.

Solution The speed of sound under these conditions is

$$u = \left[1.4\left(\frac{172 \text{ mm Hg}}{760 \text{ mm Hg}}\right)(1.01 \times 10^5 \text{ N/m}^2)(3.92 \times 10^{-1} \text{ kg/m}^3)^{-1}\right]^{1/2}$$

$$= 286 \text{ m/s} \cdot \left(\frac{1 \text{ mi}}{1.61 \times 10^3 \text{ m}}\right)\left(\frac{3600 \text{ s}}{1 \text{ hr}}\right) = 639 \text{ mi/hr}$$

At Mach 0.9 we have

$$v = (0.9)(639) \text{ mi/hr} = 575 \text{ mi/hr}$$

Exercises

3. The bulk modulus of water is 2.04×10^9 N/m². Find the velocity of sound in water.

4. When sound propagates along a thin rod, it is Young's modulus that appears in the expression for the velocity of a compressional wave, not the bulk modulus. Do sound waves travel faster or slower along a thin rod of steel when compared to the speed of sound in bulk steel? See Table 13-2.

23-4 Reflections

In Section 22-7 we saw that a wave traveling along a string was inverted upon reflection from a fixed point. We can now gain a qualitative understanding of how this occurs. When a pulse comes to a rigid support, an upward force component is exerted on the support due to the string's tension. By Newton's third law, the support exerts a force with a downward-directed component back upon the string. It is this downward-directed force that produces an *inverted* reflected pulse that then propagates back along the string.

A similar phenomenon occurs when a wave encounters a segment of the string that has an increased density. The reflection is not complete, as indicated in Fig. 23-7, but the reflected portion of the pulse is inverted as for the case of reflection from a fixed point. The coil springs of Fig. 23-7 have lower frictional losses than either rope or heavy string and make ideal "stretched strings."

The action of a free end can be understood by examining the action of a pulse passing from a heavy medium to a light one. In this case the lack of inertia of the new medium into which the pulse is moving causes an excessive upward motion of the boundary, as shown in Fig. 23-8. This upward motion, or *overshoot,* causes a pulse to propagate back along the heavy string as well as onward into the lighter string. For a completely free end, the pulse is completely reflected without inversion. An example of a mechanism for providing a completely free end and yet maintaining tension is indicated in Fig. 23-9.

Reflections also occur when sound waves enter a medium with a density differing from that in which they were propagating. When the new medium has a larger density but similar bulk modulus, a partial, inverted reflection of the displacement wave occurs. When the new medium is less dense, the reflected displacement wave retains its original polarity. Such partial reflections are used in devices that probe otherwise undetectable features. Examples include the use of sonic echoes for the location of submarines, tumors in the human body, and oil-bearing geological formations.

For sound waves, situations analogous to fixed and free ends of a stretched string are produced at the closed and open ends of organ pipes. At closed ends the displacement produced by a sound

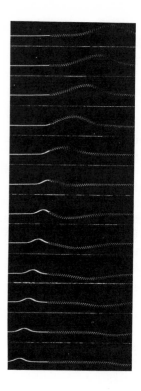

Fig. 23-7. A pulse moves from right to left along a spring. When a heavier segment of the spring is encountered, as in the fourth frame, a smaller upward motion is produced due to the greater inertia of the heavy spring. A sharp angle develops at the boundary so that a large downward force component acts upon the light spring to cause an inverted reflection. (From *PSSC Physics,* D. C. Heath and Company, Lexington, Massachusetts, © 1965) .

wave vanishes and a reflection with inversion occurs. At an open end a large displacement occurs and the wave is reflected without inversion.

Standing waves within organ pipes are caused by reflections at either the closed or open ends, just as standing waves on a stretched string result from a superposition of incident waves and the waves reflected at a fixed end. When an airstream is blown over the knife edge at the bottom end of one of these pipes, it excites the natural frequencies in that pipe. Once the oscillations within the pipe are established, it is easy to see that the pressure oscillations at the base of the pipe can cause the airstream to pass alternately outside the knife edge, then inside the pipe. Energy is thus transferred from the fluttering, moving airstream into sound energy within the pipe. The start-up, or *transient,* portion of operation when the air is first admitted to the pipe is not easily analyzed and gives the sound of a pipe organ a character not readily imitated by electronic means.

From the illustrations in Fig. 23-10, it is readily seen that the fundamental frequency for an open-ended pipe corresponds to one-half wavelength contained within the pipe. Successive harmonics each consist of one additional half-wavelength within the pipe. For a closed-end pipe, the fundamental frequency occurs when one-fourth wavelength is contained in the pipe with successive harmonics corresponding to additional half-wavelengths within the pipe. The absence of even harmonics in the closed pipe gives its sound a character different from that of the open-ended pipe.

For open-ended pipes:

$$n\frac{\lambda}{2} = L \qquad (n = 1, 2, 3, \ldots)$$

hence

$$f = \frac{u}{\lambda} = \frac{nu}{2L} \qquad (n = 1, 2, 3, \ldots)$$

Open - ended pipes   reflection without inversion

½ wavelength contained within the pipe

Post

Slip-ring

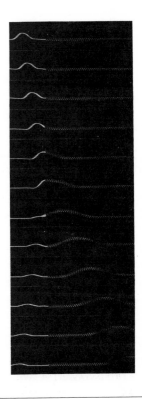

Fig. 23-8. As a pulse enters a lighter spring, it "overshoots" because of the smaller inertia of the new medium. This causes a pulse of the same sign to propagate in both directions away from the junction of the two springs. (From *PSSC Physics,* D. C. Heath and Company, Lexington, Massachusetts, © 1965)

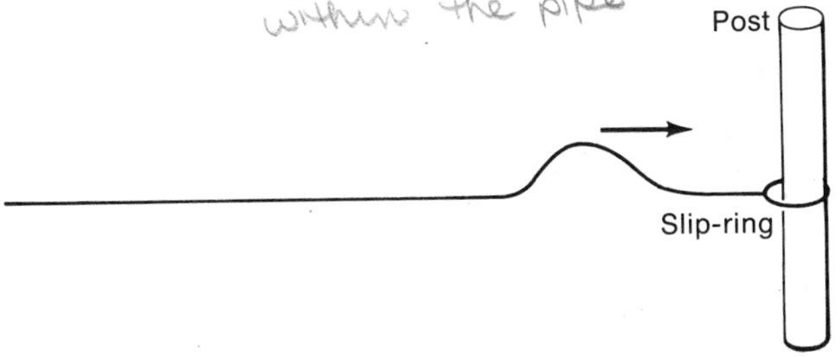

Fig. 23-9. A free end for a stretched string can be provided by a slip ring with negligible mass and friction.

For closed-end pipes: *reflection with inversion*

$$n\frac{\lambda}{4} = L \qquad (n = 1, 3, 5, \ldots)$$

hence

$$f = \frac{u}{\lambda} = \frac{nu}{4L} \qquad (n = 1, 3, 5, \ldots)$$

CLOSED END PIPES
ONE FOURTH WAVELENGTH IS
CONTAINED IN THE PIPE

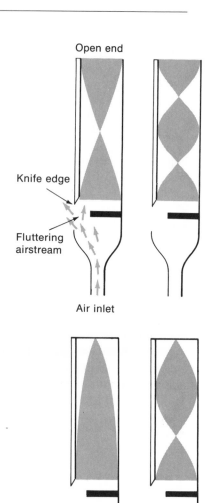

Open end

Knife edge

Fluttering
airstream

Air inlet

Exercises

5. Consider sound waves propagating through a medium of constant density. If they encounter a region of larger bulk modulus, is there a reflection? If so, is the reflected displacement wave inverted?

6. Estimate the fundamental frequency produced by blowing over the open top of a soda pop bottle by considering it to be a closed-end organ pipe.

23-5 Complex Waveforms

In Fig. 23-11 we see the sound waveform produced by a violin tuned to a concert A, which corresponds to a fundamental frequency of 440 Hz. Close inspection of the figure reveals a periodic structure that repeats every 1/440 of a second, yet apparently many higher frequencies are present simultaneously. We do not have to look far for an explanation of the presence of these higher frequencies. The violin produces sound waves by virtue of standing waves set up on a stretched string.* The allowed wavelengths for a stretched string of length L are given by

$$\lambda = \frac{2L}{n} \qquad (n = 1, 2, 3, \ldots)$$

and the allowed frequencies by

$$f = \frac{nu}{2L} \qquad (n = 1, 2, 3, \ldots)$$

where u is the propagation velocity in the string. Besides the funda-

---

*These vibrations are coupled to the wooden body of the instrument and to the air contained in the body. The end result is a reasonably efficient coupling of the vibration to the surrounding air. The interested reader can refer to "Acoustical Measurement of Violins" by Carleen M. Hutchins and Francis L. Fielding, *Physics Today*, July 1968, p. 35; or to "The Physics of Violins" by Carleen M. Hutchins, *Scientific American*, November 1962.

Fig. 23-10. The standing waves in an open-ended pipe show a maximum displacement at the open end as well as at the air inlet end. The closed-end pipes exhibit a displacement node at the closed end. The width of the shaded areas within the pipes indicate the magnitude of displacement oscillation at each point along the pipe.

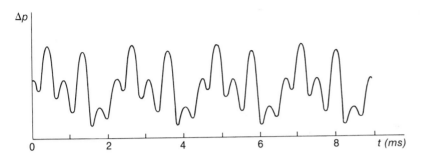

Fig. 23-11. Pressure variations due to the sound of a violin playing a concert A. Though the fundamental frequency is 440 Hz, many higher frequencies appear to be present as well.

mental frequency ($n = 1$), the various harmonics ($n = 2, 3, \ldots$) are excited by bowing the string. The simultaneous presence of various harmonics produces the richness of tone of the violin and a rather complicated-looking waveform.

The nature of the violin waveform can be presented as in Fig. 23-11 or by the bar graph shown in Fig. 23-12, where the height of each bar indicates the strength of each harmonic present. This type of bar graph is called a *spectrum* of the violin tone. The presence of strong second and fifth harmonics in the violin tone's spectrum is partly responsible for the particular quality of the sound of the violin.

The combination of many pure sine waves (the harmonics) in the violin's tone produces a rather complicated waveform. Can waveforms of even less sinusoidal appearance be composed of many sine waves? Can *any* waveform be synthesized by a sufficiently large sum of harmonic series of sinusoidal waves? In fact, it was discovered by J. Fourier that any periodic waveform can be represented by harmonic series of sine and cosine terms.* Such a series

*Jean Baptiste Joseph Fourier (1786–1830) was a French mathematician, teacher, statesman, and scientist. Today we would call him a mathematical physicist. His brilliance was noted outside the academic field as well. He accompanied Napoleon Bonaparte to Egypt in 1798 and became governor of Lower Egypt.

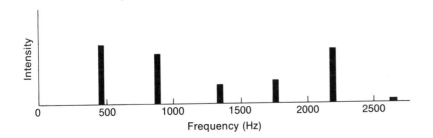

Fig. 23-12. A graph of the intensity of the various harmonics present in the complex waveform of Fig. 23-11. Note the strong fundamental tone at 440 Hz.

is called a *Fourier series*. An example is shown in Fig. 23-13, where

$$x = \sin t + \tfrac{1}{3}\sin(3t) + \tfrac{1}{5}\sin(5t) + \tfrac{1}{7}\sin(7t) + \tfrac{1}{9}\sin(9t)$$

Successive odd harmonics were added together with their amplitudes and phases chosen to produce a waveform that approximates a flat-topped wave.* In Fig. 23-14 the effect of adding each successive term in the Fourier series for a sawtooth function is shown as far as the seventh term,

$$x = \tfrac{1}{2}\pi - \sin t - \tfrac{1}{2}\sin(2t) - \tfrac{1}{3}\sin(3t) - \tfrac{1}{5}\sin(5t)$$

$$- \tfrac{1}{7}\sin(7t) - \tfrac{1}{9}\sin(9t)$$

The strength of various harmonics present in the tones of sev-

---

*The procedure for calculating the coefficients multiplying the harmonics is well defined for any given function. It is, however, a somewhat advanced mathematical topic outside our present interests.

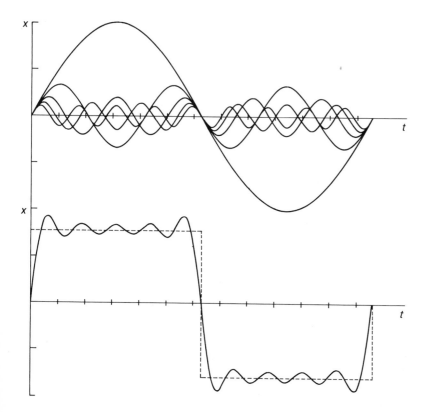

Fig. 23-13. The first five terms of the Fourier series for a square wave are shown in the upper portion of the figure. Their sum gives the result below.

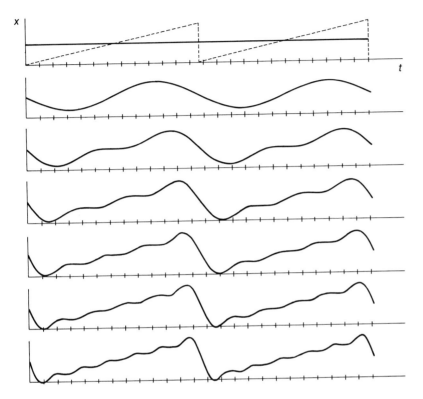

Fig. 23-14. An illustration of the effect of adding each successive term for the first seven terms of the Fourier series for the sawtooth function (shown as a dashed line).

eral musical instruments is shown in the spectra of Fig. 23-15.* Electronic music, such as that produced by a Moog Synthesizer, can be made by combining various harmonics by electronic means, then using these electronic waveforms to drive an amplifier and loudspeaker system.

The idea of harmonic analysis pervades a great deal of modern science, technology, and mathematics. This is due in part to the fact that the response of many mechanical or electrical systems to sinusoidal oscillations is easily measured in the laboratory *and* easily calculated mathematically. Since *any* periodic disturbance can be represented by a combination of these simple sinusoidal oscillations, the response of the system is easily predicted. One example is in the specifications for a high-fidelity sound-reproducing system.

---

*The strength of the various harmonics depends somewhat on which note is sounded. Further, different harmonics die away at differing rates. The more interested reader can consult John Backus, *The Acoustical Foundation of Music*, New York: W. W. Norton and Company, Inc. (1969).

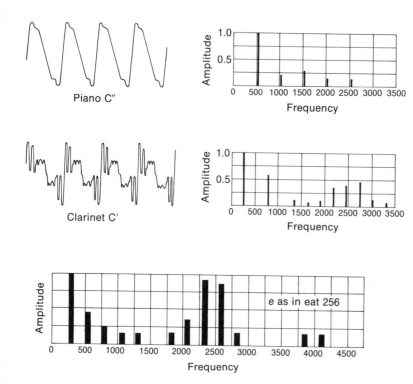

Fig. 23-15. The harmonic analysis (spectrum) of several sounds. (Adapted from Weber, R. L., Manning, K. V., and White, M., *College Physics,* 4th ed., McGraw-Hill, New York, 1965)

We could attempt to catalog the response of such a system to all the various waveforms produced by clarinets, drums, oboes, the human voice, etc. Instead, we measure the response of the system to a finite number of low-, medium-, and high-frequency sinusoidal waves, as illustrated in Fig. 23-16. Since the response is uniform over the range of audibility, the theorem of Fourier *assures* us of a faithful response to any audible complex waveform. If the relative phases of a combination of harmonics are altered, the shape of the complex waveform that they produce is also altered. However, the mechanism of human hearing is such that variations of phase have very little effect on the sound that is perceived. Thus for audible

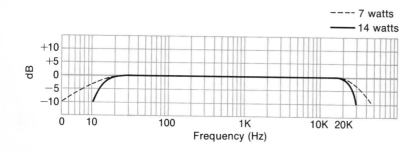

Fig. 23-16. The response of a high-fidelity amplifier to various frequencies.

sounds, the spectra in Fig. 23-15 are more significant than the waveforms.

Exercises

7. Sketch a combination of fundamental and third harmonics that gives a close fit to the triangle waveform in Fig. 23-17. Can you see how to add a bit of fifth harmonic to improve the result?

8. By inspection of Fig. 23-18, can you *estimate* the lowest-frequency and highest-frequency terms needed in a harmonic analysis of $f(t)$?

23-6   Attenuation and Dispersion

Mechanical waves, such as those propagating along a string, transport energy in the form of the kinetic energy of moving segments of the string and in the form of potential energy of distorted portions of the string. However, some of this energy is gradually dissipated by internal friction in the string and by air resistance. This causes the wave to become *attenuated* (weakened) as it propagates. Furthermore, the degree of attenuation may depend on frequency: high-frequency mechanical waves generally dissipate their energy more rapidly than low frequencies. Thus a complex waveform that contains many frequency components will gradually change its shape as it propagates, due to the more rapid loss of high-frequency components. An example is given in Fig. 23-19. The high-frequency

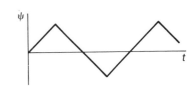

Fig. 23-17.

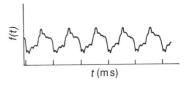

Fig. 23-18.

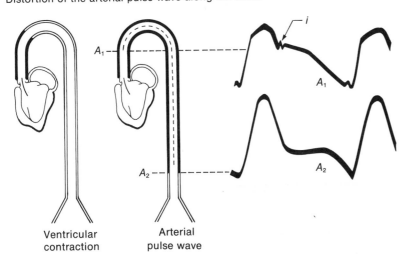

Distortion of the arterial pulse wave along the aorta

Ventricular contraction

Arterial pulse wave

Fig. 23-19. The waveform of the blood-pressure pulses produced by the pumping action of the heart is modified as it propagates through the arteries. Waveform A_1 is obtained in the aorta (near the heart) while waveform A_2 is obtained in the femoral artery (in the groin). Notice the loss of high-frequency components, labeled i in part A, due to their rapid attenuation. A further cause of change in the waveform is the reflection back from the smaller peripheral arteries. (Adapted from Robert F. Rushmer, *Cardiovascular Dynamics*, W. B. Saunders Co., Philadelphia, 1970)

components of the arterial blood-pressure pulse shown there are caused by the sudden closing of the valve between the heart and aorta. Note that these sharp features are absent some distance downstream.

Preferential attenuation of high frequencies is not the only cause of waveform modification. In some cases waves of differing frequencies travel at different velocities. This phenomenon is called *dispersion*. A pure sinusoidal wave travels at a unique velocity in a dispersive medium. It also retains its sinusoidal shape as it propagates. However, complex waveforms consist of many different frequency components and these components move at different speeds when dispersion occurs. Thus, the relative phase of the differing components changes continually as they move along. An illustration of the way dispersion can modify the waveform

$$y = \sin x - \tfrac{1}{9}\sin(3x) + \tfrac{1}{25}\sin(5x)$$

is shown in Fig. 23-20. There the low-frequency components of the complex wave continually fall behind, thus changing phase with respect to the higher frequencies and distorting the total waveform.

Since surface water waves exhibit dispersion, their waveform is continually changing as they propagate. At first glance, it seems to most observers that the bow waves from a boat move outward in an orderly way without change in shape. A careful examination, however, shows that each wave within this group of waves appears to last only a short while, gradually disappearing, only to reappear in another position. This complex behavior is due to the differing speeds of the differing frequency components. The net effect is to cause the entire *group* of bow waves to move with a well-defined velocity. This velocity is called the *group velocity*, while the velocity of individual components of the wave is called the *phase velocity*.

23-7   Surface Waves on Water

Though the word "waves" alone usually evokes images of water waves, these water waves are a rather complicated phenomenon. They contain both transverse and longitudinal components and extend their action far below the surface. A derivation of their speed from first principles is beyond the scope of this text, but some features of water waves are worth noting.

If simple sinusoidal waves of wavelength λ propagate along a canal of depth h, their wave velocity u is given by

$$u^2 = \left[\frac{g\lambda}{2\pi} + \frac{4.4 \times 10^{-4}}{\lambda}\right] \cdot \tanh\left(\frac{2\pi h}{\lambda}\right)$$

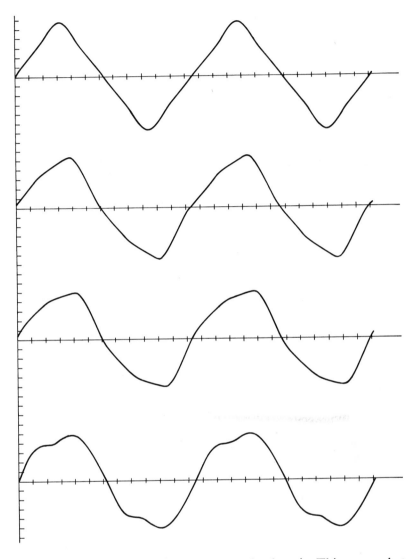

Fig. 23-20. The approximately triangular wave at the top of the figure is the sum of first, third, and fifth harmonics. If this waveform moves through a dispersive medium, so that the shorter wavelengths move steadily ahead of the longer wavelengths, the original waveform becomes progressively distorted as it moves along.

where λ is in m, g in m/s², h in m, and u in m/s. This somewhat complicated dependence of wave speed upon wavelengths leads to dispersion, as previously mentioned. However, when λ is large compared to the depth h, the second term in the brackets becomes negligible and further,

$$\tanh\left(\frac{2\pi h}{\lambda}\right) \approx \frac{2\pi h}{\lambda} \qquad \text{for } h \ll \lambda$$

In shallow water, then, the wave velocity is given by

$$u \approx \sqrt{gh} \qquad\qquad (23\text{-}12)$$

which is independent of wavelength. This absence of dispersion causes deep-water ocean waves to form the well-defined shallow-water waves called surf in the shallow water at the shore. As the depth decreases, their velocity decreases, as indicated by (23-12), until they topple over due to the inertia of the wave crests.

Though the oceans away from continental margins are about 4 km deep, there is a class of waves that have such long wavelengths that they are shallow-water waves even in deep ocean water. These are the waves produced by earthquakes, usually called tidal waves. The speed of any shallow-water wave in deep ocean water can be calculated from (23-12):

$$u = \sqrt{gh} = \sqrt{(9.8 \text{ m/s}^2)(4 \times 10^3 \text{ m})} = 198 \text{ m/s} = 443 \text{ mi/hr}$$

for an ocean of depth $h = 4$ km.

This large velocity causes earthquake-generated waves to strike distant regions so quickly that a tidal-wave warning network has been developed for the Pacific basin region. The sudden diminution of velocity when these waves strike a shore causes them to rear up to enormous heights, inflicting great damage.* Though these waves may have amplitudes of less than one foot in deep water, they can reach heights of 100 feet or more on a steep shoreline.

23-8 Summary

When the laws of mechanics are applied to the mass distributed along a stretched string, it is found that a transverse wave propagates with a velocity of

$$u = \sqrt{T/\mu}$$

TRANSVERSE WAVE

where T is the tension in the string and μ is the mass per unit length. DENSITY

Similarly, the velocity of compressional waves in a medium with bulk modulus B and density ρ is given by

$$u = \sqrt{B/\rho}$$

COMPRESSIONAL WAVE

For a gas with a ratio of specific heats $\gamma = C_p/C_v$ at a normal pressure p, this becomes:

$$u = \sqrt{\gamma p/\rho}$$

gas $\gamma = \dfrac{C_p}{C_v}$

When a transverse wave on a string encounters a segment of string with a lower mass per unit length, μ, a portion of the wave is reflected. When it encounters a segment with a larger value of μ, a portion of the wave is reflected in inverted form. The reflections

---

*See "Tsunamis" by Joseph Bernstein, *Scientific American*, August 1954.

at a fixed end of a stretched string produce an inverted reflection of the same amplitude as the original wave. Such reflections produce standing waves.

Similarly, an inverted longitudinal displacement wave is caused by a region of greater density (but similar bulk modulus). The extreme case is a closed-end tube, where all of a displacement wave is reflected but with inverted form. Such reflections produce standing waves in closed-end organ pipes. In an open-end organ pipe, reflection occurs without inversion and leads to somewhat different conditions for standing waves.

Waveforms that are not pure sinusoidal waves are called complex waveforms. Periodic waveforms of any shape can be made from a harmonic series of sinusoidal waves. The presence and relative intensity of these harmonics are responsible for the *quality* of the sound produced by various musical instruments. A bar graph of the intensity of the various harmonics is called a spectrum. Once the response of an instrument to various frequencies is known, we can predict its ability to handle complex waveforms. If, for example, a sound system is capable of handling the full range of audible frequencies without distortion, it is known that it will not distort the audible portion of *any* complex, periodic waveform.

A traveling wave is attenuated when it loses energy. If the degree of attenuation is frequency dependent, distortion of a complex waveform occurs. If the different frequencies travel with different wave velocities, dispersion occurs. This effect also causes distortion of a complex waveform as it propagates.

Water waves are an example of mechanical waves that exhibit dispersion. The longer waves tend to travel faster, but, in water that is shallow compared to the wavelength, all water waves travel with a velocity

$$u = \sqrt{gh}$$

For earthquake-produced waves in deep ocean water, this velocity is about 450 mi/hr.

Problems

23-1.   A rope with a total mass m and length L is hung vertically from an overhead support. A transverse pulse is sent from the upper end of the rope toward the bottom. Find the speed of this wave as a function of the distance, y, from the bottom end of the rope. The rope hangs freely with no additional load at the bottom.

23-2.   Utilizing data from Tables 13-1 and 13-2, which materials there have the largest and smallest sound-propagation velocities?

23-3.   The density of the earth's crust at a depth of 10 km beneath the continents is approximately 2.7 g/cm^3. The velocity of longitudinal seismic waves at this depth can be found by timing their arrival from distant earthquakes and is found to be 5.5 km/s. From this information, find the bulk modulus of the crustal material at that depth and compare it with that of steel (Table 13-2).

23-4.   The velocity of sound in air at STP is 331 m/s. Calculate the velocity of sound in air at 50°C.

23-5.   A string with a mass of 9 g is stretched between points $\frac{1}{2}$ m apart. Its third harmonic is found to have a frequency of 1000 Hz. Find the tension in the string.

23-6.   A steel cable is stretched to its yield point (Table 13-3). Calculate the speed of a transverse wave along this cable. The diameter of the cable is 7 mm and its density is 8 g/cm^3.

23-7.   The fundamental frequency and third harmonic of a stretched string are 1000 Hz apart. If the string is 0.25 m in length and under a tension of 400 N, find the mass per unit length of the string.

23-8.   A pipe is open at both ends. When air is blown past one end, the fundamental frequency produced is 400 Hz. The far end is then closed. What new fundamental frequency is produced?

23-9.   An organ designer wishes to produce the lowest possible note with an organ pipe 3 m long. Should it be open-ended or closed-ended? What frequency will be produced?

23-10. A plucked string $\frac{1}{2}$ m long with a mass density of 5 g/m emits sound waves with the string's fundamental frequency. It moves through the air at 30 m/s toward a stationary reflecting wall. The reflected sound beats with the sound of the moving string to produce a beat note of 47.7 Hz. What, in newtons, is the tension in the string?

Chapter **24**

Light as a Wave Phenomenon

24-1 Introduction Most of us think of light as an entity that streams from a source, such as the sun, reflecting or scattering from illuminated objects so that it reaches our eye to produce vision. This state of affairs is not obvious. The ancients tended to think of the *eye* as emitting something that probed the environment to discern visual details. Newton, who made such enormous strides in mechanics, also contributed heavily to *optics:* the study of light. In his view, light was a stream of tiny, bullet-like particles that left a source and eventually entered the eye to produce visual sensations. He was able to explain reflection and refraction of light with this model of light. At about the same time (1670), Christian Huygens showed that the observed laws of reflection and refraction of light also resulted if light was a wave phenomenon, although with extremely short wavelength.

About 160 years later it was finally shown that light *interferes* with itself, just as mechanical waves do when they produce standing waves. During the nineteenth century many more experiments were carried out that showed that light was a wave phenomenon. Finally, in 1873, James Clerk Maxwell showed that *electromagnetic waves* should exist. When he computed the speed of these waves he discovered it was equal to the speed of light. Maxwell guessed that light was an electromagnetic wave and his guess was later verified. Thus, by the beginning of the twentieth century, it was clear that light was a fully understood wave phenomenon, in particular, an electromagnetic wave phenomenon.

The opening of the twentieth century brought with it several rude shocks for existing scientific theories. Among these shocks was the discovery that, in some ways, light behaved as if it consisted of minute bundles of energy called *quanta,* or *photons.* In fact, some experiments by Compton in the early 1920s showed that these quanta change momentum and energy in collisions with electrons very much as though they were particles rather than continuous waves.

The modern view of light that has evolved during this century is a dualistic one in the sense that we recognize that light exhibits both wave-like and particle-like behavior. There is, however, one unified theory (called *quantum electrodynamics*) that beautifully describes the whole range of experiments and processes in which light is involved.

We will begin, as the world did in the seventeenth century, with a wave theory of light that is fully capable of explaining the operation and design of optical instruments. Unlike the world, however, we will wait only until Chapter 39 to find out just exactly *what* is "being waved." Chapter 40 will then introduce the modern, dualistic view required for an understanding of atomic effects.

24-2 Shadows

It is well known that opaque objects cast shadows. If a source of light is sufficiently small, it can produce sharp-edged shadows. An extended source of light produces shadows with fuzzy edges. These observations are compatible with the hypothesis that light travels in straight lines, as shown in Fig. 24-1. The lines followed by the light are called *rays.* Rather than showing all of the rays emanating in all directions from the light sources in Fig. 24-1, only the interesting rays that separate light from shadow are indicated. The partially illuminated region, called the *penumbra,* occurs where only a portion of the light from an extended source can reach the screen. The completely dark zone is called the *umbra.*

At the time of a total solar eclipse, the umbra of the moon's shadow touches the earth so that observers at that point see all of the sun covered by the moon's disk. The spectacular nature of this phenomenon is due to the extremely close coincidence in the apparent angular size of the moon and the sun as seen from the earth. The plane of the orbit of the moon about the earth is tipped with respect to the earth's orbital plane, so a relatively rare coincidence is required for the earth-moon-sun alignment that produces a total eclipse. Furthermore, due to the slight variations in the earth-moon

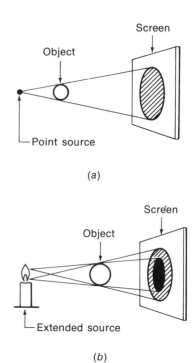

Fig. 24-1. (*a*) A point source of light casts a sharp shadow, indicating that light travels in straight lines. (*b*) The completely shadowed zone is called the *umbra.* It is the region in which no light from the extended source can reach the screen. The partially illuminated zone is the *penumbra* and consists of those portions of the screen that receive light from only a fraction of the extended source.

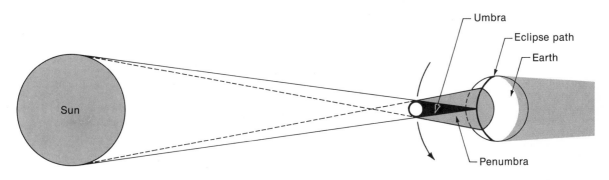

Fig. 24-2. The moon's shadow during a total eclipse of the sun.

and the earth-sun distances due to their elliptical orbits, the moon's apparent disk can totally cover the sun's disk for only about half of these rare events. Observers in the penumbra zone shown in Fig. 24-2 see a partial eclipse of the sun.

The fact that sharp-edged shadows are cast by point sources of light seems to argue against a wave theory of light, since waves spread out after they pass around an obstacle, as shown in parts *a* and *b* of Fig. 24-3. However, if the wavelength of the light is extremely small compared with the dimensions of the object casting a shadow, the edge of the shadow appears to be sharp, as indicated in Fig. 24-3*c*. Indeed, we now know that the wavelength of visible light is less than one micrometer (μm) so that sharp-edged shadows are expected in most cases.

wavelength is small compared to dimensions of the object casting a shadow → edge of shadow appears to be sharp

Exercises

1. Assuming that the sun and moon have the same apparent size when viewed from the earth (both subtend an angle of approxi-

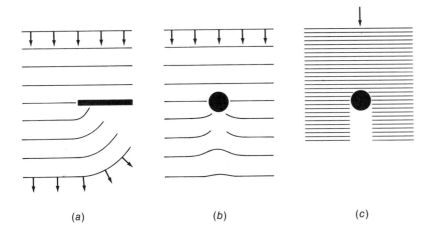

(a)   (b)   (c)

Fig. 24-3. In (a) and (b) the tendency of waves to spread into "shadow" regions is illustrated. Part (c) illustrates the sharp shadows produced when wavelengths are short compared to the obstacle's size.

mately $\frac{1}{2}°$ though the sun is 400 times larger in diameter), sketch the appearance of the sun to an observer in the penumbra of Fig. 24-2.

2.   A rectangular sheet of wood 2 ft by 3 ft is placed halfway between a point source and a wall, as shown in Fig. 24-4. What is the height and width of the shadow cast upon the wall? Is it rectangular? What happens to its shape if the top of the sheet leans toward the wall?

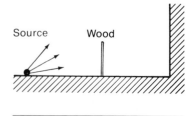

Fig. 24-4.

24-3   The Speed of Light

Light apparently moves in straight lines in empty space. What about the speed of light? How quickly does it move from one point to another? Galileo attacked this question with a simple experiment. He knew that light from a distant event, such as a cannon shot, reached an observer more quickly than sound, so that sound could not be used as a signal for the timing of light from one point to another. Instead, he and an assistant separated themselves by a large distance at night and held shutter-equipped lanterns. After locating the light of each other's lanterns, they each closed their shutters. Galileo then opened his shutter quickly. If light moved with finite velocity, he reasoned, it would take time to reach his assistant, who had been instructed to open his shutter the moment he saw light from Galileo's lantern. The light from the assistant's lantern would also take time to return to Galileo. The time delay that Galileo would observe between opening his shutter and first seeing light from the assistant's lantern thus would give a measure of the velocity of light when combined with the measured distance between lanterns. Of course, the assistant's reaction time was included in this measurement, but that could be found by doing the experiment first with a small separation then with a large separation. The *change* in the time required for the return light to appear would be the extra delay due to a finite speed of light. No perceptible change was observed.

Galileo correctly concluded that the speed of light was too great for observation by such simple means. Small wonder, since the round trip time for light with a one-mile separation is about 10^{-5} s, while human reaction times range between $\frac{1}{10}$ and $\frac{1}{5}$ s and are somewhat variable even in a given subject.

The first successful observation of effects due to the velocity of light was achieved in 1675 by Ole Roemer. He was observing the rotational motion of the moons of Jupiter about Jupiter itself and noting the times at which the moons were eclipsed behind Jupiter.

The time between two successive eclipses should be the orbital period for any given satellite. Roemer noted that this orbital period seemed to increase very slightly for six months, and then slowly decrease for the next six months. He correctly attributed this behavior to the fact that while the earth, following its own orbit, moves away from Jupiter, the news of each eclipse is delayed more and more by the additional time required for light to reach the earth.

In Roemer's time, the size of the earth's orbit was only approximately known so he could only estimate a value of 2×10^8 m/s. With modern values for the astronomical distances involved, Roemer's data would yield a value near 3×10^8 m/s which is quite close to the currently accepted value.

A terrestrial measurement was made by Hippolyte Louis Fizeau in 1849. In this measurement a beam of light was timed over a total distance of 17,260 m (over 10 mi). The arrangement used is indicated in Fig. 24-5. The light beam is "chopped" into segments by the rapidly rotating toothed wheel. The observer views the returning light beam through the toothed wheel as well. At a high enough angular velocity, a tooth has moved far enough to block the observer's view before the light flash can return. The measurement is made by adjusting the speed of the wheel until the observed light is completely extinguished. Fizeau actually used a half-silvered mirror arrangement so that the light beam and observer could use the *same* portion of the toothed wheel in order to reduce errors due to misalignment. His result was $c = 313,300$ km/s which is about $4\frac{1}{2}\%$ larger than the presently accepted value.

Because the velocity of light enters into electromagnetic theory, atomic theory, relativity theory, and is central to modern high-precision measurements of lengths, it is considered one of the most fundamental of the physical constants. It has been measured with greater and greater precision over the years by a wide variety of methods. The most recent measurement (1972) by a group at the

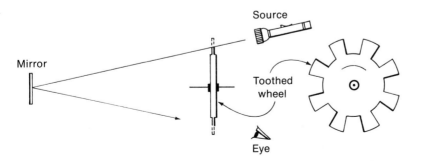

Fig. 24-5. Light passes between two teeth of a toothed wheel, to a mirror, and returns to the observer between the teeth of the toothed wheel. However, if the wheel rotates rapidly enough, the next tooth is in front of the observer before the light returns from the mirror.

National Bureau of Standards deduced c, the velocity of light in vacuum, by a simultaneous measurement of the frequency and wavelength of light from a laser. The fundamental relation $c = \lambda f$ gave the result

$$c = 299{,}792{,}460 \pm 6 \text{ m/s}$$ $3.00 \times 10^8 \, m/s$

The characteristics of laser light that made this precision possible will be discussed in a subsequent section.

Exercises

3. Modern laboratory apparatus readily measures time intervals of 1 ns (10^{-9} s). Show that light moves at a velocity of approximately 1 ft/ns.

4. How long did a flash of light take to make the round trip in Fizeau's apparatus? The wheel had 720 teeth; what was its speed in rev/min when a tooth was able to move far enough to block the returned light?

24-4   Interference and Young's Experiment

If light does consist of waves, we should be able to observe interference phenomena, such as standing waves, beats, and the cancellation created by two waves of equal intensity but opposite phase.

Standing light waves were actually observed by Wiener in 1890. The arrangement he used is sketched in Fig. 24-6. A photographic emulsion was floated on mercury. The incident light waves reflected from the mercury. The reflected waves interferred with the incident waves just as the waves at the clamped end of a string do (see Fig. 22-16). When the film was developed, the highly exposed film at the antinodes was darkened. These zones of darkening were thin layers parallel to the mercury surface but extremely close together. In

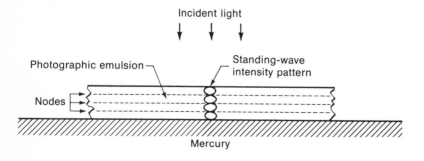

Fig. 24-6. Standing waves of light produced in a photographic emulsion are observed by the darkening produced at the antinodes.

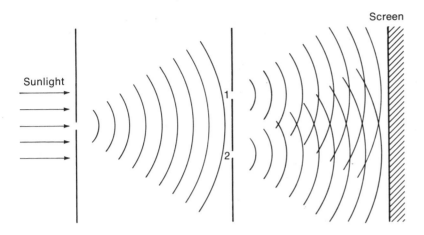

Fig. 24-7. Young's experiment demonstrated the interference of light waves from pinhole 1 with those from pinhole 2 by letting them combine on a screen.

order to measure the distance between antinodes, Wiener sliced the film at a very small angle θ so the spacing was "magnified" by the factor $1/\sin \theta$. The light wavelength inferred was less than 10^{-6} m.

The earliest experimental observation that gave clear support to a wave theory of light involved interference effects and was made by Thomas Young in 1801. The experiment is sketched in Fig. 24-7. Light from the sun illuminated a single pinhole, which produced a single source of light waves in a darkened room. Two subsequent pinholes produced two sets of light waves that were guaranteed to have well-defined phases with respect to each other, since they were merely "samples" from a common wave front. When these light waves then overlapped, they were in phase in some regions and out of phase in other regions, thus producing light and dark zones on the screen. This effect would not have been observed if light were a stream of small particles.

The analogous situation for water waves is shown in Fig. 24-8. There, water waves are set up by two vibrating balls touching the water surface. The zones of quiet water seen are the locations where the waves from one ball are always exactly out of phase with those from the other. The quiet zones are analogous to the dark regions produced by interference of light waves in Young's experiment. The appearance of these zones for light waves is indicated in Fig. 24-9, which is a photograph of a screen illuminated by the light from two thin slits in a modern version of Young's experiment. Observation of these interference effects by Young signaled the failure of Newton's corpuscular theory of light and the acceptance of Huygens' wave theory. From these observations alone there was no way of deducing the *type* of waves constituting light, but it was clear that waves were involved.

Fig. 24-8. Water waves produced by two in-phase sources in a ripple tank interfere with one another. The locus of points at which cancellation occurs are the zones of quiet water, called nodes. (Copyright © 1971 by D. C. Heath and Company. No part of the material covered by the copyright may be reproduced in any form without written permission of the publisher. Reprinted by permission of the publisher.)

exactly out of phase with each other — dark regions

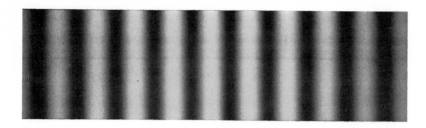

Fig. 24-9. The interference pattern produced by the light from two thin slits shows alternating regions of constructive and destructive interference. (From Cagnet, M., Francon, M., Thrien, J. C., *Atlas of Optical Phenomena,* Springer, 1962)

We can easily predict the location of the regions of constructive and destructive interference beyond two illuminated slits from the geometry shown in Fig. 24-10. Recall that the two slits are illuminated by a single source, so that the light waves are initially in phase. These waves spread in all directions toward the screen, just as ripples produced by waves beating against a slot in a breakwater spread in semicircles in a quiet harbor. If the two waves traverse distances l_1 and l_2 to reach a spot on the screen, they will be in phase only if

$$l_1 - l_2 = n\lambda \qquad (n = 0, 1, 2, 3, \ldots) \qquad (24\text{-}1)$$

in phase
light region

where λ is their wavelength. From the geometry of Fig. 24-10, we obtain:

$$l_1 = \frac{L}{\cos\theta_1} \qquad l_2 = \frac{L}{\cos\theta_2} \qquad (24\text{-}2)$$

We can also express l_1 and l_2 in terms of x, shown in Fig. 24-10, and L by using the Pythagorean theorem:

$$l_1 = \sqrt{L^2 + x^2} \qquad l_2 = \sqrt{L^2 + (d + x)^2} \qquad (24\text{-}3)$$

Either of the expressions (24-2) or (24-3) can be combined with (24-1) to find the points where the screen is bright.

A single, somewhat simpler expression is obtained if the distance to the screen is much larger than the slit separation. In this case, if a pair of rays from the two slits are destined to reach a

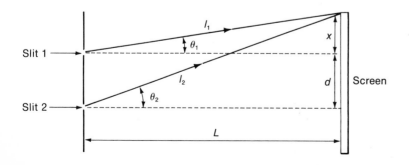

Fig. 24-10. Whenever the path lengths l_1 and l_2 differ by an integral number of wavelengths, constructive interference occurs.

common point, they are very nearly parallel. For any such pair of rays, the geometry of Fig. 24-11 is applicable and we find:

$$\Delta l = l_2 - l_1 = d \sin \theta \qquad (24\text{-}4)$$

Combining this expression with (24-1), we obtain

$$d \sin \theta = n\lambda \qquad (n = 0, 1, 2, 3, \ldots)$$

or

$$\sin \theta = n\frac{\lambda}{d} \qquad (n = 0, 1, 2, 3, \ldots)$$

bright
zones

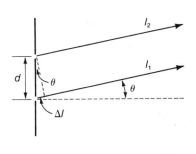

Fig. 24-11. If $d \ll L$, the emerging rays from slit 1 and slit 2 are very nearly parallel, so that they make the same angle, θ, with respect to the incident direction.

as the condition for angles producing bright zones on the screen. When the paths l_1 and l_2 differ by an odd number of half-wavelengths, destructive interference occurs and produces a dark zone. These two cases can be combined in one expression:

$$\sin \theta = n\frac{\lambda}{2d} \quad \begin{cases} n = 0, 2, 4, \ldots & \text{(constructive)} \\ n = 1, 3, 5, \ldots & \text{(destructive)} \end{cases} \quad (24\text{-}5)$$

even

odd

Since the wavelength of visible light is of the order of $\frac{1}{2}$ μm, extremely fine slits spaced close together are required to see the two-slit interference pattern.

Exercises

5. A stereo sound system utilizes two loudspeakers that can interfere with one another. The geometry involved is like that of Fig. 24-10, with the listener substituted for the point of interest on the screen. Suppose a listener stands where $l_2 - l_1 = 1$ ft. Certain frequencies in the audible range will be diminished for this listener due to destructive interference. Find these frequencies. Assume that the velocity of sound is 10^3 ft/s for simplicity.

6. Two fine slits are illuminated with light of 0.6-μm wavelength and are separated by the width of a pencil line (take this to be 0.3 mm). A screen is placed a distance $L = 3$ m away in a Young's experiment. The point on the screen equidistant from the two slits is bright and corresponds to $n = 0$ in Eq. (24-5). Find the distance x between this central bright maximum and either of the first maxima produced on each side of the central one. These maxima correspond to $n = 2$ in (24-5). Note that the approximation leading to Fig. 24-11 and Eq. (24-5) is excellent for this example, and note further that $\sin \theta \approx x/L$ is also an excellent approximation.

24-5 Huygens' Principle

The way that light waves propagate is most easily predicted by application of *Huygens' principle*. We have used this principle implicitly in our discussion of Young's experiment.

When Huygens, in the seventeenth century, proposed that light is a wave phenomenon, he was able to explain the then known behavior of light by the use of a simple guess as to how these waves propagate. His guess is now called Huygens' principle:

> *Every point of an advancing wave front can be considered as a point source of spherical secondary waves. Later positions of this wave front are given by the envelope of these secondary waves.*

A simple application of Huygens' principle is indicated in Fig. 24-12. There, secondary waves have been drawn from a sufficient number of points on the initial wavefront to allow us easily to see their envelope. The reader may wonder why the secondary waves are not also extended to the left side of the initial wave front. Indeed, if at some instant of time a disturbance is suddenly created along a line (whether the disturbance is optical or mechanical is immaterial), a wave front *will* move in both directions from it. A one-dimensional example was given in Fig. 23-1. However, once a wave disturbance is propagating at its characteristic wave velocity, no further backward-moving waves are produced. Accordingly, in a Huygens' construction we consider only the portion of secondary waves moving in the same general direction as the wave front.

Huygens' principle seems a bit arbitrary—especially since we are asked to ignore the backward portion of spherical secondary waves. Indeed, today we know that Huygens' principle is not a fundamental principle of physics; instead, it was an intuitive guess by Huygens that proved to be extremely useful. In general, the motion of a wave front can be rigorously computed in any situation by solving the three-dimensional wave equation that describes the waves.* Though this direct method of attack is quite difficult, Kirchoff was able to show that application of Huygens' principle resulted from a reasonably accurate solution of the wave equation. This proof is known as Kirchoff's theorem and allows us to use simple Huygens' constructions to solve many problems in wave motion.

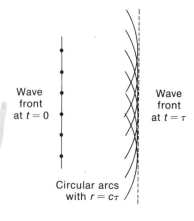

Fig. 24-12.   A plane wave moving from left to right can be considered to be the envelope of Huygen's secondary waves.

Exercise

7.   Use Huygens' principle to show that (*a*) a spherical wave front in a medium with a constant wave velocity remains spherical

---

*This equation is generally a second-order partial differential equation.

in shape; and (*b*) if a portion of this originally spherical wave enters a region of reduced wave velocity, the wave front is distorted in that region.

24-6   Diffraction

When a wave impinges on a very small opening in a screen, it spreads in all directions beyond the screen, as indicated in Fig. 24-13. The spreading of waves beyond an obstacle is called *diffraction*. If light traveled only in straight lines, all obstacles would cast sharp-edged shadows. The fact that light spreads beyond an obstacle can lead to interference effects in these "shadow" regions. An example for the case of water waves is shown in Fig. 24-14.

In general, whenever a wave front passes a sharp-edged obstacle, interference effects are visible beyond the obstacle. An illustration of this behavior for light waves is shown in Fig. 24-15. We are normally not aware of these diffraction effects because the incident light rays must be parallel to reveal these features. Light from normal, extended sources produces a penumbra region that obscures these interference features.

The diffraction of light places a definite upper limit on the useful magnification of a microscope or telescope. We will examine this point further in Chapter 27.

Fig. 24-13. Waves in a ripple tank spread in all directions after passing through a small opening. (Courtesy of Education Development Center, Inc., 39 Chapel Street, Newton, Massachusetts)

Fig. 24-14. Water waves in a ripple tank exhibit an interference pattern after passage through a single slit. (From *PSSC Physics,* D. C. Heath and Company, Lexington, Massachusetts, © 1965)

24-7   Single-Slit Diffraction

Huygens' principle can be applied to predict the locations of the regions of constructive and destructive interference in single-slit diffraction. In Fig. 24-16, secondary waves are shown emanating from two particular points separated by $w/2$, half the width of the slit. At an angle θ such that

$$\frac{w}{2} \sin \theta = \Delta l = \frac{\lambda}{2}$$

the distance Δl shown in Fig. 24-16 will be one-half wavelength. These two secondary waves will thus combine out of phase on a distant screen. Every point in the upper half of the slit opening can be paired with another point in the lower half of the slit in a similar fashion; thus at this particular angle, every such pair cancels. We conclude, therefore, that perfect cancellation results and the screen is dark at this angle. In fact, dividing the slit into *any* even number of segments will allow us similarly to pair off points of the slit for which the light waves will be one-half wavelength out of phase and thus lead to darkness on the screen. The angles at which *darkness* occurs are then given by

Fig. 24-15. Shadows of small needles illustrate diffraction caused by an obstacle. (From Rossi, *Optics,* 1957, Addison-Wesley, Reading, Massachusetts.)

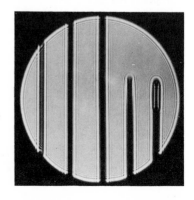

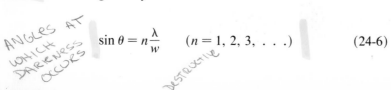

ANGLES AT WHICH DARKNESS OCCURS

$$\sin \theta = n\frac{\lambda}{w} \qquad (n = 1, 2, 3, \ldots) \qquad (24\text{-}6)$$

DESTRUCTIVE

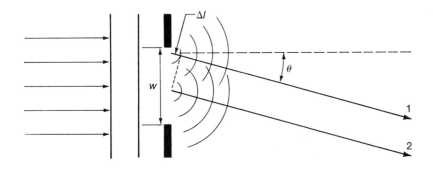

Fig. 24-16. When $\Delta l = \lambda/2$, secondary waves from all points separated by $w/2$ will be out of phase when they reach a distant screen.

However, if the slit is subdivided into an *odd* number of segments, the above argument leading to cancellation can be made for *all but the last* segment. Points in this last odd segment cannot be cancelled off with other points, since all the points in the remaining even number of segments have already been paired. We expect, then, that at angles given by

$$\sin\theta = \frac{m}{2}\cdot\frac{\lambda}{w} \qquad (m = 3, 5, 7, \ldots) \qquad (24\text{-}7)$$

ODD NUMBER OF SEGMENTS
LIGHT   CONSTRUCTIVE

we will find light on the distant screen, but each successive maximum is less bright, since the uncancelled odd segment of the screen contains a smaller fraction of the total light as m increases. This argument breaks down for $m = 1$ since in that case there are no remaining segments of the screen to be paired. Thus the central maximum in a single-slit interference pattern is twice the width of all others, as can be seen in Fig. 24-17. Furthermore, since all secondary wavelets are *exactly* in phase at $\theta = 0°$, this central maximum is far brighter than the others. The bulk of the light passing through the slit falls into the central maximum.

The single-slit pattern shown above is actually produced by *each* of the two slits in a double-slit interference experiment. In our previous discussion in Section 24-4, we assumed each slit was so narrow that light waves spread in all directions from each slit. Actually, if the slits have a width that is *not* negligible compared

Fig. 24-17. An interference pattern on a screen beyond a single slit. (From Cagnet, M., Francon, M., Thrien, J. C., *Atlas of Optical Phenomena*, Springer, 1962)

to their separation, both the double-slit pattern and the single-slit pattern will be superimposed, as shown in Fig. 24-18.

Exercise

8. If the *first* ($m = 3$) destructive interference fringe occurs at $\theta = 90°$, then the space beyond a single slit appears to be smoothly, though not uniformly, illuminated with no discrete fringes present. What fraction of a wavelength must the slit width be for this to occur?

24-8   The Diffraction Grating

Most sources of light produce light of many different wavelengths. A device that can separate and measure these various wavelengths is thus essential for the study, and eventual understanding, of light sources. The most successful such device is a *diffraction grating*. The knowledge it has produced concerning the emission of light by atoms has been invaluable in the development of the modern quantum theory of atomic structure.

One way that one might think of measuring light wavelengths involves the two-slit interference pattern discussed in Section 24-4. We could shine the light of unknown wavelength on two slits of known separation. Then we could measure the angle of deflection observed for various values of *n,* substitute them into (24-5), and solve for the unknown wavelength. There are two drawbacks to this method. First, the two-slit diffraction pattern is not very sharp, so it is difficult to obtain high precision. Second, very little light passes through two fine slits so that only bright light sources can be measured in this manner. Fortunately, a single trick cures both of these failings and leads us to the diffraction grating.

A diffraction grating consists of a great many fine slits parallel to each other. The role of these fine slits is simply to produce secondary wavelets spreading in all directions. Ideally, they are so fine

$\sin\theta = n\,\lambda/d$

Fig. 24-18. The interference pattern produced by two slits consists of the single slit diffraction pattern superimposed on the double-slit pattern. (From Cagnet, M., Francon, M., Thrien, J. C., *Atlas of Optical Phenomena*, Springer, 1962)

that each slit does not produce a multiple interference pattern but only a diffuse secondary wave. The wavelets emerging from many regularly spaced slits then combine at certain angles to produce constructive interference, as shown in Fig. 24-19. When each of the path-length differences, Δl, is an integer number of wavelengths, constructive interference results. If the spacing between slits is d, then for constructive interference,

$$d \sin \theta = \Delta l = n\lambda \quad (n = 0, 1, 2, 3, \ldots) \quad (24\text{-}8)$$

The large number of secondary waves produces an unusually bright interference pattern. Further, because many slits are involved, the total width of the grating is large, so that only a very slight change in angle is required to destroy the constructive interference between any of the widely separated slits. The resulting interference pattern is thus extremely sharp.

If light composed of many different wavelengths is incident upon a diffraction grating, the different wavelengths are deflected through different angles to their interference-pattern maxima. As indicated in Fig. 24-20, the various *orders* (values of n in Eq. (24-7)) can overlap one another.

In actual practice, gratings are made by using ruling engines to scribe fine grooves on glass or metal plates, and by a process called "blazing," the shape of the grooves is designed to throw most of the light into a chosen order. When white light is passed through a grating, it is spread out into colors, indicating that different colors correspond to different wavelengths. The dispersed array of colors is called a *spectrum*. Spectra of several different light sources are shown inside the front cover.

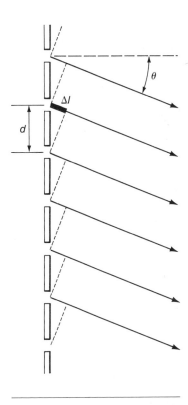

Fig. 24-19. The diffraction grating. At certain angles θ, each of the distances Δl will be an integer number of wavelengths, so that constructive interference results.

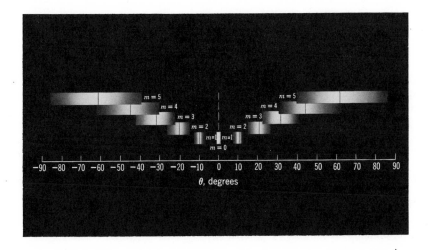

Fig. 24-20. A grating produces many spectra of a white-light source. The higher orders are displaced by a larger angle and are spread out to a greater degree. The various orders are shown here displaced vertically to avoid confusion, since the higher orders do overlap. The central line in each order is $\lambda = 550$ nm. (From Halliday and Resnick, *Fundamentals of Physics*, John Wiley & Sons, Inc., 1970)

The word *spectrum** was originally used by Newton to describe the rainbow-like spread of colors produced when white light was dispersed into its component colors. He discovered the spectrum of white light not with a diffraction grating but with a *prism*, to be discussed in Chapter 26. We now know that each color is associated with a different wavelength as indicated in Table 24-1. The wavelengths there are given in nanometers (10^{-9} m). Since Newton's time, the word spectrum is used whenever intensity as a function of wavelength (or frequency) is discussed. The spectrum of a sound source is an example. With our modern understanding of quantum mechanics, it is known that the frequencies of light produced by atomic transitions are simply related to the energy available (see Chapter 40). Thus the word "spectrum" has come to be used for intensity as a function of *energy* as well. This is the usage, for example, when we speak of the energy spectrum of neutrons produced by nuclear fission.

TABLE 24-1 The Wavelengths of Various Colors of Light

Color	Wavelength (nm)
Violet	400–450
Blue	450–500
Green	500–570
Yellow	570–590
Orange	590–610
Red	610–700

Example Suppose a light source produces two nearly equal wavelengths of light, λ_1 and λ_2. The spectrum produced by a grating will consist of repeated pairs of lines: one pair for each *order* of the spectrum. The first order corresponds to $n=1$ in (24-8), the second to $n=2$, etc. Will the angular spacing between the two members of this *doublet* of lines be the same in each order?

Solution The two wavelengths will be deflected through angles given by:

$$\sin\theta_1 = n\frac{\lambda_1}{d} \qquad \sin\theta_2 = n\frac{\lambda_2}{d}$$

For small angles, where $\sin\theta \approx \theta$, the situation is particularly simple:

$$\theta_1 - \theta_2 \approx n\left(\frac{\lambda_1 - \lambda_2}{d}\right)$$

or

$$\Delta\theta \approx n\frac{\Delta\lambda}{d}$$

so that the angular spacing of the doublet depends linearly on the order. For large angles, the angular spacing increases even more rapidly with n, since a given interval in $\sin\theta$ corresponds to greater and greater intervals in θ as $\theta \to 90°$. This is easily seen by inspection of a graph of $\sin\theta$ vs θ.

Exercises

9. A grating 1 cm wide has 1000 equally spaced rulings across its

*The word *spectrum* is related to the Latin word *spectre* for *ghost*. It describes a *shimmering pattern* or *apparition*.

width. What angular separation does it produce for the yellow sodium doublet lines in the second-order spectrum? (λ_1 = 589 nm; λ_2 = 589.6 nm.) Use the small-angle approximation $\sin \theta \approx \theta$ (in radians).

10. How far apart do the yellow sodium doublet lines in Exercise 9 fall on a screen 1 m distant? How many rulings/cm are required to increase this spacing to 5 mm?

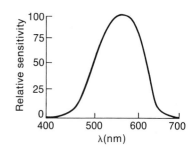

Fig. 24-21. Sensitivity of the eye to light at various wavelengths.

24-9   Characteristics of Light Sources   *NOT ON HOURLY*

A light source can be characterized by its luminous intensity, its spectral radiancy, its coherence, and the spatial distribution of the light it emits. Let us examine the definition of each of these terms.

Luminous intensity is a measure of the light output of a source, taking into account the variability of response of the eye to various wavelengths—see Fig. 24-21. Luminous intensity thus indicates how bright a source appears to the eye. The SI unit of luminous intensity is the *candela* (cd):

LUMINOUS INTENSITY →
HOW BRIGHT

> The candela is the luminous intensity of 1/600,000 of a square meter of a radiating cavity at the temperature of freezing platinum (2042K).

The *flux* of luminous intensity describes the quantity of light streaming in a given direction. The unit is the *lumen* (lm). A point source having an intensity of 1 cd radiates a net light flux of 4π lm. Thus, a sphere of 1-m radius centered on a 1-cd source has a flux of 1 lm falling on each square meter of its inner surface.

FLUX OF LUMINOUS INTENSITY
QUANTITY OF LIGHT

When a portion of the light flux from a source is spread over an area, it illuminates it. The amount of light flux striking a unit area is a measure of the *illuminance* of an object. Table 24-2 gives some representative values in lm/m². The tremendous range of sensitivity of the eye implied by the tabulated values is even greater than the six-orders-of-magnitude range of hearing sensitivity.

ILLUMINANCE →
AMOUNT OF LIGHT FLUX
STRIKING A UNIT AREA

TABLE 24-2   Illuminance

Source of Illumination	Illuminance (lm/m²)
Sunlight + skylight	100,000
Sunlight + skylight (overcast)	10,000
Interior near window (daylight)	1,000
Minimum for close work	100
Full moonlight	0.2
Starlight	0.0003

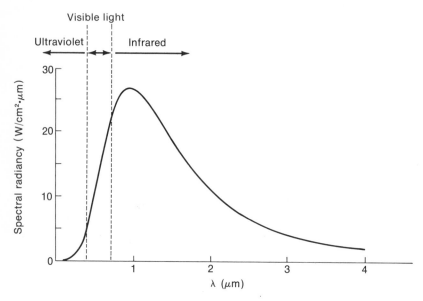

Fig. 24-22. Spectral radiancy
of tungsten at 3000K.

Spectral radiancy is a measure of the *power* radiated by a unit area of the source in each wavelength interval. Thus spectral radiancy has no relation to the ability of the eye to respond to the light. It simply gives the light intensity in terms of radiated power as a function of wavelength. Typical units for spectral radiancy are mixed: $W/cm^2 \cdot \mu m$. These units are used in Fig. 24-22, where the spectral radiancy of tungsten at 3000K is shown. Thus at the peak of the curve, each square cm of tungsten radiates about 27 W of light in a 1-μm wavelength interval.* When an electric discharge or spark passes through a gas, the emitted light occurs only at discrete wavelengths, as shown on the inside front cover. The spectral radiancy of such light sources consists of discrete sharp peaks at various wavelengths.

The light from most sources comes from atoms that have been given excess energy (either from thermal agitation or an electrical discharge), which is then given off as light. Each atom generally gives off its light in the form of a short wave train at random times. The net result from the blending of all these randomly occurring wave trains is called *incoherent light* and is illustrated in Fig. 24-23.

SPECTRAL RADIANCY
POWER RADIATED BY
A UNIT AREA OF THE SOURCE
IN EACH WAVELENGTH INTERVAL

W/cm^2 , μm

$power/cm^2$, 1-μm λ interval

---

*In order to obtain the shape of this curve, the data were obtained in much smaller wavelength intervals than 1 μm, since the spectral radiancy changes a great deal over a 1-μm range of wavelength. For example, if the data are obtained in intervals of 0.1 μm, the number of W/cm^2 in that 0.1-μm interval must be multiplied by 10 to obtain the units given in Fig. 24-22.

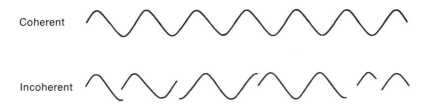

Coherent

Incoherent

Fig. 24-23. Coherent light is very nearly a pure sinusoidal wave. Incoherent light consists of short bits of wave trains with random phase relations.

Lasers are light sources capable of producing remarkably coherent light, that is, light that maintains a sinusoidal waveform over a great length. A further property of laser light is that it is highly unidirectional rather than spreading uniformly out from the source.

The *coherence length* of a given light source indicates the distance along a light ray over which the wave crests are spaced regularly, as in a sine wave. If a beam of light is split and then recombined, as in the two-slit experiment of Fig. 24-7, interference effects can be seen only if the recombined waves have a definite phase relation to each other. Thus if the path lengths, such as l_1 and l_2 in Fig. 24-10, differ by more than the *coherence length* of the light being used, the interference pattern will not appear. Some characteristics of various light sources are given in Table 24-3.

COHERENCE LENGTH → distance along a light ray over which the wave crests are spaced regularly

TABLE 24-3   Some Characteristics of Light Sources

	Spectral Distribution	Coherence Length	Spatial Distribution
Heated solids, liquids and dense gases (such as the sun)	continuous	extremely short	uniform
Electrically excited light from low-density gas.	sharp spectral lines	1 mm–1 m	uniform
Laser	one extremely sharp line or several extremely close lines	up to many km	highly collimated (unidirectional)

24-10   The Michelson Interferometer*   *NOT ON HOURLY*

An *interferometer* is a device that splits a light beam into two parts, sends them over different paths, then recombines them to produce

*"Michelson and His Interferometer" by Robert S. Shankland, *Physics Today*, April 1974.

an interference pattern. Because an extremely small change in either of the path lengths is easily detected by the change in the interference pattern, interferometers are used for precision length measurements.

The method used to split and recombine the light is shown in Fig. 24-24. The half-silvered mirror, M, reflects part of the light upward to M_2 along path l_2 while allowing part of the light to pass through along path l_1. When the light rays recombine at the observation point, O, interference will produce a bright or dark result, depending upon whether l_1 and l_2 differ by an even or odd number of half-wavelengths.*

If M_2 is moved in or out, the interference pattern changes by one *fringe* (i.e., light to dark and back to light again) for a one-wavelength change in l_2. Since l_2 is a round-trip path, each half-wavelength change in position of M_2 corresponds to a one-fringe change. If l_1 and l_2 differ by more than the coherence length of the light source, no interference effects can be seen. Thus as l_2 is steadily changed in value from l_1, the interference pattern slowly "washes out" until it disappears completely. The interferometer, in fact, provides a convenient tool for the measurement of coherence lengths. Before the advent of the laser, most light sources of reasonable intensity had coherence lengths of less than a centimeter and interferometers could be used for only short length measurements. Modern interferometers for distance measurements are equipped with lasers and a photocell coupled to an electronic register to count the number of fringes. The moveable mirror, M_2, is generally a cube-corner reflector that does not require critical alignment to return the light beam accurately. An example of this type of instrument is shown in Fig. 1-1. The fundamental definition of the meter as given in Chapter 1 is based on a light wavelength, and an interometer is used to determine the length.

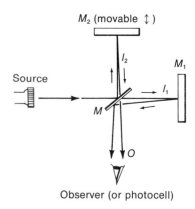

Fig. 24-24. The Michelson interferometer. The mirror M is half-silvered so that part of the light passes through, part reflects.

Exercises

11. Two light sources are available for use with an interferometer. They are equal in all respects except that one emits red light, the other blue. Which leads to the most accuracte length measurements?

12. A threaded screw with 10 threads per cm is used to move M_2

---

*Actually, the observer at O sees a "bullseye" pattern of light and dark fringes. The reader can see this is so by following an expanding cone of light through the paths of l_1 and l_2. Our simple discussion relates to the central spot in the bullseye pattern.

in an interferometer like that in Fig. 24-24. The wavelength of the light is 500 nm. How many times does the observer see the central spot go from light to dark and back to light again when the screw is turned one full turn? Estimate how rapidly the screw can be turned while allowing an observer to keep count of the alternating light and dark patterns.

24-11 Summary

The wavelength of light is so short—about $\frac{1}{2}\mu$m—that its wave-like properties escaped notice until the nineteenth century. Nonetheless, as early as the seventeenth century a wave theory of light existed along with a particle theory of light. The velocity of light, c, is extremely large. To an accuracy of better than $\frac{1}{10}\%$ it is given by $c = 3.00 \times 10^8$ m/s.

$c = 3 \times 10^8 \, m/s$

In empty space light moves in straight lines. When it encounters obstacles, its wave nature can become apparent. For example, interference between light waves emanating from two fine slits in a barrier leads to light and dark zones on a screen placed beyond the barrier. These effects are described by Eq. (24-5). Even a single slit can produce interference effects, which again lead to bright and dark zones on a screen beyond the slit. The location of these zones is given by Eqs. (24-6) and (24-7).

The analysis of single-slit diffraction was carried out by application of Huygens' principle:

Every point of an advancing wave front can be considered as a point source of spherical secondary waves. Later positions of this wave front are given by the envelope of these secondary waves.

The diffraction grating utilizes the interference of light waves scattered by many fine rulings on a glass or metal plate. Such a grating produces a spectrum of the light incident upon it, since various wavelengths are dispersed to different angles.

The spectrum of white light reveals a mixture of wavelengths. The eye responds most strongly to light waves near 550-nm wavelength, decreasing to zero at about 400 nm (violet) and 700 nm (deep red). The brightness of a light source as seen by the eye is given by the luminous intensity. The luminous intensity is proportional to the quantity of visible light *per unit area* radiated from the source. The SI unit for luminous intensity is the candela. The flux of a light source is its total light output, measured in lumens. When a portion of this flux illuminates a surface by falling upon it, we measure the illuminance of the surface in units of lumens per square meter. The spectral radiancy of a light source describes the way the light intensity varies with wavelength.

The coherence length of light waves indicates the distance over which the light waves are spaced uniformly. For a pure sinusoidal wave, the coherence length would be infinite. Most light sources are incoherent, but the light from a laser can have coherence lengths up to several km.

A Michelson interferometer splits a light beam into two parts, each of which follows a different path. When these two beams are later recombined, the interference between them can reveal relative changes between the two path lengths. In this way, distance measurements with an accuracy of a fraction of a light wavelength are possible. The fundamental definition of the meter is now made in terms of a specified light wavelength, with an interferometer as the instrument that makes the length measurement.

Problems

24-1. A rule of thumb for estimates of the distance between an observer and a lightning stroke is: *The distance in miles equals the number of seconds delay between the visible flash and the arrival of the sound divided by five.* Is this rule correct?

24-2. A pinhole camera utilizes the fact that light propagates in straight lines to form an image, as shown in Fig. 24-25. If the object is a man with a height of 1.8 m and he is 10 m away from the pinhole, what is the height of his image on the film?

24-3. Find the frequency of light waves that have a wavelength of $\lambda = 600$ nm. Take the velocity of light to be $c = 3 \times 10^8$ m/s.

24-4. Take the radius of the earth's orbit to be 1.5×10^8 km. The earth's orbital velocity is then close to 30 km/s. Con-

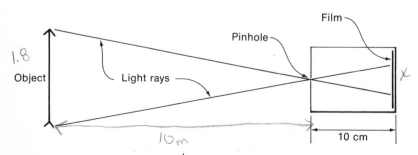

Fig. 24-25.

sider Jupiter to be very distant and the true orbital period of one of its satellites to be 40.5 hr. How long does this orbital period appear to be to an earth-bound astronomer at a time when the earth is approaching Jupiter with its full orbital velocity?

24-5. Yellow sodium light, with $\lambda = 589$ nm, is incident upon two fine slits a distance d apart. The first bright interference zones (maxima) are found to be 0.2° on either side of the central maximum. Find the distance d.

24-6. Use the conditions of Problem 24-5, except assume that the light is incident upon a single slit of width w. Find w.

24-7. Two slits of width w have a center-to-center spacing d. For $d = 2w$, how many two-slit interference maxima are contained within the central single-slit maximum?

24-8. Equation (24-8) was derived for light waves normally incident (at 90°) from the left. If, instead, light is incident as in Fig. 24-26, how is Eq. (24-8) modified?

24-9. A grating 2 cm wide has 4000 lines ruled upon it. Visible light of a single wavelength is normally incident and is diffracted to $\theta = 30°$. Find the wavelength of the light. Is there any ambiguity in your result?

24-10. Find the wavelength of light that is diffracted to 20° in first order by a diffraction grating with 6000 rulings/cm. Now find the angle through which this light is diffracted by this same grating in second order.

24-11. A point source of light radiates equally in all directions. If the total flux it emits is 125.6 lm, find the intensity of the source in cd.

24-12. A sphere is centered on a point source of light that emits light uniformly in all directions. Find the illuminance of the inside of the sphere when it has a radius of (a) 2 m; (b) 4 m. The net flux emitted by the point source is 100 lm.

24-13. Find the illuminance on a surface 4 m distant from a 10-cd source that emits uniformly in all directions.

24-14. Mirror M_2 in a Michelson interferometer is slowly moved and 443 fringes are counted. The light is provided by a He-Ne laser ($\lambda = 632.8$ nm). Find the distance through which M_2 moved.

24-15. The photocell in a commercial interferometer produces an electric current when struck by light. This current is passed through an amplifier to increase its amplitude. Each

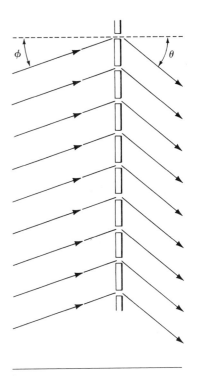

Fig. 24-26.

time an interference maximum occurs, the increase in the amplifier's output is counted on an electrical scaler. Thus, the total number of counts recorded is the net distance through which M_2 has moved, in units of half-wavelengths of the laser light. The light used is from an He-Ne laser with a wavelength of $\lambda = 633$ nm. If the amplifier's frequency response is limited to the $0-10^6$ Hz range, what is the maximum speed at which M_2 can be moved without introducing an error due to lost counts?

24-16.* The yellow light from a sodium lamp is used in a Michelson interferometer. This yellow light consists of two closely spaced wavelengths: $\lambda_1 = 589$ nm and $\lambda_2 = 589.6$ nm. As M_2 is slowly moved, the entire interference pattern slowly disappears. Eventually it reappears. This continues periodically as M_2 is moved steadily. (a) Explain this effect. (b) Find the motion of M_2 between two successive disappearances of the interference pattern.

24-17.* In a lecture demonstration, the light from a He-Ne laser is allowed to strike a steel ruler with engraved rulings spaced every $\frac{1}{64}$ in. as in Fig. 24-27. It is observed that interference from the various rulings causes a series of bright spots to appear on the blackboard beyond the ruler, in addition to the very bright spot at θ_0, which is simply the mirrored light beam reflected from the shiny portions of the steel ruler. If $\theta_0 = 2.32°$, $\theta_1 = 4.03°$, $\theta_2 = 5.17°$, and $\theta_3 = 6.09°$, find the wavelength of He-Ne light.

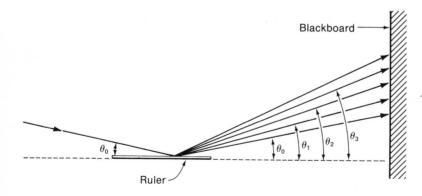

Fig. 24-27.

Chapter **25**

Reflection and Images

25-1 Introduction When a ray of light strikes the surface of an object, it may penetrate into that object or be deflected back into the medium from which it came. In the former case we speak of the *transmitted ray* and in the latter case, the *reflected ray*. We usually see objects by virtue of the light that they reflect. If a reflecting object is sufficiently smooth, the object has a shiny or mirror-like appearance. Reflection of light from such an object is called *specular* reflection. If the surface of the object is rough, different segments of the surface reflect light into different directions, and we speak of *diffuse* reflection.

The difference between the two types of reflection is responsible for the difficulty of driving a car on a rainy night. As shown in Fig. 25-1, diffuse reflection from the rough surface of a dry road

[handwritten margin notes:]
smooth → shiny
    specular reflection

rough → light reflected in
    different directions
    diffuse reflection

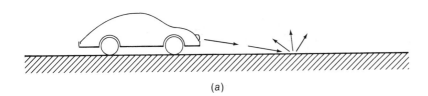

(a)

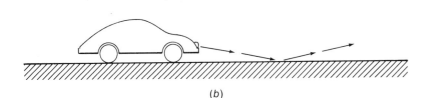

(b)

Fig. 25-1. (a) When the surface of a road is dry, diffuse reflection occurs so that the illuminated road is easily seen by the driver. (b) When the road is wet, the water surface is smooth and specular reflection occurs.

scatters light back toward the driver so that he can see the illuminated road. When the road is wet, the surface irregularities are filled
with water. It then behaves like a mirror so that the headlight
beams are reflected forward. Not only does this make it difficult for
the driver to see the road, it also reflects light from oncoming cars
directly into his eyes.

In this chapter we will use Huygens' principle to deduce the law
of specular reflection. We will then use this law to deduce the types
of images made by flat mirrors and curved mirrors.

25-2   The Law of Reflection

When light undergoes specular reflection, the angle of incidence is
observed to be equal to the angle of reflection, as indicated in Fig.
25-2. This behavior can be explained by either a particle theory of
light or a wave theory. Let us deduce this law via the application of
Huygens' principle.

Consider an extended wave front approaching a plane mirror,
as in Fig. 25-3. By use of Huygens' constructions along the wave

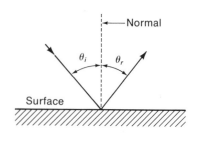

Fig. 25-2.   The angle between
the incident ray and the normal
to the surface is called the angle
of incidence, θ_i. For specular
reflection, the angle of incidence
equals the angle of reflection,
θ_r.

*specular reflection
angle of incidence =
angle of reflection*

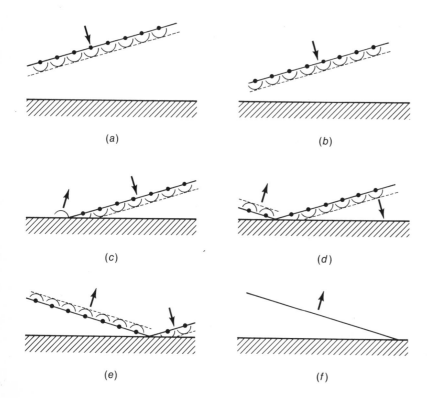

(a)

(b)

(c)

(d)

(e)

(f)

Fig. 25-3.   Subsequent positions of a plane wave front are
found by use of Huygens' principle. As the wave front encounters a plane mirror, it is
reflected bit by bit until the
wave front is entirely reflected.

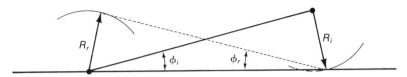

Fig. 25-4. The Huygens' wave-let from the right end of the wave front reaches the mirror in a time $T = R_i/c$. In the same time, the reflected Huygens' wavelet from the left end of the wavefront moves a distance $R_r = R_i$, so that $\phi_r = \phi_i$.

front, we see that the wave advances as a plane wave until it reaches the mirror. The wavefront is then bent bit by bit by the reflection until it is entirely reflected. The law of reflection is most easily derived by a simple two-point Huygens' construction, made at the moment one edge of the wave front has reached the mirror, as in Fig. 25-4. The wavelet emanating from the right-hand end of the wave front expands a distance

$$R_i = cT \qquad \text{right-hand end}$$

in the time T. We choose this time T such that the wavelet just touches the mirror. Meanwhile, the wavelet emanating from the left-hand edge of the mirror moves a distance

$$R_r = cT \qquad \text{left-hand end}$$

so that the left-hand end of the wavefront is now inclined at an angle ϕ_r with respect to the mirror surface, as indicated by the dotted line. From the geometry of the figure, we see that

$$\phi_r = \phi_i$$

It follows that the light rays, which are normal to these wave fronts, also form equal angles of incidence and reflection:

$$\theta_r = \theta_i$$

as was to be proved.

Exercise

1. Draw a sketch of a person looking at himself in a mirror. Prove that if the mirror is only half of his height, he can just see all of himself. Would this still be true if his eyes were located at his waist?

25-3   Images

When we see an object, we detect light coming from various points on its surface. The quantity of light emitted from each point differs in general, and the synthesis of all these points forms a complete visual representation of the object. The simplest case to consider,

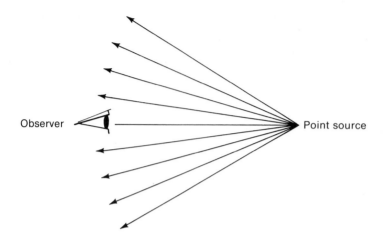

Fig. 25-5. A point source of light emits rays radially in all directions. When the observer's eye intercepts a cone of expanding light rays, he interprets the visual sensations as a bright point located at the point of convergence of the light rays.

then, is the appearance of a single light-emitting point. Light rays emanate outward from the point, as in Fig. 25-5, and when a segment of these diverging rays is intercepted by our eye, a complex series of photochemical and neurological events gives rise to the visual sensation of a point-like object. The action of the eye itself is further discussed in Section 27-3.

We can fool our eye into believing a point is present when none is actually there. In Fig. 25-6 two methods of producing this effect are shown. The first method directs many rays of light so that they cross over at a point. An observer located beyond the crossover point has all the visual sensations that would occur if a light-emitting point was actually present. He concludes that a point is present at the position where the light rays converge. The appearance of a point (when none is there) is called an *image*. When the observed light rays indeed come from a point, as in Fig. 25-6a, the image is called a *real* image. In Fig. 25-6b the rays only appear to

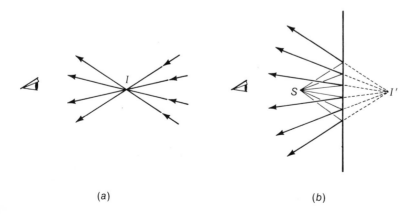

(a)   (b)

Fig. 25-6. (a) Light rays cross over at a single point I, so the observer's eye sees light rays that emanate from one point, even though they did not originate there, nor is there a material point located there. (b) Light rays from a point S reflect from a mirror. The bundle of diverging rays that reaches the eye appears to be diverging from a point I' behind the mirror, as shown by the dotted lines. The eye accordingly tells us there *is* a point behind the mirror.

come from a point and the image is called a *virtual* image. A further distinction between a real and virtual image is that when a screen is placed at the image point in Fig. 25-6a, a point of light will appear on it. The virtual image in Fig. 25-6b cannot be projected on a screen but can be seen only directly by an observer.

Since an extended object and its image can be thought of as a large collection of light-emitting points, we can expand our discussion to include extended objects. The arrow in Fig. 25-7 is an example. When the arrow is illuminated, each point on its surface scatters light into many directions. Those light rays that emanate from a point on the object and strike the mirror subsequently appear to have diverged from a point behind the mirror.

Any pair of reflected rays emanating from any point of the object enables us to locate the image of that point. Since the law of reflection results in equal angles of incidence and reflection, congruent triangles are produced (as in Fig. 25-8), so that we know that the virtual portion of the light path (behind the mirror) is the same length as the portion of the light path extending from the object to the mirror. Thus, the virtual image of a point in a plane mirror appears to be located at a distance behind the mirror that is equal to the distance from the object to the mirror. This is also true for every other point of the extended object, so that every portion of an object's virtual image in a plane mirror appears to be behind the mirror by a distance equal to its true distance from the mirror.

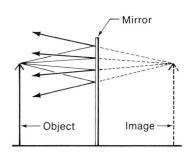

Fig. 25-7. An arrow is placed in front of a mirror. Since reflected light rays emanating from the tip of the arrow appear to emanate from behind the mirror, an observer sees a virtual image. *Each* point of the arrow images in a similar way to produce the complete image.

Exercises

2. Can you prove that the angles α and β in Fig. 25-8 are equal?
3. Sketch two slightly diverging light rays that reflect in the mirror

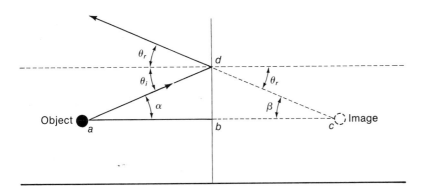

Fig. 25-8. The triangles *abd* and *cbd* are congruent triangles, since two of their angles are equal and they share one side.

for each of the three points on the object in Fig. 25-9. Locate the virtual image of each of the points and then sketch in the entirety of the image.

25-4 The Focal Point of a Curved Mirror

Consider a mirror whose shape is that of a segment of a spherical surface. This type of curved mirror is called a *spherical mirror* and the center of the complete sphere is called the *center of curvature* of the mirror. Light rays emanating from a point source placed at the center of curvature are reflected directly back to the center, as shown in Fig. 25-10*a*. If a beam of parallel light rays impinge upon it, as in Fig. 25-10*b*, they are focused toward a point halfway between the mirror and the center of curvature of the mirror. This point is called the *focal point* of the mirror.

The construction in Fig. 25-11 illustrates this behavior. The line *ABC* in Fig. 25-11 is called the *axis* of the mirror. The right triangle *APC* has an interior angle θ and sides of length *AP* and *AC*. The right triangle *APB* has an interior angle 2θ and sides given by *AP* and *AB*. We can write, then:

$$\tan \theta = AP/AC$$

$$\tan 2\theta = AP/AB$$

In the limit of small angles, $\tan \theta$ is very nearly equal to θ. For *small angles*, then:

$$\frac{AP}{AC} = \frac{1}{2}\frac{AP}{AB}$$

or

$$AB = \tfrac{1}{2}AC \qquad\qquad (25\text{-}1)$$

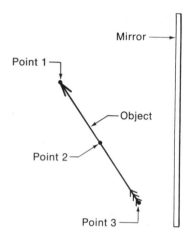

Point 1

Mirror →

Object

Point 2

Point 3

Fig. 25-9.

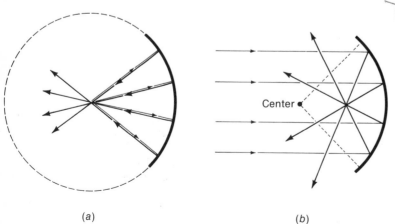

Center

(a)   *(b)*

Through center reflects back through center

Parallel reflects back through focal pt. R/2

Fig. 25-10. (*a*) Light rays emanating from the center of curvature of a spherical mirror are reflected back to the center of curvature. (*b*) Parallel light rays are focused to a point halfway between the mirror and its center of curvature.

R/2 = f

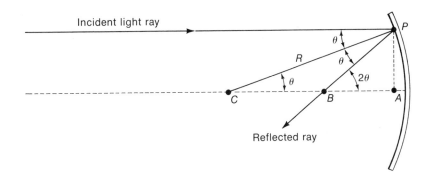

Fig. 25-11. A light ray strikes a mirror at point P. The radius R from the center of curvature of the mirror, C, makes an angle θ above the horizontal. The law of reflection tells us the reflected ray will make an angle 2θ with the incident ray.

If the angles are to be small, the mirror must be a shallow one, that is, only a small segment of a sphere. In that case the point A is nearly in contact with the surface of the mirror and the distance AC is very nearly R. Thus we conclude that the focal point B, at which all initially parallel light rays coincide, is halfway between the mirror and its center of curvature. The distance from the mirror to the focal point is called the *focal length, f,* so we have

$$f = \frac{R}{2} \qquad (25\text{-}2)$$

for a shallow spherical mirror.

If a spherical mirror is too deep, that is, if it comprises a large section of a sphere, the small-angle approximation used to derive (25-1) is not valid for rays of light striking far from the center. These light rays do not pass through the same point as rays closer to the center. This effect is called *spherical aberration* and can be remedied by using a mirror with the shape of a paraboloid (which closely resembles a sphere in its central portions), as seen in Fig. 25-12.

Exercise

4. At which points in Fig. 25-11 and in Fig. 25-12 would you place a small point of light so that its reflected rays emerging from the mirror would all be parallel? This method is used in flashlights, headlights, and searchlights.

25-5   Images Produced by a Concave Mirror

When the center of curvature of a spherical mirror is on the same side as the reflecting surface, as in Fig. 25-13, the mirror is *concave.* When it is on the opposite side, we call it *convex,* a case we will discuss in Section 25-6. Consider now the image of a point object that is produced by a concave mirror. In Fig. 25-13 the object O

[handwritten margin notes: convex — CURVATURE ON OPPOSITE SIDE as the reflecting surface; concave — CURVATURE ON SAME SIDE AS THE REFLECTING SURFACE]

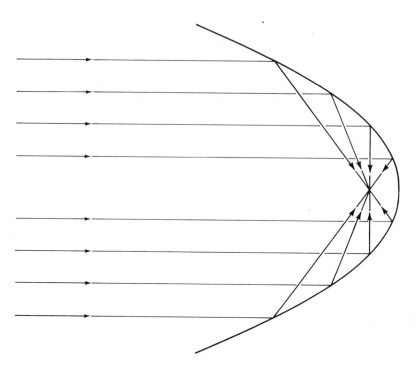

and the image O' are, respectively, distances s and s' from the mirror. Referring to Fig. 25-13b if ϕ, θ, and ϕ' are all small angles, we may write:

$$\phi = \frac{h}{s} \qquad \theta = \frac{h}{R} \qquad \phi' = \frac{h}{s'} \qquad (25\text{-}3)$$

Further, since an exterior angle of a triangle equals the sum of the two opposite interior angles, we have:

$$\theta = \psi + \phi$$

and using the law of reflection

$$\phi' = \theta + \psi$$

We can eliminate ψ between these two equations to obtain:

$$\phi + \phi' = 2\theta \qquad (25\text{-}4)$$

Substituting each of the expressions (25-3) into (25-4), we finally obtain a relation between s and s' in terms of the mirror's radius of curvature:

$$\frac{1}{s} + \frac{1}{s'} = \frac{2}{R}$$

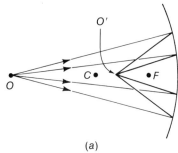

(a)

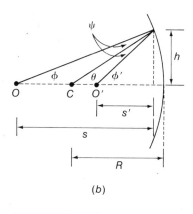

(b)

Fig. 25-13. The point object, O, is a distance s from the mirror. The image, O', is located s' away from the mirror.

Recalling that $R/2$ is the focal length, f, we have:

$$\frac{1}{s} + \frac{1}{s'} = \frac{1}{f}$$ (25-5)

Note that the result (25-5) is independent of the angles ϕ and ϕ', provided that all angles are small. Thus all rays striking the mirror from the point object at a distance s form an image at a distance s'. A collection of points constituting an extended object are similarly imaged. The size and location of such an object is easily determined by a graphical procedure such as that shown in Fig. 25-14. The particular rays shown there were chosen because their paths are easily deduced. One of them, 1, is initially parallel to the mirror's axis and hence is reflected to the focal point. Another, 2, strikes the mirror at its vertex and is reflected at an angle below the horizontal that is equal in magnitude to the initial angle above the horizontal. Still another, 3, passes through the focal point initially, so it reflects parallel to the axis. Finally, one ray, 4, passes through the center of curvature and then returns through the center of curvature. Actually, only two of these rays are required to locate the image which, in this case, is real and inverted. If the mirror is shallow, so that the approximations made in obtaining Eq. (25-5) are valid, and if the object is small enough, the image will be at the location given by (25-5). In general, when we use any two of the rays illustrated in Fig. 25-14 to locate an image, we call it a *principle ray diagram*.

If the mirror is too deep, i.e., if its diameter is comparable to f, the rays striking far off the axis will not all be focused to one point and the image will be blurred. If the object is too large, its image will not lie on a flat surface when the object itself is a plane figure. *Unless explicitly mentioned, we will assume that all mirrors discussed in subsequent sections are sufficiently large and shallow so that the approximations of a single focal point and a plane-image surface are adequate.*

(handwritten margin notes:)
① initally parallel → reflected to the focal pt
② strikes at vertex reflected at an angle below the horizontal equal to the initial angle above the horizontal
③ through the focal point reflects parallel to axis
④ through the center of curvature → returns through center of curvature

location given by
$$\frac{1}{s} + \frac{1}{s'} = \frac{1}{f}$$

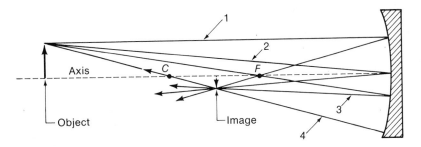

Fig. 25-14. Four rays of light emanating from the tip of the object are followed through their reflections. These particular rays were chosen because of the ease of their analysis. Note that they all cross again at one point, indicating a real image located there.

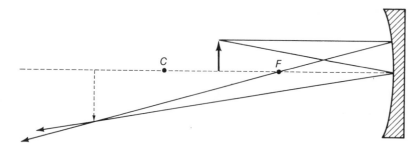

In Fig. 25-14 the object was beyond *C;* the image was formed between *C* and *F* and it was real, inverted, and smaller than the object. In Fig. 25-15 we see an object between *C* and *F*. Now the image is formed beyond *C* and is larger than the object.

Finally, in Fig. 25-16, we see the result of placing the object between *F* and the mirror. The rays reflected by the mirror do not cross anywhere. However, they appear to be diverging from a point *behind* the mirror. Thus, an observer stationed to the left of the object in Fig. 25-16 would see an enlarged, upright image of the object that appears to be behind the mirror. Such an image, we have said, is called a virtual image and is indicated by the dotted arrow in Fig. 25-16. This arrangement of object and mirror is precisely the one used when we use a curved mirror as a shaving mirror or a make-up mirror to provide a magnified view of our face.

In each of the three preceding cases, we could have used (25-5) directly without resorting to a graphical construction. In order to cover all cases, including virtual images, we need only adopt a sign convention: that the object distance *s* and the image distance *s'* are considered positive when they lie on the same side of the mirror as the focal point. The focal length, *f*, is also taken as positive. If *s'* is found to be negative, the image lies *behind* the mirror and is virtual.

Case 1: $s \geqslant 2f$ (Fig. 25-14)

By Eq. (25-5), $\dfrac{1}{s'} = \dfrac{1}{f} - \dfrac{1}{s}$, hence

$$s' = \frac{fs}{s - f}$$

Since $s \geqslant 2f$, we have

$$f \leqslant s' \leqslant 2f$$

BEYOND C —
  real, inverted, smaller than object

BETWEEN C and F
  — image larger than object

BETWEEN F & mirror
  virtual, enlarged
  upright image behind the mirror

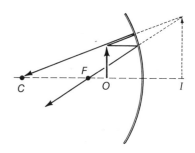

Fig. 25-16.   The image *I* is virtual, erect, and enlarged when the object *O* lies between *F* and the mirror.

The image is thus real and between C (since $R = 2f$) and F.

Case 2: $f \leqslant s < 2f$ (Fig. 25-15)

Again, $s' = \dfrac{fs}{s - f}$ but now s lies between $2f$ and f so we have

$$s' > 2f$$

That is, the image is real and beyond C.

Case 3: $s < f$ (Fig. 25-16)

Since $s' = \dfrac{fs}{s - f}$ and $s < f$, the denominator is negative and so s' is negative, hence the image is behind the mirror.

Example   We wish to form a real image of a light bulb filament on a screen. A concave mirror with a focal length of 10 cm is placed 1 m from the screen. Where should the filament be placed so that its image will fall on the screen?

Solution   We can make a graphical construction, working backwards from the image to deduce where the object should be placed. But it is quicker simply to solve (25-5) for the unknown, s:

$$\frac{1}{s} = \frac{1}{f} - \frac{1}{s'} = \frac{s'}{fs'} - \frac{f}{fs'} = \frac{s' - f}{fs'}$$

Hence
$$s = \frac{fs'}{s' - f} = \frac{10 \text{ cm} \cdot 100 \text{ cm}}{100 \text{ cm} - 10 \text{ cm}} = 11.1 \text{ cm}$$

That is, the lamp filament should be placed 11.1 cm from the mirror, as indicated in Fig. 25-17.

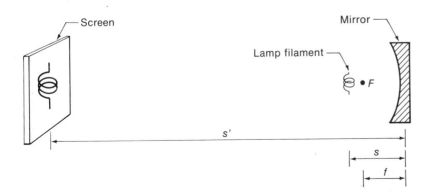

Fig. 25-17.

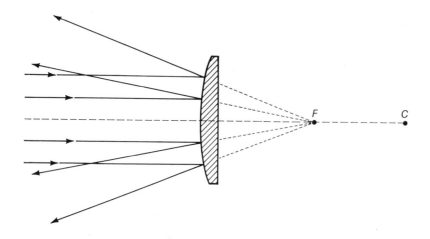

Fig. 25-18. Parallel light rays diverge after striking a convex mirror. They appear to diverge from the point F, called the focal point. The distance from the mirror to F is half the radius of curvature, as it was for a concave mirror.

Exercises

5. A concave mirror with a focal length of 10 cm is placed 6 cm behind an object. Where is the object's image located? Where would an observer stand to see it? Is it real or virtual? Is it erect or inverted?

6. A lamp filament 1 in. in height is placed 4 in. away from a concave mirror with a radius of curvature of 6 in. Use a principle ray diagram to find the location of the image. Is it erect or inverted? Is it real or virtual?

25-6 Convex Mirrors

A convex mirror, as shown in Fig. 25-18, causes incident parallel light rays to *diverge* after reflection. We can still define a focal point F, but it now lies *behind* the mirror. A proof similar to that illustrated in Fig. 25-11 shows that the distance f from the reflecting surface to F is again one-half the radius of curvature:

$$f = \frac{R}{2}$$

Several images produced by convex mirrors are shown in Fig. 25-19. The images formed are *always* virtual, erect, and reduced in size compared to the object. A familiar example is given by the view we see when we look into a reflecting, spherical Christmas tree ornament.

[handwritten margin notes:] convex mirror focal point behind the mirror

[handwritten margin notes:] images are always virtual, erect and reduced in size compared to the object

Equation (25-5) generalizes nicely to include convex mirrors if we define f to be a *negative* number when the focal point lies behind the mirror. Then we have:

$$\frac{1}{s} + \frac{1}{s'} = \frac{1}{f} \qquad \text{(where } f < 0\text{)}$$

Solving for the image distance, s':

$$s' = \frac{fs}{s - f}$$

The denominator of this expression is clearly positive, since s is positive and f is negative. The numerator is always negative, so that s' is always negative; that is, the image is always behind the mirror and, hence, virtual.

[handwritten notes in margin:]
s is positive
f is negative
s' is negative
image behind mirror
virtual

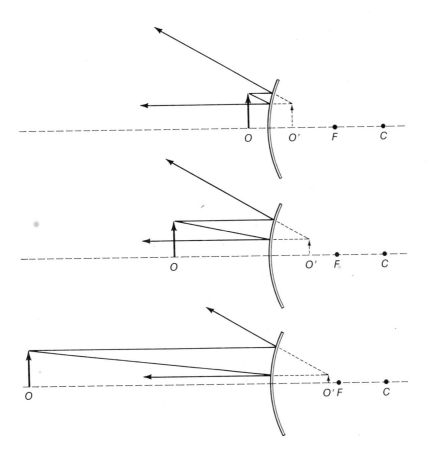

Fig. 25-19. Images formed by a convex mirror are always virtual, erect, and smaller than the object.

Exercise

7. A child's face is one foot from the center of a spherical Christmas tree ornament with a radius of 2 in. Find the location of the child's image. Is it erect or inverted?

25-7   Magnification

The images formed by curved mirrors are generally a different size than the object. The ratio of the image size, h', to the object size, h, is called the *magnification, m:*

$$m = \frac{h'}{h} \tag{25-6}$$

When the image is larger than the object, the magnification is greater than unity. A magnification less than unity indicates an image that is reduced in size from that of the object.

The dimension of an object parallel to the axis of a mirror is called its *longitudinal* dimension, while those perpendicular to the axis are the *lateral* dimensions. The magnifications of lateral and longitudinal dimensions are *not* generally equal. This inequality often leads to some interesting distortions in telephoto pictures.

To find the lateral magnification of a concave mirror, the construction in Fig. 25-20 is useful. From the law of reflection we have

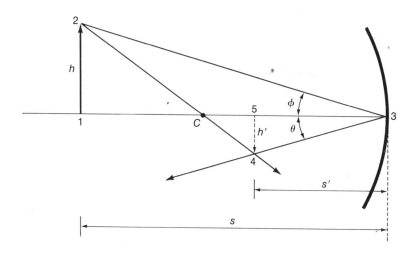

Fig. 25-20. The lateral magnification is given by the ratio of image height to object height: $m = h'/h$.

$\theta = \phi$, and from the triangles 123 and 345 we have:

$$\tan \phi = h/s$$

$$\tan \theta = -h'/s'$$

where a minus sign has been included to indicate an inverted image. Thus, since $\theta = \phi$,

$$m = h'/h = -s'/s \qquad (25\text{-}7)$$

Equation (25-7) may be regarded as being completely general if (a) the focal length of concave mirrors is taken to be positive and that convex mirrors negative; (b) the image distance is taken to be positive if the image is on the same side of the mirror as the object, negative if on the opposite side; (c) a negative magnification denotes an inverted image, a positive one an erect image. Alternatively, and perhaps more easily, one can construct a rough graphical solution to determine whether the image is erect or inverted; then use (25-7) to determine the magnitude of the magnification.

The longitudinal magnification can be found by finding the motion, $\Delta s'$, of the image parallel to the axis when the object is moved parallel to the axis a distance Δs. Then the longitudinal magnification, m_l, is given by:

$$m_l = \frac{\Delta s'}{\Delta s}$$

Both $\Delta s'$ and Δs can be approximated by differentials ds' and ds. When we take differentials of both sides of (25-5), we obtain

$$-\frac{ds}{s^2} - \frac{ds'}{(s')^2} = 0$$

Hence

$$m_l = \frac{\Delta s'}{\Delta s} \approx \frac{ds'}{ds} = -\left(\frac{s'}{s}\right)^2 = -m^2 \qquad (25\text{-}8)$$

that is, the longitudinal magnification is the *square* of the transverse magnification. The minus sign simply indicates that when s increases, s' decreases, and vice versa.

Example At the end of Section 25-3, we saw that $s' = -s$ for a plane mirror; that is, the image lies *behind* the mirror a distance equal to the

[Handwritten margin notes:]

focal length
  positive - concave
  negative - convex

$s' =$ positive
  image on same side
    as object

$s' =$ negative
  opposite side of object

negative magnification
  inverted image
positive magnification
  erect image

object distance. This can be seen from Eq. (25-5) if we let $f \to \infty$ (since R is infinite for a plane mirror):

$$\lim_{f \to \infty} \left(\frac{1}{s} + \frac{1}{s'} \right) = \lim_{f \to \infty} \frac{1}{f} = 0$$

or

$$\frac{1}{s'} = -\frac{1}{s}$$

so $s' = -s$. The lateral magnification is given by

$$m = \frac{-s'}{s} = +1$$

Thus the image is erect and the same size as the object. In this special case the longitudinal magnification is equal in size to the lateral magnification:

$$m_l = -m^2 = -1$$

where the minus sign indicates the familiar fact that the image seen in a mirror moves *longitudinally* in the opposite direction from the object.

Example   A spherical mirror with a focal length of 42 in. is used to project a real image of a distant tower onto a screen. If the tower is 200 ft tall and 1 mi away, find the height of its image on the screen.

Solution   In this case, the object distance s is so much larger than the focal length f that the image is almost exactly at the focal point. To see this, we note that

$$\frac{1}{s'} = \frac{1}{f} - \frac{1}{s} \approx \frac{1}{f} \qquad \text{if } s \gg f$$

so that $s' \approx f$ for $s \gg f$. Thus we find that

$$m = \frac{-s'}{s} \approx \frac{-f}{s}$$

Since s is one mile,

$$m = \frac{-42 \text{ in.}}{12 \cdot (5280) \text{ in.}} = -6.63 \times 10^{-4}$$

The image height is thus

$$h' = mh = -6.63 \times 10^{-4} \cdot 200 \text{ ft} = -0.132 \text{ ft} = -1.6 \text{ in.}$$

where the minus sign indicates an inverted image. In Chapter 27 we will see how a magnifying lens can be used to inspect this small image so closely that we actually see more detail than by looking directly at the distant tower. This is the basis of the reflecting telescope.

Exercise

8. A concave mirror with a focal length of 5 cm has an object of 1-cm height placed 7 cm from the mirror. Find the lateral size of the image from (25-7). Check your result by making a graphical construction and measuring the height of the image directly.

25-8   Summary

A simple empirical law describes specular reflection:

$$\theta_r = \theta_i$$

where θ_r is the angle between the reflected ray and the normal to the reflecting surface, while θ_i is the angle between the incident ray and the normal.

When light rays from an object are reflected by either plane or spherical mirrors, images are produced. In a real image, light rays actually cross over at points of the image, while in a virtual image, light rays only appear to come from points of the image. A real image can be projected on a screen while a virtual image cannot.

If a mirror constitutes a segment of a sphere of radius R, it is found that the location of an image is governed by

$$\frac{1}{s} + \frac{1}{s'} = \frac{1}{f}$$

$f = {}^{R}\!/_{2}$

where s is the mirror-object distance, s' is the mirror-image distance, and $f = R/2$ is the focal length. For a converging (concave) mirror, f is positive, while for a diverging (convex) mirror, f is negative. For a virtual image, s' is negative, while for a real image, s' is positive.

converging - concave $f = +$
diverging - convex $f = -$
virtual image $s' = -$
real image $s' = +$

The lateral magnification of a mirror is given by $m = -s'/s$. If s' is considered to be positive when on the same side of the mirror as the object (real image) and negative when on the opposite side of the mirror (virtual image), a negative magnification corresponds to an inverted image, while a positive value corresponds to an erect image.

– magnification = inverted
+ magnification = erect

The longitudinal magnification is given by $m_l = -m^2$ and is always negative. The minus sign simply means that the object and image always move in opposite directions longitudinally.

Problems

25-1. A ray of light moves from a point O, strikes a plane mirror, then reflects to the point O'. If O, O', and the plane of the mirror are fixed, the law of reflection tells us where this particular light ray strikes the mirror. Show that the actual path followed by the reflected light ray is the shortest possible path from O to O' that involves reflection from the mirror. This result was first shown by Hero of Alexandria around the year 100 A.D.

25-2. When a light ray strikes a pair of mirrors that forms a right angle, as in Fig. 25-21, the reflected ray returns along a path that is parallel to the incident light ray's path. Prove this statement.

25-3. When we look into a plane mirror, left and right are reversed. However, if we look into the vertex of a pair of plane mirrors that has been joined at right angles, our appearance is just as others see us. Explain both of the above facts by use of ray diagrams.

25-4. A barber shop customer sits at a distance $3L/4$ from a wall that is entirely covered with a plane mirror. The wall which is $L/4$ behind him is also a plane mirror, so that the two large plane mirrors are parallel to one another. Explain why he sees multiple images of himself when he looks into one mirror. What is the apparent distance of the various images he sees?

25-5. A concave mirror with a diameter of 30 cm and a focal length of 75 cm is placed at one end of a tube 30 cm in diameter and 1 m in length. The tube is then pointed at a distant $(s \gg f)$ building, so that an image is formed within tube. (a) Sketch the principle rays that locate this image. (b) Is the image real or virtual? Inverted or erect? (c) The building is 1 km distant and has a height of 50 m. Find the height of the image. (d) If this image is projected onto an opaque screen located on the axis of the tube, which is just large enough to accommodate the image, will it block a large fraction of the incident light rays?

25-6. When an object is placed 40 cm from a mirror, it produces a real image with a lateral magnification of $m = -\frac{1}{2}$. (a) Is the mirror concave or convex? (b) Is the image erect or inverted? (c) Find the focal length of the mirror.

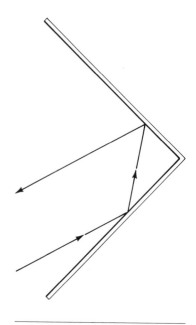

Fig. 25-21.   A light ray reflected from a pair of mirrors at right angles to one another is returned along a path parallel to the original path.

25-7. Prove that the focal length, f, of a *convex* mirror is given by $f = R/2$. You must use the usual small-angle approximations.

25-8. A penny is held 10 cm away from a silvered Christmas tree ornament with a 5-cm diameter. Find the position and size of the image.

25-9. A piece of photographic film is placed at the focal point of a concave mirror with a 2-m focal length. The real image of the moon is projected onto the film to make a photograph. If the moon has a diameter of 3,480 km and is 386,000 km distant, find the diameter of this image.

25-10. Find the type of mirror and radius of curvature that will produce an erect image one-quarter the size of an object that is 20 cm away from the mirror.

25-11. A bright lamp filament is 1 cm in length. We wish to project a 50-cm-long image of the filament onto a wall 10 m distant from a mirror. Describe the type of mirror to be used, its focal length, and the position of the lamp filament.

25-12. Draw a principle ray diagram that illustrates the use of a shaving mirror to produce an enlarged, erect, virtual image.

25-13. A spherical concave shaving mirror has a radius of curvature of 1 ft. (*a*) What is the magnification when the face (which is the object) is 4 in. from the vertex of the mirror? (*b*) Where is the image located?

Chapter **26**

Refraction and Lenses

SECTIONS 1-8

26-1 Introduction When light strikes a transparent substance, a portion of the light is reflected, as we learned in Chapter 25. The remainder of the light moves into the transparent substance. If the speed of light in this new medium differs from its value in the original medium, the phenomenon of *refraction* (bending) occurs. Since the speed of light is generally less in transparent liquids and solids than it is in air, refraction is a common phenomenon. We will see that focusing of light rays can be obtained by curving the boundary between two substances in which the speed of light differs. Devices that utilize this type of focusing are called *lenses* and will be discussed in this chapter.

refraction - speed of light in this new medium differs from its value in the original medium

26-2 The Law of Refraction

If a ray of light moves into a medium where it moves more slowly, it is deflected towards the direction of the normal to the boundary. This is shown in Fig. 26-1, where the lower medium is the one in which the light moves more slowly, i.e., $c_1 > c_2$. The change in direction is largest for large angles θ_1, and vanishes for normal incidence (when $\theta_1 = 0$). The precise way in which θ_1 and θ_2 are related is given by *Snell's law:*

$$\frac{1}{c_1} \sin \theta_1 = \frac{1}{c_2} \sin \theta_2$$

The reason for this bending is due to the fact that one side of a wave front slows before the other, as illustrated in Fig. 26-2.

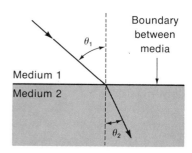

Fig. 26-1. A light ray making an angle θ_1 with the normal in medium 1 is deflected to a new angle, θ_2, in medium 2.

Snell's law can be obtained by applying Huygens' principle to these wave fronts. As in obtaining the law of reflection in the preceding chapter, a simple two-point Huygens' construction will suffice. Figure 26-3 is such a construction. The leftmost end of the wave front is just striking the slow medium in that construction. A time interval Δt is chosen so that $R_1 = c_1\Delta t$ just brings the rightmost end of the wave front into contact with the slow medium. In this same time interval the left end has moved a distance

$$R_2 = c_2\Delta t$$

Taking sines of the angles θ_1 and θ_2,

$$\sin \theta_1 = \frac{R_1}{\text{hypotenuse}} = \frac{c_1\Delta t}{\text{hypotenuse}}$$

$$\sin \theta_2 = \frac{R_2}{\text{hypotenuse}} = \frac{c_2\Delta t}{\text{hypotenuse}}$$

Now if we solve both of these equations for Δt and equate the two resulting expressions, we find

$$\frac{\sin \theta_1}{c_1} = \frac{\sin \theta_2}{c_2}$$

which is Snell's law.

Substances in which the velocity of light is much less than it is in the surrounding air are the substances in which the phenomenon of refraction is most pronounced. It is customary to define an *index of refraction n* for substances equal to the ratio of c, the velocity of light in vacuum, to c', the velocity of light in the substance:

$$n = \frac{c}{c'}$$

This index of refraction n is a measure of the substance's ability to refract a light ray, since Snell's law now becomes

$$n_1 \sin \theta_1 = n_2 \sin \theta_2 \qquad (26\text{-}1) \quad \text{Snell's law}$$

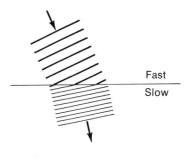

Fast

Slow

Fig. 26-2.   For all light rays that are not normally incident upon the slow medium, one end of the wave front slows before the other. This causes a bending of wave fronts and, hence, of the light ray.

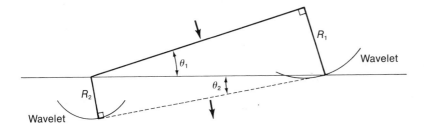

R_1

Wavelet

θ_1

θ_2

R_2

Wavelet

Fig. 26-3.   The new wave front indicated by the dashed line is the locus of all secondary wavelets emanating from the original wave front. Only those from the two endpoints are shown for clarity.

Equation (26-1) is the usual way of stating Snell's law. Table 26-1 gives values of n for a variety of substances. In the design of most optical instruments, the index of refraction with respect to *air* rather than with respect to vacuum is needed. However, since the velocity of light in air is so similar to that in vacuum, the distinction is rarely significant.

TABLE 26-1   Indexes of Refraction*

Substance	n	Substance	n
Air	1.0003	Glass, heaviest flint	1.89
Water	1.33	Glass, crown	1.52
Ethyl alcohol	1.36	Rock salt	1.54
Carbon bisulfide	1.63	Polystyrene	1.49
Magnesium fluoride	1.38	Polyvinyltoluene	1.58

*At room temperature for sodium yellow light ($\lambda = 0.589 \ \mu$m).

The speed of light in most substances varies slightly with wavelength. Thus, light waves in a material medium exhibit *dispersion*. The variation of index of refraction for several wavelengths is shown for two types of glass in Table 26-2.

TABLE 26-2   Variation of n with Wavelength

Wavelength (microns)	0.4	0.5	0.6	0.7
n (Crown glass)	1.532	1.522	1.517	1.513
n (Flint glass)	1.658	1.638	1.627	1.620

Example   A ray of light moves from within a body of water into the air, as sketched in Fig. 26-4. If θ_1 has a value of 20°, find θ_2.

Solution   Solving for θ_2 in Snell's law, we have:

$$\sin \theta_2 = \frac{n_1}{n_2} \sin \theta_1$$

$$\theta_2 = \arcsin\left(\frac{n_1}{n_2} \sin \theta_1\right).$$

From Table 26-1 we find $n_1/n_2 = 1.33$ and $\sin(20°) = 0.342$, so that

$$\theta_2 = \arcsin(0.455) = 27.1°$$

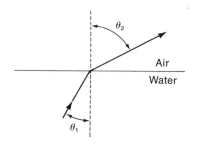

Fig. 26-4.

Exercises

1. In the preceding example, what value of θ_1 would lead to $\theta_2 = 90°$? Sketch the light ray in this situation.

2. In Fig. 26-5 both light rays make the same initial angle of incidence on the glass ($n = 1.5$) surface. Which is bent through the largest angle, that is, which has the largest value of $\theta_1 - \theta_2$?

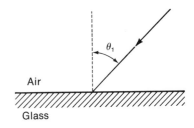

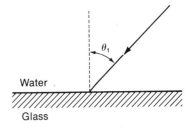

Fig. 26-5.

26-3   Total Internal Reflection

When a ray of light moves into a region with a larger index of refraction, it is bent closer to the normal. When it moves into a region with smaller index of refraction, it is bent away from the normal. Clearly, at a large enough incident angle, called the *critical angle*, a light ray can be made to emerge into the medium with smaller index with $\theta = 90°$ – that is, with the light ray moving exactly parallel to the boundary (see Exercise 1). Once this situation occurs, what is the effect of increasing the incident angle still more? Is the light ray bent so much that it reenters the first medium? If this happens, how do we apply Snell's law?

First, consider the critical angle of incidence for a light ray emerging from glass ($n = 1.5$) into air. Setting $\theta_2 = 90°$ in Snell's law, we can find the critical angle of incidence, θ_c.

$$\sin \theta_c = \frac{1}{1.50} \sin 90° = 0.67$$

$$\theta_c = 42° \quad \text{critical angle of incidence}$$

If we increase θ_1 beyond θ_c, Snell's law leads to an absurd result for θ_2:

$$\sin \theta_2 = \frac{n_1}{n_2} \sin \theta_1 = 1.5 \sin \theta_1$$

So for $\theta_1 > 42°$, $\sin \theta_2 > 1$. But *no* angle has a sine greater than unity! The physical interpretation of this result is that Snell's law is no longer valid for $\theta_1 > \theta_c$. Indeed, experimentally it is found that the light ray abruptly changes its behavior at this incident angle, as shown in Fig. 26-6. For incident angles greater than θ_c, the light ray remains in the original medium, that is, it reflects off the boundary. As with any other form of reflection, this light ray obeys the law of reflection, $\theta_1 = \theta_2$.

$\theta_1 > \theta_2$ light ray remains in the original medium – it reflects off the boundary $\theta_1 = \theta_2$

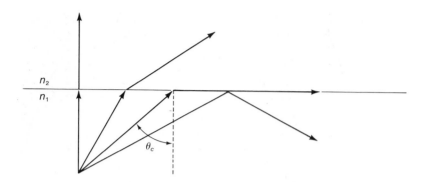

Fig. 26-6. A light ray moving from a medium of large refractive index into one of smaller refractive index. At an incident angle of θ_c, the emerging light ray is parallel to the surface. For larger incident angles, the light ray totally reflects back into the lower medium.

Since the critical angle for glass in air is less than 45°, glass prisms fabricated as shown in Fig. 26-7 can utilize the total reflection phenomenon. These devices reflect light very efficiently, since the internal reflection is indeed total. A small amount of specular reflection at the incident and exit faces of the prisms causes less than 100% of the light to follow the paths shown in Fig. 26-7. The loss at each of the two faces generally lies between 4% and 5%, so that only about 91% of the light is effectively reflected. This fraction still exceeds that for one of the best metallic reflectors, aluminum, and further is not subject to the decrease in reflectivity caused by tarnishing of a metal surface. So-called antireflection coatings (to be discussed in Chapter 27) can be applied to the faces of a prism, making the overall efficiency as high as 98%.

Total internal reflection is also responsible for the action of light pipes: devices that are used to convey light to or from relatively inaccessible areas.* Some examples are shown in Fig. 26-8. A striking example of light trapped by total internal reflection is shown in the photograph of a physics lecture demonstration in Fig. 26-9.

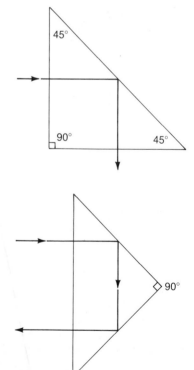

Exercises

3. Describe how Snell's law can be used to measure the index of refraction of a substance.

4. How can you use the phenomenon of total internal reflection to measure the index of refraction of a substance?

26-4   Refraction by Slabs and Prisms

When light is incident on a plane glass surface, we have seen that the light is refracted toward the normal. If a plane slab of glass is

*See "Communication by Optical Fiber" by J. S. Cook, *Scientific American,* November 1973; or "Integrated Optics" by P. K. Tien, *Scientific American,* April 1974.

Fig. 26-7. Total internal reflection in glass prisms.

Fig. 26-8. (a) A flexible light pipe partially disassembled to show the fine, flexible fibers of which it is fabricated. (Courtesy American Optical Corporation) (b) Individual fibers within this device are kept rigidly parallel to each other so that a true image is faithfully transported along its length. (Courtesy American Optical Corporation)

Fig. 26-9. A beam of laser light is trapped by total internal reflection within a falling stream of water.

involved, as in Fig. 26-10, the refraction that occurs when the ray reemerges into air precisely offsets that that occurred at entry into the glass. The net result is a parallel displacement of the light ray. However, if the second surface of the slab is tipped at an angle with respect to the first surface, the glass is in the form of a prism and the emergent light ray experiences a net change in angle, called the *deviation*.

A prism, as shown in Fig. 26-11, refracts an incident light ray through an angle α on entrance into the prism and by a further angle, β, on exit from the prism. The total deviation, γ, is equal to the sum of α and β and depends upon the index of refraction of the prism as well as the angle of incidence of the light ray. We can calculate the relation between γ and the prism's index of refraction most easily for the special case when $\alpha = \beta$. From Fig. 26-12 we then have $\theta = \phi/2$ and

$$\theta_1 = \theta + \alpha = \frac{\phi}{2} + \frac{\gamma}{2}$$

From Snell's law:

$$\sin \theta_1 = n \sin \theta$$

or

$$\sin\left(\frac{\phi + \gamma}{2}\right) = n \sin \frac{\phi}{2} \qquad (26\text{-}2)$$

which gives the desired relation between the deviation, γ, the shape of the prism, ϕ, and the index of refraction, n.

For most substances, the index of refraction is dependent upon the wavelength of the light, as indicated in Table 26-2 for two types of glass. Since the index of refraction is generally larger for the shorter wavelengths, the violet-blue end of the visible spectrum is generally deviated more by a prism than the red-orange end of the spectrum, as indicated in Fig. 26-13. In fact, Newton first discovered the nature of the visible spectrum of sunlight by use of a prism.

A similar phenomenon occurs in the formation of a rainbow. Strong sunlight shining on a region where water droplets abound (usually a rain shower) is refracted and internally reflected by each nearly spherical droplet, as in Fig. 26-14. The secondary bow that is sometimes seen outside the primary bow is due to the more complex path shown in Fig. 26-14b. Since the reflections involved in rainbow formation are within the critical angle, they are not total, so the loss of light due to the extra reflection in b causes the secondary bow to be quite faint.

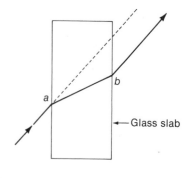

Fig. 26-10. The refraction toward the normal, at a, is balanced by the refraction away from the normal, at b.

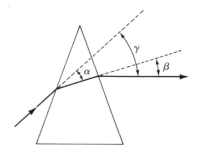

Fig. 26-11. A prism deviates a ray of light through a total angle γ.

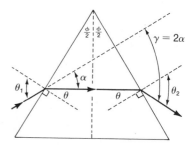

Fig. 26-12. A light ray passing through a prism with $\alpha = \beta$.

Exercises

5. Two slabs, one of crown glass and one of flint glass, are struck by a ray of light exactly as in Fig. 26-10. If the slabs are of equal thickness, which one displaces the emerging ray by the greatest distance?

6. A prism with $\phi = 30°$ is oriented as in Fig. 26-12 with $\theta_1 = \theta_2$. If the prism is made of polystyrene and the light is sodium yellow light, find γ.

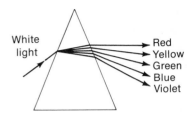

Fig. 26-13.   Due to the variation of n with wavelength, a prism disperses white light into its component colors.

26-5   Image Formation by a Plane Refracting Surface

When one looks down into a calm swimming pool, the bottom appears closer than it really is. This apparent decrease in depth is caused by refraction at the water's surface. In Fig. 26-15 we see two light rays from a point O. Their apparent origin is the point O', so that an observer in air would see a virtual image at O'. The ratio of s' to s can be calculated as follows: From the figure, $\tan \theta = b/s$ and $\tan \theta' = b/s'$. At small angles $\tan \theta \approx \sin \theta$, so we have (for small angles)

$$\frac{s'}{s} = \frac{\tan \theta}{\tan \theta'} \approx \frac{\sin \theta}{\sin \theta'} \tag{26-3}$$

Applying Snell's law,

$$\frac{\sin \theta}{\sin \theta'} = \frac{n'}{n} \tag{26-4}$$

where n' is the index of refraction of air and n that of water. Substituting (26-4) into (26-3):

$$\frac{s'}{s} \approx \frac{n'}{n} = \frac{1}{4/3} = \frac{3}{4}$$

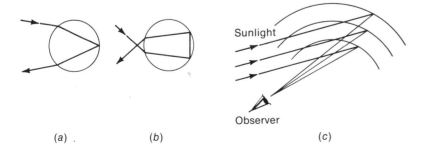

(a) .   (b)   Sunlight   (c)

Observer

Fig. 26-14.   The primary bow of a rainbow is produced by the two refractions and single internal reflection in raindrops (a). A fainter secondary bow involves the double internal reflection shown in (b). A diagram showing the three different paths followed by blue light, green light, and red light is shown in (c).

The apparent depth of an object in water viewed from air is 3/4 its actual depth.

26-6   Converging Lenses

[handwritten: CONVEX LENS ACTS LIKE CONCAVE MIRROR]

A parallel beam of light rays can be brought to a focus by refraction if the refracting substance has a curved surface. In Fig. 26-16 a spherical surface at the end of a glass rod is shown. The incident parallel rays that are at the greatest distance from the center of the rod are refracted most strongly due to their larger angle of incidence. If the angles of incidence are not too large, all the rays cross over a common focal point, *F*, inside the glass rod.

A second spherical refracting surface can be added to our glass rod, as in Fig. 26-17, to produce a *lens*. By application of Snell's law, we can find the angle of deviation of each ray at both surfaces. Ray tracing by a graphical construction then shows that an image is formed by such a lens, again subject to the restriction that the angles of incidence must be small enough so that small-angle approximations are adequate. Rather than analyzing this type of lens in detail, however, let us turn our attention to thin lenses.

The usual sort of lens we find in a telescope is close to an idealized *thin lens*. A thin lens has a thickness that is small compared to the radius of curvature of either of its surfaces and also compared to its focal length. Such is clearly not the case for the lens of Fig. 26-17. A variety of thin lenses are shown in Fig. 26-18. As we will soon see, those lenses that are thicker in their central portions (*a*, *b*, and *c* in Fig. 26-18) are converging lenses, which are analogous to concave mirrors in their focusing properties. Lenses that are thinnest in their central portions are diverging and analogous to convex mirrors.

We can show that thin lenses with spherical surfaces form images and we can also deduce their focal length by use of the con-

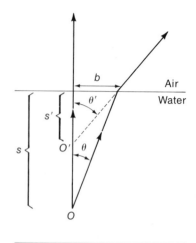

Fig. 26-15. The intersection of the two light rays is the location of the virtual image, *O'*, of the point *O*.

[handwritten: THIN LENS THICKNESS IS SMALL COMPARED TO THE RADIUS OF CURVATURE AND IT'S FOCAL LENGTH]

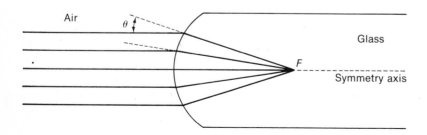

Fig. 26-16.   Light rays furthermost from the symmetry axis of a curved refracting surface have the greatest angle of incidence θ. They are thus refracted through a greater angle than rays closer to the symmetry axis, so that all rays cross at the focal point *F*.

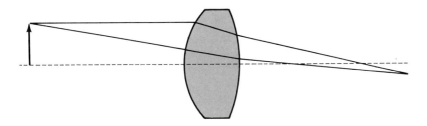

Fig. 26-17. A thick lens with two convex surfaces.

struction in Fig. 26-19. We wish to obtain a relation between object distance, s, image distance, s', and the quantities that are characteristic of the lens: its index of refraction, n, and its radii of curvature. From the geometry of Fig. 26-19, we can deduce the following exact statements:

$$\psi_1 + \phi = \alpha \qquad \psi_2 + \phi' = \delta \tag{26-5}$$

$$\beta + \gamma = \psi_1 + \psi_2 \tag{26-6}$$

For angles of incidence that are small, we may approximate the sine of an angle by the angle itself. Snell's law for the two refractions in Fig. 26-19 then becomes

$$\alpha = n\beta \qquad \text{and} \qquad n\gamma = \delta \tag{26-7}$$

Small-angle approximations also allow us to write:

$$\phi = \frac{h}{s} \qquad \phi' = \frac{h'}{s'} \qquad \psi_1 = \frac{h}{R_1} \qquad \psi_2 = \frac{h'}{R_2} \tag{26-8}$$

Combining the statements of (26-5) with (26-7), we find:

$$\psi_1 + \phi = \alpha = n\beta$$
$$\psi_2 + \phi' = \delta = n\gamma$$

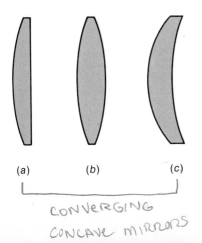

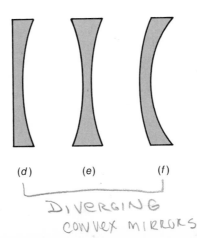

Fig. 26-18. Thin lenses. (a) Plano-convex. (b) Double convex. (c) Meniscus. (d) Plano-concave. (e) Double concave. (f) Meniscus. The first three are converging lenses, the last three are diverging lenses.

(a)   (b)   (c)   (d)   (e)   (f)

CONVERGING
CONCAVE MIRRORS

DIVERGING
CONVEX MIRRORS

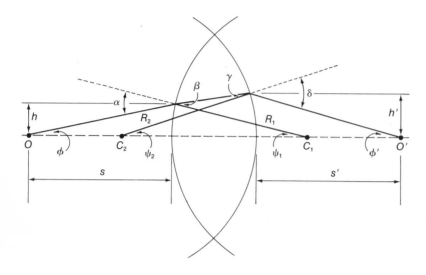

Fig. 26-19. A light ray passes through a thin lens from the object, O, to the image, O'.

Adding the information in (26-6), we obtain

$$\psi_1 + \psi_2 + \phi + \phi' = n(\beta + \gamma) = n(\psi_1 + \psi_2)$$

so that

$$\phi + \phi' = (n-1)(\psi_1 + \psi_2)$$

With the aid of (26-8), with $h' \approx h$, this becomes

$$\frac{1}{s} + \frac{1}{s'} = (n-1)\left(\frac{1}{R_1} + \frac{1}{R_2}\right) \qquad (26\text{-}9)$$

which is the desired relation between object distance, image distance, and characteristics of the lens itself.

Equation (26-9) can be rewritten in a form analogous to (25-5) for a curved mirror:

$$\frac{1}{s} + \frac{1}{s'} = \frac{1}{f} \qquad (26\text{-}10)$$

where

$$\frac{1}{f} = (n-1)\left(\frac{1}{R_1} + \frac{1}{R_2}\right) \qquad (26\text{-}11) \quad \text{LEN'S MAKER EQUATION}$$

Equation (26-10) can be used to locate the image of an extended object just as was done for the case of concave mirrors. Equation (26-11) is called the *lensmaker's equation* since it allows us to design a lens of any required focal length.

The arguments used in deriving (26-9) are equally valid if light moves from right to left in Fig. 26-19, that is, if O' is the object and O the image. The focal length of a lens thus does not depend upon which face light is incident. This is true whether the lens is symmetric (Fig. 26-18*b*) or not (Fig. 26-18*c*).

Example   A slab of glass with $n = 1.50$ is to be made into a lens with $f = 30$ cm. What values should be chosen for R_1 and R_2?

Solution   From (26-11) we find

$$\frac{1}{R_1} + \frac{1}{R_2} = \left(\frac{1}{1.5 - 1}\right)\left(\frac{1}{30 \text{ cm}}\right) = \frac{1}{15} \text{ cm}^{-1}$$

A continuous variety of choices for R_1 and R_2 exists. Some examples are:

$\frac{1}{30} + \frac{1}{30} = \frac{1}{15}$ i.e. $R_1 = R_2 = 30$ cm

$\frac{2}{45} + \frac{1}{45} = \frac{1}{15}$ i.e. $R_1 = 22\frac{1}{2}$ cm, $R_2 = 45$ cm

$\frac{5}{90} + \frac{1}{90} = \frac{1}{15}$ i.e. $R_1 = 18$ cm, $R_2 = 90$ cm

$\frac{1}{15} + 0 = \frac{1}{15}$ i.e. $R_1 = 15$ cm, $R_2 = \infty$

This last choice forms a plano-concave lens. Admitting negative values of R_2 (a meniscus lens), we can find more cases. One example is $\frac{1}{5} - \frac{2}{15} = \frac{1}{15}$, i.e., $R_1 = 5$ cm, $R_2 = -7\frac{1}{2}$ cm, which corresponds to a lens shaped like that in Fig. 26-18c.

negative values of R_2 = meniscus lens

Example   A lens with $R_1 = 15$ cm and $R_2 = 15$ cm is made of flint glass (see Table 26-2). What is the focal length of this lens for (a) red light with $\lambda = 0.7$ μm; (b) violet light with $\lambda = 0.4$ μm?

Solution   Calculating f for the two cases, we find:

$$1/f \text{ (red)} = 0.62\left(\frac{2}{15}\right) \text{cm}^{-1}$$

$$f(\text{red}) = 12.1 \text{ cm}$$

$$f(\text{blue}) = \frac{1}{0.658}\left(\frac{15}{2}\right) \text{cm} = 11.4 \text{ cm}$$

Thus parallel rays of white light are not all focused at precisely the same point. The violet component of the spectrum is focused 7 mm closer to the lens than the red component, with the remaining colors in between. This defect of a simple lens is called *chromatic aberration* and will be discussed further in Section 26-11.

Exercises

7.   A plano-convex lens is fabricated from glass with $n = 1.625$ ($= \frac{13}{8}$) and has a focal length of 8 cm. Find the radius of curvature of the convex surface.

8.   A thin lens is held in the sunlight. The sun's rays are very nearly parallel. It is observed that a sharp point of light is produced

when a screen is held 12 cm from the lens, as in Fig. 26-20a. An object is then placed 18 cm in front of the lens, as in Fig. 26-20b; where is the image located?

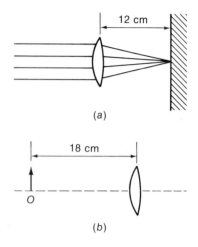

(a)

26-7   Diverging Lenses   CONCAVE LENS
ACTS LIKE CONVEX MIRROR

The thin lenses discussed in the preceding section focus or converge the light that passes through them. In this way they are quite similar to a concave mirror. A *diverging lens* causes light rays to spread out, as does a convex mirror. The focal point of a diverging lens is that point from which light rays appear to diverge when parallel light strikes the lens, as shown in Fig. 26-21.

The derivation of the lensmaker's equation for a diverging lens is essentially identical to that for a converging lens and we can, in fact, still use (26-11). It is customary to interpret the radius of curvature of a concave lens surface as negative, so that the focal length of a diverging lens becomes a negative quantity. Hence the name *negative* lens is sometimes used for a diverging lens and *positive* lens for a converging lens.

(b)

Fig. 26-20.

focal length of diverging lens is negative

Exercise

9.   A diverging lens similar to that shown in Fig. 26-18f has $R_1 = -3$ cm, $R_2 = +4$ cm, and $n = \frac{3}{2}$. Find its focal length.

26-8   Graphical Methods

The graphical methods used for locating images produced by thin lenses are similar to those used for spherical mirrors. In Fig. 26-22 the properties of certain rays are shown that are most convenient

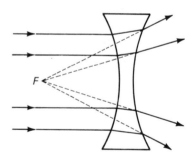

Fig. 26-21.   The focal point *F* of a diverging lens.

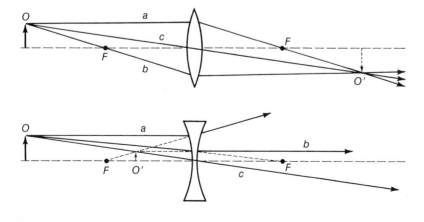

Fig. 26-22.   Rays that (a) are parallel to the axis, (b) pass through a focal point, or (c) pass through the center of a lens are easily analyzed.

for graphical constructions. (*a*) Rays parallel to the axis sub-sequently pass through a focal point. For a diverging lens they appear to come *from* a focal point. (*b*) Rays that pass through a focal point emerge parallel to the axis. For a diverging lens, rays that are directed toward the focal point beyond the lens emerge parallel to the axis. (*c*) Rays that pass through the center of the lens pass through a region where the faces are parallel and hence are not deviated. There is a parallel displacement for such a light ray, as in Fig. 26-10, but for a *thin* lens it is negligible.

There are locations of the object for a converging lens that produce real, inverted, reduced images; real, inverted, enlarged images; and virtual, erect, enlarged images, as was the case for concave mirrors. A diverging lens always produces virtual, erect, reduced images of a real object. All these cases are illustrated in Fig. 26-23. Comparison with the results of Sections 25-5 and 25-6 reveals the rather complete analogy between concave mirrors and converging lenses, as well as between convex mirrors and diverging lenses. Lateral magnification is still given by Eq. (25-7)

$$m = \frac{-s'}{s}$$ (25-7)

[handwritten margin notes:]
parallel → focal pt
focal pt → parallel
through center → not deviated

diverging lens always produces virtual, erect, reduced images of a real object

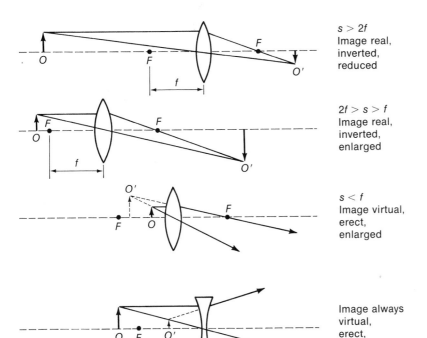

s > 2f
Image real,
inverted,
reduced

2f > s > f
Image real,
inverted,
enlarged

s < f
Image virtual,
erect,
enlarged

Image always
virtual,
erect,
reduced

Fig. 26-23. Various placements of the object lead to the different types of images indicated.

for a single lens, just as it was for images formed by a single curved mirror. Again, a sign convention may be adopted wherein (*a*) converging lenses have positive focal length and diverging lenses have negative focal length; (*b*) image distances on the opposite side of the lens from the object are taken as positive and those on the same side, negative (note the inversion with respect to the mirror convention); and (*c*) if the resulting magnification is negative, the image is inverted, while a positive magnification indicates an erect image.

Generally speaking, the number of times that a student finds need to calculate magnification is inadequate either to ensure or to justify memorization of such sign conventions. Instead, he should be able to use a graphical construction to deduce whether an image is real or virtual, erect or inverted. Equation (25-7) can then be used to calculate the absolute magnitude of the magnification. Since the longitudinal magnification for mirrors was derived from Eq. (25-5), which is identical to (26-10) for lenses, we obtain

$$m_l = -m^2 \qquad (25\text{-}8)$$

for lenses as well as for mirrors.

Example   In an opaque projector, the page to be projected is brilliantly illuminated, and a positive lens is used to cast a real image of the page on a distant screen. A page $8\frac{1}{2}$ in. by 11 in. is to be projected on a screen 20 ft away. If the final image is to be $4\frac{1}{4}$ ft $\times$ $5\frac{1}{2}$ ft, find the focal length of the required lens.

Solution   We desire a magnification of six times. The real image will be inverted, so we must have

$$m = \frac{-s'}{s} = -6$$

Since s' is 20 ft, we have $s = 20$ ft$/6 = 40$ in. The focal length is then given by (26-10):

$$\frac{1}{s} + \frac{1}{s'} = \frac{1}{f} \qquad (26\text{-}10)$$

which becomes $$\frac{1}{40 \text{ in.}} + \frac{1}{240 \text{ in.}} = \frac{7}{240 \text{ in.}} = \frac{1}{f}$$

or $$f = \frac{240 \text{ in.}}{7} = 34.28 \text{ in.}$$

Exercises

10. Construct a ray diagram to locate the image in Fig. 26-24. Use three rays from the tip of the arrow: one that passes through the center of the lens, one that passes through the

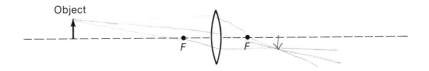

Fig. 26-24.

left focal point, and one that is initially parallel to the axis. Do all three rays (within drawing accuracy) pass through the same point?

11. Construct a ray diagram to locate the image in Fig. 26-25.

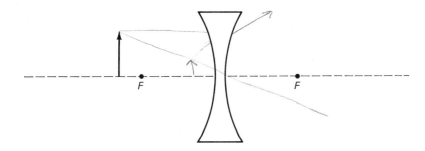

Fig. 26-25.

26-9   Aberrations   NOT ON HOURLY

In the last example of Section 26-6, we saw a defect in a lens—*chromatic aberration*, which is the inability of a simple lens to focus all colors in the same plane. A spherical mirror does not exhibit chromatic aberration, since the law of reflection is uniform for all colors. However, *both* spherical mirrors and lenses with spherical surfaces suffer from *spherical aberration:* the inability of a lens or mirror to focus to a single point *all* rays emanating from a point on the object. This aberration is related to our need for small-angle approximations in carrying out the proofs in Sections 25-4, 25-5, and 26-6. When rays strike far from the axis of a mirror or lens with spherical surfaces (violating our small-angle approximations), these rays are brought to a focus closer to the mirror or lens than the "focal point." This effect is shown in Fig.

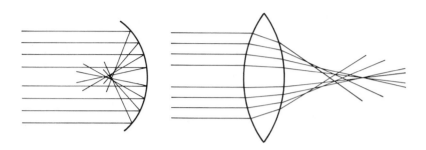

Fig. 26-26. Spherical aberration.

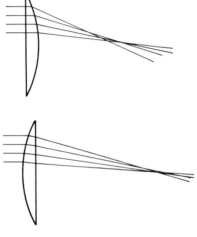

26-26 for a converging lens and for a concave mirror. Spherical aberration for a mirror can be eliminated for infinitely distant objects by using a paraboloidal rather than spherical surface (see Fig. 25-12). This is the method used in astronomical reflecting telescopes. For every other distance between the object and mirror, a different ellipsoidal shape is required. In a lens, spherical aberration can be partly removed by judicious choice of the two spherical surfaces. For example, a simple plano-convex lens has substantially less spherical aberration if light enters the curved face, as shown in Fig. 26-27. Further reduction in spherical aberration can be obtained by use of a *doublet*, two lenses composed of glasses with different indices of refraction cemented together, as in Fig. 26-28. If the two glasses not only have differing average indices of refraction but also have differing degrees of dispersion, both chromatic aberration and spherical aberration can largely be removed in such a doublet, which is then called an *achromatic doublet*.

Fig. 26-27. The degree of spherical aberration depends upon the orientation of a plano-convex lens.

Most high-quality optical instruments employ lenses that are more complicated than the simple achromatic doublet. The additional surfaces available not only allow for better correction of spherical and chromatic aberration, but they also help correct the following defects of simple lenses:

1. Curvature of the field: the image of a flat plane does not form a plane;

2. Coma: an off-axis point object is imaged in a comet-like shape;

3. Distortion: objects off-axis are magnified differently than those close to the axis.

Some examples of compound lenses for cameras are shown in Fig. 26-29.

26-10 Summary

A wave train changes its direction (is deviated) when it encounters a medium that changes the wave velocity. The deviation depends

Fig. 26-28. An achromatic doublet.

on both the change in wave velocity and the angle of incidence on the new medium and is given by Snell's law. If we write $c' = c/n$ for the velocity in a medium, we can write Snell's law in the form

$$n_1 \sin \theta_1 = n_2 \sin \theta_2$$

For a sufficiently large angle of incidence of a ray moving from a medium with index of refraction n_1 to a medium with index n_2 (where $n_2 < n_1$), Snell's law leads to an impossible conclusion: $\sin \theta_2 > 1$. In this case refraction does not occur; total internal reflection occurs instead.

When light encounters a refracting medium with two plane parallel surfaces, the emerging light ray is not deviated but is displaced. If the second plane surface is not parallel to the first, a prism is formed and a net deviation occurs. An object located within a refracting medium and viewed through a plane surface of the medium appears to be less than its true distance from an external observer.

When light encounters a curved refracting surface, focusing can occur. For a thin lens with two spherical surfaces, the focal length is given by

$$\frac{1}{f} = (n - 1)\left(\frac{1}{R_1} - \frac{1}{R_2}\right) \qquad \text{LeN's mAkeR EQUATION}$$

and, as was the case for spherical mirrors,

$$\frac{1}{s} + \frac{1}{s'} = \frac{1}{f}$$

Lenses with spherical surfaces suffer from spherical aberration: the inability to focus to a common focal point all parallel rays that strike the lens. Spherical aberration also occurs for spherical mirrors. The variation of the index of refraction with wavelength causes chromatic aberration in lenses. This type of aberration does not occur for mirrors.

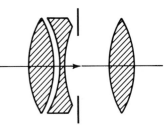

Original "Cooke triplet"

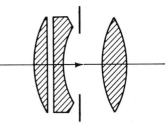

Zeiss "Tessar"

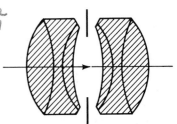

Goertz "Dagor" f/4.5

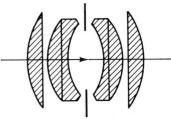

Taylor, Taylor and Hobson
Speed Pancro f/2

Fig. 26-29. Compound lenses. (Courtesy Bausch and Lomb)

Problems

26-1. The speed of light in air is 3×10^8 m/s. Find the speed of light in a medium with $n = 1.6$.

26-2. When light passes from one medium to another, its frequency is unchanged. Find λ', the wavelength of yellow sodium light, in glass with $n = 1.6$. The wavelength of sodium light in vacuum is 589.3 nm.

26-3. A ray of light strikes a plate of glass ($n = 1.5$) with an angle of incidence $\theta = 50°$. A portion of the light is reflected and part is transmitted (and refracted) into the glass. Find the angle between the reflected and refracted rays.

26-4. When a flat slab of glass ($n = 1.52$) is placed at normal incidence in one of the light paths in a Michelson interferometer, the optical path, in units of wavelengths, become longer. The light passes through the slab twice. Find a thickness (there are many) of the glass plate that does not change the interference pattern when the glass is inserted.

26-5. An aquarium is made of glass with $n = 1.6$. A ray of light in the water strikes the bottom of the tank at an angle of incidence $\theta = 45°$. Find the angle between the ray of light and the normal to the glass plate when the ray emerges into air below the aquarium.

26-6. A glass-walled cell is placed in one of the optical paths of a Michelson interferometer. The light passes through the cell twice, since it reflects back from mirror M_1 (see Fig. 24-24). An observer watches the interference pattern carefully as the air is entirely pumped out of the cell. He sees 20 fringes pass by. What is the length of the cell?

26-7. Find the critical angle of incidence for a light ray passing from glass ($n = 1.52$) into water ($n = 1.33$).

26-8. A plate of glass with $n = 1.6$ and a thickness of 2 cm is struck by a ray of light at an angle of incidence $\theta = 30°$, as in Fig. 26-10. Find the distance between the emergent ray and the path the ray would have followed in the absence of the plate.

26-9. A beam of white light is incident upon a crown glass prism, as in Fig. 26-12. The various colors are deviated by differing amounts, but assume that this variation is small enough so that (26-2) is applicable for all colors. If $\phi = 30°$ find the distance between blue light ($\lambda = 0.5\ \mu$m) and red light ($\lambda = 0.7\ \mu$m) on a screen 20 cm away from the prism.

26-10. Reverse the discussion in Section 26-5 to describe a fish's view of an object in air. How far above the water surface does an object appear to a fish if the object was 2 m above the water?

26-11. A cylindrical "light pipe" fabricated of polystyrene ($n = 1.49$) transports light from a point source, as in Fig. 26-30. The source radiates uniformly in all directions. What fraction of the light it radiates is trapped within the light pipe and piped to the right?

26-12. Light is reflected totally in the prism in Fig. 26-31a. A drop

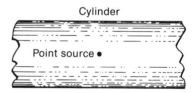

Cylinder

Point source •

Fig. 26-30. A point source of light is in the center of a polystyrene rod.

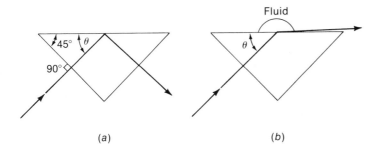

Fig. 26-31.

(a)   (b)

of fluid is added, as in Fig. 26-31b. The index of refraction of the fluid is just large enough so that the total reflection at the glass surface no longer occurs. Find the index of refraction of the fluid if n (glass) $= 1.60$.

26-13. As the sun sets, the refraction of sunlight by the earth's atmosphere causes the appearance of sunset to occur after the sun has passed below the actual horizon. Draw a ray diagram to explain this effect.

26-14. A laboratory student recalls that the real image of a distant object falls very close to the focal point of a converging lens. He finds the image of a light bulb that is 30 ft away from the lens to be $\frac{1}{2}$ ft behind the lens. He concludes that $f = \frac{1}{2}$ ft. How large is his error?

26-15. A converging plano-convex lens is to be made of plastic with $n = 1.50$. It is to have a focal length of 20 cm. Find the radius of curvature of the curved surface of the lens.

26-16. A diverging plano-convex lens is made of plastic with $n = 1.50$. It is to have a focal length of -20 cm. Find the radius of curvature of the curved surface of the lens.

26-17. A meniscus lens with $n = 1.60$ has radii of curvature $R_1 = 20$ cm and $R_2 = -30$ cm. Find its focal length. Is it diverging or converging?

26-18. A converging lens forms a real image of an object. Show that the distance between the object and its image is always greater than four times the focal length of the lens.

26-19. A diverging lens has a focal length of -16 cm. Find the location and type of image formed when the object-to-lens distance is (a) 4 cm, (b) 8 cm, (c) 16 cm, (d) 32 cm.

26-20. The dimensions of a lantern slide are 3 in. by 4 in. It is desired to project an image of the slide, enlarged to 6 ft by 8 ft, on a screen 30 ft from the projection lens. (a) Find the required focal length of the projection lens. (*b*) Where should the slide be placed?

26-21.   A slide projector forms a real image on a screen. If the image is to be cast on a screen 20 ft from the lens and the lens has a focal length of 6 in., what must be the distance from the object to the lens?

26-22.   The projector lens of the preceding problem is to be fabricated of glass with $n = 1.65$. Find values of R_1 and R_2 for this lens.

26-23.   A converging plano-convex lens made of crown glass has $R_1 = 20$ cm, $R_2 = \infty$. An object is placed 120 cm from the lens. Find the location of the image for red light with $\lambda = 700$ nm and blue light with $\lambda = 400$ nm.

Chapter **27**

Optical Instruments

SECTIONS 4, 6-9

27-1 Introduction   In this chapter we use our knowledge of image formation to discuss simple optical instruments. In the process we introduce several fundamental concepts not previously discussed: the action of lenses in combination, diffraction through circular apertures, and reflection from thin films. Human vision and correction of vision defects are also discussed.

27-2 The Camera   NOT ON HOURLY

The camera is a rather simple optical instrument that, in essence, consists of a light-tight box fitted with a lens at one end, as seen in Fig. 27-1. The real images produced by this lens fall upon light-

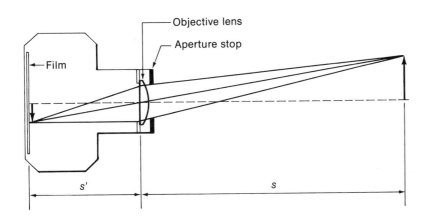

Fig. 27-1. Components of a camera.

sensitive film at the back of the camera. Since the image distance from the lens, s', depends upon the object distance, s, most cameras have provision for adjustment of the lens-film distance so that a sharply focused image can be produced on the film for various object distances.

Cameras are also equipped with shutters that allow light to enter only for a precisely timed period (typical values are between $\frac{1}{10}$ and $\frac{1}{500}$ s) so that (a) the film receives just the amount of light required to produce a satisfactory image, and (b) moving objects produce reasonably sharp images. Further control of the quantity of light reaching the film is obtained by adjustment of a diaphragm that masks off outer portions of the lens (see Fig. 27-2).

Since the objects to be photographed are typically at large distances compared with the focal length of the lens, the images are always close to the focal point. The lateral magnification is thus approximately

$$m = f/s$$

The *area* of the image is proportional to the square of the lateral size, while the lateral size itself is proportional to f. Thus the area of the image is proportional to f^2. The amount of light gathered by the lens, on the other hand, is proportional to the square of the lens diameter. The illuminance of the image formed on the film is proportional to the amount of light per unit area striking the film. Thus we find:

$$\text{illuminance} \propto \frac{\text{light}}{\text{area}} \propto \frac{D^2}{f^2}$$

where D is the lens diameter and f its focal length. It is customary to define an f-number:

$$f\text{-number} = \frac{f}{D}$$

so that illuminance $\propto 1/(f\text{-number})^2$. Since this result is independent of the focal length of the lens, a photographer can properly adjust the quantity of light striking the film by always using the same f-number for a given light level on the object, regardless of which particular camera he is using. The f-number of a camera lens is controlled by adjusting the diaphragm, which limits the active area of the lens. This adjustment is usually calibrated in terms of f-numbers. Each step changes the f-number by a factor of about $\sqrt{2}$, so that it changes the illuminance of the film by a factor of 2. The actual f-number markings on a camera lens are rounded off to two figures

Fig. 27-2. An adjustable lens diaphragm.

so that the ratios of successive *f*-numbers are not exactly equal to the square root of 2 (see Table 27-1).

TABLE 27-1 Typical *f*-Number Markings on a Camera

Increasing illuminance	*f*-numbers	Ratios of *f*-numbers (ideally = $\sqrt{2}$)
	f-16	
		16/11 = 1.45
	f-11	
		11/8 = 1.38
	f-8	
		8/5.6 = 1.43
	f-5.6	
		5.6/4 = 1.40
	f-4	
		4/2.7 = 1.48
	f-2.7	

Exercises

1. A certain type of film is correctly exposed for a sunny scene if the shutter is opened for $\frac{1}{100}$ s and the *f*-number is 16. The photographer needs a shorter exposure to prevent blurring of moving automobiles in the picture. If he chooses a shutter opening time of $\frac{1}{400}$ s, he must let four times as much light in through the lens. What *f*-number should he use?

2. A camera is fitted with a lens that has a focal length of 50 mm and a maximum *f*-number (diaphragm wide open) of 2.7. (*a*) What is the lens diameter? (*b*) Approximately what is the size of the image on the film when the object is a person 20 ft away?

27-3 The Eye NOT ON HOURLY

As can be seen in Fig. 27-3, there is a similarity between an eye and a camera. Both utilize a lens to produce an image on a light-sensitive surface, called the *retina* in the case of the eye. Both have an adjustable diaphragm, called the *iris* in the case of the eye, to help control the illuminance of the light-sensitive surface.

However, there are striking differences as well. There is no shutter in the eye (the eyelids do not serve the purpose of metering out fixed amounts of light), so the eye is continuously sensitive. Further, the sensitivity of the retina is enormously variable, so that extremely large variations in light level can be handled (see Table 24-2), even though the iris can only produce a 16-fold variation in

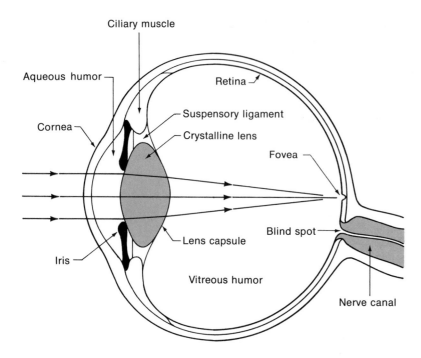

Fig. 27-3. The human eye.

retinal illuminance. Also, the eye is focused by muscular adjustment of the lens curvature (hence of its focal length) rather than by varying the lens-to-image distance, as is done with the fixed-focal-length lens of a camera.

The light receptors in the retina are called *rods* and *cones*. The cones are involved in color vision but require fairly large light levels. The rods are sensitive to extremely low light levels. At the center of the retina is the *fovea*, which contains only cones. This is the high-resolution portion of the retina. The ultimate resolution of the fovea seems to be about two cone-widths (4×10^{-6} m).* As you read this page, your eyes jump abruptly from one point to another, so that the words to be read are focused on the fovea. The region where the optic nerve leaves the eye is a blind spot. You can use Fig. 27-4 to find your blind spots.

---

*See "Floaters in the Eye" by Harvey E. White and Paul Levatin, *Scientific American*, June 1962.

Fig. 27-4. Close the right eye and focus on the circle. As you now move the book toward your eye, the X will disappear at about 8 to 10 in. The same phenomenon occurs when observing the X with the left eye closed.

A simplified sketch of the anatomy and the interconnection of nerves involved in human vision is indicated in Fig. 27-5. The interconnections between nerve fibers from adjacent receptors in the retina are arranged so as to heighten the contrast at image edges. The partial crossover of nerve fibers at the optic chiasm is involved in depth perception. Since the left eye sees a bit more of the left side of a nearby object than the right eye, the brain is able to infer

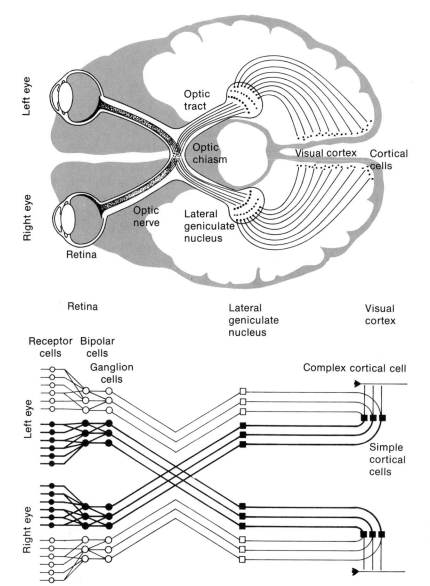

Fig. 27-5. The anatomy and a simplified "wiring diagram" of human vision. (From *The Neurophysiology of Binocular Vision*, John D. Pettigrew. Copyright © August 1972 by Scientific American, Inc. All rights reserved.)

distance information from this effect. Nerves that lead toward the cortex from the lateral geniculate nucleus already carry composite information derived from both eyes.*

By the time nerve impulses reach the "simple cells" in the visual cortex, there have been two stages of data processing that heighten contrast and give depth information. Further preprocessing is then carried out before the complex cortical cells respond. Some of these complex cells respond only when there is a light-dark edge in the field of view. Others respond only for certain orientations of the light-dark edge; still others respond to the location of edges. Finally, some respond only when edges are in motion in given directions.** The outputs of the complex cells thus contain all the logical elements required to reconstruct an image of the outside world. Many unanswered questions remain: How and where are the complex cell outputs used? What built-in prejudices for form exist in the "wiring diagram" of our visual system? How is this process influenced by infant development?

The simple type of lens in the human eye suffers from all the aberrations of such lenses and can only produce a sharp image close to its axis, which intersects the fovea. Interestingly enough, nature has evolved an extremely simple — indeed crude — optical instrument but coupled it to an astonishingly powerful and rapid computer that can actuate muscles to cause the high-resolution zone to scan rapidly over a scene. The myriad images so produced by both eyes are eventually fused by a short-term, rapid-access memory into a single three-dimensional scene.

Exercises

3. Approximately how long is the image on the retina for an object that is 4 in. high and 24 in. away? The distance from the lens to the retina is about 1 in.

4. Look at the "blind spot" experiment (Fig. 27-4). By moving the book in and out and by rotating the book, deduce the horizontal and vertical extent of the blind spot on the retina.

27-4   Combinations of Lenses

Many optical instruments use combinations of lenses to produce

---

*See "The Neurophysiology of Binocular Vision" by John D. Pettigrew, *Scientific American*, August 1972.

**See "The Visual Cortex of the Brain" by David H. Hubel, *Scientific American*, November 1963.

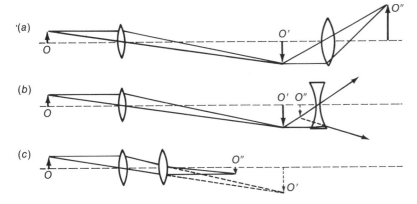

Fig. 27-6. In (a) and (b) a real image formed by a first lens is the object for the second lens. In (c) the object formed by the first lens is a virtual object for the second lens.

$$\frac{1}{s} + \frac{1}{s'} = \frac{1}{f} \qquad \text{location of the final image}$$

the desired result. Even a simple magnifying glass involves a two-lens system, the second lens being that of the eye. Accordingly, we need to examine the action of several lenses in combination. If a first lens forms a real image in front of a second lens, there is no special complication, since light emanates from a real image just as it would from an ordinary object. This is illustrated in Fig. 27-6a and Fig. 27-6b. Successive use of Eq. (26-10) for each lens can be applied to deduce the location of the final image. Successive use of Eq. (25-7) gives the overall magnification.

Example A lens with a focal length of 10 cm is placed 12 cm from an object. A second lens with a focal length of 5 cm is placed 65.2 cm from the first lens. Find the location of the final image and the overall magnification of this two-lens system.

Solution The situation is very much like that shown in Fig. 27-6a. The location of the real image formed by lens 1 is given by

$$s_1' = \frac{s_1 f_1}{s_1 - f_1} = \frac{120 \text{ cm}^2}{12 \text{ cm} - 10 \text{ cm}} = 60 \text{ cm}$$

Thus, the first real image is 5.2 cm in front of the second lens. The real image formed by the second lens is found at

$$s_2' = \frac{s_2 f_2}{s_2 - f_2} = \frac{5.2 \text{ cm} \times 5 \text{ cm}}{5.2 \text{ cm} - 5 \text{ cm}} = 130 \text{ cm}$$

so the final image is 130 cm behind the second lens. It is erect since it has undergone two inversions. The lateral magnification is:

$$m_1 = \frac{-s_1'}{s_1} = \frac{-60 \text{ cm}}{12 \text{ cm}} = -5$$

$$m_2 = \frac{-s_2'}{s_2} = \frac{-130 \text{ cm}}{5.2 \text{ cm}} = -25$$

Overall magnification is $m = m_1 \cdot m_2 = 125$.

In Fig. 27-6c we see that the first lens *would* have formed an object at O' but the second lens is placed ahead of this object. In this case, the object that would have been formed by the first lens is called a *virtual object* for the second lens. In this case we may still use (26-10) but the object distance, s_2, for the second lens is negative.

Example   Two positive thin lenses are placed in contact. If their focal lengths are f_1 and f_2, what is the focal length of a single lens that will produce the same result?

Solution   The location of the image formed by the first lens is given by

$$\frac{1}{s_1'} = \frac{1}{f_1} - \frac{1}{s_1} \tag{27-1}$$

Since the two lenses are thin and in contact, the object distance for the second lens is equal to $-s_1'$. Thus for the second lens we have

$$\frac{1}{s_2} + \frac{1}{s_2'} = \frac{1}{f_2}$$

and since $s_2 = -s_1'$,

$$-\frac{1}{s_1'} + \frac{1}{s_2'} = \frac{1}{f_2}$$

Substituting from Eq. (27-1) for $1/s_1'$:

$$-\frac{1}{f_1} + \frac{1}{s_1} + \frac{1}{s_2'} = \frac{1}{f_2}$$

or

$$\frac{1}{s_1} + \frac{1}{s_2'} = \frac{1}{f_1} + \frac{1}{f_2} \tag{27-2}$$

This last equation gives the relationship between the initial object distance, s_1, the final image distance, s_2', and the focal lengths of the lenses. Since the two thin lenses are in contact, we may drop the subscripts as follows:

$$\frac{1}{s} + \frac{1}{s'} = \frac{1}{f_e}$$

where

$$\frac{1}{f_e} = \frac{1}{f_1} + \frac{1}{f_2} \tag{27-3}$$

Equation (27-3) gives us the focal length of a single lens that will produce the same result as the two thin lenses in contact.

The preceding example showed that two thin lenses in contact add together to give a single effective focal length f_e, where

$$\frac{1}{f_e} = \frac{1}{f_1} + \frac{1}{f_2} \tag{27-4}$$

In optometry and ophthalmology, it is customary to express the action of a lens by its *power:*

$$p = \frac{1}{f} \qquad\qquad (27\text{-}5)$$

where the power, p, is expressed in *diopters* when the focal length, f, is given in meters. According to (27-4), the *powers* of thin lenses in contact add directly. Thus the power of a +3-diopter lens and a +2-diopter lens in contact is 5 diopters. A positive lens and a negative lens with equal and opposite powers give $p = 0$, which corresponds to no focusing at all.

It should be remembered that (27-4) applies only to thin lenses in contact. Consider, for example, a positive lens and a negative lens of equal but opposite powers, separated as in Fig. 27-7. In part *b* of the figure it is seen that the focusing action of the positive lens directs the rays to the central region of the negative lens, where its defocusing action is inadequate to prevent the converging rays from reaching a focus. Reversing the order of the lenses leads to initially diverging rays that then strike the positive lens at its outer portions, where it refracts more strongly and more than overcomes the initial divergence.

Exercises

5. Two lenses, each of 10 cm focal length, are separated by 5 cm. Find the image when an object is placed 20 cm ahead of the first lens. *STUDY!*

6. Find the power, in diopters, for two lenses in contact. One of them has a focal length of +10 cm, the other, −20 cm.

(a)

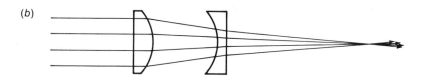

(b)

(c)

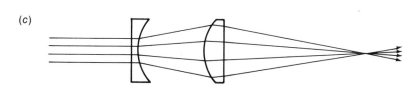

Fig. 27-7. (a) Lenses with equal but opposite powers cancel one another when in contact. (b) When the lenses are separated, a net focusing action occurs. (c) Reversing their order does not change this result.

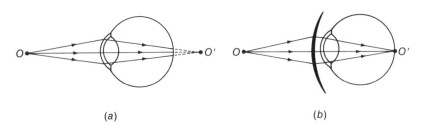

(a)   (b)

Fig. 27-8. When the lens of the eye is unable to produce a focus for nearby objects, a positive eyeglass lens is required.

27-5 Correction of Visual Defects   NOT ON HOURLY

The normal human eye is correctly focused for objects at infinity when the muscles surrounding the lens are relaxed. In order to focus on closer objects, these muscles squeeze the lens to shorten its focal length. This process is called *accommodation*. The shortest object distance that can be accommodated is called the *near point*. In normal subjects the process of accommodation becomes more difficult with age as the lens becomes less resilient. The result is to cause the near point to recede with increasing age, as shown in the following table.

We see that by age 60 practically all accommodation has been lost. When this occurs a lens of about +3 diopters is required immediately in front of the eye to be able to focus at normal reading distances, as indicated in Fig. 27-8. In some individuals, the eyeball is either too long or too short to obtain proper focus at the retina for all objects. These conditions are referred to as myopia and hypermetropia. Such individuals can obtain clear vision by the use of eyeglasses, as indicated in Fig. 27-9.

TABLE 27-2

Age (years)	Near Point (cm)
20	10
30	14
40	22
50	40
60	200

Example A person with myopia requires a $-1\frac{1}{2}$ diopter lens to see distant objects. What lens should be used (perhaps in the lower portion

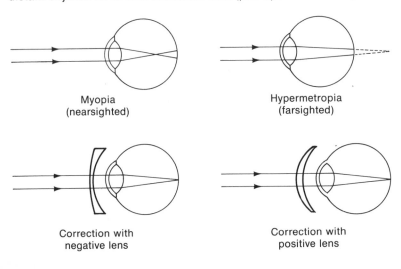

Myopia
(nearsighted)

Hypermetropia
(farsighted)

Correction with
negative lens

Correction with
positive lens

Fig. 27-9. Two common eye defects are easily corrected with eyeglasses.

of a bifocal) to see objects at 50 cm with his eye relaxed, as it is for distant vision?

Solution Since lens powers are additive for lenses in contact, we need only find the power required to bring light rays from a point 50 cm away into parallelism. This lens power added to the power required for distant vision will give the desired result, since for distant vision the incident light rays *are* essentially parallel. Bringing light rays into parallelism essentially means that s' recedes to infinity.

$$\frac{1}{s} + \frac{1}{s'} = \frac{1}{f}$$

$$\frac{1}{50 \text{ cm}} + 0 = \frac{1}{f}$$

$$f = 50 \text{ cm}; \quad p = 2 \text{ diopters}$$

Combining the power of this lens with that already required for distant vision gives

$$p = 2 \text{ diopters} - 1\tfrac{1}{2} \text{ diopters} = \tfrac{1}{2} \text{ diopter}$$

or a focal length of $+2$ m for a lens that provides clear vision at an object distance of 50 cm with a relaxed eye.

Exercises

7. What is the power, in diopters, of the lens in the human eye? The human eye is about one inch long.

8. A person with hypermetropia wears eyeglasses with a power of $+3$ diopters to view distant objects. What power lens should he use for close work with $s = 30$ cm?

27-6 The Simple Magnifier

When we wish to see more detail in an object, we bring it closer to the eye in order to enlarge the image size formed on the retina. Since the final image size is proportional to the angular size of the object, as shown in Fig. 27-10, we will define the *angular magnification, M:*

$$M = \theta'/\theta$$

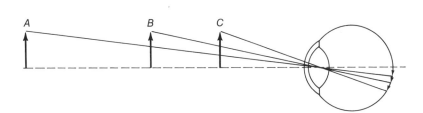

Fig. 27-10. The size of the retinal image is proportional to the angle subtended by the object.

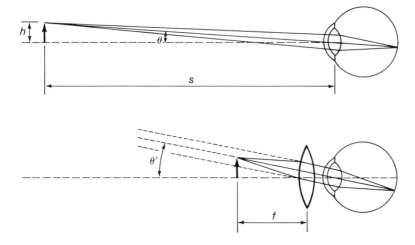

Fig. 27-11. When an object is at a comfortable viewing distance, s, it subtends an angle θ. When it is placed a distance f in front of a positive lens, light rays from it come to the eye parallel, as from an object at infinity. The eye can then be relaxed to view the object.

where θ is the normal angular size at the eye and θ' is the angular size produced by an optical system in front of the eye.

In Fig. 27-11 an object of height h is shown at a reasonable viewing distance, s. In order to bring it closer, the eye would have to strain to focus at its near point. In any event, it certainly could not be brought closer than the near point. However, by interposing a positive lens, as in Fig. 27-11, the object can be placed at the focal point of that lens to produce emerging parallel rays. The eye can comfortably focus these rays since they correspond to an object at infinity. But now the angular size of the object is approximately

$$\theta' = h/f$$

Without the lens it was

$$\theta = h/s$$

where s could be as short as 10 cm for a young person. For reasonably extended viewing times and an average person, the minimum value of s is taken to be 25 cm. The simple magnifier then produces a magnification of

$$M = \frac{\theta'}{\theta} = \frac{h/f}{h/s} = \frac{25 \text{ cm}}{f}$$

27-7 The Microscope

The simple convex lens magnifier is severely limited by lens aberrations to a magnification of 2 or 3 (often written 2× or 3×). By using compound lenses, magnifications up to 20× are obtained. The inventor of the microscope, van Leeuwenhoek, used single lenses in the form of small spherical glass beads as simple magnifiers. The image they produced was severely distorted but, nonetheless, Leeuwenhoek was able to discover red bloodcells with this instrument in 1674.

We now use the term microscope to describe a two-lens instrument that produces clear, greatly enlarged images in two steps. First, a short-focal-length lens, called the *objective* lens, is used to form an enlarged, real image. This image is then viewed through a simple magnifier to produce a final, greatly enlarged, virtual image, as shown in Fig. 27-12. The lateral magnification of the first image is simply

$$m_1 = -s'/s$$

and the magnifier gives a further magnification:

$$m_2 = \frac{25 \text{ cm}}{f_2}$$

where f_2 is the focal length of the eyepiece. The overall magnification is:

$$m = m_1 m_2$$

where m_1 is dictated by the available length of the microscope tube and the focal length of the objective lens. Since microscope tube lengths have been standardized by manufacturers, the objective and eyepiece lenses are generally marked simply with the amount of their magnification.

The object is located very close to the focal point of the objective lens in the microscope. Further, the first image is formed at a distance L that is approximately equal to the length of the microscope tube. Thus, the overall magnifying power can be approximately expressed in terms of the focal lengths of the lenses:

$$m \approx \frac{L}{f_1} \cdot \frac{25 \text{ cm}}{f_2}$$

Very short focal length lenses give the highest magnifications.

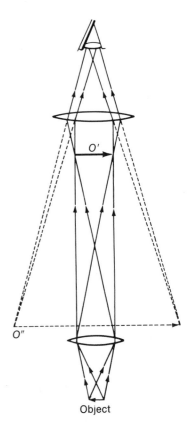

Fig. 27-12. The objective lens of a compound microscope forms an enlarged, real image at O'. The eyepiece lens is a simple magnifier to provide a further enlarged view of this real image.

Exercise

9. An objective lens is marked 10× and is used in a microscope that produces the first real image 15 cm above the objective lens. What is the focal length of this objective lens?

27-8 The Telescope

Another device that utilizes two lenses in succession to obtain an enlarged image is the telescope. Since the objects to be viewed are at great distances, we cannot use the trick of placing them very close to the focal point of an objective lens to get a greatly enlarged, real image, as was done in the microscope. Instead, a long-focal-length lens is used to produce a small, real image that is very close to the observer. This image is then viewed by a magnifier, as in Fig. 27-13.

Since the object is distant, the first image is produced essentially at the focal point of the objective lens. Its size is greatly reduced from that of the original object and is given approximately by

$$O' = O \frac{f_1}{s}$$

where O is the object size and O' the image size. The simple magnifier of focal length f_2 is placed so that the first real image is just inside its focal point. Thus, the angle θ' in Fig. 27-13 is given approximately by

$$\theta' = \frac{O'}{f_2} = \frac{O}{s} \cdot \frac{f_1}{f_2}$$

The angle seen by the unaided eye is approximately

$$\theta = \frac{O}{s}$$

so we have

$$m = \theta'/\theta = f_1/f_2$$

For high magnification, then, we desire an objective lens of large focal length and an eyepiece with short focal length. Use of a large-diameter objective lens provides much greater light collection than that of the eye, so that objects too dim to be seen with the unaided eye can be observed.

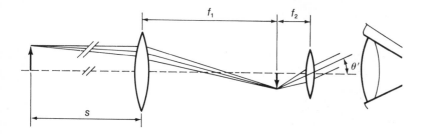

Fig. 27-13. The objective lens of a telescope forms a reduced, real image that is then viewed closely with a magnifier.

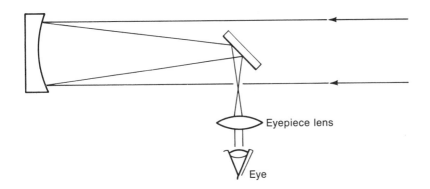

Eyepiece lens

Eye

Fig. 27-14. In the Newtonian reflector, a small plane mirror (or right-angle prism) is used to deflect the focus to one side so that the image can be viewed without obstructing the primary mirror.

In the *Newtonian reflector* telescope, the objective lens is replaced by a concave mirror, as in Fig. 27-14. Newton constructed the first telescope of this type, utilizing a metal mirror. The reflecting telescope has two great advantages over a refracting telescope. First, it is much easier to finish one surface of a mirror than it is to finish the four surfaces of an achromatic lens. Second, the entire back of a mirror can be rigidly supported, while a large lens held only at its edges can sag seriously. For these reasons the reflecting telescope has dominated astronomy since Newton's times (see Fig. 27-15).

Fig. 27-15. The Mt. Palomar 200-in. telescope. (Photo reproduced by permission of the Hale Observatories)

Exercises

10. The large astronomical telescope at Mt. Palomar, California, is a reflecting telescope with a mirror whose diameter is 200 in. and whose focal length is 660 in. Estimate the size of the real image formed at the focus of this mirror when the object is a man 1 mi distant.

11. What focal length eyepiece lens should be used with the Mt. Palomar telescope (see Exercise 10) to obtain a magnification of 330×?

27-9   Resolving Power

It would seem that arbitrarily large magnification could be obtained with either a microscope or a telescope by choosing the proper combinations of lenses. Two limitations, one technical and one fundamental, stand in the way of arbitrarily large magnification. First, highly magnifying optical systems show up even the most minute aberrations in the lenses. Modern lens-making efforts utilizing highly corrected compound lenses can surmount this difficulty, but the wave nature of light itself finally limits the usable magnification of an optical system.*

When light waves of wavelength λ are admitted into a lens or mirror of finite aperture, diffraction occurs, just as is the case with a single slit (see Section 24-7). The light waves passing through a circular hole diffract to produce a series of concentric rings of light, as shown in Fig. 27-16. The angle between the center of the pattern and the first dark ring is given by

$$\sin \theta_1 = 1.22\lambda/d$$

*See, for example, the article "Reconnaissance and Arms Control" by Ted Greenwood, *Scientific American*, February 1973.

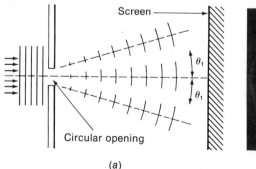

(a)

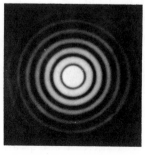

(b)

Fig. 27-16.   Diffraction of light through a circular opening. In (a) the geometrical arrangement is shown. The angle θ_1 is the angle between the center of the pattern and the first dark ring outside the central maximum. A photograph of the image on the screen is shown in (b). (From Cagnet, M., Francon, M., Thrien, J. C., *Atlas of Optical Phenomena*, Springer, 1962)

where d is the diameter of the circular opening. For our purposes it is enough to note that

$$\theta_1 \approx \lambda/d$$

If we now wish to resolve (to see separately) two point objects viewed through a telescope whose objective-lens diameter is d, the diffraction patterns produced by these points must be separated by an angle at least as large as θ_1. Hence, the objects must be separated by an angle at least as large as θ_1. Fig. 27-17 illustrates this point. The ultimate resolution of any optical system, then, is limited by the wavelength of the light and the aperture of the instrument. When the magnification of the instrument is raised to the point that diffraction effects limit the detail that can be seen, no further improvement in visible image detail is possible. Further increases in magnification enlarge the diffraction patterns as well as the image size.

Example The diameter of a microscope objective lens is 3 mm. Its focal length is 2 mm. What is the spacing of the finest detail that can be seen with such a microscope?

Solution In order to resolve the detail, it must be separated in angle by more than λ/d, where λ is the wavelength of light and d is the lens diameter. If the separation of the objects is Δx, we have approximately:

$$\theta \approx \frac{\Delta x}{f_1}$$

since the object is quite close to the focal point of the microscope objective. To resolve detail we must have

$$\frac{\Delta x}{f_1} > \frac{\lambda}{d}$$

or $$\Delta x > f_1 \lambda/d$$

Taking the average wavelength of visible light to be 5.5×10^{-7} m:

$$\Delta x > 2 \times 10^{-3}\ \text{m} \cdot \frac{5.5 \times 10^{-7}\ \text{m}}{3 \times 10^{-3}\ \text{m}} = 0.37\ \mu\text{m}$$

This microscope, regardless of its overall magnification, cannot resolve detail much finer than about one-third of a micrometer. A single virus is approximately 30 times smaller (≈ 100Å) than this limit!

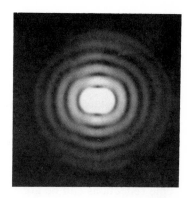

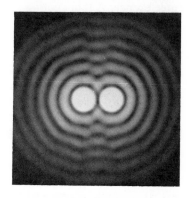

Fig. 27-17. Two point objects viewed by a telescope produce diffraction patterns, shown here highly magnified. As they are moved apart, they are reasonably well resolved when $\theta > \lambda/d$, where θ is the angle between the objects. (From Cagnet, M., Francon, M. Thrien, J. C., *Atlas of Optical Phenomena*, Springer, 1962)

Exercises

12. Find the smallest angular separation of two points that can be resolved by the Mt. Palomar telescope (see Exercise 10).

13. If the two points in Exercise 12 are 1 mi away, how far apart must they be? If instead they are two stars 10 light years away (1 light year $\approx 6 \times 10^{12}$ mi), how far apart are they?

27-10 Antireflection Coatings   NOT ON HOURLY

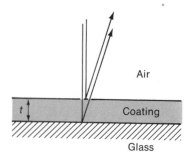

Fig. 27-18. A normally incident light ray that reflects from the glass surface travels an additional distance of 2t.

Reflection from the surfaces of lenses leads not only to decreased image brightness but also to multiple images or a generally hazy background when several lens surfaces are involved. These troublesome reflections can be greatly reduced by a technique that utilizes interference. A thin coating of a transparent material is applied over the glass surface, as shown in Fig. 27-18. For light that is normally incident upon the coated surface of thickness t, there is a path-length difference of $2t$ between the rays that reflect from the air-coating interface and those that reflect from the coating-glass interface. If t is chosen so that this path-length difference leads to destructive interference, the net reflection is greatly reduced.

The cancellation will be complete, however, only if the amount of light reflected at the two interfaces is identical. Since the amount of reflection at such an interface is proportional to the fractional change in the index of refraction, the required index of refraction n_c for the coating material will be the geometric mean of the index for air, n_a, and that for glass, n_g:

$$n_c = \sqrt{n_a n_g}$$

For $n_g = 1.5$, $\qquad n_c = 1.22$

An ideal material that is both transparent and durable has not been found, but magnesium fluoride is close enough to the ideal with $n = 1.38$. This does not give perfect cancellation of the reflected waves but does cut the reflective loss from 4% to less than 1%.

Before we can choose a value for the coating thickness, we must know whether the phase of each wave train is affected by reflection. In this connection, there is a useful analogy between light waves and mechanical waves on a string. Waves on a string are inverted upon reflection from a more dense portion of the string, while they are not inverted upon reflection from a less dense portion. Light waves are inverted when reflected from a medium with larger index of refraction, but they are not inverted upon reflection from a medium with lower index. Thus, the light waves reflecting from both the upper and lower surfaces of the coating are inverted.

In order to make both reflected wave trains be out of phase, we need only cause one path length to differ from the other by $\lambda/2$. Thus we should have

$$2t = \lambda/2$$

or $$t = \lambda/4$$

When light moves into a new medium, its frequency remains constant, so its wavelength must change:

$$\lambda' = \lambda/n_c$$

where λ is the wavelength in vacuum. Computing the wavelength of light for the center of the visible spectrum in the MgF_2 film, we find

$$\lambda' = \frac{550 \times 10^{-9} \text{ m}}{1.38} = 398 \times 10^{-9} \text{ m}$$

and $$t = \lambda'/4 \approx 10^{-7} \text{ m} = 1000 \text{ Å}$$

Such thin films are obtained by slowly evaporating MgF_2 at an orange heat ($\approx 1000°C$) in a vacuum chamber with the optical elements to be coated placed nearby.

Since the cancellation condition, $t = \lambda'/4$, is satisfied for only one wavelength, which is chosen in the center of the visible spectrum, some red and violet light is reflected from the coated surface. This mixture of red and violet light gives coated lenses a faint purple hue.

Exercise

14. What coating thickness would you choose to *enhance* the reflection of 550-nm light from a glass surface?

27-11   Summary

The simplest optical instrument is the camera, in which a single lens produces a real image of an object upon light-sensitive film. The illuminance of the film is proportional to the inverse square of the *f*-number, where the *f*-number of any lens is its focal length divided by its diameter. The eye is somewhat similar in construction to the camera.

Lenses are often combined in optical instruments. The first lens then forms an image that becomes the object of the next lens, and so on. For thin lenses in contact, their combined focal length f_e is given by

$$\frac{1}{f_e} = \frac{1}{f_1} + \frac{1}{f_2}$$

The power p of a lens is defined

$$p = \frac{1}{f}$$

so that powers of lenses in contact simply add.

Common defects of the human eye are myopia (nearsightedness) and hypermetropia (farsightedness). These defects are easily remedied by use of eyeglasses consisting of simple lenses that act with the lens of the eye to bring objects to a focus at the retina.

The simple magnifier allows an object to be brought very close to the eye for detailed examination. If the closest distance for viewing without a magnifier is 25 cm, the magnification obtained by a positive lens magnifier is

$$m = \frac{25 \text{ cm}}{f}$$

where f is expressed in cm.

In a compound microscope, a short-focal-length lens is placed very close to the object and an enlarged real image is formed. This real image is examined through a simple magnifier, called the eyepiece.

In a telescope a real image of a distant object is formed by an objective lens of long focal length. This real image is then examined at close range with the aid of an eyepiece lens that is a simple magnifier

The ultimate resolution of any optical instrument is limited by diffraction. The smallest angular separation that is resolvable is approximately given by

RESOLVING POWER

$$\theta_{min} \approx \frac{\lambda}{d}$$

$$\theta \sim \frac{\Delta x}{f_1} \qquad \frac{\Delta x}{f_1} > \frac{\lambda}{d}$$

where λ is the wavelength of light and d the aperture of the instrument.

Most lenses used in commercial optical instruments have anti-reflection coatings to minimize reflections at their surfaces. These coatings are a quarter of a wavelength thick, so that light reflected at their two surfaces destructively interferes and reduces the net reflection.

Problems

27-1. A 35-mm camera (35 mm is the film width) has an f-2 lens of 50-mm focal length. (a) Find the image size for an object with a height of 1 m that is 20 m distant from the lens. (b) How far must the lens move to focus on an object that is only 1 m from the lens?

27-2. A picture of a given scene is successfully obtained with a shutter speed of $\frac{1}{100}$ s and a lens opening of f-16. The photographer then shifts to a film that requires twice as much light and changes the shutter speed to $\frac{1}{200}$ s. Which f-number in Table 27-1 should he use?

27-3. The ultimate resolving power on the retina of the human eye is about 4×10^{-6} m. How far apart are two points that just produce this image separation if they are 1 m from the eye?

27-4. El Greco was a Renaissance artist who painted human figures in a very elongated, thin form. It has been suggested that he suffered from an eye defect (astigmatism), so that images on his retina were magnified more in the vertical dimension than in the horizontal dimension. What is wrong with this explanation?

27-5. A patient with hypermetropia requires a +2-diopter lens to focus clearly on distant objects with a relaxed eye. What lens power should he use to focus clearly on objects 25 cm away with a relaxed eye?

27-6. A patient's near point is at 200 cm. The length of his eyes is 2.5 cm. He wishes to be able to see objects at a 10-cm distance when his eyes are at their maximum accommodation ($s = 200$ cm). What power lenses should he use for eyeglasses to accomplish this?

27-7. Two positive lenses are to be used to form a magnified, erect, real image of an object. The object is 5 m from the first lens. The first lens has a 20-cm focal length, the second, 10 cm. Describe an arrangement that produces an image twice as large as the object. Where is the final image located?

27-8. A positive lens, $f = +10$ cm, is 2 m from an object. A negative lens, $f = -10$ cm, is spaced 5 cm behind the positive lens. Locate the final image. Is it real or virtual? Erect or inverted?

27-9. A microscope has an objective lens with $f = 1$ mm. Describe an instrument utilizing a second positive lens that has an overall magnification of $m = 300$.

27-10. The focal length of the eyepiece of a certain microscope is $f_1 = 2$ cm, and that of the objective lens is $f_2 = 1$ cm. The length of the microscope (distance between lenses) is $L = 22$ cm, and the final image is put at infinity. (a) What should be the distance from object to objective? (b) What is the linear magnification produced by the objective? (c) What is the overall magnification of the microscope? (d) How does your answer to c compare with the approximate result $m = 25L/f_1 f_2 = 275$?

27-11. Four positive lenses are available with focal lengths of 100 cm, 50 cm, 2 cm, and $\frac{1}{2}$ cm. Which pair of lenses give the greatest magnification when used as a telescope? What is the value of the magnification?

27-12. The human eye can resolve points 4×10^{-6} m apart on the retina. Compare this to the best possible resolution at the retina as limited by diffraction at the pupil. Assume the pupil is fully open with a diameter of 7 mm.

27-13. A microscope has an objective lens with a diameter of 1 mm. What is the minimum angular separation of two points that can be resolved by this microscope? How far apart are they if they are 2 mm from the lens?

27-14. A nation claims that its reconnaissance aircraft have photographed a missile site from an altitude of 15 mi. The camera that was used had a lens diameter of 6 in. A skeptical physicist decided to compute the best possible resolution of the lens at that distance to see whether this was physically possible. What did he conclude?

27-15. A point source of light is placed at the focal point of the McDonald Observatory's 120-in. telescope to produce a parallel beam of light. This beam is projected onto the moon. What is the minimum possible size of the spot produced on the moon? (This technique is actually used with a laser for the light source in a Lunar Laser Ranging Experiment.*)

---

*See "The Lunar Laser Ranging Experiment" by P. L. Bender et al., *Science*, vol. 182 (October 1973): p. 229.

Chapter **28**

Electricity Sections 1-9

28-1 Introduction In this chapter we begin our study of the closely related subjects of electricity and magnetism. Though these subjects seem somehow different and distinct from those we have been studying, they are not. The only new element that we will introduce into our classical mechanics is a new type of force: the electric force. Most of the phenomena that we consider to be uniquely *electrical*—the glow of a neon sign, the hum of a transformer, the snap of a spark, or the sharp smell of ozone near an electrical discharge—are ultimately due to the ability of electric charges to exert forces upon one another, even though they are separated by some distance.

This is not to say that the incorporation of this new force into our study of physics will be accomplished by a mere statement of fact. The consequences of the electrical force are far ranging, to say the least. The basic structure of matter is shaped by this force, which binds the constituents of atoms and molecules together. Our entire technology and our ability to study the universe about us is heavily indebted to the knowledge of electricity and magnetism that we have gained primarily in the last one hundred years.

The richness of topics and phenomena under the headings of electricity and magnetism will demand our attention for all but the last chapter of the remainder of this text. This attention is well deserved, for almost every force we encounter with our senses is an electric force. Every "contact" force, such as that between our feet and the floor, is due to the electric repulsion between electrons in the outer layers of atoms in each of the two contacting surfaces.

The pull of gravity is an everyday fact of life for us, but it is not so much this pull of which we are aware as it is the basically electric contact forces that act when our bodies are in equilibrium between the attraction of gravity and the contact forces supporting us.

28-2   The Electric Force

Man became aware of electrical phenomena long before the recent rapid evolution of electromagnetic theory. Indeed, the word *electricity* stems from the ancient Greek word for amber: *elektron.* It had already been known in the days of classical Greece that amber that had been rubbed by wool or cat's fur would attract small objects, like bits of lint or dust. We now say that amber, in the process of rubbing, has become *electrified,* or *electrically charged,* or simply *charged.* The ability of charged amber to attract objects some distance away places the electric force in the category of *action-at-a-distance* forces. The force of gravity is the only example we have seen previously in this book.

After a piece of amber has been electrified by rubbing with cat's fur, it is found that it attracts the piece of cat's fur. However, two pieces of electrified amber are found to repel one another. A piece of glass rubbed with silk is found to attract the silk but repel another similarly electrified glass rod. Furthermore, the electrified glass *attracts* an electrified piece of amber. Where the gravitational force can produce only attraction, the electric force can be either attractive or repulsive.

These experimental facts led early scientists to conclude that there were two types of electricity. It was supposed that opposite kinds of electricity were induced on the amber and fur (also true for the glass and silk) and that *unlike charges attracted one another.* Since electrified glass and electrified amber attracted one another, they carried unlike charges. The charge on the glass rod was called *positive;* that on the amber, *negative.* If a body had an excess neither of positive nor negative charge, it was called *neutral.*

The force between two charged objects appeared to be a *central force,* that is, its direction lay along the line joining the charges. Further, the forces between two charged bodies were found to be equal and oppositely directed, as indicated in Fig. 28-1.

28-3   Coulomb's law

The first published investigations that led to the law that describes the electrical force were carried out by Charles Augustin de Coulomb in 1785. He measured the force exerted by small charges upon

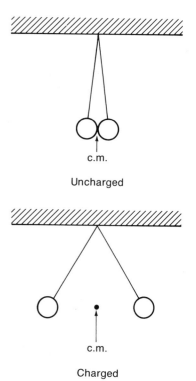

Fig. 28-1. The center of mass of two objects is not shifted to the left or right by their mutual electric repulsion, in accordance with Newton's Third Law.

one another and found how these forces depended upon the separation and magnitude of the charges. He used the torsion balance shown in Fig. 28-2 to make these measurements. Cavendish's torsional balance for the determination of G (Chapter 17) was patterned after Coulomb's balance. Coulomb reached the conclusion that the force exerted by one point charge on another point charge is given by what we now call *Coulomb's law:*

$$F_{12} = k\frac{q_1 q_2}{r_{12}^2} \qquad (28\text{-}1)$$

$$F_{12} = k\frac{q_1 q_2}{r_{12}^2}$$

minus sign
when charges have
opposite signs indicate
a force of attraction

Here F_{12} is the force on charge 2 due to the presence of charge 1, k is a constant, q_1 and q_2 are the magnitudes of the two charges, and r_{12} is their separation. The minus sign that results when the

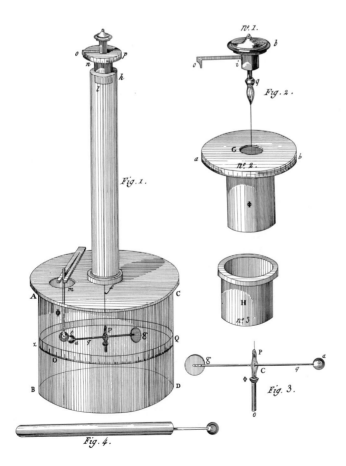

charges have opposite sign indicates a force of attraction. This law
bears an obvious similarity to the gravitational force law, (17-1):

$$F_{12} = G\frac{m_1 m_2}{r_{12}^2}$$

Two of the quantities in Coulomb's law are already well defined: force and distance. We must now decide how to define electric charge. This choice will obviously affect the value of k. One
attractive possibility is simply to set k equal to unity, use customary
units for force and distance, then let Eq. (28-1) *define* a unit of
charge. This is the approach taken in the cgs *electrostatic units*
(abbreviated esu). In these units, forces are measured in dynes,
distance in cm, k equals unity, and the magnitude of each of two
equal charges that produce a 1-dyne force at 1 cm separation is
called the *statcoulomb*.

In SI units, the unit of charge is the *coulomb* (C), but it is no
longer defined in terms of a torsion balance experiment. For practical reasons involving the ultimate accuracy of standards, the
coulomb is defined in terms of the effects of moving charges. Charges
in motion constitute an *electric current*. The SI units of current
are *amperes* and are equal to one coulomb/second. The ampere is
standardized in the SI units and the coulomb follows by definition.
In SI the force and length units are those of the MKS system of
units. Coulomb's law with SI units is

$$F_{12} = \frac{1}{4\pi\epsilon_0}\frac{q_1 q_2}{r_{12}^2} \tag{28-2}$$

where k has been rewritten as $1/4\pi\epsilon_0$. This inclusion of the factor
$1/4\pi$ in the force law is called the *rationalized* MKS system. Several
useful equations we will soon derive involve a factor of 4π in such
a way that the $1/4\pi$ in (28-2) is cancelled to give a more compact
result. We will use k when it is most convenient and $1/4\pi\epsilon_0$ when
appropriate. Important numerical values in SI or rationalized MKS
units are:

$$k = \frac{1}{4\pi\epsilon_0} = 8.99 \times 10^9 \text{ N} \cdot \text{m}^2/\text{C}^2$$

$$\epsilon_0 = 8.85 \times 10^{-12} \text{ C}^2/\text{N} \cdot \text{m}^2$$

From the SI value for k we can quickly deduce the conversion
between the SI and cgs charge units:

$$1 \text{ coulomb} = 3 \times 10^9 \text{ statcoulombs}$$

Coulomb's law can also be written in vector form:

$$\mathbf{F}_{12} = k\frac{q_1 q_2}{r_{12}^{\,3}}\mathbf{r}_{12}$$

where r_{12} is a vector that points from q_1 to q_2. Notice that if the roles of q_1 and q_2 are interchanged, we still find the same magnitude for the force but the vector reverses its direction. This indicates that the force on q_1 caused by q_2 is equal and opposite to the force on q_2 that is caused by q_1. Thus we have:

$$\mathbf{F}_{21} = -\mathbf{F}_{12}$$

in accordance with Newton's third law.

   When several different charges are near each other, the electric force on any one charge is simply the vector sum of the forces exerted by each of the other charges. This can be stated more formally by labeling the charge of interest as 1 and the remaining charges, which act upon charge 1, as 2 through N. The force on charge 1 is then:

$$\mathbf{F}_1 = \sum_{i=2}^{N} \mathbf{F}_{i1} = kq_1 \sum_{i=2}^{N} \frac{q_i}{r_{i1}^{\,3}}\mathbf{r}_{i1}$$

When resultant forces are given by such a vector sum, we say they obey a *superposition principle*. Electric forces, then, obey the superposition principle. The force between any pair of charges is not influenced by other charges and all the forces are added to obtain a net force.

Example   Two equal positive charges, q, subtend an angle θ as seen from a third positive charge, Q, and are both a distance l from the charge Q. Find the magnitude of the electric force on Q.

Solution   The situation is sketched in Fig. 28-3. The two vectors representing the two forces acting on Q are equal and separated by an angle of θ. Either we can project both force vectors onto the line bisecting the angle and then sum these components, or we can use the law of cosines:

$$(F_{total})^2 = F^2 + F^2 + 2F \cdot F \cos\theta = 2(1 + \cos\theta)F^2$$

and since $F = kqQ/l^2$,

$$(F_{total})^2 = 2(1 + \cos\theta)k^2\frac{q^2Q^2}{l^4}$$

or

$$F_{total} = \sqrt{2 + 2\cos\theta}\, k\frac{qQ}{l^2}$$

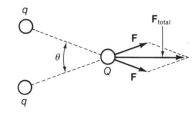

Fig. 28-3.

Exercises

1. Two bits of amber are rubbed with fur. When they are separated by 5 cm, the electric force of repulsion on each of them is 2/5 dyne. Find the charge on either bit of amber, assuming equal charges.

2. If the charge is doubled on *one* of the bits of amber in the above exercise, what happens to the electric force on it? On the other?

3. Two equal charges are separated by 30 cm. How large are these charges (in coulombs) if the electric force on either charge is one ton? You may use the approximate conversion: 1 ton $= 10^4$ N.

4. Three equal negative charges form the three vertices of an equilateral triangle with sides of length l. If the charges have a magnitude of Q, find the magnitude of the electric force on any one of them.

28-4   Continuously Distributed Charges

The Coulomb force law applies to pairs of interacting *point* charges. If an electric charge is distributed over an extended region, we may regard it as being composed of infinitesimal bits, just as was done for extended gravitating masses in Chapter 17.

Example   Find the electric force acting on the point charge q shown in Fig. 28-4. A charge Q is distributed uniformly along the line of length L.

Solution   One element of the line charge Q is indicated in the figure; it has a length dx. The amount of charge dQ contained in dx is equal to the fraction of L within dx multiplied by the total charge Q: $dQ=(dx/L)Q$. The charge dQ lies at a distance r from the point charge. Hence we may write for the magnitude of the force between dQ and q:

$$dF = kq\frac{Q}{L}\frac{dx}{r^2}$$

where the notation dF explicitly indicates that this force is an infinitesimal component of the total force. Note that the x component of $d\mathbf{F}$ is precisely cancelled by the x component of the force from the charge dQ located at $-x$. It is thus clear that the resultant force on q will be directed in the $+y$ direction, so we will need to consider only the y component of $d\,\mathbf{F}$.

$$dF_y = kq\frac{Q}{L}\frac{dx}{r^2}\cos\theta$$

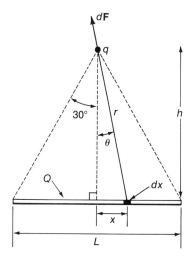

Fig. 28-4.

The total force F is the sum of all the infinitesimal components dF_y. Such a sum is an integral:

$$F_y = kq\frac{Q}{L}\int_1^2 \frac{dx}{r^2}\cos\theta \qquad (28\text{-}3)$$

where the limits of the integral must include the entire line charge. We also need to make use of the relations between x, r, and θ:

$$r^2 = h^2 + x^2$$

$$x = h\tan\theta$$

$$dx = h\sec^2\theta\, d\theta$$

since

$$\frac{d}{d\theta}(\tan\theta) = \sec^2\theta$$

Using the above relations and the identity $\sec^2\theta = 1 + \tan^2\theta$, we find:

$$F = kq\frac{Q}{hL}\int_{\theta_1}^{\theta_2}\cos\theta\, d\theta$$

where θ_1 and θ_2 represent the extreme angles subtended by elements dq at each end of the line. Evaluating the above integral we find:

$$F = kq\frac{Q}{hL}\sin\theta\Big|_{-30°}^{+30°} = kq\frac{Q}{hL}$$

Note that the dimensions in our answer are k multiplied by (charge)^2 divided by (distance)^2, as they should be.

   In Section 17-6 we saw that spherically symmetric masses exert gravitational forces on external objects as if all their mass were concentrated at their center. Inside a spherical shell of mass, no resultant gravitational force existed. Since both Coulomb's law and Newton's gravitational law are inverse square laws, we find analogous results for spherically symmetric charge distributions. For charges outside a *spherically symmetric* charge distribution, the resultant force is as if all the sphere's charge were concentrated at its center. A perfect cancellation, as in Fig. 17-12 for gravitational forces, also produces zero resultant force on a charge *within* a hollow sphere of charge.

Exercises

5.   How would you modify Eq. (28-3) if the charge Q were not uniformly distributed but varied linearly in density from zero at the center of the line to a maximum at both ends? In other words, $\lambda = \alpha|x|$, where α is a constant and $dq = \lambda dx$ gives the charge contained in each infinitesimal segment dx.

6. What is the resultant force between (*a*) a hollow sphere of charge $+Q$ and radius R and a negative point charge q at a distance r from the sphere's center $(r > R)$? (*b*) two hollow spheres of positive charge Q and radius R separated by center-to-center distance r $(r > 2R)$?

28-5 The Electrical Nature of Matter

Ordinary neutral matter is not devoid of electric charges. It is neutral only because it contains equal quantities of negative and positive electricity. These charges reside in the constituents of atoms. The nucleus of an atom contains over 99.9% of the atom's mass and is positively charged. The negative charge is distributed among the electrons held near the nucleus by electric attraction. The positive charges in the nucleus are carried by particles called *protons*, which constitute about half the particles of the nucleus. The charge residing in a proton is fixed and equal to 1.602×10^{-19} C. The remainder of the nuclear particles are *neutrons*, neutral particles that have a mass very close to that of the proton. Both neutrons and protons are called *nucleons*.

The electrons have only about 1/2000 the mass of protons but carry a negative charge of equal magnitude. The magnitude of this charge is generally abbreviated *e*. The number of charges in an atom's nucleus (and hence the number of electrons in the neutral atom) is called the *atomic number*. The chemical behavior of atoms is determined by their atomic number.

Though electrons and nucleons are particles in every sense of the word, they simultaneously exhibit a wavelike behavior. This paradoxical mixture of wave and particle aspects is at the heart of quantum mechanics and will be discussed in Chapter 40. For now, suffice it to say that the electrons surrounding a nucleus resemble a smeared-out cloud more than individual planets in a miniature solar system. Useful numbers regarding atomic structure are tabulated in Table 28-1.

$e = 1.602 \times 10^{-19} C$

TABLE 28-1 Atomic Properties

M_p (mass of proton)	1.673×10^{-27} kg
M_n (mass of neutron)	1.675×10^{-27} kg
m_e (mass of electron)	9.1×10^{-31} kg
e (elementary charge unit)	1.602×10^{-19} C
Typical nuclear radius	5×10^{-15} m
Typical atomic radius	5×10^{-11} m

Though the charge carried by each electron is small, the very large number of atoms in even a small bit of matter leads to a rather surprisingly large amount of positive and negative charge. For example, consider one gram of hydrogen. The hydrogen atom is the simplest of all atoms, its nucleus consisting of only one proton, so that its electron cloud contains only one electron. In one gram of hydrogen there are 6×10^{23} atoms. Thus the total positive charge of the protons is

no. proton = no. of electrons

$$Q = 6 \times 10^{23} \times 1.6 \times 10^{-19} \text{ C} = 9.6 \times 10^4 \text{ C}$$

There is, of course, an identical quantity of negative charge in all the electrons. To get some idea of the magnitude of this charge, let us imagine all the positive charge gathered into one point and all the negative charge at another point. If these points are separated by a distance equal to the diameter of the earth, we still find an impressive magnitude for the force of attraction between them.

$$F = k \frac{Q^2}{d_e^2} = 51.8 \times 10^4 \text{ N} \approx 60 \text{ tons}$$

The atoms in a solid are bound together by intense electric forces. The action of these forces is generally called *bonding*. One type of bond is the *covalent bond* and is schematically illustrated in Fig. 28-5. We can obtain a rough estimate of the force exerted by such a bond by calculating the force between either of the atoms and its bonding electron about 2×10^{-10} m away. This force is given by

$$F \approx k \frac{e^2}{(2 \times 10^{-10} \text{ m})^2} = 6 \times 10^{-9} \text{ N}$$

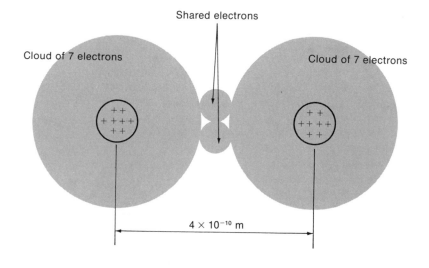

Shared electrons

Cloud of 7 electrons   Cloud of 7 electrons

4×10^{-10} m

Fig. 28-5. When two oxygen atoms combine to produce an O_2 molecule, one electron from each atom is localized between the atoms. The remainder of each atom thus carries one unit of excess positive charge.

The sum of all the bonds between adjacent planes of a crystal is responsible for its tensile strength. For a simple array, as in Fig. 28-6, the number of such bonds per unit area is $1/d^2$, where d is the spacing between atoms. The tensile strength in units of force/area is given by

$$F/\text{area} = F_{\text{bond}} \cdot (\text{bonds/area}) = 6 \times 10^{-9} \, \text{N}/d^2$$

Taking $d = 4 \times 10^{-10}$ m, we find

$$F/\text{area} \approx 4 \times 10^{10} \, \text{N/m}^2 \approx 5 \times 10^6 \, \text{lb/in.}^2$$

Our simplistic treatment of the covalent electrons as if they were points between atoms has probably made this an overestimate but strikingly illustrates the magnitude of electrical forces between atoms. In most crystalline solids, imperfections lead to tensile strengths far below our estimate. As was noted in Chapter 13, however, perfect crystals of iron do exhibit ultimate strengths of over 10^6 lb/in.^{2}.

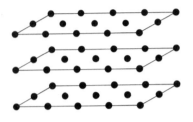

Fig. 28-6. Any given plane of a crystal is bound to the next by the sum of all the atom-atom bonds linking them.

Exercises

7. An average adult contains approximately 10^{28} elementary charges (1.6×10^{-19} C) of each sign and is very nearly electrically neutral. If there were instead 1% more negative charges than positive charges and two people stood 1 m apart, estimate the magnitude of the electric force of repulsion between them. Compare your answer to the weight of all the earth's oceans (ocean mass $\approx 10^{21}$ kg).

8. Assuming that the electron cloud and the nucleus of a typical atom are spherical, find the ratio of volumes occupied by the entire atom and by the nucleus.

28-6   Friction and Frictional Electricity

The force of friction is due to *bonds* forming between the atoms in adjacent surfaces. When two substances are in contact, only a relatively small number of atoms can touch one another, since normal surface irregularities are extremely large when viewed on an atomic scale. Those atoms that do touch form bonds that essentially weld the two surfaces together at a number of points, as in Fig. 28-7. In order to start one surface sliding over another, these bonds must

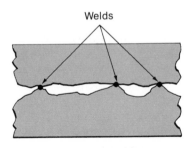

Welds

Fig. 28-7. Atomic bonds can weld two surfaces together at points where they actually touch.

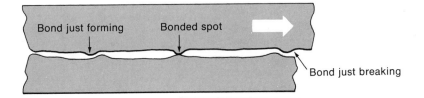

Bond just forming   Bonded spot

Bond just breaking

Fig. 28-8. The acceleration
due to the force of attraction
between bonds just forming
should offset the deceleration
due to breaking bonds. It would
seem the coefficient of *sliding*
friction should be zero!

be broken. This is the origin of the force of *static friction.** Once
one surface is sliding past another, bonds are continually being
formed between points momentarily in contact, then broken as the
surfaces move on.

As can be seen in Fig. 28-8, once one of the surfaces is in motion
there are just as many instances, on the average, when bonds are
being formed as when bonds are breaking. As bonds are first formed,
they tend to pull the sliding surface ahead, and they tend to slow it
down when they break. The complete symmetry of the situation
obviously leads us to the conclusion that the coefficient of *sliding*
friction should vanish! It is generally smaller than that of static
friction, but certainly doesn't vanish. What's wrong?

There are several levels at which we can understand the fact
that the coefficient of sliding friction is nonzero. Basically, they all
come down to the same thing: the formation and subsequent break-
age of bonds cause some energy to be used to increase the random
motion of the atoms (thermal energy). Just as a bond is formed
between opposing atoms in the two sliding surfaces, each atom is
pulled slightly away from its equilibrium position with a resultant
slight motion of the next atom deeper within the solid, and so on.
The same sort of chain reaction of one atom interacting with the
next occurs when the bond is broken. Now, given the enormous
number of jiggling atoms coupled together in a solid, it is essentially
impossible that the net result of forming and then breaking a bond
would be that all atoms are left with their original energy. Instead,
additional kinetic energy is present after such an event. It had to
come from somewhere and in this case it comes from a loss of
energy of the macroscopic moving system.

Conservation of energy makes it clear that there must be a net
retarding force that takes energy away from the moving system.

---

*For more information at an elementary level of discussion, see the book by Frank
Phillip Bowden and David Tabor, *Friction: An Introduction to Tribology;* New York:
Anchor Books Edition, Doubleday and Company Inc. (1973).

But we have not said exactly how this comes about. A simple model will give us some idea of the details. Consider the atoms bound to each other by springs. As the atoms in opposing surfaces approach one another, they are pulled slightly out of position, as shown in Fig. 28-9. The atom's inertia causes it to overshoot so that it is closer to the stationary atom during the receding phase of the motion than it was during the approaching phase. Since the atom-atom force is larger for smaller separation distances, the average retarding force on the moving atom is larger than the average accelerating force. The net result of all the atom-atom encounters between the sliding and stationary block is thus a retarding force.

The magnitude of the frictional force is known to be roughly linearly dependent upon the normal force pushing the sliding surfaces together. This is understandable on a microscopic scale. As mentioned previously, the points of actual atom-to-atom contact between surfaces are extremely few. However, as the normal force increases, the few sharp contact points are deformed by pressure, so that the contact area—and hence the number of bonded atoms— increases. The atomic interaction picture we have discussed is most easily made quantitative for clean surfaces sliding in vacuum. In everyday practice, frictional effects are often dominated by the presence of oxide films and lubricants. Some lubricants, such as the solid plastic teflon®, form only weak bonds with most materials. Liquid lubricants utilize fluid flow, as discussed in the introduction to Chapter 18.

Frictional electricity, of the sort produced when a glass rod is rubbed with silk, is produced when atoms in one of the sliding surfaces are more *electronegative* than atoms in the other surface. These atoms have larger binding energies for their electrons. When bonds are formed at the points of contact, electrons are shared between the atoms of the two sliding surfaces. The more electronegative substance then steals electrons from the other as the bonds

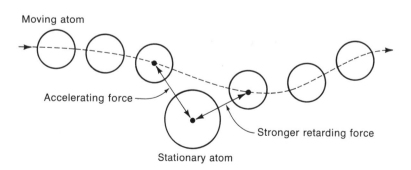

Fig. 28-9. The attractive force between atoms pulls the atoms slightly out of position in their crystal lattice as they pass by one another. The inertia of the displaced atom causes it to overshoot, so that the attractive forces are strongest as it recedes, thus producing a net retarding force. The lower atom is shown as stationary for simplicity.

break. Some charging occurs on contact, since some atoms contact each other then. The role of rubbing is simply to cause more atom-atom collisions so that more electrons are transferred.

Exercise

9. A rough value for the magnitude of atom-atom bonding forces was found in Section 28-5. Utilizing this result, estimate the number of atomic bonds linking two surfaces with a coefficient of friction $\mu = 0.5$ and a normal force loading of 600 N.

28-7   Conductors and Insulators

When an electric charge is placed upon some materials, such as amber or glass, it remains fixed. These materials are called *insulators*. A charge placed upon materials like copper or gold, called *conductors*, moves about quite readily. There is an intermediate class of materials, called *semiconductors*, in which charges move with difficulty. Examples of these classes of materials are given in Table 28-2.

insulators - electric charge remains fixed

conductors - charge moves about readily

TABLE 28-2

Insulators	Semiconductors	Conductors
Glass	Silicon	Copper
Bakelite	Germanium	Aluminum
Mylar	Carbon (diamond form)	Gold
Dry wood	Cadmium sulfide	Carbon (graphite form)
Kerosene		Salt water

Insulators are materials in which the atomic electrons are firmly bound to their parent atoms. Some conductors—the metals—contain some weakly bound electrons that easily move about to transport charge through the metal. Other conductors are *electrolytes,* such as salt water, in which positive and negative ions can move about to cause a net transport of charge.* In semiconductors, electrons or electron *vacancies* (called "holes") or both can migrate through the substance.

---

*An *ion* is an atom that has gained or lost one or more electrons, so that it is not electrically neutral.

The way in which charges are transported by conductors is illustrated in Fig. 28-10. When a positively charged conducting ball is touched to an identical but neutral ball, the charge distributes itself equally between them. In fact, this process was used by Coulomb repeatedly to obtain various subdivisions of charges during his experiments. The motion of charge from one ball to the other can also be accomplished with a conducting wire. As shown in the figure, only negative electrons move when the conductor is a metal. Electrons leave the originally neutral ball and flow to the positive ball, partially neutralizing its charge. As a result, the originally neutral ball becomes partially positively charged.

The same result occurs if positive charges are the mobile charges in the conductor. In that case, positive charges flow from the positively charged ball to the neutral ball until both are equally charged. If a string soaked with salt water is used as the conductor, both positive and negative charges flow during the equalization process. The moving charges in that case are positively charged sodium ions, Na^+, and negatively charged chlorine ions, Cl^-. These ions are produced when common salt, NaCl, is dissolved in water. The point is that charges can move from one body to another by motion of negative charges, positive charges, or both.

The ability of charges to move easily through conductors is used to advantage in the gold leave electroscope shown in Fig. 28-11. A charge applied to the metal ball at the top of the instrument quickly spreads itself over all of the conductor, including a thin, very flexible gold leaf cemented to the conductor. The charge in the nearby, rigid portion of the conductor repels the charge on the gold leaf, so that the leaf stands out at an angle, as shown. The greater the charge placed on the electroscope, the greater the angle of deflection of the gold leaf.

If a negative charge is simply brought near the ball at the top of the electroscope, free electrons in the metal rod are repelled downward in the rod and onto the gold leaf, so that the leaf will be deflected. When a positive charge is held near the ball, free electrons are attracted toward it, leaving a surplus positive charge on the bottom of the rod and on the gold leaf. Again, a deflection of the gold leaf occurs. Thus an electroscope can be used to detect charge on an object without actually transferring the charge to the electroscope, but merely by bringing it nearby.

When a neutral conductor is placed near a charged object, a net force of attraction occurs. This is because charges flow in the conductor under the influence of the charged body, as shown in

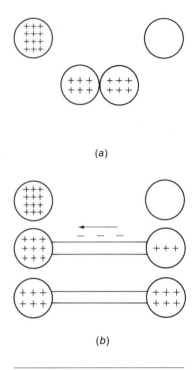

(a)

(b)

Fig. 28-10. When a metallic conductor carries charge, it is the negative electrons that actually move.

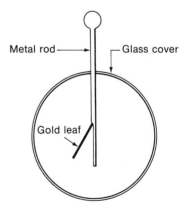

Fig. 28-11. The gold leaf electroscope.

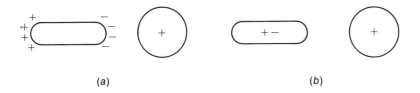

(a)   (b)

Fig. 28-12. (a) When a neutral conductor is placed near a charged object, charge separation occurs. (b) In an insulator, charges cannot flow but are slightly displaced.

Fig. 28-12a. Whatever the sign of the charge on the charged body, the opposite type of charge in the conductor moves closest to the charged body and the similar type of charge moves away. The net result is that unlike charges are closer together than like charges, so that an overall force of attraction results. When charges within a body are so separated we say the body is *polarized*.

Polarization also occurs in neutral insulating objects, as indicated in Fig. 28-12b. Charges cannot flow through an insulator but the average location of the electron clouds around its individual atomic nuclei can be slightly displaced by a nearby charged object, as indicated in Fig. 28-13. Again, the force of attraction between the unlike charges is slightly greater than the repulsion between like charges. Each atom is thus attracted toward a charged object. The result is a net force of attraction between neutral insulators and charged bodies. This is the mechanism by which a charged piece of amber or hard rubber can attract bits of paper.

28-8   Charging by Induction

The separation of charge that occurs in a neutral conductor when a charge is brought nearby can be used to charge the conductor. The process, called *charging by induction,* is indicated in Fig. 28-14. When a positive charge is brought nearby, as in Fig. 28-14b, the charges in the conductor separate. In Fig. 28-14c, connection of the far side of the sphere to any large object (the larger the better) removes the positive charges. This process is called *grounding,* since the earth is the ideal large object. A connection to the earth through a wire connected to a cold water pipe, for example, is called a *ground*

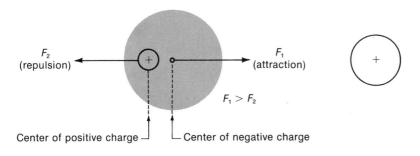

$F_1 > F_2$

Center of positive charge ┘   └ Center of negative charge

Fig. 28-13. Atoms are polarized by nearby charges.

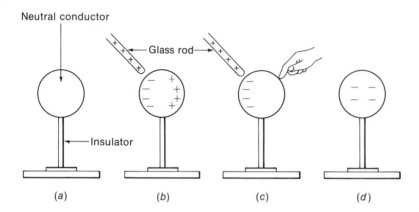

Neutral conductor

Glass rod

Insulator

(a)   (b)   (c)   (d)

Fig. 28-14.   A metal sphere on an insulating stand is charged by induction.

or a ground connection. A touch of the finger is usually adequate to ground the conductor unless it is a dry day and the experimenter is standing on a dry insulating surface. The ground (or finger) is next removed and finally the nearby charge is removed. There is then a residual charge, the *induced* charge, trapped on the conductor. The induced charge is clearly of opposite sign to that brought nearby during the process.

Induction provides a convenient method for placing small quantities of charge on an electroscope. As seen in Fig. 28-15, a positive charge may be induced on the electroscope by means of a hard rubber rod that has been charged negatively by rubbing with cat's fur. If an object with an unknown charge is then brought nearby, the electroscope deflection will increase if the unknown charge is positive and decrease if it is negative.

28-9   Atmospheric Electricity   NOT ON HOURLY

Large-scale separation of electric charge occurs within thunderstorms. Such a storm may be thought of as a large-scale heat engine

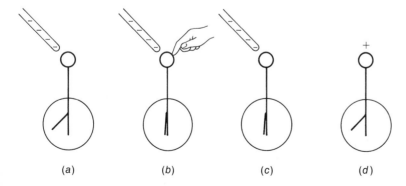

(a)   (b)   (c)   (d)

Fig. 28-15.   An electroscope can be charged by induction.

that utilizes the natural convection of warm air over sun-warmed ground to transport warm, moisture-laden air upward. As the air rises and expands, it cools until water begins to condense out of the moist air, as indicated in Fig. 28-16. The heat of vaporization given off by the condensing water drives the "engine" still faster. Eventually, the heavy droplets falling downward and the cooling produced by cool, upper-level air pulled into the updraft produces a downdraft of cold air.

The exact mechanism of electric charge separation is not well understood, but there are some natural processes that probably participate. These charging processes typically involve breakup of water droplets such that the heaviest resulting droplets are negatively charged while the lighter droplets are positively charged. The action of gravity combined with cloud updrafts then causes a charge separation so that the base of the thundercloud is negatively charged and the top is positively charged. The magnitude of the charge separation in thunder clouds is about 40 C, a large figure.

These thundercloud charges are so large that the force they exert upon electrons in the atoms of the atmosphere is sufficient to ionize some of the atoms. When enough ions have been formed between the two opposing cloud charges (or between the base of the cloud and the induced charge in the earth below), a conducting path exists and an intense electric current flows through the ionized path for a few microseconds. These paths vary from about 500 ft to 2 mi in length. This current produces intense heat and, hence, expansion of the air in its vicinity, which leads to a shock wave, the "thunderclap."*

---

*More information on atmospheric electricity can be found in *The Feynman Lectures on Physics*, Addison-Wesley Publishing Co. (1964) Volume II, Chapter 9; and "Thunder", by Arthur A. Few in *Scientific American*, July 1975.

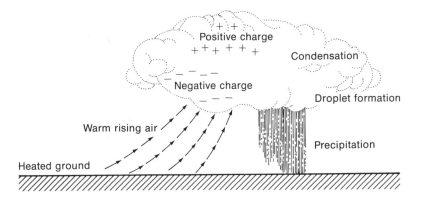

Fig. 28-16. A typical thunderstorm. (Adapted from U.S. Department of Commerce Weather Bureau Report, June 1949)

Exercise

10. The time between lightning strokes in a single thunderstorm is typically about one minute. The stroke usually lasts for 30 μs. Assuming that 40 C of charge are transferred from the bottom to the top of the cloud between strokes (and all 40 C flow back in the lightning stroke), calculate: (*a*) the average charging current between strokes; and (*b*) the average current in a lightning stroke. Express these currents in amperes, where 1 ampere = 1 C/s.

28-10 Rutherford Scattering   NOT ON HOURLY

The fact that atoms have tiny charged nuclei that contain most of the mass of an atom was first deduced in the period 1911–1913. Rutherford and his students, Geiger and Marsden, were studying the *alpha particles* produced by some radioactive substances. These particles had energies between 4 and 8 million electron volts (MeV) and carried two positive elementary units of charge:

$$q_\alpha = +2e = 3.2 \times 10^{-19} \text{ C}$$

These alpha particles were known to have a mass approximately 8000 times that of an electron. Rutherford and his coworkers were later able to show that these alpha particles were the nuclei of helium atoms. From the mass and kinetic energy of the alpha particle, we deduce that their velocity is approximately 1/20 that of light.

Rutherford decided to use these massive, rapidly moving projectiles as a probe, hoping to learn something about the structure of atoms. If the positive and negative charges were spread throughout the entire volume of atoms, then each bit of an atom would be approximately neutral. An alpha particle would not be strongly deflected by electric forces as it passed through such an atom. Alphas with energies of several MeV were known to be capable of passing through layers many tens of thousands of atoms thick. It was expected that the alphas would not be substantially deflected even then, if each atom contained uniformly distributed positive and negative charge, as in Fig. 28-17.

The atomic model most seriously under consideration in 1911 was such a uniform atom. Rutherford and his coworkers allowed a small beam of alpha particles to pass through thin metal foils (about 1000 atoms thick). Most of the alphas passed straight through

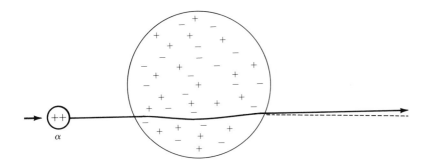

Fig. 28-17. If atoms consist of intermingled positive and nega- tive charges, alpha particles will be only slightly deflected by electric forces.

the foil, as indicated in Fig. 28-18. However, a few alphas were scattered at substantial angles. Some even came almost straight back, 180° away from their original direction. Rutherford was astonished at this result, as evidenced by his own words: "It was quite the most incredible event that ever happened to me in my life. It was almost as incredible as if you had fired a 15-inch shell at a piece of tissue paper and it came back and hit you." It was obvious that the then-accepted atomic model was wrong.

Rutherford concluded that most of the positive charge must be tied to a very small atomic component that was extremely mas- sive (compared to electrons). We *now* call this portion of the atom its *nucleus*. The reason that he thought this massive element was positively charged was that it was already known that negative atomic charges were carried on electrons—and these electrons were much lighter than alpha particles and hence unable to cause a back-scattering process. He realized that this heavy portion of the atom must be quite small, since very few alphas were scattered straight back toward the incident beam direction. Those few that were, he reasoned, must have made a rare head-on collision with this small yet massive portion of the atom.

Alphas from the radioactive source could not be purposely directed toward the center of a given atom. Instead, alphas peppered

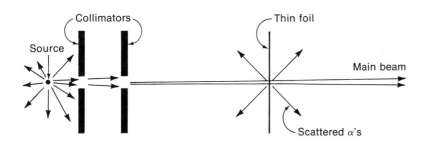

Fig. 28-18. Schematic diagram of Rutherford's alpha-scattering experiment.

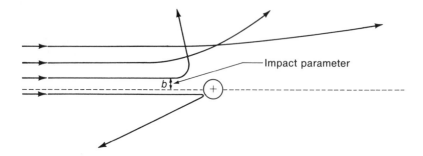

Fig. 28-19. Alphas with the smallest impact parameters are deflected through the largest scattering angles.

the target in a random fashion and, hence, had random impact parameters with respect to any given atom (see Fig. 28-19). As seen in Fig. 28-20, randomly distributed impact parameters mean that the overwhelming majority of incident alphas will have very large impact parameters. Only a very few happen to be directed close to the nucleus. Examining Fig. 28-20, we see that the number of impacts within a given range of impact parameters—say b to $b + db$—is proportional to the area of a ring of radius b and width db surrounding the nucleus. If $db \ll b$, this area is quite accurately $2\pi b \, db$ and hence is proportional to b itself. Thus the relative probability of the occurrence of two different impact parameters, b_1 and b_2, is just proportional to the ratio of the impact parameters.

$P(b_1)$ = probability of impact with $b_1 < b < b_1 + db$

$P(b_2)$ = probability of impact with $b_2 < b < b_2 + db$

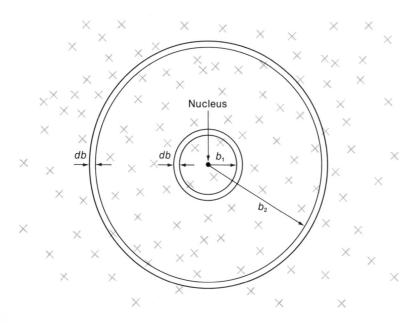

Fig. 28-20. A single target nucleus as seen from the direction of the incident alpha particles. The X's represent the line that would be followed by randomly distributed alphas if they were not deflected. Note that most alphas have large impact parameters.

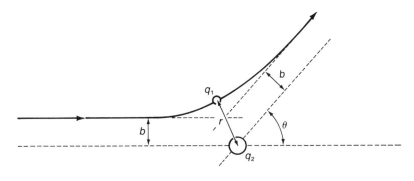

Fig. 28-21. The electric force between the alpha particle q_1 and the nucleus q_2 scatters the alpha through an angle θ. Conservation of angular momentum demands that b be the same before and after an elastic scattering event.

$$\frac{P(b_1)}{P(b_2)} = \frac{b_1}{b_2}$$

If we now relate the expected angle of scattering to the value of b, we can deduce what fraction of scattering events fall into any given angular range. From Fig. 28-19 it is qualitatively clear that the larger probability of large impact parameters will lead to a predominance of small scattering angle events. Rutherford deduced the connection between b and θ by application of Newtonian mechanics and the Coulomb force law to find the trajectory shown in Fig. 28-21. He found that:

$$b = \frac{kq_1q_2}{m_\alpha v^2} \cot\left(\frac{\theta}{2}\right)$$

With this result and the knowledge that the probability of a given impact parameter is proportional to the value of the impact parameter, he then calculated the probability of scattering into an angular range from θ to $\theta + d\theta$ and found:

$$P(\theta) \propto \left(\frac{kq_1q_2}{m_\alpha v^2}\right)^2 \frac{\sin\theta \, d\theta}{\sin^4(\theta/2)}$$

The results of the alpha-scattering experiments closely agreed with this prediction and gave birth to the idea that the atom has a small nucleus with electrons orbiting about it at comparatively enormous distances. In the hands of Niels Bohr, this "miniature solar system" model led to what is now called the "old quantum theory." We will return to this topic in Chapter 40.

Exercise

11. An alpha particle $(q = +2e)$ is deflected by a gold nucleus $(q = +79e)$. Find the magnitude of the electric force acting on the alpha particle when the center-to-center distance is 5×10^{-14} m. Find the acceleration of the alpha particle ($m_\alpha = 6.64 \times 10^{-27}$ kg).

28-11 Summary

Like electric charges repel one another while unlike charges attract one another. The magnitude of the force that point charges exert upon one another is given by Coulomb's law:

$$F = k\frac{q_1 q_2}{r_{12}{}^2}$$

We can choose to *define* our units of q by use of Coulomb's law. This is done in the esu system, where k is set equal to 1, F is measured in dynes, and r in cm. The unit of charge so defined is called the statcoulomb.

In the SI system, electric charge is defined in an independent way. This unit of charge is the coulomb (C) and flow of charge at 1 C/s defines an electric *current* of one ampere. In SI units, the constant k in Coulomb's law has the value

$$k = 8.99 \times 10^9 \text{ N} \cdot \text{m}^2/\text{C}^2$$

$9 \times 10^9 \text{ N} \cdot \text{m}^2/\text{C}^2$

For reasons of convenience, we will sometimes write k as $1/4\pi\epsilon_0$, so that

$$\epsilon_0 = 8.85 \times 10^{-12} \text{ C}^2/\text{N} \cdot \text{m}^2$$

Coulomb's law can be written in vector form:

$$\mathbf{F}_{12} = k\frac{q_1 q_2}{r_{12}{}^3}\mathbf{r}_{12}$$

and it is seen that the electric force acts in the direction of the line connecting the two charges. When several charges are present, the net electric force on any one of them is simply the vector sum of all the electric forces caused by the other charges.

We can apply Coulomb's law, which is valid for point charges, to extended, continuously distributed charges by breaking up a continuous charge into a discrete sum of infinitesimal elements. This sum becomes an integral as the elements are allowed to approach zero extent. By analogy to the case of the gravitational force, we conclude that spherical charges interact as though their charge were concentrated at their centers.

Matter is made up of atoms that are electrically neutral. However, the constituents of atoms are charged: the nucleus carries a positive charge while the surrounding electron cloud carries a negative charge. Each charged elementary particle carries a unit of charge called the elementary charge, e:

$$e = 1.602 \times 10^{-19} \text{ C}$$

The strength of atomic bonds, the energy of chemical reactions, and the strength of matter are all due to the intense electrical forces between the constituents of atoms.

The forces that bond atoms together also act between atoms in different bodies when the bodies are brought into contact. Only a very few points actually touch and the force of friction is caused by bonding at these points. If one of the two bodies is more electronegative than the other, it will gain electrons at the points of contact and give rise to frictional charging.

Conductors are those materials in which electric charges move about freely. Metals are used as conductors in most practical devices. In metals a portion of the electrons are free to move throughout the metal. In electrolytes, both positive and negative ions may move. Semiconductors conduct by motion of either electrons or electron holes or both. Insulators are those materials in which electric charges do not move freely.

Charging by induction utilizes the separation of charge that occurs when a charged body is brought close to an initially uncharged conductor. When one sign of charge is removed to ground, the conductor acquires a net charge.

Atmospheric electricity is produced in thunderstorms by charging processes involving upward-directed winds that preferentially elevate small droplets and frictional charging in collisions between droplets.

The fact that an atom contains its positive charge in a very small, central nucleus was discovered by a study of alpha-particle scattering. The intense electric force between the alpha particle and the nucleus produced a very large change in direction for those few alpha particles that, by chance, passed very close to the nucleus of the atom. Such large scattering angles would not have been produced by atoms consisting of more uniformly distributed positive and negative charge.

Problems

28-1.   Two equal, positive point charges Q are mechanically tied to one another by a steel rod of diameter d and length l. (a) Find Q_{max}, the maximum value of Q, if the yield point of the steel rod is not to be exceeded. Call this yield point YP. (b) Find the numerical value of Q_{max} when $d = 2$ cm and $l = 1$ m. The value of YP can be found in Table 13-3.

28-2.   Calculate the conversion factor between statcoulombs and coulombs.

28-3. In esu the elementary charge has a magnitude of $e = 4.8 \times 10^{-10}$ statcoulomb. (a) Find the electric force of repulsion between two alpha particles separated by 10^{-12} cm. (b) Find the acceleration of either alpha particle.

28-4. Three point charges Q, $2Q$, and $-Q$ are at the vertices of an equilateral triangle with sides of length l. Find the magnitude of the net force on the negative charge.

28-5.* Eight equal positive charges q are placed at the vertices of a cube. Find the net electric force on any one of them.

28-6. A quantity of charge Q is distributed uniformly along a straight line segment of length L. A point charge q is at a distance L away from one end of the line charge and lies on the line that contains the line segment. Find the electric force between Q and q.

28-7. Follow the procedure suggested in Exercise 5 (page 583) to find the net force acting on the point charge.

28-8.* A point charge $-q$ lies on the axis of a circular ring of charge of radius R, as shown in Fig. 28-22. The charge $+Q$ is uniformly distributed over the circular ring. Find the electric force of attraction between $-q$ and $+Q$.

28-9. A sphere with charge q and mass m hangs from a thread, as in Fig. 28-23. It is repelled by a fixed spherical charge Q, as shown. Find q in terms of Q, θ, d, m, and g.

28-10. Almost all of the mass of an atom is contained in its nucleus. Use the last two entries in Table 28-1 to estimate the density of nuclei. Assume that atoms are in contact in solids and liquids.

28-11. The radius of the nucleus of an aluminum atom is approximately 3×10^{-13} cm. This nucleus contains 13 protons and 14 neutrons. Find the density of this nucleus.

28-12. Find the charge density (C/cm^3) for the aluminum nucleus (see Problem 28-11).

28-13. All atomic nuclei have approximately spherical shapes. Their radii are closely proportional to the cube root of the number of nucleons they contain. Show that this implies a constant density for all nuclei.

28-14. Estimate the number of covalent bonds that are formed between two surfaces that have a coefficient of static friction $\mu = 0.2$ and are loaded with a normal force of 3000 N. Compare this number to the total number of atoms in either surface if they both have an area of 100 cm².

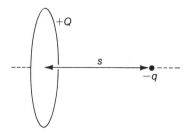

Fig. 28-22. A circular ring of charge $+Q$ attracts a point charge $-q$.

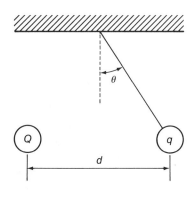

Fig. 28-23.

28-15. A copper wire carries a steady electric current of 5 am-
        peres. This current is carried by the motion of electrons
        in the copper. How many electrons per second pass a
        given point along the wire?

28-16. A glass rod is rubbed with silk, then is brought close to
        an electroscope and the electroscope is briefly touched
        with a finger. The glass rod is then removed. The electro-
        scope leaf remains deflected by the induced charge. An
        unknown charge is next brought close to the electroscope
        and the electroscope indication decreases. What is the
        sign of the unknown charge?

28-17. Find the deflection angle θ of an alpha particle that ap-
        proaches a gold nucleus with an impact parameter of 5
        $\times 10^{-12}$ cm. The number of protons in a gold nucleus is 79
        and the velocity of the alpha particle is 1/20 that of light.
        The mass of the alpha particle is 6.64×10^{-27} kg.

Electric Fields

29-1 Introduction Coulomb's law describes an action-at-a-distance force. It does not pretend to explain "how" the presence of one charge can produce a force on a distant charge; it only describes observations. Since no mention of time is made in Coulomb's law, we might presume that this force is transmitted from one charge to another instantaneously. Yet, in fact, there is a measurable time delay in the response of one charge to the presence of another. This time delay is $\tau = s/c$, where s is the separation of the charges and c is the velocity of light. If, for example, we try to deduce how two oppositely charged particles orbit about one another, we should actually compute the force on either charge as a function of the distance between the charges *earlier* in time. Yet the variation of that distance with time depends on the very force we are trying to compute. The problem of taking time delay into account is clearly not a simple one.

To make matters more complicated, the force between two charges moving at velocity v is not exactly given by Coulomb's law. Instead, there is a small correction that is v^2/c^2 smaller than the Coulomb force itself. Since c, the velocity of light, is so large, most ordinary velocities lead to a very small value for the fraction v^2/c^2. Nonetheless, this velocity-dependent term, called the *magnetic force*, can be very important, as we shall see in Chapter 34. Because of the complicated way in which the force between two *moving* charges depends upon the history of their positions and velocities, it is quite difficult to formulate the laws of dynamics with electric forces given by a time-delayed Coulomb's law with additional magnetic corrections. Instead, the laws of dynamics with electric forces taken into account, called *electrodynamics*, are generally presented in terms of electric and magnetic *fields*.

The idea of an electric field entails thinking of one charge as producing an effect (the field) in the space around it, whether another charge is present or not. Any other charge in this field responds in a simple way given by the field strength and direction at that point. Thus, we break the electric interaction into two parts: formation of a field and response to the field. It is certainly not immediately obvious that this added step will lead to a simplification of electromagnetic theory and calculations, but it does. We will begin by discussing the static situation with two charges, where no simplification is obtained by introducing the idea of the electric field. In later sections we will occasionally point out simplifications introduced by using field concepts.

29-2 The Electric Field Strength, E

Two static point charges interact in accordance with Coulomb's law. For the force exerted on charge 2 by charge 1, we write:

$$\mathbf{F}_{12} = k \, \frac{q_1 q_2}{r_{12}^{3}} \, \mathbf{r}_{12} \qquad (29\text{-}1)$$

In scalar form,

$$F = k \, \frac{q_1 q_2}{r^2}$$

F is the magnitude of force on either charge

where F is now the magnitude of the force on either charge. We can think of this force as the result of two effects. First, charge 1 creates an electric field in its surrounding space:

$$\mathbf{E}_1 = k \, \frac{q_1}{r^3} \, \mathbf{r} \qquad (29\text{-}2)$$

where $\mathbf{r}$ is a vector from charge 1 to *any* arbitrary point of interest. Then $\mathbf{E}_1$ is the *electric field vector* (or *electric intensity*, or just the *electric field*) at that point due to the presence of charge 1. The magnitude of $\mathbf{E}_1$ falls as the inverse square of the distance from the point charge producing it. It points radially outward from a positive charge and radially inward toward a negative charge. An example is shown in Fig. 29-1 for two points of interest, called *field points*. The electric field vector, **E**, is shown with its tail at the field point in both cases.

Now we wish to describe the second step: how a second charge experiences a force when acted upon by the electric field of the

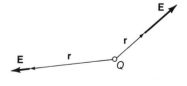

Fig. 29-1. The electric field around a given point charge is radially directed and has a magnitude inversely proportional to the square of *r*.

first. When any other charge, q_2, is placed in this electric field, it experiences a force according to the force law:

$$\mathbf{F}_2 = q_2\mathbf{E}_1 \qquad\qquad (29\text{-}3)$$

Substitution of (29-2) into (29-3) leads us back to Coulomb's law, as it should. We have simply broken the electric interaction into two parts conceptually. The first charge produces a field and the second charge responds to that field.

Equation (29-3) is a general statement concerning any charge in any electric field. In fact, it may be regarded as the *definition* of what is meant by an electric field. Dropping subscripts and considering any charge q in any field E, we have

$$\mathbf{F} = q\mathbf{E} \qquad\qquad (29\text{-}4)$$

or

$$\mathbf{E} = \mathbf{F}/q \qquad\qquad (29\text{-}5)$$

The units of electric field strength are thus force/charge; in SI units, N/C.

A device to measure electric fields could be constructed by measuring the magnitude and direction of the force (aside from non-electrical forces such as gravity) on a small body carrying a known charge q, which is called a *test charge*. If we really want to know what electric field would have been present *without* the test charge, we might worry a bit about the reaction of the charges that are the *source* of the field to the presence of the test charge. That is, if the charges producing the field of interest are not firmly constrained, they might move a bit due to the electric force caused by the test charge itself. This can be avoided, in principle at least, by considering the limiting case of a vanishingly small test charge:

$$\mathbf{E} = \lim_{q \to 0} \mathbf{F}/q \qquad\qquad (29\text{-}6)$$

However, the reader should understand that (29-5) is still *exactly* true. The need for the concept contained in Eq. (29-6) occurs only when we ask the question, "What would be the value of **E** if no perturbing charges were present?"

Electric field strengths are rarely measured, in practice, by direct observation of the force produced on a test charge. Instead, a more indirect route is used in most instruments, involving the work done by electrical forces on a moving charge. We will come back to this point in Chapter 31, but for the present we should note that the commonly used units for electric field strength are volts/meter (V/m). We shall show later that these units are identical to the units N/C in the same sense that a newton is identical to a kg · m/s². Thus an electric field strength of 1 V/m produces a force of 1 N on a 1-C charge.

Example   An electric field at a given point is directed along the positive z axis and has a strength of 10 V/m. Find the magnitude and direction of the force exerted on a charge at that point when the value of the charge is (a) 3 C; (b) —5 C.

Solution   (a) $F_x = qE_x = 0$ $F_y = qE_y = 0$

$$F_z = qE_z = 3\ C \cdot 10\ V/m = 3\ C \cdot 10\ N/C = 30\ N$$

Thus the force is directed in the positive z direction.
(b) $F_x = F_y = 0$, as above.

$$F_z = -5\ C \cdot 10\ N/C = -50\ N$$

The force is now directed in the negative z direction.

Example   Find the magnitude of the acceleration produced by a 500-V/m electric field acting on an electron.

Solution   Since $a = F/m = (q/m)E$, we see that q/m is the proportionality constant linking acceleration and applied field strength. This ratio is thus a measure of the "liveliness" of a charged object in its response to electric fields. The electron must be lively indeed, since this ratio is

$$\frac{q}{m} = \frac{e}{m_e} = 1.76 \times 10^{11}\ C/kg$$

For our 500-V/m field, we find

$$a = 1.76 \times 10^{11}\ \frac{C}{kg} \cdot 5 \times 10^2\ \frac{N}{C} = 8.8 \times 10^{13}\ \frac{m}{s^2} = 8.9 \times 10^{12}\ g$$

This astonishing value of acceleration is made even more surprising when we note that electric fields several thousand times stronger than that in our example are produced by laboratory equipment. Furthermore, nature provides far more intense electric fields within atoms. Electric field strengths very near the nuclei of atoms can be 11 orders of magnitude stronger than that in our example.

Exercises

1. The electric field due to a point charge reverses direction when the sign of the charge is changed, according to (29-2). Similarly, (29-3) indicates that the direction of the force exerted on a charge by a field depends on the sign of that charge. Show that the overall result for the force that one charge exerts upon another is in agreement with our knowledge that like charges repel, unlike charges attract.

2. A charge of 10^{-6} C is placed in an electric field. The force

acting on it is found to be 0.1 N. This charge is removed and a charged pith ball is released from the same point. If the pith ball's initial acceleration is 2 m/s², find its charge-to-mass ratio, q/m. Neglect gravitational force in this exercise.

29-3 The Superposition Principle

Since the Coulomb force obeys a superposition principle, the electric field does also. If N different charges each produce a field, the resultant at any given point is the vector sum of the N different electric fields at that point:

$$\mathbf{E}_{\text{total}} = \sum_{i=1}^{N} \mathbf{E}_i$$

In the examples shown in Fig. 29-2, the two electric fields present at point P are added to obtain the resultant field in each case.

The electric field that results at any given point due to the presence of a continuously distributed charge can be found as follows. First, the continuously distributed charge is regarded as being composed of many infinitesimal elements. Second, the field (at the point of interest) is calculated as if each element were a point charge by using (29-2). Third, each of these electric fields is vectorially summed at the field point. Finally, the infinitesimal elements are considered in the limit as they become arbitrarily small, so that they are truly point charges. The sum is now, in fact, an integral. As an example of this procedure, let us find the direction and magnitude of the electric field near an infinitely long line of charge with the charge uniformly distributed along the length of the wire. The elec-

Fig. 29-2. Electric fields add as vectors. Examples are shown for several points near a positive and negative charge.

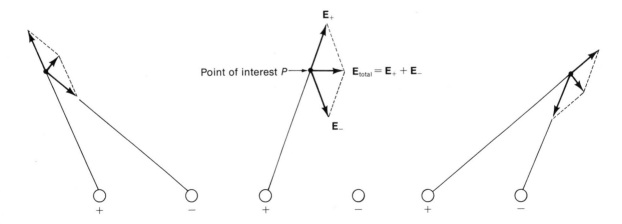

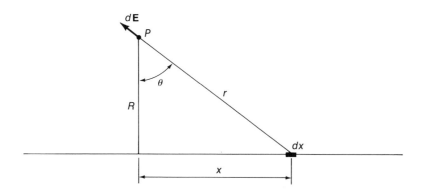

Fig. 29-3. The electric field
due to one element of length dx.

tric field at the point P due to one element of charge is shown in Fig. 29-3. This point is a distance R from the line charge (measured in the direction perpendicular to the line). The charge contained within dx is given by $dq = \lambda\, dx$, where λ is the quantity of charge per unit length of the line charge. The magnitude of the field due to this one charge element is

$$dE = k\,\frac{dq}{r^2} = k\lambda\,\frac{dx}{r^2}$$

Note that a similar element of charge located at $-x$ produces an identical field except that its horizontal component is reversed. Any given element dx at x will always be matched by another at $-x$, since the line charge extends to infinity in both directions. The total resultant field at P must therefore be directed radially outward along the direction $\mathbf{R}$. Accordingly, we need only to sum the vertical components of each dE. The magnitude of the total sum is given by:

$$E_{\text{total}} = k\lambda \int_{x=-\infty}^{x=+\infty} \frac{\cos\theta}{r^2}\,dx$$

We can express all the variables within the integral in terms of θ, r, or x, since they are all interdependent through the relations

$$r^2 = R^2 + x^2 \qquad \text{and} \qquad \cos\theta = R/r$$

In Chapter 28 we made a similar calculation for a finite length of charge and chose θ as our variable. For variety, let us now choose x as the variable:

$$E = k\lambda \int_{x=-\infty}^{x=+\infty} \frac{R/r}{r^2}\,dx = k\lambda R \int_{-\infty}^{+\infty} \frac{dx}{(R^2 + x^2)^{3/2}}$$

Since R is a constant (at any given field point), the only variable in the integral is x. Its evaluation leads to:

$$\int_{-\infty}^{+\infty} \frac{dx}{(R^2 + x^2)^{3/2}} = \frac{x}{R^2\sqrt{x^2 + R^2}} \Bigg|_{-\infty}^{+\infty} \qquad (29\text{-}7)$$

It is important to note that *at the limits*, x and r are indistinguishable since they are then parallel lines extending to infinity. Since $\sqrt{x^2 + R^2}$ is equal to r, (29-7) becomes:

$$\frac{1}{R^2}\left[\frac{x}{r}\Bigg|_{x=-\infty}^{x=+\infty}\right]$$

and since $x/r \rightarrow \pm 1$ as $x \rightarrow \pm\infty$, we find

$$\int_{-\infty}^{+\infty} \frac{dx}{(R^2 + x^2)^{3/2}} = \frac{2}{R^2}$$

Hence
$$E = \frac{2k\lambda}{R}$$

where λ is the charge per unit length of the line and R the distance to the closest point of the line. As we said previously, the direction of $\mathbf{E}$ is everywhere perpendicular to the line of charge.

At first, this might seem to be a surprising result. The electric field around a point charge falls off as the inverse square of the distance, while for an infinite line charge it falls much more slowly — only as the inverse of the distance itself. The reason is simple enough: at a point very near the line, only a small portion of the line charge contributes to the resultant radially directed field. Charge elements far to the left and right of the field point produce almost purely horizontal — and cancelling — contributions. As an observer backs away from an infinite line, he "sees" more and more of it. That is, more and more of the infinite line falls within an angular range that gives a strong contribution to the resultant. This geometrical effect amounts to one power of distance, so that the resultant field falls as $1/R$ even though each individual charge element's contribution falls as $1/R^2$.

Carrying this idea further, we might wonder if the field near an infinite plane *sheet* of charge falls off even more slowly with distance. If the field of a point charge falls as R^{-2} and a line charge as R^{-1}, perhaps the sheet of charge produces a field falling as R^{-0}, i.e., constant! We could set up the required two-dimensional integral to attack this problem directly. However, in the next chapter we will discover a simple rule governing the behavior of any electric field that will allow us to evaluate such a field very easily. Let us defer this question until then. At this point, the reader may question why anyone would care about the electric field of an infinitely long line of charge. There are no such charges in nature, so why worry about them? The reason is that any line of charge, such as a charged wire, appears nearly infinitely long if we are sufficiently close to it. For example, if the limits of the integral in (29-7) are changed from $\pm\infty$ to $\pm 10R$, the integral is changed only by $\frac{1}{2}\%$. An observer has to be $\frac{1}{6}$ the length of a line charge away from its center before the simple $1/R$ law is in error by 5%.

Exercises

3. A negative charge, $-Q$, is located at $x = -\frac{1}{2}$ m, $y = 0$, $z = 0$, and a positive charge, $+Q$, is located at $x = +\frac{1}{2}$ m, $y = 0$, $z = 0$. Find the magnitude and direction of the resultant field at (a) $x = 0$, $y = 0$, $z = 1$ m; (b) $x = 1$ m, $y = 0$, $z = 0$.

4. Check the statement at the end of the last section by substituting the limits $x = \pm 3R$ in the appropriate integral.

29-4   Spherical Charge Distributions

In Chapter 28 we noted that the force exerted on a point charge by a spherically symmetric charge distribution behaved in a simple way. The magnitude of this force was identical to what would result if all the spherical charge were located at its center. This is true as long as the point charge remains outside the spherical charge distribution. We further noted that the electric force acting on a point charge inside a hollow shell of charge is exactly zero.

Since the electric field produced by a spherical charge distribution is equal to the force per unit charge that it exerts on a test charge, we conclude:

$$E = k\frac{Q}{r^2} \quad \text{outside all spherically symmetric charge distributions}$$

$$E = 0 \quad \text{anywhere inside a hollow spherical symmetric charge}$$

This last statement refers only to the field produced by the hollow spherical charge itself. Other fields from other nearby charges may, of course, be present as well. In the next chapter we will discover that a hollow *conducting* shell behaves in a very special way so that external fields cannot reach its interior.

Exercise

5. Find the magnitude of the electric field at a point 10^{-12} m from the center of a bare (no electrons nearby) nucleus of a lead atom.

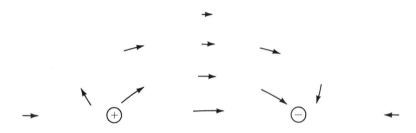

Fig. 29-4. A map of electric field vectors near opposite charges. Only field vectors lying in the plane containing the charges are shown.

This nucleus is spherical, contains 82 protons, and has a radius of about 7×10^{-15} m.

29-5   Pictorial Representations of Electric Fields

The most obvious way to construct a "picture" of an electric field is to draw, at many different field points, vectors representing the magnitude and direction of **E** at each field point. This method is illustrated for some selected points near two charges of opposite sign in Fig. 29-4. There are two difficulties with this type of field map. First, the viewer's eye is inevitably drawn along by the arrows so that it is difficult to see that each arrow gives information about the point at its *tail*. This can be remedied in part by drawing the vectors with the associated points halfway along their length. This really diminishes the difficulty by only one-half, however, since the field does *not* point along the arrow at either end in the general case.

A second difficulty occurs where the field strength changes rapidly with distance, as is the case near a point charge. As seen in Fig. 29-5, the long vectors that indicate the large field strength near the point charge reach beyond the shorter vectors that originate at a larger radius; this presents a visually confusing picture.

An alternative pictorial representation is one in which continuous lines are drawn with their local direction indicating the direction of the electric field at any point. The notion of intensity, or magnitude, of the electric field is then indicated by the density of the lines drawn. Where the lines are more tightly crowded, the field intensity is larger. An example is shown in Fig. 29-6. The lines drawn are everywhere tangent to the direction of the electric field, as illustrated at the point P. The lines drawn in this type of representation are called *lines of force*. To make the intensity representation quantitative, we make the number of lines crossing a unit area (at normal incidence) proportional to the strength of the electric field in that region.

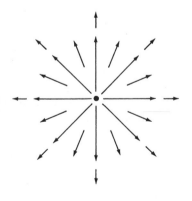

Fig. 29-5. Electric field vectors near a point charge.

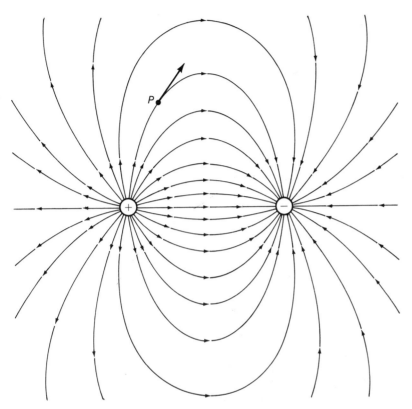

Fig. 29-6. Lines of force represent the field near two equal and opposite charges. At any given point, such as *P*, the field is tangent to the line of force.

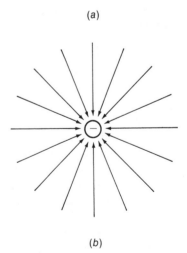

(a)

(b)

The lines of force near isolated point charges are illustrated in Fig. 29-7. If we think in three dimensions, these figures resemble pincushions with lines extending radially in all directions. Since the "crowding," or density, of lines of force is supposed to represent field intensity, let us calculate the number of lines crossing a unit of area at any given radius from the point charge. First, note that the number of lines emanating into all directions is a constant, say N_0. At any given distance r away from the point charge, these lines all pass normally through a sphere of radius r and surface area A centered on the charge. Thus the number of lines per unit area at any radius is:

$$\frac{N_0}{A} = \frac{N_0}{4\pi r^2}$$

Since the field strength is supposed to be proportional to the number of lines per unit, area we have

$$E \propto \frac{N_0}{4\pi r^2}$$

Fig. 29-7. The lines of force near a point charge are more closely crowded, indicating the larger field intensity there.

This inverse square behavior is, of course, exactly the way the electric field varies near a point charge. All we have to do is make the number of lines, N_0, proportional to the magnitude of the charge and the method of drawing lines of force gives a precise description of the fields of a point charge. It follows that the same procedure applies to all spherically symmetric charge distributions.

The fact that the $1/r^2$ dependence of the point-charge field can be represented by a fixed number of lines streaming out of the point (into the point for negative charge) is highly suggestive. Perhaps the electric field "really" consists of streams of invisible particles emanating from the charge. This view is probably a bit naive, though quantum field theory does tell us there is a germ of truth in it. Of immediate importance is the fact that this picture leads us to Gauss's law, a powerful result we will discuss fully in the next chapter. For the time being, we will simply state that when any charges are present, the number of lines of force leaving (or entering) a positive (negative) charge is proportional to the magnitude of the charge. Further, the lines are drawn so that the tangent to any field line represents the direction of the electric field at the point where the tangent is drawn.

Example   Sketch the lines of force around two equal, positive charges separated by a distance d.

Solution   A quick sketch can be made if we first obtain some idea of the direction and strength of the electric field at several representative points, as in Fig. 29-8. Having some idea of the way the field varies in strength and direction at various points, we can draw in lines of force that roughly represent the field. A carefully drawn illustration is given in Fig. 29-9. This figure, like those preceding it, is limited to two dimensions and represents the field behavior in the plane containing the two charges.

We can experimentally view something quite similar to lines of force, as shown in Fig. 29-10. This photograph was made by sus-

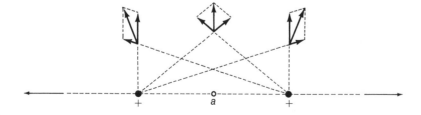

Fig. 29-8. Electric field components and their sums are indicated at several points. At point a the field is zero.

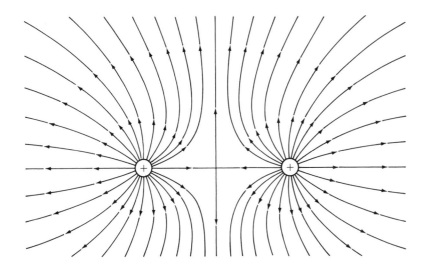

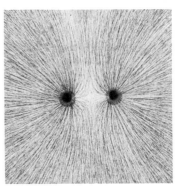

Fig. 29-9. A sketch of the field lines near two positive charges.

pending many short nylon fibers in mineral oil and then placing like charges on two small cylinders immersed in the mineral oil. The electric field polarized the fibers so that a positive charge excess appeared at one end, and a negative charge excess at the other. Each fiber then experienced a torque that tended to twist it into alignment with the local electrical field. The similarity of Fig. 29-9 and 29-10 is evident. The result for opposite charges placed on the two cylinders is shown in Fig. 29-11.

Fig. 29-10. Nylon fibers suspended in mineral oil trace out lines of force near two positively charged cylinders.

29-6 The Electric Force on a Moving Charge

When two charges are both moving, the usual Coulomb force between them is slightly altered by a velocity-dependent force called the *magnetic force*. However, when *either* of the charges is stationary, the force they exert on each other is given simply by Coulomb's law. Since *both* charges must move to produce magnetic effects we will later (Chapter 34) adopt the view that a moving charge produces a *magnetic field* with an intensity proportional to its velocity.* A second moving charge then experiences a magnetic force (over and above the usual Coulomb force) with a magnitude proportional to its own velocity and to the strength of the magnetic

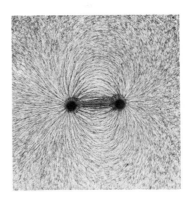

Fig. 29-11. Nylon fibers as in Fig. 29-10 except opposite charges are placed on the cylinders. Compare Fig. 29-6.

*In magnetized iron it is the continual spinning motion of atomic electrons that leads to a magnetic force.

A MOVING CHARGE PRODUCES A MAGNETIC FIELD WITH AN INTENSITY PROPORTIONAL TO IT'S VELOCITY

field. The important point for our present discussion is that no correction to the Coulomb force is required if only one of the charges of interest is moving.

Consider for example a light, negatively charged mass orbiting a massive, positively charged object, as in Fig. 29-12. Before the development of quantum mechanics, it was thought that the single electron of hydrogen atom orbited its nucleus, consisting of a single proton, in just this manner. Since the proton is nearly 2000 times as massive as the electron, it would essentially stand still while the electron orbited about it. For a circular orbit of radius r, the electric force on the electron is given by Coulomb's law:

$$F_e = k\frac{q_e q_p}{r^2} = -k\frac{e^2}{r^2}$$

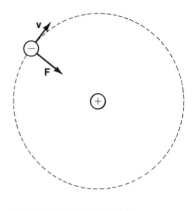

Fig. 29-12. Electric attraction leading to a circular orbit.

where e is the magnitude of the elementary charge and where the minus sign arising from the opposite charges indicates an attractive force.

This force must supply the inward centripetal force required to keep the electron in orbit:

$$F_c = -m_e\frac{v^2}{r}$$

Equating these forces, we find the equilibrium velocity:

$$k\frac{e^2}{r^2} = m_e\frac{v^2}{r}$$

$$v = \sqrt{k\frac{e^2}{m_e r}}$$

Using the experimentally known value of 5.3×10^{-11} m for the mean radius of the hydrogen atom, we obtain

$$v = \left[8.99 \times 10^9 \; \frac{\text{N} \cdot \text{m}^2}{\text{C}^2} \cdot \frac{(1.6 \times 10^{-19} \text{ C})^2}{9.11 \times 10^{-31} \text{ kg} \cdot 5.3 \times 10^{-11} \text{ m}} \right]^{1/2}$$

$$= 2.18 \times 10^6 \text{ m/s} = \frac{1}{137}c$$

This extremely large velocity—nearly 1/100 the velocity of light—illustrates the strength of electric forces at the atomic level. This simple model of the hydrogen atom has some elements of truth in it, but we will see in Chapter 40 that it cannot explain all of the behavior of hydrogen atoms.

A commonly occurring situation, in which one charge moves through the electric field of other stationary charges, is shown in

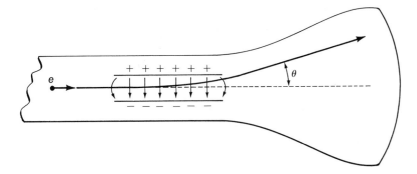

Fig. 29-13. This is one of the methods used to steer an electron beam in a *cathode ray tube* (CRT), such as that used for a television picture tube. In these vacuum tubes a beam of electrons produces a spot of light when it strikes a phosphor screen at the end of the tube. The purpose of the charged *deflecting plates*, shown in Fig. 29-13, is to position the spot of light at any desired point on the screen. Two sets of deflecting plates, at right angles to one another, provide two dimensions of steering.

The force on the moving electron is given by

$$F_e = eE$$ FORCE ON MOVING ELECTRON

and is directed upward in Fig. 29-13. Closely spaced parallel plates provide a field that is almost constant between the plates but drops quickly to zero outside the plates. Thus the electron experiences a constant upward force only during the time it traverses the plates. If the electron's velocity is v and the plate length l, this time interval is:

$$\tau = \frac{l}{v_x} = \frac{l}{v_0}$$ TIME INTERVAL $\dfrac{\text{PLATE LENGTH}}{\text{initial velocity}}$

Note that v_x will remain equal to v_0, the initial velocity, since the applied electric force is along the y direction.

The upward momentum acquired can be found from the impulse-momentum theorem:

$$\Delta m v_y = \int F_y \, dt = eE\tau = eE\frac{l}{v_0}$$

Since the initial value of v_y was zero, the final value of v_y after emerging from the plates is equal to the change in momentum divided by the mass:

$$v_y = \frac{e}{m} E \frac{l}{v_0}$$

Now the angle of deflection is found easily:

$$\tan \theta = \frac{v_y}{v_x} = \frac{e}{m} E \frac{l}{v_0^2} = \frac{1}{2} \frac{eEl}{\text{K.E.}}$$

where K.E. is the *initial* kinetic energy of the electron.

Most of the cathode ray tubes used in commercial TV sets actually use magnetic fields to deflect their electron beams because of cost considerations. We will discuss magnetic deflection in the appropriate chapter. The CRT's used in oscilloscopes and other electronic display devices generally use the electric deflection method we have just discussed. More or less typical design parameters would be K.E. $= 10{,}000$ eV, $\theta_{max} = 10°$, $l = 3$ cm.

Exercises

6. For the above parameters calculate the required electric field strength. A result in SI units (N/C or V/m) will require conversion of energy units to joules and length units to meters.

7. We calculated the velocity of an electron in a hydrogen atom by assuming that only the electrical force acted on the electron. There is, of course, a gravitational attraction as well between the proton and electron and it also varies in an inverse square fashion. Compute, in newtons, the magnitude of both the electric and gravitational forces acting between the proton and electron. Was it a good approximation to neglect the gravitational force?

29-7 Summary

In this chapter the Coulomb interaction was considered to arise in two stages. One charge produces an electric field in the space around it and a second charge is acted upon by a force proportional to the electric field strength at that point. This electric field is a vector quantity and can be determined by measurement of the electric force exerted on a test charge q:

$$\mathbf{E} = \mathbf{F}/q$$

Electric fields from several sources obey the superposition principle. The net electric field at any point is just the vector sum of the individual fields acting at that point.

The electric field due to any spherically symmetric charge distribution at any point outside that distribution is identical to that for a point charge. A hollow sphere of charge does not produce an electric field at any interior point.

Lines of force are pictorial representations of electric fields. They are drawn so that their direction is everywhere tangent to the direction of the electric field. The number density of the lines of force indicates field strength, the field being largest where lines of force are most crowded.

When an electric charge is in motion through an electric field, the electric force upon the moving charge is still given by (29-5). It is only if *both* the charge that produces the field and the charge that is acted upon by the field are in motion that (29-5) fails. In that case a magnetic force, to be discussed in a later chapter, is also present.

Problems

29-1. A test charge $q = -1.0 \times 10^{-9}$ C experiences an electric force of 5×10^{-8} N directed toward the right. Determine the electric intensity at the location of the test charge.

29-2. Find the magnitude and direction of the electric field required to hold an isolated electron stationary in the earth's gravitational field.

29-3. A proton is released from rest in an electric field of 10^6 V/m. How long does it take for its velocity to rise to a value such that relativistic effects became significant (say $v \approx c/10$)?

29-4. Two equal charges Q are separated by a distance L. Find the electric field intensity at (a) a point halfway between the charges; (b) a point $L/2$ above that halfway point; and (c) a point that is $2L$ from one charge and L from the other. Repeat for equal but opposite charges.

29-5. A 2×10^{-3} C charge is 25 cm to the right of a -4×10^{-3} C charge. Find the electric field at a point 25 cm directly above the negative charge.

29-6. A line segment has a length of 20 cm and contains 10^{-3} C distributed uniformly along its length. Find the electric field intensity at a point 20 cm above the center of the line segment.

29-7. Assume the conditions of Problem 29-6 except find the electric field intensity 40 cm from the end of the line segment. This point lies on the line that contains the line segment of charge.

29-8. A uniform sphere of charge produces an electric field of 10^5 V/m at its surface. Find the net charge contained in the sphere if (*a*) it has a 1-m radius; (*b*) it has a 1-cm radius.

29-9. One coulomb of charge is placed on a spherical shell and produces the maximum electric field strength that can be sustained in air without sparking—about 3×10^6 V/m. Find the radius of the sphere.

29-10. Two charges, $+q$ and $+4q$, are arranged as in Fig. 29-14. Find the point (or points) at which $E = 0$ and sketch the lines of force.

29-11. Assume the conditions of Problem 29-10, except assume that the smaller charge is negative.

29-12. Consider an atom of *positronium* in which an electron and positron (equal masses, opposite charges) orbit about one another following circular paths. (*a*) Find their center of mass. (*b*) Can this center of mass accelerate? (*c*) Can the radii of the two particle's orbits differ? (*d*) Find either particle's angular frequency if they are separated by $R = 10^{-10}$ m.

29-13. Find the magnitude of the centripetal acceleration of the electron in a circular orbit about a proton. Use the hydrogen atom's value for r as given in Section 29-6.

29-14. Find the magnitude of the acceleration of an electron between the deflecting plates of a CRT when the length of the plates is 3 cm, the angle of deflection is 10°, and the electron's initial kinetic energy is 10,000 eV.

29-15. When an electron is deflected in a CRT it gains kinetic energy, since it acquires a transverse velocity while maintaining a constant forward velocity. Given the conditions of Problem 29-14, find the increase in kinetic energy of the electron after it has passed the deflecting plates.

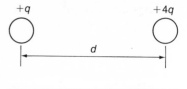

+q   +4q

$\longleftarrow\,d\,\longrightarrow$

Fig. 29-14.

Chapter **30**

Integral Properties of the Electric Field

30-1 Introduction In this chapter we will begin to see some of the advantages gained by adopting the electric field concept instead of the more direct action-at-a-distance point of view. We will find that any electric field must obey an integral law called Gauss's law. This law can often be applied in a simple way to answer questions that are not easily answered by direct use of Coulomb's law. An example of such a question is, "Exactly where do the charges reside when we charge a conductor?"

We will also discuss a second integral law that all static electric fields must obey: the circulation law, which leads us to call all static electric fields *irrotational.* This law, along with Gauss's law, aids us in making qualitative sketches of lines of force. The ability to sketch qualitative features of an electric field is often useful to an engineer who needs to design an electrostatic device or to a scientist who wishes to understand the effects of electric charges in nature.

For any given group of charges, we should be able to start with Coulomb's law, sum the electric forces for each element of charge, and finally arrive at the same conclusions as those reached by utilizing the integral properties of the electric field. The advantage of the field concept is that the field itself is found to obey simple laws regardless of the specific charge distribution that produces the field.

30-2 Flux and Flux Density

The concept of a flux first arose in the theory of fluid flow. The net transport of a fluid through a given cross-sectional area is called

NET TRANSPORT THROUGH A GIVEN CROSS-SECTIONAL AREA

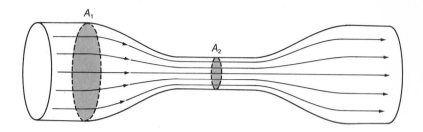

Fig. 30-1. Streamlines of fluid flow in a pipe with variable cross-sectional area.

the flux. More precisely, *flux* is the product of the fluid's velocity and an area through which it flows, where this area is at right angles to the direction of the velocity. A moment's reflection should convince us that the flux is equal to the total volume of fluid passing through that area per unit time.

In Fig. 30-1 we see streamlines of an incompressible fluid flowing through a pipe with variable cross section. Since fluid is neither created nor destroyed as it advances through the pipe, the net flux is constant through any given area, such as A_1 or A_2, that is at right angles to the flow. The amount of flux *per unit area* certainly varies as the fluid moves through the pipe. The flux per unit area is called the *flux density* and is largest in the narrow section of pipe. The net flux remains constant, however, because the cross section there is smaller. This increase in flux density is reflected by the crowding of streamlines in the narrow area. Obviously, streamlines lend themselves nicely to a pictorial representation of flux and flux density. The total number of lines is proportional to the flux, while the number of lines per unit area is proportional to the flux density.

The fact that fluid is neither created nor destroyed in the pipe means that the total number of streamlines is constant. Streamlines can arise only from *sources* and disappear only at *sinks*. In Fig. 30-2 we see an example of streamline flow with both a source and sink present. The resemblance to electric lines of force near opposite charges is made evident by comparing Fig. 30-2 and Fig. 29-11.

Electric lines of force arise at positive charges and terminate at negative charges, so that an analogy exists between streamlines and electric lines of force. We will soon find ourselves using the terms *flux* and *circulation* to describe electric fields, though these concepts were originally devised to describe fluid flow.

30-3 Electric Flux

We will define electric flux in analogy with the flux of fluid flow. As

Fig. 30-2. Streamlines are made evident in this photograph of fluid flow by dye markers that color the passing fluid. A source of fluid is on the left, a sink on the right. (Fluid mapper made and photographed by Professor A. D. Moore, University of Michigan)

we saw in the preceding chapter, the density of electric lines of force (the number passing through a unit area) is proportional to the electric field strength. Accordingly, we take the electric flux to be the product of electric field strength and a cross-sectional area that is perpendicular to the field, as in Fig. 30-3a. We generalize our definition of electric flux Φ_E by including the possibility that the lines of force pass through the area at an angle other than 90°, in which case

$$\Phi_E = EA \cos \theta \tag{30-1}$$

where θ is the angle between the field **E** and the normal to the area, as indicated in Fig. 30-3b. In fact, this direction, the normal to the area, tells us the *orientation* of that area so that we can usefully define the area as a *vector* quantity, **A**. The direction of **A** is normal to the plane containing the area and its magnitude is equal to the area.

The occurrence of products such as the one in (30-1), involving the magnitudes of two vectors and the cosine of the angle between them, is rather frequent in physics and mathematics. This type of product has been given a special name, the *dot product*, and a special notation, **A · B**, where

$$\mathbf{A} \cdot \mathbf{B} = AB \cos \theta$$

Utilizing this definition of the dot product and the vector representation of an area, we may write

$$\Phi_E = \mathbf{E} \cdot \mathbf{A}$$

(handwritten margin notes:) electric flux is product of electric field strength and cross-sectional area that is ⊥ to the field

$$\Phi_E = E \cdot A$$

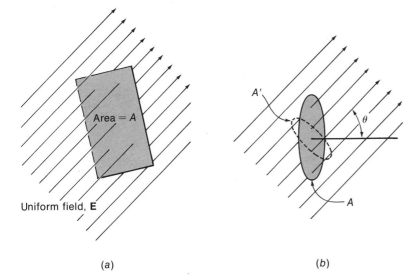

Uniform field, **E**

(a)

(b)

Fig. 30-3. (a) The electric flux through the area A is given by EA when **E** is perpendicular to the plane containing the area. (b) When **E** does not pass normally through A, the effective area is $A' = A \cos \theta$, so that $\Phi_E = EA \cos \theta$.

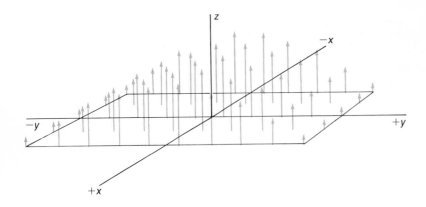

If the electric field is not constant over the area of interest, we break the area into infinitesimal area elements, so that **E** is essentially constant over any one of them, then sum all the contributions to the net flux. This sum, in the limit as the infinitesimal area approaches zero, is an integral and is written:

$$\Phi_E = \int \mathbf{E} \cdot d\mathbf{A}$$

Example An electric field is everywhere parallel to the z axis and has a magnitude that depends only on the value of y, as indicated in Fig. 30-4. The dependence of E upon y is given by $E = a - by^2$. Find the flux through a rectangle in the x,y plane with width W and length L.

Solution The rectangle is shown in Fig. 30-5. Since E does not depend on x, we may consider strips of width W in the x direction and of thickness dy in the y direction. Then we have

$$d\Phi_E = (a - by^2)\, W\, dy$$

for the flux through any strip and

$$\Phi_E = W \int_{-L/2}^{+L/2} (a - by^2)\, dy$$

gives the total flux over the entire rectangular area. Evaluating this integral, we find:

$$\Phi_E = WL\left(a - \frac{b}{12}L^2\right)$$

The sign of the flux has not yet been defined in our discussion. This question is not particularly important for a simple area such as that in the preceding example. When the surface is more compli-

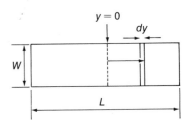

Fig. 30-5. Coordinates used to find the net flux in Fig. 30-4.

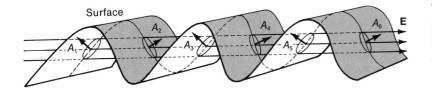

Surface

cated, however, flux can enter and then leave the surface repeatedly. In this case we need to ensure that the direction of area vectors remains on one side of the surface, as illustrated in Fig. 30-6. In that illustration, lines of force that were originally outside the surface pass in and out several times but finally emerge to the outside again, so that no *net* flux passed through the surface.

For a closed surface, such as a sphere, it is conventional to take the area element vectors as being directed outward. Clearly, then, the net flux passing through a sphere around a positive charge will be positive, while that through a sphere around a negative charge will be negative (see Fig. 30-7). As seen in Fig. 30-7c, the net flux through a sphere is zero when the charge is located outside the sphere.

Exercises

1. Show that $\mathbf{A} \cdot \mathbf{B} = \mathbf{B} \cdot \mathbf{A}$.
2. Show that $(\mathbf{A} + \mathbf{B}) \cdot (\mathbf{A} + \mathbf{B}) = A^2 + B^2 + 2\mathbf{A} \cdot \mathbf{B}$. (Remember the law of cosines.)
3. An electric field is everywhere uniform and directed parallel to the z axis. Calculate the net flux through a rectangular area

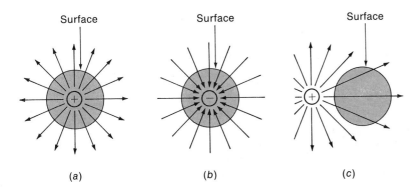

Surface   Surface   Surface

(a)   (b)   (c)

Fig. 30-7. If we take the direction of the area vectors **dA** to be outward, the net flux in (a) is positive, in (b) is negative, and in (c) is zero due to cancellation between the negative flux over the left side of the sphere and the positive flux on the right side.

of width W and length L oriented as shown in Fig. 30-8. The angle θ is 30°. What is the result for $\theta = 90°$ and $\theta = 120°$?

4. An electric field is everywhere parallel to the z axis. Its magnitude depends only on the value of x:

$$E = Cx$$

Find the net flux through the square indicated in Fig. 30-9.

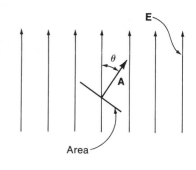

Fig. 30-8. The area A is seen edge on.

30-4   Gauss's Law

Consider now the flux passing through a spherical surface that surrounds a point charge as in Fig. 30-10a. The charge is centered, so the electric field is everywhere constant at the spherical surface and has a magnitude there given by

$$E = k\frac{q}{r^2}$$ electric field

where r is the radius of the sphere. Furthermore, the electric field is perpendicular to any infinitesimal segment of the surface of the sphere, that is:

$$\mathbf{E} \cdot d\mathbf{A} = E\, dA$$ electric field is ⊥

In order to find the total flux passing through the spherical surface, we wish to include the entire surface in our flux integral. Such an integral over a *closed* surface is denoted:

$$\Phi_{total} = \oint \mathbf{E} \cdot d\mathbf{A}$$

The limits are not stated; instead, the circle on the integral sign indicates that a closed surface is intended. For our simple case with a centered charge we obtain:

$$\Phi_{total} = \oint k\frac{q}{r^2}\, dA$$

and since r^2 is constant over the surface of the sphere, it, as well as k and q, can be factored out:

$$\Phi_{total} = k\frac{q}{r^2} \oint dA$$

Of course, the total summation of all the infinitesimal area elements, $\oint dA$, must equal the total surface area of the sphere:

$$\Phi_{total} = k\frac{q}{r^2}(4\pi r^2) = 4\pi kq$$

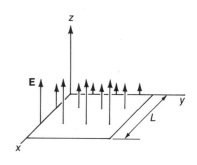

Fig. 30-9.

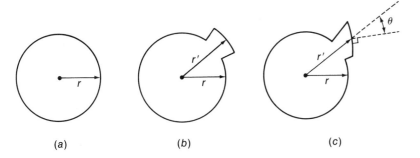

(a)   (b)   (c)

Fig. 30-10. (a) The net flux passing through the surface of the sphere depends only upon the magnitude of Q; it does not depend upon r. When the surface of the sphere is deformed, as in (b) and (c), the net flux is unchanged.

Recalling the definition of k in rationalized MKS units, $k = 1/4\pi\epsilon_0$, we obtain

$$K = \frac{1}{4\pi\epsilon_0}$$

$$\Phi_{\text{total}} = \frac{q}{\epsilon_0}$$

The cancellation of the factor 4π in the total flux integral is the reason that the force constant is chosen to be $1/4\pi\epsilon_0$ in rationalized MKS units.

Consider now the situation in Fig. 30-10b. There a segment of the surface of the sphere, ΔA, has been expanded out to a larger radius, r'. This segment contains more area, $\Delta A'$, than it did at the radius r:

$$\frac{\Delta A'}{\Delta A} = \left(\frac{r'}{r}\right)^2$$

But the electric field is weaker at this larger radius:

$$\frac{E'}{E} = \left(\frac{r}{r'}\right)^2$$

The result is that the total flux through this modified segment is unchanged. Hence, for this case we still obtain

$$\Phi_{\text{total}} = \frac{q}{\epsilon_0}$$

Finally, in Fig. 30-10c, we see a segment of the spherical surface that has been tipped at an angle θ. It now has a larger area

$$\Delta A'' = \frac{\Delta A'}{\cos\theta}$$

than it did before tipping but this increase in area is precisely offset by the factor $\cos\theta$ in the definition of flux, (30-1). Again, the final result is that we still find

$$\Phi_{\text{total}} = \frac{q}{\epsilon_0}$$

Any arbitrary shape can be given to a closed surface by changing the radius and tipping various segments of an initially spherical surface. These variations can be chosen, for example, to result in a sphere off-center from the charge q. Since any of these variations do not change the result of the total flux integral, we conclude:

net flux depends only on the magnitude of Q

$$\Phi_{\text{total}} = \oint \mathbf{E} \cdot d\mathbf{A} = \frac{q}{\epsilon_0} \quad \textit{always for q inside the surface} \qquad (30\text{-}2)$$

The need for the symbol $\oint$ is now evident. It does not matter in the least which particular closed surface is chosen, as long as the charge is inside and the surface is closed. The closed-surface integral symbol in (30-2) does not indicate limits for this reason.

If an additional charge had been placed inside the surface, its field would add vectorially to that of the first charge according to the superposition principle. Its net flux would then simply add to that already present. Generalizing to any number of charges, we see that the value of q in (30-2) should be *all* the charge contained within the surface. Note that this result depends crucially on the $1/r^2$ dependence of the field of any single point charge. In fact, (30-2) is an alternative way to state the $1/r^2$ law, since the two are logically inseparable.

Consider now a charge outside of a closed surface, as in Fig. 30-11. The flux that streams through any infinitesimal element of the surface is precisely cancelled by flux of the opposite sign at another segment of the surface. This is due to the fact that, first,

$$\frac{dA \cos \theta}{dA' \cos \theta'} = \frac{r^2}{(r')^2}$$

and second, that

$$\frac{E}{E'} = \frac{(r')^2}{r^2}$$

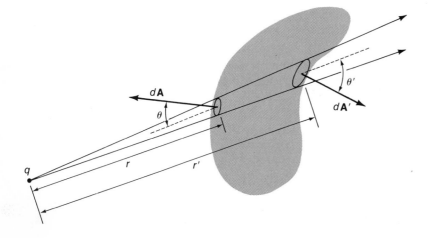

Fig. 30-11. A small portion of the lines of force from q stream into one side of a surface through the element of area $d\mathbf{A}$ and out the other side through a second element of area, $d\mathbf{A}'$. These elements are tipped at angles θ and θ' from the direction of the lines of force and are at distances r and r' from the charge q.

We thus obtain for the ratio of magnitudes of $d\Phi$ and $d\Phi'$:

$$\frac{d\Phi}{d\Phi'} = \frac{E\,dA\,\cos\theta}{E'\,dA'\,\cos\theta'} = \frac{(r')^2}{r^2} \cdot \frac{r^2}{(r')^2} = 1$$

so that these fluxes are equal in magnitude. Since flux streams in one surface and out the other, these two fluxes are of opposite sign and add to zero. This is true for every element of flux streaming from the charge, so we conclude that

$$\oint d\Phi = \oint \mathbf{E} \cdot d\mathbf{A} = 0 \qquad \text{for } q \text{ outside the surface} \qquad (30\text{-}3)$$

Combining (30-2) and (30-3), we obtain *Gauss's law:*

$$\oint \mathbf{E} \cdot d\mathbf{A} = \frac{q}{\epsilon_0} \qquad (30\text{-}4)$$

where q is all of the charge contained within the closed surface.

Note that the sign of the charge in (30-4) can also be negative. In that case the net flux is negative, since the electric field then points *inward* at the surface. If equal quantities of negative and positive charge are within the surface, the net flux is predicted to be zero by Gauss's law. This does *not* mean that the electric field is zero at the surface: it means that the net flux through the surface is zero. A simple illustration is given in Fig. 30-12. Any surface, such

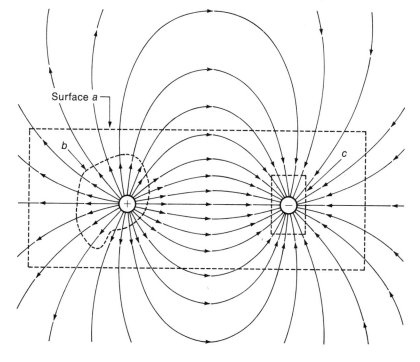

Surface a

b

c

Fig. 30-12. Two equal but opposite point charges. The net flux through the surface a is zero, through b it is positive ($+q/\epsilon_0$), and through c it is negative ($-q/\epsilon_0$).

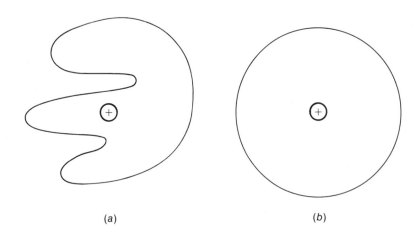

(a)   (b)

Fig. 30-13.

as b, surrounding the charge $+q$ has a net flux of q/ϵ_0. The surface c has a net flux of $-q/\epsilon_0$. The surface a has as much flux leaving it as entering it, so that $\Phi_a = 0$.

Exercises

5. According to Gauss's law, the same net flux passes through both surfaces surrounding the equal point charges in Fig. 30-13. Can you sketch lines of force that qualitatively illustrate how this happens?

6. What is the net flux passing through each of the surfaces a, b, c, and d in Fig. 30-14? Each charge shown has magnitude Q.

30-5 Applying Gauss's Law

In the last chapter we evaluated the electric field produced by an infinitely long line of charge by integrating the effects of the entire charge distribution. The result we found was

$$E = 2k\frac{\lambda}{R}$$

where λ is the (constant) charge per unit length of the line and R is the perpendicular distance from the line to the point where the field is measured (called the field point).

We can obtain this same result a bit more easily by use of Gauss's law. However, we will have to make a symmetry argument in order to do so. We need a closed surface to apply Gauss's law

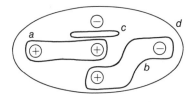

Fig. 30-14.

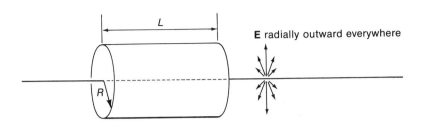

E radially outward everywhere

Fig. 30-15.

and such a surface is shown in Fig. 30-15. This surface exists only in our minds but is precisely defined. Whenever we construct a closed surface in order to apply Gauss's law, we call it a *Gaussian surface*. The Gaussian surface we have selected is a cylinder with radius R and length L, centered about the line of charge. The reason for this particular choice of closed surface will soon be apparent.

In order to obtain a simple evaluation of the flux integral in Gauss's law, we note two facts. First, the electric field is expected to vary with the distance from the line but cannot depend on which side of the line we are on. All that we see from any point around the line is a long line extending to infinity in both directions. Thus, as we examine different locations on the curved surface of the cylinder (our Gaussian surface), we expect the electric field to be constant in value. Further, the facts that the line of charge recedes to infinity in both directions and that it is everywhere uniformly charged lead us to conclude that the field cannot be tipped to the left or right in Fig. 30-15 but must simply point radially outward at right angles to the line.

Now we can calculate the flux integral:

$$\oint \mathbf{E} \cdot d\mathbf{A} = \int_{\substack{\text{left} \\ \text{end of} \\ \text{cylinder}}} \mathbf{E} \cdot d\mathbf{A} + \int_{\substack{\text{curved} \\ \text{surface of} \\ \text{cylinder}}} \mathbf{E} \cdot d\mathbf{A} + \int_{\substack{\text{right} \\ \text{end of} \\ \text{cylinder}}} \mathbf{E} \cdot d\mathbf{A}$$

Since we have argued that the field is directed radially outward, no flux passes in or out of the ends of the cylinder. Another way of saying this is that the factor $\cos \theta$ in the dot product is everywhere zero on the ends of the cylinder since $\theta = 90°$. Over the curved surface, our preceding arguments have assured us that (*a*) the field is always at right angles to the surface, and (*b*) the field $\mathbf{E}$ is constant in magnitude on that surface. Hence the $\cos \theta$ factor is unity and E can be factored out of the integral as well. We are left with:

$$\oint \mathbf{E} \cdot d\mathbf{A} = E \int_{\substack{\text{curved} \\ \text{surface of} \\ \text{cylinder}}} dA$$

FIELD IS ALWAYS AT RIGHT ANGLES TO THE SURFACE

The integral on the right is simply the total surface area of the curved surface of the cylinder, so we have:

$$\oint \mathbf{E} \cdot d\mathbf{A} = E(2\pi RL)$$

By Gauss's law we know the net flux equals q/ϵ_0, and if the line carries λ charge per unit length, the net charge contained in the cylinder is λL. Hence

$$E(2\pi RL) = \frac{\lambda L}{\epsilon_0}$$

or

$$E = \frac{1}{2\pi\epsilon_0} \cdot \frac{\lambda}{R} = 2k\frac{\lambda}{R} \tag{30-5}$$

as we found previously.

The above calculations are obviously considerably more simple than straightforward integration of Coulomb's law over the infinite line, as was done in the preceding chapter. To carry out properly the calculations using Gauss's law, however, we had to give some thought to a careful choice of the simplest possible Gaussian surface. We also had to have an idea of the basic symmetry of the electric field before we could easily evaluate the surface integral.

The result we just obtained also follows for either a hollow cylinder or a solid cylinder of charge of infinite length, as indicated in Fig. 30-16. Once outside the charge distribution, a cylindrical Gaussian surface can be utilized to deduce the behavior of the field strength. We once again obtain

$$E = 2k\frac{\lambda}{R} \tag{30-5}$$

where λ is the charge per unit length distributed within or on the cylinder. This law now holds only for values of R greater than the radius of the cylinder of charge. For smaller values of R, a Gaussian surface does not enclose *all* of the charge and the field depends on the manner in which the charge is distributed within the cylinder.

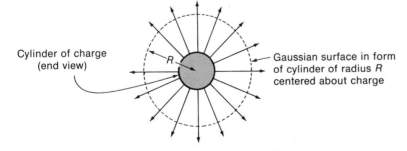

Cylinder of charge (end view)

Gaussian surface in form of cylinder of radius R centered about charge

Fig. 30-16. An end view of a cylindrical Gaussian surface surrounding a segment of a cylinder of charge of infinite length.

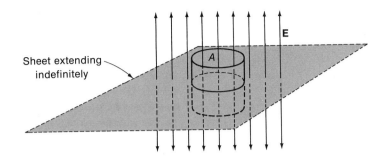

Sheet extending indefinitely

Consider now a uniform, infinite plane sheet of charge, as in Fig. 30-17. Since the sheet extends to infinity in all directions, the direction of the electric field cannot lean to the left or right in Fig. 30-17, nor can it tip into or out of the page. It is everywhere normal to the plane sheet. We see, then, that no flux will enter or leave the curved surface of the cylindrical Gaussian surface shown. Flux leaves only the ends of the cylinder. Further, all of this flux is perpendicular to the ends of the cylinder, so

$$\oint \mathbf{E} \cdot d\mathbf{A} = \int_{\substack{\text{top} \\ \text{end}}} E \, dA + \int_{\substack{\text{bottom} \\ \text{end}}} E \, dA$$

Since the sheet we are considering has a uniform distribution of charge extending to infinity, the magnitude of $\mathbf{E}$ must be constant over the ends of the cylinder, so that E can be factored out of the integrals and we obtain:

$$\oint \mathbf{E} \cdot d\mathbf{A} = EA + EA$$

where A is the area of either end of the cylinder. Finally, the amount of charge inside the cylinder is given by the product of σ, the charge per unit area, and the cross-sectional area A of the cylinder. Gauss's law then becomes:

$$\oint \mathbf{E} \cdot d\mathbf{A} = 2EA = \frac{\sigma A}{\epsilon_0}$$

from which we find

$$E = \frac{\sigma}{2\epsilon_0} \qquad\qquad (30\text{-}6)$$

The magnitude of this electric field is thus *independent* of distance from the sheet of charge. This result is coupled to the fact that the lines of force must always be perpendicular to the sheet

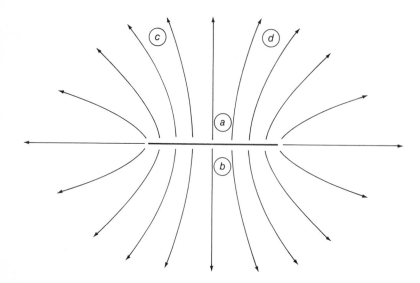

and hence cannot spread out and weaken the field strength. If the sheet is finite in extent, the field far from the sheet will eventually approach a $1/r^2$ behavior when the sheet is so distant that it appears to be a point charge. This is illustrated in Fig. 30-18.

Exercises

7. Consider two plane parallel sheets of charge of infinite extent, with equal charge densities σ but *opposite* sign. Show that the electric field they produce is zero everywhere except between the sheets and that there it has a value

$$E = \frac{\sigma}{\epsilon_0}$$

The simplest way to proceed is to superimpose the field of Fig. 30-17 having strength given by (30-6) with another due to the second parallel sheet of negative sign.

8. Construct a spherical Gaussian surface around a point charge q. Apply Gauss's law to rederive

$$E = \frac{1}{4\pi\epsilon_0} \frac{q}{r^2}$$

for the electric field due to the point charge. Note that the same result holds for either a hollow or solid spherical charge as long as the field point r is greater than the radius of the spherical charge.

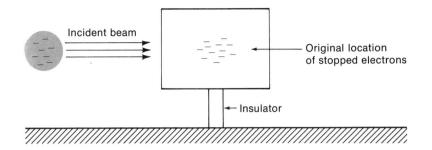

Fig. 30-19. Electrons are artificially driven into a copper block to produce an excess charge in its interior.

30-6 Location of Excess Charge on a Conductor

Suppose that a beam of electrons from a high-energy electron accelerator is momentarily directed at a copper block. The block is electrically isolated by an insulating support so that the excess negative charge of electrons will remain trapped on the copper block. If the energy of the elctrons in the beam is sufficiently large (several MeV or more), the electrons will initially come to rest deep within the block, as shown in Fig. 30-19. Since copper is an excellent conductor, however, these electrons are free to move.

Forces of electric repulsion between the electrons will next cause them to spread rapidly outward in the copper, as in Fig. 30-20*a*. Where will these excess negative charges reach equilibrium? Will their final distribution resemble Fig. 30-20*b* or *c*? The answer to this question is precisely given by application of Gauss's law.

When the original excess charges are introduced, they produce an electric field. This field, acting on the various charges, causes them to move away from one another. When the charges finally reach an equilibrium distribution, the electric forces on each and every one of them must be zero. If this were not so, they would continue to move in the conductor; that is, they would not yet have reached equilibrium. Since the electric forces must vanish at equi-

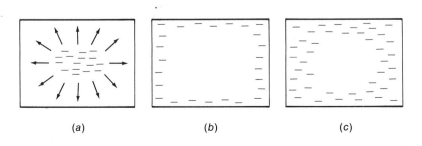

(a)   (b)   (c)

Fig. 30-20. Electric repulsion will rapidly drive the electrons away from each other, as in (a). They might get as far away from each other as possible by moving to the surface, as in (b). However, they are then crowded near the surface. Equilibrium may occur for some intermediate distribution, as in (c).

librium, the electric field also must vanish everywhere in the conductor. If the field were nonzero *anywhere* in the conductor, charges would begin to move. The absence of an electric field gives zero net flux for *any* Gaussian surface inside the conductor, such as those in Fig. 30-21. By Gauss's law we then infer that there is zero net charge within any such surface. It is of course possible that both positive and negative charges, in equal quantities, are concentrated in different portions of our chosen Gaussian enclosure. But we can shrink each Gaussian surface toward zero enclosed volume and always find zero net flux and hence zero enclosed charge. We must conclude that there are *no* excess charges anywhere within the conductor once equilibrium is attained.

Where, then, do the trapped excess charges reside? They must have moved to the surface of the conductor, for just outside the conductor we can no longer argue that the electric fields are zero. It is on the outer surface of the conductor, then, that all excess charge will be found once equilibrium is reached. In a typical conductor, the time required for charges to reach their equilibrium positions is quite negligible for most purposes.

If we now consider a hollow conductor, as in Fig. 30-22, the situation is not quite so clear-cut. The Gaussian surface shown there has no net flux through it, but it cannot be shrunk arbitrarily in size without leaving the conductor. We cannot, then, be sure that different portions of the inner surface do not have excess quantities of charge that add to zero. A more detailed analysis than we can make at present shows that no charge exists anywhere on the inner surface of a hollow conductor. Further, there is no electric field in the hollow space, regardless of the magnitude of any external fields, as illustrated in Fig. 30-23. A closed conducting surface (a wire screen cage is sufficient) is thus an excellent electrostatic shield. Michael Faraday (1791–1867) built a large metal covered box to demonstrate this fact. The box was charged from the outside by a large electrostatic generator while he was inside. He could detect no electrical effects within the enclosure, even though the outside was so highly charged that sparks were jumping from every portion of the outer surface.

Faraday's name is attached to another experiment that shows that all excess charge always moves to the outside of a hollow conductor. He used the metal pail that held his laboratory's ice supply for the hollow conductor, so the experiment is generally called the "Faraday ice-pail experiment." First, consider what occurs when a charge is placed within a hollow conductor, as in Fig. 30-24. The charge produces a field in the hollow region but this field must terminate in the inner conducting surface. Thus, the Gaussian surface

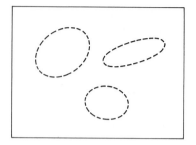

Fig. 30-21. Several Gaussian surfaces inside a charged conductor. Since $E = 0$ everywhere in the conductor, the flux is zero for each surface. There is, therefore, no net charge in any of them.

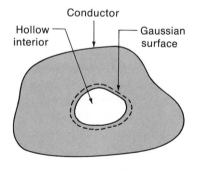

Fig. 30-22. The Gaussian surface shown here is entirely within the conductor. Since the field must be zero everywhere within the conductor, the net flux through this surface is zero. We conclude, then, that there is no *net* charge on the inner surface of a hollow conductor.

shown in Fig. 30-24 has no net flux and hence no net charge is contained within it. It must be true that an equal and opposite negative charge has been induced on the inner conducting surface by the presence of the positive charge. Motion of this induced negative charge to the inner surface finally leaves a positive charge on the outer surface.

The sequence of events in the Faraday ice-pail experiment is shown in Fig. 30-25. A metal ball on an insulating stick is used to bring a charge into the pail. The metal ball is then touched to the inside of the container so that the charge and the induced charge on the inner surface neutralize one another. The net result is a charge distributed over the outer surface of the pail. This experiment can be repeated and, even though the pail is already charged, all of the charge always flows off the metal ball touched to the inside of the pail. Actually, an open ice pail is not the completely closed hollow container for which our discussion is strictly valid. However, as long as almost all of the lines of force from the introduced charge terminate on the interior of the bucket, as illustrated in Fig. 30-26, it is a sufficiently good approximation.

30-7   The Electric Field near a Charged Conductor

In Fig. 30-27 we see electric fields near charged conductors of various shapes. The lines of force are seen to be emerging from the surfaces of the conductors at right angles to those surfaces. If this were not the case, the electric field would have a component parallel to the conducting surface. Charges would then begin flowing along the surface until they cancelled this parallel component. Once equilibrium is attained, the field must be perpendicular to the surface.

Now it is an easy matter to deduce the field strength at the surface. We construct a pillbox-shaped surface that is partially within

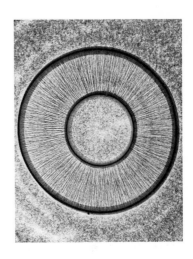

Fig. 30-23. The apparatus used to make Fig. 29-10 and 29-11 shows the field between two cylinders with opposite charges. The interior of the inner cylinder is completely shielded from the external electric field.

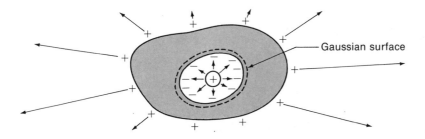

Fig. 30-24. An originally uncharged hollow conductor with a positive charge in its interior. There is a field around the charge, but it must terminate on the inner conducting surface since there cannot be an electric field within the conducting material.

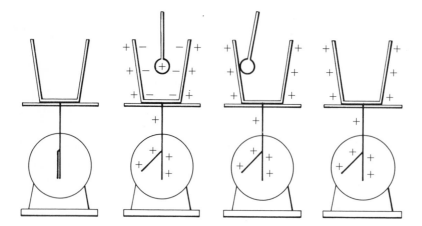

the conductor, as in Fig. 30-28. Since the field vanishes within the conductor and is parallel to the sides of the pillbox, all of the flux passes through the top area of the pillbox and we may write

$$\oint \mathbf{E} \cdot d\mathbf{A} = EA = \frac{q}{\epsilon_0}$$

where we have assumed that the pillbox is small enough so that E has a constant value over its surface. Solving for E:

$$E = \frac{q}{A\epsilon_0} = \frac{\sigma}{\epsilon_0} \tag{30-7}$$

where σ is the surface charge density. This result looks like that for an infinite sheet of charge, Eq. (30-6), except for a factor of 2. This difference arises because there is no contribution to the flux on the portion of the Gaussian surface within the conductor. Due to the nearby surface charge, the field within the conductor is just exactly cancelled everywhere in the interior by the fields from the remainder of the surface.

Our result (30-7) holds only very close to the conductor. As seen in Fig. 30-27, the electric field lines begin to fan out and change direction away from the conducting surfaces.

The preceding sections—30-5, 30-6, and 30-7—have illustrated one of the useful features of the electric field concept. One of the properties of the electric field, *independent of any particular charge distribution*, is embodied in Gauss's law. From that law we have easily deduced the behavior of electric fields near spherical charge distributions, line charges, and sheets of charge. We have also been able to determine the location of excess charge on a conductor, a task that would have been extremely difficult, if not impossible, by use of Coulomb's law alone. Further, we have found that the elec-

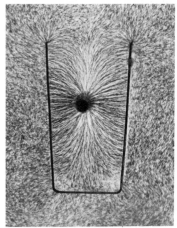

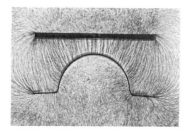

tric field (and hence the electric force on any charge) is directed perpendicular to the surface of any conductor in the region close to a conductor. These are all examples of the utility of the field description of the electrostatic interaction.

30-8 Circulation of Flux   NOT ON HOURLY

Since the idea of flux arose in connection with fluid flow, let us return to fluid flow briefly to discuss *circulation*. In Fig. 30-29 we see two fluid flow patterns. One of them has zero net circulation while the other has nonzero net circulation. To make our meaning precise, we consider an imaginary closed path in the fluid. We choose either clockwise or counterclockwise motion around the path to establish a sign convention. Then at each point along the path we find the component of the fluid velocity that is parallel to the path at that point and multiply it by an infinitesimal element of the path. We proceed around the path, summing all of these infinitesimal products until we return to the starting point. Essentially, we are finding the *average* component of fluid velocity parallel to our chosen path and multiplying by the path length. In symbols we write:

$$\text{circulation} = \oint \mathbf{v} \cdot d\mathbf{r} \qquad (30\text{-}8)$$

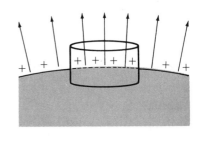

Fig. 30-28.   A pill box-shaped Gaussian surface lying partially within a conductor.

Fig. 30-29.   Two fluid flow patterns: (a) one without circulation and (b) the other with circulation. The average component of the fluid velocity parallel to the path of integration multiplied by the path length is called the *circulation*.

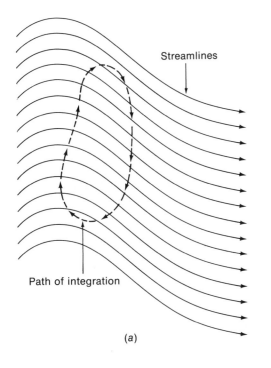

Streamlines

Path of integration

(a)

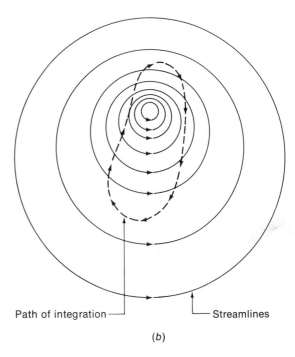

Path of integration —

Streamlines

(b)

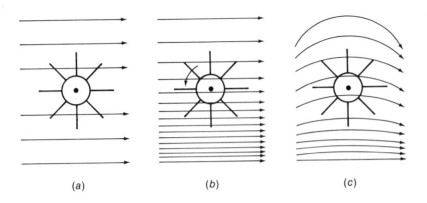

(a)   (b)   (c)

Fig. 30-30. (a) The flow is uniform and a paddle wheel will not rotate. (b) The nonuniform flow causes a paddle wheel to turn because of greater flow velocity on the lower side of the wheel. (c) The velocity is nonuniform but the weaker flow lines curl around the wheel further so that no *net* circulation exists.

where $\mathbf{v}$ is the velocity, $d\mathbf{r}$ an element of path. The dot product gives the velocity component *along* the path and the closed circle indicates a closed path is being considered.

The integral above is a *line integral*. We previously encountered line integrals when we defined mechanical work in Eq. (14-11). Flow patterns with zero net circulation, as in Fig. 30-29a, are called *irrotational flow*. *Vortex flow*, as in Fig. 30-29b, has nonzero circulation. A simple way to think about flow with circulation is to imagine placing a paddle-wheel in the fluid. If it is placed in a region with circulation it will rotate. Examples are shown in Fig. 30-30.

Example   A flow pattern such as that in Fig. 30-30b will turn a paddle-wheel and thus appears to have net circulation. Verify this by using (30-8) directly to evaluate the circulation around the path shown in Fig. 30-31. Take the velocity to be entirely in the x direction with a magnitude given by $v = a + by$.

Solution   Consider the circulation integral in four parts: from points 1 to 2, 2 to 3, 3 to 4, and back to 1 again:

$$\oint \mathbf{v} \cdot d\mathbf{r} = \int_1^2 \mathbf{v} \cdot d\mathbf{r} + \int_2^3 \mathbf{v} \cdot d\mathbf{r} + \int_3^4 \mathbf{v} \cdot d\mathbf{r} + \int_4^1 \mathbf{v} \cdot d\mathbf{r}$$

Recalling that $d\mathbf{r}$ is an element of the path and hence parallel to the path, we see that the first and third integrals are zero since $\mathbf{v}$ is perpendicular to each $d\mathbf{r}$ all along those segments. From 2 to 3 we have

$$\int_2^3 \mathbf{v} \cdot d\mathbf{r} = \int_2^3 (a + bh)\, dx$$

Since both a and bh are constant along this segment, we finally obtain:

$$\int_2^3 \mathbf{v} \cdot d\mathbf{r} = (a + bh) \int_2^3 dx = (a + bh)L$$

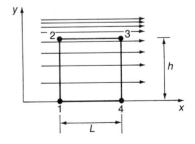

Fig. 30-31.

The last integral similarly becomes:

$$\int_4^1 \mathbf{v} \cdot d\mathbf{r} = \int_4^1 (-a)\, dx = -aL$$

The net circulation is thus nonzero and equal in value to

$$\text{circulation} = \oint \mathbf{v} \cdot d\mathbf{r} = (a + bh)\,L - aL = bhL$$

$$= b \cdot (\text{area enclosed by the path})$$

Example   A vortex consists of rotational fluid flow centered about a point, as in Fig. 30-32. The velocity is purely tangential and is a function of the radius, R. How should the velocity vary with R if it is to have constant circulation for all circular paths about the center of the vortex?

Solution   Since the velocity has no radial components (tangential only), it will be everywhere parallel to any circular path centered on the vortex. Since at any given radius the velocity is constant in magnitude, we have

$$\text{circulation} = \oint \mathbf{v} \cdot d\mathbf{r} = v \oint dr = v \cdot 2\pi R$$

For constant circulation, say C_0, we must have

$$C_0 = v \cdot 2\pi R$$

or

$$v = \frac{C_0}{2\pi R}$$

so that the flow velocity is inversely proportional to the radius in vortex flow with constant circulation.

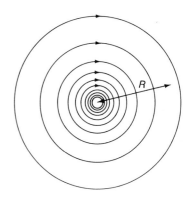

Fig. 30-32.   In vortex flow the fluid velocity decreases with increasing R.

Exercises

9. In the central parts of our galaxy, matter (stars, interstellar dust, and gas) rotates about the center very much like a rigid wheel. The tangential velocity of this matter is thus proportional to its radius. How does the circulation of this velocity pattern vary with radius?

10. A velocity field is indicated in Fig. 30-33. The velocity is parallel to the x axis and has a magnitude given by $v = a + cy^2$. Find the clockwise circulation for the path indicated.

30-9   Circulation of Electric Fields

All static electric fields must obey two integral laws. The first is Gauss's law, which we have already discussed and which involves

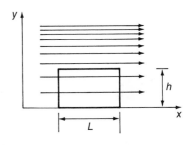

Fig. 30-33.

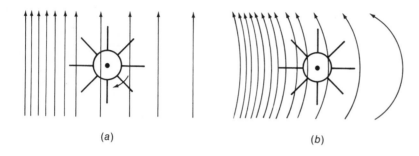

(a)   (b)

Fig. 30-34. (a) This is an impossible static field, as it has nonzero circulation. (b) If the field intensity is to decrease from left to right, it must also begin to curve rapidly enough so that the net circulation is zero.

the net flux through a closed surface. The second is the circulation law, which involves the average component of an electric field around a closed path. All *static* electric fields have zero circulation:

$$\text{circulation} = \oint \mathbf{E} \cdot d\mathbf{r} = 0 \qquad (30\text{-}9)$$

We will later discover that a changing magnetic field can produce an electric field that has nonzero net circulation. For static fields, however, there can be no circulation in any region of space.

One important consequence of the circulation law is that when electric field intensity varies over a region of space, the electric field lines must vary in curvature also. The point is best made by illustrations. In Fig. 30-34a we see an impossible static field.* The "paddle-wheel" shown would rotate, so the field has circulation. A field can, of course, diminish in strength over a region, but then it must behave as shown in Fig. 30-34b. It becomes more and more curved as it weakens, so that the net circulation is everywhere zero. The field is weaker on the right side of the paddle-wheel but it curves around the wheel for a greater path length, which offsets the greater intensity (but shorter path) of the field on the left side of the paddle-wheel.

An example of this sort of behavior is seen in the field close to two equal and opposite charges, as in Fig. 30-35. The magnitude of the field certainly varies from one region to another, but the curvature of the field lines always changes so as to maintain zero net circulation around any imaginable path.

30-10 Summary

All static electric fields obey two integral laws. The statements of both of these laws use concepts first developed for description of

---

*The reader may recall that an example and an exercise in Section 30-3 contained such fields. They were chosen for their simplicity in illustrating a flux calculation. They are not possible *static* fields.

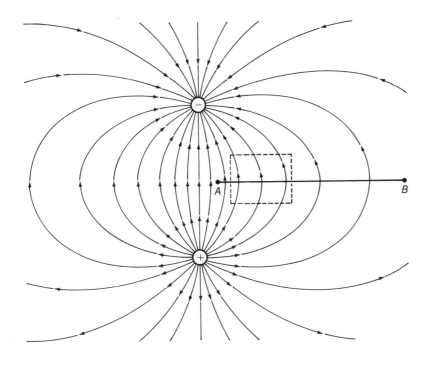

Fig. 30-35. Along the line *AB* the field strength decreases steadily with distance from *A*. The curvature of the field increases, however, so that the net circulation about *any* path, such as the dotted path, vanishes.

fluids undergoing streamline flow. The analogy between fluid flow and electric fields occurs when we compare streamlines to lines of force.

The first integral law, Gauss's law, is true for all electric fields, not just static fields. Gauss's law states that the net flux passing through any closed surface is proportional to the quantity of charge enclosed within that surface—regardless of the distribution of that charge. In rationalized MKS units, we write

$$\oint \mathbf{E} \cdot d\mathbf{A} = \frac{q}{\epsilon_0}$$

where q is the net charge within the closed surface.

We can apply Gauss's law to find the way electric field strength varies near an infinite line of charge and near an infinite plane sheet of charge. We can also use Gauss's law to deduce that all of the charge on a conductor resides on its surface, and to deduce the direction and intensity of the electric field just outside a conductor's surface.

$$E = \frac{\sigma}{2\epsilon_0}$$

All static electric fields have zero circulation:

$$\oint \mathbf{E} \cdot d\mathbf{r} = 0$$

where $d\mathbf{r}$ is an element of path length and the integral is taken around any closed path. A qualitative indication of circulation in a given region can be obtained by imagining a paddle-wheel located in a fluid flowing along streamlines given by the electric lines of force. The paddle-wheel will not turn in regions of vanishing circulation.

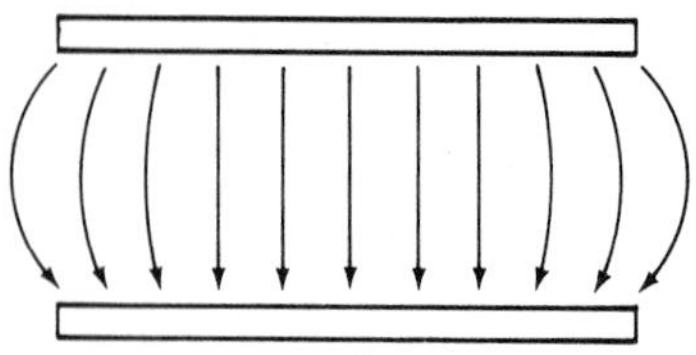

Fig. 30-36. The field both outside the plates and within the plates is zero.

Problems

30-1. Consider a nonclosed surface in the form of a hemisphere of radius R. A uniform electric field of constant intensity E and constant direction is parallel to the axis of symmetry of the hemisphere. Calculate the magnitude of the electric flux Φ_E through the hemisphere.

30-2. A point charge of 10^{-4} C is inside a hollow cube with 1-m edges. The charge is 20 cm from the center of one of the cube's faces. (a) Try to set up the integrals for each of the six faces of the cube that are required to directly calculate the electric flux Φ_E leaving the cube. (b) Use Gauss's law to find Φ_E.

30-3. Two concentric hollow spherical shells of copper have radii R_1 and R_2 ($R_1 < R_2$) and carry charges Q_1 and Q_2. Use Gauss's law to find the magnitude of $\mathbf{E}$ as a function of r measured from the center of the spheres when (a) $r < R_1$; (b) $R_1 < r < R_2$; (c) $r > R_2$. Use the result for part c to find the condition on Q_1 and Q_2 that leads to $E = 0$ for $r > R_2$.

30-4. Suppose the electric field between two metal plates is as shown in Fig. 30-36. Use Gauss's law to show that (a) no charges reside on the outer surface of either plate; (b) there must be a positive charge on the inner surface of the top plate and a negative charge on the inner surface of the bottom plate.

30-5. Assume that the charge of a proton is distributed uniformly in a sphere with a radius of 0.8×10^{-15} m. Find the electric field strength at the proton's surface.

30-6. A charge Q is uniformly distributed throughout a solid sphere of charge of radius R. Consider a spherical Gaussian surface of radius r centered on the center of the sphere of charge to prove that:

(a) $E = \dfrac{1}{4\pi\epsilon_0} \dfrac{Q}{r^2}$ for $r > R$;

(b) $E = \dfrac{1}{4\pi\epsilon_0} \left(\dfrac{Q}{R^3}\right) r$ for $r < R$.

30-7. Since Gauss's law depends only on the $1/r^2$ behavior of the electric field (and represents a central force), there should be a Gauss's law for the gravitational field **g** caused by a mass m, where

$$\mathbf{g} = G\frac{m}{r^3}\mathbf{r}$$

(*a*) Write an expression for the gravitational Gauss's law. (*b*) We used Gauss's law to show that a conductor shields charges from external fields. Gravitational fields obey Gauss's law, yet there is no such thing as a shield for gravitational fields. Explain.

30-8. A conducting sphere of radius R carries a charge Q uniformly distributed over its surface. Find the electric field immediately outside its surface by application of Eq. (30-7). Compare this result to that obtained by knowing that a spherically symmetric distribution of charge produces an external field as if all its charge were concentrated at its center.

30-9. A metal plate is charged so that there is an electric field of 10^6 V/m at its surface. Find the electric charge density σ, in C/m^2, on the surface.

30-10. Two infinitely long concentric cylinders of radius R_1 and R_2 carry uniform charge densities (charge per unit length) λ_1 and λ_2. Find the electric field as a function of r (measured perpendicular from the axis of the cylinder) when r is (*a*) between R_1 and R_2; (*b*) greater than either R_1 or R_2. (*c*) What is the condition on λ_1 and λ_2 so that the field vanishes outside the largest cylinder?

30-11. Two plane parallel infinite sheets of charge (nonconducting) are separated by 20 cm and charged so that the electric field between them is directed from sheet A toward sheet B with a magnitude of 10^5 V/m. The field is zero outside the space between the parallel sheets. Find the charge densities, including their sign, for both sheets.

30-12. A spherically symmetric charge distribution has a charge density ρ (charge/volume) given by

$$\rho = ar \quad \text{for } 0 \leqslant r \leqslant R$$

$$\rho = 0 \quad \text{for } r > R$$

Find the electric field it produces as a function of r.

30-13. An isolated charge q shown in Fig. 30-37 produces an electric field **E**. Calculate the circulation of **E** about the

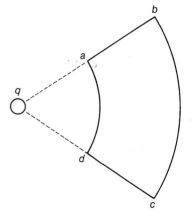

Fig. 30-37.

closed path *abcd* by carrying out the integration in four steps: radially out from *a* to *b*, along a circular arc centered on *q* from *b* to *c*, radially inward from *c* to *d*, and finally along a circular arc centered on *q* from *d* to *a*. Does your result agree with the circulation law, (30-9)? Can you find another closed path of different character than that in Fig. 30-37 for which it is immediately obvious that the circulation of **E** vanishes?

30-14.   Consider the electric field shown in Fig. 30-36. Find more than one closed path for which the circulation of **E** obviously vanishes.

30-15.   An electric field in a region of space is represented by the lines of force in Fig. 30-38. Is this a physically possible electrostatic field? Explain.

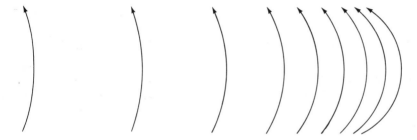

Fig. 30-38.

Chapter **31** ALL

Electric Potential

31-1 **Introduction** The intensity and direction of the electric field at any given point gives us the magnitude and direction of the electric force acting on a unit charge at that point. If that charge moves through some distance, the electric force will do work. This work will be positive if the motion is along the direction of the force and negative if the charge moves against the direction of the force.

electric force does
work
+ along direction of force
− against direction of force

The law of conservation of energy tells us that this work will appear as increased kinetic energy or as some other form of energy, such as heat. Because of the wide-ranging implications of this energy conservation law, the net work done by electric forces is often of more importance than the precise details of just how much work was done along different segments of a path. In other words, the integral quantity, work, is often more significant than the differential quantity, force.

Electric potential energy is that due to the work done on a charge by electric forces. It is often more useful to consider the work done per unit charge: this quantity is called the *electric potential*. The MKS and SI units for electric potential are *volts* (V). As remarked previously, electric potential difference (volts) is more commonly and easily measured than electric field strength at a point (V/m or N/C).

WORK/UNIT CHARGE
ELECTRIC POTENTIAL

31-2 **Electric Potential Energy**

Perhaps the reader recalls the discussions in Chapters 14 and 15 that led to the idea of potential energy. However, a brief review may be helpful. First, when a constant force acts on a mass and moves it through a distance Δx, we define the work done to be $W = F\Delta x$. The acceleration produced by the force leads to an increase in velocity, but this velocity increase is proportional only to the square root of the product of acceleration and distance.* A convenient way of summarizing the result is

$$\Delta K = W$$

where $K = \frac{1}{2}mv^2$ is called the kinetic energy and W is the work done.

If the applied force is not parallel to the direction of motion, only that component parallel to the motion will produce an increase in speed that increases the kinetic energy. The component of the force parallel to the displacement is obtained by multiplying by the cosine of the angle θ between the force, $\mathbf{F}$, and the displacement vector, $\Delta\mathbf{r}$. We now have

$$W = \mathbf{F} \cdot \Delta\mathbf{r}$$

Finally, if either F or θ varies as the object moves along its path, we must define the work by the integral

$$W_{12} = \int_1^2 \mathbf{F} \cdot d\mathbf{r}$$

where W_{12} is the work done in moving from point 1 to point 2.

This type of an integral is called a *line integral,* since some specific path or line from point 1 to 2 must be considered in order to carry out the integration it symbolizes.

Some types of force may be thought of as "storing" work. An example is the work done when we lift a weight against the force of the earth's gravity. When the weight falls back to its original elevation its kinetic energy is increased by just the amount of work required to raise it initially. Such forces are called *conservative* forces.

Other types of forces do not have this property. Frictional forces, for example, always work against motion, regardless of its direction. Thus, after we do work against a frictional force, this work will not be returned to the mechanical system by returning to the starting point. Indeed, to retrace the path of motion requires still more work to be done against frictional forces.** We categorize these forces as nonconservative or dissipative forces.

---

*The most simple illustration is the kinematic formula for the velocity of an object that starts from rest with constant acceleration: $v = \sqrt{2as}$.

**The work done against friction is not truly lost. It appears as heat. It is lost as far as the macroscopic mechanical energy of the system is concerned but is still present in the microscopic domain as increased thermal motion of the constituent atoms.

The most general distinction between conservative and non-conservative forces is made by considering the work done on an object by the force as that object follows a *closed* path. If the force is conservative, any work lost on one portion of a closed path will be regained on another as the object returns to its starting point. This is compactly stated by considering the work integral around a closed path:

$$\text{Conservative force:} \qquad \oint \mathbf{F} \cdot d\mathbf{r} = 0$$

$$\text{Nonconservative force:} \qquad \oint \mathbf{F} \cdot d\mathbf{r} \neq 0$$

Now let us apply these ideas to the forces produced by electric fields. The force is given by $q\mathbf{E}$, so we have:

$$W_{12} = q \int_1^2 \mathbf{E} \cdot d\mathbf{r}$$

The electrostatic force is obviously a conservative force, since we have seen previously that a static electric field has zero net circulation:

$$\oint \mathbf{E} \cdot d\mathbf{r} = 0$$

Hence: $\qquad\qquad \oint \mathbf{F} \cdot d\mathbf{r} = q \oint \mathbf{E} \cdot d\mathbf{r} = 0$

Since the electrostatic force is seen to be a conservative force, we may define an electrostatic potential energy. If we wish to define the potential energy so that it increases when we push against the electric force and decreases as the electric force does positive work, we require a minus sign as follows:

$$U_2 - U_1 = -W_{12} = -q \int_1^2 \mathbf{E} \cdot d\mathbf{r} \qquad (31\text{-}1)$$

where U is the electric potential energy at a given point. As was stressed in Chapter 15, only *changes* in potential energy have any physical significance.

31-3   Electric Potential

Expression (31-1) for the changes in electrostatic potential energy conveniently factors into two parts: the charge q and the integral of the electric field over the path. The magnitude of this integral is obviously equal to the work *per unit charge* done by the electric field. We give the name *electric potential* or *electrostatic potential* to the potential energy per unit charge and we denote it by the symbol V:

$$V_2 - V_1 = -\int_1^2 \mathbf{E} \cdot d\mathbf{r} \qquad (31\text{-}2)$$

The change in electric potential between two points is frequently called simply *potential difference*. The MKS and SI unit of potential difference is the *volt* (V). One joule of work is done when one coulomb of charge moves through a potential difference of one volt.

    It is important to realize that once the potential difference between two points is known, we know the work done on a charge *regardless* of which path is followed from one point to the other. A general proof of this fact can be made as follows. Referring to Fig. 31-1, we see two alternate paths between points 1 and 2. Suppose we consider paths *abc* and *def*:

$$\Delta V(abc) = -\int_1^2 \mathbf{E} \cdot d\mathbf{r}(abc)$$

$$\Delta V(def) = -\int_1^2 \mathbf{E} \cdot d\mathbf{r}(def)$$

We know that the closed loop path *abcfed* yields zero potential difference from the circulation law, so that

$$\oint \mathbf{E} \cdot d\mathbf{r}(abcfed) = \int_1^2 \mathbf{E} \cdot d\mathbf{r}(abc) + \int_2^1 \mathbf{E} \cdot d\mathbf{r}(fed) = 0 \quad (31\text{-}3)$$

Now, along any single given path, we must have

$$\int_1^2 \mathbf{E} \cdot d\mathbf{r} = -\int_2^1 \mathbf{E} \cdot d\mathbf{r} \qquad (31\text{-}4)$$

Combining (31-3) and (31-4), we see that

$$\int_1^2 \mathbf{E} \cdot d\mathbf{r}(abc) = \int_1^2 \mathbf{E} \cdot d\mathbf{r}(def)$$

thus proving that the work done per unit charge is independent of the path followed from 1 to 2. The magnitude of the total work done by the electric field is always equal to $q\Delta V$, regardless of the path.

    Since the electric potential difference between two points is found by integration of the electric field, it must be possible to find the electric field by differentiation of the electric potential. In one

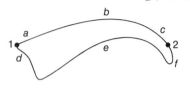

Fig. 31-1. Two paths, *abc* and *def*, lead from 1 to 2. Since the two paths together form a closed loop and the net work around that loop must be zero, the net work from 1 to 2 along either path must be the same.

dimension, x, our discussion is similar to that of Section 15-5, so we can simply write: If $V(x) = \int E\,dx$ so that

$$\Delta V = -\int_1^2 E\,dx$$

then

$$E = -\frac{dV}{dx} \tag{31-5}$$

Now we easily see from (31-5) that the units of electric field strength in SI units are indeed volts/meter.

    As a simple case, consider a constant electric field acting in the positive x direction. The potential difference between two points a distance L apart is

$$\Delta V = -\int_{x_1}^{x_1+L} E\,dx$$

(handwritten): $= -EL = V(x_1+L) - V(x_1)$ V higher at x_1 than x_1+L

$\uparrow$ higher

ALWAYS MOVE FROM A HIGHER POTENTIAL TO A LOWER

and since E is a constant, say E_0, we obtain

$$\Delta V = E_0 \cdot L$$

Thus, for $E_0 = 10^3$ V/m and $L = 1$ m, we have $\Delta V = 10^3$ V. In general, for a constant field, we have

$$V = \int E\,dx = -\left(E_0 x + \text{constant}\right)$$

Applying (31-5), we find

$$E = -\frac{dV}{dx} = -\frac{d}{dx}(E_0 x + \text{constant}) = E_0$$

which returns us to our starting point. It is also possible to generalize (31-5) to an electric field that has x, y, and z components so that V depends upon all three dimensions (see Appendix).

(handwritten): ELECTRIC POTENTIAL

$\Delta V = -\int \vec{E} \cdot \vec{dr}$

Example The potential difference between two metal plates inside a vacuum tube is 1000 V. A single proton starts from rest at the positively charged plate and finally strikes the negatively charged plate. Find its kinetic energy just as it strikes the plate.

Solution The direction of the electric field is such as to accelerate the positively charged proton from its starting point toward its final position. We do not need to know details of the electric field or the path followed by the proton, since we are given that the total potential difference is 1000 V. The kinetic energy, K, must increase as the proton moves to a lower potential:

$$\Delta K = -q\Delta V$$

The initial kinetic energy K_i is zero, so ΔK equals the final kinetic energy K_f. Hence

$$K_f = \Delta K = 1.6 \times 10^{-19} \, \text{C} \cdot 10^3 \, \text{V} = 1.6 \times 10^{-16} \, \text{J}$$

For such small energies, a much more convenient energy unit is the eV. It is equal to the energy gained by one elementary charge unit moving through a 1-V potential difference:

$$\boxed{1 \, \text{eV} = 1.6 \times 10^{-19} \, \text{J}}$$

In our example, $K_f = 10^3$ eV.

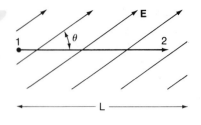

Fig. 31-2.

KINETIC ENERGY INCREASES AS MOVE TO A LOWER POTENTIAL

Exercises

1. The electric field, **E**, between two points is constant in both magnitude and direction. Consider a path of length d at an angle $\theta = 60°$ with respect to the field lines shown in Fig. 31-2. Calculate the potential difference between 1 and 2. Which point is at a higher potential? Does a positive charge gain or lose kinetic energy moving from 1 to 2? What about a negative charge?

2. In the preceding example, a proton initially at rest was accelerated through a potential difference of 10^3 V. (*a*) Find the velocity of the proton when it reaches the negative plate. (*b*) Find the velocity of an electron, released from rest at the negative plate, when it reaches the positive plate. The ratio of proton mass to electron mass is $M_p/M_e = 1836$.

31-4   Behavior of Potential Difference for Simple Cases

Zero electric field   In any region where the electric field vanishes, the potential must be constant since

$$\Delta V = \int_1^2 \mathbf{E} \cdot d\mathbf{r} = 0$$

for any points 1 and 2 within the region with zero electric field. This apparently trivial case is of some importance since there can be no static electric field within a conductor. Thus, the entirety of a conductor must be at a constant potential when charges are at rest.

Constant electric field   Consider two uniformly charged plane parallel sheets of infinite extent. The field produced by each sheet of charge is constant and perpendicular to the sheet. If the sheets are oppositely charged but with *equal* charge density σ, we can find the net electric field by superposition, as in Fig. 31-3. The two fields cancel everywhere except between the two sheets of charge, where they add. The field strength due to each sheet is given by Eq. (30-6), and we find

$$E_{total} = E_+ + E_- = \frac{\sigma_+}{2\epsilon_0} + \frac{\sigma_-}{2\epsilon_0} = \frac{\sigma}{\epsilon_0} \qquad (31\text{-}6)$$

where σ is the magnitude of the charge density of either sheet.

The electric field is constant in both magnitude and direction so that the potential difference between the two sheets of charge is easily found:

$$\Delta V = \int_1^2 \mathbf{E} \cdot d\mathbf{r} \qquad (31\text{-}7)$$

— $E_d = V_2 - V_1$ V_1 higher than V_2

where the end point 1 of this line integral is anywhere on the positive sheet and 2 is anywhere on the negative sheet. Since the potential difference is independent of the path from 1 to 2, we choose the path that makes evaluation of the integral simplest: a straight line from 1 to 2 that is parallel to the electric field. Then

$$\int_1^2 \mathbf{E} \cdot d\mathbf{r} = Ed \qquad (31\text{-}8)$$

where d is the distance between the sheets. Combining (31-6), (31-7), and (31-8), we find

$$\Delta V = \frac{\sigma d}{\epsilon_0} \qquad (31\text{-}9)$$

and also

$$E = \frac{\Delta V}{d} \qquad (31\text{-}10)$$

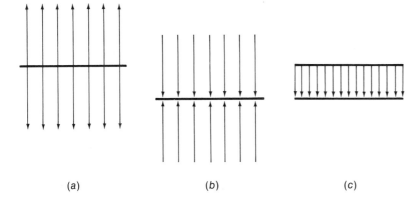

(a)   (b)   (c)

Fig. 31-3.   (a) The field near a positive sheet of charge. (b) The field near a negative sheet of charge. (c) When both are present and the charge densities are equal, the two fields cancel everywhere except between the two sheets.

A pair of oppositely charged parallel conducting plates of finite extent produce an electric field similar to that of Fig. 31-3, as long as the distance d is small compared to the width of the plates and we do not consider the region close to the edges. Near the edges the electric fields produced by each plate are no longer perpendicular to the sheets, so that their sum produces the *fringe field* seen in Fig. 31-4. The smaller the plate spacing d compared to the plate width, the smaller the extent of the fringe field. The deflecting plates of the cathode ray tube in Fig. 31-5 illustrate one use of the electric field between parallel plates.

Example   The CRT shown in Fig. 31-5 accelerates electrons through a potential difference of 10,000 V. They then pass through parallel plates that are 12 mm long and 5 mm apart. In order to deflect the electrons to the outer edge of the phosphor screen, they must be deflected through an angle of 15°. What is the magnitude of the potential difference across the deflecting plates that will accomplish this?

Solution   At the end of Section 29-6 we found that the deflection angle was given by

$$\tan \theta = \tfrac{1}{2}\frac{eEl}{\text{K.E.}}$$

where e is the electron charge, E is the field strength between the plates, l is the length of the plates, and K.E. is the kinetic energy of the electrons. If we write E in terms of the deflecting voltage, V_d, and K.E. in terms of the accelerating voltage, V_a, we have:

$$\tan \theta = \tfrac{1}{2}\frac{e\,(V_d/d)\,l}{eV_a} = \tfrac{1}{2}\frac{V_d}{V_a}\frac{l}{d}$$

where d is the plate spacing. Solving for V_d with our parameters, we find:

$$V_d = 2V_a\frac{d}{l}\tan \theta = 2 \times 10^4 \text{ V} \cdot (0.417)\,(0.268)$$

$$= 2{,}230 \text{ V}$$

Exercise

3. The fringe field outside of a pair of parallel plates decreases very rapidly with distance away from the plates. Consider the integral

$$\Delta V = -\int_a^b \mathbf{E} \cdot d\mathbf{r}$$

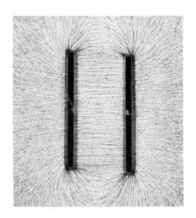

Fig. 31-4. The electric field between two oppositely charged conducting plates is uniform and perpendicular to the plates except near the edges of the plates.

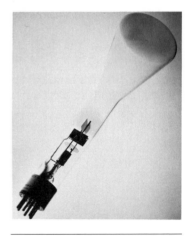

Fig. 31-5. A cathode ray tube (CRT). Electrons are liberated from a hot filament inside the small cylinder. They are accelerated through a potential difference that is maintained between the filament and the disk with a small opening. They then coast to the phosphor screen but can be steered by electric fields between the two sets of parallel plates.

along a path that goes the long way around from point a to point b in Fig. 31-6. Since ΔV is independent of path, we see that the fringe field can never completely vanish outside the plates, no matter how closely they are spaced. Can you now explain why the field between the plates is much more intense than that far from the plates?

Electric fields with 1/r dependence   We have shown that the electric field near an infinitely long line of charge is inversely proportional to the perpendicular distance from the line to the field point (Section 30-5). Furthermore, this field is directed radially outward from the line, as shown in Fig. 31-7.

Two points with the same value of r, such as points a and b in Fig. 31-8, must have the same value of electric potential. This is made clear by the path indicated in Fig. 31-8, which is everywhere perpendicular to **E.** Since $\mathbf{E} \cdot d\mathbf{r}$ then vanishes for every element of the path, ΔV must equal zero. Points such as b and c, which lie at different values of r, have differing potentials. Thus, the potential difference between points a and d must be the same as that between b and c and the potential is a function of r alone. To find the way in which the potential varies between two differing radii, we calculate:

$$\Delta V = V(r_2) - V(r_1) = -\int_{r_1}^{r_2} \mathbf{E} \cdot d\mathbf{s}$$

where $d\mathbf{s}$ is an infinitesimal element of the path. Since the integral cannot depend on the path, we consider a radial path, so that the dot product becomes simply $E \, dr$:

$$\Delta V = -\int_{r_1}^{r_2} E \, dr = -2k\lambda \int_{r_1}^{r_2} \frac{dr}{r} = -2k\lambda (\ln r_2 - \ln r_1)$$

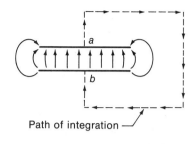

Path of integration

Fig. 31-6.

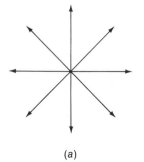

(a)

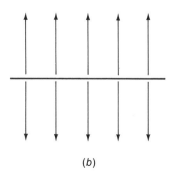

(b)

Fig. 31-7.   (a) End view of a line charge and lines of force. (b) Side view of a segment of the line and the lines of force.

where ln stands for the natural logarithm. Since the difference of two logarithms is the logarithm of the quotient, we have

$$\Delta V = 2k\lambda \ln\!\left(\frac{r_1}{r_2}\right) \qquad\qquad (31\text{-}11)$$

If point 1 at r_1 is farther away from a line of positive charge than point 2 at r_2, the natural logarithm in Eq. (31-11) is positive (see Fig. 31-9). The potential thus increases as we move closer to the line charge. If we move away from the line charge, then $(r_1/r_2) < 1$ and the potential decreases, since the natural logarithm is negative for numbers less than 1.

Vanishingly small values of r_1 could lead to trouble since

$$\ln x \to -\infty \quad \text{as} \quad x \to 0$$

However, this difficulty never actually arises since no line of charge exists that truly has *zero* radius. Expression (31-11) is still correct for a cylinder of charge with nonzero radius, but once inside the radius of the cylinder, the field does not continue to rise as $1/r$.

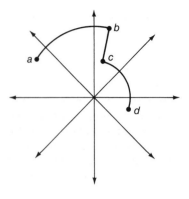

Fig. 31-8. Points *a* and *b* are at the same potential, as are points *c* and *d*. There is, however, a potential difference between points *b* and *c*.

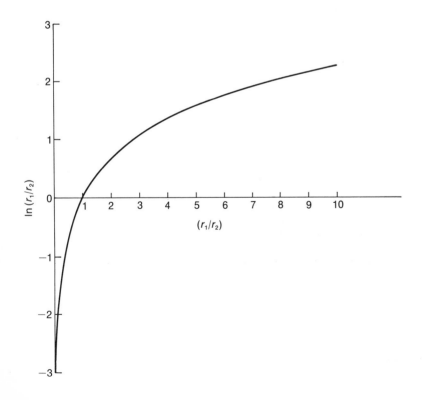

Fig. 31-9. A graph of $\ln(r_1/r_2)$.

The potential difference, then, cannot become arbitrarily large since (31-11) holds only for values of r outside the cylinder of charge.

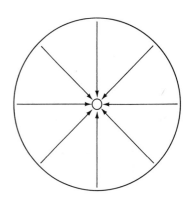

Example   A wire 1 mm in diameter is centered inside a metal cylinder with 1-cm inside diameter. A potential difference of 300 V is applied between the wire and the outer cylinder, with the wire negative. Find the charge per unit length of the wire. An electron that accelerates from the wire to the outer cylinder will gain 300 eV of kinetic energy. At what point along its path will it have gained 90% of this energy? Assume that the wire and cylinder are sufficiently long so that the results for the infinitely long case are applicable.

Solution   The electric field is directed as shown in Fig. 31-10. Its intensity varies as $1/r$ in the space between the conductors, assuming equal and opposite charges on the inner and outer conductor. The field elsewhere is zero. We know that

Fig. 31-10.

$$E = 2k\frac{\lambda}{r} \tag{31-12}$$

where λ is the charge per unit length on the inner wire, and we also know that if R_w is the radius of the wire and R_c that of the cylinder,

$$\Delta V = \int_{R_w}^{R_c} \mathbf{E} \cdot d\mathbf{r} = 300 \text{ V} \tag{31-13}$$

Rewriting (31-13) and considering a radial path:

$$\Delta V = 2k\lambda \int_{R_w}^{R} \frac{dr}{r} = 2k\lambda \ln\left(\frac{R_c}{R_w}\right) \tag{31-14}$$

Since the natural logarithm of 10 is 2.30, we have $\Delta V = 4.6k\lambda$, so that

$$\lambda = \frac{300 \text{ V}}{4.6k} = 7.25 \times 10^{-9} \text{ C/m}$$

Now we wish to find the value of r that gives 90% of the full value of 300 V. This corresponds to asking what value r of the upper limit in (31-14) gives only 90% of the result obtained there. So we require

$$\ln\left(\frac{r}{R_w}\right) = 0.9 \ln\left(\frac{R_c}{R_w}\right) = 0.9\,(2.30)$$

where r is the unknown radius. Then

$$\frac{r}{R_w} = \exp\left[(0.9)\,(2.30)\right] = 7.92$$

and $r = 7.92\,(0.5 \text{ mm}) = 3.96$ mm, which is 79% of the distance out to the 5-mm radius of the outer cylinder.

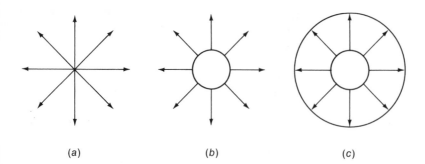

(a)   (b)   (c)

Fig. 31-11. Spherically symmetric charge distributions: (a) the isolated point charge, (b) the isolated spherical charge, and (c) the spherical charge surrounded by a spherical shell of equal and opposite charge all produce a field that varies as $1/r^2$.

Electric fields with $1/r^2$ dependence   Figure 31-11 gives examples of situations that produce electric fields whose intensities vary as $1/r^2$. Application of Gauss's law in each case leads to

$$E = \frac{1}{4\pi\epsilon_0}\frac{Q}{r^2} = k\frac{Q}{r^2}$$

where Q is the central charge. Since this field is directed in a purely radial direction, the potential difference between any two points depends only on the values of r at those points.

To evaluate the potential difference, we consider the integral

$$\Delta V = V(r_2) - V(r_1) = -\int_{r_1}^{r_2} \mathbf{E} \cdot d\mathbf{s}$$

where $d\mathbf{s}$ is an arbitrarily directed infinitesimal element of the path. Since $\mathbf{E}$ is directed radially, the dot product $\mathbf{E} \cdot d\mathbf{s}$ is equal to $E\,dr$ regardless of the path chosen:

$$\Delta V = -\int_{r_1}^{r_2} E\,dr = -kQ\int_{r_1}^{r_2} \frac{dr}{r^2}$$

$$= kQ\left(\frac{1}{r_2} - \frac{1}{r_1}\right) \tag{31-15}$$

Thus, when a charge moves from small r toward large r, the electric potential decreases. The decrease in potential energy, $q\Delta V$, produces an equal increase in kinetic energy for a freely moving charged object. If the object does not move freely, this energy appears in some other form. For example, if the charge moves through a viscous liquid that completely dissipates the kinetic energy, the lost potential energy finally appears as heat.

We define the electric potential at a given radius outside of a spherical charge distribution to be:

$$V(r) = kQ\frac{1}{r} \tag{31-16}$$

so that potential differences between differing radii are in agreement with (31-15). Since electric fields obey the superposition principle, the electric potential due to several point charges is just the sum of the electric potentials due to each separate charge.

If two closely spaced conductors are charged oppositely until a spark jumps between them, it is observed that the spark jumps much more readily from points or regions of sharp curvature. Careful measurements indicate that a certain minimum value of electric field strength must be present to produce a spark. It seems, then, that points or sharply curved surfaces on a charged conductor produce unusually large field strengths. This phenomenon is related to the fact that the electric potential outside a spherical charge varies as $1/r$, while the electric field varies as $1/r^2$.

To better understand this, let us consider a simple case: two spherical conductors with differing radii connected by a conducting wire, as in Fig. 31-12a. The entire surface of both spheres must be at exactly the same potential, since they are conductors and are connected by a conductor. The electric potential just at the surface of the spheres is given by

electric potential just at the surface of a sphere

$$V = kQ \frac{1}{r}$$

$$V_1 = kQ_1 \frac{1}{r_1} \quad \text{and} \quad V_2 = kQ_2 \frac{1}{r_2}$$

and since $V_1 = V_2$ we have:

$$\frac{Q_1}{Q_2} = \frac{r_1}{r_2} \tag{31-17}$$

The electric field strengths at the surfaces of the spheres are given by:

electric field strength at the surface of a sphere

$$E = kQ \frac{1}{r^2}$$

$$E_1 = kQ_1 \frac{1}{r_1{}^2} \quad \text{and} \quad E_2 = kQ_2 \frac{1}{r_2{}^2}$$

The ratio of these two fields strengths can be found with the aid of (31-17):

$$\frac{E_1}{E_2} = \frac{r_2}{r_1}$$

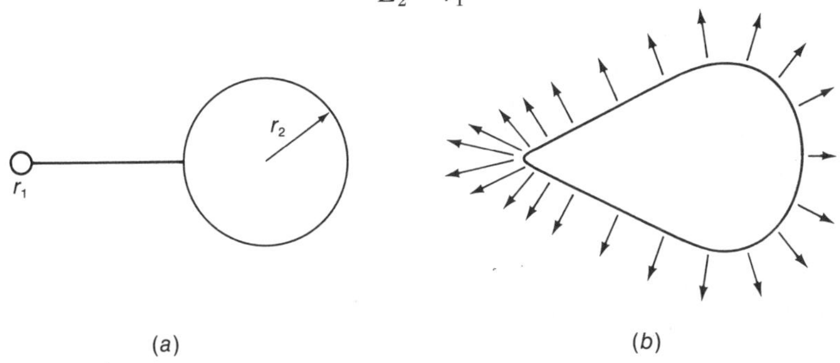

(a)   (b)

Fig. 31-12. (a) When two spheres are connected by a wire they must be at the same potential, but then the smaller sphere has the greatest electric field strength at its surface. (b) A similar phenomenon occurs for any conductor that has regions of sharp curvature.

so that the largest field exists at the surface of the smallest sphere. A similar result occurs for any sharply curved segment of a conductor, as indicated in Fig. 31-12*b* and Fig. 31-13.

Exercises

4. If 1 C of positive charge is placed on an isolated sphere of 10 cm radius, what is the value of the electric potential at its surface? What is the value of the electric potential at a great distance from the sphere ($r \rightarrow \infty$)? How much energy is gained by an electron moving from a great distance up to the surface of the sphere?

5. In air at STP, the maximum electric field strength that can be maintained without sparking is approximately 3×10^6 V/m. What, then, is the maximum charge that can be placed on the sphere of Exercise 4 if the sphere is in air? What is its electric potential then?

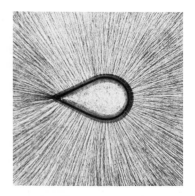

Fig. 31-13. The electric field is most intense at sharply curved regions of a charged conductor. Note also that the field is absent within the hollow conductor and that the lines of force are perpendicular to the conductor at its surface.

31-5 Electrical Discharges in Gases

Normally we think of clean, dry gases as excellent insulators. However, if an electric field of sufficient intensity is present, current can flow through such a gas. This flow can be abrupt and short-lived, as in a spark. It can also be steady, as in the electrical discharge that produces the light of a neon sign. In both cases the light comes from gas atoms that have been disrupted or excited by the bombardment of fast-moving electrons and ions. These atoms rid themselves of excess energy by giving off light—a process we will discuss in Chapter 40.

The minimum value of electric field strength required to produce an electric discharge or spark in a gas can be understood in terms of the electric potential energy available to ionize gas atoms. Generally speaking, a free electron is needed as the "seed" to start a spark. This first electron may be ejected from a metal surface by a strong electric field or by light shining on the surface.

Once a free electron is present in the gas with an electric field, this electron will be accelerated until it makes a collision with a gas molecule. If it gains enough energy, it can knock one or more electrons out of the molecule—that is, the molecule can be ionized. The newly liberated electrons, as well as the initial electron, accelerate again to produce more ionization, while the positive ions accelerate in the opposite direction. If a free electron gains 15 or more eV of energy between collisions, it can produce ionization in air. Thus, if it moves a distance L between collisions, a field strength **E** that can just produce further ionization is given by $eEL = 15$ eV. Since the distance L between collisions in air at STP is less than 10^{-6} m, a field strength

$$E > \frac{15 \text{ eV}}{e \cdot 10^{-6} \text{ m}} = 1.5 \times 10^7 \text{ V/m}$$

appears to be required to produce cumulative ionization leading to a spark. Actually, electric field strengths of about 3×10^6 V/m will produce sparks in air at STP. The reason that this lower field strength can produce an electrical discharge is that the moving electron does not lose all of its kinetic energy at each collision with a molecule. In fact, if it is below the energy required to ionize the molecule, it may simply scatter elastically and continue to gain energy from the electric field until the next collision.

The overall processes involved in electrical discharges in gases can be quite complicated, but the main features are simple enough. On the one hand, if the gas pressure is very high, the distance between collisions is very short, so that a large value of E is required to gain sufficient energy for further ionization. On the other hand, if the gas pressure is very low, there are so few molecules present that very little buildup of ionization can occur. In the intermediate region of gas pressures, electrical discharges occur readily. When large electric field strengths are required in an apparatus, the entire apparatus can be placed inside a tank filled with a high-pressure gas to prevent sparking. The opposite approach can also be used, with the tank pumped out to a very good vacuum.

Part of the process that leads to a spark between metal electrodes involves the metal surfaces themselves. Any positive ions produced near the negative electrode are accelerated toward it. When they strike the electrode, the impact can knock out several electrons, called *secondary electrons*. The impact can also knock out metal atoms—a process called *sputtering*. Both secondary electrons and sputtered atoms "feed back" into the discharge and influence its behavior. The formation of sparks in high vacuum is thus dependent upon the type of metal used for electrodes and upon the degree of cleanliness of the surfaces.

31-6   The Van de Graaff Generator

The Van de Graaff generator is a particular type of electrostatic generator developed by R. J. Van de Graaff in 1930–31. The first such devices were operated in air and could produce potential differences of 1.5 million volts. Machines of this type have been used to simulate lightning-stroke phenomena to aid in the design of electrical equipment subject to thunderstorm activity.

The Van de Graaff generator has been used extensively in nuclear physics research.* By placing the entire assembly within a tank filled with high-pressure gas, potential differences of over 10 million volts have been achieved. These potential differences are used to accelerate protons within an enclosed vacuum tube. These protons then have sufficient energy to produce nuclear reactions and are utilized to study the structure of atomic nuclei. Rather high energies are required because the electrostatic field of the nucleus is so intense that it strongly repels the positively charged protons that are directed toward it. For example, the electrostatic potential near the surface of a copper atom's nucleus is

$$V = \frac{kQ}{r} = \frac{8.99 \times 10^9 \ N \cdot m^2 \cdot C^{-2} \cdot 29 \cdot 1.6 \times 10^{-19} \ C}{5 \times 10^{-15} \ m}$$

$$= 8.34 \times 10^6 \ V = 8.34 \ MV$$

The work required to bring a proton from infinity to this potential is

$$W = qV$$

and since the charge on the proton is $e = 1.6 \times 10^{-19}$ C, we have:

$$W = eV = 8.34 \ MeV$$

This energy can be given to the proton by accelerating it in a Van de Graaff accelerator with a terminal potential of 8.34 MV.**

The operation of a Van de Graaff generator is indicated in Fig. 31-14. A positive high-voltage source is connected to a series of sharp points near the belt at the base of the machine. A voltage difference of between 10 and 50 thousand volts between the needles and the belt ionizes the gas nearby and essentially "sprays" positive charge onto the belt. The belt continuously transports this positive charge up into the hollow terminal. Inside the hollow metal terminal the electric field vanishes, so that the positive charge on the belt easily transfers to the inside of the terminal.

The operation of the machine is essentially a continuous version of the Faraday ice-pail experiment discussed in the last chapter.

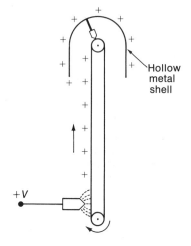

Fig. 31-14. A Van de Graaff generator. Charge is sprayed onto a belt of insulating material. It is carried up into a hollow metal shell where it readily transfers to the inside of the shell.

---

*See the article "Tandem Van de Graaff Accelerators" by Peter H. Rose and Andrew B. Wittkower, *Scientific American,* August 1970.

**Actually, energies of only a few MeV suffice to initiate nuclear reactions between protons and the nuclei in a copper target. This is partly because the strongly attractive nuclear force cancels electrostatic repulsion close to the nucleus and partly due to quantum–mechanical tunneling.

A 9-million-volt machine is shown in Fig. 31-15. Protons in this machine are accelerated inside a vacuum pipe that extends from the terminal to ground. This vacuum tube is made up of alternating segments of insulating glass and metal. The potential difference is spread evenly over the tube so that only a small fraction of the voltage appears across each insulating segment.

Exercise

6.   In a Van de Graaff generator, charge is carried into the terminal by a motor-driven belt. How much work is required to carry

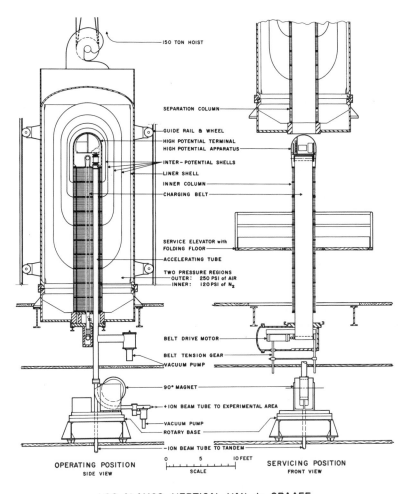

Fig. **31-15.** Drawings of the large Van de Graaff generator used for nuclear research at the Los Alamos Scientific Laboratory. The outer shell filled with air at high pressure can be lifted off for maintenance and repair. (Courtesy of Los Alamos Scientific Laboratory)

1 C of charge up to the terminal when the terminal potential is 6 MV? (In a typical machine it takes 10^3 s or more to carry 1 C of charge up the belt because of a limit in the rate at which charge can be sprayed on the belt.)

31-7   Summary

The electrostatic force is a conservative force, so we can define electrostatic potential energy. Changes in the electric potential energy U of a charge q are given by

$$\Delta U = U_2 - U_1 = -q \int_1^2 \mathbf{E} \cdot d\mathbf{r} \tag{31-1}$$

The electric potential V is the electric potential energy per unit charge:

$$V_2 - V_1 = - \int_1^2 \mathbf{E} \cdot d\mathbf{r} \tag{31-2}$$

The units of electric potential in SI units are volts. Since changes in potential have the dimensions of field multiplied by distance, the units of electric field in the SI system can be expressed as V/m.

The behavior of the electric potential for some simple cases is:

1. Zero electric field: the potential is constant;
2. Constant electric field in the x direction: the potential depends linearly on x;
3. Radially directed electric field proportional to $1/r$: the potential is proportional to $\ln r$;
4. Electric field proportional to $1/r^2$: the potential is proportional to $1/r$.

A free electron can initiate sparks in a gas if a sufficiently strong electric field is present. The gain in energy of the electron must be sufficient to ionize gas atoms with which it collides.

The Van de Graaff generator uses a mechanically driven belt to carry charge from ground (zero potential) to a high-voltage terminal. The work done to carry the charge to the higher electric potential equals the magnitude of the charge multiplied by the electric potential increase.

APPENDIX: Electric Potential in Three Dimensions; Partial Differentiation

When all three components of $\mathbf{E}$ are nonvanishing, we define changes in the electric potential V by:

$$V_2 - V_1 = - \int_1^2 \mathbf{E} \cdot d\mathbf{r}$$

Thus V itself is defined to within an arbitrary (and unimportant) additive constant. The electric potential, in general, depends upon all three coordinates; x, y, and z.

Given the electric potential $V(x,y,z)$, we can find the three components of $\mathbf{E}$ from the three equations:

$$E_x = -\frac{\partial V}{\partial x} \qquad E_y = -\frac{\partial V}{\partial y} \qquad E_z = -\frac{\partial V}{\partial z}.$$

where $\partial/\partial x$ is read, "the partial derivative with respect to x." This type of derivative is evaluated by differentiation with respect to x while holding y and z constant. A similar procedure is used for the other partial derivatives.

Example   Given $f(x,y,z) = x^2 + y^2 + z^2$, find $\dfrac{\partial f}{\partial x}, \dfrac{\partial f}{\partial y}$, and $\dfrac{\partial f}{\partial z}$.

Solution $\dfrac{\partial}{\partial x}(x^2 + y^2 + z^2) = 2x + 0 + 0 = 2x$

$\dfrac{\partial}{\partial y}(x^2 + y^2 + z^2) = 0 + 2y + 0 = 2y$

$\dfrac{\partial}{\partial z}(x^2 + y^2 + z^2) = 0 + 0 + 2z = 2z$

Example   Given $f(x,y,z) = 2xy + 4z^2$, find $\dfrac{\partial f}{\partial x}, \dfrac{\partial f}{\partial y}, \dfrac{\partial f}{\partial z}$.

Solution $\dfrac{\partial f}{\partial x} = 2y + 0 = 2y$

$\dfrac{\partial f}{\partial y} = 2x + 0 = 2x$

$\dfrac{\partial f}{\partial z} = 0 + 8z = 8z$

Exercise

7. Since $V(r) = kQ/r$ for a point charge, we can write it as

$$V(x,y,z) = kQ(x^2 + y^2 + z^2)^{-1/2}$$

Show that $\qquad E_x = kQ\dfrac{x}{r^3}$

Problems

31-1. A flat metal sheet shown in Fig. 31-16 carries a charge density of $\sigma = 10^{-5}$ C/m². Find the change in electric potential between (a) points 1 and 2, (b) points 1 and 3, (c) points 1 and 4.

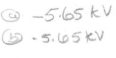

(a) -5.65 kV

(b) -5.65 kV

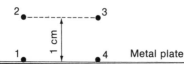

Fig. 31-16.

31-2. Two square metal plates have edges of 1-m length and are parallel and separated by 2 mm. The initially neutral plates are charged by removing 10^{-9} C of positive charge from one plate and placing it on the other. Find the potential difference between the plates.

31-3. An isolated metal sphere in air is charged until it cannot hold more charge. This occurs when the electric field at its surface equals 3×10^6 V/m. Find the magnitude of the charge for a radius of 1 m. What is the electric potential of the sphere?

31-4. A long copper cylinder with a 3-cm inner radius contains a long, coaxial copper cylinder with a 1-cm outer radius. Find the electric field strength halfway between the cylinders when the potential difference between the two cylinders is 800 V.

31-5. An electron vacuum tube starts electrons from rest at a negatively charged cathode and accelerates them into a positively charged anode. The potential difference is 300 V. The rate at which electrons are flowing from cathode to anode is such that a steady current of 100 mA (1 mA = 10^{-3} amp = 10^{-3} C/s) exists. If all the kinetic energy of the electrons produces heat in the anode, find the rate at which heat is produced.

31-6. Assume that the charge of a proton is distributed uniformly in a sphere with a radius of 0.8×10^{-15} m (a reasonable model). (a) Find the electric potential at the proton's surface. (b) Find the work required (in MeV) to bring a second proton within 1.6×10^{-15} m (center-to-center spacing) of the first proton. This is the average center-to-center distance of nucleons (both protons and neutrons) within nuclei.

31-7.* A charge Q is uniformly distributed throughout a solid, nonconducting sphere of radius R. The electric field outside this sphere is proportional to $1/r^2$, while inside the sphere it is proportional to r, as can be seen by application of Gauss's law. At the surface of the sphere, the potential is

$$V = k\frac{Q}{r}$$

Calculate the electric potential for $r < R$ and show it is

$$V = \frac{kQ}{2R^3}(3R^2 - r^2)$$

31-8. Find the electric potential halfway between two point charges q_1 and q_2.

31-9. Two conducting cylinders of differing radii are connected together, as in Fig. 31-17, and electrically charged. Find the ratio of the electric field strengths at a and b: E_a/E_b.

31-10. A sphere of 10-cm radius is connected to a sphere of 2.5 cm radius as in Fig. 31-12a. The electric field strength at the surface of the large sphere is 10^3 V/m. Find the net charge on the two spheres. Neglect any charge which is present on the connecting wire.

31-11. A Van de Graaff generator is used to accelerate a steady beam of protons that carry charge away from the terminal at a rate of 100 μA (1 μA $= 10^{-6}$ C/s). Find the power required to keep the terminal charged to a potential of 9 MV.

31-12. The x component of an electric field around a point charge is given by (see Exercise 7)

$$E_x = kQ\frac{x}{r^3}$$

Evaluate the change in potential between the points $x_1 = 0$, $y_1 = 3$, $z_1 = 0$, and $x_2 = 4$, $y_2 = 3$, $z_2 = 0$ by integration of

$$\Delta V = -\int_1^2 \mathbf{E} \cdot d\mathbf{r}$$

along a path that is a straight line along the x axis. Show that your result agrees with (31-15).

31-13.* Given that

$$V = K\frac{r\cos\theta}{r^3}$$

where θ is the angle between $\mathbf{r}$ and the x axis, find E_x (see Appendix).

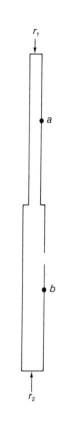

Fig. 31-17.

Chapter **32** OMIT

Capacitance

32-1 Introduction   In many applications of electricity, it is necessary to store a quantity of electrical charge for later use. In the early days of electrical research, this was done simply by placing a charge on a metal sphere mounted on an insulating stand. The quantity of charge that could be stored in this fashion was quite limited, since the electric field strength at the surface of the sphere became quite large with only a modest charge stored on the sphere. Further, the open sphere lost its charge easily and constituted somewhat of a shock hazard if heavily charged.

Devices that could store larger quantities of electricity in a small space were originally called *condensers*. These were typically fashioned of two closely spaced metal surfaces separated by a thin insulator. The capacity of any such device to store charge is called its *capacitance*. Present day usage has replaced the word condenser with the word *capacitor*.

32-2 Capacitance

When a charge is placed on an object, as in Fig. 32-1, it produces an electric field. The presence of this field causes the electric potential of the object to differ from that of its surroundings. Since the magnitude of the field is everywhere proportional to the quantity of charge, the electric potential is also proportional to the quantity of charge in the object. Thus we can write

$$Q = C\Delta V$$

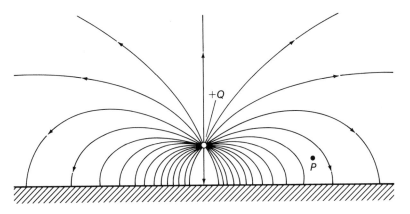

where ΔV is the potential difference between the object and its surroundings and C is a proportionality constant called the *capacitance*. As we will see shortly, C depends upon the size of the charged object, how close it is to nearby objects, and the shape of the object.

Generally speaking, any capacitor consists of two elements. When we remove charge from one element and place it on the other, as in Fig. 32-2, we *charge* the capacitor. When this separated charge flows back so that the elements are again neutral, we say the capacitor is *discharged*. Even an isolated single conductor on an insulating stand fits this description. In this case, the isolated conductor is one element of the capacitor and the remainder of the earth is the other element. When we charge the isolated conductor, we must somehow take charge from the earth and place it on the conductor.

The quantity of charge that moves from one element of the capacitor to the other as it is charged is equal to the magnitude of the charge on either element after it is charged. Calling this charge Q and the potential difference of the elements V, we have

$$C = Q/V \qquad (32\text{-}1)$$

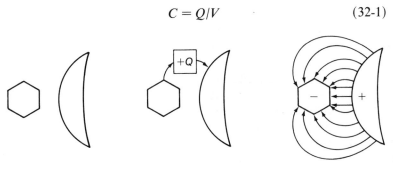

(a) Initially uncharged   (b) Transfer of charge   (c) Charged

Fig. 32-2. The quantity of charged Q transferred from one element of a capacitor to the other equals the magnitude of the charge that then exists on either element.

as a definition of the capacitance C. If Q is measured in coulombs and V in volts, the units of capacitance are *farads*, abbreviated F. Since the farad is a very large unit, both the μF (microfarad $= 10^{-6}$ F) and pF (picofarad $= 10^{-12}$ F) are widely used. The symbol used to represent any capacitor schematically depicts two parallel plates with connecting wires:

Example   A metal beaker is placed on a metal plate with a thin film of insulation between them. When a charge of 8×10^{-10} C is transferred from the beaker to the plate, their potential difference is found to be 10 V. Find the value of C for the beaker-plate capacitor.

Solution   We simply use the definition of C:

$$C = \frac{Q}{V} = \frac{8 \times 10^{-10} \text{ C}}{10 \text{ V}} = 8 \times 10^{-11} \text{ F} = 80 \text{ pF}$$

Exercise

1. We wish to store 10 μC (μC $= 10^{-6}$ C) of charge in a capacitor. The maximum voltage available for charging the capacitor is 1000 V. What is the minimum value of capacitance that could be used?

32-3   The Parallel-Plate Capacitor

Capacitors can be fabricated by placing any pair of conductors close to one another, perhaps with an insulator between them to avoid accidental discharging. The value of C can then be determined by charging the capacitor with a known charge, Q, and then measuring V. Use of the definition (32-1) then allows determination of C. However, if the two conductors have simple shapes, such as concentric spheres or parallel plates, we can use our knowledge of electric fields to calculate C directly. In this way we can design a capacitor to suit our needs without a "cut and try" procedure.

Since most commercial capacitors are in the form of closely spaced metal plates, we will calculate the capacitance of such a capacitor. Suppose we consider plates of area A and spacing d, as in Fig. 32-3. Each plate carries a charge of magnitude Q when the capacitor is charged. The electric field is given in terms of the charge density σ on either plate:

$$E = \sigma/\epsilon_0$$

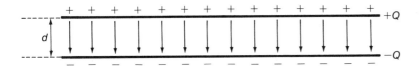

Fig. 32-3. A charged parallel-plate capacitor.

Fig. 32-3. A charged parallel-plate capacitor.

The potential difference, V, between the plates is then given by:

$$V = \frac{\sigma}{\epsilon_0} d$$

The net charge on either plate is simply the charge density multiplied by the area:

$$Q = \sigma A$$

Hence we can calculate the capacitance of a parallel-plate capacitor directly:

$$C = Q/V = \frac{\sigma A}{d\sigma/\epsilon_0} = \epsilon_0 \frac{A}{d} \qquad (32\text{-}2)$$

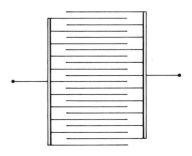

Fig. 32-4. Many parallel plates can be interleaved to increase capacitance.

If SI units are used, A in m² and d in m, then C will be given in farads by (32-2). Note that the capacitance is made larger either by increasing the area of the plates or by bringing them closer together. As the area is increased, a greater charge is stored for a given potential difference. As the plates are brought closer together, the potential difference between the plates (for a given value of Q) decreases, hence increasing the capacitance. From Eq. (32-2) we also see that the dimensions of ϵ_0 can be expressed as farads/m, as well as C²/N·m².

It would seem that C could be made arbitrarily large by making d extremely small. In practice, this possibility is limited by electrical breakdown. As d is made smaller and smaller, the maximum voltage difference at which the capacitor can be used decreases, since the field must be kept below the value that produces a spark or electrical breakdown. A thin insulating film between the plates allows higher voltages to be used and also produces an unexpected increase in capacitance. This increase of C caused by the presence of an insulator will be discussed in Section 32-5, on *dielectrics*.

The effective area can be made larger by interleaving many parallel plates together, as in Fig. 32-4. Such a design is often used for variable capacitors. In variable capacitors of the type shown in Fig. 32-5, the area of overlap of the plates is varied by rotating one set of plates into or out of the other set.

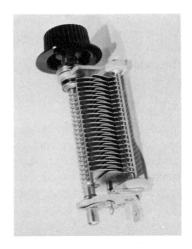

Fig. 32-5. One set of plates can be rotated to vary the overlap, and hence the capacitance, in a variable capacitor.

Example   Two rectangular copper plates, each 5 cm $\times$ 10 cm, are placed parallel to one another with a spacing of $\frac{1}{2}$ mm. Find their capacitance.

Solution   The area of each plate is

$$A = 5 \times 10^{-2} \text{ m} \cdot 10^{-1} \text{ m} = 5 \times 10^{-3} \text{ m}^2$$

We can find the capacitance with (32-2):

$$C = \epsilon_0 \frac{A}{d} = 8.85 \times 10^{-12} \frac{\text{F}}{\text{m}} \cdot \frac{5 \times 10^{-3} \text{ m}^2}{5 \times 10^{-4} \text{ m}}$$

$$= 8.85 \times 10^{-11} \text{ F} = 88\frac{1}{2} \text{ pF}$$

Exercise

2.   We wish to store the maximum amount of charge in the capacitor of the preceding example (plates 5 cm $\times$ 10 cm spaced by $\frac{1}{2}$ mm). The maximum voltage we can apply to it is limited by the breakdown strength of air:

$$E_{\text{max}} = 30{,}000 \text{ V/cm}$$

Find the quantity of charge on the capacitor when the electric field between the plates has just reached the breakdown value.

32-4   Edge Effects

In deriving (32-2) for the capacitance of parallel plates, we completely ignored edge effects. As long as the plates are close together, the field in the center of the capacitor will be given quite accurately by

$$E = \sigma/\epsilon_0$$

Since any path of integration between the plates can be chosen, our value for V is correct. The difficulty is that σ does not remain constant as we approach the edges of the capacitor. Hence Q is not exactly given by $Q = \sigma A$, where σ is the central value that was used to calculate the field strength. Nonetheless, if the plates are closely spaced, Eq. (32-2) will be reasonably accurate.

In some cases we need to be able to calculate C extremely accurately. A capacitor meant to be used as a standard is an example. In this case, *guard plates* can be used so that (32-2) is exact. These plates, as indicated in Fig. 32-6, cause the fringe field to appear at their outer edges rather than at the edges of the standard capacitor. These guard plates are maintained at the same potential

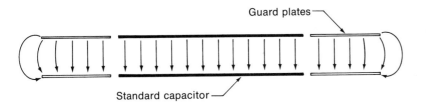

Guard plates

Standard capacitor

difference as the central capacitor, but only the charge that charges or discharges the central plates is measured.

32-5 Dielectrics

When a sheet of insulation is slipped between the plates of a capacitor, its capacitance generally increases. Such an insulator is called a *dielectric*. An illustration of this increased capacitance is seen in Fig. 32-7. For a fixed charge Q, the voltage across the capacitor actually decreases as a dielectric is placed between the plates. This is indicated by the decreased reading of the electroscope in Fig. 32-7.* Since $C = Q/V$ and V is decreased, the dielectric has *increased* the capacitance. More charge can then be added to the capacitor until the original potential difference is reached. Furthermore, a dielectric between the plates generally increases the breakdown voltage of the capacitor so that it can be used at higher working voltages.

---

*The electroscope responds to the small quantity of charge on its gold leaf. But this quantity is directly proportional to the voltage applied to the electroscope. The electroscope thus acts as a *voltmeter.*

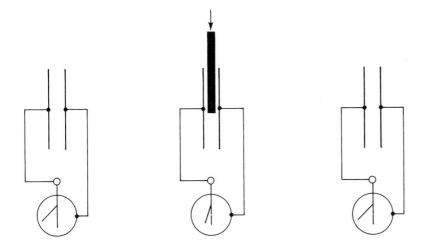

Fig. 32-7. A fixed quantity of charge is placed on a capacitor. When an insulator is inserted between the plates, the potential difference decreases. Removing the insulator restores the original potential difference.

Excess negative charge   Excess positive charge

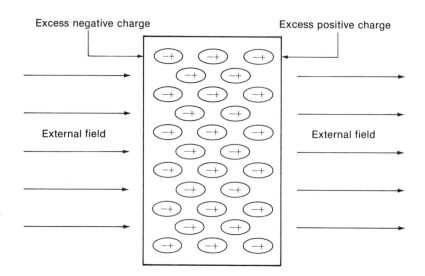

External field   External field

Fig. 32-8. An electric field causes polarization of charges in an insulator. Excess charges appear at the surfaces of the insulator.

The increased capacitance is due to polarization of the dielectric. When an insulator (the dielectric) is placed in an electric field, its atoms become slightly distorted with the negatively charged electrons displaced against the direction of the field and the positively charged nuclei displaced along the direction of the field. The conductor remains neutral overall, and within its interior the negative and positive charges still average to zero net charge, as shown in Fig. 32-8. However, at the surfaces of the insulator there is an excess layer of charge, with opposite charges on opposite faces, as shown.

This induced surface charge is still bound to its parent atoms; it cannot be drawn away by a conductor, for example. It arises from the slight displacement of the charges in each atom. It appears only at the surface because there is no nearby layer of atoms there to cancel the average charge to zero.

The two sheets of induced charge on the surfaces of the dielectric produce a field that opposes the external field that created them. The net effect, as shown in Fig. 32-9, is to produce a weaker field in the space occupied by the dielectric than will exist in that space if the dielectric is removed.

Since the field between the plates of a capacitor is reduced by a dielectric, the potential difference between the plates is also reduced. The action of a dieletric can be characterized by a dielectric constant, K. The field within the dielectric is reduced by the factor $1/K$. If all the space between the plates of a capacitor is filled with

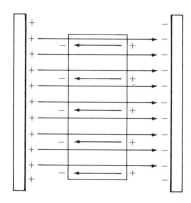

Fig. 32-9. The induced surface charges on a dielectric produce a field that opposes the applied field. The net field within a dielectric is thus weaker than the applied field.

the dielectric, the potential difference is reduced by $1/K$ so that the capacitance is *increased* by the factor K:

$$C' = KC \qquad (32\text{-}3)$$

where C is the capacitance that exists without the dielectric.

Dielectric strength indicates the maximum electric field strength that can be maintained without an electric breakdown of the dielectric. Table 32-1 gives values of K and dielectric strengths for some dielectrics.

TABLE 32-1   Values of K and Dielectric Strengths for Several Dielectrics

Substance	K	Dielectric Strength (V/m)
Air (STP)	1.00059	3×10^6
Neoprene	4.1	12×10^6
Nylon	3.4	14×10^6
Silicone oil	2.5	15×10^6
Silicone rubber	9.0	20×10^6
Water	80.0	—

Commercial capacitors are often made compact in size by making the "plates" of a thin metal foil and spacing them with paper or a plastic film, so that the entire assembly can be rolled up into a small package. This type of capacitor is the tubular capacitor shown in Fig. 32-10. The entire capacitor is then coated with a plastic film to protect it.

Fig. 32-10. Commercial capacitors.

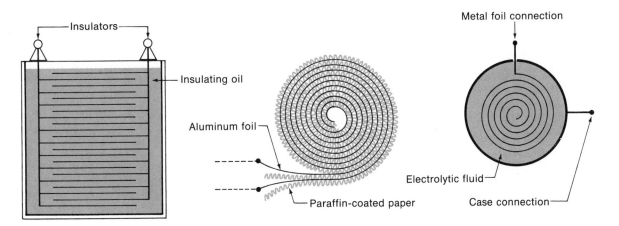

High-voltage oil-filled capacitor

Tubular capacitor

Electrolytic capacitor

Another method of getting large values of capacitance in a small space is the electrolytic capacitor, also shown in Fig. 32-10. In this device one of the conductors is a metal foil while the other is an electrolyte—a liquid that conducts electricity by the motion of ions. The metal and electrolyte are chosen so that a chemical reaction occurs on the surface of the metal to produce an insulating film. This film may be only a few atoms thick, so that the capacitance becomes enormous.

It is instructive to compare the capacitance of a capacitor made of rolled metal foils separated by a film of silicone rubber and that of an isolated metal sphere of 10-cm radius in air. The isolated sphere can be considered as being within another sphere with an infinitely large radius. The potential difference between these spheres is then given by

$$V = \int_R^\infty E\,dr = \frac{Q}{4\pi\epsilon_0} \int_R^\infty \frac{dr}{r^2} = \frac{Q}{4\pi\epsilon_0} \cdot \frac{1}{R}$$

The capacitance is then given by

$$C = Q/V = 4\pi\epsilon_0 R$$

In our case R is 10 cm and the result is

$$C = 1.1 \times 10^{-11}\ \text{F} = 11\ \text{pF}$$

For a metal foil capacitor with a silicone rubber insulator, the capacitance will be

$$C = K\epsilon_0 \frac{A}{d}$$

With a rubber film thickness of 0.001 in., our capacitor can hold up to 500 V. Taking this value for d and considering plates 2 cm $\times$ 1 m, we obtain:

$$C = 6.27 \times 10^{-8}\ \text{F} = 62,700\ \text{pF}$$

This entire capacitor can be rolled into a cylinder less than 1 cm in diameter and 2 cm long.

Exercises

3. In a given electrolytic capacitor the insulating film has a thickness of $\frac{1}{2}$ micrometer and the metal foil has an area of 40 cm². Find the value of its capacitance if $K=2.5$ for the insulating film.

4. A high-voltage capacitor is made by immersing parallel plates into silicone oil (Table 32-1). The capacitor must stand up to 15,000 V of potential difference, so the plate spacing is 1 mm. What is the area of the plates for a 50-pF capacitor?

32-6   Series and Parallel Connections

Frequently more than one capacitor is utilized in an application. In Fig. 32-11a, two capacitors are shown in what is called a *series* connection. In this configuration the overall potential difference is shared between the capacitors, so that voltages in excess of either capacitor's ratings can be sustained. In the connection of Fig. 32-11b, the capacitors are said to be in *parallel*. In this case, as we shall see, the total effective capacitance is the *sum* of the two capacitances.

First let us consider the net capacitance of several capacitors in series. As seen in Fig. 32-12, equal charges are induced on each capacitor when a charge Q is removed from point B and placed on point A. We now ask, "What single capacitance could be placed between the terminals A and B in Fig. 32-12 with the same net result?"

When the series-connected capacitors are charged, a charge Q flows into terminal A and out of terminal B. The total potential difference from A to B is the sum of the potential differences between A and B:

$$V_{\text{total}} = V_1 + V_2 + V_3 + V_4$$

Each capacitor carries the same charge Q, so that for $i = 1, \ldots, 4$:

$$V_i = Q/C_i$$

and we have

$$V_{\text{total}} = Q\left(\frac{1}{C_1} + \frac{1}{C_2} + \frac{1}{C_3} + \frac{1}{C_4}\right)$$

The net effective capacitance between A and B is then

$$C_{\text{total}} = Q/V_{\text{total}} = \left(\frac{1}{C_1} + \frac{1}{C_2} + \frac{1}{C_3} + \frac{1}{C_4}\right)^{-1}$$

A more convenient expression is

$$\frac{1}{C_{\text{total}}} = \frac{1}{C_1} + \frac{1}{C_2} + \frac{1}{C_3} + \frac{1}{C_4}$$

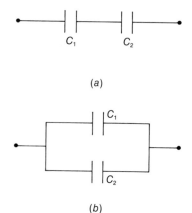

Fig. 32-11.   (a) Series and (b) parallel connected capacitors.

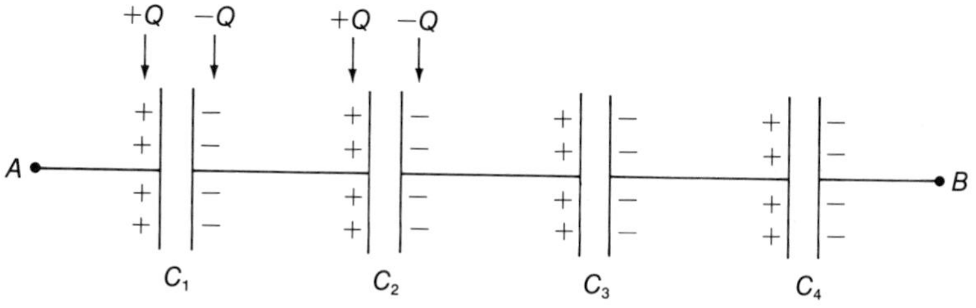

Fig. 32-12.   When a charge $+Q$ is transferred from B to A, the charge Q appears on the left plate of C_1 and it induces a charge of $-Q$ on the other plate of C_1. Since no net charge originally existed on the two connected plates of C_1 and C_2, there must now be a charge of $+Q$ on the left plate of C_2. This process continues throughout the chain of capacitors.

or, in general,

$$\frac{1}{C_{\text{total}}} = \sum_{i=1}^{N} \frac{1}{C_i} \qquad \text{(series capacitors)} \qquad (32\text{-}4)$$

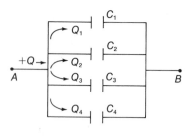

Fig. 32-13. The potential difference across each capacitor is equal to that between points A and B. The charge Q that enters terminal A is shared between the capacitors.

In a parallel arrangement of capacitors, as in Fig. 32-13, the potential difference across all the capacitors must be equal, since they are all joined by conductors and the entirety of a conductor is an equipotential. The charge Q that enters terminal A (and leaves B) is shared among the capacitors:

$$Q = Q_1 + Q_2 + Q_3 + Q_4$$
$$= C_1 V + C_2 V + C_3 V + C_4 V$$

where V is the same across each capacitor and equals the potential difference between A and B. Finally, the net effective capacitance is given by

$$C_{\text{total}} = \frac{Q_{\text{total}}}{V} = \frac{Q_1 + Q_2 + Q_3 + Q_4}{V}$$
$$= C_1 + C_2 + C_3 + C_4$$

or, in general:

$$C_{\text{total}} = \sum_{i=1}^{N} C_i \qquad \text{(parallel capacitors)} \qquad (32\text{-}5)$$

Example A 0.05-μF and 0.1-μF capacitor are connected in series with a total potential difference of 100 V. (*a*) What is the net capacitance? (*b*) What is the potential difference across each of the capacitors separately?

Solution The total capacitance for this series combination is given by

$$\frac{1}{C_{\text{total}}} = \frac{1}{C_1} + \frac{1}{C_2}$$

Solving for C_{total},

$$C_{\text{total}} = \frac{C_1 C_2}{C_1 + C_2} = \frac{(0.05)\ (0.1)\ (\mu F)^2}{0.15\ \mu F} = 0.0333\ \mu F$$

Since the same charge appears on each capacitor, the voltage across the 0.05-μF capacitor (V_1) must be twice that across the 0.1-μF capacitor (V_2):

$$V_1 = 2V_2$$

But

$$V_1 + V_2 = 100\ \text{V}$$

hence

$$3V_2 = 100 \text{ V}$$

$$V_2 = 33\tfrac{1}{3} \text{ V}$$

and

$$V_1 = 66\tfrac{2}{3} \text{ V}$$

Exercises

5.   Three capacitors, each with $C = 9 \ \mu\text{F}$, are connected (a) in series; (b) in parallel. Find their total capacitance in each case.

6.   A 1-μF and a 2-μF capacitor are connected in parallel across a 20-V potential difference. Find the charge on the 2-μF capacitor.

32-7   Energy Storage in Capacitors

As a capacitor is charged, electric charge is being raised from a low potential to a high potential. Work must be done on the charge in order to accomplish this. When we first begin to charge a capacitor, we do very little work on each element of charge since $V = Q/C$ and Q is still small. As the charging process goes on, however, Q and hence V increase steadily, so that each successive bit of charge moved from the negative plate to the positive plate requires more work than the preceding. We might guess that the total work, W, done when a capacitor is charged to a potential difference of V with a charge Q is given by

$$W = \tfrac{1}{2} V \cdot Q$$

since the average potential difference during the charging process is one-half the final value.

   To be more precise, we can consider the work done on each element of charge, dQ, as the capacitor is charged and sum these contributions to find the total. In the limit as $dQ \to 0$, this sum becomes an integral:

$$dW = dQ \cdot V = dQ \cdot \frac{Q}{C}$$

$$W_{\text{total}} = \frac{1}{C} \int_0^{Q_f} Q \, dQ = \tfrac{1}{2} \frac{Q_f^2}{C}$$

where Q_f is the final charge. Thus, when a capacitor has been charged with a total charge Q (dropping the subscript on Q_f), the net work done is:

$$W = \tfrac{1}{2} \frac{Q^2}{C} \tag{32-6}$$

We can use the relation $Q = CV$ to obtain the equivalent statements:

$$W = \tfrac{1}{2}QV \tag{32-7}$$

$$W = \tfrac{1}{2}CV^2 \tag{32-8}$$

When the capacitor is discharged, this same quantity of work is returned to the charge as it flows from the positively charged plate back to the negatively charged plate. The potential energy U stored in the capacitor is just equal to the work done in charging it:

$$U = \tfrac{1}{2}CV^2 \tag{32-9}$$

Example   An automobile spring deflects 6 in. with a 1000-lb load. Let us call it 0.2 m for a 5000-N load for ease of comparison. Compare the potential energy stored in this spring when it has been compressed by 0.2 m to the potential energy stored in a 10-μF capacitor at a potential difference of 10,000 V.

Solution   The potential energy of the spring is given by

$$U = \tfrac{1}{2}kx^2$$

where k is the spring constant and x the spring deflection. For our spring, k is given by

$$k = \frac{5000 \text{ N}}{0.2 \text{ m}} = 25 \times 10^3 \text{ N/m}$$

Hence $U = \tfrac{1}{2} (25 \times 10^3 \text{ N/m}) (0.2 \text{ m})^2 = 500 \text{ J}$

For the capacitor we have

$$U = \tfrac{1}{2}CV^2 = 5 \times 10^{-6} \text{ F} (10^4 \text{ V})^2$$

$$= 500 \text{ J}$$

A 10-μF capacitor that can withstand a potential difference of 10,000 V is certainly not a tiny device. In fact, its bulk is probably comparable to the automobile spring. A typical oil-filled 10-μF capacitor that is rated at 10,000 V takes up a space of 10 in. $\times$ 10 in. $\times$ 12 in.

Exercises

7. A charge of 30 μC is to be stored on a capacitor with the least possible expenditure of energy. Which capacitor should be used, a 15-μF or a 30-μF capacitor?

8. Compare the energy stored by a 10-μF capacitor at 1000 V to that stored by a 5-μF capacitor at 2000 V.

32-8   Summary

When charge is removed from one conductor and placed on a neighboring conductor, a potential difference is created. The quantity of charge is proportional to this potential difference and the constant of proportionality is called the *capacitance, C*:

$$Q = CV$$

The SI unit of capacitance is the *farad* and equals one coulomb per volt.

A parallel-plate capacitor with plate area A and plate spacing d has a capacitance given by

$$C = \epsilon_0 \frac{A}{d}$$

as long as edge effects are small, that is, as long as d is much smaller than the plate dimensions.

A dielectric inserted between the plates of a capacitor increases the capacitance. This is due to the decreased electric field strength within the dielectric caused by induced charges. The dielectric constant K describes the reduction in electric field within the dielectric:

$$E' = \frac{E}{K}$$

so that the potential difference is reduced by $1/K$ and the capacitance is increased by the factor K:

$$C' = KC$$

The effective capacitance of several capacitors in a series connection is given by

$$\frac{1}{C_{\text{total}}} = \frac{1}{C_1} + \frac{1}{C_2} + \frac{1}{C_3} + \cdots$$

and for a parallel connection it is given by

$$C_{\text{total}} = C_1 + C_2 + C_3 + \cdots$$

The energy stored in a capacitor is found by integrating the work done in charging it and is given by

$$W = \tfrac{1}{2}CV^2 = \tfrac{1}{2}\frac{Q^2}{C} = \tfrac{1}{2}QV$$

Problems

32-1. A charge of 0.2 μC is placed on a parallel-plate capacitor with a plate spacing of 1 mm. The potential difference is found to be 10^3 V. (*a*) Find the area of the plates. (*b*) The spacing is now increased to 2 mm with the same charge on the plates. Find the new value of the potential difference.

32-2. A parallel-plate capacitor is made with a plate spacing of 1 cm. Find the area of the plates for a capacitance of 1 F.

32-3. Suppose that the breakdown voltage of a parallel-plate capacitor is proportional to the distance d between its plates. If only a given volume V is available, how do we maximize the charge stored by such a capacitor: by making the plates of large area and d small, or by making the plates of small area and d large? Or is it independent of these factors? Assume that the capacitor is charged just below its breakdown voltage.

32-4. In Chapter 31 we saw that the potential difference between two infinitely long concentric cylinders is given by

$$\Delta V = \frac{\lambda}{2\pi\epsilon_0} \ln\left(\frac{r_2}{r_1}\right)$$

where λ is the charge per unit length on the inner cylinder of radius r_1, and r_2 is the radius of the outer cylinder. Show that the capacitance of two concentric cylinders of length L is given by

$$C = \frac{2\pi\epsilon_0 L}{\ln(r_2/r_1)}$$

if end effects are neglected.

32-5. One hollow sphere of radius r_2 encloses and is concentric with a smaller sphere of radius r_1. These two spheres constitute the two elements of a spherical capacitor. Find an expression for its capacitance in terms of the given information.

32-6. A 0.01-μF capacitor is to be fabricated with a 0.01-mm-thick dielectric that has $K = 2$ and a breakdown dielectric strength of 15×10^6 V/m. (*a*) What is the required area of the plates? (*b*) If the dielectric is to be stressed to only one-half of its dielectric strength, what is the maximum voltage that can be applied to this capacitor?

32-7. A parallel-plate capacitor with plates spaced a distance *d* apart has half of the space between its plates filled with a dielectric with dielectric constant *K*, as shown in Fig. 32-14. The field between the plates where there is no dielectric is given by σ/ϵ_0, where σ is the charge density on either plate. Within the dielectric the field is given by $\sigma/K\epsilon_0$. (*a*) Calculate ΔV between the plates for a given charge density σ. (*b*) Find the value of *C* for this capacitor with plates of area *A*.

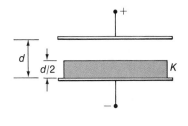

Fig. 32-14.

32-8. A 5-μF capacitor is charged to a potential difference of 20 V and then isolated. It is then connected in parallel with an initially uncharged 20-μF capacitor. Find the final charge on the 5-μF capacitor.

32-9. Show that when several capacitors of differing values are connected in series, the net capacitance is always less than that of the smallest capacitor in the group.

32-10. Four 6-μF capacitors, each with a breakdown voltage of 150 V, are to be connected either in series or in parallel. It is desired to store the maximum possible charge, and either combination may be charged up to its breakdown voltage. Which connection, series or parallel, gives the greatest possible stored charge?

32-11. Two identical parallel-plate capacitors with capacity *C* are connected in parallel. The space between the plates of one of the capacitors is then filled with oil with dielectric constant *K*. Find the total capacity of the parallel combination.

32-12. Two identical capacitors are separately charged from zero charge to a final charge of Q_f. (*a*) Find the total work required to do this. (*b*) Find the total work required to charge one of these capacitors from zero charge to a final charge of $2Q_f$.

32-13. Assume the conditions of Problem 32-3, except now we wish to maximize the stored energy.

32-14. Two 4-μF capacitors are each rated at 4 KV maximum voltage. Find the energy stored in the pair when they are (*a*) connected in parallel to a potential difference of 4 KV; (*b*) connected in series to a potential difference of 8 KV.

32-15. A capacitor is made of plates 20 cm × 20 cm that are separated by a 1-mm spacing. This capacitor is charged with $Q = 3.54 \times 10^{-9}$ C. Find the energy stored in this capacitor.

32-16. A charge *Q* is placed on a parallel-plate capacitor with plate area *A* and spacing *d*. One plate is now pulled away from the other to a spacing $2d$. (a) Find the work done against the force attracting this plate toward the other plate. (b) Use Eq. (32-9) to find the change in the energy stored in this capacitor.

32-17.   A dielectric sheet with dielectric constant K is pulled from between the plates of a capacitor, as in Fig. 32-15. (a) Consider the electrostatic forces between the charges on the capacitor plates and the charges induced on the dielectric. Must work be done to remove the dielectric? (b) For a fixed quantity of charge Q on the capacitor, does the stored energy increase or decrease when the dielectric is removed? (c) Assuming that the dielectric fills the space between the plates, calculate the change in stored energy, and hence the work done, when it is removed.

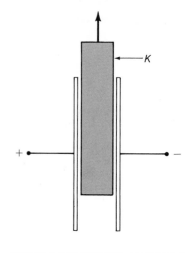

Fig. 32-15.

Chapter **33**

Charges in Motion

33-1 Introduction Thus far we have been primarily concerned with effects produced by stationary charge distributions. We have, at times, discussed moving charge elements—as in our discussion of the charging and discharging of a capacitor—but we have not yet concerned ourselves with the details of this motion. In this chapter we will quantitatively describe the motion of charges as a *current*. The energy lost by moving charges as they push their way through a conductor is due to what is called *resistance*. Finally, a source of energy that keeps charges in motion is called an *electromotive force*. We will apply these ideas to predict the behavior of simple electrical circuits.

33-2 Current

When electric charges are in motion, we say that an electrical *current* is present. If the moving charges are constrained to remain within a narrow conductor (such as a wire), the definition of current is straightforward. As in Fig. 33-1, the current is defined to be the total quantity of charge passing any fixed cross-sectional area of the conductor in one second:

$$I = \frac{\Delta Q}{\Delta t}$$

I = amp = $^{1C}/_{sec}$

$mA = 10^{-3} A$

$\mu A = 10^{-6} A$

In SI units, current I is measured in *amperes* (abbreviated A). Thus one ampere = one coulomb/second. Frequently used submultiples are the mA (10^{-3} A) and the μA (10^{-6} A). If the current varies in time, we can define the *instantaneous current:*

$$I = dQ/dt$$

The direction of the current is taken to be that in which *positive charges move.* In metallic conductors (the most commonly used conductors), the mobile charge carriers are actually electrons, which carry a negative charge. We still define the *conventional* direction of current to be that in which positive charges have to move to produce the same net result. Thus, the conventional direction for current flow happens to be in the opposite direction to that of the moving electrons in a metallic conductor. As indicated in Fig. 33-2, the *result* is the same whether negative or positive charges actually move.

Here on earth, we tend to think of metallic conductors (such as copper wires) when we discuss conductors. Actually, a great deal of the matter of the universe is in a conducting state called a *plasma.* The matter within stars and much of that distributed between stars in the tenuous interstellar medium is in the plasma phase. A plasma is a neutral mixture of positive ions, negative electrons and neutral gas. The positive ions and negative electrons quickly recombine to form neutral atoms, so that a continual supply of fresh ionization is required to maintain a plasma. This ionization can be produced by very high temperatures, which cause many ionizing collisions between gas atoms, or by electrical means, as in a gaseous discharge. Because of the plentiful supply of mobile charge carriers, plasmas are good conductors. Both the positive ions and negative electrons move when a current flows. The positive ions move in the same direction as the electric field while the negative electrons move in the opposite direction. The *current* due to each type of charge carrier is in the same direction, that of the

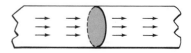

Fig. 33-1. The quantity of charge that passes through the shaded area each second is defined to be equal to the *current.*

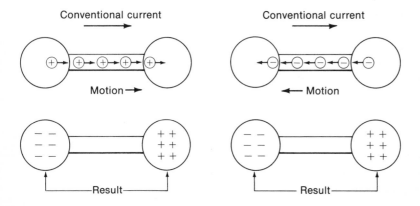

Fig. 33-2. Whether positive charges move toward the right or negative charges toward the left, the net result is the same: the conductor on the left becomes more negative, that on the right more positive. The conventional direction of current flow is that of the positive charges.

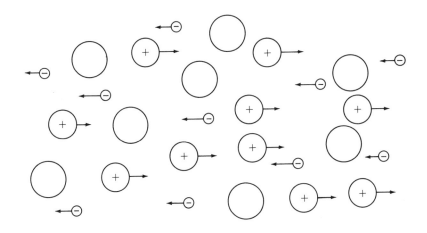

Fig. 33-3. Both plasmas and electrolytes contain mixtures of positive and negative ions. When an electric field is directed from left to right, the positive charges move toward the right, the negative charges toward the left. Both sets of moving charges contribute to the conventional current, which is directed toward the right.

electric field, as indicated in Fig. 33-3. Current is also carried through an electrolyte, such as salt dissolved in water, by motion of both positive and negative charge carriers.

Exercises

1. In a copper wire carrying an electrical current, 10^{18} electrons pass a given point every second. They are moving from left to right. (*a*) What is the direction of the conventional current? (*b*) What is the magnitude of that current?

2. An initially uncharged, 10-μF capacitor is charged for 5 s by a current of 2 mA magnitude. Find the final potential difference across the capacitor.

33-3   Resistance

Electrical currents always involve the motion of a swarm of charged particles. In metallic conduction, the metal atoms remain fixed in positions dictated by their crystal structure, while electrons move slowly through the maze of atoms. In plasmas and electrolytes, the electrons move in one direction while positively charged ions move in the other direction. It is important to note that a conductor is usually *neutral;* it contains as many negative charges as positive charges. It is the motion of one or both types of charged particles within the conductor that constitutes a current.

As the mobile charge carriers in a conductor move, they generally collide with stationary or near-stationary atoms and increase their thermal vibrations. The energy transferred to the form of heat

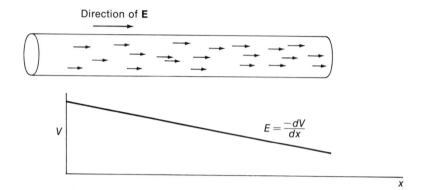

Direction of **E**

$E = \dfrac{-dV}{dx}$

V

x

Fig. 33-4. Most conductors require an electric field to keep their charge carriers in motion. The electric field along the length of the conductor implies a continually decreasing electric potential, *V*, along the conductor.

by these collisions must be continually replenished if a current is to be maintained. An exception is the *superconductor:* a conductor in which an electrical current, once established, persists (or "coasts") indefinitely without further assistance.* The known superconductors are metals or alloys and become superconductors only at temperatures approaching absolute zero.

In ordinary conductors the driving force required to keep the charge carriers in motion must be supplied by an electric field, as shown in Fig. 33-4. Unless the moving charges disappear or are created at some point (which generally is not the case), all of the current that enters one end of a conductor must leave the other end. Thus, the electric field that keeps the charge elements in motion must be present all along the length of the current-carrying conductor.

The electric field that is in the direction of the current all along a conductor causes the two ends of the conductor to be at differing electrical potentials. Each charge carrier loses potential energy as it moves along the conductor. This lost energy is the energy required to keep the charge carrier moving in spite of its repeated collisions with atoms along its path. These collisions increase the random thermal motion of the atoms, so that the lost energy appears as heat.

A measure of the difficulty of maintaining a current in a given conductor is given by its *resistance,* which is defined to be the ratio of the potential difference between the two ends of the conductor, *V,* to the current, *I,* that is present:

$$R = \frac{V}{I} \qquad \frac{\text{potential difference}}{\text{current}} \qquad (33\text{-}1)$$

*See, for example, "Applications of Superconductivity" by Theodore A. Buchold, *Scientific American,* March 1960.

where R is the symbol for resistance. In metallic conductors, the dependence of I upon V is particularly simple: it is linear. Thus for a metallic conductor, R in Eq. (33-1) is a constant, so that the current is linearly dependent upon the potential difference:

$$I = \left(\frac{1}{R}\right)V \quad (R = \text{constant}) \qquad (33\text{-}2)$$

The constancy of resistance for metallic conductors was discovered by G. S. Ohm (1789–1854) and is known as *Ohm's law*. This simple law holds over an amazing range of voltage and current (10 orders of magnitude for many metals.) The SI units of resistance are named after Ohm: one *ohm* of resistance equals one volt per ampere. The abbreviation for the ohm is the capital Greek letter omega: Ω. Because of the nearly universal use of the volt, ohm, and ampere as electrical units, it is common to refer to a potential difference simply as a *voltage*.

When the current-voltage relation in a conductor is nonlinear, the idea of resistance becomes less useful. It is convenient to express the behavior of a nonlinear conductor with a current-voltage diagram, such as that in Fig. 33-5. If we are interested only in small variations of voltage and current about some operating point, we can define a *small signal resistance* (usually denoted by a lower-case letter r) by:

$$r = \frac{dV}{dI} \qquad (33\text{-}3)$$

where the value of r now depends on the particular choice of operating point. Nonlinear dependence of I upon V is often called *nonohmic* behavior.

Current-voltage relations for various devices are indicated in Fig. 33-6. The diode is a device that passes current much more readily in one direction than another, as seen in Fig. 33-6b. The tunnel diode is a semiconductor device that, over a small region

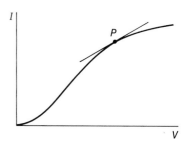

Fig. 33-5. Current in a semiconductor device is not linearly dependent upon voltage. A small signal resistance in the neighborhood of a given operating point, P, can be defined as the inverse of the slope of the tangent line shown.

$$1\,\Omega = 1\,V/Amp.$$

$$I = V/R$$

Fig. 33-6. Current-voltage relations for (a) an ohmic conductor, (b) a diode, (c) a tunnel diode, and (d) a gas discharge tube.

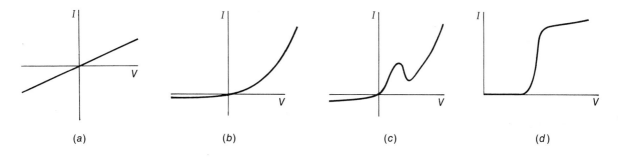

(a)   (b)   (c)   (d) of its current-voltage relation, has a *negative* resistance. That is, current *decreases* as voltage is increased. In Fig. 33-6*d* we see the characteristics of a gaseous discharge tube (such as a small, neon-filled bulb with two electrodes). At low voltages, no current flows since the gas is not a conductor. When sufficient voltage is present to cause ionization of the gas, a discharge occurs and current flows. A tiny increase in voltage produces a large change in current at this breakdown point. If a source of current that is not perfectly stable is connected to the tube, the voltage across the tube will none-theless be quite stable at this point. Accordingly, such tubes are used to provide a fixed voltage reference level in some electronic devices. A solid-state device that has similar properties over a narrow range of voltage is the *Zener diode*.

In an earlier chapter we noted that in a *static* situation, the electric potential must be uniform throughout a conductor. When charges move, however, we have seen that the potential must vary along the length of the conductor. If the resistance of the conductor is zero (a superconductor) or is so close to zero that it can be ne-glected, the length of the conductor will be at a constant potential even when a current is present.

It is often the case that a current is led along copper wires that have extremely low resistance: a value of $0.02\,\Omega$ per meter of length is common for the wires used in household wiring. If this current is then led into a conductor with much larger resistance, the bulk of the applied potential difference will exist between the ends of the high-resistance segment. The region of the conductor that has a high resistance is called a *resistor*. The symbol for a resistor is shown in Fig. 33-7. A resistor is generally a device that has been intentionally constructed to have a resistance larger than that of interconnecting wiring. Commercial resistors are readily available with values from $10^{-3}\,\Omega$ to $10^{8}\,\Omega$.

Fig. 33-7. The symbol for a resistor.

Exercises

3. A copper wire 1 m long has a resistance of $0.02\,\Omega$. The potential difference between its two ends is measured to be 2 v. What is the magnitude of the current flowing in the wire?

4. The current-voltage diagram for a device is given in Fig. 33-8. Find the values of its small signal resistance, r, near the operating points p_1 and p_2.

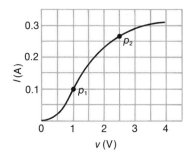

Fig. 33-8.

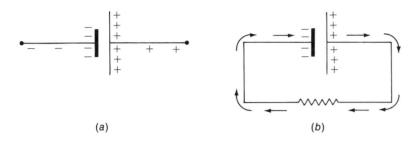

Fig. 33-9.   (*a*) The battery maintains a fixed potential difference between its terminals, even when (*b*) a current is present.

(*a*)   (*b*)

33-4   Emf and the Simple Circuit

If a steady electrical current is to flow in a conductor, two needs must be met. First, there must be a potential difference to cause the charge carriers to move through the conductor. Second, the charges that have passed through the conductor must be continually returned to their starting point so that they can pass through the conductor again. The latter requirement is provided by having the conductor close back upon itself to form an *electrical circuit.* If the charges were allowed to accumulate at an end of the conductor, they would build up until they created an electric field so large that no further charge could be forced through the conductor.

The potential difference required to keep the charges in motion can be supplied by a battery inserted in the circuit, as shown in Fig. 33-9. The symbol for a battery is indicated there, as well as a schematic diagram of a simple complete circuit in which a single battery forces current to flow through a single resistor. The solid lines indicate conductors of negligible resistance that connect the battery to the resistor.

A battery has the property that it continually maintains a potential difference between its terminals. This implies an unbalanced charge distribution, as in Fig. 33-10*a*. When a complete circuit is provided, as in Fig. 33-10*b,* charges are continually transferred through the battery and elevated in potential in order to maintain

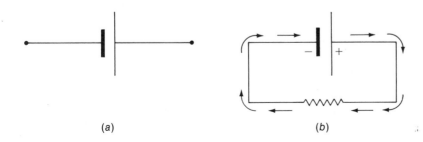

(*a*)   (*b*)

Fig. 33-10.   (*a*) The symbol for a battery. (*b*) A schematic diagram of a simple circuit.

the current. The direction of conventional current is shown in the figure. In a circuit with ordinary metallic conductors, it is the mobile electrons that actually move from the negative terminal, around the circuit, back to the positive terminal, through the battery, and back out through the negative terminal.

The battery plays a role analogous to the pump in a hydraulic circuit. Its pumping action is due to chemical forces within the battery. The positive terminal is called the *anode,* and anode material is chosen so that it has a strong chemical affinity for positive ions in the electrolytic solution, as indicated in Fig. 33-11. Thus, by attracting positive ions to itself, it becomes more and more positive. Similarly, the cathode attracts negative ions to become negatively charged. Eventually, the electrodes become so highly charged that the chemical forces are not strong enough to pull ions out of solutions against the electrostatic repulsion caused by the already-charged electrode. The occurrence of this equilibrium situation sets the *terminal voltage* of any given type of battery.

If an external circuit is connected between the terminals of a battery, the excess charges are drained off the terminals. But as soon as this begins to occur, the electrostatic field near the electrodes diminishes and chemical forces once again begin to pull ions out of solution. This process will continue until the supply of ions is exhausted — a condition aptly described by the term "dead battery."

Batteries in actual use vary widely in their chemical and physical properties. Typically, however, the potential difference they produce is set by the differences in the chemical binding energies for ions or electrons at the two terminals. Since such binding energies are generally of the order of a few eV, the potential difference for most batteries is of the order of 1 V. Higher voltages are obtained by connecting many cells in series, so that the potential energy of a moving charge carrier is increased several times. The resulting battery consisting of n cells will produce a total potential difference n times greater than that of a single cell. An example is shown in Fig. 33-12.

In some types of batteries, the chemical reactions can be reversed by forcing a current backwards through them. This then creates a fresh supply of ions, so that the battery is said to be recharged. Batteries of this type are called *storage batteries*. Table 33-1 gives the properties of several cells in wide use. These cells generally involve a more or less complicated sequence of chemical reactions in their operation. The basic idea remains, however, that chemical binding energy is utilized to elevate the potential energy of moving charges. This takes the form of an electric potential difference that is maintained between the terminals of the battery.

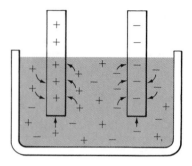

Fig. 33-11. In a battery, chemical forces between the atoms of the anode and positive ions in solution attract and bind positive ions, even though the anode is already positive. Similarly, the cathode chemically attracts negative ions from the solution.

TABLE 33-1 Characteristics of Several Battery Cell Types

Name	Anode Material	Cathode Material	Electrolyte	Reversible?	$\mathscr{E}$
Dry cell (flashlight battery)	Carbon rod	Zinc (the outer container)	Paste containing ammonium chloride	No	1.53 V
Lead-acid (storage cell used in autos)	Lead oxide	Lead	Sulfuric acid	Yes	2.2 V
Weston cell (used as a voltage standard)	Mercury	Cadmium-mercury alloy	Saturated solution of cadmium sulphate	No	1.0183 V at 20°C

Ordinarily, when a potential difference is created by piling up excess charge (as in charging a capacitor), this potential difference is lost when the charges are allowed to flow through a conductor (as in discharging a capacitor). A battery, however, maintains a potential difference between its terminals even when charges flow continually from one terminal to the other. This type of potential difference is given a special name: *electromotive force*. Electromotive force is generally abbreviated as *emf*. Since electromotive "force" actually refers to an electric potential difference (not really a force), we will simply call it emf and use the further abbreviation $\mathscr{E}$. The units are those of potential difference, i.e., volts.

The emf that is maintained between battery terminals will drive a current through a resistor connected between the terminals, as shown in Fig. 33-13. The connecting wires are assumed to

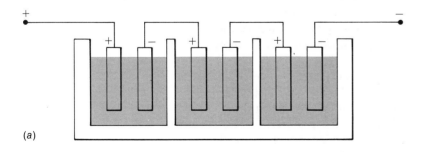

(a)

(b)

Fig. 33-12. (a) A battery consisting of three cells in series. (b) The symbol for a three-cell battery.

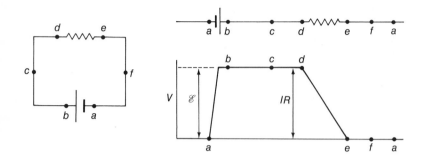

Fig. 33-13.   Changes in potential, *V*, around a simple circuit.

have negligible resistance so the potential is constant along their length. Thus, the ends of the resistor *R* have a potential difference equal to the emf of the battery, $\mathscr{E}$. From Ohm's law we then obtain the current:

$$I = \mathscr{E}/R$$

This current flows continually through the resistor and back through the battery. Each moving charge carrier is elevated in potential as it passes through the battery, moves along the wires with negligible loss in potential, then loses potential steadily as it battles its way through the resistor.

In actual practice, batteries have some resistance of their own. This resistance, called *internal resistance, R_i*, is due to ordinary resistance of the anode and cathode material as well as the resistance of the electrolyte. We can symbolize the presence of this resistance by indicating a resistance between the battery symbol and either of its terminals:

When a battery delivers a current *I*, its terminal voltage V_t is reduced by the potential difference lost across its own internal resistance:

$$V_t = \mathscr{E} - IR_i$$

TERMINAL VOLTAGE IS REDUCED BY INTERNAL RESISTANCE WHEN A CURRENT IS DELIVERED

Example   A typical "12-volt" automobile battery is made up of 6 cells in series each with an emf of 2.2 volts. The total emf of this battery is thus 13.2 volts. The average-sized automobile battery has an internal resistance of 0.01 Ω when freshly charged. What is the actual terminal voltage (the potential difference, in volts, between the battery terminals) of this battery when it is delivering a current of 20 A? When it is delivering a current of 400 A (operating the starter motor on a cold morning)?

Solution   The current passing through the battery also passes through the internal resistance. The potential lost in this resistance is given by Ohm's law and must be subtracted from the net potential difference available at the terminals. Thus we have

$$V_t = \mathscr{E} - IR_i$$

where V_t is the terminal voltage and R_i is the internal resistance. For our two cases we have

$$V_t = 13.2 \text{ V} - 20\,(0.01) \text{ V} = 13 \text{ V}$$

$$V_t = 13.2 \text{ V} - 400\,(0.01) \text{ V} = 9.2 \text{ V}$$

Exercises

5. The typical transistor portable-radio battery is nominally called a 9-volt battery. It is made up of several small dry cells connected in series. How many cells are used?

6. A dry cell has an internal resistance of $0.1\,\Omega$. What magnitude of current must it deliver in order that its terminal voltage be exactly $1\frac{1}{2}$ V?

33-5   Ammeters and Voltmeters

As their names suggest, an ammeter measures electric current (amperes) while a voltmeter measures potential difference (volts). This blurring of the distinction between a physical quantity and the units chosen for its measure is common in electrical work because of the nearly universal dominance of MKS units in electrical measurements. These instruments generally utilize the magnetic forces between moving charges and their operation will be discussed in the next chapter. For now, we will regard them as "black boxes" that simply indicate the desired quantity.

In use, an ammeter is inserted into a circuit so that the current to be measured is forced to pass through it. An ideal voltmeter will simply indicate the potential difference between any two points to which it has been connected while no current passes through the meter itself. Examples of the use of ammeters and voltmeters are shown in Fig. 33-14. For the present we will consider that the interconnecting wires have no resistance and that the ammeter is ideal, offering no resistance to a current.* Thus, the voltmeter could be placed as in either Fig. 33-14a or b, with no change in its reading.

*More realistic, nonideal meters are discussed in Section 34-8.

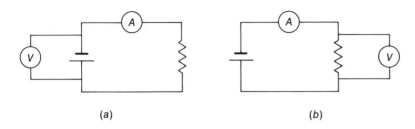

(a)   (b)

Fig. 33-14. The use of an ammeter and voltmeter in a simple circuit.

This follows because all points along a conductor with negligible resistance must be at the same potential.

Example The symbol for a switch is:

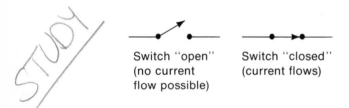

Switch "open"   Switch "closed"
(no current   (current flows)
flow possible)

Generally, if the switch can be either open or closed, it is drawn in the open position to make its presence obvious. Consider now the circuit of Fig. 33-15. The battery is a single storage cell with an emf of 2.2 V and *negligible* internal resistance. The resistor has a value of 1.1 Ω. What are the readings of V_1, A, and V_2 with the switch open and with the switch closed?

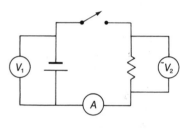

Fig. 33-15.

Solution Since V_1 is always connected to the terminals of the cell and its internal resistance is negligible; it must always read 2.2 V. When the switch is open no current flows, so the ammeter reads zero and, further, V_2 must read zero since the potential difference between the two ends of the resistor is given by Ohm's law: $V = IR$ and $I = 0$. When the switch is closed, the terminals of V_2 are electrically identical to the terminals of V_1 since there is no intervening resistance. Thus, V_2 now reads 2.2 V also. The current which flows when the switch is closed is given by Ohm's law:

$$I = \frac{V}{R} = \frac{2.2 \text{ V}}{1.1 \Omega} = 2 \text{ A}$$

The ammeter thus reads 2 A.

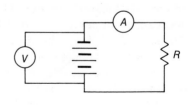

Fig. 33-16.

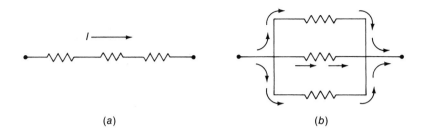

(a)   (b)

Fig. 33-17. (a) Resistors in series. (b) Resistors in parallel.

Exercises

7. A battery consisting of three dry cells in series is used to deliver a current to an unknown resistance, as in Fig. 33-16. When the ammeter reads 10 A, the voltmeter reads 4.09 V. What do you conclude is the internal resistance of the battery?

8. If a second voltmeter is now placed across R in Fig. 33-16, what does it read? What, then, is the value of R?

33-6 Series and Parallel Circuits

When resistors are placed in series, as in Fig. 33-17a, any current entering R_1 must pass successively through R_2 and R_3 as well. It seems intuitively obvious that the total resistance of this combination is larger than that of a single resistor. In Fig. 33-17b, the resistors are shown in parallel. Since several parallel paths are available to the current, we might guess that these parallel resistors offer less resistance to the total flow of current than if it were constrained to pass through only one of them.

The effective resistance of n resistors in series is easily computed. First, we note that a common current, I, passes through all the resistors. Then we ask, "what is the total potential difference between the ends of the entire string of series resistors?" It is this total potential difference, divided by I, that gives the net resistance of the combination. In Fig. 33-18 we see that the potential must decrease from point a to point b by an amount IR_1. The potential decreases from b to c by the amount IR_2. The net potential difference from a to e is just the sum of the potential differences across the individual resistors. A frequently used expression for the potential difference due to a current in a resistor is "voltage drop." Thus, the net voltage drop across the series string of resistors is the sum of the individual voltage drops, each given by IR_i, where R_i is the resistance of the ith resistor.

[handwritten margin notes:]
series → total resistance is larger than that of single resistor

parallel → offer less resistance to total flow

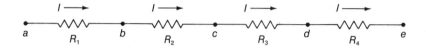

Fig. 33-18. The same current, I, passes through each of the series resistors. The net potential difference V_{ae} is the sum of the potential differences across each of the four resistors.

For N resistors in series, the current I is the same in each individual resistor, so that

$$V_{total} = IR_1 + IR_2 + IR_3 + \cdots = \sum_{i=1}^{N} IR_i$$

and

$$R_{total} = \frac{V_{total}}{I} = \sum_{i=1}^{N} R_i \qquad (33\text{-}4)$$

Series
$R_T = R_1 + R_2 + R_3 \cdots$

Thus for a series connection of resistors, the total resistance is the sum of the individual resistances.

In the case of parallel resistors, the total current is the sum of the individual currents. Further, the same potential difference exists across each resistor, since their ends share a common conductor. Thus we find that V is the same for each individual resistor, and

$$I_{total} = I_1 + I_2 + I_3 + \cdots = \sum_{i=1}^{N} I_i$$

so that

$$R_{total} = \frac{V}{I_{total}} = \frac{V}{\displaystyle\sum_{i=1}^{N} I_i} \qquad (33\text{-}5)$$

Parallel
$\frac{1}{R_T} = \frac{1}{R_1} + \frac{1}{R_2} + \frac{1}{R_3} \cdots$

This result is most easily expressed in terms of the individual resistances if we invert (33-5):

$$\frac{1}{R_{total}} = \frac{I_1 + I_2 + I_3 + \cdots}{V} = \frac{1}{R_1} + \frac{1}{R_2} + \frac{1}{R_3} + \cdots = \sum_{i=1}^{N} \frac{1}{R_i} \qquad (33\text{-}6)$$

As in Fig. 33-19, a circuit may consist of both series and parallel combinations of resistance. In the case illustrated, R_2 and R_3 can be combined, then treated as a resistance in parallel with R_4 to obtain the resistance between b and c. This resistance summed with R_5 gives the net resistance of the upper branch of the circuit. Finally, this resistance in parallel with R_1 gives the net resistance across the emf.

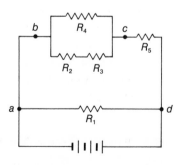

Fig. 33-19. A circuit may contain several series and parallel combinations of resistances.

Example   A $6\frac{2}{3}$-V emf is connected across the combination of 2Ω resistors illustrated in Fig. 33-20. Find the current in each resistor.

Solution   The resistance between points *a* and *b* is given by the parallel combination of a 4Ω resistance (R_2 and R_3 in series) and a 2Ω resistor:

$$\frac{1}{R_{ab}} = \left(\frac{1}{2} + \frac{1}{4}\right)\Omega^{-1} = \tfrac{3}{4}\Omega^{-1}$$

so that $R_{ab} = \tfrac{4}{3}\Omega$. The total resistance is that of R_{ab} in series with R_4:

$$R_{\text{total}} = R_{ab} + R_4 = 3\tfrac{1}{3}\Omega$$

The total current flowing through the battery is given by

$$I = \frac{\mathscr{E}}{R_{\text{total}}} = \frac{6\tfrac{2}{3}\text{ V}}{3\tfrac{1}{3}\Omega} = 2\text{ A}$$

This is the value of the current through R_4. Part of this current passes through R_1 and part through both R_2 and R_3. To find these currents, first find the potential difference between *a* and *b*:

$$V_{ab} = IR_{ab} = 2\cdot\tfrac{4}{3}\text{ V} = \tfrac{8}{3}\text{ V}$$

This potential difference exists across both R_1 and the combination $R_2 + R_3$. Hence

$$I_1 = \frac{V_{ab}}{R_1} = \frac{\tfrac{8}{3}\text{ V}}{2\Omega} = \frac{4}{3}\text{ A}$$

$$I_{23} = \frac{V_{ab}}{R_2 + R_3} = \frac{\tfrac{8}{3}\text{ V}}{4\Omega} = \frac{2}{3}\text{ A}$$

Note that the sum of these two currents is equal to that flowing in the remainder of the circuit, as it should be.

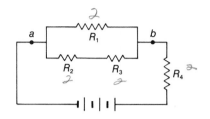

Fig. 33-20.

Exercises

9. Three 3Ω resistors are connected (*a*) in series; (*b*) in parallel. Find the total resistance in each case.

10. Three resistors with resistances of 3Ω, 4Ω, and 5Ω are available. How many different values of R_{total} can be formed by various combinations of all three resistors? What are the values obtained?

33-7   Power in Electric Circuits

When a charge Q passes through a battery in a circuit, it is elevated in potential by an amount $\mathscr{E}$. The work done on the charge is positive and equal to

$$W = Q\mathscr{E}$$

Since current is the *rate* at which charge flows through a circuit and power is the *rate* at which work is being done, we can easily express the power produced by the battery in terms of current:

$$P = \frac{dW}{dt} = \frac{dQ}{dt}\mathscr{E} = I\mathscr{E} \qquad (33\text{-}7)$$

If $\mathscr{E}$ is given in volts and I in amperes, the power given will be in SI units: *watts* (**W**).

Consider now a charge passing through a resistance. It *loses* potential in the process. If the potential difference across the resistor is V, the net loss of energy is

$$W = QV$$

The rate at which this energy is lost is the electrical power dissipated in the resistor:

$$P = \frac{dW}{dt} = \frac{dQ}{dt}V = IV \qquad (33\text{-}8)$$

This power represents the rate at which energy is lost by the moving charge carriers as they flow through the resistor. They lose this energy by repeated collisions with atoms of the resistor. The energy lost by the moving charges is given to the atoms with which they collide, thus causing these atoms to vibrate around their equilibrium locations. In other words, the electrical energy lost in the resistor appears as *heat*.

Utilizing Ohm's law, we can rewrite (33-8):

$$P = IV = I(IR) = I^2R \qquad (33\text{-}9)$$

or

$$P = IV = \left(\frac{V}{R}\right)V = \frac{V^2}{R} \qquad (33\text{-}10)$$

The heat generated within a resistor by a current is called *joule heat* but is often referred to as "I^2R heat." This heat must be taken into account when designing resistors that will carry large currents. Commercial resistors are rated in terms of the maximum power they can dissipate before their temperature rises to the point of self-destruction. Most of the small resistors seen in a TV set, for example, are in the $\frac{1}{4}$-W to 2-W range of power dissipation ratings.

Resistors intended to handle high values of power dissipation are often made by winding a high-melting-point resistive wire around a ceramic core and allowing the resistor to operate at red heat (about 900°C) so that heat can be carried off rapidly by convection and conduction. This design is often used in an electric heater that is purposely intended to convert electrical energy into

Handwritten margin notes:

$W = Q\mathscr{E}$
$P = I\mathscr{E}$

Through Resistance
$W = QV$
$P = IV$

heat. The distribution of this heat may be aided by a fan blowing air over the heater.

It is I^2R heat that limits the amount of current that can be carried by household lighting circuits. The resistance involved is the rather low resistance of the wires used to carry the current. A typical lighting circuit that utilizes #14-gauge copper wire ($r = 0.814$ mm) is limited to 15 A maximum current. This limit insures that the temperature of the wire cannot cause a fire hazard. One of the means of insuring that this limit will not be exceeded is by use of a fuse — a device containing a metal strip that is in series with the circuit. The resistance and melting point of the metal strip are chosen so that I^2R heat melts the strip and opens the circuit if the fuse rating is exceeded.

Example An electrical wall outlet provides an emf of 110 V. What value of resistance should an electrical heater have if it is to dissipate a power of 1 KW when connected to this emf?

Solution Connecting the heater to the wall outlet places the full potential difference of 110 V across the resistor. We could use Ohm's law to calculate the current and then use the product of current and voltage to obtain the power, but that was already done in obtaining (33-10):

$$P = V^2/R \qquad\qquad (33\text{-}10)$$

Hence
$$R = \frac{V^2}{P} = \frac{(110 \text{ V})^2}{1000 \text{ W}} = 12.1\Omega$$

Exercises

11. If, in the above example, the maximum current that can be drawn from the wall outlet is 15 A, what maximum power can be dissipated?

12. A given resistor cannot carry currents exceeding 20 A without exceeding its maximum power dissipation ratings. By forced-air cooling, suppose that we increase the rate at which heat can be carried away by a factor of 2. Now what is the maximum current that the resistor can carry?

33-8 Resistivity

Up until this point in our discussion, we have considered resistance as a property of an entire device, such as a segment of wire, a resistor, or an electrolytic cell. It is possible, however, to define a concept called *resistivity*, which is a property of a given *substance*. The resistance of a specific resistor made of this substance will then depend on its dimensions and the resistivity of the substance itself.

In order to define resistivity, we need the concept of a *current density*, *j*. The current density is defined to be the current per unit of area through which the current passes:

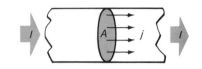

$$j = I/A \qquad (33\text{-}11)$$

Fig. **33-21**.   The current *density*, *j*, is related to the current by $j = I/A$.

The *resistivity*, *ρ*, is defined by

$$\rho = E/j \qquad (33\text{-}12)$$

where *E* is the value of the electric field strength at a point and *j* is the current density produced at the same point by that field. Equation (33-12) is Ohm's law on a point-by-point basis (if *ρ* is a constant, which it is for ohmic conductors). The notion of current density and resistivity makes it possible to discuss the behavior of currents that are not confined to a conductor of uniform cross-sectional area. We, however, will limit our discussion to that simple case.

Consider a conductor of cross-sectional area *A* and length *L*, as in Fig. 33-21. The current density is uniform everywhere along and within such a conductor, so the electric field within the conductor must also be uniform and given by

$$E = j\rho$$

The potential difference between the ends of the conductor is then given by

$$V = EL = j\rho L$$

Utiling (33-11), we have

$$V = \frac{I}{A}\rho L$$

The net resistance is then given by

$$R = \frac{V}{I} = \rho \frac{L}{A} \qquad (33\text{-}13)$$

Table 33-2 gives values of *ρ* for various materials as well as the temperature coefficient *α*. This coefficient is the fractional change in resistivity per degree Celsius:

$$\alpha = \frac{1}{\rho}\frac{d\rho}{dT}$$

TABLE 33-2   Resistivity of Various Materials at 20°C

Material	Resistivity ($\Omega \cdot m$)	Temperature Coefficient
Aluminum	2.8×10^{-8}	3.9×10^{-3}
Copper	1.7×10^{-8}	3.9×10^{-3}
Carbon	3.5×10^{-5}	-5×10^{-4}
Iron	1.0×10^{-7}	5.0×10^{-3}
Mercury	9.6×10^{-7}	8.9×10^{-4}
Nichrome®	1.0×10^{-6}	—
Silver	1.6×10^{-8}	3.8×10^{-3}
Amber	5×10^{14}	—
Beeswax	8×10^{12}	—
Glass	9×10^{11}	—

Exercises

13. When the diameter of a conductor is doubled, what happens to its resistance?

14. Household lighting circuits utilize 14-gauge copper wire (diameter = 1.63 mm). Calculate the resistance of a 4-m length of this wire.

15. We wish to fabricate a 100Ω resistor by winding 2 m of Nichrome wire around a porcelain rod. What diameter should we choose for the wire?

33-9   A Microscopic View of Resistivity   NOT ON HOURLY

In ordinary metallic conductors, the atoms are arranged in a stable crystalline array while one, two, or three (depending on the chemical valence of the metal) electrons from each of the atoms is free to wander throughout the solid. These mobile electrons are called *conduction electrons* and it is their presence that distinguishes metals from nonmetals. Thermal agitation causes the atoms of the crystal lattice to vibrate about their equilibrium positions and causes the conduction electrons to continually move from one collision to another throughout the body of the metal. Since these electrons move in random directions, no net current is produced by this thermal motion.

   The situation is changed when an electric field is present. Now the electrons drift in a direction opposite to that of the electric field (because they are negatively charged) between collisions. The combined effect of many electrons continually drifting opposite the electric field direction *does* produce a current. Note that the direction of the conventional current is the same as that of the electric field since negative charges are drifting in the opposite direction.

We can compute the net current density that this drift produces if the electrons move freely between collisions and if we know the number of conduction electrons per unit volume. If the time between collisions is τ and the electron starts from rest after each collision, the distance of electron drift between collisions is

$$s = \tfrac{1}{2}a\tau^2 = \tfrac{1}{2}\frac{F}{m}\tau^2 = -\frac{1}{2}\frac{eE}{m}\tau^2$$

where the minus sign indicates motion in the opposite direction to the field. The average drift velocity of each electron is then

$$\bar{v} = \frac{s}{\tau} = -\frac{1}{2}\frac{eE}{m}\tau \qquad (33\text{-}14)$$

Consider now a collection of such electrons with n electrons per unit volume, as in Fig. 33-22. If they are moving with a velocity v, then the number passing through one face of the unit volume (the face has unit area) is simply nv. Since each electron carries a charge $-e$, we have a current density given by:

$$j = -nve \qquad (33\text{-}15)$$

Taking v to be $\bar{v}$ in Eq. (33-14), we have

$$j = \left(\tfrac{1}{2}\frac{ne^2}{m}\tau\right)E \qquad (33\text{-}16)$$

Comparing (33-16) with (33-12), we see that the resistivity ρ is given by

$$\rho = \frac{2m}{ne^2\tau} \qquad (33\text{-}17)$$

Equation (33-17) agrees with our intuition: a large density n of charge carriers gives low resistivity, while a short time τ between collisions leads to large resistivity. Further, it gives resistivity in accord with Ohm's law, since ρ is independent of E.

The average time between collisions for copper can be found by using the tabulated value for ρ along with a value for n obtained by noting that each copper atom contributes one electron. The number of atoms per cm^3 is thus given by Avogadro's number multiplied by the density of copper divided by its atomic mass:

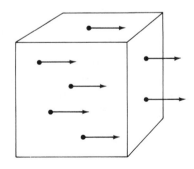

Fig. 33-22. A unit volume containing n electrons moving at velocity v. The number of electrons leaving one face is nv.

$$n = \frac{(9 \text{ g/cm}^3)(6 \times 10^{23} \text{ atoms/mole})(\text{electron/atom})}{64 \text{ g/mole}}$$

$$= 8.4 \times 10^{22} \text{ electrons/cm}^3$$

$$= 8.4 \times 10^{28} \text{ electrons/m}^3$$

Solving for τ in (33-17), we obtain

$$\tau = \frac{2m}{ne^2\rho} = \frac{2(9.1 \times 10^{-31} \text{ kg})}{(8.4 \times 10^{28}/\text{m}^3)(1.6 \times 10^{-19} \text{ C})^2(1.7 \times 10^{-8}\Omega \cdot \text{m})}$$

$$= 5 \times 10^{-14} \text{ s}$$

The average thermal velocity v_t of electrons at room temperature is 1.6×10^6 m/s, so the average distance λ between collisions is

$$\lambda = v_t\tau = 8 \times 10^{-8} \text{ m}$$

which is several hundred times the distance between atoms. This great distance between collisions cannot be understood classically but is understood in quantum mechanics. Atomic and subatomic particles, such as electrons, exhibit wavelike properties, and the matter waves that represent electron behavior spread through the crystal lattice without appreciable scattering.

It is interesting to evaluate $\bar{v}$, the average drift velocity, for electrons in a copper wire carrying a current density of 500 A/cm². Solving (33-15) for $\bar{v}$, we obtain

$$\bar{v} = j/ne = \frac{5 \times 10^6 \text{ A/m}^2}{(8.4 \times 10^{28} \text{ electrons/m}^3)(1.6 \times 10^{-19} \text{ C/electron})}$$

$$= 3.7 \times 10^{-4} \text{ m/s} = 0.37 \text{ mm/s}$$

This low velocity illustrates the magnitude of the charge carried by the extremely large number of conduction electrons in the wire.

The picture we have of a copper wire carrying a current, then, is that of a swarm of electrons moving randomly with individual velocities over 10^6 m/s but drifting steadily as a swarm with a velocity of a fraction of a mm per second. If we inquire as to the time delay between the time a current begins at one end of a wire and emerges from the other, still a third velocity is involved. The excess charges introduced at one end repel nearby charges that move and repel their neighbors and so on. The time delay involved is that required to propagate an electric disturbance via an electric field. The speed of this propagation is that of light, so the delay is short indeed. An analogous situation exists in a pipe that has been filled with water. When additional water is forced slowly into one end,

water almost instantly begins flowing out the other end. The only
time delay is that associated with the speed of a compressional disturbance in water, i.e., the speed of sound in water.

The temperature dependence of resistivity, on a microscopic
level, is understood to be due to the thermal agitation of atoms. As
the temperature rises, the amplitude of atomic vibrations increases.
This increases the likelihood of the scattering of conduction electrons. Since this decreases τ, the resistivity is predicted to increase
with temperature, as is observed for metals.

At very low temperatures, some materials exhibit *superconductivity*. In these substances, once atomic vibrations are diminished
sufficiently (temperatures near absolute zero are required), some
electrons propagate through the crystal lattice with *no* scattering.
When this occurs, resistivity vanishes and a current once started
in a loop of the material will "coast" forever.

33-10   Summary

When charge is transported past a given point at a rate $\Delta Q/\Delta t$,
we say that the current I is given by

$$I = \Delta Q/\Delta t$$

The SI units of current are amperes = coulombs/second.

The conventional direction of current is taken to be that in
which positive charge carriers move to produce the observed current. In metallic conductors, the mobile charge carriers are electrons. Electrons, carrying a negative charge, move in the opposite
direction to that we assign to conventional current.

When charge carriers pass through a substance, they lose energy in collisions with atoms. An electric field is thus required to
produce a steady current. The presence of this field causes a potential difference V between the ends of a current-carrying conductor. The resistance of the conductor is defined by

$$R = V/I \qquad \text{(33-1)}$$

Equation (33-1) is often called Ohm's law, though Ohm's discovery
was that R is a *constant* for many conductors. In cases where the
current is not strictly proportional to potential difference (voltage),
a current-voltage diagram is often useful.

In an electrical circuit with a steady current, the loss of potential energy by charge carriers moving through resistance must
continually be offset by a source of potential energy. Such a source
of electrical potential is called an emf ($\mathscr{E}$). A battery is a device
that elevates the electric potential of charge carriers by means of
chemical forces.

When N resistors are connected in series, they are equivalent to a single resistance given by:

$$R = \sum_{i=1}^{N} R_i$$

$R_T = R_1 + R_2 + R_3$

When N resistors are connected in parallel, the equivalent resistance is given by

$$\frac{1}{R} = \sum_{i=1}^{N} \frac{1}{R_i}$$

$\frac{1}{R_T} = \frac{1}{R_1} + \frac{1}{R_2} + \frac{1}{R_3}$

The power produced by an emf that creates a potential difference V is given by

$$P = VI$$

where I is the current passing through the emf. When current passes through a resistance R, power is dissipated at a rate

$$P = V_R I = I^2 R$$

The energy dissipated in a resistor appears as heat.

Current density, j, is defined to be the current passing through each unit area of an element of area perpendicular to the motion of the charge carriers. The resistivity ρ is defined to be

$$\rho = E/j$$

$E = j\rho$ electric field

A resistor of length L and cross-sectional area A has a resistance given by

$$R = \rho \frac{L}{A}$$

In metallic conductors, the current is carried by drifting conduction electrons. If the time between successive electron-atom collisions is given by τ, then

$$\rho = \frac{2m}{ne^2\tau}$$

where m is the electron mass, n the number of electrons per unit volume, and e the electron charge.

Problems

33-1.   A current of 1 mA is carried by a copper wire. How many electrons pass a given point in each second?

33-2. A 6.0-V battery is connected to a $1\,\Omega$ resistor, as in Fig. 33-9. (*a*) Find the current I in the circuit. (*b*) Find the increase in potential energy of a single electron passing through the battery. (*c*) Find the loss in potential energy of a single electron passing through the resistor.

33-3. A single lead-acid storage cell with a negligible internal resistance is connected to a light bulb that draws a current of 6.6 A. Find the resistance of the light bulb.

33-4. A dry cell is connected to a $5\,\Omega$ resistor. The potential difference across the dry cell terminals is then measured to be 1.48 V. Find the internal resistance of the dry cell.

33-5. A battery formed of three lead-acid storage cells in series is charged at a rate of 100 A. The potential difference across its terminals is measured to be 7.6 V. This same battery is now used to deliver a current of 100 A. Find its new terminal voltage.

33-6. Three identical resistors R are connected as in Fig. 33-23. The net resistance between a and b is $7.5\,\Omega$. Find R.

33-7. An ideal ammeter (zero resistance) and an ideal voltmeter (infinite resistance) are connected as in Fig. 33-24. Find their readings for $R_1 = 5\,\Omega, R_2 = 15\,\Omega, R_3 = 4\,\Omega$, and $\mathscr{E} = 20$ V.

33-8. An electrical meter called a *galvanometer* responds to small electrical currents. Suppose that a given galvanometer will give a full-scale deflection for a current of 1 mA $(10^{-3}$ A) and has a net resistance of $50\,\Omega$, as in Fig. 33-25*a*. (*a*) What value of resistance R_1 must be connected in series with this galvanometer (Fig. 33-25*b*) to convert it into a voltmeter with a full-scale reading of 100 V? (*b*) What value of resistance R_2 must be connected in parallel with the galvanometer (Fig. 33-25*c*) to convert it to an ammeter with a full-scale reading of 10 A?

33-9. A flashlight battery has an emf of 1.5 V and internal resistance $\frac{1}{4}\,\Omega$. It is connected in the circuit shown in Fig. 33-26, where all resistors have the value $3\,\Omega$. Find the current flowing through the battery.

33-10. Five equal resistors R are connected as in Fig. 33-27. Find the net resistance between points a and b.

33-11. The circuit of Fig. 33-24 is connected without the ammeter A present. The ammeter is now inserted as shown. Find the change in the voltmeter reading caused by insertion of the ammeter if $\mathscr{E} = 10$ V, $R_1 = 5\,\Omega$, $R_2 = 10\,\Omega$, $R_3 = 5\,\Omega$, and the resistance of the ammeter is $0.1\,\Omega$. Assume that the voltmeter draws no current $(R_V = \infty)$.

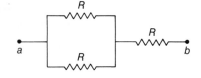

Fig. 33-23.

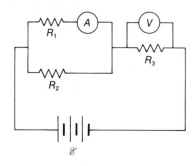

Fig. 33-24.

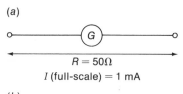

(*a*)

$R = 50\,\Omega$

I (full-scale) $= 1$ mA

(*b*)

(*c*)

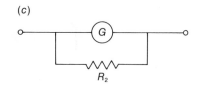

Fig. 33-25. A galvanometer (*a*) can be converted into a voltmeter (*b*) or an ammeter (*c*).

33-12. An electric heater is designed to work from a 110-V supply and produces heat at a 3-kW rate. Find its resistance.

33-13. A battery has an emf $\mathscr{E}$ and an internal resistance r. It is connected to a single resistor of value R. Show that the maximum power is dissipated in R when we set $R = r$.

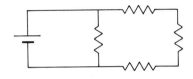

Fig. 33-26.

33-14. Three 10Ω, 2-W resistors are connected as in Fig. 33-23. Find the maximum possible voltage between points a and b without exceeding the power dissipation limits of any of the resistors.

33-15. A 10Ω resistor is connected to a 120-V emf and immersed into 100 g of water in order to heat it to the boiling point to make tea. How long does it take for the water temperature to rise from 20°C to 100°C?

33-16. An emf of 1000 V is connected by two copper wires to an electrical load of 10Ω that is 1 km distant. If the power dissipated in the wires is to be no more than 1% of that dissipated in the load, find the minimum allowable diameter of the copper wires.

33-17. A carbon resistor has a value of 1500Ω at room temperature (20°C). Under load, its temperature rises to 100°C. Find its resistance at this temperature.

33-18. Iron wire with a diameter of 1 mm is used to fabricate a resistor of 0.05Ω resistance. What length of iron wire must be used?

33-19. In an automobile starting circuit, a six-cell lead-acid storage battery with an internal resistance of 0.01Ω provides a current to the starter motor of up to 400 A. If a total of 1 m of copper connecting wire is used between the battery and motor, find the wire diameter such that no more than 20% of the *terminal voltage* of the battery is lost across the connecting wire.

33-20. Find the current density j in copper wire when the electron drift velocity equals 1 cm/s.

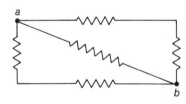

Fig. 33-27.

Chapter **34**

Magnetic Fields

SECTIONS 1-5,7

34-1 Introduction The electrostatic force between point charges obeys an inverse square law:

$$F = k\frac{q_1 q_2}{r_{12}^2} \qquad (34\text{-}1)$$

where q_1 and q_2 are the magnitudes of the two interacting charges and r_{12} is their separation. If one or the other of the two charges moves through our laboratory system of coordinates, (34-1) still describes the force that one charge experiences due to the other. However, if *both* charges are in motion, a new force appears. This is the *magnetic* force and occurs whenever *both* interacting charges are in motion. Thus, it represents a force that acts between two *currents,* as opposed to the electrostatic force, which acts between two fixed charges.

It seems strange, indeed, that the force between two charges should be modified if they move with respect to our frame of reference. If the principle of relativity is correct, this implies that the force between two stationary charges appears different to us if we move steadily through the laboratory that contains the two charges. It was in 1905 that Einstein clearly stated a new *theory of relativity* that eventually incorporated these apparent magnetic forces as a natural consequence of an observer's motion through a frame of reference.

Einstein's theory of relativity resolved difficulties with the electrodynamics of moving bodies. However, it also implied revision of older concepts of length, time, mass, and kinetic energy, as discussed in Sections 8-6 and 14-10. We will not explicitly use the theory of relativity in our discussion of magnetism, but it is interesting to reflect that magnetism is a relativistic effect that was studied long before the new theory of relativity appeared.

34-2   The Magnetic Field: A Historical View

It was known from ancient times that certain minerals found in the district of Magnesia in Asia Minor attracted one another and attracted iron as well. The observation that these mineral samples took on a specific north-south orientation when freely suspended led to the development of the magnetic compass. It was found that a *magnetized* needle or bar of iron could be made by stroking the iron with a naturally magnetized mineral sample.

When such artificially magnetized bars were allowed to pivot freely, they became aligned in a north-south direction and the end of the bar that pointed north was called a *north-seeking pole* or simply a *north pole*. By bringing bar magnets close together, it was found that like poles repel one another and unlike poles attract, in close analogy with the behavior of electric charges. One very important difference, however, was that magnetic poles were always found in opposite pairs. For example, if a bar magnet is broken in two, as in Fig. 34-1, the new pieces of the magnet will each have both a north and a south pole.

Just as an object can be electrically polarized by a nearby electric charge, certain *magnetic substances* were found to be magnetically polarized by a nearby magnet. Iron is a magnetic substance and the alignment of an iron rod near a bar magnet, as in Fig. 34-2, could be used to define a direction of the bar magnet's *magnetic field*. At the position shown in Fig. 34-2, the induced south pole of the rod is attracted toward the north pole of the bar magnet. At

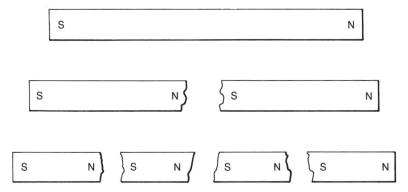

Fig. 34-1. Magnetic poles appear in opposite pairs. When a bar magnet is broken, each new piece will contain two opposite poles.

(a)

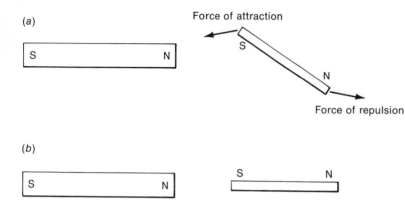

Force of attraction

Force of repulsion

(b)

S   N

S   N

Fig. 34-2. (a) Magnetic poles are induced in a small rod of iron mounted on a pivot when it is placed close to a bar magnet. (b) The torque exerted on the small rod brings it into alignment with the *magnetic field* of the bar magnet.

other positions the rod alignment is produced by the interplay of attraction and repulsion caused by both poles of the bar magnet. Thus, when a sheet covered with iron filings is placed over a bar magnet, we see the pattern of the magnetic field shown in Fig. 34-3. This much was known about magnetism in the early nineteenth century.

In 1820, Oersted made an extremely important discovery: a current-carrying wire *produces* a magnetic field. The next year Faraday found that a current-carrying wire experiences a *force* when placed in a magnetic field. The intimate connection between electricity and magnetism was thus discovered only a little over 150 years ago. We now understand that the most fundamental aspects of magnetism involve charges *in motion*. All of the experiments with magnetized iron are now understood to be due to the ceaseless motion of the atomic electrons in iron.

34-3 The Magnetic Force: Two Charges with a Common Velocity

Two stationary charges interact to produce a force. The force on either charge is given by Coulomb's law:

$$F = k\frac{q_1 q_2}{r_{12}{}^2}$$

When both charges move through our frame of reference with a common velocity, as in Fig. 34-4, the net force acting on either charge is found to be reduced. This net force between two charges with equal parallel velocities is given by

$$F_{\text{net}} = k\frac{q_1 q_2}{r^2}\left(1 - \frac{v^2}{c^2}\right)$$

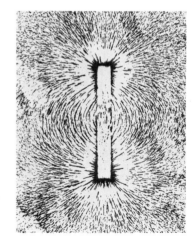

Fig. 34-3. Iron filings sprinkled over a transparent sheet resting on a bar magnet are aligned by the magnetic field. (From *PSSC Physics,* D. C. Heath and Company, Lexington, Massachusetts, © 1965)

where v is their common velocity and c is the velocity of light. The net force decreases as the velocity increases. If we factor out the velocity-dependent portion, called the *magnetic force, F_m,*

$$F_m = k\frac{q_1 q_2}{r^2}\left(\frac{v^2}{c^2}\right)$$ (34-2)

we are left with the electrostatic portion of the net force in its usual form:

$$F_e = k\frac{q_1 q_2}{r^2}$$

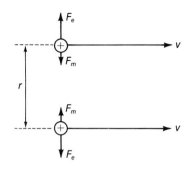

Fig. 34-4. Two moving charges experience a magnetic force, F_m, as well as an electrostatic force, F_e.

If we think of the moving charges as electrical currents, we conclude that *like currents attract,* since the magnetic force acts against the repulsion of the like charges.

It is once again useful to define a field—the *magnetic field*—which we think of as produced by one moving charge. The other moving charge then interacts with this field and experiences a magnetic force. The magnetic field will be symbolized by B. In order to describe the force given by (34-2), we could first define B by:

magnetic field produced by one moving charge

other moving charge interacts with field and experiences a magnetic force

$$B_1 = k\frac{q_1}{r^2}\frac{v}{c}$$ (34-3)

and then let the force on q_2 be given by:

$$F_2 = q_2 \frac{v}{c}B_1$$ (34-4)

Substitution of (34-3) into (34-4) then leads us back to (34-2).

Alternatively, we could place the factor $1/c^2$ entirely in the expression we choose for the magnetic field:

$$B_1 = k\frac{q_1}{r^2}\frac{v}{c^2}$$ (34-5)

with the force then given by the rule

$$F_2 = q_2 v B_1$$ (34-6)

Again, substitution of (34-5) into (34-6) leads us back to (34-2). The first approach (dividing the factor $1/c^2$ into two factors of $1/c$ shared equally between field and force) is used in a system of units called *gaussian units.* The electrostatic units of charge (esu) are used in gaussian units. Since $k = 1$ in esu, the expressions for the electrostatic and magnetic forces on charges separated by a distance r with a common velocity v are very simple in gaussian units:

$$F_e = \frac{q_1 q_2}{r^2}$$

$$F_m = \frac{q_1 q_2}{r^2}\left(\frac{v^2}{c^2}\right)$$

When esu are used in (34-3) with r in cm, the units of a magnetic field are called *gauss* (G).

In the SI units we will use, the factor $1/c^2$ that appears in the expression for F_m is entirely absorbed in the expression for the magnetic field, as in (34-5). The proportionality constant for the magnetic portion of the force is called k':

$$B = k'\frac{q}{r^2}v \tag{34-7}$$

where

$$k' = k/c^2 \tag{34-8}$$

Again, note that substitution of (34-7) into (34-6) leads back to (34-2). It is customary to include a factor of 4π in k' in the following way:

$$B = \frac{\mu_0}{4\pi}\frac{qv}{r^2}$$

so that

$$\frac{\mu_0}{4\pi} = k' = \frac{k}{c^2} = \frac{1}{4\pi\epsilon_0 c^2}$$

and

$$\mu_0\epsilon_0 = \frac{1}{c^2} \tag{34-9}$$

The nineteenth-century discovery of a relationship between the velocity of light and constants from the equations of electricity and magnetism gave the first indication that light might be an *electromagnetic wave*.

Exercises

1. Two identical charges separated by a distance s repel one another with an electrostatic force F_e. At what velocity must they move so that magnetic effects reduce the net force to $0.84F_e$?

2. Check (34-9) with the values given in the table of physical constants.

34-4   The Magnetic Field

In the preceding section we considered only two charges moving with a common velocity. The magnetic force between two charges moving with *arbitrary* velocities has a magnitude and a direction that each depend in a complicated way on the motion of the charges. The magnetic force is rather simply described, however, when we adopt the field point of view.

Let us begin by *assuming* that the *direction* of the magnetic field **B** produced by a bar magnet is reasonably described by the iron-filing pattern in Fig. 34-3. We can then experiment with moving charges in the presence of the bar magnet's field, as indicated in Fig. 34-5. At any given point it is experimentally found that:

1. The force on the moving charge is zero when its velocity is parallel to **B**.
2. The force on the moving charge is largest when **v** is perpendicular to **B**.
3. The force is always perpendicular to **v** and to **B**.
4. The force is proportional to q.
5. The force is proportional to v.

velocity parallel to field
FORCE = ZERO

velocity ⊥ to field
FORCE = MAX

FORCE ALWAYS ⊥ to
v and B

$$F = qvB$$

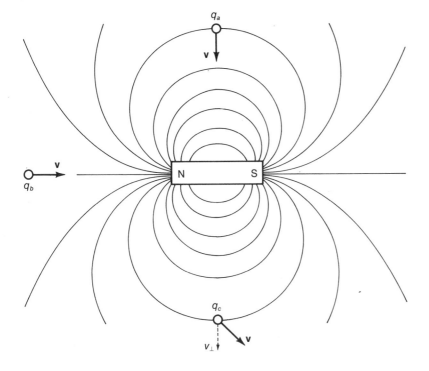

Fig. 34-5. The charge q_a moves in. a direction perpendicular to **B** and experiences a magnetic force. The charge q_b moving parallel to **B** experiences no magnetic force. The charge q_c experiences a magnetic force proportional to $v_\perp$, the component of its velocity that is perpendicular to **B**.

Recalling the properties of the vector cross product, we see that all the above observations are consistent with

$$\mathbf{F}_m = q\mathbf{v} \times \mathbf{B} \qquad (34\text{-}10)$$

where $\mathbf{F}_m$ is the force acting on the charge q when it moves with velocity $\mathbf{v}$ through a magnetic field $\mathbf{B}$. The only feature of (34-10) that is not explicitly based on the observations we have described in the dependence on the *magnitude* of $\mathbf{B}$, since we have not yet agreed on a means of assigning a magnitude to $\mathbf{B}$. Let us adopt (34-10) as a means of measuring the magnitude of $\mathbf{B}$. Just as we can use a test charge and the expression

$$\mathbf{F} = q\mathbf{E} \qquad (34\text{-}11)$$

to measure $\mathbf{E}$, we can similarly use a moving charge and (34-10) to measure the magnetic field strength, $\mathbf{B}$. We will see how this field is produced by other moving charges in the next chapter.

The vector cross product in (34-10) tells us that the magnetic force is always at right angles to both the velocity of the particle and to the direction of the magnetic field, as indicated in Fig. 34-6. If we express the force in N, the charge in C, and the velocity in m/s, the field $\mathbf{B}$ is given in units of *tesla* (T).* The tesla is the SI unit for magnetic field strength, also called *magnetic induction* or *magnetic flux density*. From (34-10) we see that the direction of $\mathbf{F}_m$ is reversed if we change the sign of the charge, or if we reverse *either* $\mathbf{v}$ or $\mathbf{B}$.

The net force on a moving particle with charge q that is acted upon by both electric and magnetic fields is given by the vector sum of (34-10) and (34-11):

$$\mathbf{F} = q(\mathbf{E} + \mathbf{v} \times \mathbf{B}) \qquad (34\text{-}12)$$

Just as we represented electric fields by electric lines of force, we can represent magnetic fields with magnetic lines of force. These lines are drawn so that they are everywhere tangent to the direction of $\mathbf{B}$ and their density indicates the magnitude of $\mathbf{B}$. The magnetic field lines between the poles of a laboratory magnet are sketched in Fig. 34-7.

Exercises

3. Consider a magnetic field whose local direction is in the plane of the page and pointing from left to right. (a) What are the two possible directions in which a charged particle can move in this

*The reader may often see magnetic field strengths expressed in *gauss* (gaussian units). One tesla $= 10^4$ gauss.

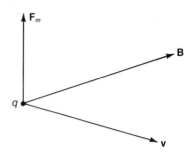

Fig. 34-6. Magnetic force is always perpendicular to the velocity of the charged particle and to the direction of the magnetic field at the position of the particle. The reader should check that the right-hand rule applied to (34-10) is correctly described by this illustration.

RT. HAND RULE

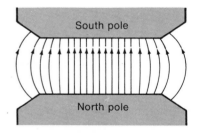

Fig. 34-7. Magnetic lines of force between the poles of a laboratory magnet indicate the direction of $\mathbf{B}$. The regions where the lines are most closely crowded are the regions where $\mathbf{B}$ has maximum magnitude.

field without experiencing a magnetic force? (*b*) With this same magnetic field present, consider a charged particle moving in the plane of the page from bottom to top. If the direction of the magnetic force acting on it is into the page, is the charge negative or positive?

4. Most laboratory electromagnets can produce a magnetic field with a magnitude of about 1 T. (*a*) What is the maximum magnetic force that can act on a proton with $v = c/4$ in a 1-T field? (*b*) What value of acceleration results if no other force acts? (*c*) How large an electric field is required to produce the same acceleration?

34-5 Circular Orbits in a Magnetic Field

A charged particle moving in a magnetic field always experiences a magnetic force that is perpendicular to its velocity. Thus, the magnitude of the velocity is unchanged by magnetic forces. Only the direction of the particle's velocity changes. If the field is uniform and the particle's initial velocity is perpendicular to the field, simple circular motion results, as shown in Fig. 34-8. To see this, recall that centripetal force F_c is described by

$$F_c = m\omega^2 R = m\frac{v^2}{R}$$

where R is the radius of the circular path. For a charged particle with charge q, mass m, and moving with velocity **v** perpendicular to a uniform field **B**, we have:

$$F_m = qvB = m\frac{v^2}{R} \qquad (34\text{-}13)$$

where R is the radius of the particle's circular path. We can cancel a factor of v in (34-13) to obtain

$$mv = qBR \qquad (34\text{-}14)$$

Equation (34-14) is often used in atomic and nuclear physics to determine the momentum, mv, of a particle with a known charge by measurement of its radius of curvature in a known field. Alternatively, the ratio q/m can be determined if we already know the velocity of the particle, as well as R and **B**.

Example A well-collimated beam of protons is directed through a magnetic field, as in Fig. 34-9. The accurately located slits ensure that the protons are deflected through an angle of 45°. The magnetic field is perpendicular to the page and uniform over the region indicated, so

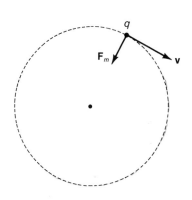

Fig. 34-8. The circular orbit of a positively charged particle in a magnetic field that is directed out of the page. The right-hand rule for **v × B** gives the direction shown for **F**$_m$.

magnitude of velocity is unchanged by magnetic forces

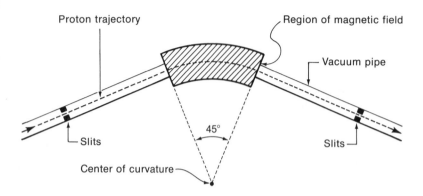

Fig. 34-9. A beam of protons within a vacuum pipe is deflected by a magnetic field.

that these protons follow a circular path with a radius of curvature of 1 m within the magnet. The magnetic field has a magnitude of 0.5 T. (a) Find the velocity of these protons. (b) What is their kinetic energy in MeV?

Solution   (a) The protons follow a segment of a circular orbit with **v** perpendicular to **B**, so we may apply Eq. (34-14):

$$v = (q/m)BR = \left(\frac{1.6 \times 10^{-19} \text{ C}}{1.67 \times 10^{-27} \text{ kg}}\right)(0.5 \text{ T}) \cdot 1 \text{ m} = 4.79 \times 10^7 \text{ m/s}$$

This velocity is about 1/6 that of light.

(b) $K = \frac{1}{2}mv^2 = \frac{1}{2}(1.67 \times 10^{-27} \text{ kg})(2.294 \times 10^{15} \text{ m}^2/\text{s}^2)$

$$= 1.92 \times 10^{-12} \text{ J}$$

$$= \frac{1.92 \times 10^{-23} \text{ J}}{1.6 \times 10^{-19} \text{ J/eV}} = 1.2 \times 10^7 \text{ eV} = 12 \text{ MeV}$$

As discussed in Section 14-10, the more accurate relativistic expression for kinetic energy differs somewhat from the classical expression, $\frac{1}{2}mv^2$, particularly at high velocities. In the present example with $v \approx \frac{1}{6}c$, the correction would only amount to 2%.

Consider now a charged particle moving in a complete circular orbit in a uniform magnetic field. Since the radius of that orbit is proportional to the velocity, the time required for each complete orbit is *independent* of velocity:

$$\tau = \frac{\text{time}}{\text{orbit}} = \frac{\text{circumference}}{v}$$

Utilizing Eq. (34-14), we find

$$\tau = \frac{2\pi R}{v} = \frac{2\pi\left(\dfrac{mv}{qB}\right)}{v} = 2\pi\left(\frac{m}{q}\right)\frac{1}{B}$$

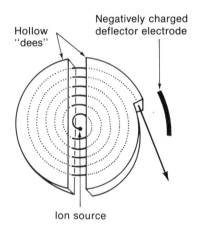

Fig. 34-10. Cyclotron operation. A magnetic field that is perpendicular to the plane of the "dees" causes positively charged ions to move in circular orbits. The dees are electrically charged with opposite polarity. The sign of this polarity is reversed repetitively so that ions are accelerated each time they cross the gap from one dee to the other.

which is independent of v. Higher-velocity particles execute proportionally larger orbits, the net result being a period of revolution that is independent of v. The number of complete orbits per time interval is the revolution frequency:

$$f = \frac{1}{\tau} = \frac{1}{2\pi} \frac{q}{m} B \qquad \text{cyclotron frequency} \qquad (34\text{-}15)$$

This frequency is called the *cyclotron frequency*, named after a nuclear particle accelerator invented by E. O. Lawrence.*

A sketch indicating the principles of the cyclotron is shown in Fig. 34-10. The entire cyclotron structure is located inside a high-vacuum enclosure, so that the accelerated particles will not collide with air molecules. The source of ions is an electric discharge in a low-pressure gas at the center of the machine. A uniform magnetic field guides the ions in circular orbits.

If protons are being accelerated, for example, ordinary hydrogen gas is continually introduced into the ion source region between two electrodes. An electric discharge ionizes the hydrogen to produce some hydrogen nuclei devoid of atomic electrons. These nuclei are protons and are attracted into a "dee" whenever it is negatively charged. High-speed vacuum pumps continually remove the excess hydrogen gas.

The polarity of the dees is repetitively reversed with a frequency exactly equal to the cyclotron frequency of the ions that are being accelerated, as indicated in Fig. 34-11. The ion orbits are segments of circles within the hollow dees, since only magnetic forces act there. This is because the dees are conductors and electric fields vanish within a hollow conductor. A photograph of a cyclotron dee is shown in Fig. 34-12.

Whenever an ion crosses from one dee to the other, it gains energy due to the electric field between the two oppositely charged dees. If the potential difference of the dees is V_0, the energy gain on a single gap crossing is

$$\Delta E = qV_0$$

where q is the charge on the ion. Since the gap is crossed twice per revolution, we have

$$\Delta E/\text{turn} = 2qV_0$$

(a)

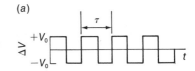

(b)

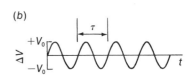

Fig. 34-11. Ideally, the potential difference between the dees would be as in (a) so that all particles liberated by the ion source would be accelerated over a full half-cycle. However, the discontinuous change in polarity implies infinite currents flowing in and out of the dees to change their polarity and cannot be achieved. The potential difference shown in (b) varies sinusoidally and is what is actually achieved. Only those ions crossing the gap at the peaks of the potential difference can gain the full energy qV_0.

*Lawrence received the Nobel Prize in Physics in 1939 for his invention of the cyclotron. An interesting history of the development of the cyclotron was written by Lawrence's graduate student whose Ph.D. thesis consisted of the development of the cyclotron into a usable instrument. It is contained in *Particle Accelerators, a Brief History* by M. Stanley Livingston; Cambridge, Massachusetts: Harvard University Press (1969).

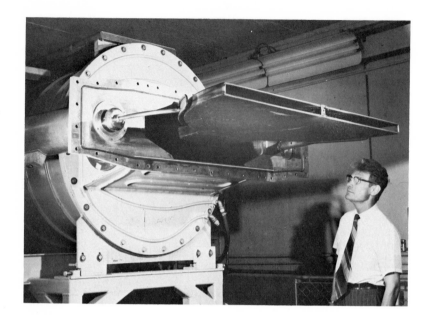

Fig. 34-12. One dee of the University of Colorado cyclotron is seen here during annual maintenance when the machine is partially disassembled. The copper block at the center of the dee is normally about 1 cm from the ion source and serves to intensify the electric field there. The protons begin their first orbit through the small hole in the block.

The role of the magnetic field is to cause the ions to cross the gap many times, so that the final energy gain is many times what could be achieved simply by letting the ions move through a potential difference of V_0. In a typical cyclotron, the ions make about 100 revolutions before reaching full energy and emerging from the machine.

Example   The University of Colorado cyclotron accelerates protons to a final kinetic energy of 28 MeV. At this energy the protons have a velocity of 7.32×10^7 m/s, which is about one-quarter the speed of light. The radius of the outermost orbit in the machine is 60 cm. The potential difference between dees is 70 kV. (a) What is the magnitude of **B**, the magnetic field? (b) Find the cyclotron frequency. (c) Find the minimum number of revolutions made before a given proton reaches full energy.

Solution   (a) The magnitude of **B** must satisfy (34-14):

$$B = \frac{m}{q} \frac{v}{R}$$
$$= \frac{1.67 \times 10^{-27} \text{ kg}}{1.60 \times 10^{-19} \text{ C}} \cdot \frac{7.32 \times 10^7 \text{ m/s}}{0.60 \text{ m}}$$
$$= 1.27 \text{ T}$$

(b) From (34-15) we have

$$f = \frac{1}{2\pi} \frac{q}{m} B$$

$$= \frac{1}{6.28} \cdot \frac{1.60 \times 10^{-19} \text{ C}}{1.67 \times 10^{-27} \text{ kg}} \, 1.27 \text{ T}$$

$$= 1.94 \times 10^7 \text{ Hz}$$

$$= 19.4 \text{ MHz}$$

(c) The maximum energy gained during each revolution is

$$\Delta E/\text{turn} = 2qV_0 = 140 \text{ keV/revolution}$$

The final energy is 28 MeV, so we find

$$N = \frac{28 \text{ MeV}}{140 \text{ keV/rev}} = 200 \text{ revolutions}$$

where N is the number of revolutions made if the protons always cross from one dee to the other at the peak of the oscillating potential difference.

Exercises

5. The nucleus of the second isotope of hydrogen is called a *deuteron*. It has a mass of 3.34×10^{-27} kg and a charge of e (1.6×10^{-19} C). A given cyclotron has a maximum orbit radius of 40 cm and a uniform magnetic field of 1.67 T. What is the cyclotron frequency of deuterons in this machine?

6. Find the maximum kinetic energy of deuterons in the above cyclotron.

34-6 The Hall Effect   NOT ON HOURLY

When a current-carrying wire is placed in a magnetic field, the individual charge carriers within the wire will, in general, experience a magnetic force. In Fig. 34-13 we see a positive charge moving

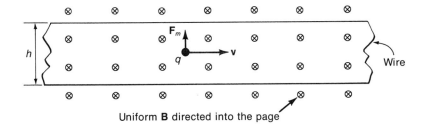

Uniform **B** directed into the page

Fig. 34-13. Each charge carrier in the wire experiences a magnetic force. For a positive charge carrier moving to the right, the force is directed upward. A negative charge moving to the *left* would also experience an upward-directed force.

to the right with a resultant upward-directed magnetic force. Note that a negative charge moving to the *left* would also be deflected upward and would still constitute a current in the same direction. Each of the many charge carriers in the wire experience this upward force.

This magnetic force causes the moving charge carriers to swerve toward the top of the wire, but the concentration of charge that develops there then repels other moving charges that try to veer toward the top. Equilibrium is quickly reached so that the electric field caused by the concentration of charge produces a force just equal to the magnetic force:

$$\mathbf{F}_e = \mathbf{F}_m$$

so that

$$q\mathbf{E} = q(\mathbf{v} \times \mathbf{B})$$

In the situation of Fig. 34-13, surplus positive charges will accumulate at the top, so that $\mathbf{E}$ will be directed downward:

$$qE = qvB \qquad \text{(since } \mathbf{v} \perp \mathbf{B}\text{)}$$

hence $E = vB$. This value of E is uniform from top to bottom of the conductor if all the moving charge carriers have the same average drift velocity.*

The potential difference between the top and bottom of a wire of thickness h is then

$$\Delta V = Eh = vBh \qquad (34\text{-}16)$$

The existence of this potential difference is called the *Hall effect* and the voltage given by (34-16) is the *Hall voltage.* We can relate the drift speed, v, to the current flowing in the conductor as was done in Section 33-9:

$$I = nqvA \qquad (34\text{-}17)$$

where n is the number of charge carriers per unit volume, q is the charge of each carrier, v is the drift velocity, and A is the cross-sectional area of the wire. Combining (34-16) and (34-17), we have

$$\Delta V = \frac{IBh}{nqA}$$

If the wire is rectangular in cross section with height h and width w, we find

$$\Delta V = \frac{IB}{nqw} \qquad (34\text{-}18)$$

---

*This is an example of a nonzero electric field within a conductor. It could not exist in a *static* situation.

Equation (34-18) gives the Hall voltage in terms of quantities that are known or can be measured – except for n, the number of charge carriers per unit volume. Furthermore, the *sign* of the Hall voltage reverses for negative charge carriers, as opposed to positive charge carriers. The Hall effect is thus quite useful in laboratory studies of the density and sign of charge carriers in materials. For typical copper conductors and field strengths, Hall voltages are small; in the μV range. For semiconductors with a small value of n, ΔV is more easily measurable – in the mV range. Such *Hall probes* are used to measure magnetic field strengths.

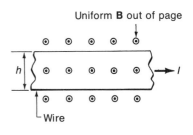

Fig. 34-14. The conventional current is toward the right. The magnetic field points out of the page.

Exercises

7. A magnetic field is directed *out* of the page in Fig. 34-14. The moving charge carriers are *electrons*. Indicate the polarity of the resultant Hall voltage.

8. A copper strip with a heighth h of 1 cm and a width of 1 mm carries a current of 200 A. A magnetic field of 1 T is applied as in Fig. 34-14. If the number density of conduction electrons is equal to that of the copper atoms (assuming one free electron/atom), find the Hall voltage generated. (See p. 703)

34-7 Magnetic Forces on Current-Carrying Conductors

In our discussion of the Hall effect, we noted that the moving charge carriers "piled up" on one side of the conductor – the same side regardless of whether the carriers were negative or positive. These carriers cannot escape the conductor entirely because of the very strong electric forces they would have to overcome. The result is that the magnetic forces acting on the charge carriers are transmitted to the entire wire.

The net force on an infinitesimal segment of wire of length dL carrying a current I in a magnetic field is the sum of the forces on the N charge carriers within dL:

$$\mathbf{F} = Nq\mathbf{v} \times \mathbf{B} \qquad (34\text{-}19)$$

where q is the charge of each carrier, $\mathbf{v}$ the drift velocity of the carriers, and $\mathbf{B}$ the magnetic field. If we define n to be the number of charge carriers per unit length, we have

$$N = n \, dL$$

and

$$Nq\mathbf{v} = I \, d\mathbf{L} \qquad (34\text{-}20)$$

where $d\mathbf{L}$ is directed along the wire in the direction of the current and hence is parallel to **v**, the drift velocity of positive charge carriers. Combining (34-19) and (34-20), we see that

$$d\mathbf{F} = I\,d\mathbf{L} \times \mathbf{B} \qquad (34\text{-}21)$$

where $d\mathbf{F}$ is the force acting on the segment $d\mathbf{L}$. We sum (or integrate over) the force all along a conductor to obtain the net force. For a straight conductor of length L in a field **B**, we find

$$\mathbf{F}_{\text{net}} = I\,\mathbf{L} \times \mathbf{B} \qquad (34\text{-}22)$$

where **L** is a vector parallel to the wire pointing in the direction of the conventional current. For negative charge carriers, (34-22) still applies since the conventional direction of current flow is then opposite to **v** and the sign of q is negative in (34-20).

Example   In a lecture demonstration apparatus, a short length of copper wire dips into two mercury pools to form a complete circuit, as in Fig. 34-15. The wire is placed between the poles of a magnet so that an upward directed force is produced when a current passes through the wire. The wire segment is projected several feet vertically by the magnetic force. For a field strength of 0.3 T acting over a 2-cm length of the wire, find **F** for the wire perpendicular to **B** and a current of 100 A.

Solution   Since **L** is perpendicular to **B**, (34-22) becomes

$$F = ILB = 100 \text{ A} \cdot (0.02 \text{ m}) (0.3 \text{ T}) = 0.6 \text{ N}$$

For a 6-g copper wire, neglecting gravity, the initial acceleration is

$$a = \frac{F}{m} = \frac{0.6 \text{ N}}{6 \times 10^{-3} \text{ kg}} = 100 \text{ m/s}^2$$

This acceleration ceases as soon as the ends of the wire leave the mercury pools.

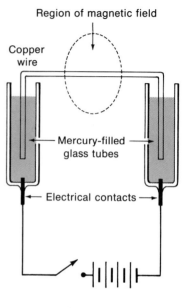

Fig. 34-15.   A segment of copper wire is in a magnetic field directed into the page. When the switch is closed the wire is driven vertically upward.

Exercise

9.   The earth's magnetic field is quite weak and directed roughly from south to north. At San Francisco, for example, the horizontal component of the earth's field is 25 μT. Suppose a wire of 2-m length is suspended vertically over a pool of mercury and a current is passed upward through the wire. In which direction will the wire move, east or west? Find the magnitude of the force exerted on this length of wire by the horizontal component of the earth's field when $I = 10^3$ A.

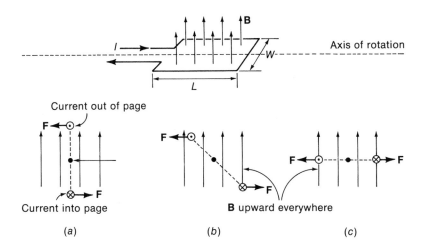

If we now consider a current-carrying *loop* in a uniform magnetic field, as in Fig. 34-16, we see that a torque acts. If the loop is free to pivot about an axis as shown, this torque will bring the *normal* to the plane of the loop into alignment with the field, **B**. If rotational inertia causes the loop to overshoot equilibrium position in Fig. 34-16*c*, the torque will then act to decrease the loop's angular velocity and to return it toward position *c*, where it is in rotational equilibrium. The reader should apply (34-22) to the loop in Fig. 34-16 to see that the directions for **F** are correct. You should also see that the right-hand rule does give Fig. 34-16*c* as the equilibrium position.

The loop shown in Fig. 34-16 has a width W and length L. Let us calculate the torque acting upon it. First, the side L of the loop experiences a force:

$$\mathbf{F} = I\,\mathbf{L} \times \mathbf{B}$$

which becomes $F = ILB$, since **L** and **B** are always at right angles as long as **B** is perpendicular to the axis of rotation, as shown. The segments along the width W of the loop produce forces parallel to the axis of rotation and, hence, no torque. The torque due to each side L is simply

$$\boldsymbol{\Gamma} = \mathbf{r} \times \mathbf{F}$$

so that

$$\Gamma = \frac{W}{2} ILB \sin\theta$$

where θ is the angle between **A**, the vector normal to the loop, and **B**, the magnetic field.

There are two forces producing a torque, one on each side of length L, so we find the net torque is

$$\Gamma = WILB \sin \theta$$

Noting that WL is the area of the loop, we write this result in vector form

$$\Gamma = I\mathbf{A} \times \mathbf{B} \tag{34-23}$$

where the magnitude of the vector $\mathbf{A}$ is equal to the area of the loop and its direction is given by the right-hand rule of Fig. 34-17. This result can be written a little more compactly if we define a *magnetic dipole moment*, $\boldsymbol{\mu}$:

$$\boldsymbol{\mu} = I\mathbf{A} \tag{34-24}$$

Then $$\Gamma = \boldsymbol{\mu} \times \mathbf{B} \tag{34-25}$$

A bar of magnetized iron contains many current loops with their planes roughly aligned. These loops of current are due to the spinning motion of atomic electrons. The result is that the iron bar has a net magnetic moment and experiences a torque when placed in a magnetic field. When such a magnet is freely suspended in the earth's magnetic field, it pivots until its own magnetic moment is aligned with the direction of the earth's field.

Exercise

10. A wire loop in the form of a 10-cm by 20-cm rectangle has a magnetic moment of $2\text{A} \cdot \text{m}^2$. (*a*) Find the current in the loop. (*b*) Find the maximum torque that can be exerted on the loop by an 0.5-T field.

34-8 The Galvanometer   NOT ON HOURLY

The torque exerted on a current-carrying loop in a magnetic field is used in an electrical instrument that detects small currents. This instrument is called a *galvanometer* and is the main element within most common electrical meters that use a pointer movement. As we will see, these meters can be connected to act as either voltmeters or ammeters.

In Fig. 34-18 we see the components of a galvanometer. The current to be measured is passed through the coil, which is placed between the poles of a permanent magnet. These poles are shaped so that the magnetic field is perpendicular to the sides of the coil regardless of its orientation (within the range of the instrument).

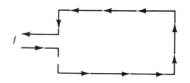

Fig. 34-17. If the fingers of the right hand are wrapped around the circumference of the loop in the direction of the current, the thumb will point up out of the page. This is the direction of **A**, the normal to the plane of the loop.

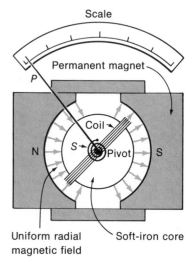

Fig. 34-18. The torque exerted on the current-carrying coil in a galvanometer advances the pointer (*P*) against the restoring torque of the spring (*S*).

A coil composed of many turns of copper wire is used rather than a single loop. The torque for n turns is thus n times larger than that given by (34-23). The spring produces a restoring torque that is proportional to the deflection angle of the pointer. Since by (34-23) the torque produced by the coil is proportional to the current in the coil, the equilibrium angle of the pointer is proportional to the current. The numerical scale shown in Fig. 34-18 is used to read the instrument.

Reasonably sturdy bearings on the coil pivots allow the instrument to be portable and rugged. The friction in the bearings limits the ultimate sensitivity of such instruments to about $\frac{1}{10}$ μA. This limit can be reduced by eliminating the bearings and hanging the coil from a very thin fiber, at the expense of portability. A more typical instrument may give a full-scale deflection for a current of 1 ma and have a net coil resistance of about 10Ω. From Ohm's law we see that such an instrument reads full-scale when the potential difference between its terminals is

$$V(\text{full-scale}) = I(\text{full-scale}) \cdot R_{\text{coil}} = 10^{-3} \text{ A} \cdot 10\Omega = 10^{-2} \text{ V}$$

Now let us convert this galvanometer into a voltmeter with a full-scale deflection of 10 V. This is accomplished by adding a series resistor R_s, as indicated in Fig. 34-19a. We wish a full-scale deflection for a potential difference of 10 V between points 1 and 2. Since full-scale deflection occurs for a current of 1 mA, we require a total resistance of

$$R_{\text{total}} = \frac{10 \text{ V}}{10^{-3} \text{ A}} = 10^4 \Omega$$

Since the galvanometer itself has a resistance of 10Ω, the resistance R_s must be $9,990\Omega$.

In Fig. 34-19b we see the conversion of a galvanometer into an ammeter using a parallel "shunt" resistor, R_p. If we wish a full-scale deflection for 1 A we must have

$$I_p + I_G = 1.0 \text{ A}$$

so that $I_p = 1 \text{ A} - 1 \text{ mA} = 0.999 \text{ A}$ when the galvanometer current is 1 mA. At this current we saw previously that the potential difference across the galvanometer is 10^{-2} V. This same potential difference exists across R_p:

$$R_p = \frac{10^{-2} \text{ V}}{0.999 \text{ A}} = 1.001 \times 10^{-2}\Omega$$

The *ideal* voltmeter would draw no current from the circuit to which it is connected. This ideal is approached as $R_s \rightarrow \infty$, but

galvanometer → voltmeter
add series resistor

galvanometer → ammeter
add parallel resistor

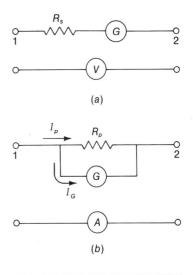

Fig. 34-19. (a) A series resistance R_s converts a galvanometer into a voltmeter; the symbol for a voltmeter is indicated. (b) A parallel resistance R_p converts a galvanometer into an ammeter; the symbol for the ammeter is shown.

large values of R_s imply an extremely sensitive galvanometer. Thus the most ideal voltmeter (one that draws a very small current) is necessarily a very delicate instrument since it uses an extremely sensitive galvanometer.

The *ideal* ammeter has zero resistance, so that it does not disturb the current to be measured when it is inserted into a circuit. However as $R_p \rightarrow 0$, the voltage available to operate the galvanometer also approaches zero. Again, an extremely delicate galvanometer is required to approach the ideal most closely.

Exercises

11. A galvanometer reads full-scale for a current of 10 mA. Its coil resistance is 1Ω. Find the value of R_s to convert it into a 100-V full-scale voltmeter.

12. The galvanometer of Exercise 11 is to be converted to an ammeter with a full-scale reading of 10 A. Find the value of R_p.

34-9 Summary

The force between charged particles is modified when they are in motion. We call the velocity-dependent portion of the net force the *magnetic* force. We adopt a field point of view in which a *magnetic field* **B** is produced by one moving charge and a second moving charge experiences a force due to the presence of the field.

In SI units the magnetic field units are *tesla* and a moving charged particle experiences a magnetic force given by

$$\mathbf{F}_m = q\mathbf{v} \times \mathbf{B}$$

$$T = \frac{N\text{-}s}{C\text{-}m}$$

where F is in N when v is in m/s and B in tesla. The direction of the magnetic force is perpendicular to both **v** and **B**. Charged particles can be held in circular orbits by uniform magnetic fields. The frequency of rotation in such an orbit is called the cyclotron frequency and is given by

$$f = \frac{1}{2\pi}\frac{q}{m}B$$

The moving charge carriers in a current-carrying conductor experience a magnetic force when the conductor is placed in a magnetic field. These forces cause the charge carriers to swerve to one side, producing a Hall voltage as well as causing a net mechanical force that is exerted on the conductor. The net force on a conductor of length L carrying a current I in a field **B** is given by

$$\mathbf{F} = I\mathbf{L} \times \mathbf{B}$$

where the sense of **L** is given by the direction of the conventional current.

The torque produced by a current-carrying loop in a magnetic field is given by

$$\boldsymbol{\Gamma} = I\mathbf{A} \times \mathbf{B}$$

where **A** is a vector equal in magnitude to the area of the loop and directed perpendicular to the plane containing the loop. Its sense is given by a right-hand rule.

The torque exerted on a current-carrying coil within a magnetic field is used to actuate a pointer in the galvanometer. Its reading is proportional to the current passing through the coil. Placing the galvanometer in series with a resistance allows it to be used as a voltmeter. A parallel shunt resistor converts the galvanometer into an ammeter.

Problems

34-1.   A proton ($q = 1.6 \times 10^{-19}$ C, $m = 1.67 \times 10^{-27}$ kg) is fired horizontally with $v = 10^7$ m/s into a magnetic field **B**. It is observed to deflect to the right, and the magnitude of its acceleration is 10^{14} m/s². Find the direction and magnitude of **B**.

34-2.   The electrons in a cathode ray tube have a kinetic energy of 15 keV. Find the magnetic force they experience when they move at right angles to a magnetic field of 500 gauss. What is the magnitude of the acceleration so produced?

34-3.   A charged particle is given an initial velocity $\mathbf{v}_0$ in a uniform magnetic field **B**, as in Fig. 34-20a. Show that the component of $\mathbf{v}_0$ along **B** is unchanged, while the component of $\mathbf{v}_0$ perpendicular to **B** swings around at the cyclotron frequency. The result is the helical path shown in Fig. 34-20b.

34-4.   An electron is fired at velocity **v** at right angles to the electric field lines between parallel plates. The plates are separated by a distance d and have a potential difference V. A uniform magnetic field **B** is directed at right angles to both the electron's velocity and the electric field. It is observed that the electron is undeflected, that is, the electric

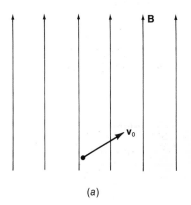

(a)

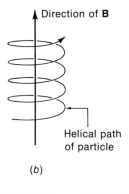

Helical path
of particle

(b)

Fig. 34-20. A charged particle with an initial velocity not perpendicular to **B** follows a helical path.

and magnetic forces precisely cancel. Find **v** in terms of the given quantities.

34-5.   A proton gains 100 keV of kinetic energy per revolution in a cyclotron. The magnetic field strength is 1.2 T. (*a*) Find the radius of the 100th revolution orbit. (*b*) Find the proton's cyclotron frequency in this machine. (*c*) Find the cyclotron frequency of a $^3\text{He}^+$ ion in this machine (same charge as the proton, three times as much mass).

34-6.   Deuterons in a cyclotron describe a circle of radius 32.0 cm just before emerging from the machine. The frequency of the applied alternating voltage is 10 MHz. Find (*a*) the magnitude of the magnetic field and (*b*) the energy of the deuterons when they emerge.

34-7.   We wish to determine the intensity of the earth's magnetic field by measuring the force exerted on a current-carrying wire 1 m in length. To see if this is feasible we ask, "How much current must we pass through the wire if we are to obtain a force of 0.01 N with the wire perpendicular to a 50-μT field?"

34-8.   A wire is to be projected vertically in the apparatus of Fig. 34-15. The wire can rise a distance x before electrical contact with the mercury is lost. A constant length l of the horizontal part of the wire is acted upon by a uniform magnetic field **B** from the time the wire begins rising until the electrical circuit is broken. The mass of the wire is m and a current I flows in the wire. Find the added height to which the wire coasts, neglecting friction and air resistance.

34-9.* Assume the conditions in Problem 34-8 except that the uniform **B** field covers a circle of radius r with $r < x$ and arranged so that the wire is initially located along a diameter of this circle.

34-10. The magnetic field of a large electromagnet is sketched in Fig. 34-21. A flexible loop of copper wire carries a current

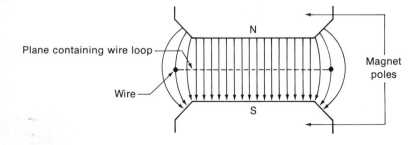

Fig. 34-21.

of 10 A and when it is placed in the magnetic field, it expands into a taut circular loop as indicated and is held at the mid-plane by a mechanical support at one point. (*a*) In which direction does current flow around the loop? (*b*) Is the wire in stable equilibrium, i.e., if the support is removed, will the wire float there? (*c*) The magnetic field strength at the wire is 1 T. Find the tension in the wire. The radius of the wire loop is 30 cm.

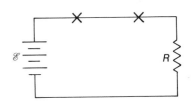

Fig. 34-22.

34-11. A loop of length 10 cm and width 15 cm is oriented as in Fig. 34-16*a* in a 0.5-T field. Find the torque Γ for a current of 10 A.

34-12. A loop with an area A and moment of inertia $\mathscr{I}$ carries a current I and is oriented as in Fig. 34-16*a* in a magnetic field **B**. The loop is released from rest and allowed to rotate freely. Find the angular velocity of the loop when it reaches the position shown in Fig. 34-16*c*.

34-13. If friction is negligible, discuss the ensuing motion of the system described in Problem 34-12.

34-14. A galvanometer coil is 2 cm × 2 cm and contains 100 turns. It is in a 0.1-T field that is directed radially everywhere, as in Fig. 34-18. The spring provides a restoring torque $\Gamma_s = k\theta$. If the deflection is to be 30° for a current of 1 mA, what is the required value of k in N · m/degree?

34-15. The coil of the galvanometer in Problem 34-14 is made of 30-gauge copper wire, which has a diameter of 0.0255 cm. Find the voltage across the galvanometer at a full-scale current of 1 mA.

34-16. Show that (34-23) applies to a plane loop of any shape.

34-17. A galvanometer with a full-scale deflection of 1 mA and a coil resistance of 10Ω is converted into an ammeter with a 5-A full-scale reading. It is inserted into the circuit of Fig. 34-22 between the points marked x. If $\mathscr{E} = 6$ V and $R = 2Ω$, find the fractional change in current, $\Delta I/I$, caused by the insertion of the ammeter.

34-18. A galvanometer with a full-scale deflection of 1 mA and a 10Ω coil resistance is converted to a voltmeter with a full-scale reading of 100 V. This voltmeter is then placed across a resistor, as in Fig. 34-23 to measure the voltage drop V_R across the resistor. If $\mathscr{E} = 100$ V and $R_1 = R_2 = 10,000Ω$, find the fractional change in V_R ($\Delta V_R/V_R$) caused by the connection of the voltmeter.

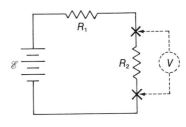

Fig. 34-23.

34-19. Bearing friction in a given galvanometer causes a frictional torque measured to be approximately 10^{-9} N · m. We choose a spring with a torque constant of 10^{-8} N · m/degree so that frictional torque will produce an error of no more than 0.1 degree. The coil frame is 2 cm $\times$ 4 cm and the field strength is 0.02 T. How many turns of wire are required to obtain a sensitivity of 30 μA for a full-scale deflection? Full-scale corresponds to 30°.

Chapter 35

Sources of
Magnetic Fields

SECTIONS 1-5

35-1 Introduction   In this chapter we will describe the production of magnetic fields by electrical currents. We will then be able to predict the magnetic forces between two current-carrying conductors. The magnetic field will be found to obey an integral law called *Ampere's law*. This law is useful in the theory of magnetism in the same way that Gauss's law is useful in electrostatics.

We will finally be able to return to the most ancient magnetic device, the bar magnet, and be able to describe its behavior from a fundamental point of view involving atomic currents. The properties of ferromagnetic materials, such as iron, will be discussed.

35-2 Oersted and Faraday

Professor Hans Christian Oersted (1777–1851) of the University of Copenhagen experimented with electric currents placed close to compass needles. He initially placed the compass needle over the wire at right angles to the wire and observed no effect. In the winter of 1819–1820, he demonstrated this in one of his lectures. Near the end of the lecture he moved the wire so that it became parallel to the needle. He was amazed to see a sudden and pronounced deflection of the compass needle, for this was the first time that anyone had observed any connection between electricity and magnetism. It was also probably one of the few times that a momentous discovery was made in the presence of a student audience.

magnet ‖ to current

The cause of the deflection of the compass needle is illustrated in the iron filing pattern of Fig. 35-1. To the surprise of the scien-

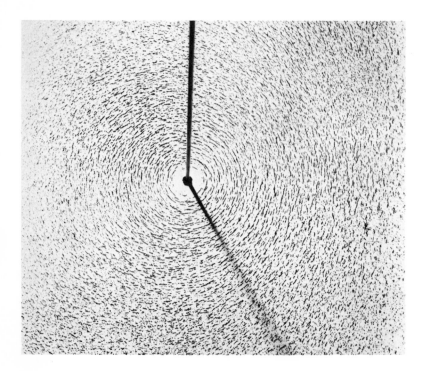

Fig. 35-1. The pattern of iron filings around a wire carrying a large current reveals the circular form of the magnetic field produced by a current. (From *PSSC Physics*, D. C. Heath and Company, Lexington, Massachusetts, © 1965)

tists of Oersted's day, there was no evidence of poles in the magnetic field produced by the current. The magnetic lines of force were apparently circles with no beginning and no end. Faraday was one of the few people at that time who was willing to take this fact at face value. He even proposed that the lines of force of a bar magnet must also be endless. He was able to show that they continued through the body of the magnet after they entered at the "south" end to reemerge at the north end of the magnet.*

The endless nature of magnetic field lines is quite different from that of electrostatic lines of force. Because the lines of **B** have no end, "Gauss's law" for magnetism is quite simple:

$$\oint \mathbf{B} \cdot d\mathbf{A} = 0 \quad \text{always} \tag{35-1}$$

The net magnetic flux passing through any closed surface must be zero. An alternative and equivalent statement is that there are no isolated magnetic charges that can provide a source or sink for magnetic field lines.

---

*He did this by use of magnetic induction, a topic we will study in the next chapter. (See Problem 36-9.)

35-3   The Magnetic Field Caused by a Current

The magnetic field produced by a current in a long, straight wire is found to obey two simple laws. The first of these gives the direction of the field by a right-hand rule illustrated in Fig. 35-2:

> When the wire is grasped in the right hand with the thumb pointing along the current, the fingers curl around the wire in the direction of the field.

The second observation is that the field intensity falls inversely as the first power of the perpendicular distance from the wire. More precisely, it is found that near a long wire carrying a current I:

$$B = \frac{\mu_0 I}{2\pi r}$$

where r is the distance, in a direction perpendicular to the wire, between the wire and the field point where **B** is measured.

The French physicist Jean Biot studied the field near conductors of various shapes and finally deduced the law for the production of a magnetic field by a current element of infinitesimal length:

$$d\mathbf{B} = k' \frac{I\, d\mathbf{l} \times \mathbf{r}}{r^3} \qquad (35\text{-}2)$$

where $k' = \mu_0/4\pi$, as indicated in Chapter 34, $d\mathbf{l}$ points in the direction of I, and the vector **r** from the current element to a field point

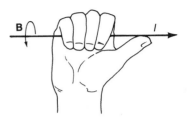

Fig. 35-2. The magnetic lines of force curl around a current-carrying wire in the same direction as the fingers of the right-hand when the thumb points in the direction of the conventional current.

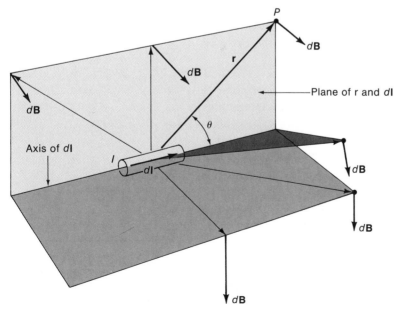

Fig. 35-3. The field $d\mathbf{B}$ produced by a current-carrying element $d\mathbf{l}$ at a distance **r** from that element is always perpendicular to the plane containing **r** and $d\mathbf{l}$.

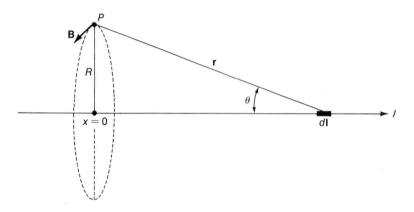

Fig. 35-4. The field **B** at the point *P* is due to the sum of contributions *d***B** generated by every element *d***l** of the wire. One element *d***l** is illustrated here. Each element *d***B**, and hence **B**, is tangent to a circle of radius *R* shown centered on the wire at *x* = 0.

is shown in Fig. 35-3. The variation of *d***B** with the magnitude of **r** is seen to go as $1/r^2$ in (35-2). Equation (35-2) cannot be directly verified since it is not possible to produce an isolated element of current. Instead, it is inferred from experiments with long wires of various shapes with the additional assumption that the total field **B** at any point is the sum of all the infinitesimal contributions *d***B**. More briefly:

$$\mathbf{B} = k' \int \frac{I\, d\mathbf{l} \times \mathbf{r}}{r^3} \tag{35-3}$$

with the integral containing all the elements *d***l** of the entire circuit. over all the contributions *d***B**, from one end of the wire to the other.

We will not attempt to evaluate (35-3) for complicated geometries but will consider two simple cases. The results we will obtain agree with experimental observations and are part of the body of facts that led to the adoption of (35-2) and (35-3).

The field near a long, straight wire   The field at a typical point *P* is illustrated in Fig. 35-4. Its magnitude is obtained by integrating over all the contributions *d***B**, from one end of the wire to the other. The integration is made simple by the fact that all the contributions *d***B** are parallel to one another and simply add as scalars:

$$\mathbf{B} = \int d\mathbf{B} = \frac{\mu_0 I}{4\pi} \int \frac{d\mathbf{l} \times \mathbf{r}}{r^3}$$

This integral should include the entire wire, hence

$$B = \frac{\mu_0 I}{4\pi} \int_{x=-\infty}^{x=+\infty} \frac{dl \sin \theta}{r^2} = \frac{\mu_0 I}{4\pi} \int_{-\infty}^{+\infty} \frac{dx\,(R/r)}{r^2}$$

This is precisely the integral we evaluated previously in finding the electric field near a long line of charge (Chapter 29, Fig. 29-3 and

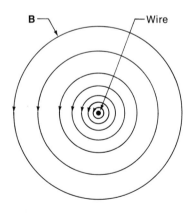

Fig. 35-5. Current flows out of the page in a long, straight wire. Magnetic lines of force encircle the wire as shown.

following discussion). The result found there gives us

$$B = \frac{\mu_0 I}{2\pi R} \qquad (35\text{-}4)$$

for the magnitude of the field near a long, straight wire. The direction of the calculated field is indicated in Fig. 35-5 and agrees with experimental observation.

The field due to a loop of current   Now consider the situation depicted in Fig. 35-6. The current flows around a circular path; what is the direction of the magnetic field? Applying the right-hand rule to any segment of the circle, we find that the direction of **B** is *out* of the interior of the loop (toward the reader) and into the page all around the loop. The resultant field of a current loop is illustrated in Fig. 35-7.

We could find the magnitude and direction of **B** at an arbitrary point, as indicated in Fig. 35-8, by integrating around the loop using (35-2) to obtain $d\mathbf{B}$ for each element $d\mathbf{l}$ of the loop. This is an extremely difficult calculation except for those field points that lie on the symmetry axis of the loop.

For a point on the axis of the loop, the geometry of Fig. 35-9 is appropriate. As indicated in Fig. 35-10, only the component of $d\mathbf{B}$ that projects onto the symmetry axis will survive in the resultant vector **B**. This is because there are as many elements $d\mathbf{B}$ above the horizontal as below, and as many out of the page as into it. Thus we need only to consider dB_h, the horizontal component of $d\mathbf{B}$:

$$dB_h = dB \cos \psi = dB \sin \theta = dB \frac{R}{r}$$

where θ and ψ are indicated in Fig. 35-9. The magnitude of $d\mathbf{B}$ is given by (35-2):

$$dB = \frac{\mu_0 I \, dl}{4\pi r^2}$$

Noting that $dl = R \, d\phi$ in Fig. 35-9, we have

$$B = \int_{\substack{\text{entire} \\ \text{loop}}} dB_h = \frac{\mu_0 I}{4\pi} \int_0^{2\pi} \frac{R \, d\phi \, (R/r)}{r^2}$$

The quantities R and r stay fixed during the integration over all the elements $d\phi$ of the loop (this is why this calculation is simple *on* the symmetry axis), so we factor them out to get

$$B = \frac{\mu_0 I R^2}{4\pi r^3} \int_0^{2\pi} d\phi = \tfrac{1}{2}\mu_0 I \frac{R^2}{r^3} \qquad (35\text{-}5)$$

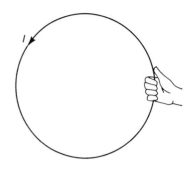

Fig. 35-6.   Current flows around a loop in the counterclockwise sense. If the loop is grasped with the right hand (thumb pointing in direction of current) at any place around the loop, the right-hand rule indicates **B** out of the page *within* the loop.

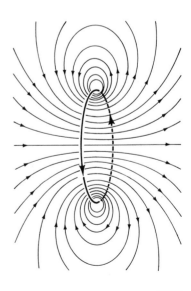

Fig. 35-7.   The magnetic field due to a single loop of current.

For a given loop, R is a constant. Thus the field strength B falls as $1/r^3$ along the axis. At large distances from the loop (when $r \gg R$), the distance r approaches the distance d in magnitude, where d is the distance from the *center* of the loop to the point P:

$$B \approx (\tfrac{1}{2}\mu_0 I R^2)\frac{1}{d^3} \qquad \text{for } d \gg R \qquad (35\text{-}6)$$

Though we will not show this explicitly, the strength of the field **B** falls off as $1/d^3$ in other directions as well, once d is much larger than R.

This type of field, which diminishes as $1/d^3$ at large distances, is called a *dipole field*. It closely resembles the *electric field* produced by two closely spaced equal and opposite charges, hence the name *dipole*. The fields produced by a current loop and by a pair of charges are illustrated in Fig. 35-11. This type of field occurs quite often in nature.

The magnetic field of the earth closely approximates a dipole field. It is thought that the central liquid core of the earth is a conductor and that the earth's field is generated by currents circulating in this core. The mechanism that drives these currents is poorly understood. The axis of the earth's field is only in rough alignment with the rotation axis of the earth. Furthermore, these currents occasionally reverse. Fossil magnetism of sedimentary and igneous rocks show that the north and south magnetic poles of the earth reverse at irregular intervals of approximately 2,000,000 years.* At present the direction of this field is such that lines of **B** emerge from the southern portion of the earth and enter the northern portion.

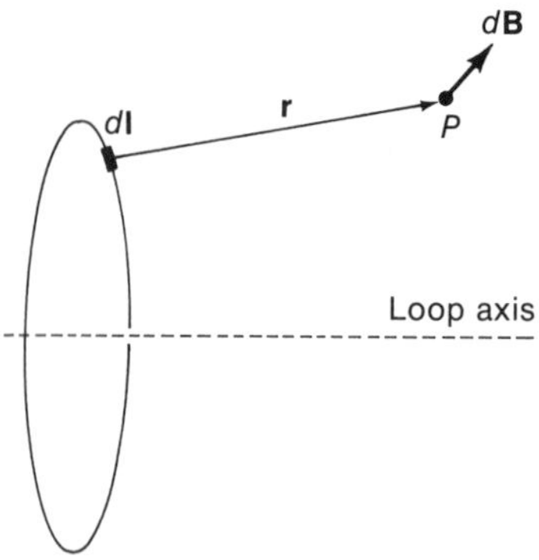

*See, for example, "Reversals of the Earth's Magnetic Field" by Allen Cox, G. Brent Dalrymple, and Richard R. Doell, *Scientific American,* February 1967.

Fig. 35-8. Summing all contributions $d\mathbf{B}$ to the total field **B** at point P is difficult due to the variation of **r** and of the direction of $d\mathbf{B}$ for each $d\mathbf{l}$ around the loop.

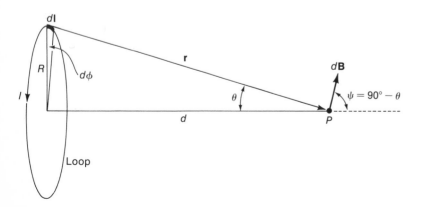

Fig. 35-9. The segment $d\mathbf{l}$ points out of the page. The vector **r** is directed toward the point P. The vector $d\mathbf{B}$ is in the direction of $d\mathbf{l} \times \mathbf{r}$ and hence lies in the plane of the page, as shown.

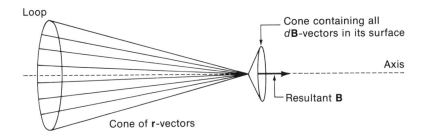

Fig. 35-10. The resultant field **B** is the vector sum of all the elements $d\mathbf{B}$. Since these are equally distributed on a cone, as shown, the resultant is horizontal and lies in the plane of the page.

Charged particles emitted by the sun (mainly protons and electrons) are trapped in cyclotronlike orbits in the earth's magnetic field, as indicated in Fig. 35-12. These energetic charged particles are concentrated somewhat toward the earth's equatorial plane at distances far outside the earth's atmosphere. This region is called the *Van Allen layer*, after James A. Van Allen, a physicist at the State University of Iowa who was responsible for some of the early artificial satellite measurements of this phenomenon.*

Fig. 35-11. (a) The magnetic field near a loop of current. (b) The electric field near a pair of equal and opposite charges. (c) Both fields are *dipole* fields at distances that are large compared to the size of the loop in (a) or the separation of the charges in (b).

*See "Radiation Belts Around the Earth" by James A. Van Allen, *Scientific American,* March 1959; or "Radiation Belts" by Brian J. O'Brien, *Scientific American,* May 1963.

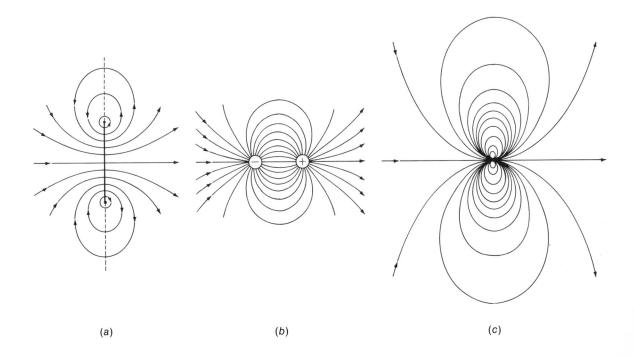

(a)   (b)   (c)

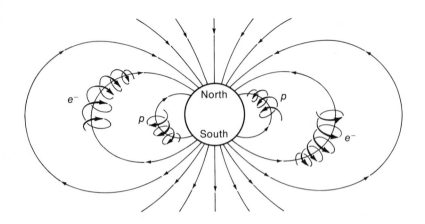

Fig. 35-12. Electrons (e^-) and protons (p) are trapped in helical trajectories in the earth's magnetic field. The size of the trajectories is not drawn to scale.

The magnetic dipole field caused by an ordinary iron bar magnet shown in Fig. 35-13 is due to the continual motion of electrons in the atoms of iron. This same type of circulating motion is common to many types of atoms, but iron is unusual in that its atoms align their individual magnetic fields so that an overall net field is produced. This phenomenon is discussed further in Section 35-6.

Exercises

1. A current of 10 A flows through a long, straight wire. Find the magnetic field strength at a distance of 30 cm from the wire.

2. The magnetic field of the earth has a strength of $\frac{1}{2}$G ($G = 10^{-4}$ T) in the vicinity of its north magnetic pole. In order to obtain an *estimate* of the size of the currents flowing within the earth, assume that the current is simply localized, as shown in Fig. 35-14. Take $R_e = 6 \times 10^6$ m and $R_c = 3 \times 10^6$ m.

35-4   Integral Properties of the Magnetic Field: Ampere's Law

All magnetic fields obey two integral laws. One of these laws concerns the flux of the magnetic field and the other concerns the circulation of the magnetic field. We have already noted that lines of **B** have neither sources nor sinks. At the end of Section 35-2 we said that this implied a simple form for the Gauss's law of magnetic fields:

$$\oint \mathbf{B} \cdot d\mathbf{A} = 0 \quad \text{always}$$

This result contrasts with the behavior of the electrostatic field. There, the net electric flux was related to the quantity of charge within the surface.

[handwritten margin notes:]
flux of magnetic field
circulation of magnetic field

FOR ELECTRIC
$\oint E \cdot dA = q/\varepsilon_0$

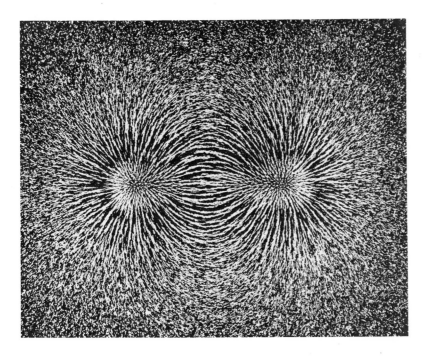

Fig. 35-13. The dipole field of a short bar magnet made visible by iron filings. Compare with Fig. 35-11. (From *PSSC Physics*, D. C. Heath and Company, Lexington, Massachusetts, © 1965)

The *circulation* of the magnetic field also behaves quite differently from that for electrostatic fields. It is obvious that the fields shown in Figs. 35-1 and 35-7 have a net circulation. For the particular case of the field of a long, straight wire, this circulation is easily calculated. We choose to evaluate the circulation along a closed path in the form of a circle of radius R centered about the wire:

$$\text{circulation} = \oint \mathbf{B} \cdot d\mathbf{s} = \int_0^{2\pi} B(R \cdot d\theta)$$

where $d\mathbf{s}$ was chosen as the name of an element of the closed path to avoid confusion with the radius. With B given by (35-4), we find

$$\text{circulation} = \frac{\mu_0 I}{2\pi R} R \int_0^{2\pi} d\theta = \mu_0 I \qquad (35\text{-}7)$$

We arrived at (35-7) for a very special case: the field of a long, straight wire and a circular closed path. This result, however, is completely general. If we had considered a noncircular path or any other current configuration, we would have arrived at the same result. This general result is called *Ampere's law:*

$$\oint \mathbf{B} \cdot d\mathbf{s} = \mu_0 I \qquad (35\text{-}8)$$

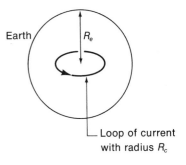

Fig. 35-14.

where ds is an element of the closed path of integration.

If we consider a path of integration that encloses more than one element carrying a current, Ampere's law is still applicable if I is taken to be the *net* current passing through the closed integration path. This is illustrated in Fig. 35-15 for several cases.

Just as it was possible to use Gauss's law to determine some electric fields (in situations with a great deal of symmetry), so is it possible to use Ampere's law to determine some magnetic fields. As an illustration, consider the field due to a long solenoid. The solenoid is a long helix of wire that may be thought of as many circular loops of current side by side. The field produced by many loops has the general behavior indicated in Fig. 35-16. To find the field near the center of the solenoid, we apply Ampere's law to the path shown in Fig. 35-17. This path, in the form of a rectangle, is located in the central portion of the solenoid, so that the field is parallel to the path within the solenoid, and at right angles to the path where the path is at right angles to the solenoid's axis. For the path shown in Fig. 35-17, the circulation integral in Ampere's law is conveniently evaluated in four parts:

$$\oint \mathbf{B} \cdot d\mathbf{s} = \int_a^b \mathbf{B} \cdot d\mathbf{s} + \int_b^c \mathbf{B} \cdot d\mathbf{s} + \int_c^d \mathbf{B} \cdot d\mathbf{s} + \int_d^a \mathbf{B} \cdot d\mathbf{s}$$

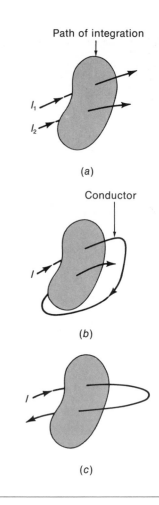

Path of integration

(a)

Conductor

(b)

(c)

Fig. 35-15. In each case the path followed in evaluating the circulation integral in Ampere's law is around the edge of the shaded area. That integral equals: (a) $\mu_0(I_1 + I_2)$; (b) $\mu_0(2I)$; (c) 0.

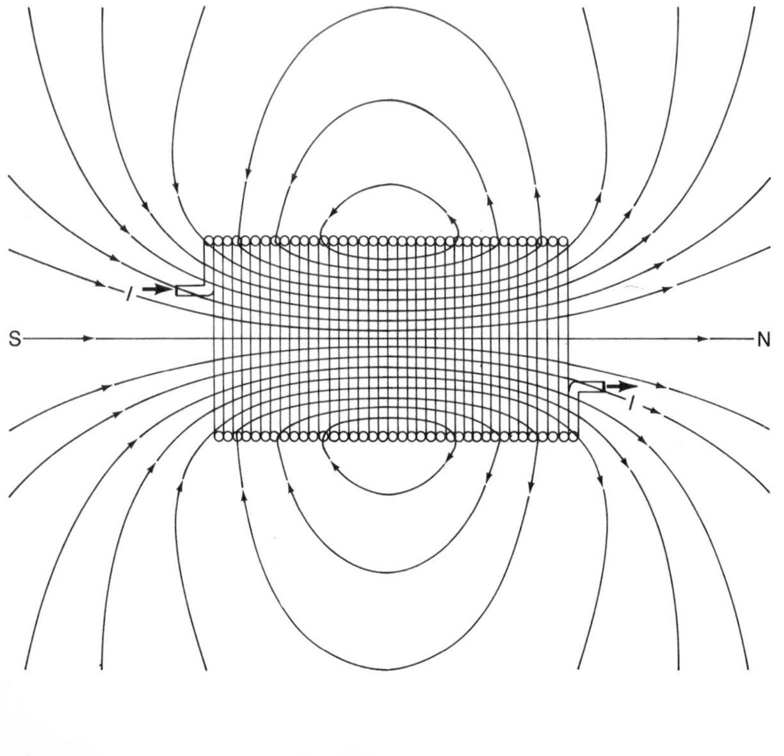

Fig. 35-16. The magnetic field due to a solenoid.

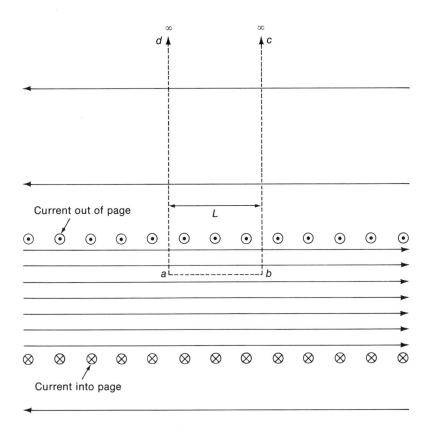

Current out of page

Current into page

Consider the first of these integrals. If the solenoid were infinitely long, B would necessarily have the same value at any location within the solenoid. Similarly, if the solenoid is *very* long, there is no reason for B to vary along its length except near the ends of the solenoid. Thus the first integral above contains a *constant* factor, B, which can be factored out. Further, **B** is parallel to the path so that the $\cos \theta$ term in the dot product equals unity:

$$\int_a^b \mathbf{B} \cdot d\mathbf{s} = B \int_a^b ds \cos \theta = B \int_a^b ds = BL$$

The two integrals along the long legs of the path (from b to c and from d to a) vanish because the field is everywhere perpendicular to those portions of the path. Finally, the integral from c to d must vanish since it is located at an infinite distance from the solenoid, so that the field strength there vanishes. We thus have:

$$\oint \mathbf{B} \cdot d\mathbf{s} = BL = \mu_0 NI$$

where N is the number of turns of the solenoid that thread through our integration path. Rearranging terms:

$$B = \mu_0 \left(\frac{N}{L}\right) I \tag{35-9}$$

where N/L is recognized as the number of turns per unit length of the solenoid.

Exercises

3. How many turns per meter should be wound on a long solenoid so that a current of 10 A will produce a field strength of $4\pi \times 10^{-3}$ T near the center of the solenoid?

4. A *toroid* is a doughnut-shaped device, as shown in Fig. 35-18. The number of turns wound on the toroid is N. From the symmetry of the toroid it should be clear that B will have a value independent of location along the path. Apply Ampere's law to the closed path of length L indicated by the dotted line to show that (35-9) also describes the magnetic field within the toroid.

35-5   The Force between Parallel Currents

In Section 35-3 we saw that the magnetic field of a long, straight wire is circular, as shown in Fig. 35-19, and has a magnitude given by

$$B = \frac{\mu_0 I_1}{2\pi r} \tag{35-10}$$

where r is the distance from the wire and I_1 is the current in the wire. A second wire carrying a current experiences a force when it is placed in this magnetic field. If this second wire is parallel to the first, as in Fig. 35-19, the force acting on a wire of length L is found from (34-22), with the field magnitude given by (35-10):

$$\mathbf{F} = I_2 \mathbf{L} \times \mathbf{B} \tag{34-22}$$

Since $\mathbf{L}$ is perpendicular to $\mathbf{B}$, we write for the force per unit length:

$$\frac{F}{L} = I_2 B = \frac{\mu_0 I_1 I_2}{2\pi r} \tag{35-11}$$

If we consider I_1 flowing in the field due to I_2, we see that (35-11) also gives the force on I_1. When the currents are in the same direction, the force is attractive. For equal currents, (35-11) becomes

$$\frac{F}{L} = \frac{\mu_0 I^2}{2\pi r} \tag{35-12}$$

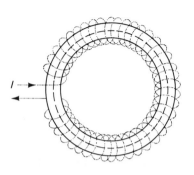

Fig. 35-18.

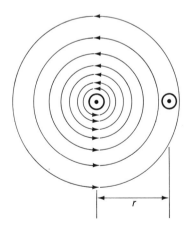

Fig. 35-19.   A current directed out of the page is encircled by lines of **B**, as shown. A second current, also directed out of the page, is attracted toward the first.

Fig. 35-20. The National Bureau of Standards' current balance. A current is passed through two coils in proximity to maximize the length of interacting currents and provide an easily measured force. The coil geometry also lends itself to an accurate calculation of the magnetic force in terms of known quantities leaving the ampere as the unknown to be determined. (Courtesy, U.S. National Bureau of Standards)

Equation (35-12) gives a direct connection between current and force. This fundamental relationship is used to define the *ampere*. In practice, an apparatus shown in Fig. 35-20 is used only as a primary standard. Calibrated emf's and resistors are then used as secondary standards. The SI definition of the ampere follows:

> The ampere is that constant current that, if maintained in two straight parallel conductors of infinite length, of negligible circular cross section, and placed 1 meter apart in vacuum, produces between these conductors a force equal to 2×10^{-7} newtons per meter of length.

For two wires carrying 50 A separated by 10 cm, we find a rather small force:

$$\frac{F}{L} = \frac{(4\pi \times 10^{-7} \text{ T} \cdot \text{m/A})(2.5 \times 10^3 \text{ A}^2)}{2\pi \cdot 10^{-1} \text{ m}} = 5 \times 10^{-3} \text{ N/m}$$

The small size of this force is due to the very low drift velocities of the electrons in the wire and the v^2/c^2 dependence of the magnetic force given by (34-2). The tremendous disparity between magnetic and electrostatic forces is numerically illustrated in Problem 35-10. The only reason we are aware of the magnetic forces between two wires is that the electrostatic forces are exactly cancelled because both conductors are neutral overall.

The drift velocity of electrons in the wires is less than 10^{-12} the velocity of light and the magnetic force is thus only about 10^{-24} of the electrostatic force, since the ratio of these forces is v^2/c^2. Thus, if the moving electrons in the wires were not neutralized by the positive charges in the stationary atoms, the electrostatic force of repulsion between the wires would be roughly 10^{24} times larger than the magnetic attraction: about 10^{21} N/m for our example above.

Exercises

5. Use the right-hand rule of (34-22) and the right-hand rule for the magnetic field about a current to show that two parallel wires carrying currents in opposite directions repel one another.

6. Two long, straight wires carrying currents pass close by one another at right angles. Show that the net magnetic force between them is zero.

35-6 Magnetic Materials   NOT ON HOURLY

In Section 32-5 we saw that some substances, called *dielectrics*, became electrically polarized when placed in an electric field. The result is that the strength of the electric field within such substances is reduced below the value that is present in their absence. This reduction in field strength is described by a *dielectric constant*, κ.

A similar phenomenon exists for magnetic fields. Many substances produce *slightly* lessened fields by their presence. These substances are said to be *diamagnetic*. The reduction of magnetic field strength in these substances is caused by changes in the motion of atomic electrons produced by the presence of a magnetic field.

Another class of substances, called *paramagnetic* materials, produce a slight *enhancement* of any externally applied magnetic field. These substances all have one thing in common: their atomic structure is such that there is a continual net circulation of charge in their electron clouds. Each such atom possesses a magnetic field, as in Fig. 35-21. Thermal agitation keeps the directions of these atomic fields randomized so that no net magnetic field is produced. However, when an external field is applied, the individual atomic fields tend to align with the external field, producing an enhancement of that field.

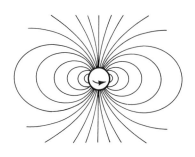

Both paramagnetic and diamagnetic substances can be characterized by their relative *permeability*, κ_m. If we construct a long solenoid and fill its interior with a magnetic substance, κ_m is the ratio of B_m, the magnetic field within the solenoid, to B_0, the magnetic field that would have been present without the substance:

$$B_m = \kappa_m B_0 = \kappa_m \mu_0 \frac{N}{L} I \qquad (35\text{-}12)$$

Fig. 35-21. The electron clouds of many types of atoms have a net rotation, so that they produce magnetic fields similar to that of a loop of current.

where μ_0 is called the *permeability of the vacuum* and $\kappa_m \mu_0$ is the permeability of the substance. The quantity $(N/L)I$ is often called the *magnetizing force* and is given in ampere-turns per meter in MKS units. The relative permeability, κ_m, is greater than unity for paramagnetic substances and less than unity for diamagnetic substances. The actual departure from unity, however, is quite small, as can be seen from Table 35-1. It is common, in fact, to express the difference between κ_m and unity as follows:

$$\chi_m = \kappa_m - 1$$

where χ_m is the magnetic *susceptibility*. Handbooks frequently tabulate the susceptibility rather than κ_m.

TABLE 35-1   Relative Permeability of Various Substances

Substance	κ_m	χ_m
Aluminum	1.0000214	2.14×10^{-5}
Bismuth	0.999833	-1.67×10^{-4}
Copper	0.9999906	-9.4×10^{-6}
Lead	0.9999831	-1.69×10^{-5}
Platinum	1.000293	2.93×10^{-4}

One class of paramagnetic substances behaves in a striking way. In these substances the interactions between individual atoms are strong enough to align their dipole moments, so that all of their individual magnetic fields are parallel in spite of thermal agitation.

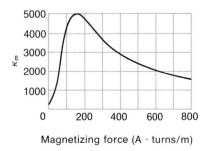

(a)

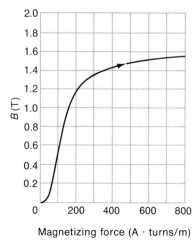

(b)

Fig. 35-22. (a) The relative permeability of a pure iron sample varies from an initial value of 200 to a peak value of 5000. (b) The field B versus magnetizing force in a long, iron-filled solenoid increases most rapidly where κ_m is largest.

Such substances are called *ferromagnetic*. The forces that cause this orderly arrangement of atoms are chemical bonding forces. The strong dependence of bonding upon relative orientation for atoms of ferromagnetic substances is a quantum mechanical effect and is understood by modern atomic theory. The number of ferromagnetic substances is limited to the five elements — Fe, Co, Ni, Gd, and Dy — and to various alloys containing these elements. At a sufficiently high temperature, the *Curie temperature,* ferromagnetic ordering breaks down and the substance becomes an ordinary paramagnet.

The spontaneous alignment of neighboring atoms in a ferromagnetic substance produces a very strong internal magnetic field. Yet, most pieces of ferromagnetic material, such as an iron nail, do not exhibit a strong magnetic field. In a typical ferromagnetic substance, very small regions — called *domains* — are highly magnetized. However, the various magnetized domains point in differing directions so that the *average* magnetic field is small.

When an external magnetic field is applied to a ferromagnetic substance, the magnitude of that field is generally enhanced by two effects. First, domains that happen to have their magnetic field in the direction of the external field grow in size at the expense of the domains that oppose the external field. Second, some domains with fields in a direction away from the external field spontaneously jump to a new alignment of their atoms with a direction closer to the external field direction. The net effect is that the external field is strengthened by the addition of many atomic fields pointing in

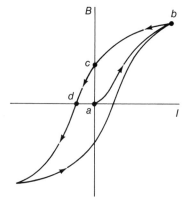

Fig. 35-23. When the current in the windings around an iron sample is increased, the field increases as from a to b. However, when the current is reduced to zero, as at c, there is still a remaining field due to residual ordering of magnetic domains. A reverse current must be applied before zero field is reached at d. The path followed as the current is alternately reversed is called a hysteresis loop (compare Fig. 15-2).

the same direction. The relative permeability of ferromagnetic substances is thus quite large.

The graph of relative permeability versus magnetizing force in Fig. 35-22a shows that κ_m is not constant for ferromagnetic materials. At low values of magnetizing force, the relative permeability is small but it grows rapidly with increasing field as more and more domains come into alignment.* At higher and higher fields, the permeability decreases until saturation is reached—the point at which all domains are aligned and no further enhancement of the field is possible. The field strength B within a long, iron-filled solenoid is graphed in Fig. 35-22b. The enormous enhancement of magnetic fields by the presence of iron has been crucial to the development of electric motors and generators.

When the magnetizing field applied to iron is reduced, the alignment of the domains returns toward its initial state but typically exhibits *hysteresis*. As seen in Fig. 35-23, the field in an iron sample does not retrace itself when the current is reduced. The point c on the graph indicates that there is a field present without a magnetizing current. This permanent magnetic field is caused by some of the domains remaining in an aligned conditions: we say the substance is *magnetized*. Mechanical shock or high temperatures will cause the specimen to return to an unmagnetized state.

In Fig. 35-24 we see a schematic illustration of the aligned atomic loops of current in a bar magnet that has been permanently magnetized by stopping at point c of the hysteresis graph in Fig. 35-23. The effect of these loops is equivalent to a single sheet of current wrapped around the cylindrical bar. This equivalent current is called an *amperian current* after Ampere, who first suggested this explanation of the mechanism of permanent magnetism. These surface currents cannot be drawn away into conductors, for they are due to endlessly revolving electrons that are *bound* to their parent atoms. Their effect is to produce a magnetic field, just as a solenoid produces a magnetic field.

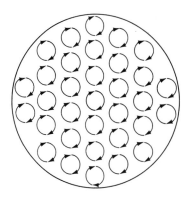

Fig. 35-24. Aligned atomic loops within a bar magnet viewed from the south-pole end of the bar. The atomic currents cancel in the interior but all rotate in the clockwise sense at the surface.

Exercises

7. A long solenoid is filled with an iron core whose characteristics are those of Fig. 35-22. The solenoid is wound with 1000 turns per meter of length. When the field in the iron core has a value of 1 T, what is the value of the current, in amperes,

*The initial relative permeability of 200 is small compared to the peak value but is *enormous* compared to nonferromagnetic substances.

flowing in the windings? Compare this to the current required to produce this field strength in the absence of iron.

8. A cylindrical bar magnet is 1 cm in diameter and 10 cm long. The value of B near the center of the magnet is 0.1 T. Find the number of ampere-turns required for an iron-free solenoid of 1-cm diameter and 10-cm length to produce the same field at its center.

35-7 Motors   NOT ON HOURLY

The torque exerted on a current-carrying loop in a magnetic field can be put to use in an electric motor of the type represented schematically in Fig. 35-25. The function of the *commutator* is to reverse the current direction every half-turn so that the torque on the loop continues to act in the same rotational sense.

A practical motor of this type generally uses an iron-core electromagnet to produce the magnetic field. The space within the loop (called the *armature*) is filled with iron to intensify the magnetic field still more in this region. Also, as indicated in Fig. 35-26, many turns of wire are wound on the armature to increase the net torque produced. Finally, as shown in Fig. 35-27, a motor utilizes many armature coils wound with their planes at differing angles, so that one coil is always in a position of maximum torque. The commutator is then arranged to deliver current only to the coil in the position of maximum torque at any given time.

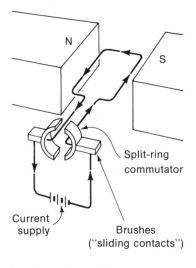

Fig. 35-25. The major elements in a simple motor are indicated in this schematic representation.

Exercises

9. Which orientation of the loop in Fig. 35-25 produces the maximum torque? (See Section 34-7.)

10. The magnetic field strength in a motor is limited by the saturation properties of iron to one tesla or less. Estimate the maximum torque produced by a 500-turn armature winding in such a field if it is wound as a rectangle with dimensions 5 cm × 10 cm. The armature current is 10 A.

35-8 Summary

In this chapter we have studied the way in which electric currents produce magnetic fields. From experiments with extended current-carrying circuits of various shapes, the following law was deduced:

$$d\mathbf{B} = k' \frac{I \, d\mathbf{l} \times \mathbf{r}}{r^3} \qquad (35\text{-}2)$$

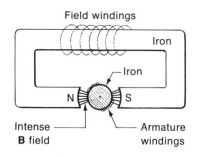

Fig. 35-26. Iron is used extensively in an electric motor to intensify the magnetic field that produces the torque on the armature.

The net field from a current of finite length is found by integration of (35-2) to include all of the circuit.

When (35-2) is applied to an infinitely long, straight wire carrying a current, (35-4) shows that the net magnetic field encircles the wire and falls off as the inverse of the distance between the wire and a field point. In fact, experimental observation of $1/r$ dependence of this field first pointed the way to (35-2).

For a current-carrying circular loop of radius R, the field on the axis was found by integration of (35-2) to be

$$B = \tfrac{1}{2}\mu_0 I \frac{R^2}{r^3} \tag{35-5}$$

where r is the distance from the loop to the on-axis field point. Experimental verification of (35-5) is further proof of the validity of (35-2).

Gauss's law for magnetic fields is

$$\oint \mathbf{B} \cdot d\mathbf{A} = 0 \quad \text{always} \qquad \text{FLUX}$$

The circulation of $\mathbf{B}$ is described by Ampere's law:

$$\oint \mathbf{B} \cdot d\mathbf{s} = \mu_0 I \qquad \text{CIRCULATION} \tag{35-8}$$

where I is the current enclosed by the closed path of integration. An infinitesimal element of that path is denoted $d\mathbf{s}$ in (35-8). Ampere's law was used to find the field of a long solenoid in Section 35-4.

The force between parallel wires is attractive when they carry currents in the same direction and repulsive for opposite currents. The force exerted by current-carrying conductors provides the primary standard for the ampere.

When substances are placed within a magnetic field, the field strength is usually slightly altered by the presence of the material. Substances that reduce the field strength are called diamagnetic while substances that enhance the field are called paramagnetic. The relative magnetic permeability, κ_m, gives the ratio of the observed field to that that would have occurred in the absence of the material

$$B_m = \kappa_m B_0 \tag{35-12}$$

Ferromagnetic substances are those paramagnetic materials in which adjacent atoms align their magnetic field to produce extremely intense fields. Alignment of the atomic magnetic fields with externally applied magnetic fields produces very large field enhancements with values of κ_m ranging into the thousands for iron.

The large enhancement of field strength by the presence of iron is put to use in electric motors, so that a large torque can be developed in a compact device.

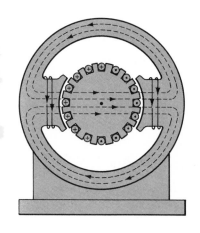

Fig. 35-27. A practical electric motor utilizes many sets of armature windings.

$$\text{FLUX} \begin{bmatrix} \oint \mathbf{B} \cdot d\mathbf{A} = 0 \\ \oint \mathbf{E} \cdot d\mathbf{A} = \dfrac{q}{\epsilon_0} \end{bmatrix}$$

$$\text{Circulation} \quad \oint \mathbf{B} \cdot d\mathbf{s} = \mu_0 I$$

$$\oint \mathbf{E} \cdot d\mathbf{s} = -\frac{d\Phi_E}{dt}$$

Problems

35-1. Two straight wires carrying equal currents I are perpendicular to one another and separated by a distance d at their smallest separation. Find the magnitude of the net magnetic field at a point halfway between the wires. You may assume the wires are infinitely long so that (35-4) is applicable for either wire.

35-2. Two parallel, straight wires carry equal currents I. Find the magnitude and direction of the magnetic field at a point halfway between them if the currents are (a) in the same direction; (b) in opposite directions. You may assume the wires are very long compared to their separation.

35-3. Two long wires carry equal currents in the same direction, as in Fig. 35-28. Find the field strength at the point P, which lies in the plane containing both wires.

35-4. Typical household electrical currents have magnitudes of about 10 A. Estimate the magnetic field strength they produce in a room and compare your result to the strength of the earth's magnetic field.

35-5. Two parallel coils are each wound with 100 turns and arranged as in Fig. 35-29. For $R = 10$ cm and $d = 10$ cm, find the on-axis magnetic field strength midway between the coils.

35-6.* Refer to Fig. 35-9 and Eq. (35-5). (a) Make a qualitative plot of the on-axis value of B as a function of d from $d = 0$ to $d = 2R$. (b) Use analytic means to find the inflection point d_0 where the second derivative of B with respect to d vanishes. (For $d < d_0$ the second derivative is negative, while for $d > d_0$ the second derivative is positive.) (c) Show that two equal coils (called Helmholtz coils) carrying equal currents can be separated by a distance $2d_0$ to produce a net on-axis field halfway between them that has a vanishing first derivative as well as a vanishing second derivative. The field is thus very nearly constant over a considerable distance between the coils.

35-7. The toroid shown in Fig. 35-30 has a thickness $d = 2$ cm and a mean radius $R = 10$ cm. It is wound with 1000 turns and carries a current of 10 A. (a) Find **B** within the windings by applying Ampere's law to the path indicated by

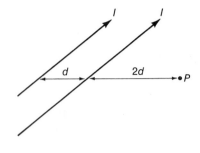

Fig. 35-28.

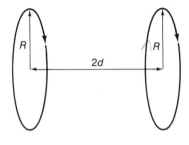

Fig. 35-29.

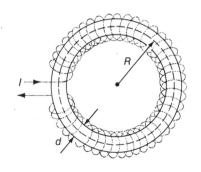

Fig. 35-30.

the dotted line with $R = 10$ cm. (*b*) Find the variation of **B** within the toroid by repeating the calculation for $R = 9$ cm and $R = 11$ cm.

35-8. A long, straight, hollow wire carries a current I. Use Ampere's law to show that (*a*) the magnetic field outside the wire is that which would be produced by the same current passing through a fine wire on the axis of the hollow wire; (*b*) the magnetic field inside the wire is zero.

35-9. A long, straight wire shown in Fig. 35-31 has a radius R and carries current out of the page with uniform current density j within the wire. A counterclockwise magnetic field is thus produced both within and without the wire, as shown. Use Ampere's law to show that the magnitude of this field is given by

$$B = \frac{\mu_0 I}{2\pi r} \qquad \text{for } r > R$$

and
$$B = \frac{\mu_0 I r}{2\pi R^2} \qquad \text{for } r < R$$

where r is the distance from the center of the wire and I is the total current in the wire.

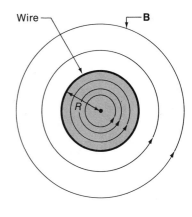

Fig. 35-31.

35-10. Two long copper wires of 1-mm diameter each carry a current of 100 A and are separated by 10 cm. (*a*) Find the magnetic force per meter of length. (*b*) Find the number of copper atoms, and hence the number of conduction electrons, per meter of length of either wire. (*c*) Find the electrostatic repulsion of these wires per meter of length if they carry the full charge of their conduction electrons. (*d*) Compute the drift velocity v of the electrons. (*e*) Compute the ratio of the answer in *a* to that in *c*, and compare to the ratio v^2/c^2.

35-11. A long solenoid with 10^2 turns/m contains an iron core with the characteristics shown in Fig. 35-22. Find the field within the core when $I = 1$ A. What value of current is required to produce this field strength in the absence of the iron core?

35-12. A toroid with an iron core as in Fig. 35-32 has a mean circumference of 40 cm and is wound with 200 turns of wire carrying a current of 1 A. What fraction of the field in the core is due to amperian currents? Use Fig. 35-22 for characteristics of the iron core.

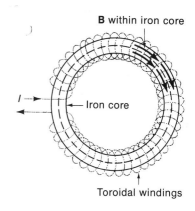

Fig. 35-32.

35-13.   Table 35-2 gives values of B versus magnetizing force for a silicon-steel specimen placed in a long solenoid. Calculate the permeability of this sample near values of H equal to (a) 10, (b) 100, (c) 1000.

TABLE 35-2

Field Strength B (T)	Magnetizing Force H (A · turns/m)
0	0
0.05	10
0.14	20
0.42	40
0.60	60
0.72	80
0.82	100
0.90	110
1.04	200
1.25	500
1.32	1000
1.37	1500

35-14.   Each coil in a motor armature carries 10 A and has 300 turns. This armature is in a magnetic field of 0.5 T. (a) Find the maximum torque of a coil with dimensions 10 cm × 10 cm. (b) If there are enough separate armature windings so that the torque is always near the maximum value, find the power produced by this motor at 3600 rev/min.

Chapter **36**

Electromagnetic Induction

electric currents induced by magnetic fields
electric field created by magnetic field

36-1 Introduction We have stressed all along that electricity and magnetism are not separate topics but part of an interrelated whole. Now we will turn our attention to one of the interrelations of electric and magnetic fields that plays a key role in the vast technological uses of electromagnetism. The phenomenon we will discuss is that of *electromagnetic induction,* whereby electric currents are induced by magnetic fields.

There are two distinct mechanisms that lead to electromagnetic induction.* One of them involves motion of a conductor through a magnetic field, so that the magnetic force on charge carriers in the conductor causes current flow. This mechanism is utilized in generators, which produce electrical energy from mechanical energy. The second mechanism is rather surprising: whenever a magnetic field *changes* in intensity, an electric field is produced. This type of *induced* electric field, in contrast to a static field, has a net circulation and is therefore capable of producing a current in a conducting closed loop. This phenomenon is utilized in the *transformer,* a widely used device that transforms current-voltage ratios in alternating current circuits.

change in magnetic field causes electric field

In Chapter 39 we will see that the inverse phenomenon—the creation of a magnetic field by a changing electric field—also occurs.

---

*The existence of two distinct phenomena in magnetic induction is clearly discussed in Chapters 16 and 17 of *The Feynman Lectures on Physics* by Richard P. Feynman, Robert B. Leighton, and Matthew Sands; Reading, Mass.: Addison-Wesley Publishing Co. (1964).

This *electric induction,* coupled with magnetic induction, is responsible for the existence of electromagnetic waves, which can carry energy over great distances.

36-2 Faraday's Observations

In 1820, Oersted showed that electric currents produce magnetic fields. It was not long before scientists were searching for the converse effect—the production of an electric current by a magnetic field. Faraday spent many years searching for such an effect. He devised many types of experiments, one of which he describes in his laboratory notebook of November 28, 1825, as involving a current-carrying wire ". . . parallel to which was another similar wire separated from it only by two thicknesses of paper. The ends of the latter wire attached to a galvanometer exhibited no action." He had previously placed coils of wire around a magnet but detected no influence by the magnet on currents flowing in the coil.

It was finally in August, 1831, that he observed an induced current. He placed two coils of wire in proximity and connected one to a wire over a compass needle. The other independent coil was connected to a battery. Whenever this battery connection was *made* or *broken,* a brief deflection of the compass needle was seen, indicating a brief passage of current in the first coil.

Faraday was quick to perceive that it was a *changing* magnetic field that was needed to produce an induced current. He verified this by plunging a bar magnet into a stationary coil connected to a galvanometer. He observed a deflection of the galvanometer when he moved the magnet into the coil and a deflection in the opposite direction when he withdrew the magnet. Further, if he reversed the orientation of the magnet poles, the galvanometer deflections also reversed. The two types of induction observations we have described are indicated schematically in Fig. 36-1.

Faraday firmly believed in the reality of magnetic lines of force and interpreted induction as being caused by the relative motion of a conductor and lines of force. Thus a magnet (and its lines of force) could be pushed toward a coil or the coil pushed toward the magnet to produce induction. When a switch was closed suddenly to create a field in a coil, induction occurred in a nearby coil as the newly born lines of the force swept over the coil as they expanded toward their steady-state configuration.

Faraday deduced that induction was dependent on the rate of the net number of magnetic lines of force sweeping through the circuit. Thus, the induced current was proportional to the rate of change of the *magnetic flux* Φ_B, where

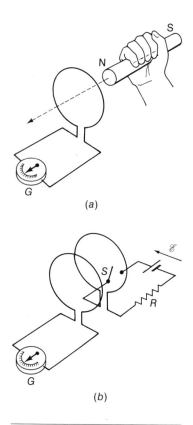

Fig. 36-1. (a) The galvanometer indicates an induced current only while the magnet moves relative to the loop. (b) A brief current is induced in the galvanometer circuit each time the switch S is opened or closed.

induced current = rate of change of the magnetic flux Φ_B

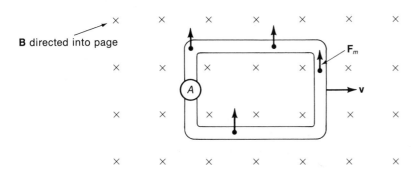

B directed into page

$\mathbf{F}_m$

A

$\mathbf{v}$

Fig. 36-2. A conducting loop moves to the right in a uniform **B** field directed into the page. The magnetic force $\mathbf{F}_m = q\mathbf{v} \times \mathbf{B}$ acts in the same direction on every mobile charge carrier and so does not generate a circulating current.

$$\Phi_B = \int \mathbf{B} \cdot d\mathbf{A}$$

FARADAY'S LAW

and the area included in the integral is that of the circuit in which induction occurs. Faraday's observations showed that

$$I(\text{induced}) \propto \frac{d\Phi}{dt}$$

We will soon state Faraday's law as an equality, but first let us turn our attention to the first of two mechanisms that lead to Faraday's law.

36-3 Motional Emf

When a conductor is moved through a nonuniform magnetic field, the magnetic force that acts upon charge carriers within the conductor can produce a net flow of current. In Fig. 36-2 we see a conductor in the form of a closed loop moving through a *uniform* magnetic field. In this case no net circulating current is produced, since all charge carriers are acted upon by a force in the same direction.*

closed loop moving through a uniform magnetic field no net circulating current produced

If we destroy the symmetry in Fig. 36-2 by making the field nonuniform, there will be a net circulation of current. This is seen readily in Fig. 36-3. There, a **B** field exists only in a well-defined region and the loop moves progressively into that region. When the leading edge of the loop enters the field, magnetic forces act on all mobile charge carriers in that segment of the loop, tending to cause

---

*Some slight *separation* of charge will occur due to the Hall effect when the conductor first begins moving. The upper portion of the loop in Fig. 36-2 will become slightly positively charged while the bottom of the loop will acquire a slight negative charge. Once the velocity is uniform, no further charging occurs.

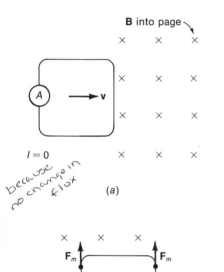

B into page

$I = 0$

(a)

because no change in flux

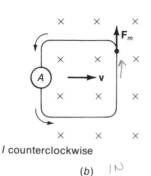

I counterclockwise

(b)   *IN*

Fig. 36-3. The loop of Fig. 36-2 moves from a region of no field through a region where there is a field directed into the page. Current flows only during those times when the left and right ends of the loop are in different field strengths, as in (b) and (d).

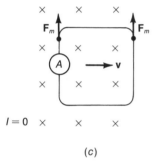

$I = 0$

(c)

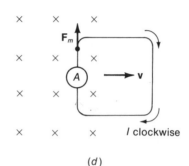

I clockwise

(d)

DIRECTION OF CURRENT OPOSES CHANGE IN FLUX

USE RT HAND RULE
WANT MORE X INTO PAGE — FINGER

DECREASING FLUX — WANTS TO GO BACK

a current directed upward, as shown. Note that the direction of the magnetic force shown is that for a positive charge carrier. A negative charge carrier would experience a force in the opposite direction, but the *current* would still be in the same direction as shown. Mobile charges in other portions of the loop do not experience a magnetic force parallel to the conductor, so the net circulation of current is driven by the forces acting on the charges along the right end of the loop.

In Fig. 36-3c, magnetic forces on mobile charges in the left and right ends of the loop cancel, as they did in Fig. 36-2. Finally, as the loop leaves the field, as in Fig. 36-3d, a current again flows but in the opposite direction to that produced in part *b*.

In Fig. 36-3b and *d*, each mobile charge within the field **B** is acted upon by a magnetic force given by

$$F_m = qvB$$

where q is the magnitude of the charge and **v** the velocity of the charge perpendicular to **B**. Because these charges are being carried

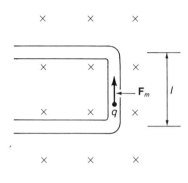

through the **B** field by physical motion of the conductor, **v** is just the velocity of the conductor itself.

The energy gained by a charge moving a distance l with a force F acting upon it is simply Fl. As seen in Fig. 36-4, when the loop enters the field, the amount of work done on a charge moving entirely along the end of length l is given by

$$W = F_m l = qvBl \qquad (36\text{-}1)$$

We call the gain in energy per unit charge a *motional emf*. In this case it is given by

$$\mathscr{E} = vBl \qquad (36\text{-}2)$$

This emf gives energy to charge carriers and causes current to flow just as surely as a battery does. In a battery, chemical forces move the charge carriers. In this device, magnetic forces move them. The energy given to the charge carriers *ultimately* comes from the force that moves the loops, but more of that in Section 36-4. Since the two long sides of the loop do not contribute to motional emf and the left side of the loop is out of the field in Fig. 36-4, the *net* emf is that given by (36-2). The induced current is then given by Ohm's law:

$$I = \frac{\mathscr{E}}{R} = \frac{vBl}{R}$$

where R is the resistance of the entire loop.

Fig. 36-4. Each mobile charge carrier has work done upon it by **F**$_m$ as it moves along the end of the loop. The net work done is $F_m l$.

Example A rectangular loop with dimensions W and l moving with constant velocity v enters and leaves a magnetic field that occupies a square area that is $2W$ on an edge, as in Fig. 36-5. Sketch a graph of the induced current in the loop as a function of time, taking counterclockwise current as positive, clockwise as negative. Assume that **B** is uniform inside the square and vanishes outside the square area. The resistance of the loop is R. The velocity of the loop is constant.

Solution The emf comes into existence the moment the leading edge of the loop enters the region of the magnetic field. This emf equals vBl and produces a counterclockwise current, as discussed above. At the moment the back end of the loop enters the field, another emf of magnitude vBl is produced that is directed so as to produce a clockwise current. The net emf now vanishes. When the leading edge of the loop leaves the field, the cancellation of the two emf's is destroyed, so that there is a clockwise current due to the emf in the trailing edge. Finally, when the loop leaves the field entirely, the emf vanishes. The time required for the leading edge to traverse the field is

$$\tau_1 = \frac{2W}{v}$$

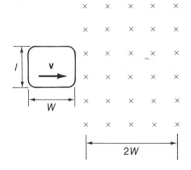

Fig. 36-5.

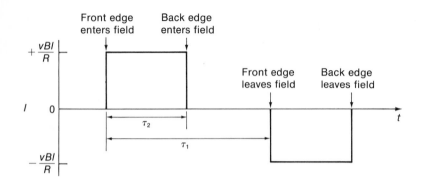

Fig. 36-6.   The current induced in the loop flows in one direction as it enters the field and in the opposite direction as it leaves.

and the trailing edge enters the field a time

$$\tau_2 = \frac{W}{v}$$

later. The induced current has a magnitude of

$$I = \frac{vBl}{R}$$

The graph of induced current versus time is shown in Fig. 36-6. For the reasonable values of $v = 10$ m/s, $B = 1$ T, $l = 10$ cm, and $R = 2\Omega$, we obtain

$$I = \frac{vBl}{R} = \frac{(10 \text{ m/s}) \cdot 1 \text{ T} \cdot 10^{-1} \text{ m}}{2\Omega} = \tfrac{1}{2}\text{A}$$

The results obtained in the preceding example can be expressed in a simple way. The net *flux* of **B** through an area is

$$\Phi = \oint \mathbf{B} \cdot d\mathbf{A} \qquad \Phi = \int \mathbf{B} \cdot d\mathbf{A} \qquad \text{not closed surface}$$

which for a *uniform* **B** is simply

$$\Phi = \mathbf{B} \cdot \mathbf{A}$$

The vector **A**, we recall, is perpendicular to the plane of the area through which the flux passes. Thus, for a uniform field parallel to **A**:

$$\Phi = \pm BA$$

depending on the direction chosen for **A**. If we curl the fingers of our right hand in the direction of the emf induced as the loop enters the field (counterclockwise in Fig. 36-5), the thumb of our right hand points against the field. Taking this direction of the induced emf as the direction of **A**, we find

$$\Phi = -BA$$

where A is the area of the loop *within* the field. This can be written as

$$\Phi = -Blx$$

where x is the length of the loop that is within the region of magnetic field. Since

$$v = \frac{dx}{dt}$$

we have

$$\frac{d\Phi}{dt} = -Blv$$

while one end of the loop is entering the field,

$$\frac{d\Phi}{dt} = +Blv$$

as it is leaving. But these are exactly the values of the induced emf that we previously found. Thus our result in the preceding example could be expressed simply as

$$\mathscr{E} = -\frac{d\Phi}{dt} \qquad (36\text{-}3)$$

where $\mathscr{E}$ is the motional emf. We will see shortly that (36-3) is an extremely general result.

Consider now a rather different situation, in which a plane rectangular coil is rotating within a uniform magnetic field, as in Fig. 36-7. If **B** is uniform and perpendicular to the axis of rotation, as shown, the force on each charge carrier along the two segments of length L in Fig. 36-7 will experience a force

$$F_m = qvB \sin \theta$$

where θ is the angle between the tangential velocity of the loop edge and the field **B**. This angle is equal in magnitude to the angle between the *normal* to the plane of the loop and the direction of **B**. If the loop is spinning with an angular velocity $\boldsymbol{\omega}$, the tangential velocity is given by

$$\mathbf{v} = \frac{W}{2}\boldsymbol{\omega}$$

and we then find

$$F_m = \tfrac{1}{2}q\omega WB \sin \theta$$

The energy gained by a mobile charge q moving along one loop side of length L is

$$E = \tfrac{1}{2}q\omega WLB \sin \theta$$

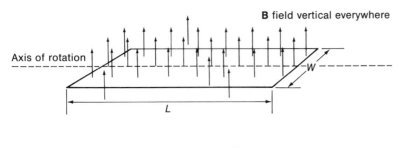

Axis of rotation

B field vertical everywhere

W

L

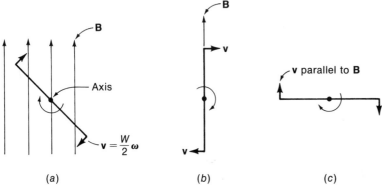

(a)

B

Axis

$v = \frac{W}{2}\omega$

(b)

B

v

v

(c)

v parallel to **B**

Fig. 36-7. A loop rotating in a uniform magnetic field. In (a) and (b) the magnetic force on mobile charges in the wire is out of the page at the top of the loop and into the page at the bottom. Thus a net emf exists around the loop. This emf is maximum in orientation (b) and vanishes in (c).

The direction of the magnetic force along the sides of length W is perpendicular to the wire and produces no net energy gain for a charge moving along those segments of the loop. The emf's generated by the two sides of length L add together, so that a charge going entirely around the loop increases its energy by

$$\Delta E = q\omega WLB \sin\theta$$

Thus, a motional emf of magnitude

$$\mathscr{E} = \omega WLB \sin\theta$$

exists around the loop. Again, if we use a right-hand rule to establish the direction of positive $\mathscr{E}$ and let the thumb of the right hand define the direction of the normal to the loop, careful analysis of Fig. 36-7 shows that

$$\mathscr{E} = +\omega WLB \sin\theta$$

and for a steadily rotating loop, $\theta = \omega t$, so that we have

$$\mathscr{E} = +\omega WLB \sin(\omega t) \tag{36-4}$$

which is plotted in Fig. 36-8. This is the type of emf produced by electric generators utilizing loops rotating in a magnetic field. The elementary components of such a generator are shown in Fig. 36-9.

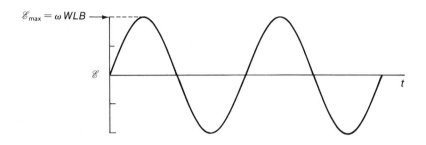

$\mathscr{E}_{max} = \omega\,WLB$

The flux of **B** through the area of the loop is given by

$$\Phi = \oint \mathbf{B} \cdot d\mathbf{A}$$

$\Phi = \int B \cdot dA$

where the integral includes all of the loop. For a uniform field this integral becomes simply

$$\Phi = \mathbf{B} \cdot \mathbf{A}$$

where **A** is the vector perpendicular to the plane of the loop and with magnitude equal to the area of the loop. Thus, for the rotating loop,

$$\Phi = BWL \cos\theta = BWL \cos(\omega t)$$

We note that

$$\left| \frac{d\Phi}{dt} = -\omega BWL \sin(\omega t) \right| \tag{36-5}$$

Comparing (36-5) to (36-4), we see that once again we can write

$$\mathscr{E} = -\frac{d\Phi}{dt}$$

Looking back over the cases of the translating loop entering a magnetic field and the loop rotating in a magnetic field, we see common features:

1. Application of the fundamental magnetic-force law

$$\mathbf{F} = q\mathbf{v} \times \mathbf{B} \tag{36-6}$$

for charge carriers in a conductor showed that charge carriers in a moving conductor gain energy.

2. The energy gained per unit charge is called a motional emf and is given by

$$\mathscr{E} = -\frac{d\Phi}{dt} \tag{36-7}$$

in both cases.

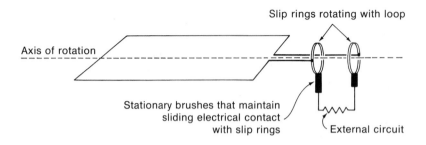

Slip rings rotating with loop

Axis of rotation

Stationary brushes that maintain
sliding electrical contact
with slip rings

External circuit

Fig. 36-9. The motional emf
induced in a rotating loop causes
an alternating current to flow in
a stationary external circuit when
sliding contacts are provided.

There are cases, however, where there is no apparent change in magnetic flux passing through a complete circuit, yet there is motional emf. An example is shown in Fig. 36-10. There, a rotating metal wheel carries its conduction electrons always in the same direction (right to left) through the fixed location of a magnetic field. Sliding contacts carry away the unidirectional current. In this device the magnetic flux is constant if we think of the circuit as passing straight along a radius from the axle contact to the rim contact. Of course, we can consider the "circuit" to contain a radius vector that, at any instant, is rotating so that $d\Phi/dt$ does not vanish. We need continually to pick a new radius vector in the vicinity of the contacts with this approach, so it is not an appealing one.

Analysis of a wide variety of devices producing motional emf leads to two general conclusions:

1. Application of the force law (36-6) to all the mobile charges of a conductor moving through a magnetic field will always correctly indicate whether or not a current will flow. Integration of this force around a circuit will always correctly give the observed motional emf.

$F = qv \times B$

2. When a motional emf is produced by a device in which it is clear that the magnetic flux passing through a circuit changes with time, the observed motional emf is given by

$$\mathscr{E} = -\frac{d\Phi}{dt}$$

where Φ is the flux *within* the circuit. However, there are some devices that are not readily analyzed in terms of a changing flux. These devices usually involve a region where the current is not confined to a well-defined conducting path in a wire, as in Fig. 36-10. In that case statement 1 is still true.

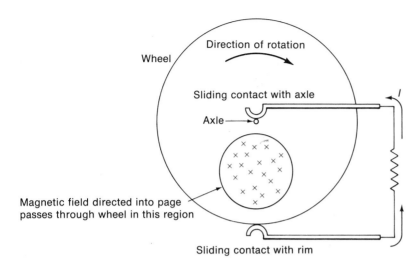

Fig. 36-10. A unipolar generator. A positive charge rotating around with the wheel experiences a downward force as it passes through the field that is directed into the page. The direction of the conventional current produced in the external circuit is indicated.

Exercise

1. A rigid conductor is moved at velocity **v** through a magnetic field, as shown in Fig. 36-11. The field is uniform in a square region that is W on an edge, as shown. The direction of the field is into the page and the rigid conductor is perpendicular to both the field **B** and its own velocity. While the conductor is moving through the region containing a field, a constant motional emf is produced that causes a current to flow through the flexible leads and through the resistor and ammeter shown. (*a*) What is the direction of this current? (*b*) Find the magnitude of the emf by integrating the magnetic force over that portion of the rod within the field. This emf will depend upon v, B, and W. (*c*) Find the emf by applying $\mathscr{E} = -d\Phi/dt$. Note that

$$\Phi_{\text{total}} = BW^2$$

and $$\Phi \text{ (within circuit)} = BWy$$

once the conductor is within the field. Further, $v = -dy/dt$.

36-4   Conservation of Energy for Motional Emf's

The devices discussed in the preceding section produce an emf and can cause current to flow. As we saw in Chapter 33, a current I

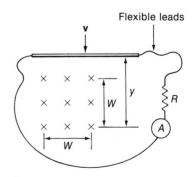

Fig. 36-11.

passing through an emf of $\mathscr{E}$ gains energy at a rate of

$$P = \mathscr{E}I$$

With I in amperes and $\mathscr{E}$ in volts, P has units of watts (J/s). When this current flows through a circuit that has a resistance R, a voltage drop exists across the resistor that opposes the current and has a magnitude of $\mathscr{E}$. The resistor thus *consumes* energy at the same rate: $\mathscr{E}I$. This energy appears as heat.

  If a wire moves at constant velocity across a magnetic field, as in the example at the end of the preceding section, electrical energy is produced, yet the law of inertia tells us that no net force is required to maintain motion of the wire at *constant* velocity. Do we then obtain the electrical energy *without* doing work? The answer is no. While it is true that no *net* force is required to move the wire at constant velocity, there is an opposing force that must be balanced by an externally applied force if we are to maintain the motion of the wire. Furthermore, the magnitude of this opposing force depends on *both* the velocity of the wire (hence the emf) *and* the current flowing in the wire, in such a way that energy is precisely conserved. This retarding force is simply the magnetic force that acts on the induced current as it flows through the magnetic field causing the induction.

  Let us carefully examine just one simple case of motional emf to see exactly how energy is conserved. In Fig. 36-12 we see again the moving conductor of Exercise 1. Since $\mathbf{v}$ and $\mathbf{B}$ are perpendicular, the magnetic force is

$$F = qvB$$

on any charge q in the conductor. The work done on a charge moving through the length, W, then increases the energy of the charge by an amount

$$\Delta E = qvBW$$

The motional emf is thus given by

$$\mathscr{E} = vBW \tag{36-8}$$

If a complete circuit exists and has resistance R, the current is

$$I = \frac{\mathscr{E}}{R} = \frac{vBW}{R} \tag{36-9}$$

Each element $d\mathbf{l}$ of the wire within the magnetic field is acted upon by a force

$$d\mathbf{F} = I\, d\mathbf{l} \times \mathbf{B} \tag{36-10}$$

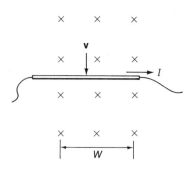

Fig. 36-12. The moving rod produces a motional emf in the direction $\mathbf{v} \times \mathbf{B}$. For $\mathbf{B}$ directed into the page, the emf is directed to the right. When a complete circuit is present, the induced current interacts with $\mathbf{B}$ to produce a force in the direction $I d\mathbf{l} \times \mathbf{B}$, which is *opposite* to $\mathbf{v}$.

Recalling that the direction of $d\mathbf{l}$ is given by I, the reader can apply the right-hand rule for the cross product in (36-10) to the situation shown in Fig. 36-12. The resultant force is upward—*opposed* to the motion!

The magnitude of this force is obtained by integrating (36-10) over the entire length of wire within the **B** field. Since this length is W, and since all elements dF are parallel and the field is uniform, we simply obtain

$$F = IWB \qquad (36\text{-}11)$$

The rate at which work is done by this force (the power consumed) is simply

$$P = Fv = IvBW \qquad (36\text{-}12)$$

But (36-8) allows us to write this as

$$P = I\mathscr{E}$$

which is *precisely* the rate of production of electrical energy. This result is independent of our choice of $\mathbf{v}$, $\mathbf{B}$, W, or R. Energy is always exactly conserved by this device.

In *any* electrical generator, mechanical energy is consumed in order to produce electrical energy. This is so because the reaction forces described above oppose the direction of motion. The retarding force that must be overcome to run a generator is produced in exactly the same way as the force driving an electric motor. In a motor, however, the force is in the same direction as the motion and mechanical energy is produced at the expense of electrical energy.

Exercise

2. In the preceding discussion we could have expressed the electrical power as $P = I^2R$. Show that this power is equal to the mechanical power given by (36-12), with I given by (36-9).

36-5   Induced Electric Fields

In the case of *motional emf*, magnetic forces cause the current to flow. Now let us examine the other type of magnetic induction, in which an *induced electric field* causes the current flow. This electric field is quite different from a static field because it has a net circulation, while a static field can never have a circulation differing from zero. This type of electric field is produced whenever a *changing* magnetic field is present.

The fundamental law that describes this field **E** is:

$$\oint \mathbf{E} \cdot d\mathbf{s} = \frac{-d\Phi}{dt} \qquad (36\text{-}13)$$

The circulation of the induced electric field around any path is equal to the negative rate of change of the magnetic flux enclosed within that path.

This law is called *Faraday's induction law*. In Fig. 36-13 we see the electric field produced by an increasing magnetic field that is directed into the page in the region shown. Since induced electric fields have nonzero circulation, a charge moving around in the direction of **E** gains energy. In fact, the circulation integral is precisely equal to the work done by electric forces per unit charge along the path of integration. To see this we write the definition of work:

$$W = \int_a^b \mathbf{F} \cdot d\mathbf{s} = \int_a^b q\mathbf{E} \cdot d\mathbf{s}$$

When the integral goes from point *a*, around a path, and back to *a*, we symbolize that by the closed-line integral sign:

$$W = \oint q\mathbf{E} \cdot d\mathbf{s}$$

Factoring *q* out of the integral and dividing, we find

$$\frac{W}{q} = \oint \mathbf{E} \cdot d\mathbf{s}$$

as was to be shown.

For electric fields produced by static charge distributions, this circulation integral is always zero, so that the electrostatic energy at a given *point* is well defined, leading to an electrostatic potential. But now we see that *induced fields can do net work on a charge following a closed path*. This work constitutes an induced emf since it is the energy gained, per unit charge, for one circuit around the path under consideration. Faraday's induction law, (36-13), can thus be restated:

$$\mathscr{E} = -\frac{d\Phi}{dt} \qquad (36\text{-}14)$$

where $\mathscr{E}$ is the emf around a path encircling the area through which the flux Φ passes. The minus sign indicates that the circulation (the emf) is in the opposite direction to that given by a right-hand rule with the thumb pointing in the direction of the *change* in **B**. Thus, in Fig. 36-13, the change in **B** is *into* the page so we point the thumb

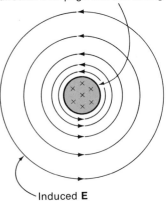

Region containing **B**, which is directed into page and increasing

Induced **E**

Fig. 36-13. A changing **B** field directed into the page produces an electric field as shown. This induced electric field has non-zero circulation.

out of the page and the induced field curls around in the direction of the fingers of the right hand.

Example A loop of wire in the form of a circle with a radius of 10 cm is placed in a uniform magnetic field that is perpendicular to the plane of the loop, as shown in Fig. 36-14c. The total resistance of the loop is 0.02 Ω. The field passing through this loop is made to increase at a rate of 0.1 T/s. Find the current in the loop due to the induced emf.

Solution The flux passing through the loop for a uniform field perpendicular to the plane of the loop is simply

$$\Phi = \mathbf{B} \cdot \mathbf{A} = \pi r^2 B$$

The rate of change of this flux is

$$\frac{d\Phi}{dt} = \pi r^2 \frac{dB}{dt}$$

The emf is then given by

$$\mathscr{E} = -\frac{d\Phi}{dt}$$

The direction of increasing flux is upward in Fig. 36-14, so the right-hand-rule direction of the induced emf will be down. Pointing the thumb of the right hand down, our fingers encircle the loop so that the emf forces a current into the page on the left side of the loop, out of the page on the right. The magnitude of this induced current is

$$I = \frac{\mathscr{E}}{R} = \frac{\pi r^2 \dfrac{dB}{dt}}{R} = \frac{3.14\,(0.1\ \text{m}^2)\,0.1\ \text{T/s}}{0.02\,\Omega}$$

$$= 1.57\ \text{A}$$

In the preceding example, the same emf would have been produced if the wire was in the form of a rectangle or any other shape — as long as its *area* was still equal to that of the loop in the example. Alternatively, if the magnetic field did not completely cover the area of the loop but was of sufficient intensity to produce the same net flux change within the loop, the induced emf would again be equal to that in our example. Faraday's induction law is thus simple and general: *whatever* the mechanism that produces a flux change within a circuit, the magnitude of the induced emf around that circuit is given by $d\Phi/dt$, where Φ is the flux within the circuit.

Example A changing magnetic flux is confined to a small cross-sectional area, as shown in Fig. 36-15. The induced electric field must extend far beyond that region, however, since the circulation integral

(a)

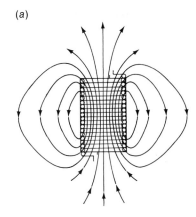

(b)

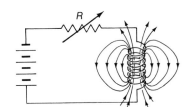

(c)

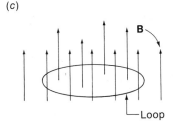

Fig. 36-14. (a) The magnetic field produced by a solenoid is reasonably uniform in a small region near its center. (b) As the variable resistor R is decreased, I, and hence B, increases. (c) A conducting loop in the uniform but increasing field near the center of the solenoid.

always equals $d\Phi/dt$, and once the path of integration is outside the region containing **B**, the value of $d\Phi/dt$ is constant. In particular, a path in the form of a circle of very large radius—*always* far from the region containing the flux—still leads to a constant value of $\oint \mathbf{E} \cdot d\mathbf{s}$, indicating that **E** does not vanish at *any* distance.

Find the magnitude of **E** as a function of r (see Fig. 36-15), assuming the direction of **E** is everywhere tangent to a circle, as indicated in Fig. 36-15.

Solution Since **E** is everywhere tangent to a circle, the circulation integral is most easily evaluated for a circular integration path. We assume that E has a constant value anywhere along such a circle since the problem, as stated, has circular symmetry.

$$\oint \mathbf{E} \cdot d\mathbf{s} = E \oint ds = E \cdot 2\pi r$$

Applying (36-13) we find:

$$E = \frac{1}{2\pi r} \frac{d\Phi}{dt}$$

for the magnitude of **E**. This $1/r$ dependence of **E** is identical, apart from a factor of μ_0, to that of a **B** field surrounding a long, straight wire. It should be, since the steps taken in using Ampere's law to find the magnetic field of a long, straight wire were identical to those in finding the present induced electric field.

If a circuit wraps around the region of changing flux more than once, the induced electric field acts to increase the energy of a moving charge over and over as the charge circles about the changing flux. Thus, for N turns of wire surrounding an area containing a flux Φ, the emf induced along the entire wire is given by:

$$\boxed{\mathscr{E} = -N\frac{d\Phi}{dt}} \tag{36-15}$$

The possibility of producing large emf's by using many turns is immediately obvious. This is the principle of the automotive ignition coil, or *spark coil*, which briefly produces potential differences of 20,000 V or more to cause a precisely timed electrical discharge within each cylinder of a gasoline engine. The heat of this electrical discharge then ignites the gasoline-air mixture to produce a power stroke in the engine.

The design of such a spark coil is indicated in Fig. 36-16. When the breaker points are closed, a current is forced through the primary windings by the automobile battery. The magnetic field that is produced is made quite large because of the highly permeable iron core. When the breaker points open, current ceases to flow in the primary and the magnetic field within the core abruptly diminishes. An emf

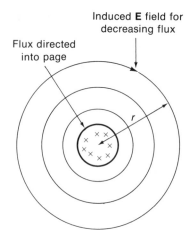

Fig. 36-15. The electric field induced by a small region of changing magnetic field extends far beyond that region.

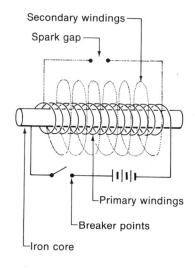

Fig. 36-16. An automotive ignition coil utilizes thousands of turns of wire to increase the induced emf to the point that an electrical discharge occurs between the terminals of a spark plug.

is induced within the secondary windings by the rapid flux change. This emf is made very large by using many thousands of turns of wire in the secondary windings. In this manner, brief potential differences of 20,000 V can be produced with a 12-V automotive battery supplying current to the primary.

Of course, we cannot get "something for nothing" in a device like the spark coil. A much larger current must flow in the primary windings than that produced in the secondary windings in order to conserve energy. We will see the way this comes about in our discussion of transformers in Chapter 38.

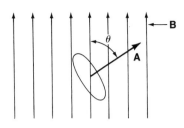

Fig. 36-17.

Exercises

3. A uniform magnetic field has an initial value of 1 T. A loop with an area of 0.01 m² is oriented with its normal vector **A** at an angle of $\theta = 60°$ away from the direction of **B**, as in Fig. 36-17. The intensity of **B** then decreases linearly with time to zero in 2 s. What is the magnitude of $\mathcal{E}$ during that time?

4. An automotive spark coil has 1000 turns in its secondary windings. When the breaker points open, a peak emf of 25,000 V is produced. (*a*) What is the maximum value of $d\Phi/dt$ that must occur? (*b*) If the diameter of the core is 2 cm, what is the maximum value of dB/dt in the core? (Assume that all of the magnetic field is confined uniformly within the cross-sectional area of the core.) (*c*) If the field falls from B_{max} to zero in 10^{-5} s, what is the value of B_{max}?

36-6 Inductance   NOT ON HOURLY

Whenever an electric current is present, a magnetic field is produced. Regardless of the shape of the electrical circuit that *produces* the field, the magnitude of that field at any given point is directly proportional to the current producing it:

$$B_1 = kI_1$$

where B_1 is the field, at any fixed point, due to the current I_1 and k is a constant. Now consider a second circuit arranged so that some of the field B_1 produces a flux within it, as in Fig. 36-18. The flux Φ_2 in the second circuit is also directly proportional to I_1:

$$\Phi_2 = MI_1 \qquad (36\text{-}16)$$

where Φ_2 is the flux passing through the second circuit and M is a constant of proportionality. The constant M has a large value if

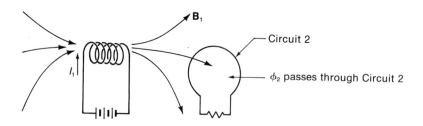

Fig. 36-18. The magnetic flux produced by one circuit can pass through another circuit. These two circuits then possess mutual inductance.

both circuits contain many turns of wire and are close together, and is small if the circuits are far apart. Now if I_1 varies, Φ_2 must vary and an emf is thus induced in the second circuit:

$$\mathscr{E}_2 = \frac{-d\Phi_2}{dt}$$

so that

$$\mathscr{E}_2 = -M\frac{dI_1}{dt} \qquad (36\text{-}17)$$

This is precisely what occurs in the automobile ignition coil: The primary current changes abruptly so that dI_1/dt is large. As a result, the flux changes quickly and an emf $\mathscr{E}_2$ is induced in the secondary.

If the two circuits have simple geometry, so that we can readily calculate the field due to I_1 everywhere within circuit 2, we can calculate Φ_2 directly in terms of I_1 and find M from Eq. (36-16). Alternatively, we can measure the emf produced when I_1 is varied at a known rate and experimentally determine M using (36-17). This coefficient M is called the *mutual inductance*. Its magnitude indicates how readily an emf is induced in one circuit by current changes in another circuit.

Example   In a particular automotive spark coil, an emf of 30,000 V is induced in the secondary when the primary current changes from 3 A to zero (uniformly) in 10 μs. Find the mutual inductance between the primary and secondary windings of this spark coil.

Solution   Applying (36-17) we immediately find

$$M = \frac{\mathscr{E}_2}{dI_1/dt} = \frac{3 \times 10^5 \text{ V}}{3 \times 10^5 \text{ A/S}} = 1 \, \frac{\text{V}}{\text{A/S}}$$

The dimensions of inductance in SI units are seen to be volts divided by amperes per second. The SI unit of inductance has been given a special name, the *henry* (H).* In our above example, the

---

*Joseph Henry (1797–1878) was an American who independently discovered many electromagnetic phenomena discovered by Faraday. Faraday's work had a far greater impact on the scientific world, partly because he was quicker in publishing his work but more importantly because of his far-reaching concept of lines of force.

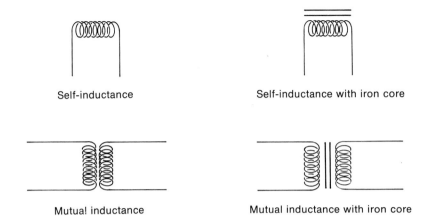

Self-inductance

Self-inductance with iron core

Mutual inductance

Mutual inductance with iron core

Fig. 36-19. The symbols for self-inductance and mutual inductance.

automotive ignition coil had a mutual inductance of 1 henry. This unit is too large for the description of most devices (the spark coil has a huge mutual inductance by virtue of the large number of turns in its secondary), so the millihenry (mH) and the microhenry (μH) are commonly used. For example, the mutual inductance of two co-axial circular loops, each with a radius of 1 cm and spaced 1 cm from one another, is 5×10^{-9} H.

Now let us turn our attention to a *single* circuit in which the current is forced to change. The flux Φ that passes through this circuit due to its own **B** field must change as I_1 changes. Hence an emf is induced *within* a circuit whenever the current in that circuit changes. Again, the flux must be proportional to the current producing it, so we have:

$$\Phi_1 = LI_1 \tag{36-18}$$

and

$$\mathscr{E}_1 = -L\frac{dI_1}{dt} \tag{36-19}$$

where L is the *self-inductance*. Symbols for mutual inductance and self-inductance are shown in Fig. 36-19. The SI unit of self-inductance is the henry.

The minus sign in (36-19) has a particularly simple meaning. When the current flowing in a circuit is changed, the emf induced in that same circuit by the changing of its own magnetic field *opposes* the change in current. If we attempt to increase the current in a circuit, the induced emf acts against the passage of current. If we attempt to decrease the current, the induced emf acts in the *same* direction as the current.

Devices can be purposely constructed to have large values of self-inductance by winding many turns of wire on an iron core. Such devices are called *inductors* and provide a sort of "inertia" in an electrical circuit. Just as a large mass tends to maintain a constant velocity, a large inductance acts to maintain a constant current by the action of its induced emf, which opposes changes in current. A particularly dramatic lecture demonstration of this property can be performed with the circuit of Fig. 36-20. A large inductor is used and a substantial current is produced by the battery with the switch closed. When the switch is abruptly opened, the induced emf acts in a direction that keeps the current flowing. This emf can be large enough to ionize the air between the switch contacts so that a rather long electrical discharge across the open contacts results.

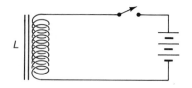

Fig. 36-20. If the inductance L is large enough, an electric discharge will occur across the switch gap as the switch is opened. The induced emf in this case acts to maintain a current in the original direction.

Exercises

5. Two circuits coupled via a mutual inductance of 50 mH are shown in Fig. 36-21. For a brief time after the switch in circuit 1 is closed, ammeter A_1 indicates a uniformly increasing current I_1, such that $dI_1/dt = 10$ A/s. While the current I_1 is increasing it is observed that the current in circuit 2 is constant at 5 mA. Find the value of R in circuit 2.

6. Under the conditions described in Exercise 5, the voltmeter V_1 reads 1 V. What is the *self*-inductance of the primary coil?

36-7 Magnetic Forces Produced by Induced Currents   NOT ON HOURLY

Consider two coaxial loops with current flowing in the same sense, as in Fig. 36-22. The magnetic forces produced by either loop upon the other pull the loops together. This can be seen simply by recalling that parallel currents in the same direction attract one another. The closest segment of one loop to a segment of the other contains a current in the same direction, so the net magnetic force on every segment of a loop is directed toward the other. This can be seen in more detail by considering the force on a given segment of a loop due to the field of the other loop, as in Fig. 36-23.

Now consider the current *induced* in a loop that is pushed toward another, as in Fig. 36-24. The stationary loop, number 1, carries a fixed current produced by a battery. The flux Φ in loop 2 increases as it moves into a stronger field. The induced emf,

$$\mathscr{E} = -\frac{d\Phi}{dt}$$

is a motional emf and produces a <u>current in the opposite direction</u> <u>to that in loop 1</u>. Thus, the force between these two loops is re-

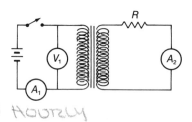

Fig. 36-21.

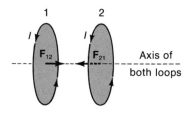

Fig. 36-22. Two loops with current flowing in the same sense attract one another.

pulsive. If we pull loop 2 away from loop 1, $d\Phi/dt$ is negative and the induced current is in the same direction as that in loop 1. Now the force between these loops is attractive.

In both cases, the magnetic force due to the induced current *opposed the mechanical motion that produced it*. Thus, an external force had to do work to move coil 2 closer to or farther away from coil 1. As noted in Section 36-4, this result indicates that energy is conserved in this process. Heat is produced by the induced current flowing in loop 2 regardless of which direction the current takes. Mechanical work occurs in both cases so that energy is conserved. If we cut loop 2 so that no current flows, this force vanishes. But since no current flows, no energy is consumed by circuit 2 in this case.

It is an extremely general result, in accordance with a generalization of the principle of the conservation of energy, that *the magnetic forces caused by induced currents always oppose the motion that produces them*. Consider now what happens if loop 2 is fixed in Fig. 36-24 and loop 1 moves toward or away from it. The induced current in loop 2 is now due to an induced electric field rather than a motional emf. Nonetheless, the same flux law (Faraday's law) describes this induced emf. This induced emf produces a current in loop 2 that now produces a magnetic field of its own. This field acts back upon the current in loop 1 and, as you may have guessed, causes a force on loop 1 that opposes *its* motion. Again, energy is conserved.

If we replace loop 1 with a bar magnet, as in Fig. 36-25, a similar result is obtained. The field produced by the magnet produces effects precisely analogous to those caused by loop 1 in Fig. 36-24.

Currents caused by induction, as in Fig. 36-26, are often called *eddy currents*. They are sometimes useful in reducing the oscillations of a system. An example is the use of eddy currents to damp out the oscillations of a chemical balance, as in Fig. 36-27.

The eddy currents we have discussed are caused by motion and produce magnetic forces that oppose that motion. Another class of eddy currents are caused by induced electric fields and produce electromagnetic effects. An example is found in the automotive ignition coil shown in Fig. 36-16. The iron core there is used to enhance the magnetic field strength in the device. However, iron is a conductor, so currents are induced in the iron core, where they are not desired, as well as in the secondary. The effect of these induced eddy currents is to *reduce* the magnetic field that would otherwise have been present. These currents are greatly reduced by making the iron core from a bundle of iron wires insulated from one another to break up the conduction path for the eddy currents.

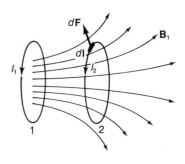

Fig. 36-23. The magnetic field **B**$_1$ due to I_1 acts upon current I_2 to produce the force on loop 2. The force on a segment of loop 2 is seen to be directed outward but also inclined toward loop 1.

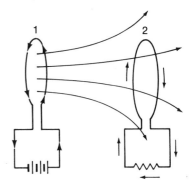

Fig. 36-24. As loop 2 approaches loop 1, the magnetic flux through loop 2 increases. The induced emf causes a current in the opposite direction to that of loop 1. A net force of repulsion results.

36-8   Lenz's Law   NOT ON HOURLY   REALLY IS

We have seen that Faraday's induction law,

$$\mathscr{E} = -\frac{d\Phi}{dt}$$

leads to induced emf's in an inductor that oppose the change in current producing them. Furthermore, we have seen that magnetic forces produced by induction oppose the motion that produced them. Lenz's law is a generalized statement embodying all such effects and is helpful in determining the direction of an induced current:

> The induced current will appear in such a direction that it opposes the change which produced it.

The following examples illustrate uses of Lenz's law.

Example   A bar magnet is pushed toward a conducting ring, as in Fig. 36-28. In order to oppose the motion, the ring must produce a "north" field on its left side, as shown, to repel the approaching north pole of the bar magnet. Using the right-hand rule for the field produced by a loop, the induced current must flow into the page at the top of the loop, out at the bottom.

Example   An observer sights along the axis of two loops, as in Fig. 36-29. In the loop closest to him, a battery causes a current to circulate clockwise so that it comes out of the page at the top of the loop. If this current is *increasing*, an emf is induced in that same loop to oppose this increase. This emf is due to an induced electric field with circulation about the axis in the counterclockwise sense. Thus, the current induced in the second loop by this same induced field will be

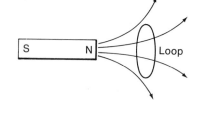

Fig. 36-25.   Motion of the loop toward or away from the bar magnet is opposed by magnetic forces. Similarly, motion of the bar magnet toward or away from the loop is opposed by the magnetic field generated by the induced current in the loop.

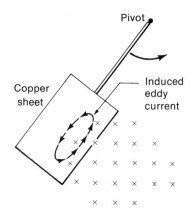

Region containing **B** field directed into page

Fig. 36-26.   The eddy current induced in this pendulum opposes its motion, so that its oscillations are quickly reduced in amplitude.

Fig. 36-27.   When the balance oscillates, eddy currents in the conducting vane produce magnetic forces that oppose the motion and stop the oscillation.

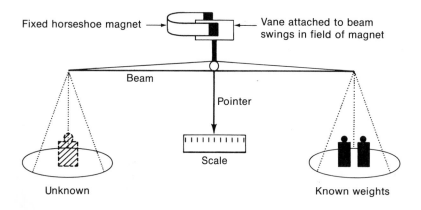

counterclockwise. If the current in the closest loop is *decreasing*, the induced current in the far loop will be clockwise.

Example A sheet of copper is pushed into the gap of a magnet, as in Fig. 36-30. Viewed from above, the induced eddy current must circulate counterclockwise so as to produce an induced north-south polarity, as shown, to be repelled by the poles of the magnet.

Exercises

7. A bar magnet is dropped toward a copper sheet, as in Fig. 36-31. Describe the currents produced in the sheet.

8. A coil is wound on a cardboard tube, as in Fig. 36-32 so that the observer "sees" the primary current I_1 circulating clockwise. The secondary coil is wound oppositely to the primary. Find the direction of the induced current (left or right through R) for (*a*) I_1 increasing, (*b*) I_1 decreasing. Now consider (*c*) I_1 flowing counterclockwise and increasing, (*d*) I_1 flowing counterclockwise and decreasing.

36-9 Relativity in Induction Phenomena   NOT ON HOURLY

The phenomenon of magnetic induction includes both *motional* emf due to purely magnetic forces and the appearance of an electric field with nonzero circulation whenever **B** changes intensity. Both effects, however, lead to the same law: Faraday's induction law. This is the result of the extremely important fact that only *relative* motion between a circuit and a source of magnetic field is important. In Fig. 36-33, an induced current results when the magnet is pushed toward the loop. It is caused by an induced electric field due to the flux change in the loop, $d\Phi/dt$. If instead the loop is moved, there is a magnetic force on each mobile charge in the loop:

$$\mathbf{F}_m = q(\mathbf{v} \times \mathbf{B})$$

which produces a current. In both cases the induced emf is equal to $d\Phi/dt$ and, hence, only *relative* motion can be detected by the presence of the induced current.

The two apparently different phenomena involved in induction, motional emf and induced electric fields, conspire to assure that the phenomena of electricity and magnetism satisfy the principle of relativity. "Absolute" motion is not detectable in induction effects; only relative motion is.

In Fig. 36-34 we see a conductor in the form of a long copper slab. If it moves as indicated, an induced current passes to the right

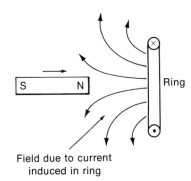

Field due to current induced in ring

Fig. 36-28. The direction of the induced current is deduced by requiring that the ring repel the magnet.

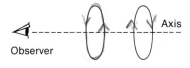

Fig. 36-29. Lenz's law gives the direction of the self-induced emf in the first loop, allowing us to deduce the direction of the emf induced in the second loop.

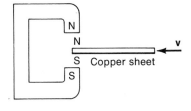

Fig. 36-30. Lenz's law demands that the force exerted on the eddy currents in the copper sheet be an opposing force.

across W and, hence, from right to left through the ammeter. We could obtain precisely the same result by keeping the slab fixed and moving the sliding ammeter circuit in the opposite direction. In both cases the magnitude of the induced emf is given by

$$\mathscr{E} = vBW$$

If, however, we move both the meter and slab together, the induced emf's in the bar and ammeter circuit are precisely equal and cancel, so that no current flows. Again, *relative* motion of slab and meter is all that is indicated by the induced current.

36-10 Energy Stored in an Inductor

When current is increasing in an inductor, the induced emf opposes the flow of current so that power must be provided by an external source, such as the battery of Fig. 36-35. Such an induced emf is often called a *back emf* for this reason. The magnitude of this power is given by

$$P = \mathscr{E} \cdot I = L\frac{dI}{dt}I$$

The net energy provided by the external source when the current increases from zero to I_f is given by the integral of the power over time:

$$W = \int_0^{t_f} P\, dt = L\int_0^{t_f} I\frac{dI}{dt}\, dt$$

$$= \tfrac{1}{2}LI_f^2 \tag{36-20}$$

If we now abruptly connect this inductor to a resistor, as by changing the switch in Fig. 36-35, the current will begin to decrease and the induced emf will act to keep the current flowing:

$$\mathscr{E} = L\frac{dI}{dt}$$

The work done by this induced emf on charges passing through the inductor is positive. Thus, when the current falls from I_f to zero, the *inductor* does a quantity of work:

$$W = \tfrac{1}{2}LI_f^2$$

which precisely equals the work done by the external source when the current was originally increased from zero to I_f.

It appears that energy is *stored* by an inductor. We call this energy the *potential energy* of the inductor:

$$U = \tfrac{1}{2}LI^2 \tag{36-21}$$

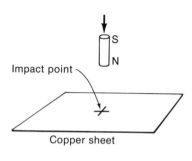

Impact point

Copper sheet

Fig. 36-31.

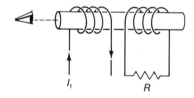

Fig. 36-32.

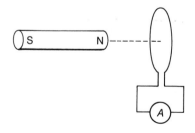

Fig. 36-33. Motion of either the magnet or the loop produces an induced current.

This result is somewhat reminiscent of the energy stored in a capacitor:

$$U = \tfrac{1}{2}CV^2$$

The units in (36-21) are those of energy (or work), which can be seen as follows:

Units of L: $\dfrac{\text{volts}}{\text{amperes/second}}$

Units of LI^2: $\dfrac{\text{volts}}{\text{amperes/second}} \cdot (\text{amperes})^2 = \text{volts} \cdot \text{coulomb}$

$$= \text{joules}$$

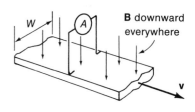

Fig. 36-34. A long copper slab moves transverse to an applied magnetic field. An induced current flows in the fixed ammeter circuit.

Exercise

9. A 10-mH inductance is carrying a constant current of 1 A. How much energy is stored in this inductor? How can we obtain this energy as heat?

36-11 Summary

A motional emf is caused by motion of a conductor through a magnetic field. If this motion results in a change of magnetic flux through a circuit, the emf is given by

$$\mathscr{E} = -\frac{d\Phi}{dt}$$

When this emf drives a current in a circuit, the energy expended is supplied by the mechanical work required to keep the conductor in motion against the retarding effects of magnetic forces.

When a circuit is stationary but the net flux passing through the circuit is changed, an induced emf appears. This emf is due to an induced electric field with nonzero circulation, a type of field that cannot occur in electrostatics. The changing magnetic field may be produced by moving a magnet near the stationary circuit or by changing the current in a second stationary circuit that produces a magnetic field. The emf induced in a circuit can be increased by wrapping the circuit into many turns so that the induced emf's add together. Rapidly changing currents can then produce very large emf's, as in an automotive ignition coil.

The magnitude of the emf induced in a circuit (2) is related to the rate of change of current in a separate circuit (1) by their mutual inductance, M:

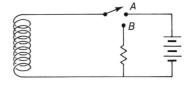

Fig. 36-35. When the switch is in position A, current increases in the inductor and its stored energy increases. When the switch is abruptly shifted to position B, this stored energy drives a current through the resistor.

$$\mathscr{E}_2 = M \frac{dI_1}{dt}$$

The self-inductance L relates the magnitude of self-induced emf to the rate of change of current in a circuit:

$$\mathscr{E} = -L \frac{dI}{dt}$$

where the minus sign indicates that the induced emf opposes the change that produces it.

When we examine the magnetic forces exerted upon induced currents by the magnetic fields that lead to those induced currents, we find that these forces always act against the motion leading to the induced currents.

Lenz's law combines the previous two statements to say that the direction of an induced current is such as to oppose the change that produced it.

Induction phenomena satisfy the principle of relativity: only relative motion is detectable.

Because of the back emf in an inductor, electrical energy is required to increase the current in an inductor. However, this energy is returned to the circuit when the current decreases. The energy used in building up the current in an inductor is thus effectively stored and we identify this energy as a potential energy:

$$U = \tfrac{1}{2}LI^2$$

Problems

36-1. A 1 m × 1 m square loop moves at a velocity of 10^2 m/s into a region containing a uniform magnetic field of 1.5 T, as indicated in Fig. 36-36. The net resistance of the loop

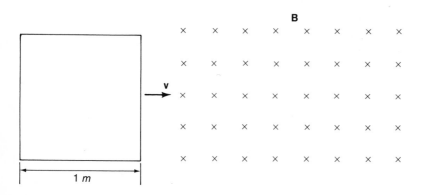

Fig. 36-36. The field **B** is directed into the page.

is 0.01Ω. Plot the induced current as a function of time starting just before the loop first reaches the region of **B** until just after the loop has entirely entered the region of **B**. Plot clockwise currents as positive, counterclockwise currents as negative.

36-2. A solenoid carrying a fixed current is oriented vertically. A copper ring with its plane horizontal is dropped from far above the solenoid so that it passes completely down the interior of the solenoid and emerges below it. Sketch a qualitative graph of the induced current in the copper ring as a function of time.

36-3. The half-loop in Fig. 36-37 is cranked at a rotational velocity $\omega = 10$ rad/s. A uniform field **B** with a magnitude of 1.2 T is perpendicular to the page. (*a*) Sketch a qualitative graph of the reading of the voltmeter V for one full revolution of the loop. Assume that good electrical connections are made at the rotating joints, that the voltmeter will indicate both positive and negative voltages, and that the response of the voltmeter is rapid enough to follow any changes present. (*b*) Calculate the maximum value of the voltmeter reading.

36-4. Repeat Problem 36-3*b* except assume that the magnitude of **B** varies uniformly from zero to 0.8 T in the 10-cm span from the left to the right edge of the half-loop.

36-5. A rigid rod moves through a square region containing a magnetic field, as in Fig. 36-11. Both the rod and its velocity are perpendicular to **B**. The net resistance of the complete circuit, including a galvanometer, is 20Ω. Find the induced current when $W = 20$ cm, $B = 0.5$ T, and $v = 4$ m/s.

36-6.* A *fluxmeter* uses a 100-turn search coil of copper wire with a cross-sectional area of 1 cm². It is placed within an unknown magnetic field with the plane containing the coil perpendicular to the field, so that the maximum possible flux passes through the coil. This coil is connected to a *ballistic galvanometer:* a device that gives a reading proportional to the net charge that passes through it. The total resistance of the circuit containing the coil and galvanometer is 10Ω. When the search coil is abruptly withdrawn to a region of zero field, the ballistic galvanometer reading indicates the unknown field strength. Find the calibration constant k in T/μC, where $B = kQ$, with B the unknown field and Q the reading of the ballistic galvanometer.

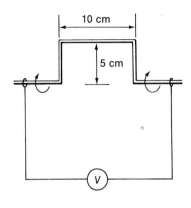

Fig. 36-37.

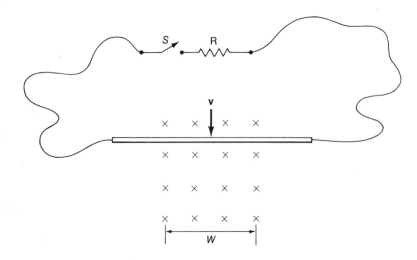

(a)

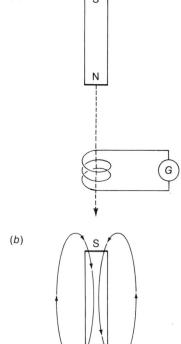

36-7. A moving rod in a magnetic field is connected by flexible leads to a resistor R, as shown in Fig. 36-38. For $B = 1$ T, $W = 10$ cm, and $R = 2\Omega$, calculate the force required to keep the rod moving at a steady 10 m/s when (a) the switch S is closed; (b) the switch S is opened. (c) Now calculate the rate at which this force does work (the power) with the switch closed. (d) Compare this rate to the power dissipated in the resistor $(P = I^2 R = \mathscr{E}^2/R)$.

(b)

36-8. A long, straight solenoid of radius r with N turns per unit length carries a current changing at a rate of dI/dt. Find the magnitude of the induced electric field immediately outside the solenoid near the central portion of the solenoid.

36-9. A bar magnet is dropped through a coil connected to a galvanometer, as in 36-39a. Describe the behavior of the galvanometer if the field of the bar magnet continues through the magnet, as in Fig. 36-39b. Describe the behavior of the galvanometer if the magnetic field lines go from the north pole to the south pole inside the magnet as well as outside, as in Fig. 36-39c.

(c)

36-10. A spark coil has an iron core with a cross-sectional area of 3 cm²; it carries essentially all of the coil's magnetic flux. If the average field strength within this core falls from 0.6 T to 0.1 T in 10 μs, how many turns are required in the secondary to obtain 30,000 V?

Fig. 36-39.

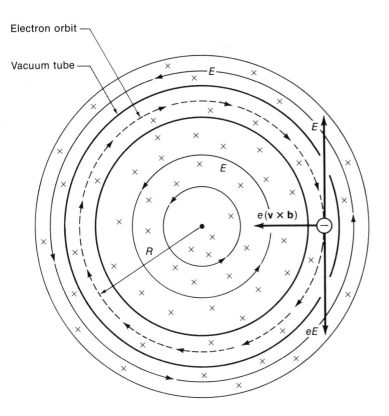

Electron orbit

Vacuum tube

Fig. 36-40. Electrons in the vacuum tube of a betatron are accelerated by the induced electric field caused by a changing magnetic flux within their orbit.

36-11. A betatron is a device that accelerates electrons in a doughnut-shaped vacuum tube by means of the electric field induced around a region of changing flux, as shown in Fig. 36-40. The net flux within the orbit is given by

$$\Phi = B_{av} \cdot \pi R^2$$

where B_{av} is the average value of B within the orbit and R is the radius of the orbit. The induced electric field is found by

$$\oint \mathbf{E} \cdot d\mathbf{s} = -\frac{d\Phi}{dt}$$

or $$-E \cdot 2\pi R = \frac{dB_{av}}{dt} \pi R^2$$

Show that the magnetic field strength at R must always be one-half of B_{av} if magnetic forces are to maintain the electrons in an orbit of fixed radius R.

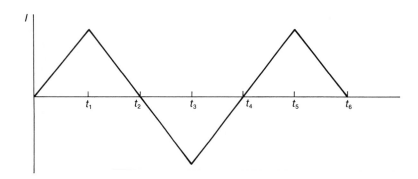

Fig. 36-41.

36-12. Two coils have a mutual inductance M. The current in the primary coil is given by the time-dependent expression

$$I_1 = a + bt + ct^2$$

The secondary coil is connected to a resistor, so that the total resistance of the secondary circuit is R. Find I_2 as a function of time.

36-13. Two solenoids are wound on a common coil-form, one on top of the other but in opposite directions. The rate at which current changes in the primary is a uniform 20 A/s. The resistance of the primary is 0.3Ω. Conventional current enters the primary at the left end of the solenoid and leaves at the right end. The voltage across the secondary is found to be negative at the left end, positive at the right end. The magnitude of the secondary voltage is $\frac{1}{2}$ V. (a) Is the current in the primary increasing or decreasing? (b) Find the value of M. (c) At the moment the primary current has an instantaneous value of 10 A, the voltage across the primary is measured to be $3\frac{1}{3}$ V. Find the value of L for the primary.

36-14. The current passing through an inductor varies with time in the manner shown in Fig. 36-41. Sketch a graph of the voltage across the inductor, assuming zero resistance.

35-15. A 50-mH inductor and 100Ω resistor are connected as shown in Fig. 36-42. At $t = 0$, a current of 100 mA exists in this circuit. The current is decreasing. (a) Find the voltage drop across the resistor. (b) Since this same voltage drop must exist across the inductor (they are connected by resistanceless wires), find the instantaneous

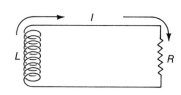

Fig. 36-42.

magnitude dI/dt at $t = 0$. (c) Find the rate at which power is dissipated in the resistor at $t = 0$. (d) Find the rate at which the energy stored in the inductor is decreasing.

36-16.   Switch S_1 in the circuit of Fig. 36-43 is closed. When a steady current I_0 has been established in L, switch S_2 is closed and S_1 is opened simultaneously. The energy stored in L now appears as heat in the resistor. How many calories of heat are produced for $I_0 = 10$ A and $L = 100$ mH?

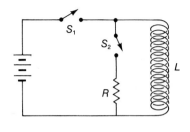

Fig. 36-43.

Circuits Containing Capacitors and Inductors

37-1 Introduction In circuits containing only resistors, there is a simple proportionality between current and the voltage drop across a given portion of the circuit. If the current varies rapidly, the voltage drop does so in a similar manner. When capacitors or inductors are present, however, a new dimension of circuit behavior appears. These two circuit elements exhibit voltage drops that are not directly dependent upon the magnitude of the current but upon its rate of change. The voltage and current in such circuits are not necessarily simply proportional to one another.

The effects of capacitors and inductors in circuits are most dramatic for rapidly varying currents. Circuits with rapidly varying currents are widely prevalent in technology, where the use of capacitors and inductors as circuit elements is common.

In this chapter we will discuss general properties of circuits with time-varying currents. In the next chapter we will concentrate our attention on one extremely important type of time variation: sinusoidally varying currents.

37-2 Voltage-Current Relations for Capacitors and Inductors

The potential difference that exists across a capacitor when it carries a charge Q is given by

$$V = \frac{1}{C} Q$$

The charge Q has been transferred from one element (or plate) of the capacitor to the other element in the process of charging the

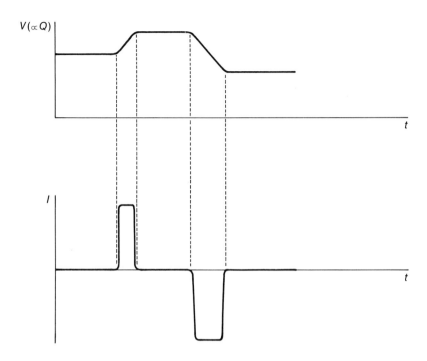

Fig. 37-1. When the charge on a capacitor is constant, no current passes into or out of the capacitor. The charge on the capacitor changes only during times when the current in non-zero. Since the capacitor voltage is proportional to Q, the current is related to the time derivative of the voltage.

capacitor. Differentiating this expression with respect to time, we obtain

$$\frac{dV}{dt} = \frac{1}{C}\frac{dQ}{dt} \qquad (37\text{-}1)$$

When a current flows from one plate of the capacitor to the other, the charge Q changes. The magnitude of the current is precisely dQ/dt, the rate at which the charge of the capacitor varies. Thus we rewrite (37-1):

$$\frac{dV}{dt} = \frac{1}{C}I \qquad (37\text{-}2)$$

We see in (37-2) that the potential difference across a capacitor is related to the current flowing from one plate to the other by a time derivative. This is quite different from the current-voltage relation of a resistor, where the current is simply proportional to the potential difference across the resistor (Ohm's law). The current-voltage relationship for a capacitor is indicated in the example graphed in Fig. 37-1.

A time derivative is also involved in the relationship between current in an inductor and the induced emf between the terminals of the inductor:

$$\mathscr{E} = -L\frac{dI}{dt} \qquad (37\text{-}3)$$

where the minus sign indicates that the induced emf opposes the change in current.

Since the voltage drop across a resistor is in the direction that opposes current, let us consider it the "normal" direction for a voltage drop and adopt the sign convention illustrated in Fig. 37-2. When current flows from A to B, the polarity of V_{AB} will be considered as positive if A is positive and B negative. Then we use Ohm's law for a resistor, (37-2) for a capacitor, and (37-3) for an inductor to find V_{AB} for each of these elements with the sign convention of Fig. 37-2:

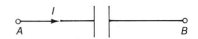

Resistor: $$V_{AB} = RI \qquad (37\text{-}4)$$

Capacitor: $$V_{AB} = \frac{1}{C} \int I\, dt \qquad (37\text{-}5)$$

Inductor: $$V_{AB} = L\frac{dI}{dt} \qquad (37\text{-}6)$$

Fig. 37-2. When we take the positive direction of current to the right, the polarity of the potential difference V_{AB} will be considered positive when A is positive and B negative.

In the remainder of this chapter we will study the way these different voltage-current relations interact when more than one type of circuit element is present in a circuit. We will begin by considering a circuit with a resistor and a capacitor.

37-3   The *RC* Circuit

Consider the circuit shown in Fig. 37-3. The capacitor is initially charged with a charge Q_0, so that an initial potential difference exists across its terminals:

$$V_0 = \frac{1}{C}Q_0$$

At a given time, say $t = 0$, the switch is closed, as in Fig. 37-3b. The terminals AB of the capacitor are now connected to the terminals CD of the resistor so that $V_{AC} = 0$ and $V_{BD} = 0$ (assuming

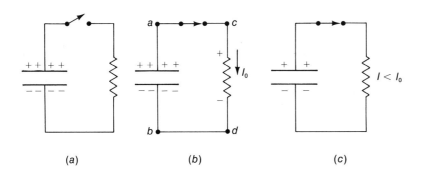

(a)   (b)   (c)

Fig. 37-3. At the instant the switch closes, the potential differences V_{ab} and V_{cd} become equal in magnitude. The current through the resistor is thus proportional to the voltage, and hence the charge, of the capacitor.

that the interconnecting wires have negligible resistance). Thus we must have

$$V_{CD} = -V_{AB}$$

where the minus sign indicates that while the voltage drop across the resistor is in the normal sense (opposing the current) given by Fig. 37-2, V_{AB} acts in the opposite direction (aiding the current). Writing V_{AB} in terms of the charge on the capacitor and V_{CD} in terms of Ohm's law, we find:

$$I_0 R = \frac{-Q_0}{C} \tag{37-7}$$

At $t = 0$ the current I_0 passes through the resistor. However, the charge on the capacitor immediately begins to decrease so that V_{AB}, and hence V_{CD}, also decreases. This decrease in voltage across the resistor causes a decrease in the current, so that the rate of discharge of the capacitor continually decreases.

At *any* particular time we may write

$$IR = -\frac{Q}{C} \tag{37-8}$$

Recognizing that $I = dQ/dt$, we have

$$\frac{dQ}{dt} R = -\frac{Q}{C}$$

and solving for dQ/dt, we obtain a differential equation:

$$\frac{dQ}{dt} = -\frac{1}{RC} Q \tag{37-9}$$

Since Q diminishes with time, Q is a function of time. The function $Q(t)$ that satisfies (37-9) is called the *solution of the differential equation*. The simple proportionality between Q and its time derivative may remind the reader of some properties of exponential functions:

If $y = e^{kt}$ then $\dfrac{dy}{dt} = ke^{kt} = ky.$

If $y = e^{-kt}$ then $\dfrac{dy}{dt} = -ke^{-kt} = -ky.$

Furthermore, simple proportionality of the exponential function and its derivative is maintained when an arbitrary constant multiplies the exponential:

If $y = Ae^{-kt}$ then $\dfrac{dy}{dt} = -kAe^{-kt} = -ky.$ \hfill (37-10)

In our application, k is equal to $1/RC$. Thus it appears that Eq. (37-9) is satisfied if the time dependence of Q is given by

$$Q = Q_0 e^{-t/RC} \tag{37-11}$$

where Q_0 is a constant. Substitution of our guess for Q, Eq. (37-11), into (37-9) leads to

$$\frac{d}{dt}(Q_0 e^{-t/RC}) = -\frac{1}{RC} Q_0 e^{-t/RC}$$

Carrying out the differentiation by use of (37-10), we find

$$-\frac{1}{RC} Q_0 e^{-t/RC} = -\frac{1}{RC} Q_0 e^{-t/RC}$$

and indeed, the differential equation (37-9) is satisfied by our solution.

The constant Q_0 that appears in our solution can be found in terms of the initial conditions. At $t=0$, Eq. (37-11) becomes $Q(t=0) = Q_0 e^{-0} = Q_0$, so that Q_0 is simply the magnitude of the charge on the capacitor at $t = 0$ (the instant the switch is closed).

As time increases, Q decreases, as shown in Fig. 37-4. When $t = RC$, Eq. (37-11) becomes

$$Q = Q_0 e^{-1}$$

so that Q has fallen to $1/e$ of its initial value. Since $e = 2.718$, this value is 0.368 (approximately 1/3) of the initial value. After another time interval of length RC, the value of Q will have fallen by still another factor of $1/e$:

$$Q = Q_0 e^{-2} \qquad \text{for} \qquad t = 2RC$$

where $e^{-2} = 0.135$. The behavior of Q given by (37-11) is sometimes called an *exponential decay law* and the time RC that controls the rapidity of the decay is called the *lifetime*, or the *decay time*, or

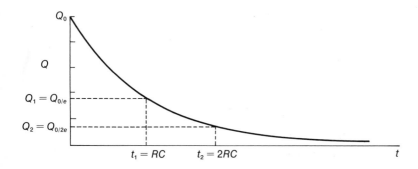

Fig. 37-4. The time dependence of Q, where $t = 0$ is the moment that the switch in Fig. 37-3 is closed.

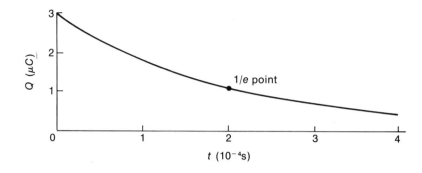

the *RC time*. To see that the product *RC* has the dimensions of time, we note that $R = V/I$ and $C = Q/V$, so that

$$RC = \frac{\text{charge}}{\text{current}} = \frac{\text{charge}}{\text{charge/time}} = \text{time}$$

For *R* in ohms and *C* in farads, the product *RC* has the units of seconds. A numerical illustration for chosen values of *R* and *C* is shown in Fig. 37-5.

Since $V = Q/C$, the magnitude of the voltage across the capacitor (and hence the resistor) is easily found from (37-11):

$$V = \frac{Q}{C} = \frac{Q_0}{C}e^{-t/RC} = V_0 e^{-t/RC} \qquad (37\text{-}12)$$

where V_0 is the initial capacitor voltage.

Exercises

1. Show that the magnitude of the current in the *RC* circuit during discharge is given by

    $$I = I_0 e^{-t/RC}$$

    where $I_0 = V_0/R$.

2. A 6-μF capacitor and 10⁴Ω resistor are arranged as in Fig. 37-3. The initial charge on the capacitor is 1 μC (10⁻⁶ C). (*a*) What is the *RC* time of this circuit? (*b*) What is the magnitude of the remaining charge at $t = 0.12$ s? (*c*) Find the potential difference at $t = 0.12$ s. (*d*) Find *I* at $t = 0$.

Consider now the *RC* circuit shown in Fig. 37-6. When the switch is closed the capacitor begins to charge. It continues to do so until its potential difference equals (and opposes) that of the battery, so that current can no longer flow. It is clear that as the

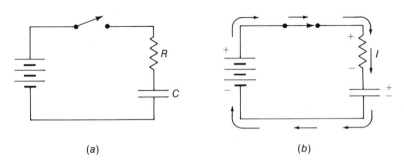

(a)   (b)

Fig. 37-6. The capacitor is initially uncharged. When the switch is closed, the capacitor begins to charge as current passes through R.

capacitor charges, the *rate* of charging will diminish, since the capacitor's potential difference increasingly opposes that of the battery. As in the previous case, we should be able to write an equation that will describe the way that this charging process depends upon time. It will again be a differential equation involving a time derivative. Its solution, hopefully, will embody all the qualitative ideas described above.

After the switch is closed in Fig. 37-6, current will flow as indicated. The voltage drop across the resistor and the potential difference across the capacitor oppose the flow of charge. At any given instant, equilibrium is maintained if the sum of the potential differences across the resistor and capacitor equals the emf of the battery:

$$\mathcal{E} = IR + \frac{Q}{C} \tag{37-13}$$

Noting that $I = dQ/dt$, we have

$$\mathcal{E} = \frac{dQ}{dt}R + \frac{Q}{C} \tag{37-14}$$

We could integrate Eq. (37-14) with respect to time. The arbitrary constants acquired in integration can then be found by demanding that the initial current (before the capacitor acquires any charge) must be $I_0 = \mathcal{E}/R$. Alternatively, we could differentiate (37-13) to obtain

$$0 = R\frac{dI}{dt} + \frac{I}{C}$$

or

$$\frac{dI}{dt} = -\frac{1}{RC}I$$

Looking back at our solution to (37-9), it is clear that now the solution is

$$I = I_0 e^{-t/RC} \tag{37-15}$$

Substituting (37-15) into (37-13), we find

$$\mathscr{E} = \mathscr{E}e^{-t/RC} + \frac{Q}{C}$$

Hence $\qquad\qquad Q = C\mathscr{E}(1 - e^{-t/RC})$ $\qquad$ (37-16)

The factor $C\mathscr{E}$ in (37-16) is the quantity of charge on the capacitor when its potential difference equals the battery emf. This, of course, occurs when the capacitor is fully charged. Thus the quantity $C\mathscr{E}$ is Q_f, the final charge on the capacitor:

$$Q = Q_f(1 - e^{-t/RC}) \qquad (37\text{-}17)$$

Graphs of current, charge, and capacitor voltage, all as functions of time, are shown in Fig. 37-7.

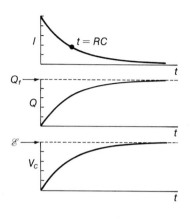

Fig. 37-7. The behavior of the circuit of Fig. 37-6 is indicated by graphs of current, charge on the capacitor, and potential difference across the capacitor. The time of closing the switch is taken as $t = 0$.

Exercises

3. Differentiate (37-16) to show that it gives the same dependence of current upon time as given by (37-15).

4. A 4-μF capacitor is connected to a 12-V battery. The total resistance in the circuit (including the internal resistance of the battery) is 0.02Ω. How long after completing the circuit is the capacitor voltage equal to $\left(1 - \dfrac{1}{e}\right)V_{\text{batt}}$?

37-4   *RC* Differentiators and Integrators

The charge on a capacitor is equal to the integral of its charging current over all past time. The voltage is given by:

$$V = \frac{Q}{C} = \frac{1}{C}\int I\,dt$$

If, then, we can produce a *current* that is proportional to the function we wish to integrate, the value of the integral of that current will always be proportional to the voltage across the capacitor. An illustration is shown in Fig. 37-8.

Since the integral over time of a positive constant is a linearly increasing function, a "linear ramp" can be produced by forcing a constant current into a capacitor, as in Fig. 37-9. The required constant current source can be approximated by a large emf in series with a large resistor. For example, if $\mathscr{E} = 1000$ V and $R = 10^6\Omega$, the current produced remains within 1% of 10^{-3} A as long as the capacitor voltage does not exceed 10 V.

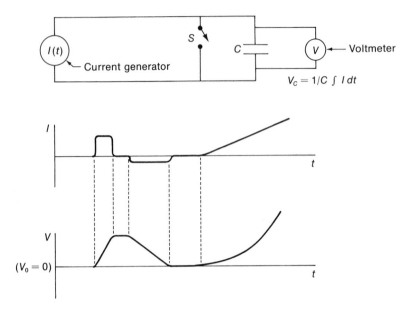

Fig. 37-8. The voltage across a capacitor is proportional to the integral of the charging current. In this circuit, the capacitor voltage rises linearly during a period of positive, constant current, falls linearly with a smaller slope during a period of negative, constant current with smaller magnitude, and finally rises quadratically during a period of linearly increasing current. The switch S clears the "answer" from any preceding integration.

However, it is often impractical to generate a current that is strictly proportional to the quantity to be integrated. For example, we could generate an *emf* proportional to the velocity of an automobile by connecting a unipolar generator to its axle. But the charging current would cease as soon as the integrating capacitor reached a voltage equal to the emf. Often, as in this case, it is desirable to obtain the integral of a *voltage* that depends upon time, not a current. The circuit of Fig. 37-10a shows such an integrator. The quantity V_{out} is only approximately proportional to the integral over

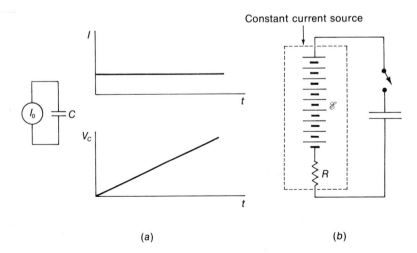

Fig. 37-9. (a) A constant current produces a capacitor voltage that increases linearly with time. (b) A constant-current source can be approximated by a large emf in series with a large resistor.

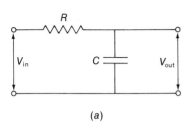

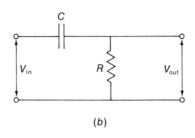

(a)   (b)

Fig. 37-10. (a) An *RC* integrator. (b) An *RC* differentiator.

time of V_{in}. As long as the voltage changes to be integrated are large compared to the capacitor voltage, it is quite accurate. However, if the capacitor voltage rises excessively (i.e., if we obtain too large an "answer" for our integration problem), the *charging rate* is no long proportional to the voltage input and thus the output voltage is no longer proportional to the integral of V_{in} (see Fig. 37-11). The differentiator of Fig. 37-10*b* gives an output closely proportional to the time derivative of the input as long as input voltage changes are slow compared to the *RC* time of the circuit. An example when this criterion is satisfied is shown in Fig. 37-12*a*, while Fig. 37-12*b* illustrates a case in which this criterion is not satisfied.

Exercises

5.   Show that the system in Fig. 37-13*b* is a hydraulic analog of the *RC* integrator shown in Fig. 37-13*a*.

6.   The waveform in Fig. 37-14 is to be differentiated with an *RC* differentiator with $C = 1 \ \mu F$. We wish to make *R* as large as possible to obtain a large signal, but *RC* must be small enough to allow proper differentiation. What value of *R* would you choose?

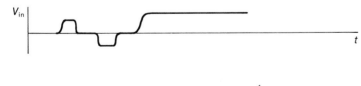

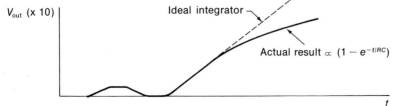

Fig. 37-11. The integrator of Fig. 37-10 is adequate as long as excessive charge does not occur, i.e., as long as $V_{out} \ll V_{in}$.

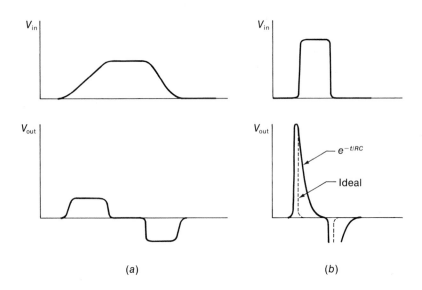

(a)   (b)

Fig. 37-12. (a) If V_{in} varies slowly in the circuit of Fig. 37-10b, the output is closely proportional to dV_{in}/dt. But if the input changes quickly, so that the time interval $\Delta t = RC$ is not negligible, the output is as shown in (b) above.

37-5 *LR* Circuits

Let us begin by considering an idealized inductor (one with no resistance) connected to an emf, as in Fig. 37-15. The voltage across the inductor must equal the emf, so we have

$$\mathscr{E} = L\frac{dI}{dt}$$

and

$$I = \int \frac{dI}{dt} = \frac{\mathscr{E}}{L}t + \text{constant}$$

The integration constant will be zero if we set $I = 0$ at $t = 0$ by keeping the switch in Fig. 37-15 open until $t = 0$. Then we find

$$I = \frac{\mathscr{E}}{L}t \qquad\qquad (37\text{-}18)$$

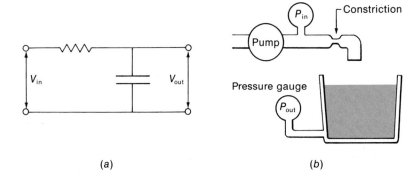

(a)   (b)

Fig. 37-13. In the hydraulic analog of the *RC* integrator, the analog of V_{in} is given by the reading of the input pressure gauge while P_{out} is analogous to V_{out}.

Resistance is included in the circuit of Fig. 37-16. When the switch is in the upper position, current begins to flow. Now as the current rises, the voltage drop across R rises, so that an ever increasing portion of $\mathscr{E}$ appears as a voltage drop across the resistor. The voltage across the inductor must decrease. The rate of increase of current thus diminishes steadily. By equating the battery emf to the sum of the voltage drops across the resistor and inductor, we can obtain a differential equation for the current:

$$\mathscr{E} = IR + L\frac{dI}{dt} \tag{37-19}$$

Solution of (37-19) proceeds in a manner similar to the solution of (37-14). When we solved (37-14) for the *RC* circuit, we found that the charge varied with time as $(1 - e^{-t/RC})$. Similarly, the current satisfying (37-19) is found to be

$$I = \frac{\mathscr{E}}{R}\left(1 - e^{-\frac{R}{L}t}\right) \tag{37-20}$$

where the quantity L/R plays the role formerly occupied by RC. We can easily see that the dimensions of L/R are those of time, since

$$L = \frac{\mathscr{E}}{dI/dt} \quad \text{and} \quad R = \frac{\mathscr{E}}{I}$$

Thus, in SI units,

$$\frac{L}{R} = \frac{\text{amperes}}{\text{amperes/second}} = \text{second}$$

A plot of the solution (37-20) is shown in Fig. 37-17.

The reader should now be able to show that if the switch is abruptly flipped to the down position in Fig. 37-16, the current will exponentially decay:

$$I = I_0 e^{-(R/L)t} \tag{37-21}$$

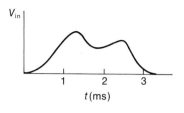

Fig. 37-14.

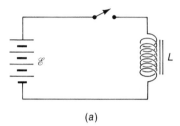

(a)

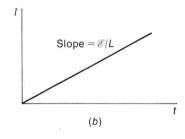

(b)

Fig. 37-15. Current increases linearly with time in an inductive circuit with no resistance.

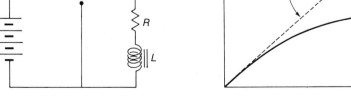

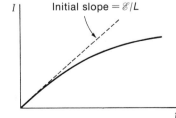

Fig. 37-16. As current increases, the voltage drop across the resistor increases so that the voltage across the inductor decreases. As $V_L \to 0$, the rate of increase of I tends toward zero.

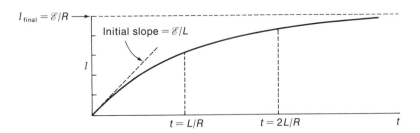

Fig. 37-17. Current as a function of time in an *LR* circuit.

where I_0 is the current at the moment the switch closed. The photographs in Fig. 37-18 were obtained with an oscilloscope arranged to display voltages that appeared across the *LR* portion of the circuit, across *L* alone, and across *R* alone, as the switch spent equal time intervals in the upper and lower positions.

An interesting puzzle arises when we compare Eqs. (37-18) and (37-20). If we let $R \to 0$ in the *LR* circuit described by (37-20), we expect to obtain (37-18). Unfortunately, it seems that allowing $R \to 0$ in (37-20) will lead to difficulty since R appears in the denominator. To resolve this difficulty we need only to examine the series expansion of the exponential function:

$$e^x = 1 + x + \frac{x^2}{2!} + \frac{x^3}{3!} + \frac{x^4}{4!} + \cdots$$

For the exponential in (37-20), this becomes

$$e^{-(R/L)t} = 1 - \frac{R}{L}t + \frac{1}{2}\left(\frac{R}{L}t\right)^2 - \frac{1}{6}\left(\frac{R}{L}t\right)^3 + \cdots \qquad (a)$$

and if $(R/L)t \ll 1$, the squared terms and all higher terms become insignificant. Thus as $R \to 0$, Eq. (37-20) becomes

$$I \to \frac{\mathscr{E}}{R}\left(1 - 1 + \frac{R}{L}t\right) \qquad \text{as} \qquad R \to 0 \qquad (b)$$

or

$$I \to \frac{\mathscr{E}}{L}t \qquad \text{as} \qquad R \to 0 \qquad (c)$$

in agreement with (37-18), which was derived *assuming* $R = 0$.

Exercises

7. The *LR* circuit of Fig. 37-16 is made up of a 1-H inductor and a resistance of 200Ω. When the switch to the battery is closed, find the time required for the current to reach the value

$$I = \frac{\mathscr{E}}{R}\left(1 - \frac{1}{e}\right) = \frac{\mathscr{E}}{R}(0.632)$$

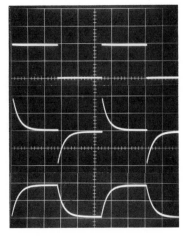

Fig. 37-18.   Oscilloscope traces that indicate voltage (upward) versus time (horizontally) in an *LR* circuit. (*a*) The voltage applied to the circuit. (*b*) The voltage across the inductor. (*c*) The voltage across the resistor. Note that (*b*) and (*c*) add to give (*a*).

8. Find the voltage across the resistor in an *LR* circuit by multiplying (37-20) by *R*. Find the voltage across the inductor by multiplying the derivative of (37-20) by *L*. Do your results agree with Fig. 37-18? Do V_L and V_R add to $\mathscr{E}$?

37-6   The *LC* Circuit

In the *LC* circuit of Fig. 37-19, a charge Q_0 is placed on *C* and the switch is then closed. As shown in the figure, the charge then oscillates from one side of the capacitor to the other. The induced emf in the inductor prevents the charge from transferring instantaneously. It also tends to keep the current moving, once established, so that the capacitor charge always "overshoots" to the opposite polarity. This process repeats indefinitely and is reminiscent of

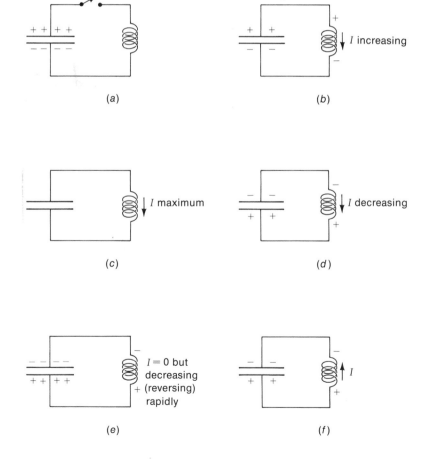

(a)

(b)

(c)

(d)

(e)

(f)

Fig. 37-19. (*a*) The initial *LC* circuit. (*b*) Charge transfers from the upper plate, through the inductor, and onto the lower plate. The induced emf in the inductor opposes the current. (*c*) The current is at a maximum with zero voltage across *C* and across *L*. (*d*) The current decreases and the induced emf in the inductor tends to maintain the current. (*e*) The current ceases (but is changing rapidly from one direction to another) and (*f*) the cycle begins anew in the opposite direction.

the endless oscillations of a harmonic oscillator. The analogy is quite accurate, as we will see by writing the differential equation that describes the behavior of this circuit.

When the capacitor and inductor are connected, there can be no voltage drop in the connecting wires (assuming $R = 0$), so following the sign convention of Fig. 37-2, we find

$$V_L = -V_C$$

or
$$L\frac{dI}{dt} = -\frac{Q}{C} \tag{37-22}$$

Noting that $I = dQ/dt$, we rewrite (37-22) to obtain

$$\frac{d^2Q}{dt^2} = -\frac{1}{LC}Q \tag{37-23}$$

The analogy with a mechanical oscillator is evident when we compare (37-23) with (16-1):

$$\frac{d^2x}{dt^2} = -\frac{k}{m}x \tag{16-1}$$

In the mass-spring system described by (16-1), the position x oscillates, while in the LC circuit the quantity Q oscillates. The inductance L is analogous to the mass in a mechanical oscillator, while the inverse of the capacitance is analogous to the spring constant in the mechanical oscillator.* Guided by our solution of (16-1) in Chapter 16, we know that Q will vary sinusoidally in time. As a check, let us assume that

$$Q = Q_0 \sin(\omega t)$$

where ω is a constant. Then

$$\frac{dQ}{dt} = \omega Q_0 \cos(\omega t)$$

and
$$\frac{d^2Q}{dt^2} = -\omega^2 Q_0 \sin(\omega t)$$

Substituting Q and d^2Q/dt^2 into (37-23), we find

$$-\omega^2 Q_0 \sin(\omega t) = -\frac{1}{LC}Q_0 \sin(\omega t)$$

---

*A large capacitor corresponds to a soft spring. Only a small voltage is produced by a large charge when C is large—just as only a small force is produced by a large displacement when k is small.

so that (37-23) is satisfied if

$$\omega^2 = \frac{1}{LC} \qquad (37\text{-}24)$$

The dimensions of ω in (37-24) must be t^{-1}, where the constant ω is the angular frequency of the simple harmonic oscillations. Let us check the dimensions of (37-24):

$$L = \frac{\mathscr{E}}{dI/dt} = \frac{\text{voltage} \cdot (\text{time})^2}{\text{charge}}$$

$$C = \frac{Q}{V} = \frac{\text{charge}}{\text{voltage}}$$

Hence

$$\frac{1}{LC} = \frac{1}{(\text{time})^2}$$

as expected. From our discussion in Chapter 16 we recall that

$$Q = Q_0 \cos(\omega t) \qquad (37\text{-}25)$$

is also a solution of (37-23). It is the solution that is appropriate if we wish $t = 0$ to occur when the switch is first closed, for then Q is maximum at $t = 0$, as shown in Fig. 37-20.

The frequency of the oscillations in Q denotes the number of complete cycles of oscillation (such as *abcde* in Fig. 37-20) per second. The units for frequency are hertz (Hz), and f is given by

$$f = \frac{\omega}{2\pi}$$

The frequency f is called the *natural frequency* or *resonant frequency* of the *LC* circuit. The period of the oscillations, T, is the time required for one complete cycle:

$$T = \frac{1}{f}$$

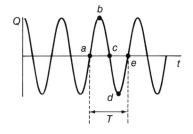

Our *LC* circuit, then, exhibits simple harmonic oscillations with frequency and period

$$f = \frac{\omega}{2\pi} = \frac{1}{2\pi} \sqrt{\frac{1}{LC}} \qquad (37\text{-}26)$$

and

$$T = \frac{1}{f} = 2\pi \sqrt{LC} \qquad (37\text{-}27)$$

Fig. 37-20. The charge on the capacitor is given by the expression in (37-25) when the switch is closed at $t = 0$ with the capacitor initially charged.

Example   A 70-μH inductor is connected to a 365-pF capacitor in an *LC* circuit. The natural oscillation frequency is given by:

$$f = \frac{1}{2\pi} \sqrt{\frac{1}{LC}}$$

$$= \frac{1}{6.28} \sqrt{\frac{1}{(7 \times 10^{-5} \text{ H}) \cdot (3.65 \times 10^{-10} \text{ F})}}$$

$$= \frac{6.26}{6.28} \times 10^6 \text{ s}^{-1} = 997 \text{ kHz}$$

Electromagnetic waves with such a frequency are near the center of the AM radio band. We will see in Chapter 39 that the *LC* circuit is one of the commonly used methods of controlling the frequency of man-made electromagnetic radiation.

Because of the lack of resistance in our idealized *LC* circuit, electrical energy is conserved. This energy resides first in the capacitor, then in the inductor, then back in the capacitor. This behavior is again reminiscent of the mechanical oscillator, where energy continually exchanges between kinetic and potential energy. For the *LC* circuit we have energy stored in the capacitor and the inductor:

$$E_C = \tfrac{1}{2}Q^2/C$$

$$E_L = \tfrac{1}{2}LI^2$$

but the values of Q and I are continually oscillating. To find I we simply differentiate (37-25):

$$I = \frac{dQ}{dt} = -\omega Q_0 \sin(\omega t)$$

Then $E_L = \tfrac{1}{2}LI^2 = \tfrac{1}{2}L\omega^2 Q_0^2 \sin^2(\omega t) = \tfrac{1}{2}\dfrac{Q_0^2}{C} \sin^2(\omega t)$

so that the energy stored in the inductor fluctuates between zero and $\tfrac{1}{2}Q_0^2/C$. The energy in the capacitor fluctuates between the same limits, but out of phase with the inductor:

$$E_C = \tfrac{1}{2}\frac{Q^2}{C} = \tfrac{1}{2}\frac{Q_0^2}{C} \cos^2(\omega t)$$

The sum of E_L and E_C gives the total energy in the circuit:

$$E_{\text{total}} = E_L + E_C = \tfrac{1}{2}\frac{Q_0^2}{C} [\sin^2(\omega t) + \cos^2(\omega t)]$$

$$= \tfrac{1}{2}\frac{Q_0^2}{C}$$

As was the case with the mechanical oscillator, the total energy is constant and equal to the initial energy of the system, as shown in Fig. 37-21. This result does not depend upon our choice of the sine or cosine form for the dependence of Q upon t (see Exercise 10).

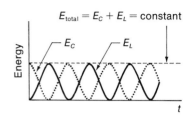

$E_{total} = E_C + E_L = \text{constant}$

Exercises

9. A variable capacitor is connected to a 70-μH inductor. What value of C is required to make this circuit oscillate with $f = 2$ MHz?

10. Prove that the total energy of an LC circuit is constant when Q is given by

$$Q = Q_0 \sin(\omega t)$$

rather than by (37-25), as was done in the text.

Fig. 37-21. The energy stored in the inductor is out of phase with that stored in the capacitor (dashed line), so that their sum is constant.

37-7   The Series *LRC* Circuit

Total energy is conserved in an oscillating LC circuit. If, however, a resistor is added, as in Fig. 37-22, electrical energy must be dissipated as the current passes repetitively through the resistor. This energy lost in the resistor appears as heat but is lost as far as electrical energy is concerned. If the resistance is sufficiently small, oscillations can still occur but their amplitude diminishes steadily, as in Fig. 37-23a. Oscillations of decreasing amplitude are called *damped* oscillations. The result of increasing the value of R in the *LRC* circuit is shown progressively in Fig. 37-23a,b,c, and d.

The differential equation describing the series *LRC* circuit is obtained by adding a term to (37-23) that represents the IR drop in the resistor:

$$\frac{d^2Q}{dt^2} + \frac{R}{L}\frac{dQ}{dt} + \frac{1}{LC}Q = 0 \qquad (37\text{-}28)$$

The solution $Q(t)$ of (37-28) is in the form of a sinusoidal, hence oscillating, function multiplied by a negative exponential so that the amplitude of the oscillations dies away as energy is dissipated in the resistor:

$$Q = Q_0 e^{-(\gamma/2)t}\sin(\omega t) \qquad (37\text{-}29)$$

where $\gamma = R/L$ and

$$\omega^2 = \omega_0{}^2 - \frac{\gamma^2}{4}$$

so that the resonant frequency is reduced by the presence of re-

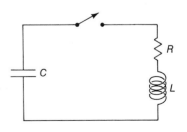

Fig. 37-22.   An *LRC* circuit.

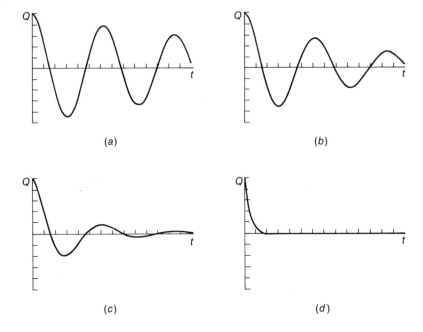

(a)

(b)

(c)

(d)

Fig. 37-23. The oscillations in an *LRC* circuit die out more rapidly as the resistance increases from (a) through (d).

sistance. The solution (37-29) decreases in amplitude more and more rapidly as R/L increases, as shown in Fig. 37-23.

This *LRC* circuit has a mechanical analog in the system shown in Fig. 37-24. For some fluids, with a proper choice of dimensions for the piston and fluid container, the force the piston exerts upon the mass m depends upon velocity in a linear manner:

$$F_p = -Rv = -R\frac{dx}{dt}$$

Such a device is called a *dashpot* or *damper* and R is its *resistance constant*. The differential equation governing the motion of m is then obtained as follows:

$$F_{net} = ma$$

$$-kx - R\frac{dx}{dt} = m\frac{d^2x}{dt^2}$$

where k is the spring constant, or

$$\frac{d^2x}{dt^2} + \frac{R}{m}\frac{dx}{dt} + \frac{k}{m}x = 0$$

which is identical in form to (37-28). The way displacement, x, can depend upon time in this mechanical system is identical with the way Q depends upon time in the series *LRC* circuit. Because of this similarity, electrical models of mechanical systems can often be used to study the effect of varying parameters of the mechanical system.

The system of Fig. 37-24 may not seem to be particularly interesting, but a great many mechanical systems are well represented by it. An example is given by the suspension system of an automobile. If the mass of the automobile itself is large compared to its "unsprung weight" (the wheels and axles that move up and down over bumps) and if the springs are not too stiff, the automobile follows a smooth course while the wheels undulate over the bumps. A given wheel plus its axle is similar to the system of Fig. 37-24, with m representing the mass of the wheel and axle, k representing the combined force constant of the spring and elasticity of the tire, and R representing the resistance to motion offered by the shock absorber (which is a sealed dashpot unit).

This analogy immediately allows us to see that if R is too small — or m too large — the wheel assembly will bounce up and down many times after striking a simple bump. On each upswing the tire pressure against the road is small and traction is easily lost. This inherently hazardous condition is eliminated by carefully tailoring the resistance constant of the shock absorber to the mass and spring constants of the wheel-axle assembly so that a response like that of Fig. 37-23d occurs. Racing cars use light-weight wheels, stiff springs and optimally adjusted shock absorbers to minimize the time required for the tire to regain traction after a bump.

The electrical circuit of Fig. 37-22 is often present in devices when we least expect it. An example is given by the circuit of Fig. 37-25. The transistor is used as a switch in this particular circuit. Its action is schematically indicated in Fig. 37-25b. Also shown are L_s and C_s, the *stray inductance* and *stray capacitance* of the

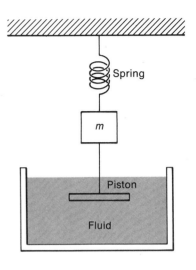

Fig. 37-24. The mechanical analog of the series *LRC* circuit.

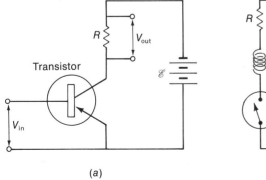

(a)

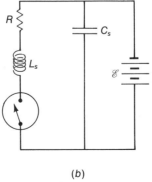

(b)

Fig. 37-25. A transistor "switch" and a crude equivalent circuit that includes stray inductance, L_s, and stray capacitance, C_s.

circuit.* Stray inductance is caused by the magnetic field surrounding *any* current-carrying circuit. Its value is typically quite small, especially when compared with values of inductance obtained in commercial inductors. Stray capacitance is caused by the proximity of one portion of a circuit to another. The adjacent sections of the circuit become the two "plates" of a capacitor—albeit of small value.

If the "switch" of Fig. 37-25 is closed abruptly (by changing the input voltage to the transistor abruptly), the *LRC* circuit can be set into damped oscillations. The occurrence of such oscillations in such a circuit is indicated by the oscilloscope traces in Fig. 37-26.

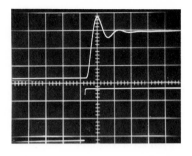

Fig. 37-26. The lower oscilloscope trace shows an abrupt increase in V_{in} in the circuit of Fig. 37-25. The upper trace shows the response, V_{out}. The slower rise time and subsequent oscillations of V_{out} are due to stray capacitances and inductances. The time scale is 10^{-8} s per major horizontal division.

Exercise

11.   The *LRC* circuit of Fig. 37-22 has $L = 100$ μH, $R = \frac{1}{2}\Omega$, and $C = 0.01$ μF. Find the time required for oscillations to die away to $1/e$ of their initial amplitude.

37-8   Summary

In a resistor, current and voltage remain in step when time-dependent variations occur. This is not the case for capacitors and inductors. The current-voltage relation for a capacitor is opposite to that for an inductor:

Capacitor:
$$V = \frac{1}{C} \int I \, dt$$

Inductor:
$$V = L\frac{dI}{dt}$$

In an *RC* circuit, charge leaks away from the capacitor through the resistor with an exponential time dependence:

$$Q = Q_0 e^{-t/RC}$$

where RC defines the characteristic time for the decay of the charge. A capacitor C charged through a resistor R also exhibits the same characteristic time RC but now

$$Q = Q_f(1 - e^{-t/RC})$$

---

*Some readers will be aware of additional stray capacitances and inductances not indicated in Fig. 37-25*b*. The point of this discussion, however, is only to indicate some of the effects they cause, not to analyze a particular circuit thoroughly.

The fact that

$$V_C \propto \int I \, dt$$

and, conversely,

$$I_C \propto \frac{dV_C}{dt}$$

is put to use in RC integrators and differentiators.

Circuits containing inductance and resistance exhibit exponential time dependences analogous to those of RC circuits. The characteristic time is given by L/R.

An LC circuit can be set into sinusoidal oscillations in a manner analogous to a mechanical oscillator composed of a mass and a spring. The characteristic frequency is given by

$$f = \frac{1}{2\pi} \sqrt{\frac{1}{LC}}$$

When a series resistance is added to form a series LRC circuit, the oscillations die away at a rate determined by the ratio L/R. A mass-spring-dashpot system is a mechanical analog to the LRC circuit.

Problems

37-1. Find the characteristic time of the circuit shown in Fig. 37-27.

37-2. A student picks up a 10-μF capacitor and touches its terminals briefly to the terminals of a dry cell battery. The internal resistance of the dry cell equals 1Ω. Find the fraction f defined by

$$V_C = f \mathcal{E}_{\text{batt}}$$

if the time of contact is 1 ms.

37-3. A computer uses high-speed transistor switches to perform digital logic operations. If the resistance of the switching circuits is 50Ω, estimate the maximum stray capacitance that can be tolerated in each switch if the desired switching rate is 10^9 Hz.

37-4. Consider a discharging RC circuit. Show that, at any given time, the rate at which energy is dissipated in the resistor equals the rate at which stored energy in the capacitor is changing.

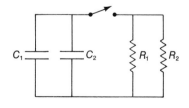

Fig. 37-27.

37-5. A capacitor C is charged to a voltage V so that its stored energy equals $\frac{1}{2}CV^2$. This capacitor is connected to a resistor so that an exponentially decreasing current flows (see Exercise 1). Integrate the rate of energy loss in the resistor from $t = 0$ to $t = \infty$ and show that the result equals the initial energy of the capacitor.

37-6. Consider the circuit of Fig. 37-6. Show that the rate at which the battery does work always equals the sum of the rate of increase of energy stored in the capacitor plus the rate of energy dissipation in the resistor.

37-7. A segment of a transistor is composed of two electrodes, each with an area of 0.01 mm² and separated by 0.1 μm of germanium (dielectric constant $K = 16$). Estimate the capacitance of these electrodes. Find the characteristic decay time when they are in a circuit with a resistance of $10^3\Omega$.

37-8. Show that the characteristic charging time, RC, of the circuit of Fig. 37-6 is also the time required for the charge on the capacitor to reach Q_f *if it were to charge steadily at the initial rate.*

37-9.* The circuit of Fig. 37-10a is wired with a 1-μF capacitor and a 1-MΩ resistor. Assume that the source of V_{in} is a generator with negligible resistance, so that charge accumulated on C leaks away through R. The input voltage V_{in} equals 1 V from $t = 0$ to $t = 0.01$ s and is zero at all other times. (*a*) Find the output voltage at $t = 0.01$ s. (*b*) A perfect integrator would maintain V_{out} equal to the value found in part *a* for all time after $t = 0.01$ s. Estimate the error $\Delta V/V_a$ at $t = 0.1$ s, where ΔV is the loss in voltage below V_a, the perfect integrator's output voltage. (*c*) This integrator will hold charge longer if R is 10 MΩ. Find the value of V_{out} at $t = 0.01$ s as in part *a* for $R = 10$ MΩ and compare with the result of part *a*.

37-10.* The RC differentiator of Fig. 37-10b is wired with $C = 1\ \mu$F and $R = 10\Omega$. The value of V_{in} is 1 V from $t = 0$ to $t = 0.01$ s and zero at all other times. (*a*) Sketch a qualitative graph of V_{out} from $t = -0.01$ s to $t = +0.02$ s. (*b*) A perfect differentiator would produce an output only when V_{in} was changing. In our case, V_{in} changes discontinuously only at $t = 0$ and $t = 0.01$ s. Find V_{out} at $t = 1$ ms and compare it to the maximum value of V_{out}.

37-11. The circuit of Fig. 37-16 is wired with $R = 25\Omega$, $L = 100$ mH and $\mathscr{E} = 10$ V. (*a*) Find the initial rate of change of current when the switch is moved to its upper position at $t = 0$. (*b*) Find the time required for the current to reach its maximum value *if the current continued increasing at the initial rate*. (*c*) Find the time required for the current to reach the fraction $(1 - 1/e) = 0.632$ of its final value. (*d*) Find the voltage across the inductor at the time found in part *c*.

37-12. The *LR* circuit in Fig. 37-16 starts with zero initial current at $t = 0$ and the switch is then thrown upward, so that current begins to increase. The circuit values are $L = 10^{-2}$ H, $R = 10\Omega$, and $\mathscr{E} = 12$ V. At $t = 10^{-3}$ s, find (*a*) the power mH produced by the emf: (*b*) the power consumed by R; (*c*) the rate at which energy is being stored in L.

37-13. A long time after $t = 0$ in Problem 37-12, the switch is flipped down. Call this time t_1. At $t_1 + 10^{-3}$ s, find (*a*) the rate at which the stored energy in L is decreasing; (*b*) the rate at which energy is consumed in R.

37-14. In a decaying *LR* circuit, the initial value of the current is I_0, so that the energy stored in the inductor is $\frac{1}{2}LI_0^2$ at $t = 0$. Integrate the power dissipated in the resistor from $t = 0$ to $t = \infty$ and show the net energy dissipated in the resistor equals the energy initially stored in the inductor.

37-15. The *LC* circuit in Fig. 37-19 has $L = 0.1$ mH and $C = 10\,\mu$F, and begins with a charge of 10^{-4} C on the capacitor. At $t = 0$ the switch is closed. (*a*) Find the value of t at which the capacitor is first charged to the initial magnitude of Q but with opposite polarity. (*b*) How much energy is stored in the capacitor at that instant? (*c*) How much energy is stored in the inductor at that instant?

37-16. The AM band of radio frequencies spans 550 kHz to 1600 kHz. You are given a variable capacitor C with a range 120 pF $< C <$ 360 pF. Can this capacitor be used in an *LC* circuit whose natural frequency can be varied over the entire range of the AM band? What value of L would you choose for this application?

37-17. The circuit of Fig. 37-19 oscillates as indicated by (37-25). Sketch four graphs with similar time scales showing the oscillations of Q, I, V_C, and V_L.

37-18. An *LC* circuit is made up of an 0.1-μF capacitor and 10-μH inductor. The initial charge on the capacitor is 1 μC and a switch is closed to complete the circuit through the inductor at $t=0$. Find the voltage across the capacitor and across the inductor at (a) $t=0$, (b) $t=\pi \times 10^{-6}$ s, (c) $t=\frac{3}{2}\pi \times 10^{-6}$ s, (d) $t=\frac{5}{2}\pi \times 10^{-6}$ s.

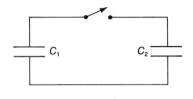

Fig. 37-28.

37-19.   In an oscillating LC circuit, the current oscillates sinusoidally. Find the maximum value of the current in terms of L, C, and the initial value of the voltage on the capacitor, V_0.

37-20.* Consider the circuit of Fig. 37-28 with $C_1=C_2$ and a charge Q_0 on C_1. (a) Calculate the energy stored in C_1. Now close the switch and wait for equilibrium, so that half the charge is on each capacitor. (b) Calculate the total of the energy stored in both capacitors. (c) Where has the lost energy gone? Remember that the interconnecting wires in a real circuit cannot have zero inductance nor can they have zero resistance.

37-21.   Show that (37-29) satisfies the differential equation (37-28).

37-22.   The circuit of Fig. 37-22 has the values $L = 100$ μH, $R = 10\Omega$, and $C = 0.01$ μF. Find its resonant frequency $f = \omega/2\pi$. Repeat for $R = 2\Omega$ and $R = 0$.

Alternating Currents

38-1 Introduction The term *alternating current* is generally used to describe currents that reverse periodically in a certain way — a *sinusoidal* dependence on time. The abbreviation for alternating current is AC. Such apparently irrational uses of this abbreviation as "AC voltage" and "AC current" abound in technical jargon. What is intended is clear enough, however. The abbreviation AC represents "sinusoidally varying" in this usage.

Sinusoidally varying currents are produced naturally by the most commonly used electric generators, of the type schematically shown in Fig. 36-9. This type of current is readily distributed over great distances, as we will see in the section on transformers. Furthermore, when applied to a suitable set of electromagnet coils, alternating currents can produce a rotating magnetic field. This type of field is the key to the operation of the *induction motor,* which is a simple, inexpensive type of motor requiring almost no maintenance. Alternating currents are extremely important in the power economy of a nation.

Alternating currents have another extremely useful property: if a collection of several such currents of different frequencies flows in the *same* conductor, an *LC* circuit can essentially "pick out" just one of these currents. This ability allows the transmission of many different information channels through the same medium — a fact that is crucial to our present communication systems.

It is not easy to predict the response of a circuit containing inductors and capacitors to an emf that varies in an arbitrary way with time. On the other hand, the response of such a circuit to a

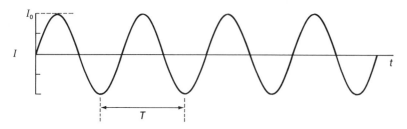

Fig. 38-1. A graph of the alternating current described by Eq. (38-1).

purely sinusoidal waveform is simple and well defined. If we recall Fourier's theorem from Section 23-5, we know that a periodic function of *arbitrary* form is equivalent to a sum of harmonically related sinusoidal functions. The response of the circuit is then easily calculated for each frequency in the harmonic spectrum and the result is then resynthesized to obtain the circuit response. We will not utilize this procedure in this text, but the reader should be aware of these wider-ranging applications of the ideas introduced in this chapter.

We could continue to list reasons for the importance of AC currents both in technology and in understanding nature. Suffice it to say they are extremely widespread.

38-2 Reactance

Consider an alternating current of the form

$$I = I_0 \sin(\omega t) \qquad (38\text{-}1)$$

shown in Fig. 38-1. The amplitude I_0 is equal to the maximum value of this time-varying current. From our previous discussions of harmonic motion, we know the frequency of complete cycles of this current is given by

$$f = \frac{\omega}{2\pi} \qquad (38\text{-}2)$$

and the duration of one cycle is equal to the period, T, given by

$$T = \frac{2\pi}{\omega} \qquad (38\text{-}3)$$

In Fig. 38-2 we see three circuits in which there is a current of the form (38-1). The alternating current generator could, for example, be of the type that utilizes a rotating loop in a magnetic field (as in Fig. 36-9). For the first circuit, where the current passes through a resistor, the current-voltage relation is one of simple proportionality. A time derivative is involved in the current-voltage relation for the other two circuits of Fig. 38-2. The voltage for each

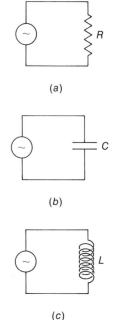

(a)

(b)

(c)

Fig. 38-2. Three alternating current circuits. The alternating current generator is symbolized by the circle with an enclosed sinusoid.

case can be found as follows. We will consider a voltage opposing the current as positive, just as indicated in Fig. 37-2. When the current is given by (38-1), we obtain

(*a*) Resistor: $V_R = IR = I_0 R \sin(\omega t)$

or $\qquad V_R = V_0 \sin(\omega t) \qquad$ where $V_0 = I_0 R$

(*b*) Capacitor: $V_C = \dfrac{Q}{C} = \dfrac{1}{C} \displaystyle\int I\, dt = \dfrac{-I_0}{\omega C} \cos(\omega t)$

or $\qquad V_C = -V_0 \cos(\omega t) \qquad$ where $V_0 = \dfrac{I_0}{\omega C}$

(*c*) Inductor: $\qquad V_L = L\dfrac{dI}{dt} = \omega L I_0 \cos(\omega t)$

or $\qquad V_L = V_0 \cos(\omega t) \qquad$ where $V_0 = I_0 \omega L$

These results are graphed in Fig. 38-3. We see that the current reaches its maximum earlier than the voltage for the capacitor. This is due to the fact that it takes time for the current to charge the capacitor. The maximum charge (and hence maximum voltage) thus occurs after the peak current has passed. The reverse is true for the inductor. There the maximum voltage is reached before the maximum current, since it is then that the current is increasing most rapidly.

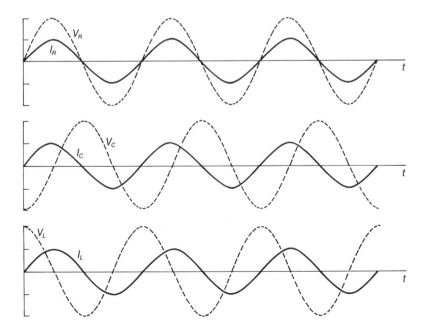

Fig. 38-3. Current and voltage for the three cases of Fig 38-2. The current is indicated by a solid line, voltage by a dotted line.

We can describe these three cases in the following way:

(*a*) Resistor—current and voltage in phase:

$$V_0 = I_0 R$$

(*b*) Capacitor—current leads voltage by a quarter-cycle:

$$V_0 = I_0 \left(\frac{1}{\omega C} \right)$$

(*c*) Inductor—current lags voltage by a quarter-cycle:

$$V_0 = I_0 (\omega L)$$

Alternatively, we can combine cases *b* and *c* in the following way:

Resistor: $I = I_0 \sin(\omega t)$ $V = V_0 \sin(\omega t)$

with $V_0 = I_0 R$ (38-4)

Capacitor and inductor:

$$V = V_0 \sin(\omega t + \phi)$$

with $V_0 = I_0 X$ (38-5)

The values of X and ϕ are given in the margin:

	X	ϕ
C	$\dfrac{1}{\omega C}$	$-\dfrac{\pi}{2} (-90°)$
L	ωL	$+\dfrac{\pi}{2} (+90°)$

The relation (38-5) for capacitors and inductors is reminiscent of Ohm's law, (38-4), for a resistive circuit with X playing the role of resistance. The quantity X is called the *reactance*.

Capacitive reactance: $X_C = \dfrac{1}{\omega C}$

Inductive reactance: $X_L = \omega L$

Reactance is analogous to resistance in that a large reactance leads to a small current for a given voltage. Capacitive reactance diminishes as the frequency increases, while inductive reactance increases with increasing frequency. Since reactance is a ratio of voltage to current, its SI unit is the *ohm*.

Example   An emf of the form

$$V = V_0 \sin(\omega t)$$

with $V_0 = 170$ V and $\omega = 377$ s^{-1} (corresponding to $f = 60$ Hz) is supplied by the wall sockets of most home lighting circuits in the United States.*

---

*Home lighting circuits are generally called "120 volts AC" for reasons to be discussed in Section 38-6.

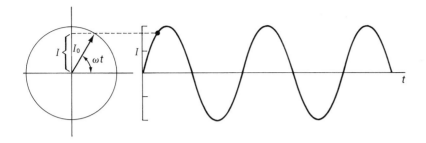

Fig. 38-4. The rotor diagram at the left represents the current $I = I_0 \sin(\omega t)$. The projection of the vector of length I_0 onto the vertical axis gives the result at the right as the vector rotates uniformly.

If we connect a 10-μF capacitor directly across this source, what is the peak value of the current that flows?

Solution   The reactance of the capacitor is

$$X_C = \frac{1}{\omega C} = \frac{1}{(3.77 \times 10^2 \text{ s}^{-1})(1 \times 10^{-5} \text{ F})}$$

$$= 265\Omega$$

The peak current, I_0, can be found by means of (38-5):

$$I_0 = \frac{V_0}{X_C} = \frac{170 \text{ V}}{265\Omega} = 0.64 \text{ A}$$

The current varies sinusoidally with a frequency of 60 Hz and leads the applied voltage by 90°.

Exercises

1. Find the reactance of a 10-mH inductor at a frequency of 60-Hz and at a frequency of 6 kHz.
2. Find the peak voltage V_0 across a 1-μF capacitor when a 10-kHz alternating current passes through it. The peak value of the current, I_0, is $\frac{1}{4}$ A.

38-3   Rotor Diagrams

In Section 16-5 we saw that simple harmonic motion is the *projection* of uniform circular motion. Similarly, we can describe *any* sinusoidally varying quantity by its *rotor diagram*. An example is shown in Fig. 38-4 for an alternating current with an amplitude I_0. A vector of length I_0 rotates uniformly counterclockwise in the rotor diagram. Its projection on the vertical axis gives the current $I = I_0 \sin(\omega t)$.

The useful nature of a rotor diagram becomes apparent when we wish to *add* two sinusoidally varying quantities that have iden-

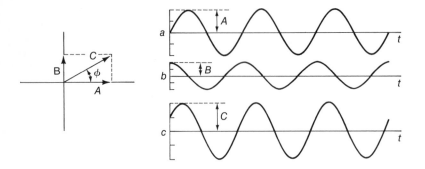

Fig. 38-5. The rotor diagram at left allows us to find the magnitude and phase of the sum c = a + b easily.

tical frequencies but are *not* in phase. In Fig. 38-5 we see the rotor diagram representing the sum of two sinusoidal quantities:

$$a = A \sin(\omega t)$$

$$b = B \sin\left(\omega t + \frac{\pi}{2}\right)$$

where $c = C \sin(\omega t + \phi)$ is the sum $c = a + b$.

As the three rotors A, B, and C rotate together counterclockwise, their vertical projections give the results at the right in Fig. 38-5. However, we need only to examine the rotor diagram at $t = 0$ (as shown) to deduce the magnitude and phase of the summed quantity c. From the geometry of the rotor diagram, we immediately see that

$$C = \sqrt{A^2 + B^2}$$

and

$$\tan \phi = B/A$$

We can apply this addition technique to find the total voltage across an *RC* circuit with an alternating current. The total voltage between points a and c in Fig. 38-6 is the sum of V_C and V_R. These

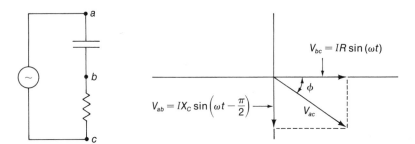

Fig. 38-6. A series *RC* circuit with a current $I = I_0 \sin(\omega t)$.

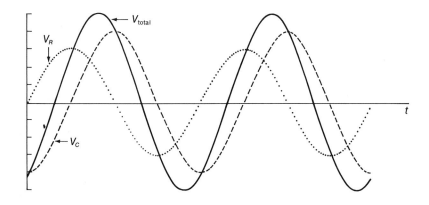

Fig. 38-7. The total voltage across the *RC* circuit of Fig. 38-6 is the sum of V_R and V_C. Its phase is intermediate between that of V_C and that of V_R.

voltages are not in phase but are easily added by means of the rotor diagram in Fig. 38-6. We find

$$V_{0ac} = \sqrt{V_{0C}^2 + V_{0R}^2}$$

Further, we can find the phase angle ϕ between V_{0ac} and I_0 from

$$\tan \phi = \frac{V_{0ab}}{V_{0bc}} = \frac{I_0 R}{I_0 X_C} = R\omega C$$

These results are indicated in Fig. 38-7.

Exercises

3. Draw a rotor diagram at $t = 0$ for the two quantities

$$a = A \sin\left(\omega t + \frac{\pi}{2}\right)$$

$$b = B \sin\left(\omega t - \frac{\pi}{2}\right)$$

Find the magnitude of their sum if $A = B$.

4. Consider the sum $z = x + y$, where $x = 10 \sin(5t)$ and

$$y = 3 \sin\left(5t + \frac{\pi}{2}\right).$$

The sum, z, is of the form $z = Z \sin(5t + \phi)$. Find the magnitude, Z, and the phase angle, ϕ.

38-4   The *LRC* Series Circuit

The *LRC* circuit of Fig. 38-8 is a series circuit, so the same current

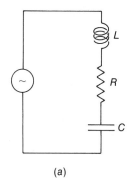

(a)

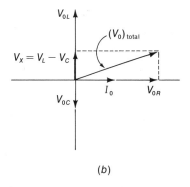

(b)

Fig. 38-8.   An *LRC* series circuit and its rotor diagram.

flows in all parts of the circuit at any instant. If this current is an alternating current, we have

$$I = I_0 \sin(\omega t)$$

and the voltages across the three circuit elements are given by

$$V_R = I_0 R \sin(\omega t) = V_{0R} \sin(\omega t)$$

$$V_L = I_0 X_L \sin\left(\omega t + \frac{\pi}{2}\right) = V_{0L} \sin\left(\omega t + \frac{\pi}{2}\right)$$

$$V_C = I_0 X_C \sin\left(\omega t - \frac{\pi}{2}\right) = V_{0C} \sin\left(\omega t - \frac{\pi}{2}\right)$$

The voltage across the entire circuit is given by the sum of these three voltages:

$$V_{\text{total}} = V_R + V_L + V_C$$

This sum is easily found by examining the rotor diagram in Fig. 38-8. First, we sum the rotors V_{0L} and V_{0C} to obtain

$$V_{0X} = V_{0L} - V_{0C}$$

where V_{0X} is the magnitude of the total *reactive* voltage. This is the voltage that is not in phase with the current. We can then add the rotor V_{0X} to that for V_R:

$$(V_0)_{\text{total}} = \sqrt{V_{0X}{}^2 + V_{0R}{}^2}$$

Substituting the values of V_{0X} and V_{0R}, we have:

$$(V_0)_{\text{total}} = \sqrt{(I_0 X_L - I_0 X_C)^2 + (I_0 R)^2}$$
$$= I_0 \sqrt{(X_L - X_C)^2 + R^2} \qquad (38\text{-}6)$$

The quantity $X_L - X_C$ is called the *reactance*, X, of the circuit, while the quantity in the radical in (38-6) is called the *impedance, Z*.

Reactance: $$X = X_L - X_C = \omega L - \frac{1}{\omega C} \qquad (38\text{-}7)$$

Impedance: $$Z = \sqrt{X^2 + R^2} = \sqrt{\left(\omega L - \frac{1}{\omega C}\right)^2 + R^2} \qquad (38\text{-}8)$$

The relationship between total voltage and current *amplitudes* in a series *LRC* circuit can now be expressed in a form similar to Ohm's law for a resistive circuit:

$$(V_0)_{\text{total}} = I_0 Z \qquad (38\text{-}9)$$

The phase difference between the total voltage and the current can be found as shown in Fig. 38-9. There the total reactive voltage,

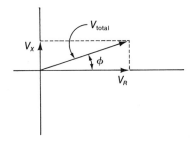

Fig. 38-9. The total reactive voltage and the resistive voltage are 90° out of phase. The total voltage has a phase intermediate between these two.

V_{0X}, and the resistive voltage, V_{0R}, are added to obtain the total voltage. The phase angle, ϕ, is given by

$$\tan \phi = \frac{V_{0X}}{V_{0R}} = \frac{X}{R} \qquad (38\text{-}10)$$

Example   The circuit of Fig. 38-8 contains a 100-pF capacitor, a 1-mH inductor, and a 5000Ω resistor. The total voltage amplitude is 100 V. The frequency is

$$f = \frac{1}{2\pi} \times 10^6 \text{ Hz.}$$

Find (*a*) the amplitude of the current; (*b*) the amplitude of V_R; (*c*) the amplitude of V_L; (*d*) the amplitude of V_C; (*e*) the phase angle between the total voltage and the current.

Solution   The reactances are:

$$X_L = \omega L = 2\pi f L = 10^6 \text{ s}^{-1} \cdot 10^{-3} \text{ H} = 10^3 \Omega$$

$$X_C = \frac{1}{\omega C} = \frac{1}{10^6 \text{ s}^{-1} \cdot 10^{-10} \text{ F}} = 10^4 \Omega$$

Thus $\qquad\qquad X = X_L - X_C = -9.9 \times 10^3 \Omega$

Then we have

$$Z = \sqrt{X^2 + R^2} = 10^3 \Omega \sqrt{(9.9)^2 + 5^2}$$

$$= 1.1 \times 10^4 \Omega$$

The amplitude of the current is

$$I_0 = \frac{V_0}{Z} = \frac{10^2 \text{ V}}{1.1 \times 10^4 \Omega} = 9.1 \text{ mA}$$

This current passes through all three circuit elements and the voltage amplitude across any of them is easily found:

$$V_{0R} = I_0 R = 9.1 \text{ mA}(5 \times 10^3 \Omega) = 45.5 \text{ V}$$

$$V_{0L} = I_0 X_L = 9.1 \text{ mA}(10^3 \Omega) = 9.1 \text{ V}$$

$$V_{0C} = I_0 X_C = 9.1 \text{ mA}(10^4 \Omega) = 91 \text{ V}$$

The direct sum of these amplitudes, of course, exceeds the amplitude of the total voltage because the individual voltages are not in phase and do not directly add. Finally, the phase angle, ϕ, is given by

$$\tan \phi = \frac{X}{R} = \frac{-9.9 \times 10^3 \Omega}{5 \times 10^3 \Omega} = -1.98$$

$$\phi = \tan^{-1}(-1.98) = -63.2°$$

Thus the voltage reaches its peak 63.2° *later* than the current. Expressed in radians:

$$\phi = \frac{-63.2°}{57.3°/\text{rad}} = -1.1 \text{ rad}$$

Since the angular frequency is 10^6 s^{-1}, the time lag between peak current and peak voltage will be 1.1 μs.

Exercises

5. Sketch a rotor diagram for the preceding example. Find the difference in phase between the voltage across the inductor and the total voltage.

6. In the preceding example, suppose that the total voltage amplitude is maintained at 100 V but the frequency is increased to extremely large values, i.e., $\omega \rightarrow \infty$. Now what are the voltage amplitudes V_R, V_L, and V_C?

38-5   Resonance

The reactive component of the total voltage in an LRC series circuit is given by $V_L - V_C$. Thus if $X_L = X_C$, the reactive voltage vanishes! Under these circumstances the impedance is at a minimum, its value equaling the value of the resistance in the circuit:

$$Z = \sqrt{(X_L - X_C)^2 - R^2} = R \qquad \text{when } X_L = X_C$$

If the frequency is now raised, X_L increases while X_C decreases, so that the term $X_L - X_C$ no longer vanishes and the impedance is greater than R. Similarly, if the frequency is decreased the impedance rises, as shown in Fig. 38-10.

The frequency at which the impedance reaches a minimum is called the *resonant frequency*, ω_r. At this frequency the current reaches a maximum value for a given applied voltage. For given values of L and C, the frequency ω_r is found as follows:

$$X_L = X_C \qquad \text{at resonance}$$

hence

$$\omega_r L = \frac{1}{\omega_r C}$$

Thus

$$\omega_r = \sqrt{\frac{1}{LC}} \quad \text{or} \quad f_r = \frac{1}{2\pi}\sqrt{\frac{1}{LC}} \qquad (38\text{-}11)$$

We previously encountered this frequency when we discussed oscillations in an LC circuit. Since ω_r is the natural oscillation frequency of the circuit, it is not surprising that the current ampli-

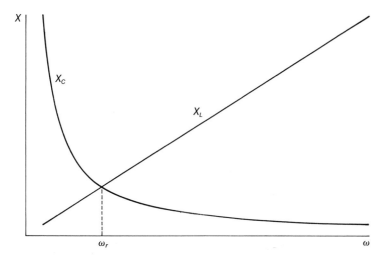

Fig. 38-10. At the resonant frequency, $X_L = X_C$. The impedance at this frequency is at a minimum: $Z = R$.

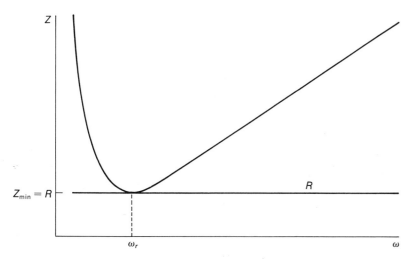

tude reaches a maximum value when the applied voltage is "in step" with this natural frequency.

If the value of the resistance in the circuit is made as small as possible, large currents will flow at resonance when the total reactance vanishes. However, the individual reactances, X_L and X_C, do *not* vanish, so that very large voltages can be developed across the reactance at resonance. For example, consider a series circuit with $L = 1$ mH, $C = 1000$ pF, and $R = 10\Omega$. We apply a total voltage with $V_0 = 100$ V. For resonance we must have

$$\omega_r = \sqrt{\frac{1}{LC}} = \sqrt{\frac{1}{10^{-3}\text{ H} \cdot 10^{-9}\text{ F}}} = 10^6\text{ s}^{-1}$$

At this frequency

$$X_L = \omega L = 10^6 \text{ s}^{-1} \cdot 10^{-3} \text{ H} = 10^3 \Omega$$

$$X_C = \frac{1}{\omega C} = \frac{1}{10^6 \text{ s}^{-1} \cdot 10^{-9} \text{ F}} = 10^3 \Omega$$

and $Z = R = 10\Omega$. At resonance we find

$$I_0 = \frac{V_0}{Z} = \frac{100 \text{ V}}{10\Omega} = 10 \text{ A}$$

Now let us find the voltages across the reactances.

$$V_{0L} = I_0 X_L = 10 \text{ A} \cdot 10^3 \Omega = 10{,}000 \text{ V}$$

$$V_{0C} = I_0 X_C = 10{,}000 \text{ V}$$

These large voltages cancel completely, since they are precisely out of phase, so that the *total* voltage is just equal to that across the resistor:

$$V_{0R} = I_0 R = 10 \text{ A} \cdot 10\Omega = 100 \text{ V}$$

If the frequency is either increased or decreased, the current falls sharply, as indicated in Fig. 38-11.

Example The tuning circuit for a radio receiver has the values $C = 3.6 \times 10^{-10}$ F, $L = 2.78 \times 10^{-5}$ H, $R = 1\Omega$. Hence

$$\omega_r = \sqrt{\frac{1}{LC}} = 10^7 \text{ s}^{-1}$$

so its resonant frequency is

$$f_r = \frac{1}{2\pi} \sqrt{\frac{1}{LC}} = 1591 \text{ kHz}$$

If a voltage with a frequency of 1580 kHz is applied to this circuit, the current is reduced compared to the current that flows at the resonant frequency. Let us see how large a reduction is obtained. At resonance we have

$$X_L = X_C = \omega_R L = 10^7 \text{ s}^{-1} \cdot (2.78 \times 10^{-5} \text{ H}) = 278\Omega$$

and $Z = R = 1\Omega$. At $f = 1580$ kHz, we have $\omega = 9.93 \times 10^6 \text{ s}^{-1}$. Then we find

$$X_L = 276\Omega \qquad X_C = 279.8\Omega$$

$$Z = \sqrt{(X_L - X_C)^2 + R^2} = \sqrt{15.4}\Omega = 3.9\Omega$$

Thus, a change in frequency of less than 1% has increased the impedance by a factor of about 4. The current induced by the nonresonant signal is accordingly reduced by a factor of more than 4 from that of

(a)

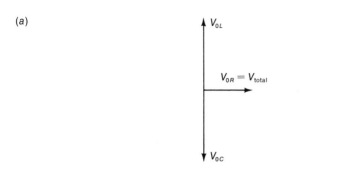

Fig. 38-11. (a) At the resonant frequency, the voltages V_L and V_C precisely cancel one another. (b) For a fixed amplitude of the applied voltage, the current amplitude is at a maximum at the resonant frequency. (c) If R is made large, the minimum in Z is not so pronounced and as a result the peak in I vs ω is also not as pronounced.

(b)

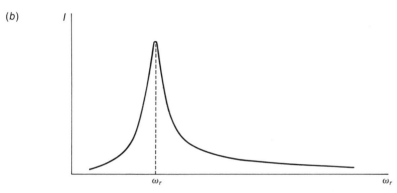

(c)

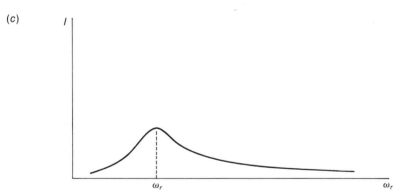

the desired signal—given equal applied voltage magnitudes. If the value of R is reduced to 0.1Ω, the change in Z is even more dramatic. Then we have:

$$Z_r(1591 \text{ kHz}) = 0.1\Omega$$

$$Z(1580 \text{ kHz}) = \sqrt{14.5}\,\Omega = 3.8\Omega$$

which is almost 40 times greater than the impedance at resonance. We see that a sharply resonant circuit is one with the smallest possible resistance.

Exercises

7. An *LRC* series circuit has the values $L = 1$ μH, $C = 20$ pF, $R = 2\Omega$. (*a*) Find the resonant frequency. (*b*) Find the value of the inductive reactance at resonance. (*c*) What is the ratio of the voltage amplitude across the inductor to that across the entire circuit at resonance?

8. Suppose the capacitor in Exercise 7 is variable and is "tuned" to a value of 40 pF. (*a*) Find the new resonant frequency. (*b*) Is the ratio of the voltage amplitude across the inductor to that across the entire circuit changed from the result in Exercise 7*c*?

38-6 Power in AC Circuits

In Section 33-7 we saw that when a current I passes through an emf $\mathscr{E}$, the charge carriers gain energy at a rate

$$P = \mathscr{E}I \qquad (38\text{-}12)$$

If the current I passes through a potential difference V that opposes the current (as in the case when a current passes through a resistor), the rate at which power is lost is given by

$$P = VI \qquad (38\text{-}13)$$

For a resistor, Ohm's law gives us $V = IR$, so we can also write

$$P = I^2R \qquad (38\text{-}14)$$

as the rate at which energy is consumed by the resistor. This energy reappears as heat.

Now let us consider an alternating current in a resistor. If

$$I = I_0 \sin(\omega t)$$

we have $\qquad P = I^2R = I_0^2R \sin^2(\omega t)$

so that the rate of energy dissipation (the power) fluctuates between zero and I_0^2R, as seen in Fig. 38-12.

The *average* rate $\bar{P}$ at which power is dissipated is that constant rate that yields the same net energy in a given time as the actual fluctuating rate. This is indicated schematically in Fig. 38-13. To find the magnitude of $\bar{P}$, then, we simply equate the two net energies:

$$\bar{P} \cdot \Delta t = \int_{t_1}^{t_2} P \, dt$$

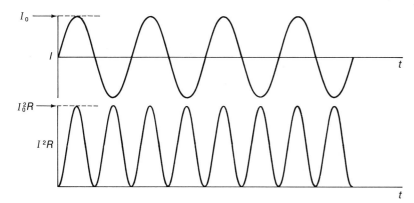

Fig. 38-12. The power dissipated by an alternating current in a resistor is always positive, since it is proportional to the *square* of the current.

where $\Delta t = t_2 - t_1$. Our result will be independent of Δt as long as it includes an integral number of cycles, so we will let $\Delta t = T = 1/f$.

$$\bar{P} \cdot T = \int_0^T I^2 R \, dt = I_0{}^2 R \int_0^T \sin^2(\omega t) \, dt$$

Utilizing a half-angle identity, we have

$$I_0{}^2 R \int_0^T \sin^2(\omega t) \, dt = I_0{}^2 R \int_0^T \tfrac{1}{2}[1 - \cos(2\omega t)] \, dt$$

$$= \tfrac{1}{2} I_0{}^2 R T - \tfrac{1}{2} I_0{}^2 R \int_0^T \cos(2\omega t) \, dt$$

The last integral is zero, since the oscillating cosine function has equal areas above and below zero. Thus we find

$$\bar{P} = \tfrac{1}{2} I_0{}^2 R$$

which is one-half the rate of power dissipation of a constant current with a value equal to our *peak* current. If we ask, "What *constant current* gives the same rate of power dissipation as the average value?" we find

$$I_{rms} = \frac{1}{\sqrt{2}} I_0 \tag{38-15}$$

where I_{rms} is the value of this *power-equivalent* constant current and I_0 is the amplitude of the alternating current. This current, I_{rms},

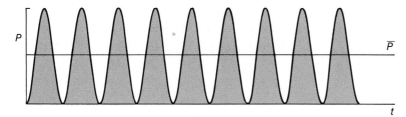

Fig. 38-13. The average power, $\bar{P}$, gives the same area under the P-vs-t curve as the actual fluctuating power, P.

is called the *root mean square* of the alternating current, since it is the square root of the mean (average) value of the square of the alternating current.

Since, in terms of *power* delivered, the rms value is the "effective" value of the alternating current, it is common to quote this value for the magnitude of an alternating current. When we say a home lighting circuit is rated at 30 amperes, we are referring to the rms value. Similarly, the rms value of the voltage is generally quoted. Most home lighting circuits use "120 volts AC," referring to the rms value. Thus, the *peak* voltage is

$$V_0 = \sqrt{2}\,V_{rms} \approx 170 \text{ volts}$$

When we wish to find the average power delivered to a resistive load by an alternating current, we use the rms values and find

$$P = I_{rms}^2 R = \frac{V_{rms}^2}{R} = V_{rms} I_{rms} \qquad (38\text{-}16)$$

in analogy to the results for a constant current.

When an alternating current flows in either a capacitor or an inductor, the power delivered by that current fluctuates in such a way that the *average* power is zero. This is seen in Fig. 38-14. The alternating sign of the power delivered to either type of reactive element indicates that energy is stored in the element during one portion of a cycle, then returned to the circuit in another portion of the cycle. Mathematically, the alternating sign occurs because V and I are 90° out of phase:

$$P = V_0 \cos(\omega t) \cdot I_0 \sin(\omega t) = \frac{V_0 I_0}{2} \sin(2\omega t)$$

$$\bar{P} = \frac{V_0 I_0}{2T} \int_0^T \sin(2\omega t)\, dt = 0$$

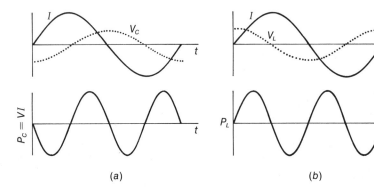

Fig. 38-14. The power delivered to (a) a capacitor or (b) an inductor alternates in sign as energy is stored then returned to the circuit.

In a circuit containing some resistance and some reactance, an intermediate situation occurs:

$$P = V_0 \sin(\omega t + \phi) I_0 \sin(\omega t)$$

where ϕ is the phase angle between voltage and current. Now we can use the identity

$$\sin(\omega t + \phi) = \sin(\omega t)\cos\phi + \cos(\omega t)\sin\phi$$

to obtain

$$P = V_0 I_0 \sin^2(\omega t)\cos\phi + V_0 I_0 \sin(\omega t)\cos(\omega t)\sin\phi$$

The first term will give an average value

$$\bar{P} = \tfrac{1}{2}V_0 I_0 \cos\phi \qquad (38\text{-}17)$$

while the second term, as we saw previously, will average to zero. The factor $\cos\phi$ is called the *power factor* and reduces the net power delivered by an alternating current (for given values of V_0 and I_0) when a reactance is present.

In a commercial power distribution system it is not economical to allow ϕ to depart far from zero. If ϕ approaches 90°, the conductors must be large to handle large average currents while relatively little power is transferred. The electromagnets in motors are quite inductive and cause serious phase shifts in most power lines. Power companies offset these as much as possible by placing large capacitors across the line or by operating some generators purposely out of phase to cancel the reactive component of the voltage.

Exercises

9. An electric toaster is a resistive load. It is connected to a 120-V (rms) outlet and draws an rms current of 10 A. What is the average power consumed by the toaster? What are the values of the peak current and peak voltage?

10. An electric motor behaves like a resistance of 10Ω in series with an inductive reactance of 5Ω when connected to a 120-V (rms), 60-Hz power source. What is the power factor for this circuit? How large is the peak current? What is the average rate of power dissipation?

38-7   Transformers

In applications of electrical power, it is sometimes desirable to utilize large values of current at low voltage (as in an arc welder) or small values of current at high voltage. The latter situation is desirable, for example, when we wish to transmit a given amount of power through a long transmission line. By using a small current, the I^2R losses in the resistance of the transmission line are minimized. However, high voltages are then required if the line is to carry a great deal of power—voltages too large for convenient use by power consumers.

Fortunately, the ratio of current to voltage is easily changed for an alternating current. This ease of tailoring the voltage-current ratio to the need at hand is precisely the reason for the dominance of alternating currents in power distribution systems. The device that accomplishes this transformation is called a *transformer.*

A transformer is nothing more than a device that provides mutual inductance between two circuits. Its symbol, as shown in Fig. 38-15, is that of a mutual inductance. If an alternating current is forced to pass through one winding of a transformer, called the *primary,* an alternating emf will be induced in the other winding, called the *secondary.* Often an iron core is utilized so that large magnetic fluxes can be generated by the primary and are efficiently guided through the secondary (Fig. 38-16).

The primary itself generates a back emf due to its self-inductance:

$$\mathscr{E}_1 = N_1 \frac{d\Phi}{dt} \qquad (38\text{-}18)$$

where N_1 is the number of turns of the primary winding. In a well-designed transformer, virtually all of the flux, Φ, will pass through the secondary winding, so we have

$$\mathscr{E}_2 = N_2 \frac{d\Phi}{dt} \qquad (38\text{-}19)$$

Comparing (38-18) and (38-19), we immediately see that

$$\frac{\mathscr{E}_2}{\mathscr{E}_1} = \frac{N_2}{N_1} \qquad (38\text{-}20)$$

Thus, the ratio of secondary voltage to primary voltage is dependent simply on the *turn ratio,* N_2/N_1. The ratio of secondary to

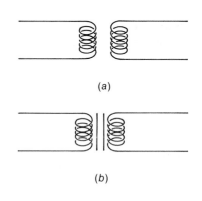

Fig. 38-15. (a) Air core and (b) iron core transformer symbols.

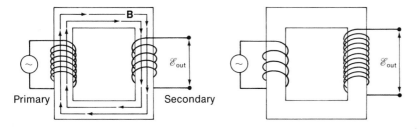

Fig. 38-16. Iron cores are often used in transformer construction. By increasing the number of turns in the secondary, the induced emf can be increased over the applied voltage at the primary.

primary currents is most easily found by applying conservation of energy and assuming a power factor of unity:

$$P_{\text{in}} = P_{\text{out}}$$

or

$$\mathscr{E}_1 I_1 = \mathscr{E}_2 I_2$$

so that

$$\frac{I_2}{I_1} = \frac{\mathscr{E}_1}{\mathscr{E}_2} = \frac{N_1}{N_2} \qquad (38\text{-}21)$$

Actually, the flux generated by I_2 is quite important and (38-21) is an approximation. Note that (38-20) and (38-21) apply to the instantaneous voltage and current. They also, then, apply to the peak value and to the rms values of voltage and current.

In actual transformers there are some power losses due to $I^2 R$ heat, heat generated by induced eddy currents, and hysteresis effects in the iron core. Most transformers, however, are 90% efficient ($P_{\text{out}} = 90\% \ P_{\text{in}}$) or better. In the following we will assume 100% efficiency and power factors of unity.

Example Electrical power is distributed at an rms voltage of 480 V. A transformer is used to "step down" to 120 V rms for household use. The required turn ratio is thus

$$\frac{N_2}{N_1} = \frac{\mathscr{E}_2}{\mathscr{E}_1} = \frac{480}{120} = 4$$

If the household consumes power at a rate of 24 kW, the total rms current at 120 V rms is

$$I_{rms} = \frac{24 \text{ kW}}{120 \text{ V}} = 200 \text{ A}$$

for a phase angle ϕ of 0°. The current flowing in the 480-V line is one-quarter of this value, 50 A.

Exercises

11. A spot welder requires an rms current of 100 A. The resistance of the load is usually quite low, so that only a few volts are required. Can a step-down transformer be utilized to power the spot welder from a 120-V (rms) lighting circuit fused at 15 A (rms)?

12. A transmission line carries 1 MW of power from a hydroelectric plant to a town. What current (rms) is required at 120 V rms? At 60,000 V? For a given resistance R of the transmission lines, what is the ratio of power losses in these two cases?

38-8 Summary

When a sinusoidally alternating current passes through a resistor, the voltage across the resistor also varies sinusoidally and is in step with the current. In a capcitor, the current leads the voltage by one-quarter cycle, and in an inductor, the voltage leads the current by one-quarter cycle.

A relationship similar to Ohm's law may be written for the peak voltage V_0 and peak current I_0 in an inductor or capacitor:

$$V_0 = I_0 X$$

where X is the *reactance*.

Capacitance reactance: $X_C = \dfrac{1}{\omega C}$

Inductive reactance: $X_L = \omega L$

Since sinusoidal oscillations result when a constant rotating vector is projected onto a fixed axis, we use rotor diagrams to describe alternating currents and voltages. By their use we find that the voltages across reactances and resistors do not add directly. If we define an *impedance*, Z, for a series LRC circuit by $Z = V_0/I_0$, we find

$$Z = \sqrt{(X_L - X_C)^2 + R^2}$$

where $X_L - X_C$ is the net reactance X of the circuit and R is its resistance. The phase angle ϕ between voltage and current in such a circuit is given by $\tan \phi = X/R$.

At a resonant frequency given by $\omega_r = \sqrt{1/LC}$, the net reactance in a series LRC circuit vanishes so that the current becomes maximum. If the resistance is small compared to the magnitude of the cancelling inductive and capacitive reactances, the voltage across both the inductor and the capacitor is much larger than the voltage applied across the entire circuit.

The average power dissipated in an AC circuit containing only resistance is given by

$$\bar{P} = \tfrac{1}{2}V_0 I_0$$

so that we define rms values for alternating voltages and currents by

$$V_{rms} = \frac{1}{\sqrt{2}}V_0 \quad \text{and} \quad I_{rms} = \frac{1}{\sqrt{2}}I_0$$

In a circuit containing pure inductance or pure capacitance, no power is dissipated on the average since power is alternately stored in the reactance then delivered back to the circuit.

In general, when voltage and current are out of phase by a phase angle ϕ, the average power is given by

$$\bar{P} = V_{rms} I_{rms} \cos \phi$$

A transformer allows voltage-current ratios to be varied to suit a given application. If the primary contains N_1 turns and the secondary N_2 turns, we find

$$\frac{\mathscr{E}_2}{\mathscr{E}_1} = \frac{N_2}{N_1} \quad \text{and} \quad \frac{I_2}{I_1} \approx \frac{N_1}{N_2}$$

Problems

38-1. A commercial inductor has a self-inductance of 250 mH and a resistance of 15Ω. Find its inductive reactance (neglect the resistance) at frequencies of 100 Hz, 1000 Hz, and 10,000 Hz. At what frequency does its reactance become ten times its resistance?

38-2. A capacitance of more than 100 μF is considered quite large since such capacitors are bulky. Find the capacitive reactance of a 100-μF capacitor at 100 Hz, 1000 Hz, and 10,000 Hz.

38-3. A pure inductance of 50 mH is in series with a 500Ω resistance. This circuit carries an alternating current with $f = 1500$ Hz. (a) Construct a rotor diagram for V_L and V_R. (b) Find the phase angle ϕ between the current and the net voltage across the circuit.

38-4. A circuit contains a 10-mH inductor and 300Ω resistor in series. The magnitude of a current I_0 is 30 mA. The voltage drop across the entire circuit has a magnitude $V_0 = 15$ V. Find the frequency of the alternating current.

38-5. A 1-μF capacitor is in series with a 1000Ω resistor. They are connected to a 100-V (peak) alternating emf. The peak voltage drop across the resistor is 30 V. Find the frequency of the alternating emf.

38-6. An LRC series circuit is made up with $L = 1$ mH, $C = 1000$ pF, and $R = 100Ω$. Find the frequency at which the voltage across the inductor is ten times that across the capacitor.

38-7. An LRC series circuit has values $L = 1$ mH, $C = 0.01$ μF, and $R = 50Ω$ and carries a current with $f = 10^5$ Hz. Find the impedance of this circuit and the phase angle between the net voltage and the current.

38-8. The *LRC* series circuit shown in Fig. 38-17 is driven by a sinusoidal AC power source. The voltage amplitude between points *a* and *d* is 100 V. The voltage between points *a* and *c* is exactly zero. (*a*) What is the frequency of the applied voltage? (*b*) How much current flows in this circuit? (Give the amplitude of the current, that is, the peak value.) (*c*) Find the amplitude of the voltage across *C* (between points *b* and *c*). (*d*) Find the rms value of the voltage between points *a* and *d*.

38-9. An *LRC* series circuit is connected to a source with V_{rms} = 170 V and $f = (1/2\pi)\,10^2$ Hz. Circuit values are $L = 0.1$ H, $C = 500\,\mu$F, and $R = 5\,\Omega$. Find (*a*) the peak current I_0, and (*b*) the average power delivered by the source.

38-10. A 100-V (rms) source at 10^6 Hz is to be connected across a series resonant circuit in order to produce a large voltage across one of the reactances. If the minimum possible value of *R* is chosen to be 1 Ω, find the values of *L* and *C* required to obtain 10,000 V (rms) across the inductor.

38-11.* Suppose we are given an ω_r and the value of *R* for a series resonant circuit. How do we maximize the voltage across both *L* and *C* for a given current? Show this is achieved by making *L* as large as possible.

38-12.* Consider a copper-wire transmission line carrying a fixed amount of power. The cross-sectional area of the wires must be proportional to the magnitude of the current to limit losses. Assuming a fixed cost per pound of copper, find the way that the cost of the transmission line wires varies with the phase angle ϕ between voltage and current.

38-13. A neon-sign transformer provides 10,000 V rms. Its primary is wound with 100 turns. How many turns are wound on the secondary if the primary is connected to a standard 120-V (rms) supply?

38-14. A 60,000-V transmission line is stepped down to 2200 V for local distribution. (*a*) What is the required turn ratio in the transformer? (*b*) Find the ratio of the current in the transmission line to that in the secondary circuit.

38-15.* A transmission line carries 100 MW of electrical power in copper wires over a distance of 10 km. What is the required copper-wire diameter if line loss is to be kept at 1% for $\phi = 0°$ and an rms voltage of (*a*) 120 V; (*b*) 60,000 V?

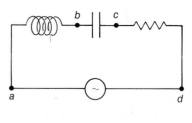

$L = 10^{-1}$ H
$C = 9 \times 10^{-9}$ F
$R = 2\,\Omega$

Fig. 38-17.

Chapter *39*

Electromagnetic Radiation

39-1 Introduction When electric charges are accelerated, as in an alternating current, a new phenomenon appears. Oscillating electric and magnetic fields appear that fall off in intensity quite slowly with distance. At large distances from the oscillating charges, the intensity of these oscillating fields diminishes only as $1/r$. It is thus possible to detect these oscillating fields at great distances from their source. This phenomenon is called *electromagnetic radiation.*

The chain of events leading to radiation involves magnetic induction and another new phenomenon: electric induction. Electric induction is the creation of a magnetic field by a changing electric field. Radiation occurs when a changing magnetic field produces an electric field, which in turn produces a magnetic field, and so on. These processes do not actually occur in separate steps in the radiation phenomenon. Electric and magnetic fields are continually changing and continually creating one another. The speed with which the fields spread away from their source is c, the velocity of light. The combined oscillating electric and magnetic fields are called *electromagnetic waves.*

When the oscillating charges move up and down the towers we see near radio stations, radio waves are produced. The frequency of oscillation for the AM radio band is about 1 MHz. When the oscillating charges have frequencies of about 10^{14} Hz, we observe *light.* Such high frequencies are possible only for objects of atomic dimensions and light is an electromagnetic wave produced by oscillating atomic charges. Still higher frequencies occur

Electric induction
creation of a magnetic
field by a changing
electric field

when charges oscillate in the even smaller atomic nucleus to produce gamma radiation.

To see the wholly new character of radiation fields that fall only inversely with distance, consider the behavior of fields we have studied to date. The electric field of a single point charge falls off as the inverse square of the distance from the charge. If we could turn charges "on and off" at will, a distant observer could conceivably detect the fluctuations in the electric field caused by the fluctuating charge. Unfortunately, electric charges can be neither created nor destroyed. We can *separate* charges to produce a negative and a positive charge where none previously existed, but the net charge remains constant. Thus, if we try to use electric fields to signal a distant observer, the best we can do is to separate charges and recombine them. In other words, we can produce a fluctuating electric *dipole*.

$$E = k\frac{q}{r^2}$$

We often pretend that we have produced an isolated charge by, for example, rubbing a rubber rod with wool and transferring some of the negative charge to a metal ball on an insulating stand. Close to the ball, we see primarily an electric field of a single charge. But if we step back far enough, the field of the positive charge on the wool becomes as important as that of our alleged single negative charge. Thus, at large distances we are aware of a dipole field. The electric field strength produced by a dipole decreases as $1/r^3$, making it difficult to detect these fields at a distance.

Electric currents *can* be switched "off and on." We can thus imagine the use of magnetic fields for signaling purposes. However, currents in a finite circuit produce fields that at large distances are, at best, dipole fields. Again, these fields strengths decrease like $1/r^3$. Without the two induction phenomena—electric and magnetic—that can produce oscillating fields that decrease only as $1/r$, there would be no electromagnetic radiation.

Another unique aspect of radiation fields is the complete detachment of the field lines from the source that was their origin. The continual induction of **E** by **B**, and vice versa, as the fields move away from the source makes these radiated fields independent entities. They continue along at the velocity of light even if the source is "switched off."

SELF-GENERATING

39-2   Electric Induction

Consider the capacitor that is being charged by a current in Fig. 39-1. The charging current produces a magnetic field surrounding the wire. We could follow the procedure of Section 35-4 by applying Ampere's law to calculate the magnitude of **B**. As in Fig. 35-15,

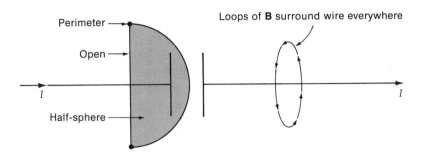

Perimeter
Open
Loops of **B** surround wire everywhere
I
I
Half-sphere

Fig. 39-1. A magnetic field surrounds a wire carrying a charging current. A peculiar surface has been chosen to apply Ampere's law to calculate **B**.

an arbitrary surface can be chosen to do this. The circulation integral around the circumference of such a surface is proportional to the net current passing through it:

$$\oint \mathbf{B} \cdot d\mathbf{s} = \mu_0 I \qquad (35\text{-}8)$$

and if we choose a surface through which the wire passes, we obtain the usual result, (35-4). But what if we choose the bizarre surface of Fig. 39-1? *No* current passes through this surface, yet there is certainly a magnetic field surrounding the wire. This field has a net circulation around the perimeter of the chosen surface. Should we dispense with Ampere's law whenever the surface happens to be between the capacitor plates but retain it for all other possible positions?

There is a way out of this dilemma. We can add a new term to the right-hand side of (35-8) to take care of the difficulty illustrated by Fig. 39-1. We will then want to see whether other predictions resulting from this new term agree with experiment.

We begin by noting that whenever a current passes through the wires in Fig. 39-1, the electric field within the capacitor is *changing.* Just as a changing magnetic field induces an electric field, it may be that a changing *electric* field induces a *magnetic* field. Defining the electric flux Φ_E by

$$\Phi_E = \int \mathbf{E} \cdot d\mathbf{A}$$

we note that for the parallel-plate capacitor of Fig. 39-1, the net electric flux passing through the surface shown in just

$$\Phi_E = EA \qquad (39\text{-}1)$$

where E is the electric field strength between the plates and A the area of the plates. The relation between charge entering (or leaving) the capacitor and the potential difference across the capacitor is

$$Q = CV = \frac{\epsilon_0 A}{d} Ed = \epsilon_0 \Phi_E$$

The charging current is now found by differentiating:

$$I = \frac{dQ}{dt} = \epsilon_0 \frac{d\Phi_E}{dt} \tag{39-2}$$

Thus, if we modify the right side of (35-8) by adding the right side of (39-2),

ADDITION OF FLUX TERM

$$\oint \mathbf{B} \cdot d\mathbf{s} = \mu_0 \left(I + \epsilon_0 \frac{d\Phi_E}{dt} \right) \tag{39-3}$$

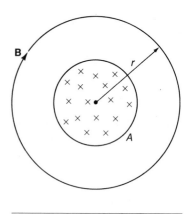

B

r

A

Fig. 39-2.

we will get a consistent answer for the value of B in Fig. 39-1 whether the surface cuts through the wire or passes between the capacitor plates. We assumed no fringing field in obtaining (39-1) and, hence, (39-2). However, a more general argument using Gauss's law gives the same result (see Problem 39-3).

James Clerk Maxwell (1831–1879) first hypothesized that (39-3) was the complete form of Ampere's law. The new term within the parentheses in (39-3) was called the *displacement current* by Maxwell, since he called the quantity $\epsilon_0 E$ the *displacement, D.*

Example   A uniform electric field, **E**, is directed into the page in Fig. 39-2. It covers an area A and is steadily increasing. Find the value of B at a radius r from the center of A. Assume from symmetry that **B** is dependent only upon r and forms circular loops about the region A.

Solution   The calculation is nearly identical to that leading to (35-7):

$$\oint \mathbf{B} \cdot d\mathbf{s} = \mu_0 I + \epsilon_0 \mu_0 \frac{d\Phi_E}{dt}$$

$\mu_0 I + \mu_0 \epsilon_0 A \frac{dE}{dt}$

Since there is no current I, we have

$$\oint \mathbf{B} \cdot d\mathbf{s} = \mu_0 \epsilon_0 A \frac{dE}{dt}$$

Since **B** is constant at a fixed value of r we have

$$\oint \mathbf{B} \cdot d\mathbf{s} = B \oint d\mathbf{s} = B \cdot 2\pi r$$

so

$$B = \frac{\mu_0 \epsilon_0 A \cdot dE/dt}{2\pi r}$$

Exercise

1. Show that the dimensions of $\epsilon_0 d\Phi_E/dt$ are those of a current.

39-3   Maxwell and Hertz

Maxwell pulled together various bits of knowledge concerning electromagnetism into a coherent, unified theory.* In 1864 he showed that electromagnetic waves were a natural consequence of the fundamental laws governing the behavior of electric and magnetic fields. Though most of the basic ideas (except for Maxwell's displacement current) were previously known, the field laws of electricity and magnetism are now generally referred to as *Maxwell's equations:*

30-2
$$\oint \mathbf{E} \cdot d\mathbf{A} = \frac{q}{\epsilon_0} \qquad (39\text{-}4) \quad \text{GAUSS'S LAW}$$

35-1
$$\oint \mathbf{B} \cdot d\mathbf{A} = 0 \qquad (39\text{-}5) \quad \text{GAUSS' FOR MAGNETIC}$$

36-13
$$\oint \mathbf{E} \cdot d\mathbf{s} = -\frac{d\Phi_B}{dt} \qquad (39\text{-}6) \quad \text{FARADAY'S LAW}$$

35-8+
$$\oint \mathbf{B} \cdot d\mathbf{s} = \mu_0 I + \mu_0 \epsilon_0 \frac{d\Phi_E}{dt} \qquad (39\text{-}7) \quad \text{AMPERE'S + ADDED MAXWELL FLUX TERM}$$

In empty space there is no charge, q, nor current, I. Maxwell was able to combine (39-6) and (39-7) with $I = 0$ to obtain a wave equation for both $\mathbf{B}$ and $\mathbf{E}$. These equations have wavelike solutions with a unique wave velocity given by

$$c = \sqrt{\frac{1}{\epsilon_0 \mu_0}} \qquad (39\text{-}8)$$

Substituting values for ϵ_0 and μ_0,

$$c = \sqrt{\frac{1}{(8.85 \times 10^{-12} \text{ C}^2/\text{N} \cdot \text{m}^2)(12.56 \times 10^{-7} \text{ N/A}^2)}}$$

$$= 3 \times 10^8 \text{ m/s}$$

which is equal to the observed velocity of light. It seemed unlikely to Maxwell that this was a mere coincidence. Instead, it seemed most likely that light waves were actually electromagnetic waves.

   Maxwell could not accept the idea that waves could exist without a medium. He believed that electromagnetic waves were carried

*See the article "James Clerk Maxwell" by James R. Newman, *Scientific American,* June 1955.

by the *ether,* a hypothetical, unobservable, weightless, yet extremely rigid medium. Electric fields and magnetic fields were thought of as stresses in this medium and it was natural enough that these stresses could propagate as waves. The modern point of view is more simple. Once changing electric and magnetic fields are produced by moving charges, they feed one another in a way compatible with Eqs. (39-6) and (39-7). A simultaneous solution of these equations predicts that the fields **E** and **B** behave like waves, spreading out at velocity c. No medium is involved, only the fields themselves.

Over 20 years after Maxwell's prediction of electromagnetic waves, Heinrich Hertz (1857–1894) was able to produce waves from obviously electrical sources. His experimental arrangement is indicated in Fig. 39-3. It utilized an *induction coil, I* (the automotive spark coil discussed in Section 36-5 is an induction coil), that produced brief surges of voltage, charging one of the spheres (Q or Q') positive, the other negative. When this voltage reached the breakdown point for the spark gap, S, a spark was initiated. Once the air in the spark gap was ionized, it became a reasonably good conductor. The discharge was then oscillatory, since the capacitance of the spheres and the inductance of the rods formed an LC circuit. The frequency was quite large: about 100 MHz. The oscillating charges produced electromagnetic waves of this frequency. The loop-shaped receiving circuit contained the capacitance of its spark-gap (R) electrodes and the inductance of the loop joining them. When suitably adjusted, it resonated with the oscillating voltage induced by the passing electromagnetic waves and produced a faint spark at its own electrodes.

With such simple apparatus for the production and detection of electromagnetic waves, Hertz was able to observe the phenomena of reflection, refraction, and interference. It became an inescapable conclusion that the "Hertzian waves" differed from light only in their wavelength and frequency.

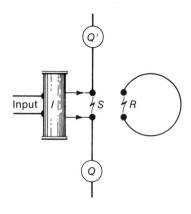

Fig. 39-3. The apparatus first used by Hertz in his observations of electromagnetic waves. The induction coil I provided a brief surge of voltage that induced an oscillatory discharge across the spark gap, S. A nearby circuit exhibited a sympathetic oscillatory discharge, R.

Exercise

2. Calculate the approximate wavelength of the Hertzian waves.

39-4 The Speed of Electromagnetic Waves

The method utilized by Maxwell in deducing the speed of electromagnetic waves involved solution of second-order partial differential wave equations that he deduced from (39-6) and (39-7). We cannot follow this elegant method in a text of this level, but we can show that traveling electromagnetic waves are *compatible* with

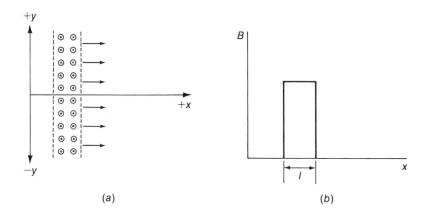

(a)

(b)

Fig. 39-4. (a) A magnetic field points along the positive z direction (out of the page) and moves in the positive x direction. It is confined to a region l along the x axis but extends indefinitely in the y and z directions. It is thus a traveling pulse in the form of a plane wave. (b) The magnitude of B vs x at a given instant.

Eqs. (39-6) and (39-7). We can also show that the speed of these waves is $c = (\epsilon_0 \mu_0)^{-1/2}$.

While the discussion that follows is necessarily somewhat contrived, it does allow us to deduce the speed of electromagnetic waves from basic principles. We will consider a traveling wave that is a rectangular pulse, much like the waveform we used to find the speed of sound in Section 23-3.

Let us suppose that there exists a traveling magnetic field pulse with the following properties. It is everywhere parallel to the z axis. It is uniform within a region l along the x axis and zero outside that region. This region moves toward positive x at velocity v. The field extends indefinitely in the directions $\pm y$, as indicated in Fig. 39-4.

A stationary observer located in the path of this traveling pulse observes a sudden increase, then decrease in **B** as it passes. Accordingly, an electric field is induced according to Faraday's induction law. In order to calculate the intensity of this induced electric field, consider a rectangular area of length L (L > l) in the x direction and width w in the y direction, as shown in Fig. 39-5. As the pulse of magnetic field strength B enters and then leaves this area, a magnetic flux exists within it. This flux begins at zero and rises linearly to a maximum value of

$$\Phi_{max} = wlB \quad \overset{?}{=} \quad \not{B\cdot A} \quad B \cdot A$$

It maintains this maximum value until the pulse reaches the other side of the area then falls linearly to zero. The time required to reach maximum value is

$$\Delta t = \frac{l}{v}$$

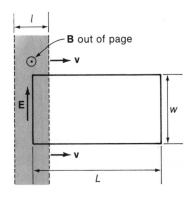

Fig. 39-5. Applying a right-hand rule to the perimeter of the rectangular area, we conclude that $\mathscr{E}$ has a clockwise sense about that path. Since the only contribution to the integral $\mathscr{E} = \oint \mathbf{E} \cdot d\mathbf{s}$ comes from the left end, **E** must point upward in the region within l.

Hence we find

$$\frac{d\Phi}{dt} = \frac{\Phi_{\text{max}}}{\Delta t} = wBv \qquad (39\text{-}9)$$

as the flux rises.

Applying Faraday's law to the perimeter of this same area, we find

$$\oint \mathbf{E} \cdot d\mathbf{s} = \frac{-d\Phi}{dt} = -wBv \qquad (39\text{-}10)$$

In Fig. 39-5 we see an overhead view (looking along z from positive z toward negative z) of the area bounded by the path of integration. Following the discussion of Section 36-5, we find that the sense of $\mathscr{E}$ is clockwise in Fig. 39-5. Assuming that $E = 0$ in the regions not yet reached by the pulse, we expect $\mathbf{E}$ to be upward (in the positive y direction) in Fig. 39-5 within the region contained by l. Thus we have

$$\oint \mathbf{E} \cdot d\mathbf{s} = Ew$$

which, combined with (39-10), gives

$$EW = wBv$$

or
$$E = Bv \qquad (39\text{-}11)$$

The minus sign in (39-10) was dropped to give a relation between only the magnitudes, E and B. That minus sign was utilized in obtaining the clockwise sense of the induced emf.

From the above, it seems reasonable that the magnetic plane wave of thickness l moving at velocity $\mathbf{v}$ has imbedded within it an electric plane wave, also of thickness l and also moving with velocity $\mathbf{v}$. The direction of $\mathbf{E}$ is perpendicular to $\mathbf{B}$ and both $\mathbf{E}$ and $\mathbf{B}$ are perpendicular to the direction of propagation, as shown in Fig. 39-6. The magnitude of $\mathbf{E}$ is v times larger than that of $\mathbf{B}$.

Now let us see if the magnetic field induced by this traveling electric pulse is consistent with our interpretation. If the traveling $\mathbf{E}$ and $\mathbf{B}$ fields continually induce one another, the $\mathbf{B}$ field induced by $\mathbf{E}$ should equal the one with which we started.

The traveling electric pulse of thickness l produces a changing electric flux within an area of width W (in the z direction) and length L ($L > l$) in the x direction. The rate of change of Φ_E as the pulse enters this area is

$$\frac{d\Phi_E}{dt} = \frac{EWl}{\Delta t} = EWv$$

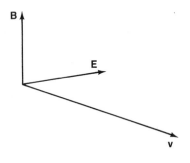

Fig. 39-6. The relative direction of **E**, **B**, and **v** for the traveling pulse of Fig. 39.5.

since $\Delta t = l/v$. Applying Ampere's law with the addition of Maxwell's displacement current and with $I = 0$ (empty space), we have

$$\oint \mathbf{B} \cdot d\mathbf{s} = \mu_0 \epsilon_0 \frac{d\Phi_E}{dt} = \mu_0 \epsilon_0 EWv$$

The entire contribution to the line integral now comes from an edge of length W:

$$\oint \mathbf{B} \cdot d\mathbf{s} = BW = \mu_0 \epsilon_0 EWv$$

or $$B = \mu_0 \epsilon_0 Ev \tag{39-12}$$

If our discussion is correct, (39-11) and (39-12) should be compatible. Further, if they are, we should be able to deduce the value of v. Combining (39-11) and (39-12), we find:

$$\frac{E}{v} = \mu_0 \epsilon_0 Ev$$

This equation is satisfied if

$$v^2 = \frac{1}{\mu_0 \epsilon_0} \tag{39-13}$$

As indicated in the preceding section, this velocity is the velocity of light. Our traveling pulse, then, moves with the velocity of light, c.

The thickness l and the dimensions W and w in our discussion dropped out in our final result. Since a sinusoidal traveling wave, as in Fig. 39-7, can be made up of a fine, "stairstep" superposition

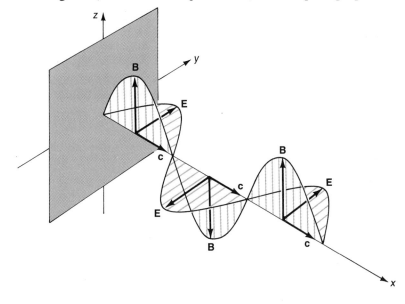

Fig. 39-7. The **E**, **B**, and **c** vectors in a sinusoidal electromagnetic wave.

of pulses, the velocity of a sinusoidal wave is also given by (39-13). This result is independent of wavelength, so that all electromagnetic waves (in empty space) travel at the same velocity, independent of wavelength. We can then infer the frequency of electromagnetic waves from their wavelength, λ:

$$f = \frac{c}{\lambda}$$

Alternatively, we can measure f and λ for electromagnetic waves to obtain the propagation velocity:

$$c = f\lambda$$

Over the range of frequencies (10^6–10^{21} Hz) where such measurements have been made, they yield values for c that are equal within the experimental precision of the measurements. The most precise such measurement was carried out recently (1972) by the National Bureau of Standards with light from a laser. Their result gives

$$c = 299{,}792{,}460 \pm 6 \text{ m/s}$$

for the speed of light (and hence for all electromagnetic waves) in vacuum. The spectrum of all electromagnetic waves that have been detected is indicated in Fig. 39-8.

Exercise

3. Consider an electromagnetic wave, similar to that of Fig. 39-4, that propagates in the $+z$ direction with an electric field strength of 1 V/m pointing in the positive y direction. Find the direction and magnitude of the magnetic field pulse that travels along with the electric field.

39-5   Radiation from Antennas   NOT ON HOURLY

In Fig. 39-9 we see a typical arrangement for the production of electromagnetic waves. The two rods are each one-quarter wavelength long at the frequency to be transmitted. The alternator may be a vacuum tube or transistor oscillator. The current shown charges the capacitance between the ends of the antenna, producing an electric field. While it exists, the current produces a magnetic field perpendicular to the electric field. The electric and magnetic fields are 90° out of phase with one another, since the current (and hence **B**) reaches a peak before the charge (and hence **E**) reaches its peak.

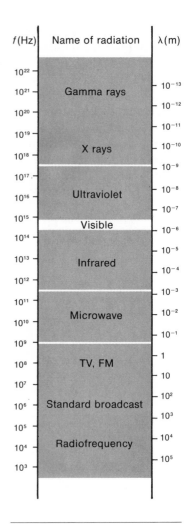

Fig. 39-8. The known spectrum of electromagnetic radiation.

These fields are both dipole fields and fall as $1/r^3$, where r is the distance from the antenna to a field point.

At large distances from the antenna, the fields produced directly by the antenna charges and currents no longer dominate. Instead, the radiation field (in which **E** and **B** fall as $1/r$) produced by the continual induction of **E** by **B** and of **B** by **E** dominates. The electric field lines at one instant are shown in Fig. 39-10. The separation of electric field lines from the moving charges is illustrated in Fig. 39-11. The intensity of the radiation field is at a maximum in directions perpendicular to the line of the antenna.

In the inverse process, currents are induced in an antenna by passing electromagnetic waves. Such an antenna is called a *receiving antenna*. A dipole also *receives* electromagnetic waves best when it is parallel to **E**. As the angle between the dipole and **E** increases, it responds less. The power transmitted (or received) as a function of angle with respect to a dipole is shown in Fig. 39-12.

Antenna *arrays* are often constructed to augment the natural directivity of a simple half-wave dipole. Such an array is shown in Fig. 39-13.

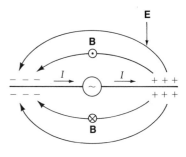

Fig. 39-9. A half-wave dipole antenna. As charges flow toward the right, a magnetic field is produced that encircles the rods. As charges accumulate at the ends of the rods, an electric field is produced that is perpendicular to the **B** field.

Fig. 39-10. Electric field lines (not to scale) of an oscillating dipole.

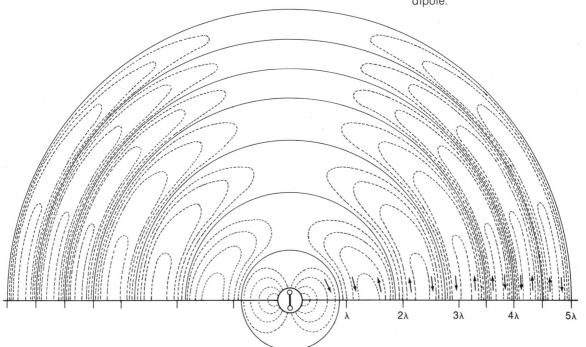

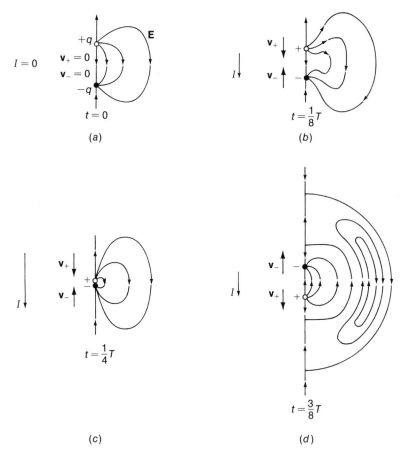

Fig. 39-11. An illustration of how radiation escapes from an oscillating electric dipole. The dipole is represented by oscillating charges $\pm q$. The lines of **E** are shown at four stages, $\frac{1}{8}$ cycle apart. The crucial instant comes in (c), when the charges pass one another in opposite directions. The electric lines of force break away and form closed loops, while a new set of electric field lines begin to appear between the separating charges in (d). (Adapted from Halliday and Resnick, *Physics*, John Wiley & Sons, Inc., 1966)

Exercise

4. Having seen rooftop TV antennas, estimate the wavelength of electromagnetic waves used for television transmission.

39-6 Radiation from Accelerated Charges   NOT ON HOURLY

A single charge moving with a constant velocity produces a $1/r^2$-dependent magnetic field and also carries its $1/r^2$-dependent electric field along with it. However, when a charge accelerates, radiation is produced. In the dipole antenna, many charges are accelerated by the alternating current. By considering a single charge only briefly accelerated, we can gain some insight into the mechanism of the $1/r$-dependent radiation fields and also see why the radiation is maximum in directions perpendicular to the direction of acceleration of the charges.

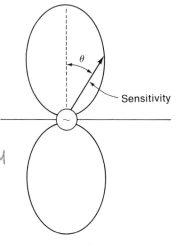

Fig. 39-12. The power transmitted (or received) by a half-wave dipole is maximum at right angles to the dipole.

In Fig. 39-14 we see the electric lines of force due to a charge Q that has been accelerated briefly. The speed of electromagnetic disturbances is c, so the lines of force at large distances from the charge have not yet responded to Q's changed position. They still appear to emanate from the old location of Q indicated by the dotted circle. The lines of force close to the charge emanate from its present position. In an intermediate zone of thickness $\Delta r = \Delta t/c$, where Δt is the time required for the charge to change positions, the lines of force are kinked and have a *transverse* component. The static lines of force are purely radial.

It is the transverse component of **E** that propagates outward as a $1/r$-dependent radiation field. It is easy to see that the magnitude of the stretching required of the field lines in the kinked zone is proportional to r. On the other hand, the field intensity falls as $1/r^2$. The net effect is that the propagating transverse component of **E** falls as $1/r$.

Figure 39-14 also makes it clear that the outward-running kink in the field lines has its maximum amplitude in a direction perpendicular to the acceleration of the charge. For the direction along the line of acceleration there is no transverse component and hence no radiation.

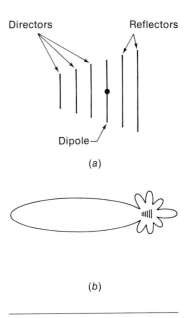

Directors   Reflectors

Dipole

(a)

(b)

Fig. **39-13.** (a) An array consisting of a half-wave dipole with several additional elements, called *reflectors* and *directors*. These elements are purposely made not equal to an integral half-wavelength, so that they reradiate induced signals out of phase with the main signal to augment or suppress reception from a given direction. (b) The pattern of sensitivity of such an antenna shows a strong lobe in the desired direction and minor lobes in other directions.

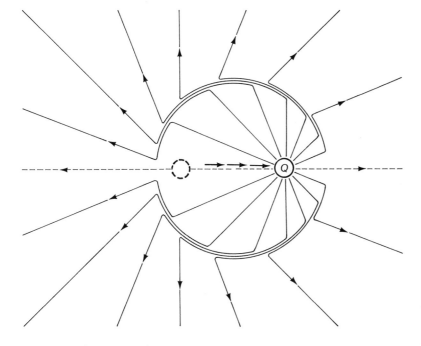

Fig. **39-14.** A charge Q accelerates abruptly from its old position, shown by the dotted circle, to a new position. The field lines at large distances still appear to emanate from the old location, while those close in emanate from the new location. The zone of disturbed field lines propagates with the velocity of light, c.

By the process described in Section 39-4, the radiated **E** field induces a **B** field that moves along with it. The intensity of this **B** field is proportional to **E** so that both fields fall as $1/r$.

The energy carried by both the **E** and **B** fields is proportional to the square of their amplitudes. The power density thus falls as $1/r^2$. The $1/r^2$ behavior of the power density assures that the same amount of energy passes through spheres of various radii that are centered on the source.

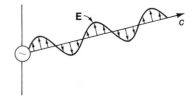

Fig. 39-15. In any given direction from the antenna, the radiated electric field oscillates in a direction perpendicular to the propagation direction and in a plane containing the line of the dipole antenna.

39-7 Polarization   *NOT ON HOURLY*

An accelerated charge produces a radiated transverse **E** field that lies in a plane containing the line of motion of the charge. Similarly, a dipole antenna produces radiated electric fields that oscillate in a plane containing the antenna, as indicated in Fig. 39-15. A receiving antenna perpendicular to this direction does not have charges accelerated back and forth along its length by the radiation field and is thus a poor receiver. If the receiving dipole is aligned parallel to the transmitting dipole, it becomes an effective receiver. We describe this situation by saying the electromagnetic waves are *polarized*.

Light waves are produced by the extremely high frequency vibrations of atomic electrons. Again, the **E** field in the radiated light wave lies in a plane that contains the direction of the atomic electron's acceleration. Most light sources contain large numbers

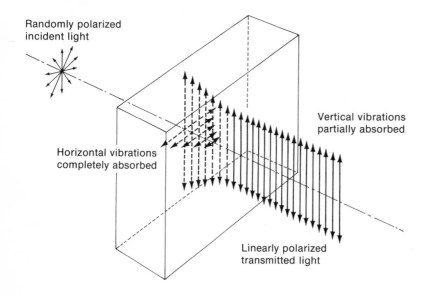

Randomly polarized incident light

Horizontal vibrations completely absorbed

Vertical vibrations partially absorbed

Linearly polarized transmitted light

Fig. 39-16. Randomly polarized light is preferentially polarized along one direction by a Polaroid filter.

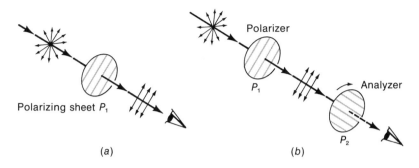

Fig. 39-17. (a) A polarizing sheet passes only light with a unique polarization direction. This direction is schematically indicated by the lines drawn on the sheet. (b) A second polarizing sheet with its direction of polarization perpendicular to the first blocks the light.

of atoms with random orientation, so that the radiated light is a mixture of polarization directions and is called *unpolarized*.

Initially unpolarized light can be polarized if it undergoes a process that selects only one plane of polarization. One way in which this can be done is used in the *Polaroid* filter. Polaroid filters, invented by Edwin H. Land, consist of sheets of long molecules of polyvinyl alcohol stained with iodine. These molecules are oriented in a single direction by stretching the sheet, which is bonded to a plastic supporting sheet.

The oriented long molecules absorb light much more strongly for one polarization direction, as indicated in Fig. 39-16. The light that does pass through is strongly polarized in the other direction. A second polarizer with its preferred polarization direction parallel to the first will pass the polarized light. If the second polarizer is rotated by 90°, it will block the polarized light, as shown in Fig. 39-17*b*.

Exercises

5. The reader has no doubt seen TV antennas on rooftops. What direction do you infer for the TV transmitter's dipole antenna: horizontal or vertical?

6. Assume that a Polaroid filter allows only the *component* of **E** in a given direction to pass through. A second Polaroid with its polarizing direction perpendicular to the first will then block the light completely, as in Fig. 39-17*b*. A third polarizer is now placed *between* the two crossed Polaroids with its axis intermediate in direction, i.e., 45° away from the first Polaroid. Does light now pass through the entire combination?

39-8 Energy and Momentum of Radiation   NOT ON HOURLY

When a radio wave strikes a receiving antenna, it gives energy to conduction electrons and sets them in motion. When infrared radiation is absorbed, it produces warmth and, again, energy is carried by the radiation. These observations and many more show that energy is carried along by any sort of electromagnetic radiation. How is the energy transported in empty space? Where does it reside?

The energy carried by electromagnetic waves must somehow be connected with the **E** and **B** fields, since these traveling fields constitute electromagnetic radiation. To find the quantity of energy that can be stored by electric and magnetic fields, let us briefly consider energy storage by static fields.

In Section 32-7 we saw that the energy stored by a charged capacitor is given by

$$U = \tfrac{1}{2}CV^2 \qquad\qquad (32\text{-}9)$$

This energy, in a very real sense, resides in the electric field between the plates. It is this field that sets the charges in motion when the capacitor is discharged and its energy is given to the moving charges. As the capacitor loses energy, the electric field decreases.

If we adopt the view that the energy given by (32-9) is stored in the field between the plates of a parallel-plate capacitor, we can find the energy per unit volume of the field as follows:

$$U = \tfrac{1}{2}CV^2 = \tfrac{1}{2}\frac{\epsilon_0 A}{d}\,(Ed)^2$$

$$= \tfrac{1}{2}\epsilon_0 dA E^2$$

and recognizing that dA is the volume containing the field, we have

$$\frac{U}{\text{volume}} = \tfrac{1}{2}\epsilon_0 E^2 \qquad\qquad (39\text{-}14)$$

This is a general result for geometry other than a parallel-plate capacitor, as well.

In Section 36-10 we saw that an inductor stores energy as given by

$$U = \tfrac{1}{2}LI^2 \qquad\qquad (36\text{-}21)$$

Again, we may consider this energy to be stored in the magnetic field. If we consider an inductor in the form of a very long solenoid, it is straightforward to calculate the energy per unit volume that must be associated with the field within the solenoid. The result is

$$\frac{U}{\text{volume}} = \frac{1}{2}\frac{B^2}{\mu_0} \qquad\qquad (39\text{-}15)$$

Again, this result is a general result, though we will not prove it.

Now we consider a sinusoidal electromagnetic wave with electric and magnetic fields given by

$$E = E_0 \sin(\omega t) \qquad (39\text{-}16)$$

and

$$B = B_0 \sin(\omega t) \qquad (39\text{-}17)$$

We have already seen (Section 38-6) that the average of the square of a sinusoidally varying quantity is one half the square of its amplitude. Thus we find the average energy per unit volume to be

$$\frac{\bar{U}_E}{\text{volume}} = \tfrac{1}{4}\epsilon_0 E_0{}^2$$

and

$$\frac{\bar{U}_B}{\text{volume}} = \tfrac{1}{4}\frac{B_0{}^2}{\mu_0}$$

Now from (39-12) and (39-13) we find that in an electromagnetic wave,

$$B = \frac{1}{c} E \qquad (39\text{-}18)$$

so that

$$\frac{\bar{U}_E}{\text{volume}} = \frac{\bar{U}_B}{\text{volume}}$$

and the total average energy per unit volume in an electromagnetic wave is

$$\frac{\bar{U}_{\text{total}}}{\text{volume}} = \frac{\bar{U}_E}{\text{volume}} + \frac{\bar{U}_B}{\text{volume}} = \tfrac{1}{2}\epsilon_0 E_0{}^2 \qquad (39\text{-}19)$$

This energy is carried along at the velocity of light, so that the energy passing through a unit area in unit time is given by

$$\frac{\bar{U}}{\text{area} \cdot \text{time}} = \tfrac{1}{2}c\epsilon_0 E_0{}^2 \qquad (39\text{-}20)$$

Example   Sunlight delivers about 1400 W/m² to the earth's surface. If this radiation were monochromatic, as given by (39-16) and (39-17), find the magnitude of E_0 and B_0.

Solution   Applying (39-20) we find

$$E_0{}^2 = \frac{2}{c\epsilon_0} \frac{\bar{U}}{\text{area} \cdot \text{time}}$$

$$= \frac{2}{(3 \times 10^8 \text{ m/s}) (8.85 \times 10^{-12} \text{ C}^2/\text{N} \cdot \text{m}^2)} \frac{1.4 \times 10^3 \text{ J}}{\text{m}^2\text{s}}$$

$$= 1.05 \times 10^6 \text{ V}^2/\text{m}^2$$

$$E_0 = 1.02 \times 10^3 \text{ V/m}$$

We then find B_0 from (39-18):

$$B_0 = \frac{E_0}{c} = \frac{1.02 \times 10^3 \text{ V/m}}{3 \times 10^8 \text{ m/s}} = 3.4 \times 10^{-6} \text{ T}$$

Maxwell predicted that electromagnetic waves carry linear momentum as well as energy. If an absorbing surface absorbs a quantity of electromagnetic energy U, the momentum imparted to the surface is given by

$$p = \frac{U}{c} \tag{39-21}$$

The force that thus acts on the absorber is called *radiation pressure* and its magnitude is very small.*

Example   A brilliant arc lamp delivers a luminous flux of 100 W to a 1-cm² absorber. Find the force due to radiation pressure.

Solution   The force is found by differentiating (39-21):

$$F = \frac{dp}{dt} = \frac{1}{c} \frac{dU}{dt} = \frac{10^2 \text{ W}}{3 \times 10^8 \text{ m/s}}$$

$$= 3.3 \times 10^{-7} \text{ N}$$

Exercises

7. At the wavelength of maximum visual sensitivity (555 nm), a radiant flux of 1 W is equivalent to a luminous flux of 680 lumens. From Table 24-2 we find that an illuminance of 100 lumens/m² is required for close work. Find E_0 for light of this intensity.

8. Show that (39-18) is dimensionally consistent.

39-9   The Michelson-Morley Experiment   NOT ON HOURLY

In the first part of this chapter we determined the speed of light by use of Maxwell's equations. Now we ask, "With respect to what particular frame of reference is this velocity measured?" To Maxwell, it seemed clear that there must be a medium for all electromagnetic phenomena and that all velocities entering into the equations of electromagnetic theory were velocities relative to

---

*See the article "The Pressure of Laser Light" by Arthur Ashkin, *Scientific American*, February 1972.

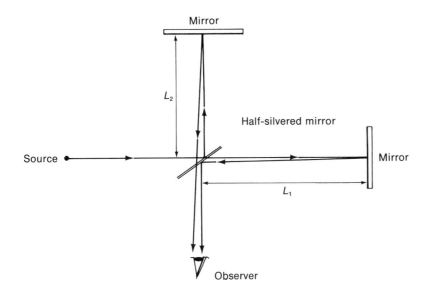

Mirror

L_2

Half-silvered mirror

Source ●

Mirror

L_1

Observer

Fig. 39-18. The Michelson interferometer. If the "ether wind" were blowing at velocity u from left to right, the velocity of the light ray would be $c + u$ on the outgoing portion of L_1 and $c - u$ on the return. Along both outgoing and returning portions of L_2, the light velocity would be $\sqrt{c^2 - u^2}$ with respect to the apparatus.

that medium. He visualized fields as stresses in the medium. This hypothetical medium was called the *ether*.

The earth's orbital velocity about the sun is so large that one might imagine being able to detect effects in the propagation of light caused by the "ether wind" generated by our motion.

In the year 1881, Albert A. Michelson attempted to measure the speed of the earth through the ether. He proposed to do this by comparing round-trip time for a light beam traveling upstream and then downstream in the "ether wind" to the round trip time for a light beam going across the wind direction and back. The earth's orbital velocity about the sun is about 30 km/s. Though this is only 10^{-4} as great as the speed of light, Michelson thought he could detect its effects by sensitive interference measurements on the light beams.

Michelson used the interferometer named for him (Section 24-10) in the arrangement shown in Fig. 39-18. One of the light paths, L_1, was made parallel to the earth's velocity while the other, L_2, was at right angles to this direction. Along path L_1 the outgoing velocity would be $c + u$, where u is the speed of the "ether wind," if the ether theory was correct. On the return it would be $c - u$. The total round-trip time would be

---

*Of course, the light-wave fronts spread in all directions. However, only those rays whose velocity relative to the apparatus is as shown in Fig. 39-18 reach the observer. It is those rays that we discuss.

$$t_1 = \frac{L_1}{c+u} + \frac{L_2}{c-u} = L_1\left(\frac{c-u}{c^2-u^2} + \frac{c+u}{c^2-u^2}\right)$$

$$= \frac{2cL_1}{c^2-u^2} = \frac{2L_1}{c}\left(\frac{1}{1-\dfrac{u^2}{c^2}}\right) \tag{39-22}$$

Along the path L_2, the velocity vector of light with respect to the ether frame of reference must be tipped in such a way that the light follows the straight path L_2 in spite of the ether wind. Using addition of velocities and the Pythagorean theorem we find the velocity with respect to the *apparatus* is

$$v^2 = \sqrt{c^2 - u^2}$$

along path L_2 (both outgoing and returning). Thus we have

$$t_2 = \frac{2L_2}{\sqrt{c^2-u^2}} = \frac{2L_2}{c}\sqrt{\frac{1}{1-\dfrac{u^2}{c^2}}} \tag{39-23}$$

Setting $L_1 = L_2$, there will still be a difference in t_1 and t_2 according to (39-22) and (39-23). This difference is a small one, since $u \ll c$. For equal paths L:

$$t_1 = \frac{2L}{c}\left(1 - \frac{u^2}{c^2}\right)^{-1}$$

$$t_2 = \frac{2L}{c}\left(1 - \frac{u^2}{c^2}\right)^{-1/2}$$

Since u^2/c^2 is much smaller than 1, we can conveniently use the binomial series:

$$y = (1+x)^n = 1 + nx + \frac{n(n-1)}{2!}x^2 + \frac{n(n-1)(n-2)}{3!}x^3 + \cdots$$

Applying this to t_1, we find:

$$t_1 = \frac{2L}{c}\left(1 + \frac{u^2}{c^2} + \frac{u^4}{c^4} + \frac{u^6}{c^6} + \cdots\right)$$

Since $u^2/c^2 \approx 10^{-8}$, we can ignore higher-order terms beyond the u^2/c^2 term and still retain high accuracy. Similarly:

$$t_2 = \frac{2L}{c}\left(1 + \tfrac{1}{2}\frac{u^2}{c^2} + \tfrac{3}{8}\frac{u^4}{c^4} + \tfrac{5}{16}\frac{u^6}{c^6} + \cdots\right)$$

The difference in time is thus accurately given by:

$$\Delta t = t_1 - t_2 = \frac{2L}{c}\left[\tfrac{1}{2}\frac{u^2}{c^2}\right]$$

Because of this difference, the two waves will no longer be in phase when they recombine. The difference in the *distances* covered is

$$\Delta s = c\Delta t = L\frac{u^2}{c^2}$$

Expressing this difference as a number N of wavelengths λ, we have

$$\Delta s = N\lambda \qquad \text{or} \qquad N = \frac{\Delta s}{\lambda} = \frac{L}{\lambda}\frac{u^2}{c^2}$$

Because λ is so small for light waves, N can approach unity for a reasonable interferometer size, even though $u^2/c^2 \approx 10^{-8}$. If $N = \frac{1}{2}$ we have destructive interference; if $N = 1$ we have constructive interference.

In the final apparatus built by A. A. Michelson and E. W. Morley, the entire instrument was built on a stone slab floated in mercury. As the instrument was slowly turned so that first one path then the other was parallel to the ether wind, the interference pattern should have shifted. It did not. The experiment was repeated at different times of the year, just in case the first observation indicated that the sun was moving through the ether and by chance the earth's orbital velocity cancelled the sun's velocity. No shift was ever observed.

It seems inconceivable that the earth, part of a rotating solar system in a rotating galaxy, could be at absolute rest with respect to the ether frame of reference. This would imply that the remaining entirety of the universe is making some fantastically immense orbits about the earth. The presently accepted point of view was clearly put forth by Albert A. Einstein in 1905.* He proposed that only *relative* motion should appear in physical laws; "absolute" motion is meaningless.

From this point of view, the observer we considered to be stationary in the derivation in Section 39-4 was as good as any other. An observer in the "stationary" frame of reference considered in Fig. 39-4 would, accordingly, measure c for the velocity of light. If Maxwell's equations are truly proper physical laws, they will agree with the principle of relativity. Thus, they will not change their form in another, uniformly moving frame of reference. An observer in this frame of reference, then, will still find $c =$

---

*The year 1905 marked the publication of his paper, "On The Electrodynamics of Moving Bodies," *Annalen der Physik*, vol. 17 (1905), in which he proposed the *Special Theory of Relativity*. The word "special" indicates that only frames of reference with constant velocity were considered.

$(\epsilon_0\mu_0)^{-1/2}$. From this point of view we would *expect* a null result in the Michelson-Morley experiment.

Since Michelson and Morley did obtain a null result, we conclude that Einstein's interpretation is correct; the equations of electromagnetism are valid in any frame of reference. Only relative velocities (as in $\mathbf{F} = q\mathbf{v} \times \mathbf{B}$) enter them; and the velocity of light is the same for all observers moving in any direction.

On the other hand, it seems *intuitively* obvious that if a given beam of light has a velocity c, as measured by one observer, that another observer moving "upstream" against the direction of light propagation should measure a larger velocity. Nonetheless, our interpretation of the result of the Michelson-Morley experiment says that all observers should measure the same velocity. Einstein's solution to this dilemma involved a reformulation of our basic concepts of space and time, as discussed in Section 8-6. Length intervals and time intervals as measured by observers in moving frames of reference are modified by factors that involve the term $\sqrt{1 - (u^2/c^2)}$, which we encountered in our discussion of the Michelson-Morley experiment.

39-10 Summary

Just as electric fields can be induced by changing magnetic fields, so can magnetic fields be induced by changing electric fields. Ampere's law is modified to include this effect:

$$\oint \mathbf{B} \cdot d\mathbf{s} = \mu_0 I + \mu_0\epsilon_0 \frac{d\Phi_E}{dt} \tag{39-7}$$

Electromagnetic radiation involves the continual induction of an $\mathbf{E}$ field by a $\mathbf{B}$ field and the reverse. The speed of propagation of electromagnetic radiation is predicted to be

$$c = \sqrt{1/\mu_0\epsilon_0}$$

in agreement with observation.

A dipole antenna radiates maximally in directions perpendicular to its own length. Similarly, a single accelerated charge radiates maximally perpendicular to the direction of its acceleration.

The electric field in a sinusoidal electromagnetic wave lies in a plane containing the line of motion of the moving source charges. Thus, electromagnetic waves from dipole antennas are linearly polarized. Most light sources produce a random mixture of polarization directions, but one of these directions may be selected by use of a polarizing filter.

The energy per unit volume stored in electric and magnetic fields is found to be

$$\frac{U_E}{\text{volume}} = \tfrac{1}{2}\epsilon_0 E^2$$

$$\frac{U_B}{\text{volume}} = \tfrac{1}{2}\frac{B^2}{\mu_0}$$

In a sinusoidal electromagnetic wave with an electric amplitude E_0, the *net* average energy density is

$$\frac{\bar{U}}{\text{volume}} = \tfrac{1}{2}\epsilon_0 E_0{}^2$$

with half of this energy carried in the magnetic field and half in the electric field.

The flux of energy carried by an electromagnetic wave is thus

$$\frac{U}{\text{area} \cdot \text{time}} = \tfrac{1}{2}c\epsilon_0 E_0{}^2$$

The momentum given to an absorber taking in a quantity of energy U is

$$p = \frac{U}{c}$$

All attempts to observe a modification of the measured velocity of light have failed. The velocity of light is always measured to be the same value, c, regardless of the state of motion of the observer or the source.

Problems

HANDOUT SHEET
OF PROBLEMS

39-1. An *LC* resonant circuit contains a 400-pF capacitor and a 100-μH inductor. It is set into oscillation and coupled to an antenna. Find the wavelength of the radiated electromagnetic waves.

39-2. A free electron is acted upon by a sinusoidal electromagnetic wave with $E_0 = 10^3$ V/m and $\omega = 10^6$ s^{-1}. (*a*) Find the maximum velocity of this electron caused by the electric force. (*b*) Now find the maximum magnitude of the magnetic force exerted on this electron by the magnetic field of the electromagnetic wave.

39-3.* Suppose that a current I passes along a wire and charges an object with a charge Q. Since charges cannot be created or destroyed, we have

$$I = \frac{dQ}{dt}$$

But Gauss's law tells us that a surface surrounding the charged object gives

$$\Phi_E = \oint \mathbf{E} \cdot d\mathbf{A} = Q/\epsilon_0$$

The current I produces a magnetic field given by Ampere's law:

$$\oint \mathbf{B} \cdot d\mathbf{s} = \mu_0 I$$

Now show that

$$\oint \mathbf{B} \cdot d\mathbf{s} = \mu_0 \epsilon_0 \frac{d\Phi_E}{dt}$$

in agreement with (39-3). This line of argument parallels that originally made by Maxwell in hypothesizing the displacement current.

39-4.* Show that the magnitude of the transverse component of $\mathbf{E}$ in Fig. 39-14 decreases with distance as $1/r$.

39-5. When incident light waves set atomic electrons into vibration, these electrons radiate as well. Thus light is reradiated (scattered) into new directions when it passes through matter. Show that when a beam of light is polarized, no light is scattered in the direction of the $\mathbf{E}_0$ vector in the incident light wave.

39-6. A He-Ne laser produces a continuous beam of light carrying a power of 10 mW. The beam is in the form of a cylinder with a diameter of 2 mm. (*a*) Find the electric field magnitude, E_0. (*b*) Find the magnetic field magnitude, B_0.

39-7. Show that the power transmitted through a pair of Polaroid filters (as in Fig. 39-17) is proportional to $\cos^2 \theta$, where θ is the angle between the polarizing directions of the two filters.

39-8. A pulsed ruby laser produces a pulse of light with a duration of 10 ns. The energy contained in this pulse is 1 J. The diameter of the beam is 1 cm. (*a*) Find the spatial length of the traveling pulse of light. (*b*) Find the energy density, U/volume, of this pulse. (*c*) Find the magnitude of the electric field amplitude, E_0.

39-9. Consider a small, spherical particle in space at a distance from the sun approximately equal to that of the earth, so that the solar flux is about 1400 W/m². The particle is completely absorbing, so that the energy it absorbs is proportional to its cross-sectional area πr^2. The density of the particle is ρ. (*a*) Show that, for a small enough radius, the force due to radiation pressure can exceed the force due to the sun's gravitational field, so that the particle is pushed away from the sun rather than attracted toward it. (This is the mechanism that causes a comet's tail to point away from the sun.) (*b*) Calculate the particle radius for which the two forces cancel when $\rho = 1$ g/cm³.

39-10. An astronaut floating freely in space decides to use his flashlight as a rocket. He shines a 10-W light beam in a fixed direction so that he acquires momentum in the opposite direction. If his mass is 80 kg, how long must he wait to reach a velocity of 1 m/s?

39-11. Let us assume that we wish to detect a hypothetical ether wind. For $u/c = 10^{-4}$, find the length L ($L_1 = L_2$) required for an interferometer that gives one full fringe shift when it is rotated by 90° in the ether wind.

Chapter 40

Quantum Physics

40-1 Introduction   The dominant theme of twentieth-century physics is atomism. We know that matter exists in discrete units: atoms and their subunits, the electrons and nucleons. Further, electric charge exists only in multiples of the smallest charge known, e, the charge carried by the electron.

When we examine the energy that resides in the motion of electrons within atoms, we find that this energy can change only in *discrete* steps. Thus the concept of *atomism,* as opposed to a smooth continuum, appears to extend into the domain of energy as well as the domain of electric charge and matter.

We have just completed a chapter of study that indicated that electromagnetic radiation is a smooth, continuous wave. Yet we will see that this radiation, too, seems to be parceled out in discrete quantities. Originally these parcels were called *quanta;* the name *quantum physics* refers loosely to the physics of matter and energy at the microscopic level of the atom.

Paradoxically, we will find that the basis for the discrete character of atomic energy levels resides in the wavelike behavior of electrons. We will see that all types of matter, and electromagnetic radiation as well, exhibit a *wave-particle duality*. This chapter will concentrate on the mainstream of the development of this dualistic concept. The chief events spanned the years 1900–1927.

The aim of this entire text has been to introduce the reader to classical physics, and a single chapter such as this is not intended to "cover" twentieth-century physics. It is included to show the reader some of the striking new ideas that arose in the physics of this century. If it whets an occasional reader's appetite to read further in modern physics, it will have served its purpose.

40-2 Planck and Cavity Radiation

In Fig. 40-1 we see a photograph of an incandescent tungsten tube with a small hole drilled in its side. The radiation escaping through this hole is a sample of the radiation within the enclosure and is called *cavity radiation*. The spectral radiance of this radiation for a temperature of 1646K is shown in Fig. 40-2. (See Fig. 24-22 and the accompanying discussion of spectral radiance.) It is found experimentally that this spectrum is *independent* of the material composing the walls of the cavity. Both the magnitude and shape of this spectrum depend only upon the temperature of the cavity.

Early attempts to arrive at the cavity radiation spectrum from purely theoretical considerations failed. The best attempts used the ideas of *statistical mechanics*, which are useful when large numbers of particles constitute a system and only average properties are desired. It is clear that the source of the electromagnetic radiation in the cavity is the accelerated charges contained in the large

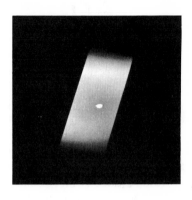

Fig. 40-1. The light that escapes through a small hole in the side of a hollow, incandescent tungsten tube is an example of cavity radiation. (From Halliday and Resnick, *Fundamentals of Physics*, John Wiley and Sons, Inc., 1970)

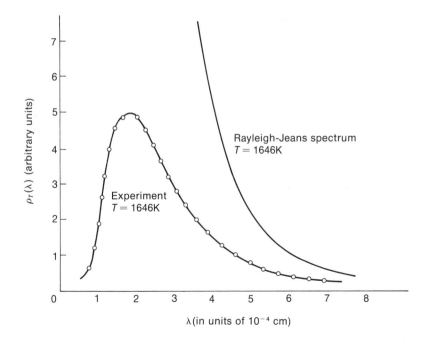

Rayleigh-Jeans spectrum
$T = 1646$K

Experiment
$T = 1646$K

$\rho_T(\lambda)$ (arbitrary units)

λ(in units of 10^{-4} cm)

Fig. 40-2. Spectrum of cavity radiation. The circles are measured points. The curves represent theoretical predictions.

number of atoms making up the cavity walls. These accelerations are caused by thermal agitation.

One of the postulates of statistical mechanics is that each and every possible state of motion of a system with many degrees of freedom is equally probable. For example, the most frequently occurring total energy of a large number of moving particles is that that corresponds to the largest number of ways that individual particles can move while adding up to this most probable energy.

Rayleigh and Jeans applied the methods of statistical mechanics to the cavity radiation problem and immediately faced a paradox. Various modes of electromagnetic waves were possible in the cavity. Each mode corresponded to an integral number of wavelengths fitting within the cavity. But as the frequency was increased and the wavelengths decreased, the number of possible modes grew without bound! The basic postulates of classical statistical mechanics said that the shortest wavelengths would then dominate the spectrum and that the spectral radiancy approached infinity as $\lambda \rightarrow 0$. This is indicated by the Rayleigh-Jeans spectrum graphed in Fig. 40-2. The data is seen to be in hopeless disagreement with the Rayleigh-Jeans prediction at short wavelengths. This difficulty was called the ultraviolet catastrophe.

A way out of this difficulty was proposed by Max Planck in 1900. If the oscillating atoms in the walls could radiate only discrete quantities of energy, and this quantity was *proportional to the frequency*, the probability of high-frequency emission could be greatly reduced below the Rayleigh-Jeans prediction. In fact, there is a complete statistical theory assuming each atomic oscillator has quantized energy states given by

$$E = nhf \qquad (40\text{-}1)$$

where n is an integer, h is a constant called *Planck's constant,* and f is the frequency; this theory leads to the curve passing through the experimental points in Fig. 40-2.

Planck's constant, h, has the dimensions of energy multiplied by time. Planck found its value by fitting his theory to experiment. The presently accepted value is

$$h = 6.63 \times 10^{-34} \text{ J} \cdot \text{s} = 4.14 \times 10^{-15} \text{ eV} \cdot \text{s}$$

Planck himself disliked the necessity for assuming that each atomic oscillator can only possess discrete quantities of energy. It seemed obvious enough that a mass on a spring, for example, could oscillate with *any* amplitude and hence any energy. Yet the beautiful agreement between Planck's theory and experiment required serious consideration of the idea of quantized energies.

Example   The hypothesis made by Planck implies that a 100-g mass attached to a spring with $k = 395$ N/m can change its energy only in steps of hf. In the present case, $f = 10$ Hz. Find the magnitude of these energy steps. Can they be discerned by experiment?

Solution   The size of the energy steps is

$$\Delta E = hf = (6.63 \times 10^{-34} \text{ J} \cdot \text{s}) \cdot 10 \text{ Hz} = 6.63 \times 10^{-33} \text{ J}$$

The total energy of this mass-spring system when it oscillates with an amplitude x_0 as small as a wavelength of light is

$$E_{\text{total}} = \tfrac{1}{2}kx_0{}^2 = \tfrac{1}{2}(3.95 \times 10^2 \text{ N/m}) (6 \times 10^{-7} \text{ m})^2 = 7.11 \times 10^{-11} \text{ J}$$

The total number of increments ΔE required to reach even this tiny amplitude is thus on the order of 10^{22}! Clearly these steps are so finely spaced for a macroscopic oscillator that they can never be noticed.

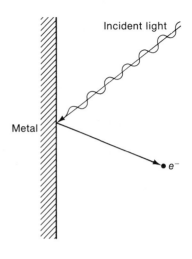

Fig. 40-3.   The photoelectric effect. Light shining on a metallic surface can eject electrons.

Exercise

1.   The frequency of visible light is about

$$f = c/\lambda = (3 \times 10^8 \text{ m/s}) (6 \times 10^{-7} \text{ m})^{-1} = 5 \times 10^{14} \text{ Hz}$$

Calculate hf for an electron oscillating at this frequency and compare it in magnitude to the typical unit of energy in atoms: the electron volt ($1 \text{ eV} = 1.6 \times 10^{-19} \text{ J}$).

40-3   Einstein and the Photoelectric Effect

When Hertz was studying electromagnetic *waves,* he stumbled upon the observation that was one day destined to lead to a *particle* model of light. He found that sparks were more easily initiated in his detection apparatus (Fig. 39-3) when ultraviolet light was illuminating the metal surfaces. Later workers found this effect was due to electrons ejected from the metal by the light, as shown in Fig. 40-3. This process is called the *photoelectric effect.* These electrons had a variety of energies up to a well-defined maximum. This seemed reasonable enough, since the electromagnetic wave had a continually varying electric field magnitude:

$$E = E_0 \sin(\omega t)$$

The apparatus sketched in Fig. 40-4 was used to find the maximum value of the electron's kinetic energy. A negative repelling voltage was slowly increased until even the fastest electrons were just prevented from reaching the collector. The kinetic energy of the most energetic electrons was then inferred to be

$$E_{\max} = eV \tag{40-2}$$

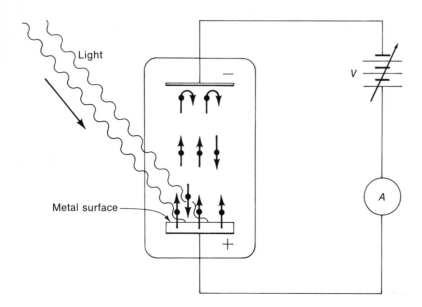

Fig. 40-4. Electrons are released from the metal surface with initial kinetic energy. The repelling voltage V is increased until no electrons can be collected (indicated by the current in the ammeter going to zero).

where e is the electron charge and V the electric potential applied to the collector.

It seemed natural enough to suppose that this maximum energy was related to the maximum electric field strength in the incident light, that is, to the light *intensity*. Surprisingly, it was found that the intensity of the light affected only the quantity of electrons liberated, not their kinetic energy.

The maximum kinetic energy of the photo-ejected electrons was found to depend linearly upon the *frequency f* of the incident light:

$$E_{max} = Af - B \qquad (40\text{-}3)$$

where A and B are constants. Einstein proposed that this dependence of energy upon frequency could be explained by an extension of Planck's hypothesis. In a paper published in 1905,* he proposed that light travels through space in bundles, or *quanta,* each carrying an energy given by

$$E = hf \qquad (40\text{-}4)$$

Annalen der Physik, vol. 17 (1905). This was a phenomenal year for Einstein. This same volume contained his paper on special relativity and a paper containing a theory of Brownian motion. The latter theory was utilized by Jean Perrin to measure Avogadro's number, N_0. It was the 1905 Einstein paper on the photoelectric effect that led to Einstein's Nobel Prize in 1921.

where h is Planck's constant and f is the frequency. Then the maximum quantity of energy that can be given to an electron in a quantum-electron collision is the full energy of one quantum. The expression (40-3) was interpreted by Einstein to be

$$E_{\max} = hf - E_0 \qquad (40\text{-}5)$$

where E_0 is the energy required to tear an electron from the metal surface. This quantity of energy is called the *work function* and is of the order of a few electron volts for most metals. In Fig. 40-5 we see the remarkable agreement between experimental points measured by Millikan and expression (40-5). The threshold frequency f_0 is that frequency that just liberates an electron with zero remaining kinetic energy:

$$E_{\max} = 0 = hf_0 - E_0$$

so

$$f_0 = E_0/h$$

In contrast with the classical prediction, light with a frequency below f_0 cannot liberate electrons, even if it is extremely intense.

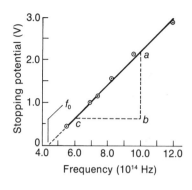

Fig. 40-5. Millikan's measurements of the stopping potential at several different frequencies of light incident upon sodium metal. The solid line is given by Eq. (40-5).

Example   According to (40-5), the slope of the data for E vs f should give Planck's constant. Does it?

Solution   The ordinate in Fig. 40-5 is the stopping potential V in volts. The maximum kinetic energy will then be given in eV by (40-2). The slope of V vs f is found from the graph:

$$\frac{\Delta V}{\Delta f} = \frac{1.6\ \text{V}}{4 \times 10^{14}\ \text{Hz}} = 4 \times 10^{-15}\ \text{V} \cdot \text{s}$$

Multiplying by e, the charge of the electron, we find

$$\frac{\Delta E_{\max}}{\Delta f} = 6.4 \times 10^{-34}\ \text{J} \cdot \text{s}$$

which, within the accuracy of these measurements, agrees with the accepted value of h.

The success of Einstein's theory of the photoelectric effect, where classical physics failed completely, meant that the concept of discrete quanta of radiation had to be taken seriously. Where Planck had proposed only that atomic oscillators had to make discrete quantum jumps of energy, Einstein proposed that the electromagnetic energy they radiated also had to travel in discrete parcels. The modern term used for these parcels is *photons*.

Exercise

2. The threshold frequency for the liberation of photoelectrons from sodium is

$$f_0 = 4.39 \times 10^{14} \text{ Hz}$$

Find the work function E_0 for sodium. Express your answer in eV.

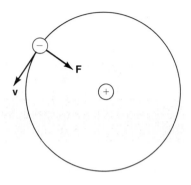

Fig. 40-6. A simple model of the hydrogen atom. The nucleus is a single proton and carries over 99.9% of the atom's mass. The electron is held in orbit by the electric attraction of the proton.

40-4   The Bohr Model of the Hydrogen Atom

In the nineteenth century it was already known that electrically excited atoms (as in a gas discharge or spark) radiated light only at certain discrete frequencies. Each element appeared to have its own characteristic spectrum, a fact that was useful for the detection of small amounts of elements in samples of unknown composition. As *useful* as the science of spectroscopy was, however, the basic mechanism responsible for radiation of only discrete frequencies was still not understood in the early part of the twentieth century.

Rutherford's discovery that the atom had a small nucleus (Section 28-10) led to the simple model indicated in Fig. 40-6. There we consider the simplest of all atoms, the hydrogen atom. If we are ever to understand any atom, this simplest case should provide a starting point.

Unfortunately, the predictions of classical physics fail badly in two respects when applied to such an atom.

1. The revolving electron is an accelerated charge and should continuously radiate electromagnetic waves. If it did so, however, it would lose energy, spiraling in toward the proton so that the ultimate state of the atom would be a tiny collapsed state with dimensions of the order of nuclear dimensions. In fact, atoms are known to remain stable with dimensions 10^4 times larger than nuclear dimensions.

2. The spiraling electron would steadily increase its frequency of revolution as it spiraled toward the proton so that all colors of light would be produced by a collection of such atoms. Instead, only sharp, discrete colors are observed.

In Fig. 40-7 we see an example of spectral lines from hydrogen. Such spectra can be observed by passing the light from electrically excited hydrogen atoms through a diffraction grating, as discussed in Section 24-8.

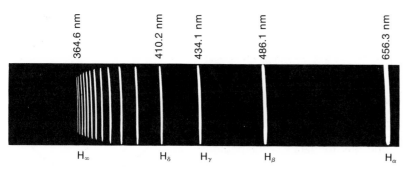

364.6 nm   410.2 nm   434.1 nm   486.1 nm   656.3 nm

H_∞ H_δ H_γ H_β H_α

Fig. 40-7. One of the series of spectral lines emitted by hydrogen atoms. This series is called the *Balmer series* and runs from visible wavelengths into the near ultraviolet. (Adapted from Holton and Brush, *Introduction to Concepts and Theories in Physical Science,* 1973, Addison-Wesley, Reading, Massachusetts)

It is also found that when light with a continuous distribution of frequencies (as that from an incandescent solid) is passed through a gas, *absorption lines* are seen at the same wavelengths at which the gas atoms would radiate if they were excited (see Fig. 40-8).

Since Einstein's photon hypothesis tells us that the energy carried by light is proportional to frequency, the observation of emission and absorption at only certain frequencies allows us to infer that the atom exists only in certain discrete energy levels. The discrete levels that are required to explain the hydrogen atom's spectrum are indicated in an *energy-level diagram* in Fig. 40-9.

When the hydrogen atom loses energy by dropping from a high-lying level to a lower level, it emits a photon with a frequency given by

$$f = \frac{\Delta E}{h} \tag{40-6}$$

where ΔE is the difference in energy between the upper and lower levels. The lowest possible level is called the *ground state,* while higher levels are called *excited states* of the atom. The various transitions that produce radiation are shown in Fig. 40-10. The Balmer series of spectral lines is made up of all transitions that *end* on the first excited state; it is the one shown in Fig. 40-7.

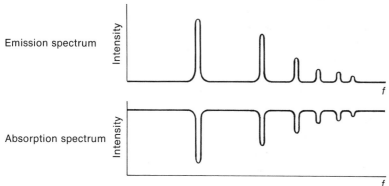

Emission spectrum   Intensity

Absorption spectrum   Intensity

f

f

Fig. 40-8. Hydrogen atoms emit and absorb radiation at only certain discrete frequencies.

The sequence of energy levels shown in Fig. 40-9 and 40-10 are described by a simple formula:

$$E_n = -\frac{E_0}{n^2} \qquad n = 1, 2, 3, \ldots \qquad (40\text{-}7)$$

where $E_0 = 13.58$ eV and $n = 1$ corresponds to the ground state of the hydrogen atom.

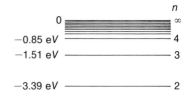

Example   Use (40-7) and (40-6) to find the frequency of the photons emitted in the Balmer series line, which correspond to the transitions $n = 3 \rightarrow n = 2$.

Solution   From (40-7) we have

$$\Delta E\,(n = 3 \rightarrow n = 2) = -E_0\!\left(\frac{1}{3^2} - \frac{1}{2^2}\right)$$

$$= \left(\frac{1}{4} - \frac{1}{9}\right)\cdot 13.58 \text{ eV}$$

$$= 1.89 \text{ eV} = 3 \times 10^{-19} \text{ J}$$

Fig. 40-9.   An energy-level diagram for the hydrogen atom.

From (40-6) we find

$$f = \frac{\Delta E}{h} = \frac{3 \times 10^{-19} \text{ J}}{6.63 \times 10^{-34} \text{ J}\cdot\text{s}} = 4.5 \times 10^{14} \text{ Hz}$$

The wavelength of this radiation is

$$\lambda = \frac{c}{f} = \frac{3 \times 10^8 \text{ m/s}}{4.5 \times 10^{14} \text{ s}^{-1}} = 670 \text{ nm}$$

which is in the visible spectrum.

The Rutherford model of the atom, indicated in Fig. 40-6, fails to explain the existence of only discrete, separate energy levels in the atom. Niels Bohr, once a student of Rutherford's, attempted to include quantum ideas into this model. He found a method to do so using Planck's constant, but he found it necessary to make seemingly arbitrary assumptions to do so.

Before we state Bohr's assumptions, let us note two facts. First, the angular frequency of emitted radiation is given by

$$\omega = 2\pi \frac{\Delta E}{h} = \frac{\Delta E}{\hbar}$$

where $\hbar$ is defined to equal $h/2\pi$. Second, we note that the dimensions in h (or in $\hbar$) are also those of angular momentum. To see this we begin with the units of h, which are energy $\times$ time, and we find:

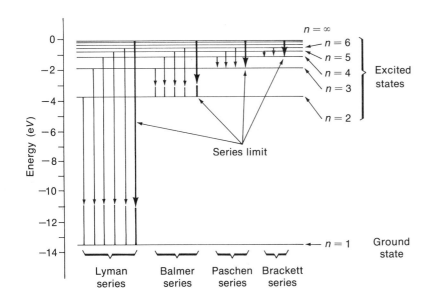

Fig. 40-10. Various spectral series found in the light emitted by excited hydrogen atoms are now understood to be due to the various transitions shown here.

$$[\text{energy}][\text{time}] = [\text{force} \cdot \text{length}][\text{time}]$$

$$= \left[\frac{\text{mass} \cdot \text{length}^2}{\text{time}^2}\right][\text{time}]$$

$$= [\text{mass}][\text{velocity}][\text{length}]$$

$$= [\text{angular momentum}]$$

The Bohr postulates are:

1. Electrons in atoms move only in certain nonradiating, stable orbits that are consistent with Coulomb's force law and Newton's second law. These orbits are specified by the *quantization of angular momentum:*

$$L = n\hbar \qquad n = 1, 2, 3, \ldots \qquad (40\text{-}8)$$

2. Photons are given off when the atom makes a transition from an upper energy state to a lower energy state, with frequencies given by Eq. (40-6).

Now let us find the energy levels given by postulate 1 above. First, applying Newton's second law to a circular orbit of radius r and with tangential velocity v, we find

$$F = m\frac{v^2}{r}$$

where m is the mass of the orbiting electron. This force is given by the electric attraction between the proton and electron, so we have:

$$F = \frac{ke^2}{r^2} = m\frac{v^2}{r}$$

or

$$\frac{ke^2}{r} = mv^2 \qquad (40\text{-}9)$$

Noting that $-ke^2/r$ is the electric potential energy U of the proton-electron system and that mv^2 is twice the kinetic energy K we have $-U = 2K$, so that

$$E_{\text{total}} = U + K = \tfrac{1}{2}U = -\frac{ke^2}{2r} \qquad (40\text{-}10)$$

Now we use the quantization postulate (40-8):

$$L = mvr = n\hbar$$

so that

$$v_n = \frac{n\hbar}{mr_n}$$

is the velocity of the electron, which depends upon n. Substituting this value of v_n into (40-9), we find

$$r_n = \frac{n^2\hbar^2}{kme^2} \qquad (40\text{-}11)$$

for the radius of the various allowed orbits. The smallest allowed radius is found by setting $n = 1$ in (40-11) and is called the *Bohr radius:*

$$r_1 = \frac{\hbar^2}{kme^2} = 5.3 \times 10^{-11} \text{ m}$$

This theoretical value is of the correct order of magnitude for the size of an atom.

The various allowed energy states are now found by substituting (40-11) into (40-10):

$$E_n = -\frac{ke^2}{2r_n} = -\frac{1}{n^2}\left(\frac{mk^2e^4}{2\hbar^2}\right) \qquad (40\text{-}12)$$

Compare this theoretical result of the Bohr theory with the experimentally derived (40-7). The form of the equation is correct but what about the theoretically calculated value of E_0? Comparing (40-12) and (40-7), we see that E_0 must be given by the quantity in parentheses in (40-12):

$$\frac{mk^2e^4}{2\hbar^2} = \frac{(9.11 \times 10^{-31} \text{ kg})(8.99 \times 10^9 \text{ Nm}^2/\text{C}^2)^2(1.6 \times 10^{-19} \text{ C})^4}{2(6.63 \times 10^{-34} \text{ J} \cdot \text{s})^2/(2\pi)^2}$$

$$= 2.17 \times 10^{-18} \text{ J}$$

$$= 13.6 \text{ eV}$$

in agreement with the empirical result. Use of four-place accuracy in the above calculation leads to 13.58 eV, as it should.

The Bohr model of the atom was a stunning success, yet no one could understand the origin of Bohr's postulates. Though Bohr's work was published in 1913, it was 1924 before the explanation of Bohr's quantization conditions was proposed by Louis de Broglie, as we will see in the next section.

Exercises

3. Show that the dimensions of the right side of (40-11) are equal to a length.

4. Find the velocity of the electron in its first Bohr orbit ($n = 1$) in hydrogen.

40-5   Louis de Broglie's Matter Waves

In 1922, Louis de Broglie published two papers on the theory of cavity radiation, in which he treated the radiation itself as a collection of particles, following Einstein's photon hypothesis. In 1923 he began to consider the novel idea that, just as electromagnetic "waves" behave as if they were particles, might not "particles" like electrons behave as if they were waves? The symmetry implicit in this suggestion was appealing and he pursued the idea, finally writing a doctoral dissertation examining the consequences of such an assumption.*

He postulated that the frequency of matter waves should be given by

$$f = \frac{E}{h}$$

as it was for photons. He then needed a way to find either the wavelength or propagation velocity of these waves. He again turned to

---

*An illuminating account of the origin and development of de Broglie's hypothesis is given in "Fifty Years of Matter Waves" by Heinrich A. Medicus, *Physics Today,* February 1974.

the successful results for photons as a guide. First, let us recall that the momentum carried by electromagnetic radiation is given by

$$p = \frac{U}{c}$$

where U is the energy carried by the radiation. Applying the same law to photons of energy E, we have

$$p = \frac{E}{c} = \frac{hf}{c} = \frac{h}{\lambda} \tag{40-13}$$

where λ is the wavelength of the electromagnetic radiation represented by the photons. We have here a blend of particle and wave concepts in one equation. The left side of (40-13) refers to the momentum of a discrete photon, while the right side refers to a wavelength.

The Einstein frequency relation for photons and the wavelength relation in (40-13) were assumed by de Broglie to apply to *both* photons and "material" particles, such as electrons. These two relations are now called the *Einstein–de Broglie relations:*

$$E = hf \qquad p = h/\lambda$$

Louis de Broglie was then able to give a reason for the arbitrary angular momentum quantization rule developed by Bohr. He proposed that the only *allowed* electron orbits were those that could exist indefinitely as standing waves. That is, the circumference of the allowed electron orbits should equal an integral number of electron wavelengths, as indicated in Fig. 40-11.

Using (40-13) for the electron wavelengths and requiring that the circumference of an orbit equal an integer number of wavelengths, we find

$$n\lambda = \frac{nh}{p} = 2\pi r_n \qquad n = 1, 2, 3, \dots$$

or

$$pr_n = n\frac{h}{2\pi} = n\hbar$$

and since $L = pr_n$, we have

$$L = n\hbar$$

which is the Bohr quantization *postulate* now derived by assuming that electrons behave like waves.

The success of the de Broglie assumption in giving a foundation for the highly successful Bohr model certainly argues strongly for the acceptance of the idea that electrons are somehow waves.

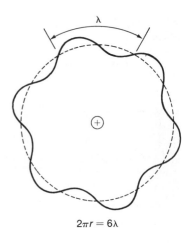

$$2\pi r = 6\lambda$$

Fig. 40-11. The Bohr quantization condition is equivalent to requiring that each allowed Bohr orbit have a circumference equal to an integral number of electron wavelengths, so that a standing wave can exist.

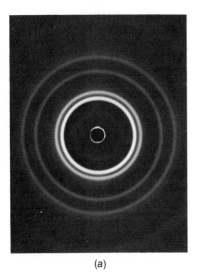

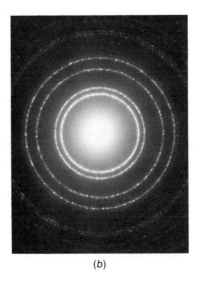

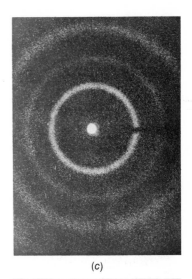

(a)   (b)   (c)

Yet many questions are unanswered, such as: "What is being waved?" "What is the meaning of the *amplitude* of the wave?" "Can interference effects be *directly* observed for electrons?" "How can electrons (and photons) simultaneously exhibit wavelike and particlelike properties?" The answers to these questions appeared in the three years spanning 1925–1927 in one of the most intensely creative periods in the history of physics.

Einstein was struck by the novelty of de Broglie's hypothesis and interested many others, so that de Broglie's thesis was widely circulated and read. A few attempts were made over the next several years to observe interference effects in the propagation of electrons. However, the first published (1927) observation of interference effects in the behavior of electrons resulted from an accident.

In 1925, Davisson and Germer had been studying the way in which electrons scattered from a nickel surface in vacuum. A cracked vacuum tube had led to a badly oxidized nickel surface, so the nickel was heated in vacuum to clean it up. When they again began recording data, they found a maximum in the scattered electron intensity at a specific angle. They finally realized that the heating of the nickel target had produced crystals of nickel and the ordered rows of atoms had acted as a diffraction grating for the electron-matter waves.

This direct observation of interference effects in the motion of electrons was quickly followed by G. P. Thomson's publication of his electron scattering experiments, which produced diffraction rings confirming the de Broglie wavelength formula (see Fig. 40-12a). These direct experimental observations actually followed

Fig. 40-12. Diffraction rings produced by (a) x rays with = 0.71Å. (Courtesy Education Development Center, Inc., 39 Chapel Street, Newton, Massachusetts), (b) 600-eV electrons (Courtesy Education Development Center, Inc., 39 Chapel Street, Newton, Massachusetts), and (c) 0.0568-eV neutrons scattered by polycrystalline metal samples (Courtesy Dr. C. G. Shull, Massachusetts Institute of Technology).

the highly successful expansion of de Broglie's theory by Erwin Schroedinger in 1926. Nonetheless, these direct observations were required finally to drive home the *necessity* of a dualistic wave-particle theory of matter and radiation.

Exercises

5. Calculate the momentum of low-energy ($E = 1$ eV) electrons. Now use (40-13) to find the wavelength. How close would two slits have to be placed to see an interference pattern with its first minimum at 30°?

6. From (40-10) we infer that the kinetic energy of an electron in its first Bohr orbit equals 13.58 eV. Show that the circumference of the first Bohr orbit equals one wavelength for a 13.58-eV electron.

40-6   The Meaning of the Wave Function

Observation of interference effects for matter as well as for light waves makes it clear that both matter and radiation are wavelike phenomena. Nonetheless, the photoelectric effect makes it clear that light behaves like a collection of particles, called photons. Electrons certainly behave as if they were particles, yet they exhibit interference effects as well. How do we reconcile these apparent conflicts?

In 1926, Max Born suggested a unifying picture that gives meaning to the *amplitude* of a matter wave. First, let us recall that the energy density of an electromagnetic wave is proportional to the square of its amplitude:

$$\frac{U}{\text{volume}} = \tfrac{1}{2}\epsilon_0 E^2 \tag{40-14}$$

In the photon picture, this energy is carried by n discrete photons:

$$\frac{U}{\text{volume}} = \frac{n}{\text{volume}}(hf) \tag{40-15}$$

From (40-14) and (40-15) we see that the number of photons per unit volume is proportional to the square of the classical electromagnetic wave amplitude:

$$\frac{n}{\text{volume}} \propto E^2 \tag{40-16}$$

In analogy with (40-16), Born assumed that a *matter wave* function $\psi(x,t)$ had an amplitude such that its square was equal to the number of particles per unit volume:

$$\frac{n}{\text{volume}} = \psi^2 \qquad (40\text{-}17)$$

Both (40-16) for electromagnetic radiation and (40-17) for matter waves provide a marriage between the wave and particle concepts. The *probability* that a particle will be observed is given by the square of the associated wave amplitude. An illustration of this probabilistic interpretation is given by the steady growth of the two-slit interference pattern indicated in Fig. 40-13. This pattern could be caused *either* by light waves or by electron waves. In either case, each distinct grain of exposed film indicates the arrival of a discrete particle.

The spatial and temporal dependence of a matter wave for a given energy E and momentum p is assumed to be sinusoidal of the form* discussed in Section 22-3:

$$\psi(x,t) = \psi_0 \sin(kx - \omega t) \qquad (40\text{-}18)$$

where

$$k = \frac{2\pi}{\lambda} = \frac{p}{\hbar}$$

and

$$\omega = 2\pi f = \frac{E}{\hbar}$$

so that (40-18) can be written

$$\psi(x,t) = \psi_0 \sin \frac{1}{\hbar}(px - Et)$$

In 1926, Erwin Schroedinger proposed a wave equation for these matter waves. This equation indicated the way in which matter waves change in space and time (e.g., how they refract and reflect) in a physical system where the effect of forces could be represented by a potential energy. The action of a varying potential energy was found to be similar to a varying index of refraction for light waves.

With Schroedinger's fully three-dimensional wave equation, the problem of the hydrogen atom was finally solved correctly and a difficulty with the older Bohr theory was properly resolved. The Schroedinger wave equation is a partial differential equation that is beyond the expected level of mathematical preparation for the

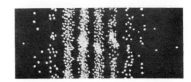

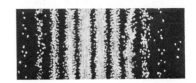

Fig. 40-13. A piece of film is exposed in a two-slit interference experiment. The interference maxima correspond to a large value of the wave function, where the discrete events that cause film exposure have the maximum probability. We see sketches of the film as it appears (*a*) after 28 electrons have passed through the apparatus, (*b*) after 1000 electrons, and (*c*) after 10,000 electrons. Part (*d*) is an actual photograph (different scale) after exposure to millions of electrons. (From Elisha R. Huggins, *Physics* I, copyright © 1968, W. A. Benjamin, Inc., New York)

---

*It is actually found necessary to include *two* sinusoidal components in the matter-wave function, one of which is multiplied by $i = \sqrt{-1}$. These two components are always 90° out of phase so that the net probability density in a standing matter wave is *truly* stationary.

reader of this text. In what follows we will qualitatively describe some of its predictions.

The Bohr theory claimed that the orbital angular momentum of the hydrogen atom was given by

$$L = n\hbar \qquad n = 1, 2, 3, \ldots$$

where $n = 1$ corresponds to the ground state. What is actually true is that the ground state has *zero* angular momentum and higher-lying states can be in various discrete states of angular momentum as indicated below:

$n = 1$ (ground state) $L_z = 0$

$n = 2$ (1st excited state) $L_z = 0$ or $1\hbar$

$n = 3$ (2nd excited state) $L_z = 0$, $1\hbar$, or $2\hbar$

$n = 4$ (3rd excited state) $L_z = 0$, $1\hbar$, $2\hbar$, or $3\hbar$

and so on, where L_z is the maximum z-component of **L**.

These results are in direct conflict with the Bohr theory. The de Broglie picture only partially resolves this conflict. The standing waves he pictured must consist of two interfering waves moving clockwise and counterclockwise around the Bohr orbit so that the net angular momentum is zero for *all* levels. Nonetheless, it is experimentally found that the hydrogen atom can possess some angular momentum, though not as much as suggested by the Bohr model.

The three-dimensional solutions of Schroedinger's wave equation with a potential energy given by

$$U = -\frac{ke^2}{r}$$

agree with the actual experimental findings. The lowest state of the hydrogen atom is found not to be a "racetrack" orbit for the matter waves but a "breathing mode" of vibration that produces a standing wave in the in-out radial direction.

A computer-generated dot plot for this state and for the first excited state with $L = 0$ is shown in Fig. 40-14. There, the density of the dots is made proportional to ψ^2 so that the regions of high probability for finding the electron are where the dot density is largest. Note the radial node in the first excited state.

The electron waves can refract around the hydrogen atom with a net angular momentum, as well as move purely radially in and out. The combination of both types of wave motion leads to nodes in azimuthal angle as well as in the radial direction. The lowest

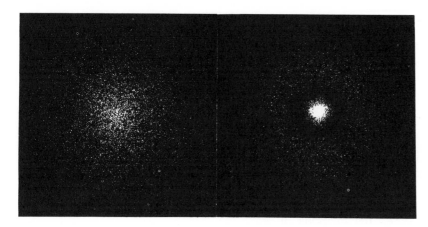

Fig. 40-14. Computer-generated probability-density plots for the ground state (1 S) and first excited state (2 S) of the hydrogen atom. The curves below are plots of ψ as a function of r, with a indicating the size of the first Bohr radius. (From Ashby and Miller, *Principles of Modern Physics,* Holden-Day, Inc., San Francisco, 1969)

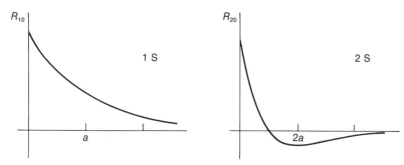

energy state where this occurs is the first excited state that has $L = 1\hbar$. The probability density for this state is shown in Fig. 40-15 and for a higher excited state in Fig. 40-16.

Though the old Bohr theory and the racetrack standing wave pattern of de Broglie were unable to explain both the energy levels and the angular momentum of the hydrogen atom, the Schroedinger theory does so beautifully. The Schroedinger wave equation has also been successfully applied to many other microscopic systems, including the motion of nucleons within the nucleus of the atom. The mechanics of atomic and subatomic particles is thus well described by the matter wave hypothesis of de Broglie and the Schroedinger equation.* This mechanics is called *wave mechanics.*

---

*The Schroedinger equation is limited to the nonrelativistic domain, where the classical particle velocities are much smaller than that of light. A relativistic treatment of electron matter waves is successfully given by the Dirac equation.

40-7   Classical and Quantum Physics

The quantum physics we have briefly described in this chapter rests upon strange new assertions about the nature of matter and radiation. The reader may well wonder, at this point, why we spend so much time on classical physics if modern physics is so vastly different from it. The answer to this question is twofold.

First, the concepts we used in developing the new ideas of quantum physics are *classical* concepts. The new quantum physics did not spring newborn from a vacuum; it grew out of a steady evolution of ideas beyond classical physics but yet based upon the older concepts.

Second, there is a beautiful correspondence between classical physics and quantum physics. When the methods of wave mechanics are applied to systems with masses and energies well above the atomic level, the predictions of wave mechanics are indistinguishable from those of classical Newtonian mechanics. Just as relativistic mechanics becomes indistinguishable from classical mechanics when $v \ll c$, so does wave mechanics blend into classical mechanics when $\lambda \ll x$ (λ is the de Broglie wavelength and x is a typical dimension of the mechanical system). Classical physics is thus appropriate for systems with velocities much less than that of light and dimensions much larger than atomic dimensions.

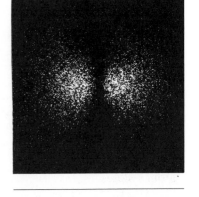

Fig. 40-15. Computer-generated probability density for the first excited state of the hydrogen atom with $l = 1\hbar$. The electron waves in this case circulate about the proton at the same time they reverberate in and out. (From Ashby and Miller, *Principles of Modern Physics*, Holden-Day, Inc., San Francisco, 1969)

40-8   Summary

Max Planck found it necessary to assume quantized energy states for atomic oscillators in the walls of an enclosure filled with thermally excited cavity radiation. He assumed that the energy of each oscillator was quantized and given by

$$E = nhf$$

where n is an integer. He was only then able to predict correctly the observed spectrum of cavity radiation.

Einstein carried the quantization idea further in explaining the photoelectric effect. He postulated that radiation with a frequency f consisted of individual photons, each with $E = hf$.

Bohr was able to describe the spectrum of hydrogen correctly by quantizing the angular momentum of the electron in the atom according to the rule

$$L = n\hbar$$

The result was a discrete series of allowed energy levels whose energy agreed with experiment.

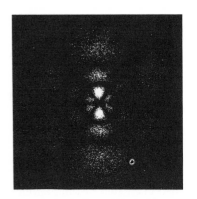

Fig. 40-16. Computer-generated probability density for a more complicated excited state. It represents the fifth excited state when it has three units ($l = 3\hbar$) of angular momentum. (From Ashby and Miller, *Principles of Modern Physics*, Holden-Day, Inc., San Francisco, 1969)

The origin of the Bohr quantization rule was hinted at by de Broglie's discovery that his proposed matter-wave picture led to each Bohr orbit having a circumference equal to an integer number of wavelengths, so that there could be a standing wave. Neither the Bohr theory nor de Broglie's conjecture correctly described the angular momentum of the hydrogen atom, however.

Finally, Schroedinger was able to deduce a form for a wave equation of matter waves that predicted that three-dimensional electron standing waves could exist in the region around a proton. These standing waves correctly corresponded, both in energy and angular momentum, to the observed properties of the hydrogen atom's states.

Use of the Schroedinger equation approach to atomic mechanics is called wave mechanics. Wave-mechanical predictions are strikingly different from the predictions of classical mechanics in the atomic domain. For systems much larger than atoms, the two types of mechanics are indistinguishable and use of classical mechanics is appropriate.

Appendix

1. General Physical Constants*

Name	Symbol	Value	Standard deviation (ppm)
Speed of light	c	299792458 ms^{-1}	0.004
Permittivity of vacuum	ϵ_0	8.85418782 × 10^{-12} C^{2}J^{-1}m^{-1}	0.008
Permeability of vacuum	μ_0	4π × 10^{-7} Hm^{-1}	exactly
Gravitational constant	G	6.6720 × 10^{-11} Nm^{2}kg^{-2}	615
Elementary charge	e	1.6021892 × 10^{-19} C	2.9
Planck's constant	h	6.626176 × 10^{-34} Js	5.4
Boltzmann's constant	k	1.380662 × 10^{-23} JK^{-1}	32
Stefan-Boltzmann constant	σ	5.67032 × 10^{-8} Wm^{2}K^{-4}	125
Avogadro's constant	N_A	6.022045 × 10^{23} mol^{-1}	5.1
Proton mass	m_p	1.6726485 × 10^{-27} kg	5.1
Atomic mass unit	u	1.6605655 × 10^{-27} kg	5.1
Electron mass	m_e	9.109534 × 10^{-31} kg	5.1
Standard volume of ideal gas		22.41383 × 10^{-3} m^{3}mol^{-1}	31
Gas constant	R	8.31441J mol^{-1} K^{-1}	31
Bohr radius	a_0	0.52917706 × 10^{-10} m	0.8

* Three-significant-figure values suitable for most computational work are given inside the front cover. The values above are based on the analysis of E. R. Cohen and B. N. Taylor, *Journal of Physical and Chemical Reference Data*, vol. 2, no. 4 (1973).

2. The International System of Units (SI)
A. SI BASE UNITS*

Quantity	Name	Symbol
length	meter	m
mass	kilogram	kg
time	second	s
electric current	ampere	A
thermodynamic temperature	kelvin	K
amount of substance	mole	mol
luminous intensity	candela	cd

* These units and their definitions are discussed at the appropriate points in the text (see index).

B. SI PREFIXES

Factor	Prefix	Symbol	Factor	Prefix	Symbol
10^{12}	tera	T	10^{-1}	deci	d
10^{9}	giga	G	10^{-2}	centi	c
10^{6}	mega	M	10^{-3}	milli	m
10^{3}	kilo	k	10^{-6}	micro	μ
10^{2}	hecto	h	10^{-9}	nano	n
10^{1}	deka	da	10^{-12}	pico	p
			10^{-15}	femto	f
			10^{-18}	atto	a

C. SI SUPPLEMENTARY UNITS

Quantity	Name	Symbol
plane angle	radian	rad
solid angle	steradian	sr

D. SI DERIVED UNITS WITH SPECIAL NAMES

Quantity	Name	Symbol	Expression in other units	Expression in base units
frequency	hertz	Hz		s^{-1}
force	newton	N		$m \cdot kg \cdot s^{-2}$
pressure	pascal	Pa	N/m^2	$m^{-1} \cdot kg \cdot s^{-2}$
energy, work	joule	J	$N \cdot m$	$m^2 \cdot kg \cdot s^{-2}$
power	watt	W	J/s	$m^2 \cdot kg \cdot s^{-3}$
electric charge	coulomb	C	$A \cdot s$	$s \cdot A$
electric potential, potential difference, electromotive force	volt	V	W/A	$m^2 \cdot kg \cdot s^{-3} \cdot A^{-1}$
capacitance	farad	F	C/V	$m^{-2} \cdot kg^{-1} \cdot s^4 \cdot A^2$
resistance	ohm	Ω	V/A	$m^2 \cdot kg \cdot s^{-3} \cdot A^{-2}$
magnetic flux	weber	Wb	$V \cdot s$	$m^2 \cdot kg \cdot s^{-2} \cdot A^{-1}$
magnetic flux density	tesla	T	Wb/m^2	$kg \cdot s^{-2} \cdot A^{-1}$
inductance	henry	H	Wb/A	$m^2 \cdot kg \cdot s^{-2} \cdot A^{-2}$
luminous flux	lumen	lm		$cd \cdot sr$
illuminance	lux	lx		$m^{-2} \cdot cd \cdot sr$

E. UNITS IN USE WITH SI

Name	Symbol	Value in SI units
minute	min	1 min = 60 s
hour	h	1 h = 60 min = 3600 s
day	d	1 d = 24 h = 86400 s
degree	°	$1° = (\pi/180)$ rad

E. UNITS IN USE WITH SI (*continued*)

Name	Symbol	Value in SI units
minute	′	$1' = (1/60)° = (\pi/10800)$ rad
second	″	$1'' = (1/60)' = (\pi/648000)$ rad
litre	l	$1l = 1$ dm³ $= 10^{-3}$ m³
tonne	t	1 t $= 10^3$ kg

3. Conversion Factors

Length

1 in. = 2.54 cm (exactly)
1 meter = 39.4 in. = 3.28 ft
1 mile = 1.61 km = 5280 ft
1 parsec = 3.26 light years = 1.92×10^{13} mi = 3.08×10^{13} km
1 angstrom = 10^{-10} m

Volume

1 liter = 10^3 cm³ = 1.06 quarts = 61.02 in.³ = 0.035 ft³

Density

1 kg/m³ = 10^{-3} g/cm³ = 1.94×10^{-3} slug/ft³ = 6.24×10^{-2} lb/ft³

Time

1000 s = 16.64 min
1 day = 86,400 s
1 year = 3.16×10^7 s

Mass

1 oz = 28.4 g
1 lb = 454 g
1 kg = 2.21 lb
1 slug = 32.2 lb = 14.6 kg

Velocity

60 mi/hr = 88 ft/s (exactly)
1 mi/hr = 0.447 m/s = 1.47 ft/s
1 km/hr = 0.621 mi/hr

Angle

1 rad = 57.3° = 0.159 rev

Angular velocity

1 rad/s = 9.54 rpm (rev/min)

3. Conversion Factors (*continued*)

Force

1 N = 10^5 dyne = 0.225 lb (force)
1 oz (force) = 0.278 N = 2.78×10^4 dyne

Pressure

1 atm = 76 cm Hg = 10.33 m H_2O = 1.01×10^5 N/m^2 = 2117 lb/ft^2 = 14.7 lb/in.^{2}
1 N/m^2 = 1.45×10^{-4} lb/in.^{2}

Energy and power

1 J = 10^7 erg = 0.738 ft · lb
1 eV = 1.60×10^{-19} J = 1.60×10^{-12} erg
1 cal = 4.18 J
1 hp = 550 ft · lb/s = 746 W

Temperature

T(Celsius) = T(Kelvin) − 273.15° C
T(Fahrenheit) = $\frac{5}{9} T$(Celsius) + 32° F

Electricity and magnetism

1 C = 3×10^9 statcoulombs
1 statvolt = 300 V
1 W/m^2 = 1 T = 10^4 G

4. Some Approximate Values Suitable for Making Estimates

Value	$\left(\dfrac{\text{true value} - \text{approximate value}}{\text{true value}}\right) \times 100$
1 year = $\pi \times 10^7$ s	1%
1 meter = 1 yard	10%
1 liter = 1 quart	6%
1 U.S. ton = 1000 kg	10%
Density of water = 60 lb/ft^3	4%
Density of iron = 500 lb/ft^3	10%
1 newton = $\frac{1}{5}$ pound	12%
1 horsepower = $\frac{3}{4}$ kw	$\frac{1}{2}$%
1 rev/min = $\frac{1}{10}$ rad/s	5%
velocity of earth's equator due to rotation = 1000 mi/hr	3%
$\pi = 3$	5%
$e = 3$	8%
$\sqrt{10} = 3$	5%

5. Values of Trigonometric Functions

Radians	Degrees	Sine	Cosine	Tangent	Cotangent			
.0000	0	.0000	1.0000	.0000	∞	90	1.5708
.0175	1	.0175	.9998	.0175	57.29	89	1.5533
.0349	2	.0349	.9994	.0349	28.64	88	1.5359
.0524	3	.0523	.9986	.0524	19.08	87	1.5184
.0698	4	.0698	.9976	.0699	14.30	86	1.5010
.0873	5	.0872	.9962	.0875	11.430	85	1.4835
.1047	6	.1045	.9945	.1051	9.514	84	1.4661
.1222	7	.1219	.9925	.1228	8.144	83	1.4486
.1396	8	.1392	.9903	.1405	7.115	82	1.4312
.1571	9	.1564	.9877	.1584	6.314	81	1.4137
.1745	10	.1736	.9848	.1763	5.671	80	1.3963
.1920	11	.1908	.9816	.1944	5.145	79	1.3788
.2094	12	.2079	.9781	.2126	4.705	78	1.3614
.2269	13	.2250	.9744	.2309	4.332	77	1.3439
.2443	14	.2419	.9703	.2493	4.011	76	1.3265
.2618	15	.2588	.9659	.2679	3.732	75	1.3090
.2793	16	.2756	.9613	.2867	3.487	74	1.2915
.2967	17	.2924	.9563	.3057	3.271	73	1.2741
.3142	18	.3090	.9511	.3249	3.078	72	1.2566
.3316	19	.3256	.9455	.3443	2.904	71	1.2392
.3491	20	.3420	.9397	.3640	2.748	70	1.2217
.3665	21	.3584	.9336	.3839	2.605	69	1.2043
.3840	22	.3746	.9272	.4040	2.475	68	1.1868
.4014	23	.3907	.9205	.4245	2.356	67	1.1694
.4189	24	.4067	.9135	.4452	2.246	66	1.1519
.4363	25	.4226	.9063	.4663	2.144	65	1.1345
.4538	26	.4384	.8988	.4877	2.050	64	1.1170
.4712	27	.4540	.8910	.5095	1.963	63	1.0996
.4887	28	.4695	.8829	.5317	1.881	62	1.0821
.5061	29	.4848	.8746	.5543	1.804	61	1.0647
.5236	30	.5000	.8660	.5774	1.732	60	1.0472
.5411	31	.5150	.8572	.6009	1.664	59	1.0297
.5585	32	.5299	.8480	.6249	1.600	58	1.0123
.5760	33	.5446	.8387	.6494	1.540	57	0.9948
.5934	34	.5592	.8290	.6745	1.483	56	0.9774
.6109	35	.5736	.8192	.7002	1.428	55	0.9599
.6283	36	.5878	.8090	.7265	1.376	54	0.9425
.6458	37	.6018	.7986	.7536	1.327	53	0.9250
.6632	38	.6157	.7880	.7813	1.280	52	0.9076
.6807	39	.6293	.7771	.8098	1.235	51	0.8901
		Cosine	Sine	Cotangent	Tangent	Degrees	Radians

5. Values of Trigonometric Functions (*continued*)

Radians	Degrees		Cosine	Tangent	Cotangent		
.6981	40	.6428	.7660	.8391	1.192	50	0.8727
.7156	41	.6561	.7547	.8693	1.150	49	0.8552
.7330	42	.6691	.7431	.9004	1.111	48	0.8378
.7505	43	.6820	.7314	.9325	1.072	47	0.8203
.7679	44	.6947	.7193	.9657	1.036	46	0.8029
.7854	45	.7071	.7071	1.0000	1.000	45	0.7854
		Cosine	Sine	Cotangent	Tangent	Degrees	Radians

6. Mathematical Formulas

Quadratic formula

If $ax^2 + bx + c = 0$, then $x = \dfrac{-b \pm \sqrt{b^2 - 4ac}}{2a}$

Trigonometric functions in terms of lengths of sides of a right triangle

$$\sin\theta = \frac{\text{side opposite}}{\text{hypotenuse}} \qquad \cos\theta = \frac{\text{side adjacent}}{\text{hypotenuse}}$$

$$\tan\theta = \frac{\text{side opposite}}{\text{side adjacent}} \qquad \cot\theta = \frac{\text{side adjacent}}{\text{side opposite}}$$

$$\sec\theta = \frac{\text{hypotenuse}}{\text{side adjacent}} \qquad \csc\theta = \frac{\text{hypotenuse}}{\text{side opposite}}$$

Trigonometric identities

$$\sin^2\theta + \cos^2\theta = 1 \qquad \sec^2\theta - \tan^2\theta = 1 \qquad \csc^2\theta - \cot^2\theta = 1$$

$$\sin(a \pm b) = \sin a \cos b \pm \cos a \sin b$$

$$\cos(a \pm b) = \cos a \cos b \mp \sin a \sin b$$

$$\tan(a \pm b) = \frac{\tan a \pm \tan b}{1 \mp \tan a \tan b}$$

$$\sin 2\theta = 2 \sin\theta \cos\theta$$

$$\cos 2\theta = \cos^2\theta - \sin^2\theta = 2\cos^2\theta - 1 = 1 - 2\sin^2\theta$$

$$\sin\theta = \frac{e^{i\theta} - e^{-i\theta}}{2i} \qquad \cos\theta = \frac{e^{i\theta} + e^{-i\theta}}{2}$$

$$e^{\pm i\theta} = \cos\theta \pm i \sin\theta$$

Expansion of $\sin\theta$ and $\cos\theta$

$$\sin\theta = \theta - \frac{\theta^3}{3!} + \frac{\theta^5}{5!} - \frac{\theta^7}{7!} + \cdots$$

$$\cos\theta = 1 - \frac{\theta^2}{2!} + \frac{\theta^4}{4!} - \frac{\theta^6}{6!} + \cdots$$

6. Mathematical Formulas (*continued*)

Expansion of exponential function

$$e^x = 1 + x + \frac{x^2}{2!} + \frac{x^3}{3!} + \frac{x^4}{4!} + \cdots$$

Binomial series

$$(1 + x)^n = 1 + nx + \frac{n(n-1)}{2!}x^2 + \frac{n(n-1)(n-2)}{3!}x^3 + \cdots$$

Derivatives and antiderivatives

The letters u and v stand for any functions of x; and a, b, and m are constants. Here x is taken to be the independent variable. In the early chapters of this text we often considered t to be the independent variable, in which case formula 4 below (for example) would read

$$\frac{d}{dt}t^m = mt^{m-1} \quad \text{and} \quad \int t^m dt = \frac{t^{m+1}}{m+1} \quad (m \neq -1)$$

1. $\dfrac{dx}{dx} = 1$
2. $\dfrac{d}{dx}(au) = a\dfrac{du}{dx}$
3. $\dfrac{d}{dx}(u + v) = \dfrac{du}{dx} + \dfrac{dv}{dx}$
4. $\dfrac{d}{dx}x^m = mx^{m-1}$
5. $\dfrac{d}{dx}\ln x = \dfrac{1}{x}$
6. $\dfrac{d}{dx}(uv) = u\dfrac{dv}{dx} + v\dfrac{du}{dx}$
7. $\dfrac{d}{dx}e^x = e^x$
8. $\dfrac{d}{dx}\arctan x = \dfrac{1}{1 + x^2}$
9. $\dfrac{d}{dx}\arcsin x = \dfrac{1}{\sqrt{1 - x^2}}$
10. $\dfrac{d}{dx}\operatorname{arcsec} x = \dfrac{1}{x\sqrt{x^2 - 1}}$
11. $\dfrac{d}{dx}\cos x = -\sin x$
12. $\dfrac{d}{dx}\sin x = \cos x$
13. $\dfrac{d}{dx}\tan x = \sec^2 x$
14. $\dfrac{d}{dx}\cot x = -\csc^2 x$
15. $\dfrac{d}{dx}\sec x = \tan x \sec x$
16. $\dfrac{d}{dx}\csc x = -\cot x \csc x$

1. $\int dx = x$
2. $\int au\, dx = a \int u\, dx$
3. $\int (u + v)dx = \int u\, dx + \int v\, dx$
4. $\int x^m\, dx = \dfrac{x^{m+1}}{m+1} \quad (m \neq -1)$
5. $\int \dfrac{dx}{x} = \ln |x|$
6. $\int u\dfrac{dv}{dx}dx = uv - \int v\dfrac{du}{dx}dx$
7. $\int e^x\, dx = e^x$
8. $\int \dfrac{dx}{1 + x^2} = \arctan x$
9. $\int \dfrac{dx}{\sqrt{1 - x^2}} = \arcsin x$
10. $\int \dfrac{dx}{x\sqrt{x^2 - 1}} = \operatorname{arcsec} x$
11. $\int \sin x\, dx = -\cos x$
12. $\int \cos x\, dx = \sin x$
13. $\int \tan x\, dx = \ln |\sec x|$
14. $\int \cot x\, dx = \ln |\sin x|$
15. $\int \sec x\, dx = \ln |\sec x + \tan x|$
16. $\int \csc x\, dx = \ln |\csc x - \cot x|$

6. Mathematical Formulas (*continued*)

Additional antiderivatives

$$\int (a + bx)^m dx = \frac{(a + bx)^{m+1}}{(m + 1)b} \qquad (m \neq -1)$$

$$\int \frac{dx}{a + bx} = \frac{1}{b}\ln(a + bx)$$

$$\int \frac{dx}{(a + bx)^2} = -\frac{1}{b(a + bx)}$$

$$\int b^{ax} dx = \frac{b^{ax}}{a \ln b}$$

$$\int \ln x \, dx = x \ln x - x$$

Answers to Exercises and Odd-Numbered Problems

CHAPTER 1
Answers to exercises

1. (*a*) 1.6 cm (*b*) 0.016 m (*c*) 0.000016 km
   $= 1.6 \times 10^{-5}$ km
2. 0.01 μs; 10 ns
3. 26.8 m/s; 96.6 km/hr
4. yes
5. 10,000; 45; 3.333; 3.3
6. 59.3; 59.33; 55.01; 45

CHAPTER 2
Answers to exercises

1. (*a*) 2 m/s (*b*) 0 (*c*) $\frac{8}{3}$ m/s (*d*) 4 m/s
   (*e*) 4 m/s (*f*) 4 m/s (*g*) 2 m/s (*h*) 1 m/s
2. Values obtained are approximate:
   (*a*) −0.8 m/s (*b*) −1 m/s (*c*) 0 (*d*) 4 m/s
3. $\bar{v} = \dfrac{\Delta x}{\Delta t} = \dfrac{x_2 - x_1}{t_2 - t_1} = \dfrac{c(t_2 - t_1)}{t_2 - t_1} = c$
4. See Fig. *E2-4*(*below*).

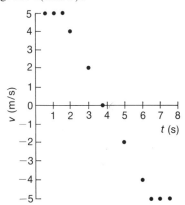

5. because derivatives involve only *slopes*, not magnitudes; see Fig. *E2-5*(*below*).

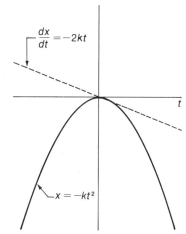

6. $x = ct$ and $x + \Delta x = c(t + \Delta t)$, so
   $\Delta x = ct + c\Delta t - x$ or $\Delta x = ct + c\Delta t - ct$. Thus
   $\Delta x/\Delta t = c$ *always*, even when $\Delta t \to 0$.
7. $x = -kt^2$ and $x + \Delta x = -k(t + \Delta t)^2$
   $= -kt^2 - 2kt\Delta t - k\Delta t^2$. Proceed as on page 26:
   $\dfrac{\Delta x}{\Delta t} = -2kt - k\Delta t$ and $\lim\limits_{\Delta t \to 0} \dfrac{\Delta x}{\Delta t} = -2kt$. Yes;
   length/(time)^2. See Fig. *E2-7*(*below*).

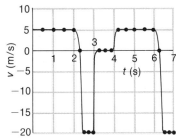

8. $\dfrac{dx}{dt} = 80t^4$

9. $\dfrac{dx}{dt} = b + 2ct$

10. $v = \dfrac{dx}{dt} = 10\dfrac{m}{s^2} \cdot t;\; v(t = 3\,s) = 30\ m/s$

11.   (a) $x = 10\dfrac{mi}{hr} \cdot t$ (b) $x = -10\ mi + 10\dfrac{mi}{hr} \cdot t$

     (c) $x = 10\ mi - 10\dfrac{mi}{hr} \cdot t$. See Fig. E2-11(*below*).

(a)

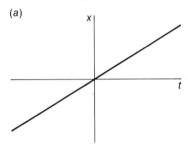

(b)

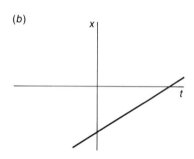

(c)

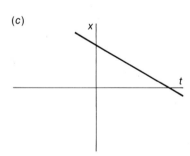

12.   (a) yes   (b) yes, missing points could give different slopes   (c) 8 ft/s

13.   (a) $t_p = \dfrac{x_0}{v_2 - v_1}$ (b) 1 hr

14. $f(t) = \frac{1}{4}t^4 + C$

15. $x = \int v\, dt = \dfrac{A}{3}t^3 + \dfrac{B}{4}t^4 + C$

16. $x(t = 3\ hr) = 65\ km$

17. $\frac{1}{2}K(t_f^2 + t_i^2) = 36\ ft$

18. $19\frac{1}{2}$ squares = 49 m; deduce $v = 4\dfrac{m}{s^2} \cdot t$ and

     integrate: $\Delta x = 2\dfrac{m}{s^2}(t_f^2 - t_i^2) = 48\ m$

Answers to odd-numbered problems

2-1. $\bar{v}$ (m/s): 9.09, 9.23, 10.1, 10.3, 8.99, 7.67, 7.34.

2-3.   1 m/s; $2\frac{1}{2}$ m/s

2-5.   (a) k seconds   (b) $3k$ seconds
     (c) $5k$ seconds   (d) length/seconds²

2-7.   no, as long as the function is smooth and has its maxima and minima *within* the given interval

2-9. $K = 10\ m/s^2$

2-11. The *slopes* are identical, even though the absolute values differ. See Fig. P2-11 (*below*).

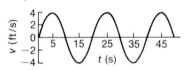

2-13. $t = 3$ hr

2-15. $f(t) = \dfrac{A}{2}t^2 + Bt + C$

2-17. 219

2-19. $c_1 t + \frac{1}{2}c_2 t^2 + \frac{1}{3}c_3 t^3 + \frac{1}{4}c_4 t^4 + c_5$

2-21. 9.08×10^7 neutrons

CHAPTER 3
Answers to exercises

1.   (a) $v = b + 3ct^2$ (b) $v(t = 0) = 10$ m/s; $v(3\ s) = 91$ m/s

2.   78 m

3.   The magnitudes graphed are approximate. The sequence of positive and negative accelerations should be understood by the reader. See Fig. E3-3(*top left, page 887*).

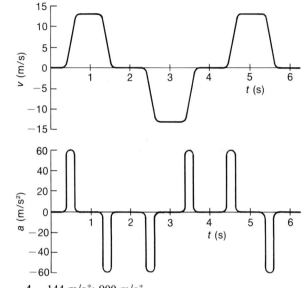

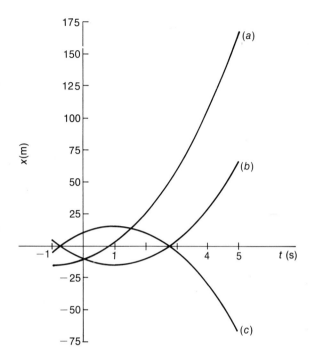

4. 144 m/s²; 900 m/s²
5. 40 m/s (= 89½ mi/hr)
6. (a) 50 m/s   (b) 250 m
7. a car braking to a stop
8. $y = y_0 + v_0t - \frac{1}{2}gt^2$
9. 0.1 yr
10. $v^2 = v_0^2 - 2ad$
11. 21.2 s
12. 10 m/s²; yes, to $16\frac{2}{3}$ m/s²
13. Approximately $19\frac{1}{2}$ m/s
14. 2962.5 m

3-13. Yes, 189 m from place where brakes applied
3-15. $6\frac{3}{4}$ m
3-17. At $t = 4$ s
3-19. 57.9 ft/s²; 48.5 ft/s². The acceleration is non-uniform but (3-15) and (3-17) *assume* constant acceleration.

Answers to odd-numbered problems

3-1. See Fig. P3-1 (*top right, page 887*).
3-3. (a) $t = 9$ s   (b) 88.5 m/s
3-5. (a) $t = \dfrac{-2B \pm \sqrt{4B^2 - 12AC}}{6A}$

 (b) (1) $4B^2 - 12AC > 0$
 (2) $4B^2 - 12AC = 0$ (3) $4B^2 - 12AC < 0$
3-7. (a) $v = 3At^2 + 2Bt + C$; $a = 6At + 2B$
 (b) $B^2 = 3AC$
3-9. (a) $t = 14\frac{2}{3}$ s; $x = 1936$ ft   (b) 90 mi/hr and 180 mi/hr

3-11. $\Delta x = \dfrac{b}{12}\left(t_3^4 - t_2^4\right) + \dfrac{c}{2}\left(t_3^2 - t_2^2\right)$

 $+ \left(v_1 - \dfrac{b}{3}t_1^3 - ct_1\right)\left(t_3 - t_2\right)$

CHAPTER 4
Answers to exercises

1. $\frac{1}{2}$ m/s
2. About 1.3 kg
3. 0.06 m/s
4. 40 N
5. 10³ N; 9 × 10³ N
6. 33 lb
7. 10.2 m/s²
8. 25.8 ft; 12.9 ft
9. 30 m/s; 33 m/s; 27 m/s
10. 3.46 s
11. 2 N

Answers to odd-numbered problems

4-1. 35 kg

4-3. 64.25 km/hr

4-5. (a) 8.8 ft/s   (b) 352 ft/s²   (c) 1750 lb
      (d) 437 lb

4-7. (a) $\dfrac{m}{M}v$ (b) $\dfrac{m}{M}L$ (c) $\dfrac{mv}{m+M}$

4-9. $V = \dfrac{mv_s}{m+M}$

4-11. 4.46 N/lb

4-13. 784 N

4-15. (a) 50 m/s   (b) 150 m/s

4-17. (a) $x' = a + v't$ (b) $x = a + b + v't + Vt$

4-19. (a) 5000 dynes   (b) v = constant
      (c) $a = -100$ cm/s²   (d) Seatback presses
      forward against occupant.

4-21. The block moves 0.476 m with respect to the
      water (not 0.50 m).

CHAPTER 5
Answers to exercises

1. 400 lb

2. 0.43 s; same

3. $v_{earth} = \dfrac{m_b}{m_e} v_{book}$, hence $\Delta y_{earth} = \dfrac{m_b}{m_e} \Delta y_{book}$
    $= 6.53 \times 10^{-26}$ m (actual distance book falls in
    0.4 s is 0.784 m and there is a difference in
    round-off errors)

4. Paddle on water, water on paddle; arms on
    paddle, paddle on arms; man on boat, boat on
    man; boat on water, water on boat

5. (a) 384 lb (mass)   (b) 96 lb (force)

6. 98 lb

7. $a = \dfrac{m_1 g - T}{m_1} = g - \dfrac{m_2 g}{m_1 + m_2} = \dfrac{m_1 g}{m_1 + m_2}$

8. 12000 N

9. Less of rocket's mass is fuel, more can be
    payload

10. no

11. 139 m/s

12. 0.47 g

13. 7.4 metric tons

Answers to odd-numbered problems

5-1. 2.7 m/s² upward

5-3. 2.7 m/s²

5-5. (a) 39.2 N, 19.6 N, 0   (b) 59.2 N, 29.6 N, 0
      (c) 19.2 N, 9.6 N, 0   (d) 0, 0, 0

5-7. 1333 N; 667 N

5-9. $T_1 = \dfrac{m_3(m_1 + m_2)g}{m_1 + m_2 + m_3}$;

      $T_2 = m_1 g\left(1 - \dfrac{m_1 + m_2}{m_1 + m_2 + m_3}\right)$;

      $a = \dfrac{(m_1 + m_2)g}{m_1 + m_2 + m_3}$

5-11. 100 lb; yes, then $T = 150$ lb

5-13. 1.93 v_e

5-15. 3.48 tons

5-17. 10 slugs

5-19. $7.39M$; $54.6M$; $403M$; $2981M$. The massive
      structural components that would be a useless
      part of the final payload can be left behind
      when a second stage is fired, so that less
      total mass need by brought to the final
      velocity.

CHAPTER 6
Answers to exercises

1. 1.37×10^5 lb · s

2. 1000 N

3. $6\frac{1}{2}$ N

4. $8\frac{1}{3}$ N

5. 18 kg · m/s

6. $\bar{x} = 18$; no, x is not a linear function of t.

7. Results are very similar if the cars are identical.

8. The car with smaller mass experiences the
    greatest acceleration (see Problem 4-5); thus
    greater forces must be exerted to restrain the
    occupants of the smaller car.

9. Assuming a crumpling distance of 3 ft for each
    car, the acceleration is 40 g's and $F = 6,400$ lb.

10. 2500 N

11. 100 N; 400 N; mass rate also doubles to
    20 kg/s.

Answers to odd-numbered problems

6-1.   approximately 3×10^5 lb

6-3.   three times his weight; his weight

6-5.   approximately 2.5×10^3 lb

6-7. $\bar{g} = 44\frac{1}{3}$; $g(1) = 9$, $g(3) = 93$

6-9.   Relative velocity falls from 20 m/s to 0 during

collision. Thus $\Delta t = \dfrac{\text{total crumpling distance}}{\text{average relative velocity}}$

$= \dfrac{1 \text{ m}}{10 \text{ m/s}} = 0.1$ s. Belch-Fire V8: 3500 N;

Tofu 4: 10500 N; $\frac{2}{3}$ of these values with additional $\frac{1}{2}$-m motion.

6-11. 1.5×10^{-6} N

CHAPTER 7
Answers to exercises

1. $A_x = A \cos \theta$ becomes $A_x = A \cos(\theta + \phi)$;
    $A_y = A \sin \theta$ becomes $A_y = A \sin(\theta + \phi)$.
    The same result occurs if the vector is rotated;
    no.

2. $\sqrt{32}$ m $= 5.66$ m; $45°$

3. $L = 5$; $\theta = 126.87°$

4. $A_x = 8.66$; $A_y = 0.5$

5. R is horizontal and 2.8 units long

6. R is 16.3 cm long and $71°$ above the horizontal

7. Yes, see Fig. E7-7(below).

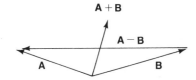

A + B

A − B

A   B

8. 14.2 units

9. $R_x = 14$; $R_y = 20$; $R_z = 0$

10. (a) $R_x = 6$ m, $R_y = 8$ m   (b) $S_x = 10$ m,
    $S_y = -2$ m   (c) $I_x = -10$ m, $I_y = 2$ m
    (d) $R = 10$, $S = 10.2$, $S = 10.2$

11. 31.3 ft/s

12. $v_R = \sqrt{v_B^2 + v_C^2}$; $\theta = \tan^{-1} \dfrac{v_B}{v_C}$ where θ is the

    angle between v_R and v_C

13. 10 m/s; $-53°$

14. $x = 8$ m, $y = -6$ m

15. Initial value of v_x remains constant for both.

16. (a) $v_{0x} = 87$ ft/s; $v_{0y} = 50$ ft/s
    (b) $R = 2v_{0x}v_{0y}/g$ (c) $R = 270$ ft

17. $v_y^2 = v_{0y}^2 - 2gy$ and when $y = h$, $v_y = 0$. Thus
    $2gh = v_{0y}^2$ or $h = v_{0y}^2/2g = 38.8$ ft.

Answers to odd-numbered problems

7-1. $R_x = 1$, $R_y = 2$

7-3. $\mathbf{A} + \mathbf{B}$ is $15\frac{3}{4}$ units long, $\theta = 39°$; $\mathbf{A} - \mathbf{B}$ is $24\frac{1}{2}$
       units long, $\theta = 24°$; see Fig. P7-3(below).

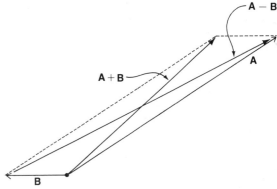

A − B

A + B

A

B

7-5. $R_x = 6.23$, $R_y = -0.53$; or $R = 6.25$,
       $\theta = -4.86°$

7-7.   14.9 units

7-9.   208.8 mi/hr; $16.7°$ west of due north

7-11. (a) $0°$ and $90°$ (b) $45°$

7-13. Proceed as in Exercise 16 to find $R = 2v_{0x}v_{0y}/g$
      $= 2v_0^2 \sin \theta \cos \theta/g = v_0^2 \sin 2\theta/g$; $\theta = 45°$
      gives $\sin 90°$, which is maximum value.

CHAPTER 8
Answers to exercises

1. 5.66 m/s

2. 2740 m/s

3. The pool table + earth. The table is sufficient if
    it rests on a frictionless surface.

4. $F = 2070$ lb southwest

5. $90°$

6. Note that $\mathbf{W}_\perp$ is perpendicular to the plane
    while $\mathbf{W}$ is perpendicular to the horizontal.
    Thus both sides, which include the angle θ,
    have been rotated by $90°$ so that they still
    include an angle θ.

7. 2.9 m/s^2
8. $T_1 = W/\cos\theta$
9. The axes of the new coordinate system are parallel to those of the old system, and the origin of the new system has coordinates in the old system given by
    $X = x_0 + v_{0x}t$; $Y = y_0 + v_{0y}t$;
    $Z = z_0 + v_{0z}t$
10. 51 ft/s; 43 ft
11. 105 N; 686 N; 694 N
12. The mass increases by an amount $\Delta m = 7.5 \times 10^{-12}$ kg
13. $x = 0.73$ m, $y = 2.27$ m, $z = -0.36$ m
14. $v_x = -2.4$ m/s, $v_y = 1.2$ m/s; or $v = 2.7$ m/s, $\theta = 153°$
15. Taking the origin at the lower left corner of the figure:
    $$x = \frac{\frac{1}{2}a^2 d + abc + \frac{1}{2}b^2 c}{ad + bc}; \qquad y = d/2$$
16. With the center of the coin at $x = 0$, $y = 0$, note that *any* strip parallel to the y axis has $y_{cm} = 0$, so the entire coin has $y_{cm} = 0$. Note that *every* strip to the left of $x = 0$ is balanced by one an equal distance to the right of $x = 0$, so $x_{cm} = 0$.
17. $x = \frac{2}{3}c$

Answers to odd-numbered problems

8-1. 2.5 m/s at $36.9°$ away from initial direction.
8-3. no
8-5. 9.9×10^{-20} kg · m/s; $-60°$
8-7. 4.6×10^{-30} kg
8-9. $2Nmv \cos\theta$
8-11. 64.8 m/s or 28.4 m/s
8-13. *no*
8-15. On the sailboat, the trajectory is a straight line. A stationary observer sees a parabolic trajectory given by Eq. (7-3).
8-17. $a = hg/d$
8-19. mL/M; no
8-21. $x_{cm} = -R/8$
8-23. (*a*) $\theta = \tan^{-1}(a/g)$ (*b*) $\theta = \tan^{-1}(a/g)$ (*c*) 0   (*d*) accelerator pedal
8-25. $1\frac{1}{3}$ in. from left side

CHAPTER 9
Answers to exercises

1. 7040 ft/s$^2 = 219$ g's
2. 0.1 ft/s^2 (= 250 mi/hr^2)
3. to increase r and thus decrease centripetal force
4. 2730 g's; 170 lb
5. 13.9 m (= $45\frac{1}{2}$ ft)
6. 3.1 m/s
7. 0.25 s (travels 0.13 m)
8. 91.7 m; 45.9 m; no
9. no; 5 N
10. $0.2g$
11. $0.124g$
12. accelerates at 1.68 m/s^2

Answers to odd-numbered problems

9-1. 352 rev/s (= $21{,}150$ rpm)
9-3. $4.5°$
9-5. $v_{min} = \sqrt{gR}$
9-7. 15.6 ft (15.45 ft from point of release)
9-9. $\mu = 0.86$
9-11. 0.6
9-13. $v = \sqrt{2gR}$
9-15. 1.3 m/s^2
9-17. $a = \left(\sin\theta - \mu\cos\theta - \dfrac{m_2}{m_1}\right)\dfrac{m_1 g}{m_1 + m_2}$;
      $m_2 = m_1(\sin\theta - \mu\cos\theta)$

CHAPTER 10
Answers to exercises

1. 8 m
2. $\omega = 20t$
3. 10 rad/s^2
4. (*a*) 0.39 s   (*b*) 97 rad/s   (*c*) 0.06 s
5. (*a*) $\alpha = 2bt$ (*b*) $\theta = a(t_2 - t_1) + \frac{1}{3}b(t_2{}^3 - t_1{}^3)$
6. 660 lb
7. (*a*) perpendicular to radius, 22.5 N   (*b*) $31°$ away from radius, 43.7 N   (*c*) $4.5°$ away from radius, 284 N
8. yes
9. yes, no

10. $\sqrt{2}\ V$; 45° above horizontal (for a stationary observer)

11. 15.15 rad/s; horizontal and to the left of the direction of motion

12. $A^2 \sin 0 = 0$

13. $A = 7\frac{1}{2}$ in.; $B = 9\frac{1}{2}$ in.; $C = 71\frac{1}{4}$ in.² out of the page

Answers to odd-numbered problems

10-1. 20 s

10-3. (a) -1800 rev/min² ($=-3.14$ rad/s²)
        (b) 3600 rev   (c) 18.9 m/s   (d) 3560 m/s²

10-5. 8 times per minute

10-7. $v = r_0[4k^2t^2 + c^2(1 + kt^2)^2]^{1/2}$

10-9. yes

10-11. 0; AB, into page; 0; AB, out of page; 0 (assuming that θ is swept out by clockwise motion of A); $C = -D$

CHAPTER 11
Answers to exercises

1. $f' = u\dfrac{dv}{dt} + v\dfrac{du}{dt} = 3abt^2 + 5act^4$

   $f' = \dfrac{d}{dt}(uv) = 3abt^2 + 5act^4$

2. $\dfrac{d}{dt}\left(\mathbf{r} \times \dfrac{d\mathbf{r}}{dt}\right) = \mathbf{r} \times \dfrac{d^2\mathbf{r}}{dt^2} + \left(\dfrac{d\mathbf{r}}{dt} \times \dfrac{d\mathbf{r}}{dt}\right) = \mathbf{r} \times \dfrac{d^2\mathbf{r}}{dt^2}$
    since cross product of identical vectors equals zero.

3. (a) $\Gamma = rF$ into page   (b) $I = 2mR^2$
    (c) $\alpha = \dfrac{rF}{2mR^2}$ (d) $\omega = \dfrac{rF\tau}{2mR^2}$

4. 150 rad/s²

5. perpendicular to radius; 0.9 rad/s²

6. 8 rev/s

7. Net angular momentum remains constant. When atmosphere moves with rotation, earth's angular velocity decreases slightly (seasonal effect is about 5 parts per billion).

8. 30.8 rad/s

9. (a) Precession reverses direction.
    (b) Precession reverses direction.

10. 0.17 rad/s (37 s for one full precessional revolution)

11. (a) forward, in the direction of bicycle's velocity   (b) **L** points to left so **L** swivels clockwise, as seen from above.   (c) less difficult, since bicycle turns into direction of incipient fall

12. 19.6 N · m

13. $\frac{1}{2}mgl \cos \theta$

14. beans: $a = \frac{2}{3}g \sin 45°$; juice: $a = g \sin 45°$

15. Initially accelerates (without rotation) with $a = g \sin \theta$, acceleration then drops to $\frac{2}{3}g \sin \theta$, and rotation begins with $\alpha = \frac{2}{3}g \sin (\theta/R)$

Answers to odd-numbered problems

11-1. $v = \omega R = \alpha t R = \Gamma t R/I = 2.25$ m/s

11-3. $a = \dfrac{(m_3 - m_2)g}{\frac{1}{2}m_1 + m_2 + m_3}$ (positive a means m_3 falling)

11-5. Treating the man as a cylinder with a *diameter* just over 1 ft, we have $I_m = \frac{3}{4}$ slug · ft². If the weights go from 3-ft radius to 1-ft radius, we find $\omega = 3\frac{1}{2}$ rev/s.

11-7. 155.8 rad/s

11-9. Presently, the angular momentum of the rotation of the earth on its axis is one-quarter that of the moon's orbital angular momentum. If the orbital period were to remain at 27 days, the orbital radius would increase by $\frac{5}{4}$. The actual effect will be larger, since the orbital period will decrease (see Section 17-5).

11-11. Thrusts will be perpendicular to desired shifts of axis. Reader should sketch $\boldsymbol{\omega}$ and indicate thrust directions for a variety of changes in orientation.

11-13. The key point is that both the expression for ΔL and for Γ are modified by a common factor, $\sin \phi$.

11-15. at a point $\frac{2}{3}$ of the stick's length from the pivot

11-17. (a) $h = 1\frac{2}{5}R$ (b) $R + \frac{2}{5}R\left(1 - \dfrac{\mu mg}{F}\right)$

   $\leqslant h \leqslant R + \frac{2}{5}R\left(1 + \dfrac{\mu mg}{F}\right)$,

   where F is the magnitude of the force. Note that (b) reduces to (a) as $F \to \infty$.

CHAPTER 12
Answers to exercises

1. $c = 2\frac{1}{3}$ ft

2. $T = \dfrac{Wb}{a \sin \theta}$; $T = \dfrac{Wb + \frac{1}{2}Ab}{a \sin \theta}$, where A is the
    weight of the forearm; for $b = 1\frac{1}{2}$ ft,
    $a = 0.1$ ft, $A = 5$ lb, $\theta = 120°$, we find
    $T = 390$ lb

3. 57.6 lb

4. no

5. both $\Sigma F_y = 0$ and $\Sigma \Gamma = 0$

6. (a) $\mu \geqslant 0.18$ (b) 35 N

Answers to odd-numbered problems

12-1. $T_1 = 1.97W$; $T_2 = 1.88W$

12-3. 78 kg

12-5. $x = w$

12-7. (a) $\mu = 0.577$ (b) $\mu = 0.321$

12-9. $T_1 = 277$ lb; $T_2 = 212$ lb; $\theta_2 = 49°$

12-11. $T(\text{left}) = 987$ N; $T(\text{right}) = 494$ N

12-13. (a) 14.14 lb (b) 10 lb

CHAPTER 13
Answers to exercises

1. 33 mg

2. 180 lb (20 ft × 14 ft × 8 ft room)

3. 4.09×10^{22}

4. 1.99×10^{-23} g

5. 16.6 μm

6. (a) 1.43×10^{-4} (b) 1.43×10^{-4} rad

7. 2.5×10^{-3} cm^3

8. 1.37×10^3 N · m/rad

Answers to odd-numbered problems

13-1. 3.02 kg

13-3. 6.24×10^5 lb (ocean water is a little more
       dense than fresh water; would give
       6.4×10^5 lb)

13-5. 2.4×10^{-10} m

13-7. 2.75×10^6 N/m; 6.87×10^5 N/m;
       1.37×10^6 N/m

13-9. $9\frac{1}{2}°$ (using small-angle approximations,
       $\sin \theta \approx \tan \theta$)

13-11. 2.54×10^3 N · m

13-13. $\Delta V/V = 5 \times 10^{-3}$

13-15. 0.24 in.

13-17. 9×10^3 N · m

CHAPTER 14
Answers to exercises

1. (a) 2.25×10^3 J (b) 1.425×10^4 J
    (c) 75.5 m/s

2. $d = \frac{1}{2}v^2/\mu g = 280$ ft

3. (a) $8\frac{1}{2}$ m/s (b) 22 m/s (c) $24\frac{1}{2}$ m/s

4. positive; negative

5. ~~100 N/m~~   ⋅ 01 N/m

6. (a) 11.1 ft · lb (b) $75\frac{1}{2}$ ft/s

7. $W = FC$

8. (a) 0 (b) $Kh^3/3$ (c) $0.12h^3$

9. Such a value of y_f is not reached.

10. Same calculations are involved.

11. 76 ft/s; 5 ft/s

12. If v falls to zero at peak height h (projectile
     fired vertically), $h = v_0^2/2g$

13. 1.44×10^5 s = 40 h

14. $K = \Gamma\theta = 377$ J

15. $v = \sqrt{gh/0.7}$

16. 6.26 m/s

17. 18.2 s

18. 399 lb

19. 10.4 N · m($= 7.7$ lb · ft)

Answers to odd-numbered problems

14-1. $W = F \cdot d = \frac{1}{2}mv^2$; $d = v^2/2(0.1 - 0.04)g$
       $= 104$ ft

14-3. 2.04 m

14-5. (a) 6.67×10^3 hp (b) 1.33×10^4 hp

14-7. 358 kw ($= 480$ hp)

14-9. 9.6×10^4 ft · lb just to overcome inertia and
       reach 30 mi/hr; 2.5×10^4 ft · lb to cruise one
       block

14-11. 1.9×10^{12} W ($=$ current power consumption
        of U.S.)

14-13. $h = 2.7R$

14-15. When the engine turns 2.48 times faster than the wheels, the maximum possible speed (161 mi/hr) is developed at 4800 rpm.

14-17. $2\sqrt{h/g}$; $1.83\sqrt{h/g}$; $2.12\sqrt{h/g}$

CHAPTER 15
Answers to exercises

1.   30 m/s
2.   12 ft/s
3.   1.5 m/s
4.   0.4 m
5.   Approximately 40 J
6. $F = -3C_1x^2$
7.   9.7 ft/s
8.   (a) $v_0 = \sqrt{2gh}$ (b) maximum height $= h/4$ (see Fig. E15-8)(below).

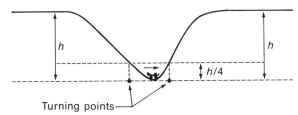

Turning points

9. 2.31×10^3 m/s; at 2 nm, $U \approx \frac{3}{4}$eV so $v \approx 1.2 \times 10^3$ m/s
10.   (a) 3 eV   (b) 2.89×10^5 J
11.   1024 m/s
12.   Take a series of measurements, starting with a large value of separation s; decrease s until an apparent rise in muzzle velocity is seen for small s. Extrapolate the large s values back to $s = 0$.
13. $v_f = v_i \cos 45°$; $K_f = \frac{1}{2}K_i$
14. $v_p/v_h = 11.5$

Answers to odd-numbered problems

15-1.   (a) 750 J   (b) −343 J   (c) 407 J   (d) 407 J

15-5. $d = \sqrt{2mg(h + d)/k}$

15-7.   1.14 eV

15-9. $d = l \sin\theta/\mu$

15-11.   (a) $\Delta U = \frac{1}{2}a(x_1{}^2 - x_2{}^2) + \frac{1}{3}b(x_1{}^3 - x_2{}^3)$ (b) $U = -(\frac{1}{2}ax^2 + \frac{1}{3}bx^3)$

15-13.   2500 J; 9.96 J

15-15.   60°; 0.0866c and 0.05c

15-17.   (a) $\Delta K = \left(\dfrac{r_1{}^2}{r_2{}^2} - 1\right)K_1$, where

$K_1 = \frac{1}{2}mr_1{}^2\omega_1{}^2$

    (b) By direct evaluation of $\int F \cdot dr$ with

$F = m\omega^2r$, we find $W = \left(\dfrac{r_1{}^2}{r_2{}^2} - 1\right)K_1$ also.

CHAPTER 16
Answers to exercises

1.   See Fig. E16-1(below).

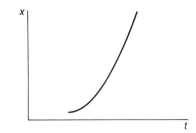

2.   See Fig. E16-2(below).

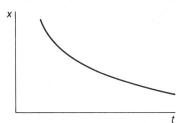

3. $\dfrac{dx}{dt} = 2At \sin(\omega t) + \omega At^2 \cos(\omega t)$

4. $\dfrac{d^2x}{dt^2} = -\dfrac{k}{m}x$ becomes $-\omega^2 A \sin(\omega t)$

$= -\dfrac{k}{m} A \sin(\omega t)$

5. $\sqrt{\dfrac{k}{m}} = \sqrt{\dfrac{F/l}{m}} = \sqrt{\dfrac{mlt^{-2}/l}{m}} = \sqrt{t^{-2}} = t^{-1}$

6. $f = \dfrac{4}{\pi} \text{ s}^{-1} = 1.273 \text{ s}^{-1}$

7.   (a) $-\pi/2$ (b) 0   (c) $\pm\pi$ (d) $\pi/2$

8.   1474 lb

9.   (a) $W < 0$; K decreases   (b) $W > 0$; K increases   (c) $W < 0$; K decreases   (d) $W > 0$; K increases

10. $v_x = -\omega r \sin(\omega t)$ and $v_x = \omega r \cos(\omega t)$; at $t = 0$, $v_x = 0$ and v_y positive. As t first increases, v_x becomes negative; $\mathbf{v}$ turns left.

11. For $\phi = \pi/2$ we have $x = A \cos(\omega t)$ and $y = A \sin(\omega t)$, so that $x^2 + y^2 = A^2[\cos^2(\omega t) + \sin^2(\omega t)] = A^2$, which is the equation of a circle of radius A. There is one more value of ϕ that also gives a circle.

12. 3.98 m or 13.05 ft

13. $f = \dfrac{1}{2\pi} \sqrt{\dfrac{3g}{2l}}$

14. $\sqrt{g/l} = \sqrt{lt^{-2}/l} = t^{-1}$; $\sqrt{mgd/I} = \sqrt{mlt^{-2}l/ml^2} = t^{-1}$; $\sqrt{K/I} = \sqrt{F \cdot l/ml^2} = \sqrt{mlt^{-2}l/ml^2} = t^{-1}$

15. See Fig. E16-15(below).

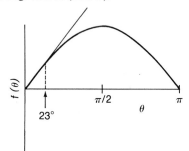

16. 1.58 min

Answers to odd-numbered problems

16-3. $C = \sqrt{A^2 + B^2}$; $\tan \phi = B/A$

16-5. 0.8 s

16-7. (a) 63 m/s (b) 4×10^{13} m/s²

16-9. 6 mm; 2.45 Hz

16-15. 0.99 m ($= 3.26$ ft)

16-17. 0.78 s

16-19. $x = 3.6$ cm; $v = 94$ cm/s; $a = 1800$ cm/s²

CHAPTER 17
Answers to exercises

1. Yes. Since the area of a sphere is proportional to the square of its radius, the number of vortices per unit area is proportional to $1/r^2$.

2. (a) $0.250G \dfrac{m_1 m_2}{l^2}$ (b) $0.262G \dfrac{m_1 m_2}{l^2}$

   (c) $0.267G \dfrac{m_1 m_2}{l^2}$

3. 2.7×10^{-3} rad (neglecting effects of the large mass on the more distant of the small masses)

4. approximately 2×10^{-6} N

5. $R^3 = \left(\dfrac{24}{1.4}\right)^2 R_e^3$; $R = 6.7R_e (= 4.2 \times 10^7$ m$)$

6. $a = \dfrac{(3.7)^2}{81} g = 0.17g$

7. $F = \left(\dfrac{3960 \text{ mi}}{4960 \text{ mi}}\right)^2 W = 102$ lb

8. 1.9 hr

9. $K = \Delta U = mgR_e/2 = 3.1 \times 10^7$ J

10. $v = \sqrt{2gR_e} = 1.1 \times 10^4$ m/s

Answers to odd-numbered problems

17-1. Earth-moon: 2×10^{20} N; Sun-moon: 4.4×10^{20} N

17-3. 0.22 m/s²; 2.36 hr

17-5. 2 hr

17-9. $W = 0.95$ mg

17-11. 6.37×10^6 J

17-13. $v(\text{escape}) = \sqrt{2}\, v(\text{orbit})$

CHAPTER 18
Answers to exercises

1. 9.8×10^4 dyne/cm² $= 9.8 \times 10^3$ N/m²

2. The pressure must be identical at the base of each vessel. If this were not so, the liquid in the horizontal pipe at the bottom would be in motion. Since pressure is proportional to depth, the depth must also be identical for each vessel.

3. For a given force, pressure is proportional to $1/A$, so the small piston more easily produces high pressure, although at the expense of moving a smaller volume of air per stroke. Since high-pressure tires have a small volume anyway, we conclude that the salesman needed the money.

4. Connect the right-hand side of the diaphragm to a vacuum.

5. 1.01×10^6 dyne/cm²

6. 6650 lb

7. 1.77×10^3 kg/m³

8. The difference in pressure causes an upward force $F = \rho g(h_2 - h_1)A$, but the quantity in parentheses is the height of the block, so we

have $F = \rho g V$, in agreement with Archimedes' principle.

9. 103 cm/s

10. 7.47×10^4 N $(= 8.4$ U.S. tons)

11. Bullets fired horizontally out of the left side fall faster than their normal trajectory in still air. On the right side they fall more slowly than the normal trajectory.

12. See Fig. *E18-12(below)*.

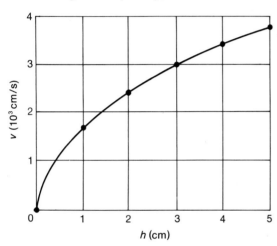

Answers to odd-numbered problems

18-1. $\Delta p = 624$ lb/ft^2 (approximately 1500 lb on chest)

18-3. $d = 9$ mm

18-5. 69.7 kg; no, the weight of the air is offset by the buoyancy of the surrounding air.

18-7. Total mass thrown out, M; density of stones, ρ_s; density of water, ρ_w; cross-sectional area of boat, a; cross-sectional area of lake, A; height h of boat above lake bottom is given by:
$$h = \frac{M}{\rho_w a}\left(1 - \frac{a}{A}\right) + \frac{M}{\rho_s A}$$
Note that the lake level falls slightly.

18-9. gold/silver $= 17/83$

18-13. Air flows more rapidly over top of wave and reduces pressure there. This would seem to be a small effect, since a difference in air velocities of 100 mi/hr and 80 mi/hr supports a column of water only $7\frac{1}{2}$ in. high.

18-15. 774 lb

CHAPTER 19
Answers to exercises

1. 37°C

2. 717.9K $(= 444.7°C)$

3. 6.75 cm^3

4. 0.036 ft $(= 0.43$ in.) (remember $\alpha = \beta/3$)

5. 60 kilocalories (Calories)

6. 152.5K

7. 10.5 cal/s

8. 32°C

9. 23.5 W

10. For $\epsilon = 1$, $A = 1.5$m^2: 368 W; 90 W

Answers to odd-numbered problems

19-3. 0.33 mm

19-5. 1.27 cal/g°C

19-7. 19.1K

19-9. 106g

19-11. 650K

19-13. 5760K

CHAPTER 20
Answers to exercises

1. 7.61×10^4 cal/s; 427 hp

2. (*a*) 335 s (*b*) 3 g/s (*c*) 668 s

3. 9.63 atm $(= 9.75 \times 10^5$ N/m$^2)$

4. 5 liters

5. 1905 J

6. 15.7 atm

Answers to odd-numbered problems

20-1. 2.02 kg

20-3. 9×10^6 cal

20-5. m/27

20-7. 48°C

20-9. 11 cm^3

20-11. 51.4 atm; 902K

20-13. 462 m/s; 2580 m/s

20-15. 831 J

20-17. $W = \dfrac{K}{1 - \gamma}\,[V_2^{1-\gamma} - V_1^{1-\gamma}]$

CHAPTER 21
Answers to exercises

1. 9,970 J (= 1 hp for $13\frac{1}{3}$ s)
2. Yes. 2500 J = 597 cal but the engine cannot produce more than 400 cal.
3. (a) 1124.8 cal   (b) 804.8 cal
4. 957 J
5. 3000K
6. 62%
7. All. The refrigerator becomes a room heater with an output equal to the power supplied to it.
8. $P = 6.75$; 403 cal/s
9. 6.28×10^{25} J. No low-temperature reservoir available.
10. 16.82 cal/K
11. Heat lost by water is gained by another body increasing its entropy. Net entropy increases or, at best, is constant.

Answers to odd-numbered problems

21-1.   (a) 596 cal   (b) $2\frac{2}{3}$ atm   (c) 24.6l
21-3.   450K
21-5. 1.7×10^4 W = 22.8 hp
21-7. $\text{Eff} = 1 - \left(\dfrac{V_1}{V_2}\right)^{\gamma-1}$
21-9.   (a) 400 cal/s   (b) 219 W
21-11. $P = 6.825$; 245 W
21-13. $\Delta S = +2.425$ cal/K

CHAPTER 22
Answers to exercises

1. $y + (5 \text{ m/s})t$
2. For $|\xi| \leq 1$, $y = 1 - |\xi|$; for $\xi < -1$ and $\xi > 1$, $y = 0$.
3. For $kx - \omega t = $ constant, we have $x = \dfrac{\omega}{k}t$

   + constant/k, so $u = \dfrac{dx}{dt} = \omega/k$
4. (a) 2 m   (b) 0.5 Hz   (c) 10 m
   (d) 0.628 m^{-1} (e) 5 m/s
5. 4.4 s
6. 10^{10}

7. 434 Hz
8. $f' = f_0\left(\dfrac{u}{u+w}\right)\left(\dfrac{u+w}{u}\right) = f_0$
9. 443 Hz and 437 Hz; larger mass reduces frequency.
10. The reflected wave's frequency is mixed with the frequency of the transmitted wave so that the beat frequency gives an accurate measure of the Doppler shift.
11. 0.25 m
12. 40 Hz
13. Segments of the string are moving rapidly at the instant its displacement vanishes. The energy is in the form of kinetic energy and the inertia of the string carries it to position d.

Answers to odd-numbered problems

22-1. See Fig. P22-1(*below*).

(a)

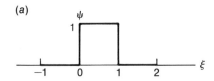

(b)

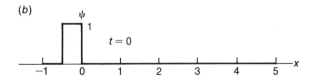

(c)

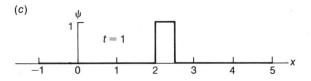

(d)

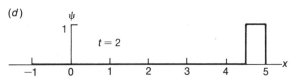

22-3.   0.039
22-5.   31.6 m
22-7.   See Fig. P22-7(*top left, page 897*).

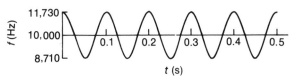

22-9.   739 mi/hr

22-11. 440 Hz

22-13. (a) $v = \left(\dfrac{u}{f_0}\right) \cdot \text{beats/s}$

(b) $d = \displaystyle\int v\, dt = \dfrac{u}{f_0}\int (\text{beats/s})\, dt = \dfrac{u}{f_0} \cdot \text{beats}$

CHAPTER 23
Answers to exercises

1. The rope first accelerates upward on the right side of the pulse, where positive curvature provides an upward force component. At the left side of the pulse, a similar upward-directed acceleration brings the downward-moving rope to rest.
2. m/s; slugs/ft and lb
3. 1.43×10^3 m/s
4. faster, since $Y > B$
5. yes; no inversion
6. For $L \approx 10$ in.; $u \approx 1100$ ft/s; $f = 330$ Hz
7. First, add some third harmonic with the phase shown in Fig. E23-7a to sharpen the point of the fundamental. The addition of fifth harmonic, as in Fig. E23-7b sharpens the point still more, as indicated in Fig. E23-7c (*below & top right*).

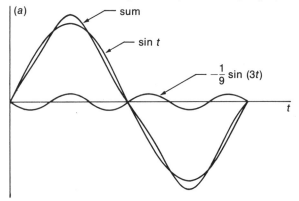
(a) sum, sin t, $-\dfrac{1}{9}\sin(3t)$

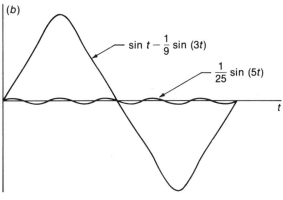
(b) $\sin t - \dfrac{1}{9}\sin(3t)$, $\dfrac{1}{25}\sin(5t)$

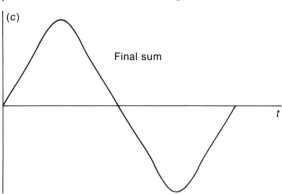
(c) Final sum

8.   1000 Hz and 8000 Hz

Answers to odd-numbered problems

23-1. $u = \sqrt{gy}$

23-3. $B = 8.2 \times 10^{10}$ N/m² (about half that of steel)

23-5. 2000 N

23-7. 6.4 g/m

23-9. 28 Hz

CHAPTER 24
Answers to exercises

1. See Fig. E24-1 (*below*).

2. Twice as high, twice as wide: 4 ft × 6 ft;
    yes; Height decreases; width of top decreases
    so that shape becomes trapezoidal.
3. $(3 \times 10^8 \text{ m/s})(3.28 \text{ ft/m})(10^{-9} \text{ s/ns}) = 0.98 \text{ ft/ns}$
4. 5.75×10^{-5} s; 725 rev/min
5. $n \cdot 500$ Hz $(n = 1, 3, 5 \ldots)$
6. 6 mm
7. See Fig. E24-7(*below*).

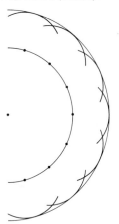

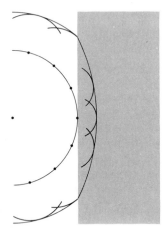

8. $w = \dfrac{3}{2}\lambda$
9. 1.2×10^{-4} rad
10. 0.12 mm; 4.17×10^4 lines/cm
11. blue
12. 4×10^3; at 4 counts per second,
    $\omega_{max} = 10^{-3}$ rev/s $(= 6 \times 10^{-2}$ rev/min$)$.

Answers to odd-numbered problems

24-1. yes
24-3. 5×10^{14} Hz
24-5. 0.169 mm
24-7. three
24-9. 417 nm or 500 nm or 625 nm
24-11. 1.58×10^3 cd
24-13. 0.625 lm/m²
24-15. 31.65 cm/s
24-17. Average over all data gives $\lambda = 645$ nm.

CHAPTER 25
Answers to exercises

1. yes
2. The line *ac* is parallel to the upper dotted
    line, so $\theta_i = \alpha$ and $\theta_r = \beta$. Given $\theta_i = \theta_r$, we
    have $\alpha = \beta$.
3. See Fig. E25-3(*below*).

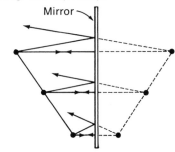

Mirror

4. the focal points
5. 15 cm behind mirror; in front of mirror;
    virtual; erect
6. inverted; real
7. 0.923 in. behind surface; erect
8. 2.5 cm

Answers to odd-numbered problems

25-5.   (*b*) real, inverted   (*c*) 3.75 cm   (*d*) no
25-9.   1.803 cm
25-11. A concave mirror with a focal length
        of 19.6 cm. The lamp filament is 20 cm
        from the mirror.
25-13. (*a*) $m = 3$ (*b*) 12 in. behind mirror

CHAPTER 26
Answers to exercises

1. 48.75°; see Fig. 26-6 for the case θ_c.
2. air-glass
3. If n_2 is known, the unknown n_1 can be determined by measuring θ_1 and θ_2: $n_1 = n_2 \sin \theta_2 / \sin \theta_1$.
4. One method, used for liquids, is to place a drop of the unknown substance on a medium with a known, and larger, index of refraction (such as flint glass). The angle at which total internal reflection first occurs gives the unknown index. Commercial instruments are calibrated so that the unknown index is read directly from a scale.
5. flint
6. 15.37°
7. 5 cm
8. 36 cm to the right of the lens
9. −24 cm

Answers to odd-numbered problems

26-1. 1.88×10^8 m/s
26-3.   99.3°
26-5.   70.1°
26-7.   61°
26-9.   1 mm
26-11. 0.164
26-15. 10 cm
26-17. 100 cm; converging
26-19. s' always negative, image virtual
         (a) 3.2 cm   (b) 5.33 cm   (c) 8.0 cm
         (d) 10.67 cm
26-21. 6.15 in.
26-23. red: 57.75 cm; blue: 54.74 cm

CHAPTER 27
Answers to exercises

1. $f = 8$
2. (a) 18.5 mm   (b) 0.6 in.
3. 0.17 in.
5. 6 cm beyond second lens
6. 5 diopters
7. approximately 40 diopters
8. $6\frac{1}{3}$ diopters
9. 1.36 cm
10. $\frac{3}{4}$ in.
11. 2 in.
12. 1.08×10^{-7} rad
13. 6.9×10^{-3} in.; 6.5×10^6 mi
14. 199 nm

Answers to odd-numbered problems

27-1.   (a) 2.5 mm   (b) 2.6 mm
27-3.   0.16 mm
27-5.   6 diopters
27-7.   Lenses 31 cm apart; the image is 5.1 m beyond the second lens.
27-11. 100 cm (objective) and $\frac{1}{2}$ cm (eyepiece); $m = 200$
27-13. 5.5×10^{-4} rad; 1.1 μm
27-15. approximately 70 m

CHAPTER 28
Answers to exercises

1. $\sqrt{10}$ statcoulomb
2. Both forces double.
3. 3.16×10^{-4} C
4. kQ^2/l
5. $F_y = 2kq\alpha \displaystyle\int_0^{L/2} \frac{x\,dx}{r^2} \cos \theta$
6. (a) $F = -k\dfrac{Qq}{r^2}$ (b) $F = k\dfrac{Q^2}{r^2}$
7. $F \approx 2 \times 10^{24}$ N; weight of oceans $\approx 10^{22}$ N
8. 10^{12}
9. Approximately 10^{11} bonds
10. (a) $\frac{2}{3}A$ (b) $1.33 \times 10^6 A$
11. 14.5 N; 2.2×10^{27} m/s²

Answers to odd-numbered problems

28-1.   (a) $Q = \sqrt{\dfrac{\pi YP}{4k}}\, d$

         (b) $Q = 4.58 \times 10^{-3}$ C
28-3.   (a) 9.2×10^5 dyne   (b) 1.39×10^{29} cm/s²
28-5. $F = 3.29kq^2/l^2$

28-7. $F = 0.268 \, kq\alpha$

28-9. $q = d^2 mg \tan \theta / kQ$

28-11. $4 \times 10^{17} \, \text{kg/m}^3$

28-15. 3.125×10^{19} electrons/s

28-17. $52°$

CHAPTER 29
Answers to exercises

1. Substituting (29-2) into (29-3), we find attraction $(F < 0)$ if charges are opposite and repulsion $(F > 0)$ for similar charges.

2. $2 \times 10^{-5} \, \text{C/kg}$

3. (a) $kQ/(1.25)^{3/2} \, \text{m}^2$ in the $-x$ direction
   (b) $3.55 kQ/\text{m}^2$ in the $+x$ direction

4. Evaluation of (29-7) now gives $1.9 R^{-2}$ rather than $2 R^{-2}$.

5. $1.18 \times 10^{17} \, \text{V/m}$

6. $1.18 \times 10^{5} \, \text{V/m}$

7. $8.2 \times 10^{-8} \, \text{N}; \; 3.6 \times 10^{-47} \, \text{N}$

Answers to odd-numbered problems

29-1. 50 V/m (to the left)

29-3. about $\frac{1}{3} \mu s$

29-5. 4.84×10^8 V/m pointed downward but tipped 12.1° to the left of vertical

29-7. 3.75×10^7 N/C

29-9. 54.7 m

29-11. d to the left of $-q$

29-13. $9 \times 10^{22} \, \text{m/s}^2$

29-15. 313 eV

CHAPTER 30
Answers to exercises

1. $AB \cos \theta = BA \cos \theta$

2. Magnitude of $\mathbf{A} + \mathbf{B}$ is $[A^2 + B^2 + 2AB \cos \theta]^{1/2}$ (law of cosines). Since a vector dot-multiplied by itself gives the square of its magnitude, the result follows.

3. $0.866 EWL; \; 0; \; -0.5 EWL$

4. $CL^3/2$

5. Lines of force passing through the reentrant portions in (a) produce cancelling contributions to net flux, so that result is same as (b).

6. Applying Gauss's law we find: (a) $2Q/\epsilon_0$
   (b) 0   (c) 0   (d) Q/ϵ_0

7. Between plates, $E = \sigma/2\epsilon_0 + \sigma/2\epsilon_0 = \sigma/\epsilon_0$; outside plates, $E = \sigma/2\epsilon_0 - \sigma/2\epsilon_0 = 0$. See Fig. 31-3.

8. $\oint \mathbf{E} \cdot d\mathbf{A} = 4\pi r^2 E = q/\epsilon_0$ so $E = \dfrac{1}{4\pi\epsilon_0} \dfrac{q}{r^2}$

9. circulation $\propto r^2$

10. $ch^2 L$

Answers to odd-numbered problems

30-1. $\pi R^2 E$

30-3. (a) 0   (b) $E = \dfrac{1}{4\pi\epsilon_0} \dfrac{Q_1}{r^2}$

   (c) $E = \dfrac{1}{4\pi\epsilon_0} \dfrac{Q_1 + Q_2}{r^2}$ (d) $Q_1 = -Q_2$

30-5. 2.25×10^{21} V/m

30-7. (a) $\oint \mathbf{g} \cdot d\mathbf{A} = 4\pi Gm$ (b) Only one sign of gravitational "charge," m, exists.

30-9. $1.77 \times 10^{-5} \, \text{C/m}^2$

30-11. $\sigma = 8.85 \times 10^{-7} \, \text{C/m}^2$ on sheet A;
   $\sigma = -8.85 \times 10^{-7} \, \text{C/m}^2$ on sheet B

30-13. Contributions from b to c and from d to a are both zero, while those from a to b and from c to d cancel; yes; a circular path centered on the charge so that $\mathbf{E} \cdot d\mathbf{r} = 0$ everywhere because $\mathbf{E} \perp d\mathbf{r}$.

30-15. No. The circulation is nonzero since a paddle wheel would rotate counterclockwise.

CHAPTER 31
Answers to exercises

1. $\Delta V = EL \cos 60°$; point 1 is higher; gains energy; loses energy.

2. (a) 4.38×10^5 m/s   (b) 1.88×10^7 m/s

3. As the path length covered by the integration increases, E must decrease to give the same net result.

4. 8.99×10^{10} V; 0; 8.99×10^{10} eV ($= 1.44 \times 10^{-8}$ J)

5. 3.34×10^{-6} C; 3×10^5 V

6. 6×10^6 J

7. $E_x = -\dfrac{\partial V}{\partial x} = -kQ(-\frac{1}{2})(x^2 + y^2 + z^2)^{-3/2} \cdot 2x$

   $= kQx/r^3$

Answers to odd-numbered problems

31-1. (*a*) 5.65 kV   (*b*) 5.65 kV   (*c*) 0
31-3. 3.34×10^{-4} C; $3 \times x\ 10^6$ V
31-5. 30 W
31-9. $E_a/E_b = (r_b \ln r_b)/(r_a \ln r_a)$
31-11. 900 W (= 1.2 hp)
31-13. $3K \dfrac{x^2}{r^5} - \dfrac{K}{r^3}$

CHAPTER 32
Answers to exercises

1. 0.01 μF
2. 0.133 μC
3. 0.177 μF
4. 22.6 cm^2
5. (*a*) 3 μF   (*b*) 27 μF
6. 40 μC
7. 30 μF (see Eq. (32-6))
8. 5 J and 10 J

Answers to odd-numbered problems

32-1. (*a*) 2.26×10^{-2} m^2 (*b*) 2000 V
32-3. *A* large, *d* small
32-5. $C = 4\pi\epsilon_0 \left(\dfrac{1}{r_1} - \dfrac{1}{r_2} \right)^{-1}$
32-7. (*a*) $\dfrac{\sigma d}{2\epsilon_0} \left(1 + \dfrac{1}{K} \right)$ (*b*) $\dfrac{2\epsilon_0 A}{d} \left(1 + \dfrac{1}{K} \right)^{-1}$
32-11. $(K + 1)C$
32-13. All combinations equivalent; $U \propto$ volume
32-15. 1.77×10^{-8} J
32-17. (*a*) yes   (*b*) increases
$\qquad$ (*c*) $\Delta U = \frac{1}{2} Q^2 \left(\dfrac{1}{C} - \dfrac{1}{KC} \right)$

CHAPTER 33
Answers to exercises

1. (*a*) right to left   (*b*) 160 mA
2. 1000 V
3. 100 A
4. p_1: 5 Ω; p_2: 18 Ω

5. 6 cells
6. 0.3 A
7. 0.05 Ω
8. 4.09 V; 0.409 Ω
9. (*a*) 9 Ω (*b*) 1 Ω
10. 12 Ω; $\frac{47}{7}$ Ω; $\frac{47}{8}$ Ω; $\frac{47}{9}$ Ω; $\frac{60}{47}$ Ω
11. 1.65 kW 6.7   5.9   5.2
12. 28.28 A
13. one-fourth of original resistance
14. 3.26×10^{-2} Ω
15. 0.16 mm

Answers to odd-numbered problems

33-1. 6.25×10^{15} electrons/s
33-3. $\frac{1}{3}$ Ω
33-5. 5.6 V
33-7. 10.32 V and 1.94 A
33-9. 0.6 A
33-11. $\Delta V = 3.16 \times 10^{-2}$ V
33-15. 23.26 s
33-17. 1440 Ω
33-19. 2.17 mm

CHAPTER 34
Answers to exercises

1. 2.75×10^8 m/s
3. (*a*) left to right or right to left in the plane of the page   (*b*) positive
4. (*a*) 1.2×10^{-11} N   (*b*) 7.19×10^{15} m/s^2
$\qquad$ (*c*) 7.5×10^7 V/m
5. 12.7 MHz
6. 10.7 MeV
7. top positive
8. 14.9 μV
9. west; 5×10^{-2} N
10. (*a*) 100 A   (*b*) 1 N $\cdot$ m
11. 9,999 Ω
12. 1.001×10^{-3} Ω

Answers to odd-numbered problems

34-1. up; 0.0104 T
34-5. (*a*) 38.1 cm   (*b*) 18.3 MHz   (*c*) 6.1 MHz

34-7.   200 A

34-9. $\dfrac{\pi I B r^2}{mg} - r$

34-11. 0.075 N · m

34-15. 2.66 mV

34-17. 0.1%

34-19. 625 turns

CHAPTER 35
Answers to exercises

1. 6.67×10^{-6} T

2. 2.7×10^{9} A

3. 10^{3} turns/m

7.   0.16 A; 796 A

8.   7958 A · turns

9.   as shown

10. 25 N · m

Answers to odd-numbered problems

35-1. $\sqrt{2}\,\mu_0 I / \pi d$

35-3. $\dfrac{5}{12}\dfrac{\mu_0 I}{\pi d}$

35-5. $4.445 \times 10^{-4} I$

35-7.   (a) 0.02 T (= 200 gauss)   (b) 222.22 gauss, 181.18 gauss

35-11. 0.54 T; 4300 A

35-13. (a) 3.98×10^{3} (b) 6.525×10^{3}
        (c) 1.05×10^{3}

CHAPTER 36
Answers to exercises

1.   (a) left to right   (b) $\mathscr{E} = vBW$ (c) $\mathscr{E} = vBW$

3.   2.5 mV

4.   (a) 35 Weber/s (Weber = T · m²)
     (b) 7.96×10^{4} T/s   (c) 0.796 T

5.   100 Ω

6.   100 mH

7.   counterclockwise as seen from above

8.   (a) left   (b) right   (c) right   (d) left

9. 5×10^{-3} J; connect a resistor across L

Answers to odd-numbered problems

36-1.   See Fig. P36-1(below).

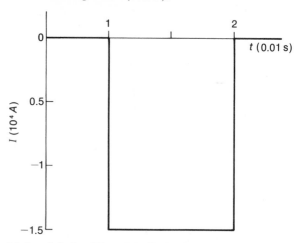

36-3.   (a) See Fig. P36-3(below).   (b) 0.06 V

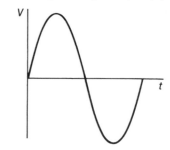

36-5.   20 mA

36-7.   (a) 0.05 N   (b) zero   (c) 0.5 W
        (d) 0.5 W

36-9.   See Fig. P36-9(top left, page 903).

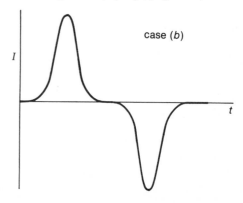

case (b)

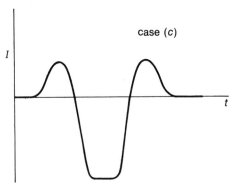

case (c)

36-13. (a) increasing   (b) 25 mH   (c) 16.7 mH
36-15. (a) 10 V   (b) 200 A/s   (c) 1 W   (d) 1 W

CHAPTER 37
Answers to exercises

2. (a) 6×10^{-2} s   (b) $Q = Q_0/e^2 = 0.135\ \mu C$
   (c) 0.0226 V   (d) 16.7 μA

4. 8×10^{-8} s

6. Want $\tau \leqslant 0.5$ ms so $R = 500\ \Omega$

7. 5 ms

9. 90.47 pF

11. 4×10^{-4} s

Answers to odd-numbered problems

37-1. $\tau = \dfrac{R_1 R_2 (C_1 + C_2)}{R_1 + R_2}$

37-3. approximately 20 pF

37-7. 14.2 pF; 14.2 ns

37-9. (a) 10^{-2} V   (b) 10% error   (c) 10^{-3} V

37-11. (a) 100 A/s   (b) 4 ms   (c) 4 ms
   (d) 3.68 V

37-13. (a) 1.94 W   (b) 1.94 W

37-15. (a) 99.35 μs (b) 5×10^{-4} J   (c) zero

37-17. See Fig. P37-17(below).

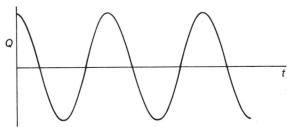

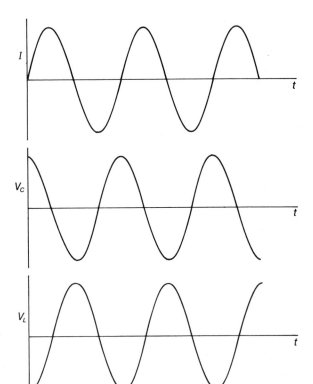

37-19. $I_{max} = V_0 \sqrt{C/L}$

CHAPTER 38
Answers to exercises

1. 3.77 Ω; 377 Ω

2. 25 V

3. See Fig. E38-3(below); sum = zero

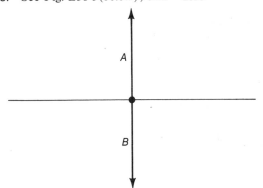

4. $Z = 10.44 \ \Omega$; $\phi = 16.7°$

5. See Fig. *E38-5*(*below*); 153.2°

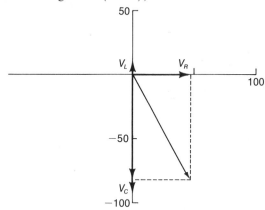

6. $V_R = 0$, $V_L = 100$ V, $V_C = 0$

7. (*a*) 3.56×10^7 Hz   (*b*) $224 \ \Omega$ (*c*) 112:1

8. (*a*) 2.52×10^7 Hz   (*b*) yes, 79:1

9. 1.2 kW; 14.14 A, 170 V

10. 0.894; 15.15 A; 1.15 kW

11. Yes, the secondary could deliver up to 900 A.

12. 8.33×10^3 A; 16.67 A; 2.5×10^5

Answers to odd-numbered problems

38-1. 157.1 Ω; 1571 Ω; 15710 Ω; 95.5 Hz

38-3. (*a*) See Fig. *P38-3*(*below*).   (*b*) 43.3°

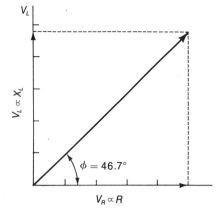

38-5. 50 Hz

38-7. 472 Ω; 83.9°

38-9. (*a*) 15.2 A   (*b*) 1.29 kW

38-13. 8333 turns

38-15. (*a*) 12.26 cm   (*b*) 0.245 mm

CHAPTER 39
Answers to exercises

1. $\epsilon_0 \dfrac{d\phi_E}{dt} = \left[\dfrac{C^2}{N \cdot m^2} \right] \cdot \left[\dfrac{N}{C} \cdot \dfrac{m^2}{s} \right] = \dfrac{C}{s} = A$

2. 3 m

3. 3.33×10^{-9} T in $-x$ direction

4. approximately 2 m

5. horizontal

6. yes

7. 10.5 V/m

Answers to odd-numbered problems

39-1. 377 m

39-9. (*a*) $F_g \propto r^3$ while $F_{rad} \propto r^2$, so $F_{rad}/F_g \propto 1/r$ (*b*) 5.9×10^{-7} m

39-11. 50 m

CHAPTER 40
Answers to exercises

1. 3.315×10^{-19} J ($= 2.07$ eV)

2. 1.82 eV

4. 2.19×10^6 m/s ($= c/137$)

5. 2.23×10^{-9} m

Index